READINGS IN THE MANAGEMENT OF INNOVATION

Second Edition

Edited by

Michael L. Tushman

William L. Moore

READINGS IN THE MANAGEMENT OF INNOVATION

Second Edition

Edited by

Michael L. Tushman

Columbia University
Graduate School of Business

William L. Moore

University of Utah
Graduate School of Business

BALLINGER PUBLISHING COMPANY
Cambridge, Massachusetts
A Subsidiary of Harper & Row, Publishers, Inc.

International Standard Book Number: 0-88730-244-0

Library of Congress Catalog Card Number: 88-24199

Printed in the United States of America

Library of Congress Cataloging-in-Publication Data

Tushman, Michael L.
  Readings in the management of innovation.

  Includes bibliographical references and index.
  1. Technological innovations—Management. I. Tushman, Michael.
II. Moore, William L., 1943–
HD45.R3   1988   658.5′77   88-24199
ISBN 0-88730-244-0 (pbk.)

Table of Contents

Preface

The ability to develop successful innovations, both in the product or service offered and in the way of producing it, is crucial to the health of individual firms, industries, and the larger economy. As the rate of change of technology accelerates and as the degree of international competition increases, firms must enhance their ability to develop and introduce new products, services, and processes over time. Even with this heightened emphasis on innovation, however, managers are simultaneously pushed to increase efficiency and productivity. One generic challenge, then, is to create organizations that have the capability to operate in two time frames simultaneously—capable of organizing to accomplish today's work today, even as the firm develops products, services, and/or processes to address tomorrow's challenges.

Ironically, however, organizing for today's efficiency often sows the seeds for stunted innovation. Throughout this collection, we ask the reader to consider the balance between short-term efficiency and long-term adaptability. This duality—the need to build different kinds of organizations to accomplish different kinds of innovation—is a central managerial challenge that pervades the book.

Several themes recur throughout this second edition of *Readings in the Management of Innovation:*

1. We continue our emphasis on *managerial problem solving.* The vast majority of innovation problems deal not with a lack of technology, but with a mismatch between technological possibilities and market demands. Even with the right technology and clear market demand, innovation frequently suffers from a host of organizational pathologies. Our focus is on problem solving: helping managers develop appropriate strategies and, in turn, appropriate structures, human resources, and cultures to facilitate sustained innovation.

2. *Different kinds of innovation are appropriate at different stages of a product life cycle.* Incremental innovation, as well as major product or process innovation, is important. While the relative importance of different kinds

of innovation changes over time, managers must develop appropriate leadership styles and organizational configurations to facilitate each type. The particular innovation profile should be linked to the organization's strategy, which should, in turn, be driven by an assessment of external opportunities and threats.

3. The development of an innovation from idea generation to commercialization demands *close collaboration across several functional areas* (and perhaps across different divisions and/or geographic locations). Effective innovation is built on strong, competent functional areas that are effectively integrated. Managers must develop strong discipline areas even as they shape structures, rewards, and informal systems to achieve collaboration.

4. As product class conditions change, different kinds of innovation are required. For example, major product or service innovation is primary early in a product class, but incremental innovation is considerably more important after the emergence of a dominant design (or industry standards). Because different organization forms are required to develop different types of innovation, *organization adaptation* is an important managerial issue. Managers need to extend incremental innovation even as they build organizations to develop fundamentally new products, processes, and/or services. Organization adaptation involves actions taken within the firm as well as a range of extra-organization efforts including venturing, licensing, joint ventures, and alliances.

5. As innovation involves disruption in the status quo, *managing change* is always an important managerial concern. In the innovation business, managers must know both what to do as well as *how* to implement required organization changes to facilitate innovation.

6. Since organization change is always involved in innovation management, required changes take place in the context of the *firm's history.* History reflects how the organization has worked in the past. As innovation frequently involves changes in methods, structures, and leadership styles, history frequently stifles innovation. Effective managers need to build on the firm's history, even as they develop systems, structures, and processes to facilitate new product, process, and/or service development.

7. Managing innovation involves mediating between external forces for change and internal forces for stability. Effective innovation over time involves developing the *leadership styles and executive team* that can create the conditions to facilitate both short-term efficiency and long-term adaptability. The manager and his or her team must develop their own learning abilities and, in turn, facilitate the organization's ability to adapt. Organization learning is at the heart of managing innovation. This learning can be shaped only by an executive team that is itself flexible and adaptive over time.

Readings in the Management of Innovation provides a unified and comprehensive view of core issues in managing innovation. It presents a multidisciplinary as well as temporal approach to the topic through research-based articles from a variety of sources.

This book will meet the needs of several audiences. It was initially developed for a graduate course in managing corporate innovation. While texts were available in both product management and individual entrepreneurship, we found no book that focused on the special managerial, strategic, and functional concerns of managing innovation within corporations over time (or over the product life cycle). Since 1982, several books on innovation management have been published. Our book offers a unique focus on managerial problem solving and detailed attention to an array of managerial issues through a diverse set of research-based articles from a range of disciplines. Reflecting the rate of change in the innovation literature, this edition includes significant updates throughout and a new section on executive leadership and managing change. Approximately 60 percent of the articles are new.

Given the book's interdisciplinary and integrative approach, it can also be used as a complementary text for product management, operations management, or business strategy courses that emphasize product, service, or process innovation. A third target audience is managers and executives who want an integrated exposure to the diverse literature on managing innovation and change.

As in the first edition, the information in this new edition is not limited to those interested only in technology-based innovation or high-technology products. Rather, its ideas and principles fit a wide range of situations including industrial and consumer products and services, high and low technology, as well as small and large organizations. Finally, the book examines not just revolutionary innovations that change the way we live, but also on the series of smaller, more evolutionary innovations that follow, and that cumulatively have enormous effects on individuals and organizations.

Readings in the Management of Innovation is organized into eight sections. The introduction, a discussion of the VCR industry, sets the tone for the book by focusing on technological, market, organization, and leadership factors that affect firm performance. Section I examines the nature of innovation and the innovation process. Readings in this section introduce some basic concepts on technological change including dominant design; incremental and discontinuous innovation; and product, process, and service innovation. It also discusses the concept of history and organizational context as an important determinant of innovation. Section II explores organizational issues affecting the innovation process. Chapter 2 presents the congruence model—an approach to managerial problem solving—as well as several generic approaches to managing innovation. Chapter 3 focuses on history, precedent, commitment processes, politics, and careers as organizational factors that can impede innovation.

Section III presents several readings linking technology forecasting and technology strategy to business unit strategy. Section IV is made up of three chapters on building competence in functional areas. Chapter 6 discusses R&D, Chapter 7 covers marketing, and Chapter 8 explores manufacturing operations. Section V focuses on managing linkages and building structures, controls, and informal processes to achieve effective integration among diverse areas in the firm. Section VI discusses venturing, joint ventures, alliances, and acquisitions as internal and extra-organizational mechanisms to enhance organizational learning. Section VII highlights regulatory and governmental affectors that affect innovation in firms, industries, and countries. Finally, Section VIII looks at the role of the manager and his or her team in shaping innovation, managing strategic change, and building organizations capable of adapting to changing environmental conditions.

Several colleagues and institutions have been most helpful in giving us feedback, ideas, and support for this edition. The Graduate School of Business and the Innovation and Entrepreneurship and Strategy Research Centers at Columbia have provided support for research and pedagogical experimentation. Several colleagues have provided ideas and enthusiasm, including Philip Anderson, Robert Drazin, Ralph Katz, Donald Lehman, Charles O'Reilly, and Richard Nelson. Joy Glazener, as usual, provided outstanding support in all phases of this book. Carol Franco and Marjorie Richman at Ballinger furnished the support, encouragement, and pressure required for us to complete the revision. Finally, our students, both MBAs and executive audiences, have tolerated our experimentation and have been a source of insight and learning.

MLT
WM

INTRODUCTION

The Managerial Challenge

Our initial reading sets the stage for the articles to follow. Rosenbloom and Cusimano present rich detail on competitive performance in the VCR industry. Their article tracks the fates of various firms in the industry, tracing the roots of success and failure not to cultural differences between Japan and the United States, but to generic strategy, organization, and leadership issues in managing innovation.

Technological Pioneering and Competitive Advantage: The Birth of the VCR Industry

Richard S. Rosenbloom

Michael A. Cusumano

This article examines a significant example of "technological pioneering"—the development of an emerging technology in pursuit of future commercial opportunity. In this case, the pioneers' efforts resulted in the birth of a major industry, the manufacture of videocassette recorders (VCRs) for the mass consumer market. We compare the actions of six firms that were "pioneers" in development of home video recording technology—RCA and Ampex in the U.S. and the Victor Company of Japan (JVC), Matsushita, Sony, and Toshiba in Japan.[1] Three firms—Matsushita, JVC, and Sony—emerged in the late 1970s as the big "winners" in the growth of this new industry. But although all three are Japanese, this is not another case in which international differences in "competitiveness" were decisive. The real story of their success in the VCR industry lies in the manage- ment practices of three pioneers, who just hap- pened to be Japanese.

Sony, JVC, and Matsushita succeeded where others failed primarily because of the way they managed development of video tech- nology. Over a period of nearly two decades, they repeatedly focused on more rewarding op- portunities, positioned their technical efforts more productively, and executed those efforts more effectively to master the development of diverse constituent technologies. The success of all three leaders in VCR innovation was rooted in distinctive capabilities that are broadly relevant to the development of major new technologies:

- *Learning.* Each firm proved able, step by step, to solve the diverse set of technologi- cal challenges posed by the requirements of a mass-market videorecorder; and be- yond that, each was able to couple product design and manufacturing to create low- cost mass-production capabilities.
- *Strategic Clarity and Consistency.* Each firm proved able to conduct long-term develop- ment efforts consistently targeted on the

©1987 by the Regents of the University of Califor- nia. Reprinted from the *California Management Review*, Vol. 29, No. 4. By permission of The Regents.

The authors gratefully acknowledge helpful com- ments contributed by Robert Burgelman, Mel Horwitch, and David Teece, and by Alfred Chandler, William Lazo- nick, and other members of the Business History Seminar at Harvard Business School.

most challenging and also, potentially, the most lucrative market—the consumer; and each based those commitments on sound technical choices, reflecting an understanding of the technology on the part of both managers and engineers.

Strategic Management of Technology

The strategic management of technology is a topic that has been addressed actively in recent years in writings about business strategy and about technological innovation, two literatures that had long progressed largely independently.[2] The central issue for managers is how to translate technological capability into competitive advantage. Leadership in technology does not always bring substantial economic return. Yet, in numerous cases, a firm that led in development of an important new technology, has gained a decisive and enduring advantage.

Most discussions of technology in the literature of competitive strategy focus on innovation—that is, on determining when and how best to introduce a novel product or process to commercial use. The *timing of market entry* (first mover or follower), *product positioning* (segmentation, pricing, etc.), and *the organization of production and distribution* (make or buy, channels, etc.) are the most salient strategic issues in the management of innovation. Michael Porter discusses the relationship between these choices, industry structure, and a firm's existing competitive strategy.[3] David Teece extends the analysis by focusing on three principal situational factors: the character of the technology (especially the degree to which it can be protected against its appropriation by others), the state of industry evolution (in relation to appearance of a dominant design), and the character of complementary (non-technological) assets required to complete the innovation.[4]

Technological Pioneering. But these three factors are themselves, at least in part, conse-

quences of choices made earlier by an aspiring innovator and its rivals. In other words, the benefits that a firm can realize from alternative strategies for innovation—the consequences of choices about timing, positioning, and implementation—are shaped substantially by steps taken years earlier during development of the technology. For an emerging technology, a firm's success in the management of innovation depends on its prior conduct as a technological pioneer.

Every innovator in a new technical field must first be a pioneer. When executives perceive that a development in technology presents a major opportunity (or threat), their firm (at least implicitly) develops a "posture" toward it.[5] Its stance may be primarily passive, letting others shape the course of events, becoming, in effect, a "monitor" in the sense of "keeping watch" over developments elsewhere. Alternatively, a firm may become a pioneer, investing in development aimed at achieving some commercially relevant product or process. A first mover logically must have been among the pioneers in a new technology, but each pioneer retains the option to be a "follower" in commercialization.

The strategic issues in development parallel those of innovation: *defining a strategic role* (pioneer or monitor); *technological positioning* (product/market focus, technical agenda, time horizon, level of commitment); and *implementation* (internal development, joint venture, purchase technology). "Technological positioning," analogous to market positioning, establishes the product/market targets, the time horizon for accomplishment, and the level of resources committed. An important aspect is the definition of a "technical agenda," the hierarchy of salient problems to be resolved in development. Managers shape the distinctive capabilities necessary for competitive advantage by the ways in which they implement the tasks of development and innovation. For example, they can establish proprietary know-how by investing in internal development of technology. Firms also acquire necessary complementary assets by development, purchase, or through alliances.[6]

An Overview of VCR Development and Innovation

In broad outline, the story of videorecording presents a familiar pattern in the history of technology. A major innovation, the introduction of television after the Second World War, created a need for a way to make recordings that would offer the fidelity on playback and ease of use that radio broadcasters had attained with magnetic tape recorders. Firms already in the broadcast equipment business attempted to invent a device suitable for this well-defined application. In the early 1950s, before anyone had solved the problem of inventing a practical videotape recorder (VTR), fertile technical work was underway in the laboratories of RCA and Ampex in the United States and Toshiba in Japan. The Ampex team was first to succeed and their machine took the broadcasting market by storm. Within a few years, RCA (through cross-license with Ampex) and Toshiba (in a joint venture with Ampex) also became important suppliers of VTRs for use in broadcast studios.

A Mass Market: Visions and Realization. Although some engineers in all three companies perceived the possibility that videorecording technology could be developed to serve a mass market, all three organizations elected not to pursue that opportunity. At the same time, three other firms (JVC, Sony, and Matsushita), stimulated by the inventive and commercial successes of the other trio, soon began pioneering technical efforts of their own. The latter trio of pioneers embarked on programs of technical development and commercial enterprise that they sustained right up to the birth of the VCR mass market in 1975. Persistent technical effort, informed by the lessons of failed innova-

tions, ultimately yielded a design synthesis well matched to the needs of the mass market. Rival innovators then vied to establish competitive advantage for the highly profitable growth stage of a new industry.

This new industry, annually yielding its participants billions of dollars of revenues in the mid-1980s, began with the commercialization of Sony's Betamax[7] in 1975. JVC, withstanding pressure to adopt a common standard, introduced the rival VHS[8] system in 1976. By 1977, seven other Japanese firms were marketing VCRs under license from Sony or JVC. Worldwide consumer demand passed the million-unit level in 1978 and doubled annually during the five years 1976–81.[9] When growth began to slow in the mid-80s, demand exceeded 30 million units annually.

TECHNOLOGICAL ORIGINS AND STRATEGIC INITIATIVES

In 1951, David Sarnoff, Chairman of RCA, had challenged his engineers to develop a video recorder within five years.[10] The RCA team focused on a machine that moved a narrow tape at very high speeds past fixed magnetic heads. Meanwhile, in the central research laboratories of Toshiba, one of Japan's largest electrical equipment and electronics firms, Dr. Norikazu Sawazaki, a young engineer assigned to work with NHK, the Japanese broadcasting company, also undertook (with scant resources) to invent a videorecorder. He tried a different approach, in which the recording head rotated at high speed while the tape moved past at a relatively slow pace. Because the tape path resembled a helix, the device came to be called a "helical scanner." In 1954, Sawazaki filed the first patents for his new scanner.[11]

Innovation. But the first commercially important videorecorder was created by a small team

5

at Ampex Corporation in Redwood City, California.[12] This revolutionary machine employed a "transverse scanner," in which four recording heads on a rapidly rotating drum scanned across a two-inch-wide tape.[13] The first public demonstration of the Ampex VTR, in March 1956, stunned a convention of television broadcasters. The machine embodied patented technology without which it appeared impractical to build a competitive VTR. But Ampex's management soon began to share its technical monopoly. The desire to offer a color recorder led to a cross-licensing agreement with RCA, which controlled color TV patents. Fearful then that the giant RCA would be first to introduce a transistorized VTR, Ampex began to collaborate with Sony in development of solid-state circuitry for videorecording, intending thereby also to gain access to the Japanese market.[14]

At that time, demand from broadcasting for VTRs was booming, despite $50,000 price tag on the machines.[15] But there was no sign that other users would be interested in products based on the transverse scanner, with its inherent complexity and cost. Nonetheless, some speculated about the prospects for a mass market, anticipating uses of VTRs that would become commonplace in the 1980s. For example, in the aftermath of Ampex's first public demonstrations, one reporter quoted George Long, then president of Ampex, suggesting that

> eventually they [VTRs] might be mass-produced for home use by persons who want to see a program over and over again or want it recorded during their absence.[16]

A few days later a reporter elaborated:

> A more visionary project is the thought of a home recorder and playback for taped TV pictures. Why not pick up the new full-length motion picture at the corner drugstore and then run it through one's home TV receiver?[17]

Defining Strategic Roles

Engineers at Ampex and in Japan opened the path to development of a VTR for the masses when they combined the helical scanner concept with other elements of the successful broadcast recorder. Some engineers at RCA, Toshiba, and Ampex saw the potential for developing that technology, but top managers in each firm decided to focus video efforts on the broadcast market. The opportunity inherent in the new design synthesis was most clearly perceived, instead, by three other firms—JVC, Sony, and Matsushita—that consequently became video pioneers after 1957.

RCA. At RCA in 1956, the engineer who had headed their unsuccessful attempt to build a broadcast VTR believed it was possible to reduce the size of the fixed-head machine and build a "home television magnetic-tape player."[18] But senior corporate executives were preoccupied instead by challenges from firms such as Zenith, Texas Instruments, and IBM in businesses far more important to the company's welfare at the time.[19] RCA's leadership in television receiver manufacture was under challenge by a creative Zenith marketing strategy. Meanwhile, the new semiconductor technology was eroding RCA's large vacuum tube business. At the same time, the sale of electronic computers was becoming a major industry, catching RCA by surprise. Management terminated work on consumer applications of videorecording in 1958. It was not until the early 1970s that RCA attempted (abortively) to innovate again in VTR design.

Toshiba. In Tokyo in 1959, Sawazaki of Toshiba demonstrated a working prototype of a VTR with helical scanner, stimulating widespread interest.[20] Soon afterward, a Japanese newspaper concluded that "the development of the VTR may change the world. Perhaps we should say that completion of the initial prototype is the first step."[21] Sawazaki tried to get support within Toshiba for a VTR development program to bring his invention to commercial use. But Toshiba's senior executives failed to understand the potential of Sawazaki's inventions and assigned him to other responsibilities in research management, effectively cutting off video development. In 1964, Toshiba and Ampex formed a joint venture that gave Ampex access to Japanese buyers and allowed Toshiba to enter the broadcast VTR market, while ceding control of product development to Ampex.[22]

Ampex. At Ampex, once the first successful VTR design was developed, Alex Maxey began experimenting with simpler methods of scanner design.[23] His work led to a helical-scan, transistorized VTR marketed by Ampex for closed-circuit use in 1962. Ampex, at that time, had produced some 75% of all television recorders in use throughout the world and seemed well-positioned to extend its dominant share into institutional and then consumer markets. But top managers wanted to reduce dependence on VTR revenues and instead pursued new markets in consumer audio products, computer peripherals, and businesses further afield from the company's core capabilities in magnetic recording. The Professional Products Division, which sold equipment to broadcasters, understandably gave top priority to assuring continued dominance in the studio equipment market.

New Pioneers Emerge

While Ampex, RCA, and Toshiba concentrated on broadcast users of VTRs and on other priorities in their diversified businesses, three Japanese firms specializing in consumer electronics began to focus on new applications for the VTR. The catalyst was the Japan National Broadcasting Corporation (NHK), which imported its first Ampex VTR in 1958. NHK invited engineers from Sony, Toshiba, Matsu-

shita, JVC and other Japanese firms to examine the VTR. At the same time, the Ministry of International Trade and Industry (MITI), which wanted to limit imports of the $50,000 Ampex machine, organized a VTR research group and began offering modest subsidies to firms willing to develop domestic versions. NHK engineers provided valuable technical data to those companies. At the same time, Sawazaki's achievements in design of the helical scanner became known to the technical leaders of JVC and Sony, with whom he had ties from school days.[24]

Sony. Sony Corporation rapidly became the leader in development of VTR technology for the mass market. Still a small firm in 1958, Sony had been established in 1946 by a college-educated electrical engineer, Masaru Ibuka, and a physicist, Akio Morita. Sony's founders from the beginning intended to apply "a mix of electronics and engineering to the consumer field."[25] They did not believe in market research. As Ibuka proclaimed in a 1980 interview, "You can't research a market for a product that doesn't exist."[26] These two strategies—applying "high technology" to the consumer market, and developing innovative products based on ambitious but specific targets and their personal convictions as to what would sell, rather than what existing markets seemed to demand—guided Sony's approach to development of video technology.

In retrospect, Sony in the 1950s was already assembling the skills that would prove important to success in home video: design and manufacturing know-how for magnetic recording equipment, television receivers, and analog semiconductor circuitry. Sony had introduced Japan's first sound magnetic recorder and tape in 1950, and followed with a TV camera in 1953 as well as the first Japanese stereo tape recorder for home use in 1955. Ibuka expressed interest in video recording as early as 1950 and, in 1951, encouraged Sony engineers to experiment with a fixed-head machine. In 1953, Nobutoshi Kihara, a talented mechanical engineer who had joined Sony in 1947, also began studying video recording technology. But Ibuka decided to concentrate on more immediate applications for the transistor and set aside VTR research until NHK's Ampex VTR reached Tokyo in 1958.[27]

In cooperation with NHK's central laboratory, Sony built a replica of the four-head Ampex VTR in only three months. The speed with which Sony engineers accomplished this feat led Ibuka to give Kihara a new challenge—to develop a home VTR, leaving the broadcasting market to Ampex. Ibuka, who had learned about Sawazaki's helical scanner in 1958, first set a target price of $5,000 and later reduced this to $500—1% of the price of Ampex's VTR.[28] Ampex provided some technical guidance to Sony in 1960 in exchange for transistor know-how, although the two companies ended their cooperation after a few months, due to a dispute over royalty payments for a critical Ampex patent. Sony demonstrated the world's first fully-transistorized VTR in 1961, using a two-head helical scan system that Ampex engineers still believed would never succeed commercially.[29] In 1962, Sony marketed its first VTR product, the PV-100.

The Victor Company of Japan (JVC).

JVC was another company with expertise in magnetic recording and television technologies, as well as a commitment to the consumer market. Initially a producer of audio equipment and phonograph records, JVC moved into television after World War II and introduced an audio tape recorder in 1956.[30] JVC began VTR research in 1955, on the suggestion of Managing Director Kenjiro Takayanagi, who had pioneered television broadcasting in Japan during the late 1920s. Takayanagi had worked in the NHK labs before joining JVC in 1946 to develop commercial television; with him came about 20 other TV engineers from NHK, creat-

ing JVC's considerable expertise in television technology.[31]

Takayanagi had become interested in video recording after hearing of Sarnoff's challenge to RCA engineers in 1951. He started collecting information on video and studying audio magnetic recording. In 1955, he asked a young engineer, Yuma Shiraishi, to study the concept of a "video disc" (because tape was of such poor quality), but he directed a return to magnetic video recording work in 1956, after a visit to RCA following the debut of Ampex's VTR.[32] Like Ibuka, Takayanagi learned about Sawazaki's helical scanners in 1958.

JVC's engineers familiarized themselves with the Ampex technology by copying NHK's VTR for in-house study in 1958–1959.[33] The four-head Ampex system, Takayanagi concluded, was too expensive and complicated for home use.[34] Ampex was then refusing to grant patent rights, further encouraging JVC to work on a two-head helical scanner and electronic circuitry that would bypass the key Ampex patents. In January 1960, JVC was the first to announce the design of a two-head VTR. The company brought its first helical-scan VTR to market in April 1963, selling the first units to the Tokyo police.

Matsushita.

The third Japanese consumer electronics company to initiate VTR technical development in the late 1950s, Matsushita Electrical Industries, was destined to become the dominant producer in the global mass market of the 1980s. Matsushita, which dates back to an electric-plug factory established in 1918, gradually diversified into a variety of electrical equipment and home appliance markets, including audio tape recorders in 1958 and television receivers in 1960. In 1953, Matsushita Electric acquired a 50% ownership of JVC, but elected to operate it as a fully-independent enterprise.

In 1957, the company was still headed by its founder, Konosuke Matsushita, who, like

Ibuka and Takayanagi, displayed an early and avid interest in video recording. He encouraged Tetsujiro Nakao, then head of the central research laboratories, to support research on VTR design. When NHK asked the Matsushita company to develop a video recording head, Nakao gave the task to Dr. Hiroshi Sugaya, a young physicist who had joined the firm three years earlier. Sugaya had been studying audio heads and had already developed a prototype for NHK. MITI also provided him with a grant in 1958 to subsidize research on the video head. To test it, Sugaya needed a VTR; this prompted Matsushita to build a prototype modeled after the Ampex design.

But Matsushita's orientation as a consumer-products company led Sugaya and other company engineers to work on cheaper models, potentially for home use. MITI again provided modest subsidies in the early 1960s to help fund this effort. After surveying the available technologies, Sugaya decided in 1960 that a two-head design employed at JVC was best. Matsushita produced its first commercial VTR in 1964, to coincide with the Tokyo Olympics.[35]

The Pioneers' Divergent Paths

By the early 1960s, Sony, Matsushita, and JVC had established VTR development programs focused on consumer markets and based on what proved to be the winning technical concept—the 2-head helical scanner. In all three firms, top executives proved ready to invest in a technology whose commercial benefits were distant in time and highly uncertain. In contrast, the continuing technical effort at Ampex lacked a market focus, RCA had abandoned video technology development relevant to a mass market, and Toshiba had ended VTR technical development entirely.

How can these diverging paths be explained? Both the prevailing strategies of the firms and the dispositions and capabilities of top management appear to have been impor-

tant. For JVC, Matsushita, and Sony, the prospect of an innovative mass-market product based on magnetic recording fit well with existing commitments, strategies, and aspirations. And leaders in all three were ready to bet on the long-term. While RCA and Ampex had earlier demonstrated willingness to persist in risky long-term developments, in the late 1950s managers in both firms were preoccupied with other strategic challenges.

Those firms ready to make a long-term commitment were also well-equipped to do so wisely. In Sony, Matsushita, and JVC, extremely capable technical leaders operated at or close to the top of the management hierarchy. Thus general managers were able to base their choices on sound technical judgments—which sometimes were lacking in RCA and Ampex in the 1960s. Technical leaders, furthermore, were in tune with corporate strategies and could guide development efforts that effectively served the firm's long-term commercial interests.

RCA, Ampex, and Toshiba returned to pursue the vision of a mass market for VTRs before 1975, but in each case it was both too late, and too little. The choices made in 1959–61 to set direction for VTR technical work in all six companies had a major impact on their differing fates in the 1980s.

DEVELOPING PRODUCTS FOR THE CONSUMER MARKET

By 1963, competitors had breached the barriers protecting Ampex's VTR know-how, but its technological leadership was still pronounced. A decade later, Sony was the technological leader in the VTR marketplace, closely followed by JVC and Matsushita, sharing the leading-edge technology of the U-format ¾ inch videocassette recorder.[36] Among the main ingredients of this reversal of fortunes was a clear difference in the ways in which the rival pioneers managed development of the technology.

Market Opportunities

In the early 1960s, rival firms targeted VTR technical development on one or more of the three main markets: broadcasters, "closed circuit television" (CCTV) users, and households.[37] VTR producers had a finely-detailed sense of user needs and of the character of demand in the broadcast market, but only a vague sense of the needs of other users. Broadcasters adopted Ampex's transverse-scan two-inch format as their standard for studio equipment, creating a profitable market that Ampex dominated.[38]

The CCTV market began to open up following the introduction of Sony's PV-100 and the comparable Ampex VTR in 1962, but the real nature of the market remained to be discovered. Sony claimed in 1962, for example, that the PV-100 was applicable in "various fields—technological, industrial, educational, medical, sports, etc." and "a forerunner to consumer VTRs." But, in truth, like Sony's first audio recorder, the first-generation helical scan VTR was a technical wonder in search of users.

VTRs for the Home. The first attempt to develop the consumer market came in 1965 when Sony introduced its "CV" model (using half-inch tape) and Ampex offered a new one-inch design, both priced below $1,000. Both Sony and Ampex merchandised their new products as a means for "instant" home movies. Other applications were in prospect as well; a feature story in *LIFE* magazine pointed out their use to record TV programs, while noting "this would not be cheap since an hour's tape now costs up to $60." *LIFE* also reported "talk about a time when taped recordings of Broadway musicals and plays will be made and either sold or rented."[39]

While some journalists were enthralled, consumers remained apathetic; few bought the new VTRs. But the innovators learned that other users liked these machines. Schools, companies, hospitals, and other users of 16 mm. films—the "audio-visual market"—began to buy VTRs for training, sales promotion, and similar uses.

The experience of failure in 1965 taught Sony and Ampex valuable lessons about the latent consumer market. A successful home VTR, they concluded, would have to be no more expensive than the 1965 models, yet easier to use and more reliable, and also capable of recording color images.[40] All of these characteristics were embodied in the next generation of VTRs, first marketed in the early 1970s. More compact than the CV and its brethren, they used tape packaged in easy-loading cartridges or cassettes, and they produced color images.

Birth of the VCR. In April 1969, Sony, again the first-mover, announced the development of a "magazine-loaded" VTR using one-inch tape. JVC soon disclosed that it had designed a cartridge-loaded machine using half-inch tape. Meanwhile, Ampex, which was being pushed out of the low-end audio-visual markets by Japanese rivals, took a bold step. Attempting to leapfrog the evolving VTR products, Ampex took aim at a potential high-volume market for an "audio-visual scratchpad"—a high-performance, light-weight, battery-operated recorder and camera dubbed "Instavideo." Within a year, Ampex engineers had produced a prototype which drew rave reviews at showings to the trade and press in 1970–71. William Roberts, Ampex CEO, doubted the company's capabilities in mass production and turned, instead, to Toamco, the Toshiba-Ampex joint venture in Japan, to produce Instavideo. But Toamco was also inexperienced in high-volume production, having previously made only studio equipment for broadcasters. Because the two companies never solved the problem of transferring the design technology across the Pacific, Ampex never did ship Instavideo to customers.[41]

With the introduction of the new generation of VCRs, the trade and business press anticipated the imminent appearance of the "next big consumer electronics market." They dubbed this prospective industry "cartridge television," because all the equipment employed "cartridges" or "cassettes"[42] to package the medium on which programs were recorded. Enthusiasts forecast that millions of households would soon buy videoplayers and buy or rent program material distributed in "cartridges."[43] But this formulation of the innovation embodied a "chicken and egg" problem that would prove difficult to solve: consumers would not buy players unless there was attractive "software" (programs); producers would not invest substantially in software until there was a large base of installed equipment. Nevertheless, in 1970–72, dozens of firms in Europe, North America, and Japan announced plans to offer magnetic videoplayers using cassettes or cartridges. In addition, RCA and CBS were promoting rival electro-optical playback technologies, called "Selectavision" (RCA) and "Electronic Video Recording" (CBS), while Teldec and Philips in Europe added to the variety by introducing videodisc technologies.

VCR Standards. Sporadic attempts to agree on an international standard for home video foundered on the diversity of technologies and corporate interests. But Sony, seeking to strengthen its hand against foreign rivals, invited JVC and Matsushita to join in establishing a standard for cassettes using ¾ inch tape. By December 1970, the three firms had agreed on a format now familiar as the "U-Matic." Sony marketed its version in late 1971 at a price of $1,000.[44]

As had happened in 1965, none of the many products offered as home videoplayers in the early 1970s attracted enough buyers to be viable as consumer products. But the ease of use and improved picture quality of new VCRs like the U-Matic stimulated growth in the audio-visual market, thus adding to the attractiveness of, and availability of resources for, further development of VCR technology.

Design for the Mass Market

In 1970, engineers at RCA, Ampex, and the pioneering Japanese firms had access to all the technologies required to design a VCR for the mass market. The remaining tasks were to synthesize the various design elements and to create a manufacturing facility capable of low-cost volume production. Because of a financial crisis brought on by heavy losses in other businesses, Ampex killed the troubled Instavideo program and withdrew from the race. RCA, Matsushita, Toshiba, Sony, and JVC renewed their efforts to produce a new generation of home VCR designs, although only Sony and JVC would achieve products truly successful in the marketplace.

RCA. Top management at RCA had committed the company in 1969 to leadership in "the next big innovation in consumer electronics," a home videoplayer they dubbed "Selectavision." The technical vehicle they chose initially was "Holotape," a laser-based marvel from the Princeton Laboratories. In 1971, with the Holotape project in deep technical trouble and the threat of imminent commercialization of Japanese VCRs built to a common standard, RCA pressed forward with videodisc development and attempted simultaneously to commercialize its own innovative VCR design.[45]

Low-cost mass production of precision scanner mechanisms was certain to be a key factor in any successful consumer VCR innovation. RCA's senior managers doubted their company's ability to perform that task. The VCR design that RCA engineers proposed in 1971 addressed that concern with a scanner designed to eliminate tape threading by scanning inside the cassette. All the precision mechanical elements were contained in a single scanner-transport module to be produced by Bell and Howell. But like Ampex and Toamco, RCA and Bell and Howell never could bridge the gap between design and manufacture. In 1974, RCA scrapped its VCR project and went looking for a source in Japan.[46]

Matsushita. During the same period, 1972–74, Matsushita failed in repeated efforts to build a consumer video market in Japan. In 1973, after consumers had shown no interest in a low-cost version of the ¾-inch U-format VCR, Matsushita introduced a model using half-inch tape on a one-reel self-threading cartridge. A new plant in Okayama, dedicated entirely to VTR production, was staffed with 1,400 workers and equipped to produce 120,-000 units in 1974. But the product was expensive—$1,300—and recordings were limited to 30 minutes. Sales were a fraction of forecasts, and Matsushita had to cut back employment drastically. But management chose to keep the main cadre of manufacturing engineers intact, assigning them to "research," thereby preserving an important resource that would be mobilized in 1976 to start production of machines built to JVC's VHS design.[47]

Toshiba. After the collapse of the Instavideo partnership with Ampex, Toshiba joined forces with Sanyo to design and market a VCR for home use. Their product, based on a 3-head helical scanner, reached the market in Japan in September 1974. A second-generation design, extending playing time to two hours, was introduced in May 1976. But even the improved design was surpassed in cost-effectiveness by rival 2-head machines and was soon withdrawn from the market.

Sony. Unlike the opportunistic innovations that RCA, Matsushita, and Toshiba attempted, the VCRs next designed at Sony and JVC resulted from systematic efforts in which they carefully considered user needs. While Sony moved to establish the innovative U-Matic in the institutional (audio-visual) market, Chairman Ibuka pressed for the design of a smaller and cheaper VCR for home use. According to the project chief, Kihara, Ibuka "gave us the actual target and the dimensions of the target, and we knew we had the technology to achieve this." Kihara felt there were "at least ten" different ways of building a home VCR, so his team tried each one, working in small groups operating in parallel. After about 18 months, Ibuka and Morita helped to select the best prototype and the design team spent another 18 months working with production engineers to prepare for manufacture. Most of Kihara's group then followed the project into the production engineering department to insure a smooth transition into mass production.[48]

JVC. JVC, like Sony, had been aiming at the home market for years. It priced a model introduced in 1967 at about $630, although the open-reel format was awkward and the machine did not record in color. A cartridge model and a smaller open-reel VTR were shown in 1969, but these met the same fate as similar VTRs offered by Sony and Matsushita and flopped as consumer products.[49] Throughout this period there was a close connection between design engineers and manufacturing operations. At first, JVC produced VTRs in a small manufacturing section within its central R&D Department. When a separate VTR Division was organized in 1967, design and production continued on one site.[50]

In 1971, while JVC's Video Division joined Sony and Matsushita in adopting the U-Matic standard for the audio-visual market, it also launched a new R&D project that, five years later, would result in the VHS. Yuma Shiraishi, then manager of VTR development for JVC, at first assigned only two of his ten engineers to the new project. He provided broad guidelines: Develop a VCR to sell at around $500 and use as little tape as possible, without degrading picture quality. He also kept the effort secret from the board of directors, including Takayanagi, then the head of R&D for the company. The Video Division was operating at a loss and some senior executives responsible for finances wanted to curtail new product development. Only after engineers had completed a promising prototype in 1973 did Shiraishi reveal the VHS project to Takayanagi and other top executives.[51]

The first prototype, ready in 1972, used 1/2-inch tape, as Shiraishi wanted, but recorded for only one hour and produced a poor picture. Shiraishi added a third engineer in 1973 and then more personnel as the team came closer to a commercial product, carefully selecting staff with expertise in design and manufacturing, as well as marketing. He asked his team to determine exactly what consumers wanted in a home VTR and then develop the necessary technology. The members viewed this as a "problem in logic" and wrote down a matrix of technical difficulties and solutions. This matrix served as a roadmap for the project and as the proposal that Shiraishi used to get top management to back the effort.[52]

Moving to Market. While engineers at JVC worked in secret on the VHS, Sony worked on two fronts simultaneously to prepare to launch the Betamax. Anticipating a mass market for VCRs, Sony had built a "greenfield" facility in Kohda. To build toward a volume production capability at Kohda, Sony trained staff first to produce the established U-Matic product and then began experimental production of the Betamax in mid-1974.

In parallel, Sony sought to persuade other firms to adopt the Betamax design. In December 1974, Sony management showed the Betamax prototype to Matsushita executives. Matsushita ignored this appeal; not only did top managers not want to acknowledge Sony's

leading position, but they felt the one-hour recording time of the Betamax was too short to guarantee wide consumer acceptance. JVC and RCA also declined to adopt the Beta format.[53]

Having invested already in production tooling, Sony executives were unwilling to change the design. The first Betamax model was offered for sale in April 1975. A few days later, JVC revealed to its parent firm, Matsushita, that a small team of engineers had been secretly developing a two-hour "Video Home System" that was near completion.[54] JVC began marketing the VHS in fall 1976 and Matsushita adopted it in early 1977 as its standard for the consumer market.[55]

LEARNING AND THE STRATEGIC MANAGEMENT OF TECHNOLOGY

In the late 1970s and early 1980s, Matsushita, JVC, and Sony were the world's leading producers of VCRs for the home, accounting for 80% of output in 1978 and 57% in 1984 (see Table 1). Ampex, RCA, and Toshiba, although they had been better positioned in 1960 to lead both in technical development and commercialization, profited only modestly or not at all from the success of the VCR.[56] These contrasting outcomes had a major impact on the prosperity of those corporations in the 1980s (see Table 2).

Matsushita's astute actions to market VHS machines under its own brands and as a supplier to other OEMs made it the leader in VCR production, despite the failure of earlier attempts to market home VTRs. But the boom in home video had an even greater impact on JVC and Sony, where video products became the dominant source of revenue in the late 1970s and early 1980s. JVC grew from a relatively small firm to become one of the global leaders in consumer electronics, with some $3 billion in annual revenues during the mid-

TABLE 1. *Half-Inch VCR Production Shares (1976–1983)*

	1976	1978	1980	1983
VHS Group:				
Matsushita	28.7	35.8	28.7	28.7
JVC	8.7	18.8	18.3	16.4
Hitachi				11.1
Sharp				8.5
Mitsubishi	1.4	5.1	18.8	3.4
Tokyo Sanyo				3.6
Others				3.3
VHS Total:	38.8	59.7	65.8	74.9
Beta Group:				
Sony	55.9	27.9	22.4	11.8
Sanyo				7.7
Toshiba	5.3	12.5	11.1	3.6
Others				2.0
Beta Total:	61.2	40.4	33.5	25.1
TOTAL:	100%	100%	100%	100%
Units (1000)	286	1,470	4,441	18,217

Sources: *Nikkei Business*, June 27, 1983. *Nihon Keizai Shimbun*, December 21, 1984.

TABLE 2. *Revenues of the Six Firms: 1960–1985 (millions of dollars)[a]*

	1960	1970	1980	1985
Ampex:	68	296	469	480[d]
Japan Victor[b]				
total:	55[d]	293	1,810	2,941
video:	nil	2	1,068	1,912
Matsushita[c]				
total:	256[d]	2,588	13,690	17,120
video:[d]	nil	6	800	3,000
RCA	1,495	3,326	8,011	8,972
Sony[b]				
total:	52	414	4,321	5,357
video:	nil	17	973	1,982
Toshiba[b]				
total:	390[d]	1,667	7,738	12,598
video:[d]	nil	n.a.	n.a.	400[d]

a. Yen values converted at the following rates:
    1960 and 1970: 360 = $1.00
    1980 and 1985: 200 = $1.00
b. Does not include unconsolidated subsidiaries
c. Excludes Japan Victor and certain other subsidiaries
d. Authors' estimates
Sources: Annual reports: *Nihon Keizai Shimbun, Nikkei Sangyo Shimbun;* Matsushita Video Division, Ministry of International Trade and Industry, *Kikai tokei nenkan* (annual report on machinery statistics); Toyo Keizai, *Kaisha shikiho* (company quarterly reports).

1980s. In contrast, revenues were stagnant in the 1980s at RCA and Ampex. By 1986, both companies had lost their independence, having been taken over by larger conglomerates.

These dramatic differences were rooted in differences in the management of technological development, rather than in "first-mover advantages" or other aspects of the management of innovation. Within the span of a "window of opportunity" lasting three or four years, the precise sequence in which firms entered the market was not consequential. What did matter was the technological capacity possessed by the entrants in 1976 when the window opened. That, in turn, was a result of steps taken years earlier, actions in which differences in established corporate strategies and differences in the firms' capacities to develop and produce innovative products led them to possess such sharply differing capabilities by 1976.

Learning by Trying

The first commercial VTRs using helical scanners, introduced in 1962 and 1963 by Sony, Ampex, and JVC, had marked the first steps on a path that would lead to a product for the masses. But a lengthy journey still remained. Table 3 illustrates the magnitude of technical progress required: an eleven-fold increase in recording density and 71% reduction in equipment weight, combined with 88% reduction in product prices, while enhancing video performance and user convenience. Ampex had achieved its original VTR innovation as a "breakthrough" invention, seemingly in one giant leap.[57] The VHS and Betamax came, in-

TABLE 3.   *VTR Technical Progress 1962–75*

	Sony PV-100 1962	Sony Betamax 1975
Tape Utilization (Sq. meters/hour)	19.8	1.8
Video	Monochrome	Color
Tape Loading	Manual threading (reel-to-reel)	Automatic threading (cassette)
Price to User	$12,000	$1,400
Weight (pounds)	145	41

stead, as the culmination of an evolutionary synthesis of solutions to diverse design problems. They were the tangible results of fifteen years of "learning by trying."

A "System" of Technology. The VCR is a "systems" product, its performance dependent on a combination of diverse technologies in electronic circuitry, magnetic materials, and electromechanics. The pioneers enhanced product performance by skillfully integrating novel component technologies and persistently seeking new increments of performance in available technologies. Given these characteristics, product performance was not only a question of design, but was intimately bound up in manufacturing technique. Cost reduction depended on developing new techniques for high-volume and high-precision assembly of mechanisms, the mass production of circuit boards, and the precision fabrication of components of the magnetic and mechanical subsystems. The path to competitive advantage lay in incremental design improvements and then the integration of design and manufacturing.

Four generations of helical-scan VTRs reached the market between 1962 and 1976. Each new design reflected continuity with earlier VTRs. Yet, in each generation, companies incorporated novel elements (in circuitry, scan-

ner designs, head fabrication, etc.) that enhanced performance. Achieving far greater recording density was a central element in this record of technical progress. Not only did it bring economies in use—Sony's 1962 model used a dollar's worth of tape every minute—but it reduced the size and cost of the machine itself.[58] Higher density, like other advances in the product, was achieved gradually, through a combination of improvements in scanner design, head fabrication, and tape manufacture.

Sources of Technology. The necessary innovations in constituent technologies came from many sources. Vendors of magnetic materials contributed important solutions, such as high-energy recording tape, developed using different technologies at DuPont, Fuji Film, and TDK. Each of the pioneering VTR producers gradually acquired experience in fabricating magnetic materials and mechanical parts—skills vital for meeting the increasingly precise requirements of heads and scanners.[59] And they borrowed from each other. Ampex's split-drum scanner (invented in 1964) was later licensed by all but a few VCR producers. Exchange of manufacturing information between Philips and Sony led the latter to change its method of head fabrication.

In addition to techniques developed specifically for videorecording, companies borrowed from the design of other consumer products. For example, they adapted the concept of the tape cassette, plus some details of tape transport design, from the audio recorder. They transferred techniques of solid-state circuit design and assembly from TV receiver manufacture, in which the Japanese were becoming world leaders.

In summary, learning was the central task of pioneering in video technology. Each pioneer had to learn what functional performance consumers required as well as how to achieve it in design. What may have seemed to some, in 1976, to be a dramatic and sudden innovation was really the result of a series of steps. The successful designs thus emerged from an iterative process by which the pioneers came to understand not only the full range of capabilities inherent in the new technical synthesis, but also how to realize those capabilities in commercial-scale production and what value the market would place on them once achieved. Each successful organization learned these lessons largely by experience, that is, by offering products to users, studying their responses, and trying again until they got it right—in effect, "learning by trying."

A Key Difference: Strategic Management of Technology

"Learning by trying" was made possible by the clarity of focus and the consistency of strategic management at Sony, JVC, and Matsushita. The conduct of VCR development activities in all three firms was broadly similar in style: persistent yet pragmatic, and highly focused yet flexible. Over two decades, the three Japanese companies pushed forward, step by step, in design and manufacturing technology. In contrast, lacking consistent strategic direction from the top, the leading American rivals—Ampex and RCA—failed to sustain a focus for technical development and fell short.

Strategic Consistency. Strategic consistency, then, differentiated the successful VCR pioneers from the others. Leadership provided by general managers was essential to achieving consistent direction over a lengthy period of development. At Sony, top managers explicitly and consistently guided the direction of VTR development efforts. At Ampex, top management direction was sporadic and inconsistent over time; engineers on their own failed to develop a fruitful direction for technical development.

By the early 1960s, JVC, Sony, and Matsushita all had realized that consumer use could become the primary application of the helical-scan machines they were developing. As a result, the technical agenda that shaped their efforts included specific, ambitious goals for reductions in machine size, cost, and tape consumption. In contrast, Ampex (by then paired with Toshiba) and RCA gave priority in video development to products for broadcasting companies and government uses. Because management took no initiative to define technical agendas clearly targeted on the mass market, development work was more opportunistic, seeking "breakthroughs" that remained elusive because of the need to develop manufacturing skills in tandem with advances in product design.

Clarity of commitment to pursue the potential consumer market shaped critical choices among uncertain technical alternatives. As a result, by 1965 Sony, JVC, and Matsushita all had converged on the basic design of the 2-headed scanner that emerged as the dominant design in the marketplace after 1975.[60] Ampex, like RCA earlier, flirted with fixed-head scanner designs, before committing to the helical approach in 1965. But the Ampex engineers persisted in believing that inherent cost factors would favor single-head designs, which they continued to promote.

Thus, in contrast to the steady incremental progress of the Japanese, Ampex's efforts at developing a mass-market VTR progressed with leaps and lags. Bold technical steps, in 1964 and again with the Instavideo design in 1971, were insufficient, in themselves, to break open this different marketplace. A similar pattern emerged at RCA, which looked for the grand technical solution, finally fixing on the ill-fated Videodisc.[61]

Perhaps most important, Ampex managers were influenced by an erroneous conception of how to gain competitive advantage in the mass market. A "breakthrough" mentality prevailed, as engineers and managers looked for ways to replicate the dramatic success of the broadcast VTR in another bold technical stroke. In the 1950s, Ampex had protected the first VTR design by a strong patent long enough to establish an enduring lead in design capability for studio equipment. Manufacturing prowess was irrelevant to that low-volume market niche. But, while Ampex repeatedly sought to use high-performance designs to set *de facto* standards in other markets, the company failed to build the manufacturing capability that clearly would be called for in a mass production regime.

Underlying the persistence of the winning Japanese pioneers was a conviction that VTRs would be the next major consumer electronics product after color television.[62] Managers and engineers confidently planned their work in relation to a distant time horizon. In 1958, when Ibuka articulated Sony's ultimate goal—a compact, inexpensive VTR suitable for the home—Sony's engineers were confident they would achieve it. In a 1980 interview, Kihara recalled the assessment he had made at the outset:

> By transistorization, the size of the unit could be reduced dramatically; also, through our own experience and experiments, we felt that the two-head system was simpler than the four-head system and therefore the mechanism could be made simpler; . . . also very important was the factor of recording density. We were able to get high density in a narrow track even at this stage. Expecting further developments in the videotape as well as the recording head, the recording density [could be increased], thereby reducing . . . the width of the tape. All of these factors contributed together to give us the conviction that we could certainly work toward a consumer product.[63]

Technical teams, confident that they could reach the ultimate goal, pragmatically seized opportunities that developed along the way. Sales to the audio-visual market served as a stepping-stone toward developing a consumer product, which they continued to pursue.

Engineers at Sony and JVC chose to view home-VCR development as a "problem in logic"—to be solved by setting down all the performance specifications a product needed to succeed in the home market and then identifying and mastering the necessary technologies, using R&D resources to solve technical problems related to specific design paramaters. Ampex and RCA, in contrast, pursued too many different product ideas and treated shortcomings as "failures" rather than as indicators of targets for the next stage of work. The more focused but still adaptable Japanese approach provided direction for their VTR projects and, in the case of JVC, helped to persuade skeptical members of top management that a home VCR was an achievable goal.

Organizational Underpinnings. An important underpinning of the strategic consistency achieved by JVC, Sony, and Matsushita was the high degree of organizational stability during the lengthy period of development. All three firms maintained stable technical teams with good connections to senior executives. For example, Kihara led Sony's design group for every new VTR generation. Sony's VTR manufacturing was centered in one factory until 1974 and then management took care to transfer skills to a new workforce by production of the proven U-Matic design. Moreover, at Sony, the same two top managers, Ibuka and Morita, were closely involved in all critical decisions.

Instability characterized Ampex's VTR organization. Alex Maxey, designer of Ampex's earliest helical scanners, left the company in 1964, largely unappreciated. The locus of helical-scan VTR activity moved to Chicago in 1965 and back to California in 1970. Manufacturing operations moved from California to Illinois to Japan. Four different general managers were responsible for the helical-scan VTR business between 1965 and 1971. George Long, CEO of Ampex at the birth of the VTR, had been fired in 1961 and his successor was replaced in 1971—both during periods when rival pioneers were making fateful decisions about product development and innovation.

Facilitating a persistent and flexible development of the technology was the fact that the Japanese firms contained senior executives with high levels of theoretical understanding of both television and magnetic recording technologies.[64] This gave managers and engineers the confidence to strive for continual improvement in their products, with better video quality, smaller mechanisms, and lower prices. Because the resources that companies committed to development were modest, the efforts proved to be sustainable over a long period and in the face of occasional financial adversity. Despite the "imperfection" of designs produced at the first, second, third, or even fourth tries, Sony, JVC, and Matsushita repeatedly opted to risk failure with probes of the market that produced valuable design and marketing insights and also built up manufacturing experience.

CONCLUDING COMMENTS

In the case of the VCR, as in some other industries of current concern, Japanese firms achieved outstanding success in a global marketplace for products founded on basic technologies originated in the U.S. or Europe. But the Japanese leaders in VCR production did more than simply copy Western technology or adapt it to low-cost mass production; as we have shown, they were technological pioneers who had entered the field early and persisted. Sony and JVC, especially, exhibited great insight and imagination as they pushed a difficult technology far beyond the limits thought practical by most U.S. and some Japanese competitors.

The story of the VCR demonstrates the magnitude of the reward available to firms ready to pioneer in a new technology for uncertain markets. But it also illustrates how limited is the "window of opportunity" for a firm seeking to participate in the development of a complex and novel technology. The foundations of future competitive advantage are created by the timing of the decision to pioneer, the choice of direction for development, and the skill applied to manage the learning process that ensues.

There is nothing distinctively Japanese about successful pioneering in technology. Japan has no monopoly of companies who followed the invention of a robust technology into an industry and eventually dominate it with superior engineering or manufacturing capabilities. Yet it is worth calling attention to the unusual frequency with which the current success of Japanese firms rests on their role as early technological pioneers. Examples include products as diverse as small cars (the Datsun line dates back to before World War I), digital computers (Toshiba and Tokyo University began research in 1950 and Japanese engineers completed the world's first transistorized computer in 1956), the numerical-control machine tool (Hitachi, Seiki, and Fujitsu developed one jointly in 1958), small-scale computer printers (Epson introduced its first model in the late 1960s), and semiconductor memory (Japanese firms began domestic shipments of high-quality chips in the early 1970s). It was also a Japanese company, Hitachi, that made a historic (and, apparently, a successful) attempt to apply disciplined engineering and manufacturing practices to large-scale software development by opening the world's first "software factory" in 1969.[65]

Thus, the story of pioneering in VCRs is not an exceptional case. Strategic experimentation and disciplined learning at critical, early stages in the evolution of particular technologies, combined with stable teams of managers and engineers, has no doubt affected, if not determined, the current performance of a large number of firms in Japan—perhaps more than elsewhere, although we leave the exploration of that question to further research.

REFERENCES

1. These six do not exhaust the list of those who attempted to bring videorecording into the home, but they shall suffice for purposes of our analysis. Other firms, mainly in North America, also sought to develop videorecording for consumer use in the early 1960s, but none of those opportunistic ventures persisted for more than a few years. See Richard S. Rosenbloom and Karen J. Freeze, "Ampex Corporation and Video Innovation," in Richard S. Rosenbloom, ed., *Research on Technological Innovation, Management and Policy,* Vol. 2 (Greenwich, CT: JAI Press, 1985). In the late 1960s, N.V. Philips, the European electronic giant, and Cartridge Television Inc. (CTI), an American venture capital start-up, began to pursue the same goal. Philips was the innovator in the European VCR market in 1972, but failed to keep pace with the advances made in Japan in the next 5–7 years. Cartridge Television was the only American firm to produce a VCR for consumer use, but its product was commercialized too early and CTI fell into bankruptcy and was liquidated in 1974.

2. For reviews of the treatment of technology and strategy in these literatures a decade ago, see Richard S. Rosenbloom, "Technological Innovation in Firms and Industries: An Assessment of the State of the Art," in P. Kelly and M. Kranzberg, eds., *Technological Innovation: A Critical Review of Current Knowledge* (San Francisco, CA: San Francisco Press, 1978); and Alan Kantrow, "The Strategy-Technology Connection," *Harvard Business Review* (July/August 1980), pp. 16–21. Pavitt and Graham illustrate the current interest in "strategic" aspects of technology management; see, Keith Pavitt, "Technology, Innovation, and Strategic Management," in J. McGee and H. Thomas, eds., *Strategic Management Research* (London: John Wiley & Sons,

1986); and Margaret B. W. Graham, *The Business of Research: RCA and the Videodisc* (Cambridge: Cambridge University Press, 1986). Similarly, the strategy literature exhibits an increasing concern with issues of technology management; see John Friar and Mel Horwitch, "The Emergence of Technology Strategy: A New Dimension of Strategic Management," in Mel Horwitch, ed., *Technology in the Modern Corporation* (New York, NY: Pergammon Press, 1986). For example, in his most recent book, Michael Porter devotes a chapter to "Technology and Competitive Advantage," a topic treated sparingly in his previous *Competitive Strategy;* see, Michael E. Porter, *Competitive Advantage* (New York, NY: The Free Press, 1985), and Michael E. Porter, *Competitive Strategy* (New York, NY: The Free Press, 1980).

3. Michael E. Porter, *Competitive Advantage,* op. cit., chapter 5.

4. David Teece, "Profiting from Technological Innovation," *Research Policy,* 15 (1986).

5. The character of these choices is discussed in Richard S. Rosenbloom, "Managing Technology for the Longer Term: A Managerial Perspective," in Kim B. Clark, Robert H. Hayes, and Christopher Lorenz, *The Uneasy Alliance* (Boston, MA: Harvard Business School Press, 1985), especially pp. 311–314. Hamilton discusses the strategic options offering increased focus and commitment for a firm addressing an emerging technology. His "window" strategy corresponds to the idea of a "monitor's role" advanced here. See William F. Hamilton, "Corporate Strategies for Managing Emerging Technologies," in Mel Horwitch, ed., *Technology in the Modern Corporation* (New York, NY: Pergammon Press, 1986), pp. 111–113. Useful managerial discussions are found in E.A. Gee and C. Tyler, *Managing Innovation* (New York, NY: Wiley Interscience, 1976); and George Pake, "From Research to Innovation at Xerox," in Richard S. Rosenbloom, ed., *Research on Technological Innovation, Management and Policy,* Vol. 2 (Greenwich, CT: JAI Press, 1986).

6. See Rosenbloom, (1985), op. cit.

7. Betamax is a trademark of the Sony Corporation.

8. VHS is a trademark of Japan Victor Corporation.

9. For comparison, it should be noted that total output of VTRs in Japan ranged between 114,000 and 134,000 units annually from 1972 through 1975, averaging 124,000.

10. A VTR is a system combining magnetic, electro-mechanical, and electronic components. At the core are the magnetic elements—the recording medium (coated tape) and the transducer (head); unless they perform well, the system will not. Also important are the scanner and associated transport mechanism, through which the head scans the medium to "read" or "write," and the electronic circuitry, which both processes the signals to and from the head and controls the mechanical actions of the scanner and transport.

11. The patent was issued in 1959.

12. Ampex had been founded in 1944 to manufacture motors and generators for naval radar systems. After the war, the company began to make audio tape recorders and military instrumentation equipment. For a description of Ampex and the invention of videorecording. See Rosenbloom and Freeze, op. cit., pp. 115–122.

13. The Ampex VTR, its circuits designed entirely with vacuum tubes, was massive, complex, and expensive, filling a large console and two equipment racks the size of refrigerators. But it sold like the proverbial hotcakes to broadcasters.

14. Sony, in 1959, was smaller than Ampex but had already developed substantial expertise in semiconductor circuit design for both audio and video.

15. Television broadcasters used the VTR to replace the expensive and awkward kinescope recording method, generating cost savings that repaid the price of the VTR in 11 months.

16. *Wall Street Journal*, April 16, 1956, p. 16. Long was fired in 1960 when Ampex experienced a drop in revenues and a net loss for the year.

17. *New York Times*, April 22, 1956, II. p. 13.

18. RCA Industry Service Laboratory, RB-98, April 12, 1957.

19. John Burns, an "outsider," was appointed President of RCA in 1957 and set about to reorient R&D in the company and diversify its businesses. He was forced out by David Sarnoff within 4 years. Graham, op. cit., pp. 81–83.

20. Sawazaki excited interest in helical scanning among parts of the American technical community with a paper presented at an SMPTE conference in California in May 1960. Norikazu Sawazaki, et al., "A New Videotape Recording System," *Journal of the SMPTE*, 60 (December 1960): 868–871. Following this, RCA initiated a small program for development of helical-scan VTRs, but it was aimed entirely at an emerging government market for compact recorders that could be used in space.

21. NHK, Nippon Hoso Kyokai (Japan National Broadcasting Corporation), ed., *Nihon hosho shi* (History of Japanese Broadcasting), Tokyo, 1965.

22. *Nihon Keizai Shimbunsha*, ed., *"Gekitotsu! Soni tai Matsushita—bideo ni kakeru soryokusen"* (Crash! Sony versus Matsushita—the all-out war waged on video), (Tokyo, Nihon Keizai Shimbun, 1978), pp. 29–30, 158–161; and Rosenbloom-Cusumano interview with Norikazu Sawazaki, September 7, 1985.

23. A self-educated engineer who had joined Ampex in 1954 to work on the VTR project, Maxey initiated and championed work on helical-scan technology, with no knowledge of Sawazaki's invention. The prototype he built in late 1957 was the first video recorder using the helical format to produce adequate television images. See Rosenbloom and Freeze, op. cit., pp. 130–142.

24. NHK, op. cit., pp. 475, 645, 748.

25. Nick Lyons, *The Sony Vision* (New York, NY: Crown Publishers, 1976), p. 5.

26. Rosenbloom interview with Masaru Ibuka, July 24, 1980.

27. Ibid.; Masaru Ibuka, "How Sony Developed Electronics for the World Market," *IEEE Transactions on Engineering Management*, 22 (February 1975): 17; Masaru Ibuka, *"Watakushi no rireki-sho"* (My career), *Nihon Keizai Shimbunsha*, 18 (1963): 42–43.

28. Ibuka and Sawazaki interviews. This critical period is discussed also in Lyons, op. cit., pp. 150–151.

29. Lyons, op. cit., pp. 203–204; Rosenbloom interview with Masahiko Morizono, Sony Senior Managing Director, July 1980.

30. Nihon Bikuta Kabushiki Kaisha (Japan Victor Company—JVC), *Nihon Bikuta 50-nen shi* (A 50-year history of Japan Victor), (Tokyo: JVC, 1977), pp. 107–112. Established in 1927 as a subsidiary of the Victor Talking Machine Company of America, JVC had once been a subsidiary of RCA, which had acquired its parent (Victor) in 1929. Majority ownership was acquired by Toshiba in 1937 and was taken over by a bank in 1951 when the Victor Company was in deep financial difficulty.

31. Masaru Terakado, *Nihon Bikuta no jinzai keiei* (Personnel management at Japan Victor), (Tokyo: Purejidento-sha, 1980), pp. 103, 112–118, 123.

32. Cusumano interview with Kenjiro Takayanagi, December 28, 1984.

33. *Nikkei shimbun* (Nikkei Newspaper), January 10, 1960.

34. Takayanagi interview.

35. Hiroshi Sugaya, *Katei-yo VTR no kanosei* (Feasibility of a home-use VTR), Matsushita Denki Sangyo K.K., Chuo Kenkyusho Denki-bu, Daisan ka, Project No. SE-002, March 1960; Rosenbloom interview with Sugaya, August 4, 1980; Cusumano interview, January 8, 1985. For a history of the company through the late 1960s, see Matsushita Denki Sangyo Kabushiki Kaisha, *Matsushita Denki 50-nen no ryaku shi* (A short 50-year history of Matsushita Electric), (Osaka: Matsushita, 1968). A survey of the video area is Matsushita Denki Sangyo Kabushiki Kaisha, *Bideo Jigyo-Bu no go-annai* (An introduction to the Video Division), (Osaka: Matsushita, 1980).

36. N. V., Philips introduced a VCR in 1972 that was technologically comparable (perhaps superior) to the Japanese designs, but failed to invest in the further extension of the technology until after the Japanese had introduced their half-inch designs.

37. American firms were also addressing an aerospace market for highly specialized satellite-borne equipment for secret application. But this market and technology had little influence on the path of commercial development.

38. Although both JVC and Ampex had developed helical-scan recorders designed for studio use, both firms pulled back from commercialization of those products in 1962 and turned their helical-scan developments toward the CCTV market.

39. *LIFE*, September 17, 1965, pp. 55–60.

40. Sales of color receivers had just begun rapid growth in the U.S. market.

41. This story appears in greater detail in Rosenbloom and Freeze, op. cit., pp. 165–171.

42. In technical usage, a cassette is a package containing two reels (supply and take-up)—usually in the same plane. Because early designs required larger reels (a consequence of lower recording density), programs were packaged on an enclosed single reel (called a "cartridge") and players designed for automatic threading. Tape from a cartridge had to be rewound fully before it could be removed from the player.

43. See, for example, "The Greatest Thing Since the Nickelodeon?" *Forbes*, July 1, 1970, p. 14; and Lawrence Lessing, "Stand by for the Cartridge Revolution," *Fortune* (June 1971), p. 82.

44. In Japan, 358,000 yen for a VCR with built-in TV tuner for recording broadcasts. Matsushita and JVC introduced similar products simultaneously, as provided in the agreement between the companies. Not all Japanese companies were convinced this would become the dominant video technology. Hitachi and Mitsubishi, for example, took licenses from CBS for Electronic Video Recording, believing, at that time, that it would become the "center" of the video-entertainment field. See Sony Corporation, "Table of Sony VTR History," August 16, 1977; and *Nihon Keizai Shimbun*, March 24, 1970, p. 7.

45. Graham, op. cit., chapter 6.

46. Ibid.

47. Rosenbloom interview with Akio Tanii, Manager of Matsushita VTR Division, 1980.

48. Rosenbloom interview with Kihara, 1980.

49. JVC, op. cit., p. 129.

50. JVC, op. cit., pp. 116–120; and JVC internal memorandum, September 29, 1981.

51. Cusumano interviews with Takayanagi, December 28, 1984, and with Yuma Shiraishi, December 20, 1984.

52. See Yuma Shiraishi, "VHS: bideo kasetto rekoda" (VHS: Video cassette recorder), in Egawa Akira, *Seihin senryaku no jidai* (The age of product strategies), (Tokyo: Dobunsha, 1980), pp. 154–166; and Kunio Yanagida, "VHS kaihatsu dokyumento" (A document on VHS development), *Shukan gendai* (May 1980).

53. A demonstration for senior RCA technical staff failed to deter the latter's commitment to Videodisc, but it did convince them that the RCA VCR design had been outmoded.

54. *Nihon Keizai Shimbunsha,* op. cit., pp. 13–22.

55. JVC also licensed the VHS to Hitachi, Sharp, and Mitsubishi in Japan (which sold machines supplied by Japan Victor); to RCA, Magnavox, Sylvania, GE, and Curtiss in the U.S. (which sold machines produced by Matsushita); and to Thomson in France (supplied by JVC). Toshiba and Sanyo (which had their own VCR design) lined up behind the Betamax standard, and were followed by Pioneer and Aiwa (a Sony subsidiary) by spring 1977. Zenith and Sears, Roebuck in the U.S. also bought Betamax models on an OEM basis. Eventually, while 11 firms in Japan, the U.S. and Europe supported the Betamax standard, 42 adopted the VHS. See *Nihon Keizai Shimbunsha,* op. cit., p. 73; and Nomura Management School, *VTR sangyo noto* (VTR industry note), Tokyo, 1984, p. 47.

56. Toshiba captured a small share of the mass market as licensee of Sony's technology. RCA, buying VHS machines from Japanese suppliers, built a profitable but limited business marketing VCRs in the North American market. Ampex, the source of the most basic inventions in VTR technology, failed to establish any position in the VCR markets.

57. More precisely, there was a series of "breakthroughs" on critical elements of the design, achieved during four years of inventive effort. But all of them were achieved in secrecy within the small team working at Ampex. See Rosenbloom and Freeze, op. cit.

58. Smaller reels of tape could be driven by smaller, less-expensive transport mechanisms, and the whole machine fitted in a smaller case.

59. The rotating drum on the Betamax scanner, for example, had to be machined to tolerances of $+/-.01$ millimeter. The tiny magnetic heads mounted on the drum had a thickness, initially, of only .06 millimeter and had to be positioned on the drum to accuracies measured in thousandths of a millimeter.

60. JVC's first VTR product used 2 heads, in contrast to the earlier Sony and Ampex machines with a single head. By 1965, Sony had switched to the 2-head design.

61. See Graham, op. cit.

62. See, for example, *Sankei shimbun,* April 27, 1965.

63. Rosenbloom interview with Kihara (1980).

64. Specifically, Takayanagi, Ibuka, and Nakao.

65. For the auto story, see Michael A. Cusumano, *The Japanese Auto Industry: Technology and Management at Nissan and Toyota* (Cambridge, MA: Harvard University Press, 1985). For computers, see Sigeru Takahashi, "Early Transistor Computers in Japan," *Annals of the History of Computing,* 8/2 (1986); and Noburo Minamisawa, *Nihon Konpyuta hattatsu shi* (History of the development of Japanese computers), (Tokyo: *Nihon Keizai Shimbunsha,* 1978). For machine tools, see Harvard Business School, "Hitachi Seiki (A)," Boston, MA, Harvard Case Services, #9-686-1004, 1986. For Epson, see Ken-ichi Imai, Ikujiro Nonaka, and Hirotaka Takeuchi, "Managing the New Product Development Process: How Japanese Companies Learn and Unlearn," in Kim B. Clark, Robert H. Hayes, and Christopher Lorenz, eds., *The Uneasy Alliance* (Boston, MA: Harvard Business School Press, 1985). For semiconductors, see Franklin B. Weinstein, Michiyuki Uenohara, and John G. Linvill, "Technological Resources," in Daniel I. Okimoto, Takuo Sugano, and Franklin B. Weinstein, eds., *Competitive Edge: The Semiconductor Industry in the U.S. and Japan* (Stanford, CA: Stanford University Press, 1984). For software, see Michael A. Cusumano, "Pioneering the 'Factory' Model for Large-Scale Software Development," M.I.T. Sloan School of Management Working Paper, 1987.

SECTION
I

Innovation Over Time
and in Historical Context

This section introduces the basic issues of technology and patterns in technology-based innovation. Abernathy and Utterback trace the linkages between product (or service) and process innovation, as well as discuss the importance of both incremental and discontinuous innovation. Abernathy and Utterback also introduce the idea of dominant design in a product class, which Gould and Wise illustrate in detail with examples from the typewriter and computer industries. Abernathy and Clark discuss the importance of different types of manufacturing, marketing, and product innovations, with particular emphasis on discontinuous or substitute product and/or process innovation. Abernathy and Clark also link innovation type to organization characteristics. The Marquis article picks up the organization-innovation theme as he focuses on managing incremental innovation. Marquis also presents a framework for the innovation process that emphasizes the need to link technological potential with market needs. Finally, Smith's history of the early telephone industry illustrates the impact of prior success on an organization's and executive team's ability to learn. The interrelated themes of technology, innovation, organization, leadership, and history will be discussed throughout this collection.

1

INNOVATION OVER TIME AND IN HISTORICAL CONTEXT

Patterns of Industrial Innovation

William J. Abernathy

James M. Utterback

A new model suggests how the character of its innovation changes as a successful enterprise matures; and how other companies may change themselves to foster innovation as they grow and prosper.

How does a company's innovation—and its response to innovative ideas—change as the company grows and matures?

Are there circumstances in which a pattern generally associated with successful innovation is in fact more likely to be associated with failure?

Under what circumstances will newly available technology, rather than the market, be the critical stimulus for change?

When is concentration on incremental innovation and productivity gains likely to be of maximum value to a firm? In what situations does this strategy instead cause instability and potential for crisis in an organization?

Intrigued by questions such as these, we have examined how the kinds of innovations attempted by productive units apparently change as these units evolve. Our goal was a model relating patterns of innovation within a unit to that unit's competitive strategy, production capabilities, and organizational characteristics.

This article summarizes our work and presents the basic characteristics of the model to which it has led us. We conclude that a productive unit's capacity for and methods of innovation depend critically on its stage of evolution

The research reported in this article was supported by the National Science Foundation Division of Policy Research and Analysis.

from a small technology-based enterprise to a major high-volume producer. Many characteristics of innovation and the innovative process correlate with such an historical analysis; and on the basis of our model we can now attempt answers to questions such as those above.

A SPECTRUM OF INNOVATORS

Past studies of innovation imply that any innovating unit sees most of its innovations as new products. But that observation masks an essential difference: what is a product innova-

TABLE 1.

	Fluid Pattern	Transitional Pattern	Specific Pattern
Competitive emphasis on	Functional product performance	Product variation	Cost reduction
Innovation stimulated by	Information on users' needs and users' technical inputs	Opportunities created by expanding internal technical capability	Pressure to reduce cost and improve quality
Predominant type of innovation	Frequent major changes in products	Major process changes required by rising volume	Incremental for product and process, with cumulative improvement in productivity and quality
Product line	Diverse, often including custom designs	Includes at least one product design stable enough to have significant production volume	Mostly undifferentiated standard products
Production processes	Flexible and inefficient; major changes easily accommodated	Becoming more rigid, with changes occurring in major steps	Efficient, capital-intensive, and rigid; cost of change is high
Equipment	General-purpose, requiring highly skilled labor	Some subprocesses automated, creating "islands of automation"	Special-purpose, mostly automatic with labor tasks mainly monitoring and control
Materials	Inputs are limited to generally-available materials	Specialized materials may be demanded from some suppliers	Specialized materials will be demanded; if not available, vertical integration will be extensive
Plant	Small-scale, located near user or source of technology	General-purpose with specialized sections	Large-scale, highly specific to particular products
Organizational control is	Informal and entrepreneurial	Through liaison relationships, project and task groups	Through emphasis on structure, goals, and rules

tion by a small, technology-based unit is often the process equipment adopted by a large unit to improve its high-volume production of a standard product. We argue that these two units—the small, entrepreneurial organization and the larger unit producing standard products in high volume—are at opposite ends of a spectrum, in a sense forming boundary conditions in the evolution of a unit and in the character of its innovation of product and process technologies.

One distinctive pattern of technological innovation is evident in the case of established, high-volume products such as incandescent light bulbs, paper, steel, standard chemicals, and internal-combustion engines, for examples.

The markets for such goods are well defined; the product characteristics are well understood and often standardized; unit profit margins are typically low; production technology is efficient, equipment-intensive, and specialized to a particular product; and competition is primarily on the basis of price. Change is costly in such highly integrated systems because an alteration in any one attribute or process has ramifications for many others.

In this environment innovation is typi-cally incremental in nature, and it has a gradual, cumulative effect on productivity. For example, Samuel Hollander has shown that more than half of the reduction in the cost of producing rayon in plants of E. I. du Pont de Nemours and Co. has been the result of gradual process improvements which could not be identified as formal projects or changes. A similar study by John Enos shows that accumulating, incremental developments in petroleum refining processes resulted in productivity gains which often eclipsed the gain from the original innovation. Incremental innovations, such as the use of larger railroad cars and unit trains, have resulted in dramatic reductions in the cost of moving large quantities of materials by rail.

In all these examples, major systems innovations have been followed by countless minor product and systems improvements, and the latter account for more than half of the total ultimate economic gain due to their much greater number. While cost reduction seems to have been the major incentive for most of these innovations, major advances in performance have also resulted from such small engineering and production adjustments.

Such incremental innovation typically results in an increasingly specialized system in

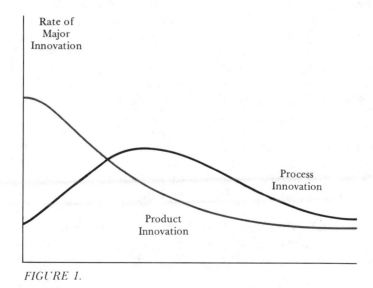

FIGURE 1.

which economies of scale in production and the development of mass markets are extremely important. The productive unit loses its flexibility, becoming increasingly dependent on high-volume production to cover its fixed costs and increasingly vulnerable to changed demand and technical obsolescence.

Major new products do not seem to be consistent with this pattern of incremental change. New products which require reorientation of corporate goals or production facilities tend to originate outside organizations devoted to a "specific" production system; or, if originated within, to be rejected by them.

A more fluid pattern of product change is associated with the identification of an emerging need or a new way to meet an existing need; it is an entrepreneurial act. Many studies suggest that such new product innovations share common traits. They occur in disproportionate numbers in companies and units located in or near affluent markets with strong science-based universities or other research institutions and entrepreneurially oriented financial institutions. Their competitive advantage over predecessor products is based on superior functional performance rather than lower initial cost, and so these radical innovations tend to offer higher unit profit margins.

When a major product innovation first appears, performance criteria are typically vague and little understood. Because they have a more intimate understanding of performance requirements, users may play a major role in suggesting the ultimate form of the innovation as well as the need. For example, Kenneth Knight shows that three-quarters of the computer models which emerged between 1944 and 1950, usually those produced as one or two of a kind, were developed by users.

It is reasonable that the diversity and uncertainty of performance requirements for new products give an advantage in their innovation to small, adaptable organizations with flexible technical approaches and good external com-

munications, and historical evidence supports that hypothesis. For example, John Tilton argues that new enterprises led in the application of semiconductor technology, often transferring into practice technology from more established firms and laboratories. He argues that economies of scale have not been of prime importance because products have changed so rapidly that production technology designed for a particular product is rapidly made obsolete. And R. O. Schlaifer and S. D. Heron have argued that a diverse and responsive group of enterprises struggling against established units to enter the industry contributed greatly to the early advances in jet aircraft engines.

A TRANSITION FROM RADICAL TO EVOLUTIONARY INNOVATION

These two patterns of innovation may be taken to represent extreme types—in one case involving incremental change to a rigid, efficient production system specifically designed to produce a standardized product, and in the other case involving radical innovation with product characteristics in flux. They are not in fact rigid, independent categories. Several examples will make it clear that organizations currently considered in the "specific" category—where incremental innovation is now motivated by cost reduction—were at their origin small, "fluid" units intent on new product innovation.

John Tilton's study of developments in the semiconductor industry from 1950 through 1968 indicates that the rate of major innovation has decreased and that the type of innovation shifted. Eight of the 13 product innovations he considers to have been most important during that period occurred within the first seven years, while the industry was making less than 5 per cent of its total 18-year sales. Two types of enterprise can be identified in this early period of the new industry—established units that came into semiconductors from vested positions in vacuum tube markets, and new entries such as Fairchild Semiconductor, I.B.M., and Texas Instruments, Inc. The established units responded to competition from the newcomers by emphasizing process innovations. Meanwhile, the latter sought entry and strength through product innovation. The three very successful new entrants just listed were responsible for half of the major product innovations and only one of the nine process innovations which Dr. Tilton identified in that 18-year period, while three principal established units (divisions of General Electric, Philco, and R.C.A.) made only one-quarter of the product innovations but three of the nine major process innovations in the same period. In this case process innovation did not prove to be an effective competitive stance; by 1966 the three established units together held only 18 per cent of the market while the three new units held 42 per cent. Since 1968, however, the basis of competition in the industry has changed; as costs and productivity have become more important, the rate of major product innovation has decreased, and effective process innovation has become an important factor in competitive success. For example, by 1973 Texas Instruments which had been a flexible, new entrant in the industry two decades earlier and had contributed no major process innovations prior to 1968, was planning a single machine that would produce 4 per cent of world requirements for its integrated-circuit unit.

Like the transistor in the electronics industry, the DC-3 stands out as a major change in the aircraft and airlines industries. Almarin Phillips has shown that the DC-3 was in fact a cumulation of prior innovations. It was not the largest, or fastest, or longest-range aircraft; it was the most economical large, fast plane able to fly long distances. All the features which made this design so completely successful had been introduced and proven in prior aircraft.

And the DC-3 was essentially the first commercial product of an entering firm (the C-1 and DC-2 were produced by Douglas only in small numbers).

Just as the transistor put the electronics industry on a new plateau, so the DC-3 changed the character of innovation in the aircraft industry for the next 15 years. No major innovations were introduced into commercial aircraft design from 1936 until new jet-powered aircraft appeared in the 1950s. Instead, there were simply many refinements to the DC-3 concept—stretching the design and adding appointments; and during the period of these incremental changes airline operating cost per passenger-mile dropped an additional 50 per cent.

The electric light bulb also has a history of a long series of evolutionary improvements which started with a few major innovations and ended in a highly standardized commodity-like product. By 1909 the initial tungsten filament and vacuum bulb innovations were in place; from then until 1955 there came a series of incremental changes—better metal alloys for the filament, the use of "getters" to assist in exhausting the bulb, coiling the filaments, "frosting" the glass, and many more. In the same period the price of a 60-watt bulb decreased (even with no inflation adjustment) from $1.60 to 20 cents each, the lumens output increased by 175 per cent, the direct labor content was reduced more than an order of magnitude, from 3 to 0.18 minutes per bulb, and the production process evolved from a flexible job-shop configuration, involving more than 11 separate operations and a heavy reliance on the skills of manual labor, to a single machine attended by a few workers.

Product and process evolved in a similar fashion in the automobile industry. During a four-year period before Henry Ford produced the renowned Model T, his company developed, produced, and sold five different engines, ranging from two to six cylinders. These

---

THE UNIT OF ANALYSIS

As we show in this article, innovation within an established industry is often limited to incremental improvements of both products and processes. Major product change is often introduced from outside an established industry and is viewed as disruptive; its source is typically the start-up of a new, small firm, invasion of markets by leading firms in other industries, or government sponsorship of change either as an initial purchaser or through direct regulation.

These circumstances mean that the standard units of analysis of industry—firm and product type—are of little use in understanding innovation. Technological change causes these terms to change their meaning, and the very shape of the production process is altered.

Thus the questions raised in this article require that a product line and its associated production process be taken together as the unit of analysis. This we term a "productive unit." For a simple firm or a firm devoted to a single product, the productive unit and the firm would be one and the same. In the case of a diversified firm, a productive unit would usually report to a single operating manager and normally be a separate operating division. The extreme of a highly fragmented production process might mean that several separate firms taken together would be a productive unit.

For example, analysis of change in the textile industry requires that productive units in the chemical, plastics, paper, and equipment industries be included. Analysis involving the electronics industry requires a review of the changing role of component, circuit, and software producers as they become more crucial to change in the final assembled product. Major change at one level works its way up and down the chain, because of the interdependence of product and process change within and among productive units. Knowledge of the production process as a system of linked productive units is a prerequisite to understanding innovation in an industrial context.
W.J.A., J.M.U.

---

were made in a factory that was flexibly organized much as a job shop, relying on trade craftsmen working with general-purpose machine tools not nearly so advanced as the best then available. Each engine tested a new concept. Out of this experience came a dominant design—the Model T; and within 15 years 2 million engines of this single basic design were being produced each year (about 15 million all told) in a facility then recognized as the most efficient and highly integrated in the world. During that 15-year period there were incremental—but no fundamental—innovations in the Ford product.

In yet another case, Robert Buzzell and Robert Nourse, tracing innovations in processed foods, show that new products such as soluble coffees, frozen vegetables, dry pet foods, cold breakfast cereals, canned foods, and precooked rice came first from individuals and small organizations where research was in progress or which relied heavily upon information from users. As each product won acceptance, its productive unit increased in size and concentrated its innovation on improving manufacturing, marketing, and distribution methods which extended rather than replaced the basic technologies. The major source of the latter ideas is now each firm's own research and development organization.

The shift from radical to evolutionary product innovation is a common thread in these examples. It is related to the development of a dominant product design, and it is accompanied by heightened price competition and increased emphasis on process innovation. Small-scale units that are flexible and highly reliant on manual labor and craft skills utilizing general-purpose equipment develop into units that rely on automated, equipment-intensive, high-volume processes. We conclude that changes in innovative pattern, production process, and scale and kind of production capacity all occur together in a consistent, predictable way.

Though many observers emphasize new-product innovation, process and incremental innovations may have equal or even greater commercial importance. A high rate of productivity improvement is associated with process improvement in every case we have studied. The cost of incandescent light bulbs, for example, has fallen more than 80 per cent since their introduction. Airline operating costs were cut by half through the development and improvement of the DC-3. Semiconductor prices have been falling by 20 to 30 per cent with each doubling of cumulative production. The intro-

duction of the Model T Ford resulted in a price reduction from $3,000 to less than $1,000 (in 1958 dollars). Similar dramatic reductions have been achieved in the costs of computer core memory and television picture tubes.

MANAGING TECHNOLOGICAL INNOVATION

If it is true that the nature and goals of an industrial unit's innovations change as that unit matures from pioneering to large-scale producer, what does this imply for the management of technology?

We believe that some significant managerial concepts emerge from our analysis—or model, if you will—of the characteristics of innovation as production processes and primary competitive issues differ. As a unit moves toward large-scale production, the goals of its innovations change from ill-defined and uncertain targets to well-articulated design objectives. In the early stages there is a proliferation of product performance requirements and design criteria which frequently cannot be stated quantitatively, and their relative importance or ranking may be quite unstable. It is precisely under such conditions, where performance requirements are ambiguous, that users are most likely to produce an innovation and where manufacturers are least likely to do so. One way of viewing regulatory constraints such as those governing auto emissions or safety is that they add new performance dimensions to be resolved by the engineer—and so may lead to more innovative design improvements. They are also likely to open market opportunities for innovative change of the kind characteristic of fluid enterprises in areas such as instrumentation, components, process equipment, and so on.

The stimulus for innovation changes as

31

a unit matures. In the initial fluid stage, market needs are ill-defined and can be stated only with broad uncertainty; and the relevant technologies are as yet little explored. So there are two sources of ambiguity about the relevance of any particular program of research and development—target uncertainty and technical uncertainty. Confronted with both types of uncertainty, the decision-maker has little incentive for major investments in formal research and development.

As the enterprise develops, however, uncertainty about markets and appropriate targets is reduced, and larger research and development investments are justified. At some point before the increasing specialization of the unit makes the cost of implementing technological innovations prohibitively high and before increasing cost competition erodes profits with which to fund large indirect expenses, the benefits of research and development efforts would reach a maximum. Technological opportunities for improvements and additions to existing product lines will then be clear, and a strong commitment to research and development will be characteristic of productive units in the middle stages of development. Such firms will be seen as "science based" because they invest heavily in formal research and engineering departments, with emphasis on process innovation and product differentiation through functional improvements.

Although data on research and development expenditures are not readily available on the basis of productive units, divisions, or lines of business, an informal review of the activities of corporations with large investments in research and development shows that they tend to support business lines that fall neither near the fluid nor the specific conditions but are in the technologically-active middle range. Such productive units tend to be large, to be integrated, and to have a large share of their markets.

A small, fluid entrepreneurial unit requires general-purpose process equipment which is typically purchased. As it develops, such a unit is expected to originate some process-equipment innovations for its own use; and when it is fully matured its entire processes are likely to be designed as integrated systems specific to particular products. Since the mature firm is now fully specialized, all its major process innovations are likely to originate outside the unit.

But note that the supplier companies will now see themselves as making product—not process—innovations. From a different perspective, George Stigler finds stages of development—similar to those we describe—in firms that supply production-process equipment. They differ in the market structure they face, in the specialization of their production processes, and in the responsibilities they must accept in innovating to satisfy their own needs for process technology and materials.

The organization's methods of coordination and control change with the increasing standardization of its products and production processes. As task uncertainty confronts a productive unit early in its development, the unit must emphasize its capacity to process information by investing in vertical and lateral information systems and in liaison and project groups. Later, these may be extended to the creation of formal planning groups, organizational manifestations of movement from a product-oriented to a transitional state; controls for regulating process functions and management controls such as job procedures, job descriptions, and systems analyses are also extended to become a more pervasive feature of the production network.

As a productive unit achieves standardized products and confronts only incremental change, one would expect it to deal with complexity by reducing the need for information processing. The level at which technological change takes place helps to determine the extent to which organizational dislocations take

DESIGN AS A MILESTONE OF CHANGE

The milestone in all the examples of transition in the accompanying article is a dominant new product synthesized from individual technological innovations introduced independently in prior products. This dominant design has the effect of enforcing standardization so that production economies can be sought. Then effective competition begins to take place on the basis of cost as well as of product performance.

Similar product design milestones can be identified in other product lines: sealed refrigeration units for home refrigerators and freezers, effective can-sealing technology in the food canning industry and the standardized diesel locomotive in the locomotive and railroad industry. In each case the milestone signals a significant transformation, affecting the type of innovation which follows it, the source of information, and the size, scope, and use of formal research and development.

In an earlier article in this series, George R. White (*see his "Management Criteria for Effective Innovation," February, pp. 14–23*) contends that dominant designs can be recognized in the early stages of their development. His analysis suggests that dominant designs will more likely display one or more of the following qualities:

- Technologies which lift fundamental technical constraints limiting the prior art while not imposing stringent new constraints.
- Designs which enhance the value of potential innovations in other elements of a product or process.
- Products which assure expansion into new markets.
  W.J.A., J.M.U.

place. Each of these hypotheses helps to explain the firm's impetus to divide into homogeneous productive units as its products and process technology evolve.

The hypothesized changes in control and coordination imply that the structure of the organization will also change as it matures, becoming more formal and having a greater number of levels of authority. The evidence is strong that such structural change is a characteristic of many enterprises and of units within them.

FOSTERING INNOVATION BY UNDERSTANDING TRANSITION

Assuming the validity of this model for the development of the innovative capacities of a productive unit, how can it be applied to further our capacity for new products and to improve our productivity?

We predict that units in different stages of evolution will respond to differing stimuli and undertake different types of innovation. This idea can readily be extended to the ques-

33

WHEN TRANSITION IS INVISIBLE—OR EVEN ABSENT

Identifying the evolutionary transition from product to process innovation is sometimes troublesome. In some cases the transition may have occurred so rapidly as to be unrecognized; this appears to be the case with some continuous-flow processes where advanced, elaborate, and large-scale equipment is necessary to make a new product virtually from its initial introduction. Rapid transition is also characteristic of certain products with low unit values, such as cigarettes and simple plastic and metal parts, where the availability of a process technology may have made the product feasible in the first place.

More interesting cases are those where the transition from product to process innovation and from unit production to mass production, though predicted, has not come about. Examples include home construction, nuclear power, and some other energy alternatives. In each of these examples, experimental programs to stimulate cost reductions, greater standardization, or other aspects of transition have been undertaken under government and private sponsorship; but none has had long-run impact. These cases are of special interest because the model may help in identifying barriers and pinpointing appropriate responses.

W.J.A., J.M.U.

---

tion of barriers to innovation; and probably to patterns of success and failure in innovation for units in different situations. The unmet conditions for transition can be viewed as specific barriers which must be overcome if transition is to take place.

We would expect new, fluid units to view as barriers any factors that impede product standardization and market aggregation, while firms in the opposite category tend to rank uncertainty over government regulation or vulnerability of existing investments as more important disruptive factors. Those who would promote innovation and productivity in U.S. industry may find this suggestive. *(See "Why Innovations Fail," by Sumner Myers and Eldon Sweezy, March/April, pp. 40-46.)*

We believe the most useful insights provided by the model apply to production processes in which features of the products can be varied. The most interesting applications are to situations where product innovation is competitively important and difficult to manage; the model helps to identify the full range of other issues with which the firm is simultaneously confronted in a period of growth and change.

CONSISTENCY OF MANAGEMENT ACTION

Many examples of unsuccessful innovations point to a common explanation of failure: certain conditions necessary to support a sought-after technical advance were not present. In such cases our model may be helpful because it describes conditions that normally support advances at each stage of development; accordingly, if we can compare existing conditions with those prescribed by the model we may discover how to increase innovative success. For example, we may ask of the model such questions as these about different, apparently independent, managerial actions:

·   Can a firm increase the variety and diversity of its product line while simultaneously realizing the highest possible level of efficiency?
·   Is a high rate of product innovation consistent with an effort to substantially reduce costs through extensive backward integration?
·   Is government policy to maintain diversified markets for technologically active industries consistent with a policy that seeks a high rate of effective product innovation?
·   Would a firm's action to restructure its work environment for employees so that tasks are more challenging and less repetitive be compatible with a policy of mechanization designed to reduce the need for labor?
·   Can the government stimulate productivity by forcing a young industry to standardize its products before a dominant design has been realized?

The model prompts an answer of "no" to each of these questions; each question suggests actions which the model tells us are mutually inconsistent. We believe that as these ideas are further developed they can be equally effective in helping to answer many far more subtle questions about the environment for innovation, productivity, and growth.

FURTHER READINGS

For readers who wish to explore this subject in greater detail, the authors recommend:

Abernathy, W. J. and P. L. Townsend, "Technology, Productivity and Process Change," *Technological Forecasting and Social Change,* Vol. 7, No. 4, August, 1975, pp. 379-396.

Abernathy, W. J. and K. Wayne, "Limits of the Learning Curve," *Harvard Business Review,* Vol. 52, No. 5, September-October, 1974, pp. 109-119.

Utterback, James M., "Innovation in Industry and the Diffusion of Technology," *Science,* Vol. 183, February, 1974, pp. 620-626.

Utterback, James M. and W. J. Abernathy, "A Dynamic Model of Process and Product Innovation," *Omega,* Vol. 3, No. 6, 1975, pp. 639-656.

For the sources of the examples cited and some others, see:

Abernathy, William J., *The Productivity Dilemma: Roadblock to Innovation in the Automobile Industry;* Baltimore: Johns Hopkins University Press, 1978.

Bright, James R., *Automation and Management;* Boston: Division of Research, Graduate School of Business Administration, Harvard University, 1958.

Buzzell, Robert D. and Robert E. Nourse, *Product Innovation in Food Processing: 1954-1964;* Boston: Division of Research, Harvard Graduate School of Business Administration, 1967.

Clarke, R. "Innovation in Liquid Propelled Rocket Engines," doctoral dissertation, Stanford Graduate School of Business, 1968.

Enos, J., *Petroleum Progress and Profits;* Cambridge: M.I.T. Press, 1967.

Hogan, W. T., *Economic History of the Iron and Steel Industry in the United States,* Vol. 3; Lexington, Mass.: D.C. Heath Books, 1971, p. 1011.

Hollander, S., *The Sources of Increased Efficiency;* Cambridge: M.I.T. Press, 1965.

Jenkins, Reese V., *Images and Enterprise: Technology and the American Photographic Industry, 1839 to 1925;* Baltimore: Johns Hopkins University Press, 1975.

Knight, Kenneth E., "A Study of Technological Innovation: The Evolution of Digital Computers," doctoral dissertation, Carnegie Institute of Technology, Pittsburgh, 1963.

Little, Arthur D., Inc., *Patterns and Problems of Technical Innovation in American Industry: Report to the National Science Foundation,* PB 181573, U.S. Department of Commerce, Office of Technical Services; Washington, D.C.: U.S. Government Printing Office, September, 1963.

Miller, R. E., and D. Sawers, *The Technical Development of Modern Aviation;* Praeger Publishers, New York, 1970.

Schlaifer, R. O. and S. D. Heron, *Development of Aircraft Engines and Fuels;* Cambridge: Harvard University Press, 1950.

Stigler, G. J., *The Organization of Industry,* Homeward, Illinois: Richard D. Irwin, 1968.

Tilton, John E., *International Diffusion of Technology: The Case of Semiconductors;* Washington, D.C.: Brookings Institution, 1971.

The Panda's Thumb of Technology

Stephen Jay Gould

We have an example of the principle of imperfection at our fingertips

The brief story of Jephthah and his daughter (Judg. 11:30–40) is, to my mind and heart, the saddest of all biblical tragedies. Jephthah makes an intemperate vow, yet all must abide by its consequences. He promises that if God grant him victory in a forthcoming battle, he will sacrifice by fire the first living thing that passes through his gate to greet him upon his return. Expecting (I suppose) a dog or a goat, he returns victorious to find his daughter, and only child, waiting to greet him "with timbrels and with dances."

Handel's last oratorio, *Jephtha,* treats this tale with great power (although his librettist couldn't bear the weight of the original and gave the story a happy ending, with angelic intervention to spare Jephthah's daughter at the price of her lifelong chastity). At the end of part 2, while all still think that the terrible vow must be fulfilled, the chorus sings one of Handel's wonderful "philosophical" choruses. It begins with a frank account of the tragic circumstance:

> How dark, O Lord, are they decrees! . . .
>   No certain bliss, no solid peace,
>   We mortals know on earth below.

With permission from NATURAL HISTORY, Vol. 96, No. 1; Copyright the American Museum of Natural History, 1986.

Yet the last two lines, in a curious about-face, proclaim (with magnificent musical solidity as well):

> Yet on this maxim still obey:
> WHATEVER IS, IS RIGHT

This odd reversal, from frank acknowledgement to unreasonable acceptance, reflects one of the greatest biases ("hopes" I like to call them) that human thought imposes upon a world indifferent to our suffering. Humans are pattern-seeking animals. We must find cause and meaning in all events (quite apart from the probable reality that the universe both doesn't care much about us and often operates in a random manner). I call this bias "adaptationism"—the notion that everything must fit, must have a purpose, and in the strongest version, must be for the best.

The final line of Handel's chorus is, of course, a quote from Alexander Pope, the last statement of the first epistle of his *Essay on Man,* published just thirteen years before Handel's oratorio. Pope's text contains (in heroic couplets to boot) the most striking paean I know to the bias of adaptationism. In my favorite lines, Pope chastises those people who may be unsatisfied with the senses that nature bestowed upon us. We may wish for more acute vision,

37

hearing, or smell, but consider the consequences.

If nature thunder'd in his op'ning ears
And stunn'd him with the music of the spheres
How would he wish that Heav'n had left him still
The whisp'ring zephyr, and the purling rill!

And my favorite couplet, on olfaction:

Or, quick effluvia darting thro' the brain,
Die of a rose in aromatic pain.

What we have is best for us—whatever is, is right.

By 1859, most educated people were prepared to accept evolution as the reason behind similarities and differences among organisms—thus accounting for Darwin's rapid conquest of the intellectual world. But they were decidedly not ready to acknowledge the radical implications of Darwin's proposed mechanism of change, natural selection, thus explaining the brouhaha that the *Origin of Species* provoked—and still elicits (at least before our courts and school boards).

Darwin's world is full of "terrible truths," two in particular. First, when things do fit and make sense (good design of organisms, harmony of ecosystems), they did not arise because the laws of nature entail such order as a primary effect. They are, rather, only epiphenomena, side consequences of the basic causal process at work in natural populations—the purely "selfish" struggle among organisms for personal reproductive success. Second, the complex and curious pathways of history guarantee that most organisms and ecosystems cannot be designed optimally. Indeed, to make the statement even stronger, imperfections are the primary proofs that evolution has occurred, since optimal designs erase all signposts of history.

This principle of imperfection has been the main theme of these essays for several years. I call it the panda principle to honor my favorite example, the panda's false thumb, subject of an old essay (*Natural History*, November 1978) that reemerged as the title to one of my books. Pandas are the herbivorous descendants of carnivorous bears. Their true anatomical thumbs were, long ago during ancestral days of meat eating, irrevocably committed to the limited motion appropriate for this mode of life and universally evolved by mammalian Carnivora. When adaptation to a diet of bamboo required more flexibility in manipulation, pandas could not redesign their thumbs but had to make do with a makeshift substitute—an enlarged radial sesamoid bone of the wrist, the panda's false thumb. The sesamoid thumb is a clumsy, suboptimal structure, but it works. Pathways of history (commitment of the true thumb to other roles during an irreversible past) impose such jury-rigged solutions upon all creatures. History inheres in the imperfections of living organisms—thus we know that they had a different past, converted by evolution to their current state.

We can accept this argument for organisms (we know, after all, about our own appendixes and aching backs). But is the panda principle more pervasive? Is it a general statement about all historical systems? Will it apply, for example, to the products of technology? We might deem it irrelevant to the manufactured objects of human ingenuity—and for good reason. After all, constraints of genealogy do not apply to steel, glass, and plastic. The panda cannot shuck its digits (and can only build its future upon this inherited ground plan), but we can abandon gas lamps for electricity and horse carriages for motor cars. Consider, for example, the difference between organic architecture and human buildings. Complex organic structures cannot be reevolved following their loss; no snake will redevelop front legs. But the apostles of so-called post-modern architecture, in reaction to the sterility of so many glass-box buildings of the international style, have juggled together all the classical forms of history in a cascading effort to rediscover the virtues of ornamentation. Thus, Philip Johnson could place a broken pediment atop a New York skyscraper and raise a medieval castle of plate glass in downtown Pittsburgh. Organisms cannot recruit the virtues of their lost pasts.

Yet I am not so sure that technology is exempt from the panda principle of history, for I am now sitting face to face with the best example of its application. Indeed, I am in most intimate (and striking) contact with this object—the typewriter keyboard.

I could type before I could write. My father was a court stenographer, and my mother is a typist. I learned proper eight-finger touch-typing when I was about nine years old and still endowed with small hands and weak, tiny pinky fingers. I was thus, from the first, in a particularly good position to appreciate the irrationality of the distribution of letters on the standard keyboard, called QWERTY by all aficionados in honor of the first six letters on the top letter row.

Clearly, QWERTY makes no sense (beyond the whiz and joy of typing QWERTY itself). More than 70 percent of English words can be typed with the letters DHIATENSOR, and these should be on the most accessible second, or home, row—as they were in a failed competitor to QWERTY introduced as early as 1893. But in QWERTY, the most common English letter, E, requires a reach to the top row, as do the vowels U, I, and O (with O struck by the weak fourth finger), while A remains in the home row but must be typed with the weakest finger of all (at least for the dexterous majority of right handers)—the left pinky. (How I struggled with this as a boy. I just couldn't depress that key. I once tried to type the Declaration of Independence and ended up with: th t ll men re cre ted equ l.)

As a dramatic illustration of this irrationality, consider the keyboard of an ancient Smith-Corona upright, identical with the one (my Dad's original) that I used to type these essays (a magnificent machine—no breakdown in twenty years and a fluidity of motion unmatched by any manual typewriter since). After more than half a century of use, some of the most commonly struck keys have been worn right through the surface into the soft pad below (they weren't solid plastic in those days). Note that E, A, and S are worn in this way—and note also that all three are either not in the home row or are struck with the weak fourth and pinky fingers in QWERTY.

This claim is not just a conjecture based on idiosyncratic personal experience. Evidence clearly shows that QWERTY is drastically suboptimal. Competitors have abounded since the early days of typewriting, but none have supplanted or even dented the universal dominance of QWERTY for English typewriters. The best-known alternative, DSK, for Dvorak Simplified Keyboard, was introduced in 1932. Since then, virtually all records for speed typing have been held by DSK, not QWERTY, typists. During the 1940s, the U.S. Navy, ever mindful of efficiency, found that the increased speed of DSK would amortize the cost of retraining typists within ten days of full employment. (Mr. Dvorak was not Anton of the *New World Symphony,* but August, a professor of education at the University of Washington, who died disappointed in 1975. Dvorak was a disciple of Frank B. Gilbreth, pioneer of time and motion studies in industrial management.)

Since I have a special interest in typewriters (my affection for them dates to those childhood days of splendor in the grass and glory in the flower), I have wanted to write such an essay for years. But I never had the data I needed until Paul A. David, Coe Professor of American Economic History at Stanford University, kindly sent me his fascinating article, "Understanding the Economics of QWERTY: The Necessity of History" (in *Economic History and the Modern Economist,* edited by W.N. Parker. New York: Basil Blackwell Inc., 1986, pp. 30–49). Virtually all the nonidiosyncratic data in this essay come from David's work, and I thank him for this opportunity to satiate an old desire.

The puzzle of QWERTY's dominance resides in two separate questions: Why did QWERTY ever arise in the first place? And why has QWERTY survived in the face of superior competitors?

My answers to these questions will invoke analogies to principles of evolutionary theory. Let me, then, state some ground rules for such a questionable enterprise. I am convinced that comparisons between biological evolution and human cultural or technological change have done vastly more harm than good—and examples abound of this most common of all intellectual traps. Biological evolution is a bad analogue for cultural change because the two systems are so very different for three major reasons that could hardly be more fundamental.

First, cultural evolution can be faster by orders of magnitude than biological change at its maximal Darwinian rate—and questions of timing are of the essence in evolutionary arguments. Second, cultural evolution is direct and Lamarckian in form: the achievements of one generation are passed by education and publication directly to descendants, thus producing the great potential speed of cultural change. Biological evolution is indirect and Darwinian, as favorable traits do not descend to the next generation unless, by good fortune, they arise as products of genetic change. Third, the basic topologies of biological and cultural change are completely different. Biological evolution is a system of constant divergence without any subsequent joining of branches. Lineages once distinct, are separate forever. In human history, transmission across lineages is, perhaps, the major source of cultural change. Europeans learned about corn and potatoes from Native Americans and gave them smallpox in return.

So, when I compare the panda's thumb with a typewriter keyboard, I am not attempting to derive or explain technological change by biological principles. Rather, I ask if both systems might not record common, deeper principles of organization. Biological evolution is powered by natural selection, cultural evolution by a different set of principles that I understand but dimly. But both are systems of historical change. There must be (perhaps I now only show my own bias for intelligibility in our complex world) more general principles of structure underlying all systems that proceed through history—and I rather suspect that the panda principle of imperfection might reside among them.

My main point, in other words, is not that typewriters are like biological evolution (for such an argument would fall right into the nonsense of false analogy), but that both keyboards and the panda's thumb, as products of history, must be subject to some regularities governing the nature of temporal connections. As scientists, we must believe that general principles underlie structurally related systems that proceed by different overt rules. The proper unity lies, not in false applications of these overt rules (like natural selection) to alien domains (like technological change), but in seeking the more general rules of structure and change themselves.

THE ORIGIN OF QWERTY

True randomness has limited power to intrude itself into the forms of organisms. Small and unimportant changes, unrelated to the working integrity of a complex creature, may drift in and out of populations by a process akin to throwing dice. But intricate structures, involving the coordination of many separate parts, must arise for an active reason—since the bounds of mathematical probability for fortuitous association are soon exceeded as the number of working parts grows.

But if complex structures must arise for a reason, history may soon overtake the original purpose—and what was once a sensible solution becomes an oddity or imperfection in the altered context of a new future. Thus, the panda's true thumb permanently lost its ability to manipulate objects when carnivorous ancestors found a better use for this digit in the limited motions appropriate for creatures that run and claw. This altered thumb then becomes a constraint imposed by past history upon the panda's ability to adapt in an optimal way to its new context of herbivory. The panda's thumb, in short, becomes an emblem of its different past, a sign of history.

Similarly, QWERTY had an eminently sensible rationale in the early technology of typewriting but soon became a constraint upon faster typing as advances in construction erased the reason for QWERTY's origin. The key (pardon the pun) to QWERTY's origin lies in another historical vestige easily visible on the second row of letters. Note the sequence: DFGHJKL—a good stretch of the alphabet in order, with the vowels E and I removed. The original concept must have simply arrayed the letters in alphabetical order. Why were the two most common letters of this sequence removed from the most accessible home row? And why were other letters dispersed to odd positions?

Those who remember the foibles of manual typewriters (or, if as hidebound as yours truly, still use them) know that excessive speed or unevenness of stroke may cause two or more keys to jam near the striking point. You also know that if you don't reach in and pull the keys apart, any subsequent stroke will produce a repetition of the key leading the jam—as any key subsequently struck will hit the back of the jammed keys and drive them closer to the striking point.

These problems were magnified in the crude technology of early machines—and too much speed became a hazard rather than a blessing, as key jams canceled the benefits of celerity. Thus, in the great human traditions of tinkering and pragmatism, keys were moved around to find a proper balance between speed and jamming. In other words—and here comes the epitome of the tale in a phrase—QWERTY arose in order to slow down the maximal speed of typing and prevent jamming of keys. Common letters were either allotted to weak fingers or dispersed to positions requiring a long stretch from the home row.

This basic story has gotten around, thanks to short takes in *Time* and other popular magazines, but the details are enlightening, and few people have the story straight. I have asked nine typists who knew this outline of QWERTY's origin and all (plus me for an even ten) had the same misconception. The old machines that imposed QWERTY were, we thought, of modern design—with keys in front typing a visible line on paper rolled around a platen. This leads to a minor puzzle: key jams may be a pain in the butt, but you see them right away and can easily reach in and pull them apart. So why QWERTY?

As David points out, the prototype of QWERTY, a machine invented by C.L. Sholes in the 1860s, was quite different in form from modern typewriters. It had a flat paper carriage and did not roll paper right around the platen. Keys struck the paper invisibly from beneath, not patently from the front as in all modern typewriters. You could not view what you were typing unless you stopped to raise the carriage and inspect your product. Keys jammed frequently, but you could not see (and often did not feel) the aggregation. Thus, you might type a whole page of deathless prose and emerge only with a long string of E's.

Sholes filed for a patent in 1867 and spent the next six years in trial-and-error efforts to improve his machine. QWERTY emerged from this period of tinkering and compromise. As another added wrinkle (and fine illustration of history's odd quirks), R joined the top row as a last-minute entry, and for a somewhat capricious motive according to one common tale—for salesmen could then impress potential buyers by smooth and rapid production of the brand name TYPE WRITER, all on one row. (Although I wonder how many sales were lost when TYPE EEEEEE appeared after a jam!)

THE SURVIVAL OF QWERTY

We can all accept this story of QWERTY's origin, but why did it persist after the introduction of the modern platen roller and front-stroke key? (The first typewriter with a fully visible printing point was introduced in 1890.) In fact, the situation is even more puzzling. I thought that alternatives to keystroke typing only became available with the IBM electric ball, but none other than Thomas Edison filed a patent for an electric print-wheel machine as early as 1872, and L.S. Crandall marketed a writing machine without typebars in 1879. (Crandall arranged his type on a cylindrical sleeve and made the sleeve revolve to the required letter before striking the printing point.)

The 1880s were boom years for the fledgling typewriter industry, a period when a hundred flowers bloomed and a hundred schools of thought contended. Alternatives to QWERTY were touted by several companies, and both the variety of printing designs (several without typebars) and the improvement of keystroke typewriters completely removed the original rationale for QWERTY. Yet during the 1890s, more and more companies made the switch to QWERTY, which became an industry standard by the early years of our century. And QWERTY has held on stubbornly, through the introduction of the IBM Selectric and the Hollerith punch card machine to that ultimate example of its nonnecessity, the microcomputer terminal (Apple does offer a Dvorak option with the touch of a button but emblazons QWERTY on its keyboard and reports little use of this high-speed alternative).

To understand the survival (and domination to this day) of drastically suboptimal QWERTY, we must recognize two other commonplaces of history, as applicable to life in geological time as to technology over decades—contingency and incumbency. We call a historical event—the rise of mammals or the dominance of QWERTY—contingent when it occurs as the chancy result of a long string of unpredictable antecedents, rather than as a necessary outcome of nature's laws. Such contingent events often depend crucially upon choices from a distant past that seemed tiny and trivial at the time. Minor perturbations early in the game can nudge a process into a new pathway, with cascading consequences that produce an outcome vastly different from any alternative.

Incumbency also reinforces the stability of a pathway once the little quirks of early flexibility push a sequence into a firm channel. Suboptimal politicians often prevail nearly forever once they gain office and grab the reins of privilege, patronage, and visibility. Mammals waited 100 million years to become the dominant animals on land and only got a chance because dinosaurs succumbed during a mass extinction. If every typist in the world stopped using QWERTY tomorrow and began to learn Dvorak, we would all be winners, but who will bell the cat or start the ball rolling? (Choose your cliché, for they all record this evident truth.) Stasis is the norm for complex systems: change, when it happens at all, is usually rapid and episodic.

QWERTY's fortunate and improbable ascent to incumbency occurred by a concatenation of circumstances, each indecisive in itself, but all probably necessary for the eventual outcome. Remington had marketed the Sholes machine with its QWERTY keyboard, but this early tie with a major firm did not secure QWERTY's victory. Competition was tough, and no lead meant much with such small numbers in an expanding market. David estimates that only 5,000 or so QWERTY machines existed at the beginning of the 1880s.

The push to incumbency was complex and multifaceted, dependent more upon the software of teachers and promoters than upon the hardware of improving machines. Most early typists used idiosyncratic hunt-and-peck, few-fingered methods. In 1882, Ms. Longley, founder of the Shorthand and Typewriter Institute in Cincinnati, developed and began to teach the eight-finger typing that professionals use today. She happened to teach with a QWERTY keyboard, although many competing arrangements would have served her purposes as well. She also published a do-it-yourself pamphlet that was widely used. At the same time, Remington began to set up schools for typewriting using (of course) its QWERTY standard. The QWERTY ball was rolling but this head start did not guarantee a place at the summit. Many other schools taught rival methods on different machines and might have gained an edge.

Then a crucial event in 1888 probably added the decisive increment to QWERTY's small advantage. Longley was challenged to prove the superiority of her eight-finger method by Louis Taub, another Cincinnati typing teacher, who worked with four fingers on a rival non-QWERTY keyboard with six rows, no shift action, and (therefore) separate keys for upper and lower case letters. As her champion, Longley engaged Frank E. McGurrin, an experienced QWERTY typist who had given himself a decisive advantage that, apparently, no one had thought of before. He had memorized the QWERTY keyboard and could therefore operate his machine as all competent typists do today—by what we now call touch-typing. McGurrin trounced Taub in a well-advertised and well-reported public competition.

In public perception, and (more importantly) in the eyes of those who ran typing schools and published typing manuals, QWERTY had proved its superiority. But no such victory had really occurred. The tie of McGurrin to QWERTY was fortuitous and a good break for Longley and for Remington. We shall never know why McGurrin won, but reasons quite independent of QWERTY cry out for recognition: touch-typing over hunt-and-peck, eight fingers over four fingers, the three-row letter board with a shift key versus the six-row board with two separate keys for each letter. An array of competitions that would have tested QWERTY were never held—QWERTY versus other arrangements of letters with both contestants using eight-finger touch-typing on a three-row keyboard or McGurrin's method of eight-finger touch-typing on a non-QWERTY three-row keyboard versus Taub's procedure to see whether the QWERTY arrangement (as I doubt) or McGurrin's method (as I suspect) had secured his success.

In any case, the QWERTY steamroller now gained crucial momentum and prevailed early in our century. As touch-typing by QWERTY became the norm in America's typing schools, rival manufacturers (especially in a rapidly expanding market) could adapt their machines more easily than people could change their habits—and the industry settled upon the wrong standard.

If Sholes had not gained his tie to Remington, if the first man who decided to memorize a keyboard had used a non-QWERTY design, if McGurrin had a bellyache or drank too much the night before, if Longley had not been so zealous, if a hundred other perfectly possible things had happened, then I might be typing this essay with more speed and much greater economy of finger motion.

But why fret over lost optimality. History always works this way. If Montcalm had won a battle on the Plains of Abraham, perhaps I would be typing *en français*. If a portion of the African jungles had not dried to savannas, I might still be an ape up a tree. If some comets had not struck the earth (if they did) some 60 million years ago, dinosaurs might still rule the land, and all mammals would be rat-sized creatures scurrying about in the dark corners of their world. If *Pikaia,* the only chordate of the Burgess Shale, had not survived the great sorting out of body plans after the Cambrian explosion, mammals might not exist at all. If multi-cellular creatures had never evolved after five-sixths of life's history had yielded nothing more complicated than an algal mat, the sun might explode a few billion years hence with no multicellular witness to the earth's destruction.

Compared with these weighty possibilities, my indenture to QWERTY seems a small price indeed for the rewards of history. For if history were not so maddeningly quirky, we would not be here to enjoy it. Streamlined optimality contains no seeds for change. We need our odd little world, where QWERTY rules and the quick brown fox jumps over the lazy dog.[1]

NOTE

1. I must close with a pedantic footnote, lest nonaficionados be utterly perplexed by this ending. This quirky juxtaposition of uncongenial carnivores is said to be the shortest English sentence that contains all twenty-six letters. It is, as such, *de rigueur* in all manuals that teach typing.

I.B.M.'s 5,000,000,000 Gamble

T. A. Wise

The decision by the management of the International Business Machines Corp. to produce a new family of computers, which it calls the System/360, has emerged as the most crucial and portentous—as well as perhaps the riskiest—business judgment of recent times. The decision committed I.B.M. to laying out money in sums that read like the federal budget—some $5 billion over a period of four years. To launch the 360, I.B.M. has been forced into sweeping organizational changes, with executives rising and falling with the changing tides of the battle. Although the fact has largely escaped notice, the very character of this large and influential company has been significantly altered by the ordeal of the 360, and the way it thinks about itself has changed, too. Bob Evans, the line manager who had the major responsibility for designing this gamble of a corporate lifetime, was only half joking when he said: "We called this project 'You bet your company.'"

Evans insists that the 360 "was a damn good risk, and a lot less risk than it would have been to do anything else, or to do nothing at all," and there is a lot of evidence to support him. But despite the fact that I.B.M. claims that the 360 is a great success, and hails it as a commercial triumph, it is still far too early to say whether the decision to press on with it was a thoroughly sound one. A long stride ahead in the technology of computers in commercial use

was taken by the 360. So sweeping are the implications that it may be ten years before there is enough data to evaluate the wisdom of the whole undertaking.

The new System/360 was intended to obsolete virtually all other existing computers—including those being offered by I.B.M. itself. Thus the first and most extraordinary point to note about this decision is that it involved a challenge to the marketing structure of the computer industry—an industry that the challenger itself had dominated overwhelmingly for nearly a decade. It was roughly as though General Motors had decided to scrap its existing makes and models and offer in their place one new line of cars, covering the entire spectrum of demand, with a radically redesigned engine and an exotic fuel.

The computer is recognized as the most vital tool of management introduced in this generation. It increasingly affects not only business corporations, but government and education as well. There are now perhaps 35,000 computers in use, and it has been estimated that there will be 85,000 by 1975. I.B.M. sits astride this exploding market, accounting for something like two-thirds of the worldwide business—i.e., the dollar value of general-purpose computers currently installed or on order. I.B.M.'s share of this market last year represented about 77 percent of the company's $3.6-billion gross revenues, and probably accounted for about the same share of I.B.M.'s profits, which totaled $477 million.

Several separate but interrelated steps

were involved in the launching of System/360. Each one of the steps involved major difficulties, and taking them all meant that I.B.M. was accepting a staggering challenge to its management capabilities. First, the 360 depended heavily on microcircuitry, an advanced technology in the field of computers. (See "In Electronics, the Big Stakes Ride on Tiny Chips," FORTUNE, June.) In a 1952 vacuum-tube model of I.B.M.'s first generation of computers, there were about 2,000 components per cubic foot. In a second-generation machine, which used transistors instead of tubes, the figure was 5,000 per cubic foot. The System/360 model 75 computer, using hybrid microcircuitry, involves 30,000 components per cubic foot. The old vacuum-tube computer could perform approximately 2,500 multiplications per second; the 360 model 75 was designed to perform 375,000 per second. The cost of carrying out 100,000 computations on the first-generation model was $1.38; the 360 will reduce the cost to 3½ cents.

The second step was the provision for compatibility—that is, as the users' computer requirements grew, they could move up from one machine to another without having to discard or rewrite already existing programs. Limited compatibility had already been achieved by I.B.M., and by some of its competitors too, for that matter, on machines of similar design but different power. But it had never been achieved on a broad line of computers with a wide range of powers, and achieving this compatibility depended as much on developing compatible programs or "software" as it did on the hardware. All the auxiliary machines—"peripheral equipment" as they are called in the trade—had to be designed so that they could feed information into or receive information from the central processing unit; this meant that the equipment had to have timing, voltage, and signal levels matching those of the central unit. In computerese, the peripheral equipment was to have "standard interface." The head of one competing computer manufacturing company acknowledges that at the time of the System/360 announcement he regarded the I.B.M. decision as sheer folly and doubted that I.B.M. would be able to produce or deliver a line that was completely compatible.

Finally—and this was the boldest and most perilous part of the plan—it was decided that six main units of the 360 line, originally designated models 30, 40, 50, 60, 62, and 70, should be announced and made available simultaneously. (Models at the lower and higher ends of the line were to be announced later.) This meant that all parts of the company would have to adhere to a meticulous schedule.

UP IN MANUFACTURING, DOWN IN CASH

The effort involved in the program has been enormous. I.B.M. spent over half a billion dollars on research and development programs associated with the 360. This involved a tremendous hunt for talent: by the end of this year, one-third of I.B.M.'s 190,000 employees will have been hired since the new program was announced. Between that time, April 7, 1964, and the end of 1967, the company will have opened five new plants here and abroad and budgeted a total of $4.5 billion for rental machines, plant, and equipment. Not even the Manhattan Project, which produced the atomic bomb in World War II, cost so much (the government's costs up to Hiroshima are reckoned at $2 billion), nor, probably, has any other privately financed commercial project in history.

Such an effort has already changed I.B.M.'s nature in several ways:

· The company, which was essentially an assembler of computer components and a business-service organization, has now also become a major manufacturing concern as well. It is the world's largest maker of integrated circuits, producing an estimated 150 million of the hybrid variety annually.

• After some ambivalence, I.B.M. has abandoned any notion that it is simply another American company with a large foreign operation. The view now is that I.B.M. is a fully integrated international company, in which the managers of overseas units are presumed to have the same capabilities and responsibilities as those in the U.S. The company's World Trade subsidiary has stopped trying to develop its own computers; instead, it is marketing the 360 overseas, and is helping in the engineering and manufacturing of the 360.

• The company's table of organization has been restructured significantly at least three times since 1960. Several new divisions and their executives have emerged, while others have suffered total or partial eclipse. An old maxim of the I.B.M. organization was that few men rose to line executive positions unless they had spent some time selling. Now a new group of technically oriented executives has come to the forefront for the first time, diluting some of the traditional power of the marketing men in the corporation.

• I.B.M.'s balance sheet has a new look. In 1963, about 27 percent of the company's assets were in the form of cash or marketable securities. By the end of 1965 assets of that kind were down to 18 percent, and the balance sheet showed $173 million less cash than it had two years earlier. Last spring the company called on its shareholders for $371 million of equity capital to help finance the new computer family.

THE MISSIONARIES AND THE SCIENTISTS

Oddly enough, the upheaval at I.B.M. during the past two years went largely unnoticed. The company was able to make itself over more or less in private. It was able to do so partly because I.B.M. is so widely assumed to be an organization in which the unexpected simply doesn't happen. Outsiders viewing I.B.M. presume it to be a model of rationality and order—a presumption related to the company's products, which are, of course, instruments that enable (and require) their users to think clearly about management.

This image of I.B.M., moreover, has been furthered over the years by the styles of the two Watsons. Tom Watson Sr. combined an intense devotion to disciplined thinking with formal, rather Victorian attitudes about conduct, clothes, and courtesy. The senior Watson's hostility toward drinking, and his demand that employees dedicate themselves totally to the welfare of the corporation, created a kind of evangelical atmosphere. When Tom Watson Jr. took over from his father in 1956, the manner and style shifted somewhat, but the missionary zeal remained—now overlaid by a new dedication to the disciplines of science. The overlay reinforced the image of I.B.M. as a chillingly efficient organization, one in which plans were developed logically and executed with crisp efficiency. It was hard to envision the company in a gambling role.

The dimensions of the 360 gamble are difficult to state precisely. The company's executives, who are men used to thinking of risks and payoffs in hard quantitative terms, insist that no meaningful figure could ever be put on the gamble—i.e., on the odds that the program would be brought off on schedule, and on the costs that would be involved if it failed.

OUTSAILING THE BOSS

At the time, it scarcely seemed that any gamble at all was necessary. I.B.M. was way out ahead of the competition, and looked as if it could continue smoothly in its old ways forever. Below the surface, though, I.B.M.'s organization didn't fit the changing markets so neatly any more, and there really was, in Evans' phrase, a risk involved in doing nothing.

No one understood this more thoroughly, or with more sense of urgency, than one of the principal decision makers of the company, T. Vincent Learson. His entire career at I.B.M., which began in 1935, has been concerned with getting new products to market. In 1954 he was tapped by young Tom Watson as the man to spearhead the company's first big entry into the commercial computer field—with the 702 and the 705 models. His success led to his promotion to vice president and group executive in 1956. In 1959 he took over both of the company's computer development and manufacturing operations, the General Products Division and the Data Systems Division.

Learson stands six foot six and is a tough and forceful personality. When he is managing any major I.B.M. program, he tends to be impatient with staff reports and committees, and to operate outside the conventional chain of command; if he wants to know why a program is behind schedule, he is apt to call directly on an executive at a much lower level who might help him find out. But he often operates indirectly, too, organizing major management changes without his own hand's being visible to the men involved. Though he lacks the formal scientific background that is taken for granted in many areas of I.B.M., Learson has a reputation as a searching and persistent questioner about any proposals brought before him; executives who have not done their homework may find their presentations falling apart under his questions—and may also find that he will continue the inquisition in a way that makes their failure an object lesson to any spectators. And Learson is the most vigorous supporter of the company's attitude that a salesman who has lost an order without exhausting all the resources the company has to back him up deserves to be drawn and quartered.

Learson's personal competitiveness is something of a legend at I.B.M. It was significantly demonstrated in this year's Newport-to-Bermuda yacht race, in which Learson entered his own boat, the *Thunderbird.* He boned up on the history of the race in past years, and managed to get a navigator who had been on a winning boat three different times. He also persuaded Bill Lapworth, the famous boat designer, to be a crewman. Learson traveled personally to California to get one of the best spinnaker men available. All these competitive efforts were especially fascinating to the people at I.B.M. because Tom Watson Jr. also had an entry in the Bermuda race; he had, in fact, been competing in it for years. Before the race Watson good-humoredly warned Learson at a board meeting that he'd better not win if he expected to stay at I.B.M. Learson's answer was not recorded. But Learson won the race. Watson's *Palowan* finished twenty-fourth on corrected time.

When Learson took over the computer group he found himself supervising two major engineering centers that had been competing with each other for some time. The General Products Division's facility in Endicott, New York, produced the low-priced 1401 model, by far the most popular of all I.B.M.'s computers—or of anyone else's; to date something like 10,000 of them have been installed. Meanwhile, the Data Systems Division in Poughkeepsie made the more glamorous 7000 series, of which the 7090 was the most powerful. Originally, I.B.M. had intended that the two centers operate in separate markets, but as computer prices came down in the late 1950's, and as more versions of each model were offered, their markets came to overlap—and they entered a period in which they were increasingly penetrating each other's markets, heightening the feeling of rivalry. Each had its own development program, although any decision to produce or market a new computer, of course, had to be ratified at corporate headquarters. The rivalry between the two divisions was to become an element in, and be exacerbated by, the decision to produce the 360.

Both the 1401 and the 7000 series were selling well in 1960. But computer engineers and architects are a restless breed; they are apt to be thinking of improvements in design or circuitry five minutes after the specifications of their latest machines are frozen. In the General Products Division, most such thinking in 1960 and 1961 was long term; it was assumed that the 1401 would be on the market until about 1968. The thinking at the Data Systems Division concerned both long-range and more immediate matters.

A $20-MILLION STRETCH

One of the immediate matters was the division's "Stretch" computer, which was already on the market but having difficulties. The computer had been designed to dwarf all others in size and power, and it was priced around $13,-500,000. But it never met more than 70 percent of the promised specifications, and not many of them were sold. In May, 1961, Tom Watson made the decision that the price of Stretch should be cut to $8 million to match the value of its performance—at which level Stretch was plainly uneconomic to produce. He had to make the decision, it happened, just before he was to fly to California and address an industry group on the subject of progress in the computer field.

Before he left for the coast, an annoyed Watson made a few tart remarks about the folly of getting involved in large and overambitious projects that you couldn't deliver on. In his speech, he admitted that Stretch was a flop. "Our greatest mistake in Stretch," he said, "is that we walked up to the plate and pointed at the left-field stands. When we swung, it was not a homer but a hard line drive to the outfield. We're going to be a good deal more careful about what we promise in the future." Soon after he returned the program was quietly shelved; today only seven of the machines are in operation. I.B.M.'s over-all loss on the program was about $20 million.

The Stretch fiasco had two consequences. One was that the company practically ignored the giant-computer field during the next two years—and thereby enabled Control Data to get a sizable headstart in the market. (See "Control Data's Magnificent Fumble," FORTUNE, April.) The market's customers are principally government and university research centers, where the most complex scientific problems are tackled and computers of tremendous power are required. Eventually, in 1963, Watson pointed out that his strictures against overambitious projects had not been meant to exclude I.B.M. from this scientific market, and the company is now trying to get back into it. Its entry will be the 360-90, the most powerful machine of the new line.

A second consequence of the Stretch fiasco is that Learson and the men under him, especially those in the Data Systems Division, were under special pressure to be certain that the next big project was thought out more carefully and that it worked exactly as promised. As it happened, the project the division had in mind in 1960–61 was a fairly ambitious one: it was for a line of computers, tentatively called the 8000 series, that would replace the 7000 series, and would also provide a limited measure of compatibility among the four models projected. The 8000 series was based on transistor technology, and therefore still belonged to the second generation; however, there had been so much recent progress in circuitry design and transistor performance that the series had considerably more capability than anything being offered by I.B.M. at that time.

The principal sponsor of the 8000 concept was Fred Brooks, head of systems planning for the Poughkeepsie division. An imaginative, enthusiastic twenty-nine-year-old North Carolinian with a considerable measure of southern charm, Brooks became completely dedicated to the concept of the new series, and beginning in late 1960 he began trying to enlist support for it. He had a major opportunity to make his case for the 8000 program at a briefing for the division's management, which was held at Poughkeepsie in January, 1961.

By all accounts, he performed well: he was relaxed, confident, informed on every aspect of the technology involved, and persuasive about the need for a change. Data Systems' existing product line, he argued, was a mixed bag. The capability of some models overlapped that of others, while still other capabilities were unavailable in any model. The 8000 series would end all this confusion. One machine was already built, cost estimates and a market forecast had been made, a pricing schedule had been completed, and Brooks proposed announcing the series late that year or early in 1962. It could be the division's basic product line until 1968, he added. Most of Brooks's auditors found his case entirely persuasive.

ENTER THE MAN FROM HEADQUARTERS

Learson, however, was not ready to be sold so easily. The problems with Stretch must have been on his mind, and probably tended to make him look hard at any big new proposals. Beyond that, he was skeptical that the 8000 series would minimize the confusion in the division's product line, and he wondered whether the concept might not even *contribute* to the confusion. Learson had received a long memorandum from his chief assistant, Don Spaulding, on the general subject of equipment proliferation. Spaulding argued that there were already too many different computers in existence, and that they required too many supporting programs and too much peripheral equipment; some drastic simplification of the industry's merchandise was called for.

With these thoughts in mind, Learson was not persuaded that Brooks's concept was taking I.B.M. in the right direction. Finally, he was not persuaded that the company should

again invest heavily in second-generation technology. Along with a group of computer users, he had recently attended a special course on industrial dynamics that was being given at the Massachusetts Institute of Technology. Much of the discussion had been over his head, he later recalled; but from what his classmates were saying he came away with the clear conviction that computer applications would soon be expanding rapidly, and that what was needed was a bold move away from "record keeping" and toward more sophisticated business applications.

There was some direct evidence of Learson's skepticism about the 8000 series. Shortly after the briefing Bob Evans, who was then manager of processing systems in the General Products Division, was dispatched to Poughkeepsie as head of Data Systems' planning and development. He brought along a number of men who had worked with him in Endicott. Given the rivalry between the two divisions, it is not very surprising that he received a cool welcome. His subsequent attitude toward the 8000 concept ensured that his relations with Brooks would stay cool.

Evans made several different criticisms of the concept. The main one was that the proposed line was "nonhomogeneous"—that is, it was not designed throughout to combine scientific and business applications. Further, he contended that it lacked sufficient compatibility within the line. It would compound the proliferation problem. He also argued that it was time to turn to the technologies associated with integrated circuits.

BLOOD ON THE FLOOR

For various reasons, including timing, Brooks was opposed, and he and Evans fought bitterly for several months. At one point Evans called him and quietly mentioned that Brooks was getting a raise in salary. Brooks started to utter a few words of thanks when Evans said flatly, "I want you to know I had nothing to do with it."

In March, 1961, Brooks had a chance to make a presentation to the corporate management committee, a group that included Tom Watson, his brother, A. K. Watson, who headed the World Trade Corp., Albert Williams, who was then president of the corporation (he is now chairman of the executive committee), and Learson. Brooks made another effective presentation, and for a while he and his allies thought that the 8000 might be approved after all.

But early in May it became clear that Evans was the winner. His victory was formalized in a meeting, at the Gideon Putnam Hotel in Saratoga, of all the key people who had worked on the 8000. There, on May 15, Evans announced that the 8000 project was dead and that he now had the tough job of reassigning them all to other tasks. In the words of one participant, "There was blood all over the floor."

Evans now outlined some new programs for the Data Systems Division. His short-term program called for an extension of the 7000 line, both upward and downward. At the lower end of the line there would be two new models, the 7040 and the 7044. At the upper end there would be a 7094 and a 7094 II. This program was generally noncontroversial, except for the fact that the 7044 had almost exactly the same capabilities as a computer called Scamp, which was being proposed by another part of I.B.M. It would obviously make no sense to build both computers; and, as it happened, Scamp had some powerful support.

Scamp was a small scientific computer developed originally for the European market. Its principal designer was John Fairclough, a young man (he was then thirty) working in the World Trade Corp.'s Hursley Laboratory, sixty miles southwest of London. The subsidiary had a sizable stake in Scamp. It had been trying for

many years to produce a computer tailored to the needs of its own markets, but had repeatedly failed, and had therefore been obliged to sell American-made machines overseas.

But Scamp looked especially promising, and the subsidiary's executives, including Fairclough and A. K. Watson, were confident that it would meet American standards. It had previously tested well and attracted a fair amount of attention in I.B.M.'s American laboratories. Evans himself came to Hursley to look at it, and was impressed. But its similarity to the 7044 finally took Fairclough and some associates to the U.S. to test their machine against a 7044 prototype.

MERE EQUALITY WON'T DO

As things turned out, Scamp did about as well as the 7044—but, also as things turned out, that wasn't good enough. Evans and Learson were resolved to stretch out the 7000 line, but opposed to anything that would add to proliferation. In principle, A. K. Watson, who had always run World Trade as a kind of personal fiefdom, could have stepped in and ordered the production of Scamp on his own authority. In practice, he decided the argument against proliferation was a valid one. And so, in the end, he personally gave the order to drop Scamp. Fairclough got the news one day soon after he had returned to England, and he found himself with a sizable staff that had to be reassigned. He says that he considered resigning, but instead worked off his annoyance by sipping Scotch and brooding much of that night.

Evans and Learson had also agreed that Data Systems should try its hand at designing a computer line that would blanket the market. The General Products Division was asked to play a role in the new design, but its response was lukewarm, so the bulk of the work at this stage fell to Data Systems. The project was

51

dubbed NPL, for new product line; the name System/360 was not settled on until much later. To head the project, Evans selected his old adversary Brooks—a move that surprised a large number of I.B.M. executives, including Brooks himself.

Still smarting over the loss of the 8000 project, and suspicious that the NPL was just a "window-dressing" operation, Brooks accepted the job only tentatively. To work with him, and apparently to ensure that NPL did not end up as the 8000 under a new name, Evans brought Gene Amdahl, a crack designer whom the company had called on to work on several earlier computers. However, Amdahl's influence was offset by that of another designer, Gerrit Blaauw, a veteran and past supporter of the 8000 project. Brooks's group received enough money to show that the company took NPL seriously (the first-year appropriation was $3,800,000), but Amdahl and Blaauw disagreed on design concepts, and the project floundered until November of 1961.

Even to the trained eye I.B.M.'s main divisions appeared to be in excellent health in the summer of 1961. The General Products Division, according to Evans, was "fat and dumb and happy" in the lower end of the market, selling the 1401 at a furious rate, and still feeling secure about its line through about 1968. The World Trade Corp. was growing rapidly, although it had suffered its third major setback on getting a computer line of its own. The Data Systems Division was extending its old 7000 line to meet the competition, and working on the NPL.

THE PROLIFERATING PRODUCTS

But it was around this time that Tom Watson and Learson—then a group executive vice president, and nominally at least working under Albert Williams, the company president—developed several large concerns. There was the company's persistent difficulty in grappling with the new technology and with the expanding demands of the market. There was the absence of any clear, over-all concept of the company's product line; fifteen or twenty different engineering groups scattered throughout the company were generating different computer products, and while the products were in most cases superior, the proliferation was putting overwhelming strains on the company's ability to supply programing for customers. The view at the top was that I.B.M. required some major changes if it expected to stay ahead in the computer market when the third generation came along.

Between August and October, 1961, Watson and Learson initiated a number of dialogues with their divisional lieutenants in an effort to define a strategy for the new era. By the end of October, though, neither of them believed that any strategy was coming into focus. At this point Learson made a crucial decision. He decided to set up a special committee, composed of representatives from every major segment of the company, to formulate some policy guidance. The committee was called SPREAD—an acronym for systems programing, research, engineering, and development. Its chairman was John Haanstra, then a vice president of the General Products Division. There were twelve other members, including Evans, Brooks, and Fairclough.

HAANSTRA THE HAMMER

The SPREAD committee was conducted informally, but with a good amount of spirited discussion. For some purposes it broke up into separate committees, such as one on programing compatibility. Haanstra, as one member put it, acted as a hammer on the committee anvil, forcing ideas into debate and demanding definitions. Still, there was some feeling that Haanstra was bothered by the fact that the group was heavily represented by "big machine" oriented men.

The progress of the committee during November was steady, but it was also, in Learson's view, "hellishly slow." Suddenly Haanstra found himself promoted to the presidency of the General Products Division and Bob Evans took over as chairman of SPREAD. The committee meetings were held in the New Englander Motor Hotel, just north of Stamford, Connecticut. In effect, although not quite literally, Learson locked the doors and told the members that they couldn't get out until they had reached some conclusions.

While Evans accelerated the pace of the sessions somewhat, Fred Brooks increasingly emerged as the man who was shaping the direction of the committee recommendations. This was not very surprising, for he and his group had had a headstart in thinking out many of the issues. By December 28, 1961, the SPREAD committee had hammered out an eighty-page statement of its recommendations. On January 4, 1962, the committee amplified the report for the benefit of the fifty top executives of the corporation.

Brooks was assigned the role of principal speaker on this occasion. The presentation was split into several parts and took an entire day. The main points of the report were:

- There was a definite need for a single, compatible family of computers ranging from one with the smallest existing core memory, which would be below the 1401 line, to one as powerful as I.B.M.'s biggest—at that time the 7094. In fact, the needs were said to extend beyond the I.B.M. range, but the report expressed doubt that compatibility could be extended that far.
- The new line should not be aimed simply at replacing the popular 1401 or 7000 series, but at opening up whole new fields of com-

puter applications. At that time compatibility between those machines and the new line was not judged to be of major importance, because the original timetable on the appearance of the various members of the new family of computers stretched out for several years.

- The System/360 must have both business and scientific applications. This dual purpose was a difficult assignment because commercial machines accept large amounts of data but have little manipulative ability, while scientific machines work on relatively small quantities of data that are endlessly manipulated. To achieve duality the report decided that each machine in the new line would be made available with core memories of varying sizes. In addition, the machine would provide a variety of technical and esoteric features such as "floating point arithmetic," "variable word length" and a "decimal instruction set" to handle both scientific and commercial assignments.
- Information input and output equipment, and all other peripheral equipment, must have "standard interface"—so that various types and sizes of peripheral equipment could be hitched to the main computer without missing a beat. This too was to become an important feature of the new line.

Learson recently recalled the reaction when the presentation ended. "There were all sorts of people up there and while it wasn't received too well, there were no real objections. So I said to them, 'All right, we'll do it,' The problem was, they thought it was too grandiose. The report said we'd have to spend $125 million on programing the system at a time when we were spending only about $10 million a year for programing. Everybody said you just couldn't spend that amount. The job just looked too big to the marketing people, the financial people, and the engineers. Everyone recognized it was a gigantic task that would mean all our resources were tied up in one project—and we knew that for a long time we wouldn't be getting anything out of it."

Considering the fact that shipments of System/360 machines have been moving well—some 4,000 to date, with an estimated 20,000 on order—Learson's "go ahead" decision, which led to so huge a commitment, would seem indisputably to have been the wise one. But the validity of some of the critics' points was proved with company-shaking accuracy when I.B.M. turned from planning to production, financing, and selling.

Innovation:
Mapping the Winds of Creative Destruction

William J. Abernathy

Kim B. Clark

This paper develops a framework for analyzing the competitive implications of innovation. The framework is based on the concept of transilience—the capacity of an innovation to influence the established systems of production and marketing. Application of the concept results in a categorization of innovation into four types. Examples from the technical history of the US auto industry are used to illustrate the concepts and their applicability. The analysis shows that the categories of innovation are closely linked to different patterns of evolution and to different managerial environments. Special emphasis is placed on the role of incremental technical change in shaping competition and on the possibilities for a technology based reversal in the process of industrial maturity.

1. INTRODUCTION

Technological innovation has been a powerful force for industrial development, productivity growth and indeed, our rising standard of living throughout history, but intense study of its

"Innovation: Mapping the Winds of Creative Destruction" by William J. Abernathy and Kim B. Clark, from *Research Policy*, Number 14, 1985, pp. 3–22. Copyright 1985 by Elsevier Science Publishers, B.V. (North Holland). Reprinted by permission.

Support for research underlying this paper was provided by the Division of Research, Harvard University Graduate School of Business Administration, whose contribution is gratefully acknowledged. Appreciation is also expressed to those colleagues who provided encouragement and ideas, especially George White, Alan Kantrow, Richard Rosenbloom, and Robert Hayes. Special appreciation is expressed to John Corcoran for extraordinary research assistance.

industrial role and influence is a relatively recent phenomenon. In traditional economics, it has long been customary to treat technological innovation as something that happened to the economic system but was not determined within it[1]. Some recent work has examined the determinants of innovation, with emphasis on the role of market demand and the influence of market structure. The focus has not been on the process of innovation itself, but rather on those aspects of the firm's (or industry's) environment that spur or retard technical advance[2].

In contrast to the bulk of economic analysis, the work of technologists and behavioral scientists has focused on what goes on inside the black box of technology. In this line of work, innovation is viewed as a sequence of activities involving the acquisition, transfer and utilization of information[3]. Although the importance of activities outside the firm (or project) is often recognized, the orientation of these studies is internal. Of principal concern are personality traits of individuals, the origin of innovative ideas and the way that administrative practices and organization structure influence their development. In this work, however, the internal traits of the firm have not been well linked to the competitive requirements of firms or industries.

More recently studies by Porter, Rosenbloom, Rosenberg, Nelson and Winter, and others, have begun to illuminate some of the important aspects of the relationship between innovation and competition[4]. While earlier work tended to deal with the effect of structural characteristics (i.e. levels of hierarchy, firm size, concentration) and administrative practices on innovation, these new studies have begun to ask questions about innovation's role in shaping the competitive environment.

Regrettably however, the conclusions Rosenbloom drew in his comprehensive review of the literature on innovation are largely as applicable today as they were ten years ago.[5] Notwithstanding some interesting findings, he concluded that results from this field offered little guidance to either the public policy maker or the business manager who must contend with practical aspects of technology investment and use.

As a remedy for this plight, Rosenbloom called for the development of a conceptual framework that would integrate knowledge concerning technology to other policy arenas of the firm (i.e. marketing, finance, operations, etc.). Without a schema that managers might apply to develop and communicate perspectives on technology throughout their organizations, it is not surprising that technology and its management have received little systematic attention in the formulation and implementation of policy.

The challenge of formulating such a framework for technology policy marks the point of departure of the present work. Our purpose in this paper is to develop a descriptive framework that may be useful in categorizing innovations and analyzing the varied role they play in competition. The framework recognizes that innovation is not a unified phenomenon: some innovations disrupt, destroy and make obsolete established competence; others refine and improve. Further, the effects of innovation on production systems may be quite different from their effects on linkages to customers and markets. The framework reflects these differences, as well as the important notion (developed in the work of Burns and Stalker, and others) that different kinds of innovation require different kinds of organizational environments and different managerial skills. Our intent is thus to develop concepts that may prove useful in the effort to incorporate technological considerations into business strategy, and perhaps in developing appropriate public policies.

We explicate the model through intensive analysis of the technical and competitive history of the US auto industry. An in-depth look at a particular industry provides a level of detail that seems essential in making the kinds of distinctions we are after. We thus do not offer the auto story as a representation of the whole economy, but rather see it as a source of useful examples. Further, the use of an historical perspective in explaining the framework is deliberate. We are interested in the pattern of technological development and competitive rivalry over time. In our emphasis on the pattern of development over time and our conception of competition as a contest among rivals (actual and potential) with different capabilities, the model we develop is evolutionary in spirit.

The paper is divided into three parts. In section 2 we develop criteria for categorizing innovation in terms of its competitive significance. The criteria are based on the notion that competitive advantage depends on the acquisition or development of particular skills, relationships and resources. How innovation affects those requirements can be used to gauge its significance. Building on these concepts, section 2 further identifies four different "modes" of innovation, and uses examples from the auto industry to illustrate them. In section 3 we relate the different kinds of innovation to the pattern of industry evolution. Special emphasis is placed on the role of incremental technical change in shaping competition, and on the possibilities for a technology based reversal in the process of industrial maturity. The paper concludes (section 4) with observations on implications for practice and further research.

2. IDENTIFYING THE ROLE OF INNOVATION IN COMPETITION: THE TRANSILIENCE MAP

The first step in developing a categorization of innovation is to get straight the question of perspective. Technological innovation may influence a variety of economic actors in a variety of ways, and it is this variety that gives rise to differing views of the significance of changes in technology. What may be a startling breakthrough to the engineer, may be completely unremarkable as far as the user of the product is concerned. In this paper we shall evaluate innovation in terms of its implications for the success (or failure) of the innovating firm in its rivalry with competitors. We are thus concerned with how, and to what extent, innovation affects the relative advantages of actual and potential competitors.

The notion of competitive advantage that we use here is broader than the differences in costs among competitors. Without minimizing the importance of cost in competition, we propose to consider the competitive position of a firm in terms of a variety of dimensions. We assume therefore that products are not homogeneous, and that firms compete by offering products that may differ in many aspects: performance, reliability, availability, ease of use,

aesthetic appearance, and image (among others), as well as initial cost. A firm gains a competitive advantage when it achieves a position in one of these featured dimensions, or a combination of them, that is both valued by customers and superior to that of its competitors.

It is important to note that the product features themselves, and the firm's position with them, are not in and of themselves the fundamental source of advantage. Such a position is the immediate, outward manifestation of a more fundamental, internal reality. The foundation of a firm's position rests on a set of material resources, human skills and relationships, and relevant knowledge. These are the competencies or competitive ingredients from which the firm builds the product features that appeal to the marketplace. Thus, the significance of innovation for competition depends on what we shall call its "transilience"—that is, its capacity to influence the firm's existing resources, skills and knowledge[6].

Table 1 presents a list of the major competitive ingredients divided into two groups. In the top half of the table we have placed the factors that determine the capabilities of the firm in technology and production. For ease of notation, we shall refer to this set as the "Technology" side of the firm, but mean it to include production and operations as well. The resources, skills and knowledge within this domain are linked to competition through their effect on the physical characteristics of the product—its performance, appearance, quality, and so on—and through its cost. The list includes traditional factors of production like materials, people, building and equipment, as well as knowledge relevant to design and production. This not only includes links to scientific, engineering and design disciplines, but it also includes the knowledge embedded in the systems and procedures used to organize production.

In this formulation we make a distinction between the skills and the knowledge embodied in individuals and the collective under-

TABLE 1. *Innovation and Firm Competence*

Domain of Innovative Activity	Range of Impact of Innovation		
I. *Technology/Production*			
Design/embodiment of technology	improves/perfects established design	⟷	offers new design/radical departure from past embodiment
Production systems/ organization	strengthens existing structure	⟷	makes existing structure obsolete demands new system, procedures, organization
Skills (labor, managerial, technical)	extends viability of existing skills	⟷	destroys value of existing expertise
Materials/supplier relations	reinforces application of current materials/ suppliers	⟷	extensive material substitution; opening new relations with new vendors
Capital equipment	Extends existing capital	⟷	extensive replacement of existing capital with new types of equipment
Knowledge and experience base	builds on an reinforces applicability of existing knowledge	⟷	establishes links to whole new scientific discipline/destroys value of existing knowledge base
II. Market/Customer			
Relationship with customer base	strengthens ties with established customers	⟷	attracts extensive new customer group/creates new market
Customer applications	improves service in established application	⟷	creates new set of applications/ new set of customer needs
Channels of distribution and service	builds on and enhances the effectiveness of established distribution network/service organization	⟷	requires new channels of distribution/new service, after market support
Customer knowledge	uses and extends customer knowledge and experience in established product	⟷	intensive new knowledge demand of customer; destroys value of customer experience
Modes of customer communication	reinforce existing modes/methods of communication	⟷	totally new modes of communication required (e.g. field sales engineers)

standing shared among groups of employees, and partly incorporated into teamwork routines, procedures, practices, and so forth. We make a further distinction between the factors of production—labor, capital and materials— and their organization and deployment. We assume that a given set of inputs can be combined and organized in a variety of ways to achieve different results, either technically in terms of the sequencing of operations and factor combinations, or organizationally in terms of systems for acquiring and processing information. The infrastructure of production—e.g. organization, system, procedures—thus merits separate consideration.

The second half of table 1 is devoted to

markets and linkages to customers. Of central importance in this domain is the relationship with the customer base. We include in this category both the strength of the relationship, as well as the composition of the customer group. The other items in this domain affect the way in which customers relate to the product. We make a distinction between the applications the product serves, and the knowledge and experience required in the product's use. Further, we have included both the way in which customers obtain the product and related services, and the way in which they obtain information about its characteristics.

Each item listed in the table is accompanied by a scale that depicts the range of effects an innovation might have. In each case the range is defined by polar extremes, the one conservative, the other radical. On the conservative end of the scale are those innovations that serve to enhance the value or applicability of the firm's existing competence. Clearly, all technological innovation imposes change of some kind, but change need not be destructive. Innovation in product technology may solve problems or eliminate flaws in a design that makes existing channels of distribution more attractive and effective. Further, innovation in process technology may require new procedures in handling information, but utilize existing labour skills in a more effective way. Such changes conserve the established competence of the firm, and if the enhancement or refinement is considerable, may actually entrench those skills, making it more difficult for alternative resources or skills to achieve an advantage. Such innovation may have an effect on competition by raising barriers to entry, reducing the threat of substitute products, and making competing technologies (and perhaps firms) less attractive.

On the radical end of the scale, the effect of innovation is quite the opposite. Instead of enhancing and strengthening, innovation of this sort disrupts and destroys. It changes the technology of process or product in a way that imposes requirements that the existing resources, skills and knowledge satisfy poorly or not at all. The effect is thus to reduce the value of existing competence, and in the extreme case, to render it obsolete. This kind of change is at the heart of Schumpeter's theory of innovation and economic development in which "creative destruction" is the vehicle of growth[7]. Its effect on competition works through a redefinition of what is required to achieve a competitive advantage. In strong form, where disruption is both deep and extensive, such innovation creates new industries.

In the framework laid out in table 1 the significance of an innovation for competitive advantage depends on more than technical novelty or scientific merit. In an age of gene splicing and other scientific marvels that do indeed create new industries and destroy old ones, it is easy to develop a stereotype of innovation that may obscure judgments about technical change and its significance for competition. One need only take the effects listed on the right hand side of table 1 to obtain the following exaggeration:

> An innovation is the initial market introduction of a new product or process whose design departs radically from past practice. It is derived from advances in science, and its introduction makes existing knowledge in that application obsolete. It creates new markets, supports freshly articulated user needs in the new functions it offers, and in practice demands new channels of distribution and aftermarket support. In its wake it leaves obsolete firms, practices, and factors of production, while creating a new industry.

This is a stereotype of an ideal that is both rarely encountered in practice and misleading. Novelty and connection with scientific advance may have little to do with an innovation's competitive significance. The entry of the

59

Timex Corporation into the watch industry provides a useful example.[8] Its success in the market was based on refinements of an old technology (pin lever movement), that was applied in upgraded styling and offered through new channels of distribution (drug stores, discount houses). Using standardization of parts, mechanization, low skilled labor and precision tooling, Timex produced a consistent, durable product, at very low cost. Its use of hard alloy bearings avoided the need for jewels, and its simple design made complex adjustments unnecessary. Further, the company employed a cadre of engineers and technicians to design and build its own tooling and production equipment. In combination with modern styling and an aggressive marketing strategy, the refinements in product and process design gave Timex a significant competitive advantage.

The refinements in technology undertaken by Timex in conjunction with an appropriate business plan provided competitive leverage out of proportion to the technical changes involved. In so doing, they provided the basis for an assault on established barriers to entry that had been built through franchise sales, service and repair, and status connected imagery. Thus, what were a series of relatively mundane changes in technology, came to have major ramifications in the market.

The Timex example reinforces the notion that the competitive significance of an innovation depends on what it does to the value and applicability of established competence— that is, on its transilience. But the example also illustrates the importance of distinguishing between effects on markets and effects on technology or systems of production. A given innovation may affect the two domains in quite different ways. It is the particular combination or pattern of technology and market transilience that is important in determining competitive impact. One way to depict the pattern of effects is to use composite transilience scales for each domain as the axes of a two-dimen-

sional diagram. In fig. 1 we have positioned the market transilience scale in the vertical dimension, and the technology transilience scale in the horizontal. This creates a "transilience map", with four quadrants representing a different kind of innovation. Working counter clockwise from the upper righthand corner, the categories of innovation are: Architectural, Niche, Regular, and Revolutionary. We shall illustrate each category using examples from the US automobile industry; specific innovations in each group are presented in table 2. Further, in section 3 we show that the categories are closely linked to patterns of industry development, and that the four quadrants represent phases of innovative development. Moreover, this categorization of innovative effects is linked to other differences in the evolution of firms and industries, so that the four quadrants also represent different managerial environments.

2.1. Architectural Innovation

New technology that departs from established systems of production, and in turn opens up new linkages to markets and users, is characteristic of the creation of new industries as well as the reformation of old ones. Innovation of this sort defines the basic configuration of product and process, and establishes the technical and marketing agendas that will guide subsequent development. In effect, it lays down the architecture of the industry, the broad framework within which competition will occur and develop. We have thus labelled innovation of this sort "Architectural"; it is graphed in the upper right hand quadrant of the transilience map.

Whether it creates a new industry like xerography or radio, or whether it reformulates an established industry as with photo typesetting in the printing industry, architectural innovation seems to involve a process and an organizational climate that is distinctive. The Charpie report, the work of Burns and Stalker, and re-

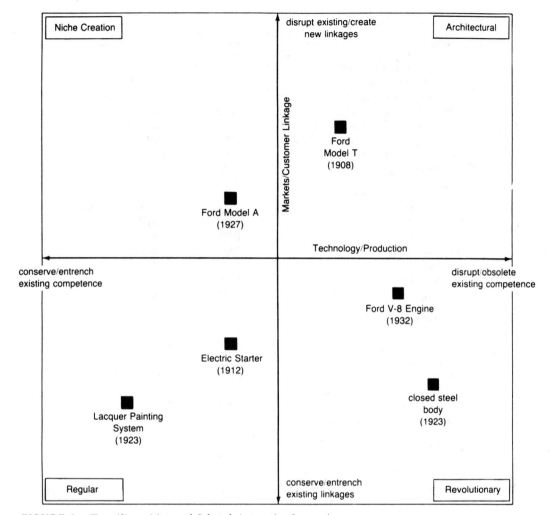

FIGURE 1.   Transilience Map and Selected Automotive Innovations

search by Jewkes, Sawyers and Stillerman, suggest that entrepreneurial action occurs in a unique managerial climate and with firms whose organizational structure is not bureaucratic and rigid[9]. The potential for stimulating architectural innovation seems to hinge on the juxtaposition of individuals with prior experience in relevant technologies and new user environments latent with needs. These insights are well illustrated in the case of the early development of the US automobile industry[10].

The early years of this century marked a turbulent era in the technological development of the motorless carriage. Early cars were made by bicycle manufacturers and wagon makers, and the designs reflected each prototype's origin. Technologies from these industries as well as locomotive manufacturing and the electrical industry were competing to serve an emerging market for personal transportation. What developed out of the contest among diverse design concepts was a creative synthesis that es-

TABLE 2.   Selected Automotive Innovations

Innovation	Year	Company
I. Architectural innovations		
Left hand steering	1890	G and J Jeffrey
Front mounted engine	1900	most producers
Planetary transmission	1902	Northern
Unitary engine and transmission	1902	Northern
Magneto integrated into flywheel	1908	Ford
Removable cylinder heads	1908	Ford
Vanadium crankshaft	1908	Ford
II. Market Niche innovation		
Safety glass	1926	Rickenbacker, Stutz
Streamlined bodies	1934	Chrysler
Station wagon	1923	Star
Hardtop convertible	1946	Chrysler
Bucket seats	1959	GM
Wide-track chassis	1959	GM
Low-priced sports car (Mustang)	1965	Ford
III. Regular innovation		
Electric starter	1912	GM
Moving assembly line	1913	Ford
Lacquer finish (DUCO-pyroxolin)	1924	GM
Rubber engine mounts	1922	Nash
Constant temp. inspection room	1924	Ford
Automatic welding	1925	Budd
Thin wall gray cast iron engine	1959	Ford
IV. Revolutionary innovation		
Closed steel body	1922	Hudson
V-8 engine cast en-bloc	1934	Ford
Automatic transmission	1940	GM
Cast steel cam and crankshaft	1934	Ford
Independent suspension	1934	GM
Unit body construction	1936	Ford

Source: W.J. Abernathy et al., *Industrial Renaissance* (Basic Books, New York, 1983), Appendix D, pp. 150–179.

tablished the architecture of product and process. The dominant producer in that development was Ford, and the dominant product, the Model T. The first panel in table 2, presents a list of some of the major technological innovations in the architectural era with special emphasis on the design concepts embodied in the Model T.

Three themes are evident in this architectural pattern of innovation. The first is the importance of breaking the grip of the prior industries on the technological structure of the new industry. Second is the durability of the concepts. The designs that constitute the creative synthesis in this period stand for a very long time in the industry's future. The third theme is the role of science. Although science-based innovations are apparent as underpinnings of the dominant design, the overall design itself is not stimulated by science. These observations find support in the specific innovations in table 2. From 1903 to 1907, the spark advanced the longitudinal mounting of engines, torque tubes for drive shafts and bevel gear systems were introduced in successive Ford models. They were supported by a series of innovations in manufacturing processes: multiple machining, new assembly methods and so forth. Many of these concepts remained the industry standard for decades, and part of their success lies in their departure from the design rules of wagon makers and locomotive manufacturers. The spark advance and the torque tube, for example, marked a mating of technologies—electrical control concepts to the thermodynamics of engine design, and the dynamics of torque in motored propulsion to chassis design—that advanced and displaced the conventions of carriage makers.

Breaking out of the confines of established industries not only influenced technical development, but also opened up new possibilities in linkages to markets and customers. The market implications of the innovations of this period are well illustrated in the case of the flywheel magneto and vanadium steel alloy. The flywheel magneto provided an electrical power source for firing the spark plugs that was built into the engine. This ingenious integration of new electrical technology with engine design meant that the car could be used in remote locations and would not be vulnerable to the life of the dry cell or other batteries for ignition. In similar fashion, the application of vanadium steel alloy in engines and chassis components was an important part of the development of a lightweight vehicle. When Ford and his engineers found an alloy that afforded three times the design strength of traditional materials, it freed them to apply new concepts of lightweight design to many parts. As a result they were released from many old constraints that had been adopted from carriage technologies. The engine and the chassis came to be based on concepts of flexible lightweight structural design as opposed to the more traditional

designs that sought durability through stout rigidity.

Improved reliability and lightweight design were important elements in the creation of a rugged, durable vehicle, able to withstand the rigors of rural operation, and yet sufficiently low cost to permit the development of a mass market. Technological innovation thus gave Ford the kind of product he needed to open up a new market and establish a new set of applications. Design concepts like the integrated flywheel and new alloys were essential to the development of a rural market for basic transportation—for a people's utilitarian vehicle that could go anywhere and was relatively easy to maintain. This was a product concept that made possible new distribution channels and new types of aftermarket support, and thus broke many of the marketing conventions of the day.

Our location of the Model T in the Architecture quadrant of fig. 1 reflects the fact that its transilience was greater in the market dimension. Though there were some disruptive elements in its technology, its genius lay more in a creative synthesis of technology innovated by its diverse predecessors. The Model T experience thus suggests that architectural innovations stand out as creative acts of adapting and applying latent technologies to previously unarticulated user needs. It is the insight and conception about fresh roles for existing inventions and technologies that mark this kind of innovation. Scientific work plays a part in freeing thought, and relaxing old rules of thumb. The challenge lies in linking understanding of technical possibilities to insights about unarticulated needs.

2.2. Innovation in the Market Niche Phase

Using new concepts in technology to forge new market linkages is the essence of architectural innovation. Opening new market opportunities through the use of existing technology is central to the kind of innovation we have labelled "Niche Creation", but here the effect on production and technical systems is to conserve and strengthen established designs. There are numerous examples of niche creation innovation, ranging from the Timex example referred to earlier, to producers of fashion apparel, and consumer electronics products. The mating of light weight earphones and a portable radio or casette player in Sony's Walkman, used established technologies to create a new niche in personal audio products. Makers of women's apparel have traditionally used changes in ornamentation, color, configuration, fabrics and finishes to create profitable if transitory market niches. Innovation of this sort represents what Utterback has called "sales maximization", in which an otherwise stable and well specified technology is refined, improved or changed in a way that supports a new marketing thrust. In some instances, niche creation involves a truly trivial change in technology, in which the impact on productive systems and technical knowledge is incremental. But this type of innovation may also appear in concert with significant new product introductions, vigorous competition on the basis of features, technical refinements, and even technological shifts. The important point is that these changes build on established technical competence, and improve its applicability in the emerging market segments.

It is clear that successful niche creation innovation requires the matching of customer needs with refinements in technology. But the evidence suggests that such an alignment is in and of itself not sufficient to establish a long term competitive advantage. Innovation that helps to create a niche may be important, even essential, to the continued existence of the innovating firms. But if the innovation is readily copied, its competitive significance may be greatly diminished. Such was the experience of Ford with the introduction of the Model A in 1927.

By 1926 Ford had sold more 15 million Model Ts, and had driven the price as low as $290 on some models. Yet the Model T was a 20-year-old product whose basic design could not economically accommodate the range of new features and levels of performance demanded in the marketplace. The market was no longer dominated by rural buyers interested in rugged durability. Emphasis in the market was moving towards urban customers and the Model T's competitors offered improved power, increased comfort and convenience, and easier operation. Though Ford had created the world's largest industrial complex and had reduced costs significantly, demand for the Model T declined in the mid-1920s. With no new products in the pipeline, Ford decided in 1926 to shut down the giant Rouge River facility for nine months and develop the Model A.

With little time for development, and with a need to meet the demands of an emerging new market, it was almost inevitable that Ford would rely on established technology and thus move into a niche creation mode of innovation. Where GM had approached the changing market with new concepts in design and in product development (i.e. constantly upgraded product technology), Ford responded with a major model change, but not with a technologically dynamic strategy (see fig. 1).

The Model A was introduced in late 1927 and was a great success. Ford's first completely redesigned model in 20 years aroused great public interest. Its appeal stemmed from the combination of features, the refinements and improvements in existing design concepts, and major advances in performance and styling. The new engine was light, but powerful. The car was capable of high speed, yet offered a smooth and quiet ride. Craftsmanship was evident in the fit of the body and the appearance of the interior and trim. And as far as styling was concerned, the Model A was lower and sleeker than the Model T, and color coordinated paint and fabrics were offered.

64

In its basic design, the Model A was a synthesis and refinement of concepts that had been introduced by other manufacturers. There was a margin of innovation in technology (e.g. mushroom valve stems, laminated safety glass, resistance welding), and a creative improvement and packaging of existing elements. Much of the improved engine performance, for example, came from improvements in existing machining processes. Smoothness and quietness of ride came from design changes that eliminated joints, the liberal use of sound deadening material, and the introduction of hydraulic shock absorbers.

In its overall configuration, the Model A gave definition to an emerging market segment (the moderately priced family car—good performance, modern styling, comfortable, convenient) through incremental innovation. It was sold through existing channels of distribution, but forged links to a new customer base, and defined new applications. These changes in market segment composition and definition has a further effect in changing methods of customer communication and influenced the delivery of aftermarket services. The Model A was thus moderately transilient in the market dimension, while building on and strengthening technical competence. Its introduction was critical to Ford's survival and enabled the company to regain market leadership from GM. But the triumph was short lived. By 1929, Chevrolet had developed new models that were a little bigger, a little faster, and a little more comfortable and stylish. That pattern continued in the 1930s as GM consolidated its market position.

The model A's market share gains were not durable because competitors were able to copy and even advance the design quite easily. Unlike the Model T, the new design was not based on company developed innovations: The technological advances in the car involved either the application of new materials developed by others, small changes to existing components or features, or manufacturing advances

that could be copied or that failed to lead to unique product features of value in the market. The Model A offered a technical configuration that was on target in the market, but that gave Ford no unique competitive strength on which to build a sustainable advantage. A similar conclusion applied to the other market niche innovations listed in the second panel of table 2.

The experience of the Model A seems to be characteristic of niche creation innovation. Such changes in technology may be associated with highly visible and transilient changes in the market, but any competitive gains from one particular innovation are likely to be transitory. No matter how well the new design meets the current demands of the market, the lasting significance of an innovation will be greatly reduced if the new technology is insufficiently unique to defy ready acquisition by competitors. But that does not imply that innovation is of no importance in markets characterized by niche creation. Rather, it suggests that the advantage derived from a given innovation will be temporary, and that long term success in this mode will require a sequence of new products and processes to counter the moves of rivals. It appears that in niche creation innovation, timing and quick reaction are everything.

2.3 Regular Innovation

The creation of niches and the laying down of a new architecture involve innovation that is visible and after the fact apparently logical. In contrast, what we call "Regular" innovation is often almost invisible, yet can have a dramatic cumulative effect on product cost and performance. Regular innovation involves change that builds on established technical and production competence and that is applied to existing markets and customers. The effect of these changes is to entrench existing skills and resources.

Research on rocket engines, computers and synthetic fibers has shown that regular innovation can have a dramatic effect on product cost, reliability, and performance. Although the changes involved may be minor when examined individually, their cumulative effect often exceeds the effect of the original invention. This same pattern is evident in the dramatic declines in price and the improved reliability of the early Model T. From 1908 to 1926, the price of the car fell from $1200 to $290, while the productivity of labor and capital increased markedly. These reductions in cost were the result of numerous changes in the process, most of which Ford himself thought to be too insignificant to recount. While improvements in casting, welding and assembly, and material substitution helped to reduce cost, they also interacted with changes in the product to improve reliability and performance. Electric lights, enamel finishes on the body, rubber engine mounts and an integral brake drum and hub, are examples of the kinds of changes in product design that improved the Model T's appeal in the market.

Regular innovation can have a significant effect on product characteristics and thus can serve to strengthen and entrench not only competence in production, but linkages to customers and markets. It is important to note that these effects tend to take place over a significant period of time. They require an organizational environment and managerial skills that support the dogged pursuit of improvement, no matter how minor. The effects of a given regular innovation on competition are thus of less concern than the cumulative effects of a whole series of changes.

Some of these effects are quite direct and involve advantages due to improvements in the product's existing technology. Other effects, however, are more subtle and indirect; it is these effects that we explore in detail in section 3. Here it suffices to note that incremental change in process technology tends to both raise productivity, and increase process capacity, often through mechanization. This has the effect of increasing economies of scale and the

capital required to compete. In addition, refinements in product design and in processes reinforce increases in scale economies by enlarging the amount of product variety that a given technology can support. Though the changes imposed by a given innovation in the regular mode may not be dramatic, a sustained pattern of such change can transform the business, altering substantively what must be done well to achieve competitive advantage.

2.4 Revolutionary Innovation

Innovation that disrupts and renders established technical and production competence obsolete, yet is applied to existing markets and customers, is the fourth category in the transilience map and is labelled "Revolutionary". The reciprocating engine in aircraft, vacuum tubes, and mechanical calculators are recent examples of established technologies that have been overthrown through a revolutionary design. Yet the classic case of revolutionary innovation is the competitive duel between Ford and GM in the late 1920s and early 1930s.

While Ford's competitive moves with the Model A were based on imitative use of technology, the behavior of competitors was a different case entirely. For the industry as a whole, the mid-1920s marked the beginning of a revolutionary phase of innovation. Ford was focused on volume production of its established design, while GM began investing in new concepts in suspensions, body forming and transmissions. Studebaker and Chrysler contributed in important ways to advanced body, suspension and engine technology. In contrast to Ford's pursuit of volume and lower cost through the Model T, GM, Chrysler and other producers developed new designs in suspensions, bodies, and transmissions that redefined the nature of the automobile. The innovation that contributed more than any other to this change in competitive and technical emphasis was the closed steel body.

First marketed by Hudson in its 1921 Essex, the closed body made of steel was a clear departure from the open (no solid top or sides) wooden bodies then dominant in the market Chevrolet's Model K perfected the concept and GM introduced process changes that made the closed steel body an affordable feature in mass production vehicles. The innovation raised new criteria for automotive design—passenger comfort, room, heating and ventilation—and deepened and broadened the appeal of the product to the American consumer by making it more convenient, enjoyable and useable.

The closed steel body strengthened market linkages, but its impact on manufacturing was disruptive. Steel bodies depended on sheet metal forming technology rather than the craft skills of the wooden body maker, or the metal removal technologies used in engine and transmission production. What was required was new machinery, new skills in labor and management, and new relationships with suppliers. Moreover, the new technology increased minimum economies of scale, as giant presses and expensive dies were used to form the metal parts.

The closed steel body came to dominate the industry, and in so doing substantially altered the nature of competition. Along with other changes in technology it formed the basis for Chevrolet's sustained attack on Ford and the Model T. It weakened the relative position of small firms, at the same time that it changed the product characteristics on which competition had been pursued. Convenience, performance and comfort became the central theme in subsequent competition and technical innovation.

Not all innovations that fall in the revolutionary quadrant have a profound competitive impact. Some fail to meet market needs, while others encounter problems in production. And others, like Ford's 1932 V-8 engine, are poorly timed. In 1932 Ford introduced the Model 18 with a new V-8 engine. Through a stunning engine design and unique manufacturing process based on Ford's own casting and machining technology, the Model 18 offered a high performance engine in a popular price range. Here was an example of an innovative design with high technological transilience applied to an existing market position. In contrast to the Model A, product features were dependent on a technology in which Ford played a leadership role and in which the company had a sustainable lead. But the launch of the product was not well timed. In the depression era of 1932, a performance engine for the working-man was not a concept for the times. To make matters worse, the extra engine performance brought with it extra stress on reliability. Problems with knocking, thrown rods and burning oil were more visible in a period of tight budgets.

While the Model 18 enjoyed some market success, it did not captivate the market; its power for change was moderate. It thus seems clear that the power of an innovation to unleash Schumpeter's "creative destruction" must be gauged by the extent to which it alters the parameters of competition, as well as by the shifts it causes in required technical competence. An innovation of the most unique and unduplicative sort will only have great significance for competition and the evolution of industry when effectively linked to market needs.

3. THE TRANSILIENCE MAP AND INDUSTRY EVOLUTION

Our application of the transilience map to the history of the auto industry shows that all four kinds of innovation have shaped the industry's development in subtle and diverse, but powerful ways. A similar conclusion as to the role of innovation emerges from detailed studies of a variety of other products and markets. The historical evidence suggests further, that innovations of a given type appear in clusters, and that the temporal pattern of innovation is closely linked to the overall evolution of the industry. The transilience map is thus much more than a simple categorization of technical change; it provides a framework within which one can examine the relationships among innovation, competition and the evolution of industries, as well as develop insight about the strategies of specific competitors.

Existing models of industry evolution posit a life cycle of development in which new products (and industries) emerge, are developed, defined, and mature[11]. Framed in terms of the transilience map, models based on the product life cycle, or the "fluid-to-specific" stage model of Abernathy and Utterback, are dominated by the transition from architectural to regular innovation. In fact, it is useful to conceive of the traditional life cycle as a development vector describing the firm's transition from one innovative phase to another. Our discussion of the different types of innovation, however, suggests that vectors of industry development may be richer and more varied than simple life cycle notions might suggest. In particular, the implicit (or sometimes explicit) biological life cycle metaphor seems to be misleading; the reversal of an older industry to embrace the emergence of revolutionary or architectural innovation may serve as the basis for renewal in its pattern of industry development. Kuhn's *The Structure of Scientific Revolutions* suggests that the advancement of science is characterized by long periods of regular development, punctuated by periods of revolution[12]. Historical evidence suggests that a similar pattern characterizes the development of technology. Furthermore, even within the traditional architecture-regular pattern, the role of innovation in competition may be more important, albeit subtle and indirect, than traditional approaches have assumed.

In this section of the paper we re-examine the regular phase of innovation, and its im-

plications for the evolution of the industry's competitive environment. Our focus is on how regular innovation contributes to the embodiment of labor and managerial skills in capital equipment; to increasing rigidity in processes and products; and, somewhat paradoxically, to increased versatility in established designs. We then explore transitions out of the regular phase of innovation. Such moves result in technological ferment and form a varied set of complex but important strategic vectors of industrial development.

3.1. Regular Innovation, Capital Embodiment and Technical Rigidity

The transition from architectural to regular innovation is often associated with the emergence of a dominant design in the product. With this the focus of innovation shifts from meeting emerging needs with new concepts, to refining, improving and strengthening the dominant design and its appeal in the market. It is important to recognize that this transition is but the search for a strategic advantage over competitors. Where advantage in the architectural phase rests on enhanced product performance that may be gained through creativity in linking new technology to latent needs, exploiting the advantage inherent in a dominant design demands a change in strategic orientation. It is for this reason that the transition to regular innovation can thus be seen as a "strategic vector" in the transilience map, leading out of architecture into a phase of refinement and improvement. The transition is thus not a move from one well defined "state" to another; it is more like charting a new path through an emerging environment.

In section 2 we noted some of the more obvious effects of regular innovation. But there are other subtle effects which a strategic vector in the regular direction may create. Consider, for example, the effects of typical improvements in processes. Innovation of this sort is often little more than the act of taking the skill that workers or managers use in performing tasks and embodying it into the design of a machine. The innovation may replace elements of the task, or eliminate the need for the worker (or manager) entirely. Examples of such embodiment are prevalent in the three innovations in the early auto industry as listed in table 2: mechanized welding, moving assembly lines, and enamel finishes.

Electric welding of metal parts was used before welding was mechanized. However, reliable performance in high-volume applications was not achieved until Ford developed a mechanical seam welder that could be operated by an unskilled operator. This was an important early step in the development of extensive welding in the body-building process. It is important to note, however, that the automatic welding device became linked to a particular model. In order to promote efficiency, its design was so specialized that, for example, an automatic welding machine of Model A vintage was not usable on the next model. Thus, the embodiment but not the concept of the technology became vulnerable to change.

Mechanized welding illustrates the embodiment of labor skills and the "shaping" effect that accrues from incremental innovation as it renders the process for a given product more specific and rigid in nature. The moving assembly line illustrates the same phenomenon in managerial skills. Just before the moving assembly line was adopted in January of 1914, Ford had developed a method of assembly involving team scheduling with stationary product and roving teams that offered the same degree of work force specialization as the moving assembly line. The team method, however, required much more careful supervision and a close eye on inventory control. The moving assembly line in contrast, simplified the supervision and inventory management problems in one stroke through the imposition of a novel conveyor technology. This made the task of management much less demanding although not necessarily more efficient—if the quality of management was otherwise excellent. With less demanding tasks, the range of experience to which the supervisor was exposed became limited. The moving assembly line decreased opportunities for training managers to cope with nonstandardized production such as small-scale stall build systems. Small lot or stall build became an anathema to the industry. As with the welding innovation, the innovation left the industry more rigid and limited in the range of unanticipated change it could accommodate.

A third case, the adoption of new finish coatings on steel body parts, illustrates the role of regular innovation in achieving mass production. The Model T was initially offered only in black because a satisfactory colored finish required repeated sanding, rubbing and polishing operations between many successive coats of finish. By one estimate, 106 days were required to produce a car body with a superior color finish. Most of this time was in drying. Such skill, care and time were required that mass production was unthinkable. The innovative new application and finish technologies of the 1920s changed all of this by requiring only that operators perform short duration tasks that were quite teachable, that demanded less training, that were machine paced, and that involved much less inventory. The innovation made possible and reinforced mass production of colored bodies. The new technique could not be scaled down very well to handle low production rates and in this sense, the innovation raised minimum economies of scale.

What Ford gained from capital embodiment was an ability to more rapidly expand its capacity in the face of burgeoning demand and a limited supply of skilled labor. It was easier to duplicate machines than skilled labor and management in a tight labor market. Furthermore, the demand for new equipment embodying labor and management was not satisfied through internal development alone. Outside suppliers played a significant role in producing the volume of machinery that Ford required.

Recent experience in the semiconductor industry underscores the close link between capital embodiment and appropriability. In new industries that are based upon innovative technologies, the skills which are critical to competitive success initially are usually not capital embodied. Entry in this phase involves much more than the acquisition of capital assets; witness the experience of many large firms who tried to enter the semiconductor industry during its formative years. There is evidence that scarcity of technical and management talent was the single most important reason that large vacuum tube producers failed to buy their way into a successful semiconductor business during the 1960s and 1970s[13]. These skills in management and technology were not appropriable through traditional means of merger or capital acquisition. Logically under these conditions the most important mode of competitive industrial propagation was by spin-off.

3.2 Regular Innovation and Market-niche Versatility

Regular innovation that embodies skills in equipment serves to increase minimum economies of scale and to make established processes more specific and rigid. Yet, somewhat paradoxically, regular innovation may also increase the versatility of technology. Once again, the development of the Model T provides a fruitful example.

At the end of its life, the Model T was a far different car than its early predecessor. The car of 1926 started electrically, had electric lights, a closed colored body, and sold for less than one half the earlier price. Furthermore, it was much easier to use and appealed to a much wider range of market segments. As the grand old product evolved it simultaneously became more versatile and appealing to a broad range of market segments, while the supporting productive unit became more rigid and vulnerable to unanticipated change. This expanded versatility is not just a definitional distinction. It may be observed in the very specific detail of engineering data on the product and process technology. The evidence shows that Ford's engine line was functionally broadened through innovation so that it could more robustly accommodate a variety of market segments with less and less actual mechanical variety[14].

The technology of early automotive engines was not well developed. In order to power a large, expensive car a large elaborate engine was needed. Since Ford in the early years served many market segments (Model K was the largest and most expensive line; Models A, B and C were the less expensive and smaller products) a line of engines was needed that could meet a wide range of customer demands. Engine variety as a function of market segmentation is illustrated in fig. 2 which compares the range of cubic inches of displacement and delivered horsepower in Ford's engine line from 1903 to 1908. The data show that large variation in engine size (353 CID), yielded the small, but critical difference of 32 horsepower in the market.

Over time, incremental advances in both product and process technology not only improved the engine's performance, but also increased its versatility in meeting market needs. Incremental innovations like dynamic balancing reduced vibration and prolonged operating life; constant temperature inspection rooms led to greater precision in manufacturing and therefore a wider dynamic range of performance in aspects like speed (r.p.m.), compression and horsepower per unit size. The rubber engine mount, widely adopted in the 1930s, broadened the appeal of small cylinder count engines. Before its use, smoothness of ride was achieved largely by high cylinder count engines like the V-12, or even the V-16. Such engines were associated with speed, power and prestige, but they also served to guarantee smooth

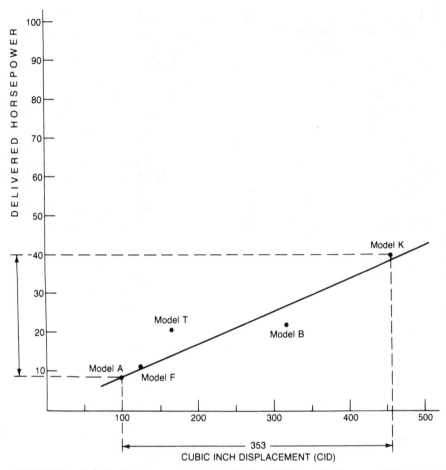

FIGURE 2. *Relationship between Mechanical Variety (CID) and Performance (HP) in Ford Engines 1903–1908*

operation. The advent of the simple rubber engine mount in a V-8 design eliminated the need for great mechanical variety since it isolated the transfer of vibration to the chassis, without adding more cylinders.

Increased technical versatility is an important dimension of regular phase innovation, but it may also be the basis for a strategic thrust into niche creation. The essence of increased versatility is greater understanding of the possibilities in the technology and the implications of alternative refinements. Enhanced technical insight and understanding may provide the de-

sign change required to offer a new product configuration, or a different mix of performance characteristics that meets an identified new set of customer needs. This seems to have been the case with the Model A in 1927, with the creation of "muscle" cars in the 1950s, and affordable sports cars (e.g. Mustang) in the 1960s.

Niche versatility was characteristic of automotive development through periods of regular innovation and later in the creation of market segments and niches. The cumulative effect of numerous innovations on Ford's engine line from 1900 to 1958 is depicted in fig. 3. In this figure the earlier trend in market variety versus mechanical variety is extended for groups of major Ford engines by decades. The increasing market-niche versatility is evident in the counter clockwise rotation of the lines indicating the relationship between mechanical variety and horsepower. By the 1950s a difference of 245 horsepower was provided by a range in mechanical variation of a little over a 200 cubic inches of displacement. This compared to a variation of 32 horsepower for a large 353 cubic inch displacement in the early 1900s.

The significance of a trend toward greater market-niche versatility for our purposes lies in its implication for innovation. As the versatility increases there is less need for novelty and technical variety in meeting a variety of market needs. In a sense, incremental innovation is self-limiting, just as Gilfillan, an early scholar on the subject observed many decades ago[15]. As long as market demands are anticipated, as long as they are consistent with embedded experience, a technology that has been refined and improved will be relatively robust in meeting them.

These observations on evolutionary adaptation in an industrial unit are related to eco-

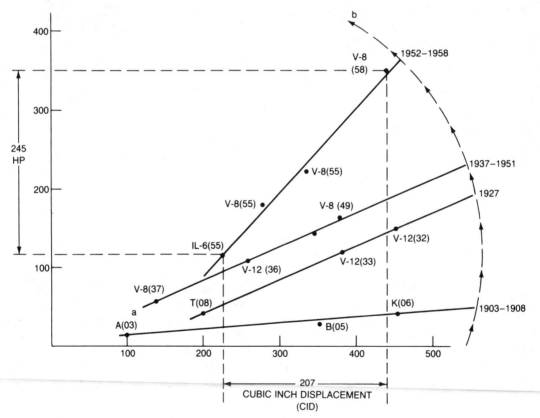

a. The number in parenthesis is the year in which the engine model was introduced.
b. Rotation (the dotted line) shows a decrease in the ratio of mechanical variety needed to deliver a given range of performance in horsepower from the 1903–1908 era to the 1952–1958 era.
Source: W.J. Abernathy, *Managerial Dimension of Technological Trends* (unpublished, 1975).

FIGURE 3.   *Trends in Market-niche Versatility for the Ford Engine Line 1903–1958*

logical concepts about development and dominance of biological species within their "niche" in the environment.[16] As the product and process technologies evolve and develop, they become more robust in the way they accommodate the full range of variety in the existing environment. Like the tree that develops an extensive root system to weather the dry seasons it must occasionally face, management refines and perfects a product over time to better accommodate the range of variation in the market. Yet a product and process technology that becomes more highly organized and efficient in the way it meets established requirements, it also becomes more vulnerable to sudden and unanticipated variations in the environment. The highly productive, efficient and developed product unit is also more vulnerable to economic death.

3.3 Innovation and Reversals in Development

Sustained periods of regular innovation are predicated on a stable relationship between the needs and preferences of customers and the design concepts in the technology. Regular innovation thus follows the emergence of a dominant design. But it also reinforces the dominance of that design through improved performance, reliability and lower cost. As the technology develops in the regular phase, the preference-technology nexus inherent in the dominant design strengthens as it grows more complex. If, however, the relationship between customer demands and technical characteristics begins to break down, if new technical options emerge, or if the range of demands begins to strain the ability of the existing designs to meet them, firms may find a move away from regular innovation advantageous.

We have in mind something more than a move to exploit new market niches. What we want to consider is the possibility that changes in the environment may create the opportunity (or the necessity) for a strategic vector out of the regular mode of innovation, into the revolutionary or architectural phase. A strategic thrust of this sort implies the re-emergence of a kind of innovation—new concepts, departures from existing designs—and a degree of technical variety and ferment that is more like the early stages of an industry's development. In contrast to the typical "birth-growth-maturity-decline" pattern of development, the transilience map thus suggests the possibility of "de-maturity".[17]

There are three kinds of changes in the industrial environment that may create the conditions for de-maturity. The first is new technical options that open up possibilities in performance or new applications that the existing design concepts could meet only with great difficulty or not at all. These options may come through research and development from within the industry, or they may be the basis for an invasion by competitors from a related field. The second impetus for dematurity may come from changes in customer demands. Whether through changes in tastes, or through changes in prices of substitutes or complements, new customer demands may impose requirements that can best be met with new design approaches. The third source is government policy. Regulations imposed on an established industry, for example, may set technical requirements or demand performance standards that favor revolutionary or architectural strategic development. De-regulation may have the same effect.

The impact of the closed steel body on the auto industry of the 1920s illustrates the possibilities of de-maturity and the critical elements of a successful strategy in that environment. In section 3 we noted the emphasis in technical development on comfort, convenience and ease of operation that followed the new body technology. One of the first that exploited the opportunities that a new direction in technology presented was the Chrysler Corporation. The company was founded in the 1920s and in its rise one can see the consequences of a well executed strategic thrust into revolutionary and architectural innovation.

Figure 4 shows the position of Chrysler innovations from 1924 to 1949 on two transilience maps; the first covers the period 1924–1939, while the second covers 1940–1949. The position of each point on the map is derived from a qualitative assessment of the impact of the innovation. The assessment was based on historical evidence of the changes required to implement the innovation. The first map (top half of figure) shows Chrysler's departure from the pattern of innovation that had characterized the industry in the Model T era. From 1930 to 1939 Chrysler introduced several innovations in carburetion, body design, transmissions and chassis construction that departed from established practice. These changes in technology were embodied in models designed for existing customer groups and existing channels, and thus have been placed in the revolutionary quadrant. Chrysler also introduced refinements in noise and harshness characteristics, seats, and dashboards that appear in the regular innovation category. Chrysler's regular innovation and those that were more revolutionary reinforced one another. The emphasis in both cases was on comfort and convenience, power and smoothness of ride. Both kinds of innovation helped to build and then to strengthen Chrysler's appeal in the marketplace, but it was the significant revolutionary changes that gave Chrysler its distinctive character, and its competitive advantage.

The contrasts between Ford and Chrysler, and between the Chrysler of the 1930s and the Chrysler of later years are instructive. Faced with similar market developments and technical possibilities, Ford's strategic thrust (as evidenced in the Model A) was to define the family car on the basis of imitative product technology, and to push forward with innovations in production processes. Almost all of Ford's innovations in the 1930s involved processes, and few of those resulted in product characteristics valued in the market. Thus while GM and Chrysler (and others) were at work on new concepts in suspensions, brakes, transmissions, and bodies. Ford's development efforts were largely focused on the cost of manufacture.

Chrysler's strategy of new design concepts and thus of flexibility in product characteristics was highly successful in the decade of the 1930s; by the end of the decade Chrysler had passed Ford in market share. In the 1940s, however, the second transilience map (bottom of Fig. 4) shows clearly that Chrysler shifted its focus to incremental technical change in both the regular and market niche modes. This shift was consistent with the emergence of a new dominant design—the "all-purpose road cruiser"—in the early 1940s, and subsequent efforts to segment the market. Although Chrysler was a significant participant in subsequent years, its market share declined as the rate of technological change in the product diminished. Product engineering remained a central element in the Chrysler approach, but the kinds of changes introduced offered little that was unique or distinctive. In contrast to its heritage as a pioneer in new design concepts, the Chrysler of the 1950s and 1960s was caught between two strategies: too small and inefficient to compete on cost; not innovative enough to create a technology-based differentiation.

4. MANAGERIAL IMPLICATIONS AND CONCLUSIONS

This paper has presented a new way of assessing the competitive significance of an innovation—the transilience map. Each quadrant in the map represents a different kind of innova-

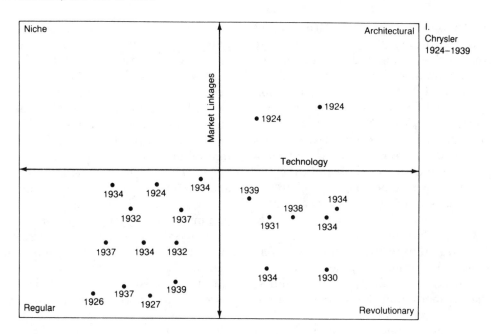

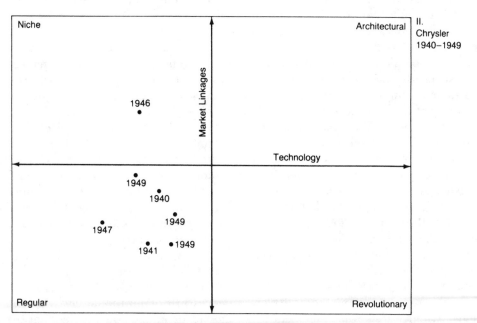

a. For a definition of the dimensions of the map see fig. 1.
Source: W.J. Abernathy et al., *Industrial Renaissance* (Basic Books, New York, 1933), Appendix D. pp. 155–77.

FIGURE 4.   *Transilience Maps of Chrysler Innovations*[a]

tion, and tends to be associated with a different competitive environment. Moreover, because the nature of change imposed on the firm is so different, the framework implies that the successful persuit of different kinds of innovation will require different kinds of organizational and managerial skills. The transilience map may thus illuminate the managerial environments required to nurture innovation and technical progress in each mode. Much further research remains to be done, but our work to this point suggests a number of working hypotheses.

In the architectural phase management must encourage the creative synthesis of information and new insight into user needs with information about technological possibilities. Architectural innovation thus demands attention to the management of creativity with a keen insight into business risk. Unique insights about user needs, usually accrued through personal experience must be combined with an ability to see the application of technologies in a new way. The task is one of constantly scanning for technological developments and unmet market needs, and orchestrating the creative, first-time combination of resources.

In contrast, timing is the essence of management in the niche creation phase. The technology is generally available; the key skill is sizing up new market opportunities, and developing a product package that exploits them. Management must nurture quick-footed capability for getting into the market before competitors enter the same niche and destroy profitability. Under these conditions, manufacturing must be quick and responsive, insuring timely delivery, responsive service and adequate capacity for a quick buildup.

In the regular mode, methodical planning and consistency are key management factors. Stability of product design and sources of materials are needed to support directed technological progress, engineering improvements, market refinement and continuing pro-

cess development. Management may need to buffer the organization against supply disruptions or other environmental changes. The idea is to achieve volume production and use scale economies to lower costs and improve products. Every opportunity must be taken to advance quality, improve product features, break bottlenecks in production, and foster process innovations that reduce scrap and increase yields. This is the work of the administrator, and the functionally oriented engineer.

The revolutionary mode of innovation is dominated by "technology push". Management must be capable of sustaining a consensus about long-term goals through investments in new technology and innovation. Here the task is to focus possible unruly technical talent toward specific markets and to marshal the financial resources for this purpose. Good technical insight is needed to break established conventions and foster close collaboration between product designers, process designers, and market planners. The climate must be one that encourages a sense of competitive assault.

The hypotheses about managerial environments and the historical analysis of the auto industry suggest a number of questions and directions for further research. The transilience map itself, of course, requires further study through application in other industries. Detailed examination of the technological and competitive development of other industries should focus both on testing the ideas discussed here, and refining and extending them. Work along these lines is underway in aircraft and semiconductors.

A particularly fruitful area for further analysis is the notion of strategic vectors, and associated industry transitions. The differences in organizational environments in each mode of innovation implies that a transition from one phase to another may pose a significant challenge to established firms. This suggests the need to examine the impact of innovation on market structure during periods of transition, as well as implications for management. Historical evidence suggests that the creation and development of technology-based industries leads the industry from quadrant to quadrant. It is at these points of transition that originating firms exit and are replaced by new firms better able to manage in the new mode. On the other hand, firms like Ford and Douglass Aircraft have managed to make the transition. The nature of transitions, and the determinants of survival deserve additional attention and analysis.

The differences across types of innovation also have implications for the management of technology at a given point in time. While a firm may have a dominant orientation, it is likely that the firm will face the task of managing different kinds of innovation at the same time. While one part of the product line may be in the regular phase of development, the firm may try to introduce a revolutionary development in another, and may try to develop new niches in a third. Whether firms ought to manage in that way, and if so, how to do it seems an important area for further work. Pursuit of these issues will require an examination of the impact of market structure and competitive rivalry on the different kinds of innovation.

NOTES AND REFERENCES

1. See for instance: M. Abramovitz, Resource and Output Trends in the United States since 1870, *American Economic Review* 46 (1956) 4–23; Robert Solow, Technical Change and the Aggregate production Function, *Review of Economics and Statistics* 39 (1957) 312–30. S. Kuznets, *Secular Movements in Production and Prices* (Houghton Mifflin, Boston, MA, 1930).

2. For example, F.M. Scherer, *Industrial Market Structure and Economic Performance,* (2nd edition, Rand McNally, Chicago, IL, 1980); Joe S. Bain, *Barriers to New Competition,* (Harvard University Press. Cambridge, MA, 1956).

3. See, for example, T. Burns and G. Stalker, *The Management of Innovation* (Tavistock Publications, London, 1961); U.S. Department of Commerce, *Technological Innovation; Its Environment and Management.* A report by the panel on innovation, Robert Charpie, Chairman (U.S. Government Printing Office, Washington, DC, January 1967); Also John Jewkes et al., *The Sources of Invention* (Macmillan, London, 1960).

4. Michael E. Porter, The Technological Dimension of Competitive strategy, in: R. Rosenbloom (ed.) *Research on Technological Innovation, Management and Policy* Vol. 1, (JAI Press, Inc., Greenwich CT, 1983) pp. 1–34. Richard S. Rosenbloom, Technological Innovation in Firms and Industries: An Assessment of the State of the Art, in: P. Kelly and M. Kranzberg et al. (eds.) *Technological Innovation: A Critical Review of Current Knowledge* (San Francisco Press, San Francisco, 1981). pp. 215–30. Nathan Rosenberg, *Inside the Black Box* (Cambridge University Press, Cambridge, 1982). Richard R. Nelson and Sidney G. Winter, *An Evolutionary Theory of Economic Change,* (The Belknap Press of the Harvard University Press, Cambridge, MA, 1982).

5. Rosenbloom, Technological Innovation in Firms and Industries. See 4.

6. Although the three types of competence are closely related, it seems useful to make distinctions among them. In the first place, they may play somewhat different roles in determining the firm's position, and innovation may affect them in different ways. Furthermore, the competencies are acquired and developed in very different ways. Material resources, and some human skills are often available through established markets. But some skills and some types of knowledge are specific to the firm and must be developed internally through experience. Whether through market transactions or through internal development, the acquisition of resources, skills and knowledge is costly and especially in the case where internal development is required, may take much time.

7. Joseph A. Schumpeter, *The Theory of Economic Development* (Harvard University Press, Cambridge, MA. 1934).

8. *Timex Corporation.* Harvard University Graduate School of Business. Case No. 9-373-080 (1977).

9. W.J. Abernathy and J.M. Utterback. Patterns of Industrial Innovation, *Technology Review* 50 (7) (June-July, 1978). T. Burns and G. Stalker, *The Management of Innovation* (Tavistock Publications, London, 1961). *Ibid.,* p. 19. Also John Jewkes et al., *The Sources of Invention* (Macmillan Co., London, 1960).

10. Examples cited in this section as they relate to the automobile industry are not referenced individually. The background of this section draws on underlying research published in William J. Abernathy. *The Productivity Dilemma: Roadblock to Innovation in the Automobile Industry* Johns Hopkins University Press, Baltimore, 1978); and W.J. Abernathy, K.B. Clark and A.M. Kantrow. *Industrial Renaissance: Producing a Competitive Future for America* (Basic Books, New York, 1983).

11. We refer to the "technology life cycle" concept as a representation of the changes in infrastructure within the firm that support a given state of technology, in particular, see W.J. Abernathy, *The Productivity Dilemma* [10]. Closely related is the International Product Life Cycle, see L.T. Wells (ed.) *The Product Life Cycle in International Trade* (Division of Research, Harvard Graduate School of Business Administration, Boston, 1972).

12. T.S. Kuhn, *The Structure of Scientific Revolutions* University of Chicago Press, Chicago, 1962).

13. J.E. Tilton, *The International Diffusion of Technology: The Case of Semiconductors* (Brookings Institution, Washington, DC, 1971); and D.C. Mueller and J.E. Tilton, R and D Cost as a Barrier to entry, *Canadian Journal of Economics* 2 (4) (November 1969) 570–579.

14. These trends were developed and reported in an early draft of the *Productivity Dilemma* by W.J. Abernathy, Managerial Dimension of Technological Change: Innovation and Process Change in the U.S. Automobile Industry (1975).

15. C.S. Gilfillan, *The Sociology of Invention,* (MIT Press, Cambridge, MA. 1935).

16. The analogy with ecological models is interesting; see for instance S.J. McNaugton and L.L. Wolf, Dominance and the Niche in Ecological Systems, *Science* 167 (3915) (January 9, 1970) 131–139.

17. The nation of "de-maturity" is discussed in W.J. Abernathy, K.B., Clark and A.M. Kantrow, *Industrial Renaissance* (Basic Books, New York, 1983). See also K.B. Clark, Competition, Technical Diversity and Radical Innovation in the U.S. Auto Industry in: Richard S. Rosenbloom (Ed.) *Research on Technological Innovation, Management and Policy* (JAI Press, Greenwich, Conn., 1983).

The Anatomy of Successful Innovations

Donald G. Marquis

*What characteristics make a significant technical advance a success? Marquis analyzes
more than 500 innovations of the nuts and bolts kind and reports his conclusions.*

Whenever I hear someone begin talking about innovation, I immediately try to answer two questions: Does he understand the distinction between innovation and invention? If so, what *kind* of innovation is he talking about?

The answers are important, for without them it's difficult to make sense out of such discourse. Take the matter of definitions. People tend to use the words innovation and invention interchangeably when, in fact, they are different though related terms.

I shall use the distinction drawn by the economist Jacob Schmookler: "Every invention is (a) a new combination of (b) pre-existing knowledge which (c) satisfies some want."

Innovation is a more subtle concept, but Schmookler's articulation of it is clear and useful:

"When an enterprise produces a good or service or uses a method or input that is new to it, it makes a technical change. The first enterprise to make a given technical change is an innovator. Its action is innovation." He goes on to add, "Another enterprise making the same technical change is presumably an imitator and its action, imitation."

Thus an innovation can be thought of as the unit of technological change; an invention—if present—is part of the process of innovation.

The small exception I want to make to Schmookler's concept of innovation is to wash away his distinction between true innovation (a technical change new to both the enterprise and the economy) and imitation (a change that has diffused into the economy, but is picked up and used by the firm).

I do this simply because many enterprises have been profoundly changed by innovations from other organizations and even from other fields (textile firms influenced by synthetic fibers, or typesetting firms changed by computers, for example). Thus to insist on too narrow a conception of innovation would mean ignoring important industry transformations, as well as a vital mechanism for change—imitation or adoption of a technical idea.

Now I must come back to that second question: What *kind* of innovation am I going to talk about?

As we look over the panorama of technological change since, say, the turn of the century, it seems to me that we can discern three distinct types of innovation:

First, there is the complex system, such as communications networks, weapons systems, or moon missions, that take many years and many millions of dollars to accomplish.

Such innovation is characterized by thorough, long-range planning that assures that the requisite technologies will be available and that they will all fit together when the final development stage is reached. Success tends to turn on the skill of managers to sort out good approaches from bad ones on a very large scale indeed. It is not a common type of innovation in most industrial firms as yet, simply because few enterprises, by themselves, face the kind of systems problem that requires it.

Then there is the kind of innovation represented by the major, radical breakthrough in technology that turns out to change the whole character of an industry. The jet engine, stereophonic sound, xerography, and the oxygen converter would be typical examples. Such innovation is quite rare and unpredictable, and is predominantly the product of independent inventors or of research by firms outside the industry ultimately influenced by it.

The reason that, by and large, innovation of this type comes from the outside is simply that technical people within an industry are apt to be preoccupied with short-term concerns. They see their problems as essentially those of product improvement, cost cutting, quality control, expanding the product line, and the like—all of them problems which they can cope with quite naturally through their own technical competence.

This is, in fact, the third kind of innovation—what I call "nuts and bolts" innovation. Modest as it is, such innovation is absolutely essential for the average firm's survival. So long as your competitors do it, so must you. If your competitor comes out with a better product,

you must make a technical change in your own—innovate—to get around the advance in his. Thus, this sort of innovation is more intimately paced by economic factors than is innovation of the systems type or of the breakthrough type.

This article is concerned with the third kind of innovation—the ordinary, everyday, within-the-firm kind of technological change without which industrial firms can, and do, perish. In what follows, I shall be attempting to answer the question: What are the characteristics of successful innovations in this category? The answer turns out to be a complex mosaic of factors, but within it you may be able to discern a pattern against which you can measure the innovativeness (or non-innovativeness) of your own organization.

In answering the question posed above, I shall be drawing on the results of a rather interesting study of actual innovations that Sumner Myers and I conducted for the National Science Foundation. Our exploration covered more than 500 innovations in products or processes that occurred in the last five to ten years in 121 companies whose interests encompassed five manufacturing industries—railroad companies, railroad suppliers, housing suppliers, computer manufacturers, and computer suppliers. The innovations studied were judged by responsible executives as most important in their companies and so, presumably, commercially significant. Hence the emphasis in my remarks here is on the characteristics of *successful* innovations.

Innovation—the nuts-and-bolts type I'm speaking of—may be carried out from conception to implementation within a single organization. But more commonly it draws on contributions from other sources at different times and places. Thus in the illustration you see two broad arrows representing the major sources of inputs for the process, which is shown taking place schematically between them. Beneath ev-

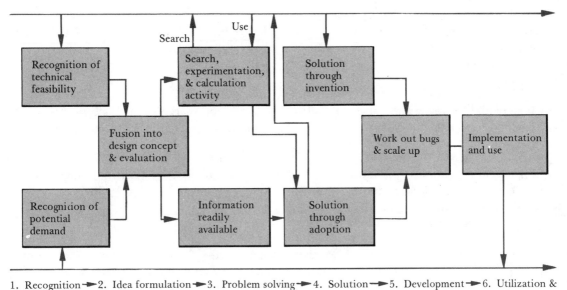

Donald G. Marquis

1. Recognition ➤ 2. Idea formulation ➤ 3. Problem solving ➤ 4. Solution ➤ 5. Development ➤ 6. Utilization & diffusion

MODEL OF THE PROCESS OF INNOVATION

erything I have put a temporal scale which shows the principal stages in the process. The events in these stages will not, of course, always occur in the linear sequence implied by the sketch. Now let's examine these stages a bit more closely.

Successful innovation begins with a new idea which involves the *recognition* of both technical feasibility and demand. At this point in time, there exists a current state of the art, or inventory of technical knowledge, of which the innovator is more or less aware, and on which his estimate of technical feasibility is based. This is the topmost broad arrow. At the same time, represented by the bottom arrow, there is a current state of social and economic utilization in which the innovator can recognize an existing or potential demand.

That word "or" is vitally important, because there's a great difference between recognizing an existing demand and recognizing a potential demand. A company may try stimulating a demand, if it feels that there's a potential one lurking there, by advertising, promotion, and demonstration. But many inventions are ahead of their time, since demand depends on the customer's judgments of the value of a new item in relation to its cost.

A classic example is the reluctance of the railroads to adopt diesel locomotives because of the heavy commitment already made to steam. General Motors had to stimulate demand, in the end, by lending a diesel to a railroad for use in its switching yard. With this successful demonstration, railroads eventually bought diesels for yard use and, later on, for long-haul trains—but only when replacement by the new technology appeared to offer economic advantages.

The next stage is *idea formulation,* which consists of the fusion of a recognized demand and a recognized technical feasibility into a design concept. This is truly a creative act in which the association of both elements is essential. If a technical advance alone is considered, it may or may not result in a solution for which there will be a demand. Similarly, a search for a response to a recognized demand may or may not be successful, depending on the technical feasibility in the current state of technical knowledge.

Part of the stage of innovation I've called idea formulation is really evaluation. It naturally comes after fusion of demand recognition and feasibility recognition into the design concept. Evaluation will recur, of course, all along the line as the process of innovation is managed. But a strong judgmental input must be made here so that the firm can decide whether to commit resources to the next stage.

The design concept is only the identification and formulation of a problem worth committing resources to work on. Then comes the *problem-solving* stage. In some happy instances, the information necessary for the solution is readily at hand in the state of the art; in others, R & D and inventive activity are called for. Unanticipated problems usually arise along the way, and new solutions and trade-offs are sought. And, in many instances, the obstacles are so great that a solution cannot be found, and work is terminated or deferred.

If problem-solving activity is successful, a *solution*—often in the form of an invention—is found, and this knowledge passes into the state of the art once patent protection is assured. Alternatively, the problem may be solved by the adoption of an invention or other input from this pool of technical art. In this case the ultimate technical change becomes, simply, an innovation by adoption—or, in Schmookler's phrase, by imitation.

Whether the solution—invented or adopted—verifies the technical feasibility and demand which were originally recognized, or focuses on a modified problem with somewhat different objectives, uncertainty still remains. Here is where the *development* stage comes in. The innovator attempts to resolve uncertainties with respect to market demand and the problems of scaling up production. Innovation

is never really achieved until the item is introduced into the actual market or production process, and sales or cost reductions are achieved.

Finally, we come to the stage in the process where the solution is first *utilized* and *diffused* in the marketplace. As anyone who has been through it can tell you, this stage is by no means guaranteed. Only one or two new products out of five achieve sales whose profits provide a break-even return on the investment in the innovation. The dollars seep away quickly at this point, too, since the costs of manufacturing start-up, market promotion, and distribution commonly far exceed the costs of achieving the solution. To be sure, some of the uncertainties present at the design-concept stage have been reduced, but the risks—in terms of investment—have increased. In the case of the adopted innovation, the uncertainties are less and the risks can be more accurately evaluated, which, of course, accounts for the popularity of this form of innovation.

Now, with this model before us, let's go back to that question I raised earlier concerning what factors make for successful nuts-and-bolts innovations. To answer the question, I shall cite four examples. This will solidify what I mean by this type of innovation and, I believe make the overall results more vivid. The names used, of course, are fictitious.

The first innovation comes from the housing supply industry, and was but one of nearly 200 technical advances we examined in more than 50 companies therein.

The Janske Asbestos Company wanted to expand its position in the home supply field and began looking for ways to improve its product line. It looked as though a better windowsill had a good market potential, since conventional ones made of wood or cement either required too much maintenance or were too expensive. A different material might solve both problems, but the company wanted to use a material and a method which would not require a significant change in its manufacturing processes.

Janske decided to tackle the problem by extruding windowsills out of asbestos cement. This process and others like it were already widely used in the industry for other products. Janske itself was not unfamiliar with the process, having recently begun to extrude asbestos-cement pipe for commercial use. It proved relatively easy to adapt the process to make the required windowsill cross section, and the total cost of the innovation was less than $100,000.

A simple story, yet one replete with experiences that recurred over and over in our broad sample of innovations in this industry and the four others we explored.

Take the matter of cost. Many companies shrink from innovating out of fear that coming up with a sound technical change in their product or process means investing an enormous sum of money. That certainly may be true for the big breakthrough or the systems type of innovation, but for the incremental type I'm talking about here, it's patent nonsense, as you can see if you glance at table 1. Here are summarized the costs of all of the innovations we looked at. Fully three-quarters of them cost less to accomplish than what Janske Asbestos spent to develop its new windowsill.

Not shown in the table, but revealed by our more detailed analysis, were the following characteristics of the successful innovations we studied: One-fourth of them required little or no adaptation of information readily obtainable

from some source, and one-third were modifications of existing products or processes rather than new items. Almost half of the innovations required little or no change in the firm's production processes.

What all this adds up to is what I'll call Lesson No. 1: *Small, incremental innovations contribute significantly to commercial success.* I conclude, therefore, that management ought to back sustained support for innovative activities so as to maintain the competence, experience, and personal contacts of its professional technical staff.

There are some other lessons, so let me turn to the second of my four examples.

Industrial engineers at Mid-North Railroad Co. found that the water coolers the company was required, by union rules, to have in each of its locomotive cabs were costing far too much to maintain. Upon learning of this, the head of the locomotive equipment department asked the director of Rail Lab Inc., the company's wholly owned R & D subsidiary, to find a water cooler that would stand up better under the shock and vibration in a locomotive cab. The director assigned a group to the problem, but none of the alternative designs could stand the gaff.

Impasse.

Then one of the group's engineers remembered his school days. Why not, he suggested, use a modification of the Hilch-Ranque tube? (This is a laboratory curiosity that uses a vortex of compressed air to separate "cold" gas

TABLE 1. *Cost of Innovations*

Cost of Innovation	No. of Cases	%	
Less than $25,000	187	33	→ 2/3 cost less
$25,000–$100,000	180	32	than $100,000
$100,000–$1,000,000	132	23	
More than $1,000,000	68	12	
	567	100	

Innovations aren't as expensive as you think. Of the 567 innovations examined by the author, two-thirds cost less than $100,000 each and fully one-third cost less than $25,000 each.

molecules from "hot" gas molecules in a fluid stream. The idea is about 40 years old and is commonly used to demonstrate the theory of Maxwell's demon in physics and engineering classes.)

This engineer worked with the tube, raising its cooling efficiency from 4 percent to about 26 percent. Applied to the water cooler problem after a few months' work, the tube idea produced a new cooler, with no moving parts, which could function in the locomotive cab without breaking down. The only power it required was compressed air, and that was readily available on the engine. Christened the *Whirl-Cool*, the novel cooler saved the railroad company nearly $250,000 per year—the maintenance cost of the old-style coolers on some 2,000 locomotives. In addition, the device was offered for sale to other users.

The key point about this innovation is the fact that it was initiated (stage 1 in our model) by the recognition of a need, rather than by the recognition of the potential of a technical idea. This came later, during design-concept formulation (stage 2). The need came out of a prosaic maintenance situation.

In our analysis of the whole set of 567 innovations, we found that the vast majority of them—three-quarters, in fact, as you can see from table 2—were stimulated by a market demand or a production need. Only one-fifth arose from someone saying, aha, maybe we can find a use for *this* technical idea.

This brings me straight to Lesson No 2:

Recognition of demand is a more frequent factor in successful innovation than recognition of technical potential. It seems to me, therefore, that management ought to concentrate on any and all ways of analyzing such demands and needs. For example, more effective communication should be established among specialists in sales, marketing, production, and R & D to see that such opportunities are not overlooked. Some companies do this as a matter of course. They are the innovative companies.

My third innovation was found in the railroad supply industry. Here, the production manager of the Miles Engine and Gear Co., a man wise in the ways of spotting out-of-line costs, noted that the material cost of cylinder-head inserts for one of the company's line of V-8 diesel engines was much too high. It turned out that the inserts were made of stainless steel and, according to the company metallurgist, the cheapest grade of stainless steel at that.

Then the metallurgist began to wonder why stainless steel was being used at all. Other engines in the line did not, nor did competitors' engines. Where inserts were used, they were of cast iron. It was a case of the seventh soldier standing at attention (because years ago he held the mule-reins) while the artillery was fired; the reason for stainless inserts was lost in the mists of history.

The upshot was that the production manager ordered that some inserts be cast out of valve-guide material, an inexpensive grade of iron. When several hundred engines were

TABLE 2.   *Sources of Innovations*

Innovation Initiated by:	No. of Cases	%	
Technical feasibility	120	21	
Market demand	257	45	¾ based on
Production need	169	30	demand or need
Administrative change	21	4	
	567	100	

Spotting technical opportunities plays a surprisingly minor role in sparking innovations. In the author's sample, three-quarters stemmed from recognizing a market potential or a need in a production process.

tested with the experimental inserts over a period of a year, no difference in performance could be detected. The changeover saved $200 per day in materials cost, based on the production rate at the time the problem was noted.

Here, again, is an innovation arising from a production need, not a technical potential. But the key information input was the production manager's noting, from his experience, that costs were out of line, and the metallurgist, from his experience, concluding that stainless steel wasn't needed at all.

This influence of the training and experience of the innovator in stimulating successful innovations is clearly shown in table 3, which covers the whole gamut of innovations we analyzed. In the actual study, we looked separately at the primary information inputs which

evoked the basic idea and those which led to its solution—stages 2 and 4 of our model of the innovative process. The influences were so similar in both cases that I have combined the data into just one table, as shown.

I've annotated the table to indicate the key results. The main one is that in half the cases, the innovators' training and experience—either on that job or previous ones—provided the key information input. If we add to experience the innovators' personal contacts both in and out of the firm, the percentage rises to fully three-quarters. Note that printed materials and R & D were minor sources of information input.

Thus we come to Lesson No. 3: *The training and experience of the people right in your own firm are the principal sources of information for successful*

TABLE 3. *Key Information Inputs*

Innovator Got Key Input from:	No. of Cases	%	
Inside the firm			
Printed materials	9	2	
Personal contacts	25	4	
Own training and experience	230	41	
Formal courses	1	0	
Experiment or calculation	40	7	
	305	54	over ½ from inside sources
Outside the firm			
Printed materials	33	6	
Personal contacts	120	21	
Own training and experience	39	7	R & D
Formal courses	8	2	
	200	36	
Multiple sources	62	11	over ½ from innovator himself
	567	101*	

*Exceeds 100% because of rounding.

You don't necessarily need to look outside the firm for innovative ideas. Most of the major information inputs which evoked the basic idea or led to its solution came from inside the firm. A healthy half of them arose from the innovator's education and experience. Surprisingly, printed materials and R & D work turned out to be minor information sources.

innovations. Thus it's fair to say that competent people within the firm are an invaluable resource. Management should have as its primary responsibility, therefore, the selection, development, retention, and effective utilization of technical personnel—including the facilitation of personal contacts both inside and outside the firm. Almost a cliché, you may say, but the results drive the point home.

The final innovation I've chosen comes from a company we've met before—Mid-North Railroad—and stems from some trouble they were having with their computer system for keeping track of cars, billing other railroads for their use, and so on.

The situation was this: Car-routing and billing data were handled on IBM cards for local tabulating purposes, but to transmit the data elsewhere the cards were fed into an IBM 047, which converted the Hollerith code on the cards into Baudet teleprinter code punched into paper tape. The paper tape was then used for the transmission over teleprinter lines. At the receiving end, the information was punched into paper tape. This was fed into an IBM 063, which repunched the data into IBM-card form.

An engineer at Mid-North, hired by the president specifically to modernize the railroad's communications, decided to eliminate the cumbersome card-to-tape, tape-to-card conversion process. Although the engineer knew from previous experience that equipment existed which could convert card data into an audio tone code transmittable over voice circuits, the combination was not applicable to teleprinter networks and would have required considerable development to make it so.

He searched among manufacturers of computer peripheral equipment, but could not find a conversion device available of the type needed. So he asked for specifications on what such a code-conversion device should be able to do and began looking around for help. One promising approach envisioned a combination of modular transistorized building blocks, and at a trade show the engineer was impressed with what he saw of modules exhibited by the Navigation Computer Co. After discussion of the problem, Navigation gave him a price for a prototype, which was built and eventually developed into a final model. Mid-North purchased the first 60 code-conversion devices for less than the $300,000 cost of one year's rental of the old card-type conversion equipment and, in the process, got some unusual operating features that need not concern us here.

In this case history, as in the previous three, we see the factors of need and previous experience of the innovator playing important roles. But crucial to the success of the code-conversion innovation was the role played by adoption of a technical idea developed by another organization—the alternative way of solving a problem that I talked about under stage 4 of our model.

Table 4 suggests just how common inno-

TABLE 4.   Original vs. Adopted Innovations

Type of Innovation	Original		Adopted		Totals		
	No.	%	No.	%	No.	%	
Product	263	60	65	51	328	58	product
Component	83	19	16	13	99	18	innovation
Process	93	21	47	37	140	25	predominant
	439	100	128	100	567	100	

23% adopted
process innovation
more likely

vation by adoption is. In 128 of the 567 innovations we studied 23 percent, or nearly one out of every four—the problem was solved by going outside the firm for a key product or process. Adopted innovations proved more likely to be process advances than was the case with original innovations from within the firm, which were usually product advances.

There are some other subtleties about adopted innovations that the limited data here do not reveal. For example, compared with the original innovations, we found that fewer of the adopted ones required a lot of adaptation of the major information input or a major change in the production process of the firm. The cost of implementing adopted innovations was about the same as that of developing original innovations, although the uncertainties were undoubtedly less, as discussed earlier. And, as you might expect, vendors proved to be solid sources leading to adoption of technical advances.

All this leads me to set out Lesson No. 4: *Don't overlook adopted innovations; they, as well as those originated within the firm, contribute significantly to commercial success.* Of course applying this lesson brings management smack up against the not-invented-here sort of resistance to technical change that is widespread in many companies. But since no one firm can perform more than a very small proportion of the worldwide innovative activity in any area of technology, it behooves managers to pay serious attention to technology sources outside the firm. Then a deliberate and intelligent trade-off can be made between advancing by original and adoptive innovation. The management of innovation is a corporatewide task that is too important to be left to any one specialized functional department. The R & D department can make its full contribution to the *total process* of innovation not only by effective problem solving, but also by building its competence, knowledge, and personal contacts so as to contribute to the generation of new ideas and to the evaluation of proposed adoptive innovations. Only in this way can it participate effectively in the overall corporate strategy for technical innovation.

SUGGESTED READING

Technological Innovation: Its Environment and Management. U.S. Department of Commerce, 1967. Available from the Superintendent of Documents, Washington, D.C.

Jacob Schmookler, *Invention and Economic Growth* (Cambridge, Mass.: Harvard University Press, 1966).

The Bell-Western Union
Patent Agreement of 1879:
A Study in Corporate Imagination

George David Smith

Telegraphy was a technological innovation of the first order. Save for the railroad, the telegraph did more to revolutionize the basic structure and pattern of American commercial life than any other of the proliferating new technologies of the nineteenth century. The telegraph hastened the process of urbanization and significantly reduced the costs of information while facilitating commercial and managerial transactions in virtually every sector of the economy. By offering high-speed communications over thousands of miles, the telegraph made possible the expansion of coordinated enterprise on a large scale over greatly expanded markets at a nonprohibitive cost. The development of the large, national, integrated corporation would have been inconceivable without it.[1]

One such corporation was the Western Union Telegraph Company itself. In the mid-1870s Western-Union was the country's largest single corporate enterprise. It was a highly sophisticated, well-managed, and innovative monopoly, constantly expanding its business and constantly improving its technology. By virtue of its size and economies of scale and through its ongoing introduction of ever more efficient means for transmitting messages over electrical wire, Western Union was able to sustain high profits, lower its costs, and ward off threats from smaller, less efficient, less profitable competitors.[2]

Yet by 1880, Western Union had forgone an opportunity to take control of one new technological device, the "speaking telephone," that had demonstrated its promise for rendering electrical communication *over short distances* more efficiently than the telegraph. Western Union's failure to develop the telephone, which would ultimately displace the telegraph as the primary mode of telecommunications in the United States, was not merely an oversight. It was a calculated decision made on reasonable business grounds. Only in retrospect do we know that it was also a failure of corporate imagination.

In May 1877 the telephone was brought to market by a small, unincorporated association of four men who held the patent rights to Alexander Graham Bell's now famous invention—a small, clumsy, rectangular box that could receive and transmit sounds replicating human speech electrically over a grounded iron telegraph wire.[3] The Bell patent holders had few resources, but they were determined to exploit what they thought was a unique and saleable invention. A few months earlier they had failed to interest the president of Western Union, William Orton, in their patents for $100,000;[4] and so having no other plausible buyer, they were compelled to establish the value of the telephone themselves. They advertised the telephone as a substitute for short-distance, point-to-point telegraph service and created the Bell Telephone Company to handle the business transactions. By the end of the year, the company succeeded in renting more than 5,500 telephone instruments at rates of $40 per annum for business customers and at $20 per annum for "social," or residential, users.[5]

The success of the Bell telephone, and the potential threat it offered to part of Western Union's own business, brought that company into the field with its own patent claims by the end of the year. Western Union was determined either to eliminate or to absorb the Bell Company, a small but significant new source of competition.[6] For nearly two years, the Bell interests and Western Union struggled for control of the nascent telephone business and the technology that supported it. After the first successful commercial telephone exchange switchboard was introduced by a Bell agent in Connecticut in January 1878, the battle lines formed around the emerging local central telephone office, where any one telephone subscriber could be electrically connected to any one of many other subscribers within a fifteen-mile radius. If the Bell telephone were left unchallenged, it would have competed for much of the traffic of Western Union's district telegraph offices—systems which dealt primarily with the multi-point interchange of local business messages via telegraph and messenger in conjunction with Western Union's long-distance, nationwide telegraph network.[7]

The competition was vigorous. Throughout, Western Union wielded advantages in financial strength, productive capacity, technical resources, and distribution outlets. It acquired an impressive array of patents and controlled the long-distance wires over which Bell's message traffic had to be transferred, if Bell's customers wished to transmit anything beyond their immediate neighborhoods. And yet on November 10, 1879, in a decision that has intrigued students of business and technology ever since, Western Union pulled out of the telephone business. In a contract that served as a settlement to a patent infringement suit brought by Bell against Western Union a year earlier, Western Union conveyed under license to the Bell Company all of its eighty-four patents on telephones and telephone apparatus (i.e., signalling and switching devices), 56,000 telephones in fifty-five cities, valuable assets in plant and equipment, and a commitment not to reenter telephone until 1896.[8]

Western Union did not give the business away. In return for the assignment of its telephone rights, Western Union extracted from Bell a 20 percent royalty of the rental income from every Bell telephone leased in the United States. Western Union received a number of other concessions from Bell that would help protect its (Western Union's) control of the more lucrative elements of its telegraph business, most importantly the long-distance transmission of business traffic. Telephone exchanges were generally limited to a fifteen-mile radius and any telephonic connection between such exchanges was limited to "personal conversation" only. Such connections were, in other words, "not to be used for the transmission of general business messages, market quotations, or news for sale or publication in competition with the [main] business of Western Union. . . ."[9]

The Bell Company agreed, moreover, to transfer to Western Union, exclusively, all extra-exchange telegraphic messages received over its telephone wires. This would protect Western Union against a possible alliance of the telephone with other telegraph companies that inevitably emerged from time to time.[10]

In structural and functional terms the agreement amounted to Western Union's assigning to the Bell interests the problems of managing the local loop, voice-grade communications systems attached to Western Union's nationwide telegraph network. As a captive feeder service, Bell's business would enhance Western Union's telegraph traffic and revenues while not competing for the latter's most important market: interexchange business communications. Western Union was freed from financing and managing local telephone exchange operations with their relatively expensive, capital intensive, and technically complex switchboard and wire plant. With the Bell interests taking on the burden of providing local feeder services, Western Union increased the overall value of its long-distance operations in a way roughly analogous to contemporary arrangements struck by major railroads and smaller lines—arrangements that left to the smaller, weaker companies the worries and costs of managing the less profitable, short-haul traffic.[11]

Thus Western Union arrived at a solution to its rivalry with the Bell Company that allowed it to reap a handsome profit from telephone rentals without absorbing the overhead, to eliminate a source of competition for interexchange business traffic, and to gain a captive ally in competition with other telegraph companies. From a contemporary standpoint this all looked perfectly sensible.[12]

Yet in the long run we know that the agreement was a strategic disaster for Western Union. By giving up its telephone interests, Western Union lost control of the technology that would ultimately relegate the wire telegraph to a secondary and diminishing role in telecommunications. By 1896, when the Bell-Western Union agreement expired, telephones

were already generating more revenues than the nation's telegraph system and were well on the way to becoming the preferred mode of telecommunication in both the business and the newly emerging residential markets.[13] Why, then, did Western Union commit what in retrospect looks like such an obvious strategic blunder?

The question is more compelling when we consider the disparity in size and in the financial, technological, and organizational strength of the two competitors. Initially the Bell enterprise consisted of Alexander Graham Bell, a teacher of the deaf and a novice inventor with no taste for business; Bell's youthful laboratory assistant, Thomas Watson; and Bell's two financial backers, Gardiner G. Hubbard, a Boston patent attorney, and Thomas Sanders, a Lowell, Massachusetts, leather merchant. These four Bell patent holders had little capital, no manufacturing facilities, no distribution outlets, no sales force, and (save for Watson and an office clerk) no full-time personnel to manage the development of the business or the technology on which it was based. When these men formed the Bell Telephone Company in July 1877, they licensed local agents on a geographically exclusive basis to provide simple, point-to-point telephone lines to subscribers along with a pair of telephones.

It was the licensees, the forerunners of the modern Bell System operating companies, who strung the wire, installed the telephones, collected the rentals, and maintained the service. To make the hardware, the Bell entrepreneurs licensed an electrical manufacturing concern in Boston operated by Watson's former employer, Charles Williams, Jr. At Williams's shop Watson supervised the production of telephones and auxiliary apparatus, tinkered with improvements on the hardware, and purchased finished equipment which he then shipped to the operating licensees on Bell's behalf. The Bell interests earned no income on either the provision of service or on the production of equipment. They made their profits entirely from leasing the telephone to their licensees.[14]

This loose set of vertical relationships made it possible for the Bell patent holders to bring the telephone to market. But because these organizational relationships were held together entirely by the Bell Company's claim to "controlling" patents—which, if defensible, gave Bell a legal monopoly of the telephone market for seventeen years—it was crucial for the company to protect these claims as fully as possible against would-be competitors who could easily copy the mechanically simple telephone device. Threats of infringement litigation worked well enough against other small concerns wanting to enter the market, but such threats were of no avail against a corporation like Western Union, whose superiority in financial, technological, and organizational strength might easily overwhelm the foundling Bell enterprise before it could become firmly established.[15]

Western Union was largely owned and controlled by the great nineteenth-century industrialist William H. Vanderbilt, his family, and his allies. The company was capitalized at $40,000,000, a staggering number for the time, and it controlled a vast network of wire, offices, and agents that extended to nearly every established settlement in the United States. Western Union owned a third interest in and effectively controlled the Western Electric Manufacturing Company of Illinois, the world's largest and most advanced developer and producer of electrical equipment. Western Union commanded a large cadre of electricians and inventors who marched in the vanguard of research and development in electrical technology.[16] By mid-1878 the company had acquired patents on telephone receivers, transmitters, and apparatus. Two of these patents were the potentially precedential claims, pending adjudication, of Elisha Gray—whose caveat on the telephone had been filed on the same day as Bell's first patent in 1876—and of Thomas Edison—whose microphonic transmitter was different in principle and superior in performance to Bell's magneto model.[17]

Western Union used its advantages in economies of scale and scope in competing for prime telephone markets in the country's major urban centers. It tried, sometimes successfully, to buy out Bell licensees. It offered potential agents—many of whom were already operating district telegraph offices—access to existing wire plant, lenient terms of credit, indemnification against possible losses in patent suits, and lower priced telephone equipment, all easily subsidized from the company's $3 million a year in net profits. The Bell Company, which struggled even to pay its manufacturing costs through 1878, could not hope to match Western Union's terms. Western Union, moreover, could effectively discriminate among whom it allowed to convey messages from local telephone facilities to its long-distance wires. And midway through 1878, the introduction of the Edison transmitter gave Western Union a clear technological lead. By 1879 Western Union had prevailed in several important cities, including Chicago and New York, and its rate of growth in telephony exceeded that of Bell's.[18]

Early in the struggle, the Bell patent holders attempted to negotiate a merger of their telephone rights with Western Union's. Bell treasurer Thomas Sanders, who was rapidly exhausting his personal fortune to finance the Bell Company's operations, particularly favored a negotiated settlement. Otherwise, he feared, Western Union would "crush us by fair means or foul." But because of the stubborn desire of Bell Company president, Gardiner Hubbard, to preside over a consolidated telephone company in which the Bell patent holders would hold half the stock (demands which Western Union found utterly excessive), negotiations fell through. That course having failed, the Bell interests had no choice but to compete.[19]

They competed on several fronts. In September 1878, the Bell Company sued Peter Dowd, a Western Union agent in Massachusetts, for infringement of Alexander Graham Bell's pair of basic telephone patents—one on the principle of telephony and one on the hardware. From that point the Bell Company stressed its patent claims both as a way of keeping its own agents in line and as an inducement to potential licensees. Bell refrained from trying to compete with Western Union on price, emphasizing instead the quality of its service and equipment in its marketing strategy, which proved successful.[20] To keep abreast of technological developments, the company began its own program of research and development, hiring expert electricians to augment the work of Thomas Watson while searching aggressively for new patents from outside sources. By the end of the year, the company was able to develop a microphonic transmitter superior to Edison's in performance.[21] And throughout, Bell constantly increased its rate of production and recruitment of operating agents in order to keep pace with the demand that was now being fanned by the very process of competition.[22]

To manage and afford all this expanded activity, the Bell interests placed their affairs on a more formal administrative footing and sought wider sources of capital. Gardiner Hubbard hired Theodore Vail away from the superintendency of the National Railway Telegraph Service to fill the position of general manager. Vail initiated formal contracts and systems of accounting controls over the licensees and established a system of traveling agents to extend and monitor the licensee business. In the spring of 1879 the patent holders reorganized their holdings (for the third time since May 1877) as the National Bell Telephone Company with $850,000 in capitalization supported by a greatly enlarged group of stockholders from the Boston financial community. One of these financiers, William Forbes, took over the presidency of the firm from Gardiner Hubbard,

whose part-time, informal, and unsystematic direction of the firm had become a drag on its ability to grow and to coordinate its affairs efficiently.[23]

Forbes and Vail turned the National Bell Company into a more recognizably modern managerial corporation. They developed goals and strategies for the business on a long-term as well as an annual basis. They developed plans for the expansion of the technological capability of the telephone service, particularly through the financing and construction of experimental, intercity toll lines. They devised contracts that would allow the Bell Company to acquire equity in its operating licensees over time. They embarked on a program for shoring up the company's patent position—"to surround the business in patents," as Vail put it.[24] They licensed additional manufacturers for auxiliary apparatus to relieve the strain on Charles Williams, Jr., who was increasingly unable to increase his capacity rapidly enough to meet growing demand. And, perhaps most important, to defend their patents, they renewed negotiations with Western Union in order to find a resolution to their destabilizing competitive war.[25]

In June 1879 the conflation of two events exogenous to the Bell Company brought Western Union to the bargaining table. In the first instance, Western Union was advised by its chief attorney, George Gifford, that he had become legally "convinced that [Alexander Graham] Bell was the first inventor of the telephone . . . that the defendant [Western Union] had infringed. . . , and . . . that the best policy for them was to make some settlement for the complainants."[26]

This alone would unlikely have daunted Western Union. Both sides were well aware that Western Union had both the resources and the will to outlast Bell in protracted litigation and was probably able to delay the outcome for years. Indeed, some Bell officials were terrified at the prospect.[27] Moreover, a single patent rarely has the potential to control the business development of an industry, especially one whose technology was as complex as telephony with its manifold requirements for transmission, switching, and signaling. J.J. Storrow, Bell's chief attorney, knew that Bell and Western Union patents had to be pooled "in order to enable everybody to use good telephones," an opinion echoed by a pair of Bell System attorneys who argued in 1890 that both Bell and Western Union had had sufficient patent strength "to practically exclude [each] other from the telephone business." At the very least, under these circumstances, Western Union could have bargained for a consolidation of patent claims on very favorable terms to itself.[28]

However, just as Gifford was advising his client to settle, William H. Vanderbilt's notorious rival in industrial finance, Jay Gould, was organizing the American Union Telegraph Company in his second attempt in four years to disrupt the telegraph monopoly. In the first attempt Gould had been bought out by Western Union for the nuisance value; but this time he was tenacious. He bought up scattered independent telegraph companies, forged alliances with key foreign concerns and American railway operators (who controlled the rights of way for telegraph lines), and cut rates on competing wires that were being rapidly constructed by some 5,000 men in his employ. Gould also attacked the credit of Western Union in his newspaper, the *New York World.* Within eighteen months Gould would take substantial market share from Western Union, driving down its revenues and stock price until he was finally able to take control of his rival by buying up its depreciated stock as it was dumped onto the market by anxious stockholders. Although Western Union officials could not have foreseen this stunning outcome at the commencement of Gould's attack, they did not fail to perceive the threat of a possible alliance between the American Union Telegraph Company and National Bell Telephone.[29]

The threat was clear enough. In 1866 Western Union had achieved its control of the nation's telegraph business through the merger and absorption of major competing companies. It subsequently protected its monopoly through its willingness to buy out well-financed competitors and through its aggressive program of technical research and development and patent acquisition. The first serious technological innovation to challenge its broad patent base arose with the advent of the Bell telephone. But the Bell telephone had very limited technical capacity, and the small group of entrepreneurs who were bringing it to market would not have seemed so problematic in the absence of a significant rival in telegraphy. Now, should Gould's telegraph provide a long-distance outlet for Bell's telephone (the prospect of which the Bell interests were cheerfully aware), or should Gould gain control of Bell licensees (he was purchasing interests in several in Connecticut and New York), a combined telephone-telegraph concern with great capital resources would be able to sustain a protracted battle in the courts and in the marketplace. Within fifteen days of Gould's entry, Western Union offered Bell terms for a "consolidation" of interests.[30]

Negotiations proceeded in earnest throughout the summer and fall. The upshot was, of course, not a merger but the creation of what one writer has termed "a duopoly in telecommunications with the market divided according to local service (Bell telephone) and long distance service (Western Union telegraph)." In the end, Western Union was delighted to be rid of what its president referred to as "a bitter and wasteful competition."[31] By turning over its telephone rights to Bell in exchange for royalties and other commitments, it could achieve a reasonable settlement to its competitive problems at no cost.

Just why Western Union chose to settle its conflict with Bell on these terms, rather than by merger or acquisition, was to some extent due to the stubbornness of Bell's management, which became more confident over time that it could survive to manage the telephone business independently.[32] More fundamentally, Western Union simply failed to imagine two plausible developments: (1) that the potential market for telephones was broader than their initial use indicated; and (2) that the telephone might achieve a long-distance capability to render it technologically independent from the telegraph network.

The market for the telephone was conceptually circumscribed by three assumptions. First, based on a generation's experience, it appeared that the most important application for electrical wire communication lay in commercial and financial correspondence and in the transmission of news. For these purposes long-distance communication was vital. Western Union's best customers were business firms that had distant agents or correspondents, banks, brokerage houses, and news wire services. Western Union explicitly excluded Bell from these markets in cases beyond a fifteen-mile radius in the 1879 agreement. Second, it was assumed that there was no great value in prolonged voice-grade conversation beyond what was required for the exchange of brief transactions between business entities or between branches or departments of a single firm. There was, of course, a small demand for "social" telephones from the beginning, but even the Bell Company regarded this essentially luxury market with only modest interest until the twentieth century, when per capita income rose to a level to spark demand for residential telephones. The telephone's main value in the latter part of the nineteenth century was its ability to transfer a *message* (a telegraphically defined commodity) without the need for an experienced key operator.[33]

Third, the market was limited by the technology. It would be years before the telephone became anything less than a strenuous

exercise of the subscriber's volume and elocution in speech and his acuteness of hearing. The earliest instruments transmitted weakly, and grounded iron telegraph wires were susceptible to induction noise from the earth, making the telephone less than reliable over distances of a few miles. Conventional habits of written discourse—by mail, messenger, or electrical wire—were transformed only gradually as the technology of telephony improved and as cultural habits adapted to the improvements.[34]

By 1879, improvements on the telephone had made it possible to have reasonably good service up to forty miles, more than enough to connect business offices in Boston with their factories in Lowell.[35] But the elements of telephone technology that would make feasible the interexchange transmission of voice-grade conversation over very long lines—such developments as the copper wire metallic circuit, the common switchboard battery, and the loading coil[36]—were as yet unforeseen. And the development of long-distance telephony as a substitute for the telegraph was largely a matter of faith.

Western Union had no such faith. It had always undervalued the telephone. when William Orton rejected Gardiner Hubbard's offer of the Bell patents in the winter of 1876–77, he scorned the device as a mere "scientific curiosity", a "toy." Western Electric's electrician, the great professional inventor Elisha Gray, agreed. Even though Elisha Gray's own insight into the principle of the telephone had actually preceded Alexander Graham Bell's, he did not rush to patent. He pressed his claims belatedly, at the behest of Western Union, only after the Bell Company had begun to establish a market. Gray, and Edison too, for that matter, had been far more absorbed in trying to develop improvements on the message-carrying capacity of the telegraph (the payoffs for which were sure to be high) than to grasp the significance of telephonic discoveries they made as by-products of their research.[37] In other words, the men for whom telegraphy was the primary focus of their business activities and technical expertise were committed too deeply to the systems which had become their bread and butter to appreciate fully what might be a radical break in the technology.

On the other hand, the Bell patent holders, whose orientation to telegraphy was at the margin's of its development, saw things differently. In the first place, Alexander Graham Bell was a rank amateur compared to the likes of Gray, Edison, and others whom Western Union had under retainer or employ. Bell, too, was interested in the lucrative reward that would go to anyone who could develop a "multiple" telegraph, but he was obsessed with the idea of reproducing speech electrically, an interest related to his own professional concerns as an elocutionist.[38] He believed, naively perhaps, that a latent demand for telephony existed; and though he quickly withdrew from active participation in the business that was created to bring his invention to market, he continued to press his peculiar vision of the telephone's potential on his partners and the public. In a famous letter to a group of London capitalists in 1878, Bell predicted

> that cables of Telephonic wires could be laid underground or suspended over head communicating by branch wires with private dwellings, counting houses, shops, manufactories, etc., etc., uniting them through the main cable with a central office . . . connected together . . . establishing direct communication between any two places in the City. . . . Not only so but I believe that in the future wires will unite the head offices of Telephone Companies in different cities and a man in one part of the Country may communicate with another in a distant place.[39]

Bell himself characterized his vision as somewhat "Eutopian," and, to be sure, his backers' view of the technology was more tightly bound by a telegraphic mind-set. Early Bell advertising, in 1876, represented the telephone as nothing more than a better short-haul telegraph. In a letter to an unknown recipient in 1878, Gardiner Hubbard surmised that the telephone might well eventually overcome the technical constraints on long-distance transmission but "ultimately," he thought, "the chief use of the Telephone on long lines will be for the transmission of *telegraphic* messages."[40] This conception of the telephone as handmaiden to the telegraph faded only incrementally over the years in the minds of Bell officials, even as Hubbard's successors moved ahead with telephonic experiments in the "connecting of cities and towns."[41]

Even though the 1879 agreement prohibited the Bell Company from competing for telegraph business over long wires, the firm did not give up that line of technological development. On the eve of the settlement, Bell and Western Union were hung up on a final resolution of the problem of interexchange, or toll line, connections. Bell Company owners and managers were divided on whether even to press the issue, but Theodore Vail's more expansive view prevailed. In its final negotiations with Western Union the Bell Company insisted on, and won, the right to develop its telephone business between exchanges, even under the prohibitions against business messages that were included in the contract. To yield the toll line altogether, Vail reflected some thirty-five years later, would have "meant the curtailment of our future—the absolute interdiction of anything like a [telephone] *system.*"[42]

It would be tempting for a historian to dismiss Vail's words as *ex post facto* rationalization, if the records did not reveal that the Bell Company was alert to the opportunity presented to it at the actual time of the settlement. To be sure, the company's view of its technological future was still framed largely in telegraphic terms. In the spring of 1880, William Forbes explained his long-term strategy in the following way: once the Bell Company was able

95

to purchase control of its operating licensees, it would have "it in [its] power to own the terminal facilities, as well as the lines connecting the Cities." This, then, he predicted, would allow Bell "to secure a large control of the Telegraph business of the country, as well as hold the Telephone business probably without competition."[43]

Nevertheless, the Bell company now had a long-distance strategy and the resources to pursue it. With its most natural and powerful competitor out of the way, its stock price soared.[44] With the acquisition of Western Union's telephone rights, the company was able to withstand more than 600 legal challenges to its patent base before the expiration of the original Bell patents by 1894. Such patent strength made it possible for Bell to insist upon equity positions in its licensees, equity positions it could now afford. In 1882 the new and more heavily capitalized American Bell Telephone Company integrated backwards into production by buying out Western Union's share of Western Electric into which Bell's other manufacturing interests were merged and acquired. Three years later, Bell established the American Telephone and Telegraph Company, a New York corporation, to develop intercity telephone service, which was proving to require substantial investment in both plant and new, non-telegraphic technologies.

Over time, technological improvements unforeseen in 1879 made it possible for the telephone to displace the telegraph as the primary mode of long-distance telecommunications. When the Bell-Western Union agreement expired in 1896, the Bell System of manufacturing and operating companies had become so large, so organizationally efficient, so technologically complex, and so far advanced in the enormous capital investment and novel techniques required to operate long-distance lines that the company was able to withstand a torrent of pent-up telephone competi-

tion that surfaced at the turn of the century. While many new companies carved out lucrative shares of local and regional markets in places where the Bell System was unestablished or weak, no one was able to muster the capital or technology to challenge the intercity network being developed by AT&T by which it established its dominance of the nation's wire communications system.[45]

If this case means anything, it is not that the managers of the National Bell Telephone Company were any smarter than their counterparts at Western Union. Indeed, given the state of telecommunications technology in 1879, Western Union appears to have used its power vis-à-vis its smaller rival to perfectly logical advantage. Nor can it be said that the Bell interests had that much clearer a view of the practical potential of the telephone than their technologically sophisticated rivals at Western Union. What can be said is that Western Union failed to imagine the opportunities posed by the telephone, which they were prone to view, just as the Bell managers did, as an *adjunct* to telegraphy. The point is that the smaller Bell Company's orientation to the technology—its necessarily more specialized focus on the telephone—gave it a subtle but decisive shift in vision as it came to terms with its larger, more powerful competitor.

Seen in this light, we can understand the Bell-Western Union case as something more than an idiosyncratic event, an amusing David and Goliath story in the annals of corporate strategy. The case is of a type: it is one powerful example of what can happen to a dominant company whose view of its technology has become so committed to a certain path of development that it cannot imagine the alternatives that lie within its technological purview and control.

Consider a couple of familiar examples. In 1877, the Cowles Company of Cleveland, Ohio, the nation's leading producer of aluminum-bronze alloys, gave up on the development of Charles Martin Hall's patent on an inexpensive way of smelting pure aluminum, so apparently difficult was it for them to conceive of applications for the light metal the technologies for which they did not already control. Hall found backing elsewhere from the founders of what was to become Alcoa. In our own time, Steven Jobs and Steven Wozniak launched Apple after it had become clear that there was no market for their idea of a personal desk-top computer among the established corporate masters of computer technology.

Such cases are in fact legion, and it would be useful for historians of business and technology to examine them each in some detail. In that way we will be able to build a more comprehensive record of the ways in which dominant companies can become locked into mind-sets that inhibit their ability to grasp some of the more subtle threats and opportunities posed to their mainstream technologies by seemingly ancillary developments. In the meantime, it is enough to suggest that corporate executives, scientists, and engineers keep these cautionary tales in mind.

NOTES

1. Some standard discussions of this point are Thomas C. Cochran, *200 Years of American Business* (New York, 1977), pp. 77–79; Glenn Porter, *The Rise of Big Business, 1860–1910* (Arlington Heights, Illinois, 1973), p. 43; Elisha P. Douglas, *The Coming of Age of American Business: Three Centuries of Enterprise, 1600–1900* (Chapel Hill, 1971), p. 487; Alfred D. Chandler, Jr., and Hermann Daems, eds., *Managerial Hierarchies* (Cambridge, Mass., 1980), p. 15; Joel A. Tarr with, Thomas Finholt, and David Goodman, "The City and the Telegraph: Urban Telecommunications in the Pre-Telephone Era," *Journal of Urban History,* XIV (November, 1987), pp. 38–80.

2. The standard work on the organization of Western Union is Robert L. Thompson, *Wiring a Continent: The History of the Telegraph Industry in the United States, 1832–1866* (Princeton, 1947). A useful supplement for the market history of the telegraph after 1866 is Richard B. Du Boff, "Business Demand and the Development of the Telegraph in the United States, 1844–1860," *Business History Review,* LIV (Winter, 1980), pp. 459–79. See also, Alvin F. Harlow, *Old Wires and New Waves* (New York, 1936), pp. 255ff.

3. The origins of the Bell Patent Association are detailed in numerous histories. Accessible accounts are Robert V. Bruce, *Bell: Alexander Graham Bell and the Conquest of Solitude* (Boston, 1973) and John Brooks, *Telephone: The First Hundred Years* (New York, 1976 and 1977); Robert W. Garnet, *The Telephone Enterprise: The Evolution of the Bell System's Horizontal Structure, 1876–1909* (Baltimore, 1985); George David Smith, *The Anatomy of a Business Strategy: Bell, Western Electric, and the Origins of the American Telephone Industry* (Baltimore, 1985).

4. Bruce, *Bell,* pp. 258–59.

5. "The Telephone." May 1877, AT&T Historical Archives, 195 Broadway, New York City (hereafter cited as AT&T Archives), Box 1097.

6. The story of the Bell-Western Union competition has become almost legendary and forms an important part of the folklore of the Bell System. Particularly interesting analyses of the struggle not found in the many standard histories of the industry are "Early Competition for Financial Control of the Telephone Industry," Federal Communications Commission, *Investigation of the Telephone Industry* [1939], Exhibit No. 2096f; Michael F. Wolff, "The Marriage that Almost Was," *IEEE Spectrum* (February, 1976), pp. 41–51. I have written in great detail on the subject in *The Anatomy of a Business Strategy,* esp. pp. 36–38, 76–80, 154–55.

7. For Bell's challenge to the district telegraph offices, see Bell Telephone Company, "Instructions to Agents, No. 3," February 1, 1878, AT&T Archives, Box 1001. See also J.E. Kingsbury, *The Telephone and Telephone Exchanges, Their Invention and Development* (London, 1915), chap. VIII; Smith, *Anatomy of a Business Strategy,* p. 36.

8. I have relied on the printed version of the agreement entitled Contract, November 10, 1879, AT&T Archives, Box 1006. (The Western Union telephones referred to throughout this paper were actually the property of the Gold and Stock Telegraph Company, a Western Union subsidiary through which its telephones were marketed. Another subsidiary, The American Speaking Telephone Company, held the Western Union patents in telephony. While these subsidiaries were parties to the agreement, I have, for convenience' sake, lumped them under the rubric of "Western Union." Telephone policy of these subsidiary companies was dictated by Western Union in any case.)

9. *Ibid.*

10. *Ibid.* For a good analysis of this point, see Gerald W. Brock, *The Telecommunications Industry* (Cambridge, Mass., 1981), pp. 97–98.

11. Smith, *Anatomy of a Business Strategy, p. 80. Cf. Brock, The Telecommunications Industry,* p. 99.

12. One contemporary commentator in the *Cincinnati Daily Enquirer,* October 18, 1879, upon hearing of the terms of the pending settlement, thought the Bell Company was being placed at a serious disadvantage.

13. The Bell Telephone System generated more than $24 million in revenues in 1895 compared with Western Union's nearly $21 million for the fiscal year ending in mid-1896. (This does not account for other telegraph companies, however.) By that time the telephone was employed in more than 2,100,000 *daily* conversations. Western Union, by comparison, handled about 160,000 telegraph messages per diem. See the *Historical Statistics of the United States,* bicentennial edition (Washington, D.C., 1977), II, pp. 786–87.

14. These arrangements are detailed in Smith, *Anatomy of a Business Strategy,* chaps. I and II.

15. The patents were Alexander Graham Bell's March 7, 1876, grant—U.S. Patent No. 174,465—on the basic principle of telephony and his January 30, 1877, grant—U.S. patent No. 186,787—on the telephone hardware. These patents, if deemed to "control" the use of telephone technology, would legally secure the patent holders from commercial competition until 1894. Others could use the patents for commercial purposes only if licensed to do so by the patent holder. Those trying to make or sell a patented device without a license were liable for "infringement" and subject to serious financial penalties.

16. Bruce, *Bell,* pp. 260–61; Brooks, *Telephone,* pp. 61–62; Charles G. DuBois, "A Half Century of Western Electric Achievement," *Western Electric News,* VIII (November, 1919), pp. 1–6; Report on Western Electric Corporate Structure," FCC, *Investigation* (1939), Exhibit 1952, pp. 3ff.

17. See Brooks, *Telephone,* pp. 61–62; James D. Reid, *The Telegraph in America* (New York, 1879), pp. 629–33.

18. Smith, *Anatomy of a Business Strategy,* chap. 2, passim.

19. Thomas Sanders to Gardiner G. Hubbard, December 5, 1877, AT&T Archives, Box 1006; Sanders to Hubbard, February 23, 1878, ibid., Box 1193; Sanders to Hubbard, January 30, 1878, ibid., General Manager's Letterbook. Sanders thought that the Bell Company should "make the best terms we can with this powerful combination [Western Union and its subsidiaries]," even if it meant selling the business.

20. Thomas Watson to Theodore Vail, February 19, 1879, ibid., Box 1205.

21. Frederic William Wile, Emile Berliner: Maker of the Microphone (Indianapolis, 1926), pp. 107ff.; Frederick Leland Rhodes, *Beginnings of Telephony* (New York, 1929), pp. 76–79.

22. Smith, *Anatomy of a Business Strategy,* chap. 3.

23. *Ibid.,* pp. 43–49; Garnet, *The Telephone Enterprise,* chap. 3. See also J. Warren Stehman, *The Financial History of the American Telephone and Telegraph Company* (Cambridge, Mass., 1925), pp. 12–15.

24. Theodore Vail is quoted in N.R. Danielian, *AT&T: The Story of Industrial Conquest* (New York, 1939), p. 96.

25. Smith, *Anatomy of a Business Strategy,* chap. 3; Garnet, *The Telephone Enterprise,* chap. 4.

26. George Gifford testified to this in a subsequent patent suit involving the telephone in which he appeared as a witness for the American Bell Telephone Company. His remarks are printed in the "Brief for the American Bell Telephone Company," Supreme Court of the United

States, October Term, 1886, *The Telephone Appeals*, pp. 2–3, 39–43.

27. See, for example, Charles Cheever to Gardiner G. Hubbard, December 3, 1877, AT&T Archives, Box 1006. Cheever warned Hubbard that a legal challenge to Western Union would become an "extremely tedious" process. Thomas Watson in his Diary of a Trip, 1878, ibid., Box 1069, reported his encounter with Western Union vice president, Anson Stager, in Chicago during which Stager "told a horrible story of how the W.U. Co. could keep the matter in the courts five years." Theodore Vail wrote to Oscar Madden on May 29, 1879, ibid., General Manager's Letterbook, surmising that Western Union could afford to "lose" $500,-000 on the telephone fight.

28. [J.J. Storrow], Memorandum, January, 1880, ibid., Box 1326; "Argument of Charles L. Buckingham and Edward G. Bradford for the defendant at Wilmington, Delaware, April 24 and 25, 1890," in their petition for Mandamus, pp. 1–8, U.S. Circuit Court, District of Delaware, June Term, 1890: The Postal Telegraph-Cable Company vs. The Delaware and Atlantic Telegraph and Telephone Company as quoted in FCC, *Investigation*, Exhibit 2096f, p. 25.

29. On the Jay Gould attack on Western Union see *ibid.*, pp. 13 and 19–21; Julius Grodinsky, *Jay Gould* (Philadelphia, 1957), p. 206; Harlow, *Old Wires and New Waves*, p. 381; Matthew Josephson, *The Robber Barons: The Great American Capitalists, 1861–1901*, pp. 205ff.

30. William Forbes to Gardiner G. Hubbard, May 26, 1879, AT&T Archives, President's Letterbook, 1G. The initial proposal from Western Union suggested the creation of a new company in which the Bell interests and Western Union were each to hold stock, the allocation of which was to be settled by arbitration.

31. Brock, *The Telecommunications Industry*, p. 98; Norvin Green (who replaced Orton as Western Union president after the latter's death in the spring of 1878) to William Forbes, September 3, 1879, ibid., Box 1006.

32. The growing confidence of Bell's management becomes apparent to anyone who follows the day-to-day correspondence of the firm through 1879. The change of management, greater sources of funds and improvements in technology all seemed to have contributed to the company's growing resolve.

33. Cf. for the telegraph market DuBoff, "Business Demand," esp. pp. 471ff.

34. On the primitive state of the technology see Rhodes, *Beginnings of Telephony*, chaps. I–V, and the interesting brief discussion in Brooks, *Telephone*, pp. 85ff. Thomas Watson, *The Birth and Babyhood of the Telephone* (New York, 1913), wrote that the earliest commercial telephones required that a subscriber have "a voice with the carrying capacity of a steam calliope" to be practically audible. The earliest telephone advertisement noted that "conversation can be easily carried on after slight practice and the occasional repitition of a word or sentence. On first listening to the Telephone, though the sound is perfectly audible, the articulation seems to be indistinct; but after a few trials the ear becomes accustomed to the peculiar sound. . . ." See "The Telephone", AT&T Archives, Box 1097.

35. The Boston-Lowell line was the first experimental long-distance line attempted by the National Bell Company. Correspondence relating to its development can be found in ibid., Boxes 1126 and 1185. It was conceived as a telegraph line on which telegraph and telephone instruments could be used interchangeably. See esp. Charles J. Glidden to George L. Bradley, November 11, 1878, ibid., Box 1126.

36. The copper wire metallic circuit was introduced gradually throughout the Bell operating companies during the 1880s and 1890s. It was the *sine qua non* of long-distance service. A metallic circuit was a twisted pair of wires, one wire used for the return of the circuit. The elimination of the ground circuit relieved the serious problem of "earth-noise" induction. The common battery eliminated the need for each telephone to be provided with a power source and eliminated the technical problem of variable local battery efficiency which could disrupt the operation of the switchboard. The loading coil, developed at the turn of the century, reduced the attenuation of the electrical impulse as it traveled over a long wire or through an underground cable. See *A History of*

Engineering & Science in the Bell System (Bell Telephone Laboratories private printing, 1975); Neil H. Wasserman, *From Invention to Innovation: Long-Distance Telephone Transmission at the Turn of the Century* (Baltimore, 1985).

37. See a pair of articles by David A. Hounshell, "Elisha Gray and the Telephone: The Disadvantages of Being an Expert," *Technology and Culture,* 16 (April, 1975), pp. 144 and 151ff. and "Bell and Gray: Contrasts in Style, Politics and Etiquette," *Proceedings of the IEEE,* 64 (September, 1976), pp. 1312–15.

38. Hounshell, "Elisha Gray", p. 151, notes that Hubbard thought the most important part of Bell's work was related to the multiple telegraph and was discouraged by Bell's enthusiasm with telephony.

39. Alexander Graham Bell to the Capitalists of the Electric Telephone Company, March 25, 1878, copy in AT&T Archives, uncatalogued. Bell's enthusiasm for the telephone did not extend to the business. He effectively withdrew from its affairs after the formation of the National Bell Telephone Company when his (by then) father-in-law, Gardiner G. Hubbard, was removed from its presidency. Bell had participated little in the company's affairs prior to that, although he held the title of company electrician. Bell sold most of his stock in 1883 and retired to a life of leisure.

40. Gardiner G. Hubbard to [unknown] post-dated 1878 (possibly 1879), AT&T Archives, Box 1115.

41. Vail to T.E. Cornish, June 20, 1879, ibid., Box 1054; Vail to G.S. Glen, May 29, 1879, ibid., General Manager's Letterbook.

42. Address of Theodore Vail to the National Geographic Society, March 7, 1916; Talk by Theodore N. Vail over Telephone from Morristown, N.J., to Providence, R.I., November 15, 1915, ibid., uncatalogued.

43. William H. Forbes to H.L. Higginson, April 24, 1880, ibid., Box 1055.

44. Stehman, *Financial History,* p. 19; Brooks, *Telephone,* p. 72. In the immediate aftermath of the settlement National Bell's stock price soared from $300 per share to $1,000 per share and then settled to $600 per share in March 1880. A year earlier the company's stock had been trading for around $50 per share.

45. Garnet, *The Telephone Enterprise,* pp. 55–108; Smith, *Anatomy of a* Business Strategy, pp. 111–38; Brock, *Telecommunications Industry,* pp. 109–25.

SECTION II

Organization and Innovation

This section focuses on the interrelations between organizations and innovation. In Chapter 2, Van de Ven's review illustrates several generic problems in the management of innovation, and the Quinn and Peters articles provide broad concepts for effective innovation management. These three articles all reinforce the notion that organizations must be able to achieve an adaptive linkage between strategy, structure, people, and informal processes in order to be innovative over time. Given these general innovation-organization ideas, Nadler and Tushman present a specific framework for organization analysis and problem solving. Their congruence model provides an integrative approach to organization diagnosis and design for innovation. The idea of building internally congruent social and technical systems to be effective in the short term and innovative over time will be reinforced throughout this book.

Chapter 3 introduces several inherent organizational forces that block innovation. Morrison's classic history of continuous-fire guns illustrates the stultifying effects of organization inertia, history, and limited identifications on innovation. Salancik and Pfeffer focus on power and the politics of stability, while Katz traces the roots of organization inertia to commitment and learning processes within groups and individuals. As innovation always involves breaking with the status quo, managers must understand the roots of this inertia. The following sections build on Section II in designing strategies, structures, cultures, and executive teams both to take advantage of as well as to break through organization inertia.

2

APPROACHES TO INNOVATION AND ORGANIZING

Central Problems in the Management of Innovation

Andrew H. Van de Ven

Innovation is defined as the development and implementation of new ideas by people who over time engage in transactions with others within an institutional order. This definition focuses on four basic factors (new ideas, people, transactions, and institutional context). An understanding of how these factors are related leads to four basic problems confronting most general managers: (1) a human problem of managing attention, (2) a process problem in managing new ideas into good currency, (3) a structural problem of managing part-whole relationships, and (4) a strategic problem of institutional leadership. This paper discusses these four basic problems and concludes by suggesting how they fit together into an overall framework to guide longitudinal study of the management of innovation.

INTRODUCTION

Few issues are characterized by as much agreement as the role of innovation and entre-

Reprinted by permission of the publisher from "Central Problems in the Management of Innovation" by Andrew H. Van de Ven, from *Management Science*, May 1986, Volume 32, Number 5. Copyright 1986 by The Institute of Management Sciences.

preneurship for social and economic development. Schumpeter's (1942) emphasis on the importance of innovation for the business firm and society as a whole is seldom disputed. In the wake of a decline in American productivity and obsolescence of its infrastructure has come the fundamental claim that America is losing its innovativeness. The need for understanding and managing innovation appears to be widespread. Witness, for example, the common call for stimulating innovation in popular books by Ouchi (1981), Pascale and Athos (1981), Peters and Waterman (1982), Kanter (1983), and Lawrence and Dyer (1983).

Of all the issues surfacing in meetings with over 30 chief executive officers of public and private firms during the past few years, the management of innovation was reported as their most central concern in managing their enterprises in the 1980's (Van de Ven 1982). This concern is reflected in a variety of questions the CEOs often raised.

1. How can a large organization develop and maintain a culture of innovation and entre-preneurship?
2. What are the critical factors in successfully launching new organizations, joint ventures with other firms, or innovative projects within large organizations over time?
3. How can a manager achieve balance between inexorable pressures for specialization and proliferation of tasks, and escalating costs of achieving coordination, cooperation, and resolving conflicts?

Given the scope of these questions raised by CEOs, it is surprising to find that research and scholarship on organizational innovation has been narrowly defined on the one hand, and technically oriented on the other. Most of it has focused on only one kind of organizational mode for innovation—such as internal organizational innovation (Normann 1979), or new business startups (e.g., Cooper 1979)—or one stage of the innovation process—such as the diffusion stage (Rogers, 1981)—or one type of innovation—such as technological innovation (Utterback 1974). While such research has provided many insights into specific aspects of innovation, the encompassing problems confronting general managers in managing innovation have been largely overlooked.

As their questions suggest, general managers deal with a set of problems that are different from and less well understood than functional managers. We concur with Lewin and Minton's (1985) call for a general management perspective on innovation—one that begins with key problems confronting general managers, and then examines the effects of how these problems are addressed on innovation effectiveness. The purpose of this paper is to present such a perspective on the management of innovation. Appreciating these problems and their consequences provides a first step in developing a research program on the management of innovation.

The process of innovation is defined as the development and implementation of new ideas by people who over time engage in transactions with others within an institutional context. This definition is sufficiently general to apply to a wide variety of technical, product, process, and administrative kinds of innovations. From a managerial viewpoint, to understand the process of innovation is to understand the factors that facilitate and inhibit the development of innovations. These factors include ideas, people, transactions, and context over time. Associated with each of these four factors are four central problems in the management of innovation which will be discussed in this paper.

First, there is *the human problem of managing attention* because people and their organizations are largely designed to focus on, harvest, and protect existing practices rather than pay attention to developing new ideas. The more successful an organization is the more difficult it is to trigger peoples' action thresholds to pay attention to new ideas, needs, and opportunities.

Second, *the process problem is managing ideas into good currency* so that innovative ideas are implemented and institutionalized. While the invention or conception of innovative ideas may be an individual activity, innovation (inventing and implementing new ideas) is a collective achievement of pushing and riding those ideas into good currency. The social and political dynamics of innovation become paramount as one addresses the energy and commitment that are needed among coalitions of interest groups to develop an innovation.

Third, there is *the structural problem of managing part-whole relationships,* which emerges from the proliferation of ideas, people and transactions as an innovation develops over time. A common characteristic of the innovation process is that multiple functions, resources, and disciplines are needed to transform an innovative idea into a concrete reality—so much so that individuals involved in individual transactions lose sight of the whole innovation effort. How does one put the whole into the parts?

Finally, the context of an innovation points to *the strategic problem of institutional leadership.* Innovations not only adapt to existing organizational and industrial arrangements, but they also transform the structure and practices of these environments. The strategic problem is one of creating an infrastructure that is conducive to innovation.

After clarifying our definition of innovation, this paper will elaborate on these four central problems in the management of innovation. We will conclude by suggesting how these four problems emerge over time and provide an overall framework to guide longitudinal study of innovation processes.

INNOVATIVE IDEAS

An Innovation is a new *idea,* which may be a recombination of old ideas, a scheme that challenges the present order, a formula, or a unique approach which is perceived as new by the individuals involved (Zaltman, Duncan, and Holbek 1973; Rogers 1982). As long as the idea is perceived as new to the people involved, it is an "innovation," even though it may appear to others to be an "imitation" of something that exists elsewhere.

Included in this definition are both technical innovations (new technologies, products, and services) and administrative innovations (new procedures, policies, and organizational forms). Daft and Becker (1979) and others have emphasized keeping technical and administrative innovations distinct. We believe that making such a distinction often results in a fragmented classification of the innovation process. Most innovations involve new technical and administrative components (Leavitt 1965). For example Ruttan and Hayami (1984) have shown that many technological innovations in agriculture and elsewhere could not have occurred without innovations in institutional and organizational arrangements. So also, the likely success of developments in decision support systems by management scientists largely hinges on an appreciation of the interdependence between technological hardware and software innovations on the one hand, and new theories of administrative choice behavior on the other. Learning to understand the close connection between technical and administrative dimensions of innovations is a key part of understanding the management of innovation.

Kimberly (1981) rightly points out that a positive bias pervades the study of innovation. Innovation is often viewed as a good thing because the new idea must be useful—profitable, constructive, or solve a problem. New ideas that are not perceived as useful are not normally called innovations; they are usually called mistakes. Objectively, of course, the usefulness of an idea can only be determined after the innovation process is completed and implemented. Moreover, while many new ideas are proposed in organizations, only a very few receive serious consideration and developmental effort (Wilson 1966; Maitland 1982). Since it is not possible to determine at the outset which new ideas are "innovations" or "mistakes," and since we assume that people prefer to invest their energies and careers on the former and not the latter, there is a need to explain (1) how and why certain innovative ideas gain good currency (i.e., are implemented), and (2) how and why people pay attention to only certain new ideas and ignore the rest. These two questions direct our focus to problems of managing ideas into good currency and the management of attention.

THE MANAGEMENT OF IDEAS

It is often said that an innovative idea without a champion gets nowhere. *People* develop, carry, react to, and modify ideas. People apply different skills, energy levels and frames of reference (interpretive schemas) to ideas as a result of their backgrounds, experiences, and activities that occupy their attention. *People become attached to ideas over time through a social-political process of pushing and riding their ideas into good currency,* much like Donald Schon (1971) describes for the emergence of public policies. Figure 1 illustrates the process.

Schon states that what characteristically precipitates change in public policy is a disruptive event which threatens the social system. Invention is an act of appreciation, which is a complex perceptual process that melds together judgments of reality and judgments of value. A new appreciation is made as a problem, or opportunity is recognized. Once appreciated, ideas gestating in peripheral areas begin to surface to the mainstream as a result of the efforts of people who supply the energy necessary to raise the ideas over the threshold of public consciousness. As these ideas surface networks of individuals and interest groups gravitate to and galvanize around the new ideas. They, in turn, exert their own influence on the ideas by further developing them and providing them with a catchy slogan that provides emotional meaning and energy to the idea.

However, Schon indicates that ideas are not potent to change policy unless they become an issue for political debate and unless they are used to gain influence and resources. The debate turns not only on the merits of the ideas, but also on who is using the ideas as vehicles to gain power. As the ideas are taken up by people who are or have become powerful, the ideas gain legitimacy and power to change institutions. After this, the ideas that win out are im-

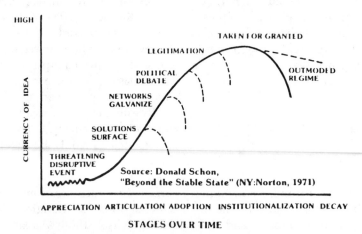

FIGURE 1.   *Managing Life Cycle of Ideas in Good Currency*

plemented and become institutionalized—they become part of the conceptual structure of the social system and appear obvious, in retrospect. However, the idea remains institutionalized for only as long as it continues to address critical problems and as long as the regime remains in power.

Schon's description of the stages by which ideas come into good currency is instructive in its focus on the social-political dynamics of the innovation process. The description emphasizes the *centrality of ideas as the rallying point around which collective action mobilizes*—organizational structures emerge and are modified by these ideas. Moreover, it is the central focus on *ideas* that provides the vehicle for otherwise isolated, disconnected, or competitive individuals and stakeholders to come together and contribute their unique frames of reference to the innovation process. Schon (1971, p. 141) states that these stages characteristically describe the process features in the emergence of public policies "regardless of their content or conditions from which they spring." Analogous descriptions of this social-political process have been provided by Quinn (1980, especially p. 104) for the development of corporate strategies, and by March and Olsen (1976) for decision making in educational institutions.

However, there are also some basic limitations to the process that lead to inertia and premature abandonment of some ideas. First, there tends to be a short-term problem orientation in individuals and organizations, and a facade of demonstrating progress. This has the effect of inducing premature abandonment of ideas because even if problems are not being solved, the appearance of progress requires moving on to the next batch of problems. Thus, "old questions are not answered—they only go out of fashion" (Schon 1971, p. 142). Furthermore, given the inability to escape the interdependence of problems, old problems are relabeled as new problems. As a result, and as observed by Cohen, March and Olsen (1972),

decision makers have the feeling they are always working on the same problems in somewhat different contexts, but mostly without results.

Except for its use in legislative bodies, the idea of formally managing the socio-political process of pushing and riding ideas into good currency is novel. However, as Huber (1984, p. 938) points out, the decision process is similar to project management and program planning situations. Thus, Huber proposes the adoption of proven project management and program planning technologies (e.g., PERT, CPM and PPM) for managing the production of ideas into good currency. For example, based upon a test of the Program Planning Model, Van de Ven (1980a, b) concluded that the PPM avoids problems of decision flight and falling into a rut that are present in March and Olsen's (1976) garbage can model of anarchical decision making. This is accomplished by the PPM's three-way matching of phased tasks with different decision processes and with different participants over time in a program planning effort.

A second limitation of the process is that the inventory of ideas is seldom adequate for the situation. This may be because environmental scanning relevant to an issue does not uncover the values and partisan views held by all the relevant stakeholders. Gilbert and Freeman (1984) point out that with the general concept of environmental scanning, current models of strategic decision making gloss over the need to identify specific stakeholders to an issue and to examine their underlying values which provide reasons for their actions. Viewing the process from a game theoretic framework, they state that "effective strategy will be formulated and implemented if and only if each player successfully puts himself or herself in the place of other players and engages in trying to see the situation from the others' viewpoints" (Gilbert and Freeman 1984, p. 4).

A third, and even more basic problem is

the management of attention—how do individuals become attached to and invest effort in the development of innovative ideas? Human beings and their organizations are mostly designed to focus on, harvest, and protect existing practices rather than to pave new directions. This is because people have basic physiological limitations of not being able to handle complexity, of unconsciously adapting to gradually changing conditions, of conforming to group and organizational norms, and of focusing on repetitive activities (Van de Ven and Hudson 1985). One of the key questions in the management of innovation then becomes how to trigger the action thresholds of individuals to appreciate and pay attention to new ideas, needs and opportunities.

THE MANAGEMENT OF ATTENTION

Much of the folklore and applied literature on the management of innovation has ignored the research by cognitive psychologists and social-psychologists about the limited capacity of human beings to handle complexity and maintain attention. As a consequence, one often gets the impression that inventors or innovators have superhuman creative heuristics or abilities to "walk on water" (Van de Ven and Hudson 1985). *A more realistic view of innovation should begin with an appreciation of the physiological limitations of human beings to pay attention to nonroutine issues, and their corresponding inertial forces in organizational life.*

Physiological Limitations of Human Beings

It is well established empirically that most individuals lack the capability and inclination to deal with complexity (Tversky and Kahneman 1974; Johnson 1983). Although there are great individual differences, most people have very short spans of attention—the average person can retain raw data in short-term memory for only a few seconds. Memory, it turns out, requires relying on "old friends," which Simon (1947) describes as a process of linking raw data with pre-existing schemas and world views that an individual has stored in long-term memory. Most individuals are also very efficient processors of routine tasks. They do not concentrate on repetitive tasks, once they are mastered. Skills for performing repetitive tasks are repressed in subconscious memory, permitting individuals to pay attention to things other than performance of repetitive tasks (Johnson 1983). Ironically as a result, what most individuals think about the most is what they will do, but what they do the most is what they think about the least.

In complex decision situations, individuals create stereotypes as a defense mechanism to deal with complexity. For the average person, stereotyping is likely to begin when seven (plus or minus two) objects or digits are involved in a decision—this number being the information processing capacity of the average individual (Miller 1956). As decision complexity increases beyond this point, people become more conservative and apply more subjective criteria which are further and further removed from reality (Filley, House, and Kerr 1976). Furthermore, since the correctness of outcomes from innovative ideas can rarely be judged, the perceived legitimacy of the decision *process* becomes the dominant evaluation criterion. Thus, as March (1981) and Janis (1982) point out, as decision complexity increases, solutions become increasingly error prone, means become more important than ends, and rationalization replaces rationality.

It is generally believed that crises, dissatisfaction, tension, or significant external stress are the major preconditions for stimulating people to act. March and Simon (1958) set forth the most widely accepted model by arguing that dissatisfaction with existing conditions stimulates people to search for improved conditions, and they will cease searching when a satisfactory result is found. A satisfactory result is a function of a person's aspiration level, which Lewin et. al. (1944) indicated is a product of all past successes and failures that people have experienced. If this model is correct (and most believe it is), then scholars and practitioners must wrestle with another basic problem.

This model assumes that when people reach a threshold of dissatisfaction with existing conditions, they will initiate action to resolve their dissatisfaction. However, because individuals unconsciously adapt to slowly changing environments, their thresholds for action are often not triggered while they adapt over time. In this sense, individuals are much like frogs. Although we know of no empirical support for the frog story developed by Gregory Bateson, it goes as follows.

> When frogs are placed into a boiling pail of water, they jump out—they don't want to boil to death.
> However, when frogs are placed into a cold pail of water, and the pail is placed on a stove with the heat turned very low, over time the frogs will boil to death.

Cognitive psychologists have found that individuals have widely varying and manipulable adaptation levels (Helson 1948, 1964). When exposed over time to a set of stimuli that deteriorate very gradually, people do not perceive the gradual changes—they unconsciously adapt to the worsening conditions. Their threshold to tolerate pain, discomfort, or dissatisfaction is not reached. As a consequence, they do not move into action to correct their situation, which over time may become deplorable. Opportunities for innovative ideas are not recognized, problems swell into metaproblems, and at the extreme, catastrophes are sometimes necessary to reach the action threshold (Van de Ven 1980b).

These worsening conditions are sometimes monitored by various corporate planning and management information units and distributed to personnel in quantitative MIS reports of financial and performance trends. However, these impersonal statistical reports only increase the numbness of organizational participants and raise the false expectation that if someone is measuring the trends then someone must be doing something about them.

When situations have deteriorated to the point of actually triggering peoples' action thresholds, innovative ideas turn out to be crisis management ideas. As Janis (1982) describes, such decision processes are dominated by defense mechanisms of isolation, projection, stereotyping, displacement, and retrospective rationalizations to avoid negative evaluations. As a result, the solutions that emerge from such "innovative" ideas are likely to be "mistakes."

Group and Organizational Limitations

At the group and organizational levels, the problems of inertia, conformity, and incompatible preferences are added to the above physiological limitations of human beings in managing attention. As Janis (1982) has clearly shown, groups place strong conformity pressures on members, who collectively conform to one another without them knowing it. Indeed, the classic study by Pelz and Andrews (1966) found that a heterogeneous group of interdisciplinary scientists when working together daily became homogeneous in perspective and approach to problems in as little as three years. Groups minimize internal conflict and focus on issues that maximize consensus. "Group Think" is not only partly a product of these internal conformity pressures, but also of external conflict—"out-group" conflict stimulates "in-group" cohesion (Coser 1959). Consequently, it is exceedingly difficult for groups to entertain threatening information, which is inherent in most innovative ideas.

Organizational structures and systems serve to sort attention. They focus efforts in prescribed areas and blind people to other issues by influencing perceptions, values, and beliefs. Many organizational systems consist of programs, which create slack through efficient repetitive use of procedures believed to lead to success (Cyert and March 1963). But as Starbuck (1983) argues, the programs do not necessarily address causal factors. Instead, the programs tend to be more like superstitious learning, recreating actions which may have little to do with previous success and nothing to do with future success. As a result, the older, larger, and more successful organizations become, the more likely they are to have a large repertoire of structures and systems which discourage innovation while encouraging tinkering. For example, strategic planning *systems* often drive out strategic thinking as participants "go through the numbers" of completing yearly planning forms and review cycles.

The implication is that without the intervention of leadership (discussed below), structures and systems focus the attention of organizational members to routine, not innovative activities. For all the rational virtues that structures and systems provide to maintain existing organizational practices, these "action generators" make organizational participants inattentive to shifts in organizational environments and the need for innovation (Starbuck 1983). It is surprising that we know so little about the management of attention. However, several useful prescriptions have been made.

Ways to Manage Attention

At a recent conference on strategic decision making (Pennings 1985), Paul Lawrence reported that in his consulting practice he usually focuses on what management is *not* paying attention to. Similarly based on his observations in consulting with large organizations, Richard Normann observed that well-managed companies are not only close to their customers, they search out and focus on their *most demanding*

customers. Empirically, von Hippel (1977) has shown that ideas for most new product innovations come from customers. Being exposed face-to-face with demanding customers or consultants increases the likelihood that the action threshold of organizational participants will be triggered and will stimulate them to pay attention to changing environmental conditions or customer needs. In general, we would expect that *direct personal confrontations with problem sources* are needed to reach the threshold of concern and appreciation required to motivate people to act (Van de Ven 1980b).

However, while face-to-face confrontations with problems may trigger action thresholds, they also create stress. One must therefore examine the effects of stress on the innovative process. Janis (1985) outlines five basic patterns of coping with stress, and states that only the vigilance pattern generally leads to decisions that meet the main criteria for sound decision making. Vigilance involves an extended search and assimilation of information, and a careful appraisal of alternatives before a choice is made. Janis proposes that vigilance tends to occur under conditions of moderate stress, and when there may be sufficient time and slack resources to make decisions. Under conditions of no slack capacity or short-time horizons (which produce stress) the decision process will resemble crisis decision-making—resulting in significant implementation errors (Hrebiniak and Joyce 1984).

Argyris and Schon (1982) focus on single loop and double loop learning models for managing attention that may improve the innovation process. In single loop learning, no change in criteria of effective performance takes place. Single loop learning represents conventional monitoring activity, with actions taken based on the findings of the monitoring system. Because it does not question the criteria of evaluation, single loop learning leads to the organizational inertia which Starbuck (1983) indicates must be unlearned before change can occur. Double loop learning involves a change in the criteria of evaluation. Past practices are called into question, new assumptions about the organization are raised, and significant changes in strategy are believed to be possible.

While double loop learning can lead to change, it can also lead to low trust, defensive behavior, undiscussibles, and to bypass tactics. Thus, the management of attention must be concerned not only with triggering the action thresholds of organizational participants, but also of channeling that action toward constructive ends. Constructive attention management is a function of how two other central problems are addressed: part-whole relations and institutional leadership—which we will now discuss.

THE MANAGEMENT OF PART-WHOLE RELATIONSHIPS

Proliferation of ideas, people, and transactions over time is a pervasive but little understood characteristic of the innovation process, and with it come complexity and interdependence—and the basic structural problem of managing part-whole relations.

The proliferation of ideas is frequently observed in a single individual who works to develop an innovation from concept to reality. Over time the individual develops a mosaic of perspectives, revisions, extensions, and applications of the initial innovative idea—and they accumulate into a complex set of interdependent options. However, as the discussion of managing ideas into good currency implies, innovation is not an individual activity—it is a collective achievement. Therefore, over time there is also a proliferation of people (with diverse skills, resources, and interests) who become involved in the innovation process. When a single innovative idea is expressed to others, it proliferates into multiple ideas because peo-

ple have diverse frames of reference, or interpretive schemas, that filter their perceptions. These differing perceptions and frames of reference are amplified by the proliferation of transactions or relationships among people and organizational units that occur as the innovation unfolds. Indeed, management of the innovation process can be viewed as managing increasing bundles of transactions over time.

Transactions are "deals" or exchanges which tie people together within an institutional framework (which is context). John R. Commons (1951), the originator of the concept, argued that transactions are dynamic and go through three temporal stages: negotiations, agreements, and administration. Most transactions do not follow a simple linear progression through these stages. The more novel and complex the innovative idea, the more often trial-and-error cycles of renegotiation, recommitment, and readministration of transactions will occur. Moreover, the selection of certain kinds of transactions is always conditioned by the range of past experiences and current situations to which individuals have been exposed. Therefore, people have a conservative bias to enter into transactions with parties they know, trust, and with whom they have had successful experiences. As a consequence, what may start as an interim solution to an immediate problem often proliferates over time into a web of complex and interdependent transactions among the parties involved.

There is an important connection between transactions and organizations. Transactions are the micro elements of macro organizational arrangements. Just as the development of an innovation might be viewed as a bundle of proliferating transactions over time, so also, is there proliferation of functions and roles to manage this complex and interdependent bundle of transactions in the institution that houses the innovation.

The prevailing approach for handling this complexity and interdependence is to divide the labor among specialists who are best qualified to perform unique tasks and then to integrate the specialized parts to recreate the whole. The objective, of course, is to develop synergy in managing complexity and interdependence with an organizational design where the whole is greater than the sum of its parts. However, the whole often turns out to be less than or a meaningless sum of the parts because the parts do not add to, but subtract from one another (Hackman 1984). This result has been obtained not only when summing the products of differentiated units within organizations, but also the benefits member firms derive from associating with special interest groups (Maitland 1983, 1985). Kanter (1983), Tushman and Romanelli (1986), and Peters and Waterman (1982) have shown that this "segmentalist" design logic is severely flawed for managing highly complex and interdependent activities. *Perhaps the most significant structural problem in managing complex organizations today, and innovation in particular, is the management of part-whole relations.*

For example, the comptroller's office detects an irregularity of spending by a subunit and thereby eliminates an innovative "skunkworks" group; a new product may have been designed and tested, but runs into problems when placed into production because R & D and engineering overlooked a design flaw; the development of a major system may be ready for production, but subcontractors of components may not be able to deliver on schedule or there may be material defects in vendors' parts. Typical attributions for these problems include: lack of communication or misunderstandings between scientific, engineering, manufacturing, marketing, vendors and customers on the nature or status of the innovation; unexpected delays and errors in certain developmental stages that complicate further errors and rework in subsequent stages; incompatible organizational funding, control, and re-

ward policies; and ultimately significant cost over-runs and delayed introductions into the market.

Peters and Waterman (1982) dramatized this problem of part-whole relationships with an example of a product innovation which required 223 reviews and approvals among 17 standing committees in order to develop it from concept to market reality. Moreover, they state that

> The irony, and the tragedy, is that each of the 223 linkages taken by itself makes perfectly good sense. Well-meaning, rational people designed each link for a reason that made sense at the time. . . . The trouble is that the total picture as it inexorably emerged . . . captures action like a fly in a spider's web and drains the life out of it. (Peters and Waterman 1982 pp. 18–19).

This example clearly illustrates a basic principle of contradictory part-whole relationships— *impeccable micro-logic often creates macro nonsense*, and vice versa.

Is there a way to avoid having the whole be less than or a meaningless sum of its parts? Perhaps a way is needed to design the whole into the parts, as Gareth Morgan (1983a, b, 1984) has been pursuing with the concept of a *hologram*. He concluded that the brain, with its incredible complexity, manages that complexity by placing the essential elements of the whole into each of its parts—it is a hologram.

Most organizations, however, are not designed with this logic, but if possible ought to be. The hologram metaphor emphasizes that organization design for innovation is not a discrete event but a process for integrating all the essential functions, organizational units, and resources needed to manage an innovation from beginning to end. It requires a significant departure from traditional approaches to organizing innovation.

Traditionally the innovation process has

been viewed as a sequence of separable stages (e.g., design, production, and marketing) linked by relatively minor transitions to make adjustments between stages. There are two basic variations of this design for product innovation. First, there is the technology-driven model where new ideas are developed in the R & D department, sent to engineering and manufacturing to produce the innovation, and then on to marketing for sales and distribution to customers. The second, and currently more popular, design is the customer or need-driven model, where marketing comes up with new ideas as a result of close interactions with customers, which in turn are sent to R & D for prototype development and then to engineering and manufacturing for production. Galbraith (1982) points out that the question of whether innovations are stimulated by technology or customer need is debatable.

"But this argument misses the point." As reproduced in Figure 2, "the debate is over whether [technology] or [need] drives the downstream efforts. This thinking is linear and sequential. Instead, the model suggested here is shown in Figure [2b]. That is, for innovation to occur, knowledge of all key components is simultaneously coupled. And the best way to maximize communication among the components is to have the communication occur intrapersonally—that is, within one person's mind. If this is impossible, then as few people as possible should have to communicate or interact. (Galbraith 1982, pp. 16–17).

As Galbraith implies, with the hologram metaphor the innovation process is viewed as consisting of interactions of inseparable and simultaneously-coupled stages (or functions) linked by a major ongoing transition process. Whereas the mechanical metaphor of an assembly line of stages characterizes most current views of the innovation process, the biological metaphor of a hologram challenges scholars and practitioners to find ways to place essential characteristics of the whole into each of the parts.

Although very little is known about how to design holographic organizations, four interrelated design principles have been suggested by Morgan (1985) and others: self-organizing

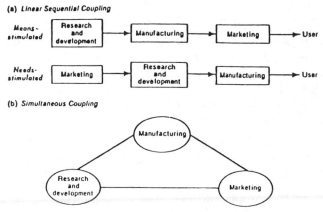

Source: J.R. Galbraith "Designing the Innovating Organization," *Organizational Dyanmics,* Winter 1982, pp. 3–24.

FIGURE 2. *Linear Sequential Coupling Compared with Simultaneous Coupling of Knowledge*

units, redundant functions, requisite variety, and temporal linkage.

First, the hologram metaphor directs attention to identifying and grouping together all the key resources and interdependent functions needed to develop an innovation into one organizational unit, so that it can operate as if it were an *autonomous unit*. (Of course, no organizational unit is ever completely autonomous.) The principle of autonomous work groups has been developed largely by Trist (1981), and is consistent with Thompson's (1967) logical design principle of placing reciprocally-interdependent activities closely together into a common unit in order to minimize coordination costs. By definition, autonomous groups are self-organizing, which implies that management follows the "principle of minimum intervention" (Hrebiniak and Joyce 1984, p. 8). This allows the group to self-organize and choose courses of action to solve its problems within an overall mission and set of constraints prescribed for the unit by the larger organization.

Second, flexibility and a capacity for self-organizing is needed by creating *redundant functions,* which means that people develop an understanding of the essential considerations and constraints of all aspects of the innovation in addition to those immediately needed to perform their individual assignments. Redundant functions does not mean duplication or spare parts as may be implied by the mechanistic metaphor, nor does it eliminate the need for people to have uniquely-specialized technical competencies. It means that all members of an innovation unit develop the capacity to "think globally while acting locally." The principle of redundant functions is achieved through training, socialization, and inclusion into the innovation unit so that each member not only comes to know how his or her function relates to each other functional specialty, but also understands the essential master blueprint of the overall innovation. The former is needed for inter-

dependent action; the latter is essential for survival and reproduction of the innovative effort.

Third, following Ashby's (1956) principle of *requisite variety,* learning is enhanced when a similar degree of complexity in the environment is built into the organizational unit. This principle is a reflection of the fact that any autonomous organizational unit at one level is a dependent part of a larger social system at a more macro level of analysis. Requisite variety means placing critical dimensions of the whole environment into the unit, which permits the unit to develop and store rich patterns of information and uncertainty that are needed in order to detect and correct errors existing in the environment. The principle of requisite variety is not achieved by assigning the task of environmental scanning to one or a few boundary spanners, for that makes the unit dependent upon the "enactments" (Weick 1979) of only one or a few individuals whose frames of reference invariably filter only selective aspects of the environment. Requisite variety is more nearly achieved by making environmental scanning a responsibility of all unit members, and by recruiting personnel within the innovation unit who understand and have access to each of the key environmental stakeholder groups or issues that affect the innovation's development.

Whereas the principles of redundant functions and requisite variety create the slack needed to integrate members of the unit and between the unit and its environment (respectively), the principle of *temporal linkage* integrates parts of time (past, present, and future events) into an overall chronology of the innovation process. While innovations are typically viewed as making additions to existing arrangements, Albert (1984c) proposes another arithmetic for linking the past, present and future. Given a world of scarcity, Albert (1984a, b) notes that the implementation of innovations often results in eliminations, replacements, or transformations of existing arrangements. As a consequence, the management of innovation must also be the management of termination, and of transitioning people, programs, and investments from commitments in the past toward the future. In common social life, funerals and wakes are used to commemorate and bereave the passing of loved ones and to make graceful transitions into the future. As Albert suggests, there is a need to create funerals, celebrations, and transitional rituals that commemorate the ideas, programs, and commitments falling out of currency in order to create opportunities for ushering in those that must gain good currency for an innovation to succeed.

INSTITUTIONAL LEADERSHIP AND INNOVATION CONTEXT

Innovation is not the enterprise of a single entrepreneur. Instead, it is a network-building effort that centers on the creation, adoption, and sustained implementation of a set of ideas among people who, through transactions, become sufficiently committed to these ideas to transform them into "good currency." Following holographic principles, this network-building activity must occur both within the organization and in the larger community of which it is a part. *Creating these intra- and extra-organizational infrastructures in which innovation can flourish takes us directly to the strategic problem of innovation, which is institutional leadership.*

The extra-organizational context includes the broad cultural and resource endowments that society provides, including laws, government regulations, distributions of knowledge and resources, and the structure of the industry in which the innovation is located. Research by Ruttan and Hayami (1983) and Trist (1981) suggests that innovation does not exist in a vacuum and that institutional innova-

tion is in great measure a reflection of the amount of support an organization can draw from its larger community. Collective action among institutional leaders within a community becomes critical in the long run to create the social, economic, and political infrastructure a community needs in order to sustain its members (Astley and Van de Ven 1983). In addition, as Aldrich (1979) and Erickson and Maitland (1982) indicate, a broad population or industry purview is needed to understand the societal demographic characteristics that facilitate and inhibit innovation.

Within the organization, institutional leadership is critical in creating a cultural context that fosters innovation, and in establishing organizational strategy, structure, and systems that facilitate innovation. As Hackman (1984, p. 40) points out, "an unsupportive organizational context can easily undermine the positive features of even a well-designed team." There is a growing recognition that innovation requires a special kind of supportive leadership.

This type of leadership offers a vision of what could be and gives a sense of purpose and meaning to those who would share that vision. It builds commitment, enthusiasm, and excitement. It creates a hope in the future and a belief that the world is knowable, understandable, and manageable. The collective energy that transforming leadership generates, empowers those who participate in the process. There is hope, there is optimism, there is energy (Roberts 1984, p. 3).

Institutional leadership goes to the essence of the process of institutionalization. It is often thought that an organization loses something (becomes rigid, inflexible, and loses it ability to be innovative) when institutionalization sets in. This may be true if an organization is viewed as a mechanistic, efficiency-driven tool. But, as Selznick (1957) argued, an organization does not become an "institution" until it becomes infused with value; i.e., prized not as a tool alone, but as a source of direct personal gratification, and as a vehicle for group integrity. By plan or default, this infusion of norms and values into an organization takes place over time, and produces a distinct identity, outlook, habits, and commitments for its participants—coloring as it does all aspects of organizational life, and giving it a social integration that goes far beyond the formal command structure and instrumental functions of the organization.

Institutional leadership is particularly needed for organizational innovation, which represents key periods of development and transition when the organization is open to or forced to consider alternative ways of doing things. During these periods, Selznick emphasized that the central and distinctive responsibility of institutional leadership is the creation of the organization's character or culture. This responsibility is carried out through four key functions: defining the institution's mission, embodying purpose into the organization's structure and systems, defending the institution's integrity, and ordering internal conflict. Selznick (1957, p. 62) reports that when institutional leaders default in performing these functions, the organization may drift. "A set of beliefs, values and guiding principles may emerge in the organization that are counterproductive to the organization's mission or distinctive competence. As institutionalization progresses the enterprise takes on a special character, and this means that it becomes peculiarly competent (or incompetent) to do a particular kind of work" (Selznick 1957, p. 139). Organization drift is accompanied by loss of the institution's integrity, opportunism, and ultimately, loss of distinctive competence.

Lodahl and Mitchell (1980, pp. 203–204) insightfully apply Selznick's perspective by distinguishing how institutional and technical processes come into play to transform innovative ideas into a set of guiding ideals—see Figure 3. First there are the founding ideals for an innovation or an enterprise, followed by the recruitment and socialization of members to serve those ideas. Leadership and formalization guide and stabilize the enterprise.

When viewed as a set of technical or instrumental tasks, the process is operationalized into setting clear goals or ends to be achieved; establishing impersonal and universal criteria for recruitment, developing clear rules and procedures for learning and socialization; analytical problem solving and decision making; and routinizing activities in order to reduce uncertainty. Institutional processes are very different from this well-known technical approach.

As Figure 3 illustrates, institutional processes focus on the creation of an ideology to support the founding ideals; the use of personal networks and value-based criteria for recruitment; socialization and learning by sharing rituals and symbols; charismatic leadership; and the infusion of values as paramount to structure and formalize activities.

Lodahl and Mitchell (1980, p. 204) point out that an innovation is an institutional success to the degree that it exhibits authenticity, functionality, and flexibility over time. Authenticity requires that the innovation embodies the organization's ideas; functionality requires that the innovation work; and flexibility requires that the innovation can incorporate the inputs and suggestions of its members. If these tests are met, organizational members will make a commitment to the innovation. In contrast, if institutional skills are not used while technical skills are in operation, the innovation may be an organizational success but an institutional failure. In that case, there will be evidence of drift and disillusionment. Such a result will be characterized by individual self-interest, differentiation, and technical efficiency.

These distinctions between institutional and technical processes have three significant implications for addressing the problems of

INSTITUTIONAL PROCESSES	IDEA	TECHNICAL PROCESSES
CREATION, ELABORATION OF IDEOLOGY	FOUNDING IDEALS	STATEMENT OF ORGANIZATIONAL GOALS
USE OF PERSONAL NETWORKS; SELECTION BASED ON VALUES AND IDEALS	RECRUITMENT	BROAD SEARCH: USE OF UNIVERSALISTIC CRITERIA
FACE-TO-FACE CONTACT WITH FOUNDERS: SHARING RITUALS, SYMBOLS	SOCIALIZATION	RULES AND PROCEDURES LEARNED THROUGH COLLEAGUES
CHARISMATIC, MYTHIC IMAGES (TRANSFORMING)	LEADERSHIP	PROBLEM SOLVING AND CONSENSUS MAKING (TRANSACTIONAL)
IDEALS PARAMOUNT: STRUCTURE TENTATIVE	FORMALIZATION	EARLY ROUTINIZATION: UNCERTAINTY REDUCTION

Source: T. Lodahl and S. Mitchell (1980).

FIGURE 3.   *Institutional and Technical Processes*

managing attention, ideas, and part-whole relations discussed in previous sections. These implications draw upon cybernetic principles and the hologram metaphor, as Morgan (1983b, 1984) proposes.

First, organizational members can develop a capacity to control and regulate their own behavior through a process of *negative feedback*, which means that goals are achieved by avoiding not achieving the goal. In other words, deviations in one direction initiate action in the opposite direction at every step in performing an activity so that in the end no error remains. In order for learning through negative feedback to occur, an organization must have values and standards which define the critical limits within which attention to innovative ideas is to focus. Whereas technical processes focus attention on clear-cut goals and targets to be achieved, institutional pro-

cesses define the constraints to avoid in terms of values and limits. Institutional leadership thus involves a choice of limits (issues to avoid) rather than a choice of ends. As Burgelman (1984, p. 1349) points out, "top management's critical contribution consists in strategic recognition rather than planning." As a result, a space of possible actions is defined which leaves room for innovative ideas to develop and to be tested against these constraints.

Second, whereas single loop learning involves an ability to detect and correct deviations from a set of values and norms, double loop learning occurs when the organization also learns how to detect and correct errors in the operating norms themselves. This permits an institution to adjust and change the ideas considered legitimate or to have good currency.

From an institutional view legitimate

error stems from the uncertainty inherent in the nature of a situation. The major problem in dealing with uncertainty is maintaining a balance on organizational diversity and order over time (Burgelman 1984). Diversity results primarily from autonomous initiatives of technical units. Order results from imposing standards and a concept of strategy on the organization. Managing this diversity requires framing ideas and problems so that they can be approached through experimentation and selection. The process of double-loop learning is facilitated by probing into various dimensions of a situation, and of promoting constructive conflict and debate between advocates of competing perspectives. Competing action strategies lead to reconsideration of the organization's mission, and perhaps a reformulation of that mission.

Finally, although technical processes of formalization press to reduce uncertainty, institutional processes attempt to preserve it. Just as necessity is the mother of invention, preserving the same degrees of uncertainty, diversity, or turbulence within an organization that is present in the environment are major sources of creativity and long-run viability for an organization. Embracing uncertainty is achieved by maintaining balance among innovative subunits, each designed according to the holographic principles of autonomous groups, requisite variety, and redundant functions discussed above. Application of these principles results in mirroring the turbulence present in the whole environment into the decision processes and other activities of each of the organization's parts. As a consequence, innovation is enhanced because organizational units are presented with the whole "law of the situation."

CONCLUDING DISCUSSION

Innovation has been defined as the development and implementation of new ideas by peo-ple who over time engage in transactions with others within an institutional context. This definition is particularly relevant to the general manager for it applies to a wide variety of technical, product, process, and administrative kinds of innovations that typically engage the general manager. From a managerial viewpoint, to understand the process of innovation is to be able to answer three questions: How do innovations develop over time? What kinds of problems will most likely be encountered as the innovation process unfolds? What responses are appropriate for managing these problems? Partial answers to these questions can be obtained by undertaking longitudinal research which systematically examines the innovation process, problems, and outcomes over time. Undertaking this research requires a conceptual framework to guide the investigation. The main purpose of this paper has been to develop such a framework by suggesting what key concepts, problems, and managerial responses should be the guiding focus to conduct longitudinal research on the management of innovation.

As our definition of innovation suggests, four basic concepts are central to studying the innovational process over time: ideas, people, transactions, and context. Associated with these four concepts are four central problems in the management of innovation: developing ideas into good currency, managing attention, part-whole relationships, and institutional leadership. Although these concepts and problems have diverse origins in the literature, previously they have not been combined into an interdependent set of critical concepts and problems for studying innovation management.

An invention or creative idea does not become an innovation until it is implemented or institutionalized. Indeed by most standards, the success of an innovation is largely defined in terms of the degree to which it gains good currency, i.e., becomes an implemented reality and is incorporated into the taken-for-granted assumptions and thought structure of organizational practice. Thus, a key measure of innovation success or outcome is the currency of the idea, and a basic research question is how and why do some new ideas gain good currency while the majority do not? Based on work by Schon (1971), Quinn (1980), and others, we think the answer requires longitudinal study of the social and political processes by which people become invested in or attached to new ideas and push them into good currency.

But what leads people to pay attention to new ideas? This is the second major problem to be addressed in a research program on innovation. We argued that an understanding of this issue should begin with an appreciation of the physiological limitations of human beings to pay attention to nonroutine issues, and their corresponding inertial forces in organizational life. The more specialized, insulated, and stable an individual's job, the less likely the individual will recognize a need for change or pay attention to innovative ideas. It was proposed that people will pay attention to new ideas the more they experience personal confrontations with sources of problems, opportunities, and threats which trigger peoples' action thresholds to pay attention and recognize the need for innovation.

Once people begin to pay attention to new ideas and become involved in a social-political process with others to push their ideas into good currency, a third problem of part-whole relationships emerges. A common characteristic in the development of innovations is that multiple functions, resources, and disciplines are necessary to transform innovative ideas into reality—so much so that individuals involved in specific transactions or parts of the innovation lose sight of the whole innovative effort. If left to themselves, they will design impeccable micro-structures for the innovation process that often result in macro nonsense. The hologram metaphor was proposed for de-

signing the innovation process in such a way that more of the whole is structured into each of the proliferating parts. In particular, application of four holographic principles was proposed for managing part-whole relationships: self-organizing groups, redundant functions, requisite variety, and temporal linkage.

However, these holographic principles for designing innovation units simultaneously require the creation of an institutional context that fosters innovation and that links these self-organizing innovative units into a larger and more encompassing organizational mission and strategy. The creation of this macro context for innovation points to the need to understand and study a fourth central problem, which is institutional leadership. Innovations must not only adapt to existing organizational and industrial arrangements, but they also transform the structure and practices of these environments. The strategic problem for institutional leaders is one of creating an infrastructure that is conducive to innovation and organizational learning.

Three cybernetic principles were proposed to develop this infrastructure. First, the principle of negative feedback suggests that a clear set of values and standards are needed which define the critical limits within which organizational innovations and operations are to be maintained. Second, an experimentation-and-selection approach is needed so that the organization develops a capacity for double-loop learning, i.e., learning how to detect and correct errors in the guiding standards themselves. Third, innovation requires preserving (not reducing) the uncertainty and diversity in the environment within the organization because necessity is the mother of invention. Embracing uncertainty can be achieved at the macro level through the principles of requisite variety and redundancy of functions.

It should be recognized that this has been a speculative essay on key problems in the management of innovation. Little empirical evidence is presently available to substantiate these problems, their implications, and proposed solutions. However, the essay has been productive in suggesting a core set of concepts, problems, and propositions to study the process of innovation over time, which is presently being undertaken by a large group of investigators at the University of Minnesota. A description of the operational framework being used in this longitudinal research is available (Van de Ven and Associates 1984). As this research progresses we hope to provide systematic evidence to improve our understanding of the central problems in the management of innovation discussed here.[1]

NOTE

1. The author wishes to gratefully recognize the stimulation of ideas for this paper from faculty and student colleagues involved in the Minnesota Innovation Research Program. Helpful comments on earlier drafts of this paper were provided by George Huber, William Joyce, Arie Lewin, Kenneth Mackenzie, and Donald Schon. This research program is supported in part by a major grant from the Organizational Effectiveness Research Programs, Office of Naval Research (Code 4420E), under Contract No. N00014-84-K-0016. Additional research support is being provided by 3M, Honeywell, Control Data, Dayton-Hudson, First Bank Systems, Cenex, Dyco Petroleum, and ADC Corporations.

REFERENCES

Albert, S., "A Delite Design Model for Successful Transitions," in J. Kimberly and R. Quinn (Eds.), *Managing Organizational Transitions*, Irwin, Homewood, Il., 1984a, Chapter 8, 169–191.

————, "The Sense of Closure," in K. Gergen and M. Gergen (Eds.), *Historical Social Psychology*, Lawrence Erlbaum Associates, 1984b, Chapter 8, 159–172.

————, "The Arithmetic of Change," University of Minnesota, Minneapolis, unpublished paper, 1984c.

Aldrich, H., *Organizations and Environments*, Prentice Hall, Englewood Cliffs, N.J., 1979.

Argyris, C. and D. Schon, *Reasoning, Learning, and Action*, Jossey-Bass, San Francisco, 1983.

Ashby, W. R., *An Introduction to Cybernetics*, Chapman and Hall, Ltd., London, 1956.

Astley, G. and Van de Ven, "Central Perspectives and Debates in Organization Theory," *Admin. Sci. Quart.*, 28 (1983), 245–273.

Burgelman, R. A., "Corporate Entrepreneurship and Strategic Management: Insights from a Process Study," *Management Sci.*, 29, 12 (1983), 1349–1364.

Cohen, M. D., J. G. March and J. P. Olsen, "A Garbage Can Model of Organizational Choice," *Admin. Sci. Quart.*, 17 (1972), 1–25.

Commons, J., *The Economics of Collection Action*, MacMillan, New York, 1951.

Cooper, A., "Strategic Management: New Ventures and Small Business," in D. Schendel and C. Hofer (Eds.), *Strategic Management*, Little, Brown and Company, Boston, 1979.

Coser, L., *The Functions of Social Conflict*, Routledge and Kegan Paul, New York, 1959.

Cyert, R. M. and J. G. March, *A Behavioral Theory of the Firm*, Prentice-Hall, Englewood Cliffs, N.J., 1963.

Daft, R. and S. Becker, *Innovation in Organization*, Elsezier, New York, 1978.

Erickson, B. and I. Maitland, "Healthy Industries and Public Policy," in Margaret E. Dewar (Eds.), *Industry Vitalization: Toward a National Industrial Policy*, Elmsford, N.Y., 1982.

Filley, A., R. House and S. Kerr, *Managerial Process and Organizational Behavior*, Scott Foresman, Glenview, Il., 1976.

Galbraith, J. R., "Designing the Innovating Organization," *Organizational Dynamics*, (Winter 1982), 3–24.

Gilbert, D. and E. Freeman, "Strategic Management and Environmental Scanning: A Game Theoretic Approach," presented to the Strategic Management Society, Philadelphia, October 1984.

Hackman, J. R., "A Normative Model of Work Team Effectiveness," Yale School of Organization and Management, New Haven, Conn., Research Program on Group Effectiveness, Technical Report #2, 1984.

Helson, H., "Adaptation-Level as a Basis for a Quantitative Theory of Frames of Reference," *Psychological Rev.*, 55 (1948), 294–313.

———, "Current Trends and Issues in Adaptation-Level Theory," *American Psychologist*, 19 (1964), 23–68.

Huber, G., "The Nature and Design of Post-Industrial Organizations," *Management Sci.*, 30, 8 (1984), 928–951.

Janis, I., *Groupthink*, 2nd edition, Houghton Mifflin, Boston, 1982.

———, "Sources of Error in Strategic Decision Making," in J. Pennings (ed.), *Strategic Decision Making in Complex Organizations*, Jossey-Bass, San Francisco, 1985.

Johnson, Paul E., "The Expert Mind: A New Challenge for the Information Scientist," In M. A. Bemmelmans (Ed.), *Beyond Productivity: Information Systems Development for Organizational Effectiveness*, North Holland Publishing, Netherlands, 1983.

Kanter, R., *The Change Masters*, Simon and Schuster, New York, 1983.

Kimberly, J., "Managerial Innovation," in Nystrom, P. and W. Starbuck (Eds.), *Handbook of Organizational Design*, Volume 1, Oxford University Press, Oxford, 1981, 84–104.

Lawrence, P. and P. Dyer, *Renewing American Industry*, Free Press, New York, 1983.

Leavitt, H. J., "Applied Organizational Change in Industry: Structural, Technological, and Humanistic Approaches," Chapter 27 in J. March (Ed.), *Handbook of Organizations*, Rand McNally, Chicago, 1965, 1144–1170.

———, "Applied Organizational Change in Industry: Structural, Technological, and Humanistic Approaches," Chapter 25, in J. March (Ed.), *Handbook of Organizations*, Rand McNally, Chicago, 1965, 1144–1170.

Lewin, Artie Y., and John W. Minton, "Organizational Effectiveness: Another Look, and an Agenda for Research," *Management Sci.*, 32, 5 (May 1986).

Lewin, K., T. Dembo, L. Festinger, and P. Sears, "Level of Aspiration," Chapter 10 in J. McV. Hunt (Ed.), *Personality and the Behavior Disorders*, Vol. 1, Ronald Press, New York, 1944.

Lodahl, T. and S. Mitchell, "Drift in the Development of Innovative Organizations," in J. Kimberly and R. Miles (Eds.), *The Organizational Life Cycle*, Jossey-Bass, San Francisco, 1980.

Maitland, I., "Organizational Structure and Innovation: The Japanese Case," in S. Lee and G. Schwendiman, *Management by Japanese Systems*, Prager, New York, 1982.

———, "House Divided: Business Lobbying and the 1981 Budget," *Research in Corporate Social Performance and Policy*, 5 (1983), 1–25.

———, "Interest Groups and Economic Growth Rates," *J. Politics*, (1985).

March, James G., "Decisions in Organizations and Theories of Choice," In A. Van de Ven and W. F. Joyce (Eds.), *Perspectives on Organizational Design and Behavior*, Wiley, New York, 1981.

——— and J. P. Olsen, *Ambiguity and Choice in Organizations*, Universitetsforlaget, Bergen, 1976.

——— and H. Simon, *Organizations*, Wiley, New York, 1958.

Miller, G. A., "The Magical Number Seven, Plus or Minus Two: Some Limits on our Capacity for Processing Information," *Psychological Rev.*, 63 (1956), 81–97.

Morgan, G., "Action Learning: A Holographic Metaphor for Guiding Social Change," *Human Relations*, 37, 1 (1983a), 1–28.

———, "Rethinking Corporate Strategy: A Cybernetic Perspective," *Human Relations*, 36, 4 (1983b), 345–360.

———, "Images of Organizations," York University, Downsview, Ontario, Prepublication manuscript, 1986.

Normann, R., *Management for Growth*, Wiley, New York, 1977.

———, "Towards an Action Theory of Strategic Management," in J. Pennings (Ed.), *Strategic Decision Making in Complex Organizations*, Jossey-Bass, San Francisco, 1985.

Ouchi, W., *Theory Z*, Addison-Wesley, Reading, Mass., 1981.

Pascale, R. and A. Athos, *The Art of Japanese Management,* Warner Books, New York, 1981.

Pelz, D. and F. Andrews, *Scientists in Organizations,* Wiley, New York, 1966.

Pennings, J., *Strategic Decision Making in Complex Organizations,* Jossey-Bass, San Francisco, 1985.

Peters, T. and R. Waterman, *In Search of Excellence: Lessons from America's Best-Run Companies,* Harper and Row, New York, 1982.

Quinn, James Brian, *Strategies for Change: Logical Incrementalism,* Irwin, Homewood, Ill., 1980.

Roberts, N., "Transforming Leadership: Sources, Process, Consequences," presented at Academy of Management Conference, Boston, August 1984.

Rogers, E., *Diffusion of Innovations,* 3rd Ed., The Free Press, New York, 1982.

Ruttan, V. and Hayami, "Toward a Theory of Induced Institutional Innovation," *J. Development Studies,* 20, 4 (1984), 203–223.

Schon, D., *Beyond the Stable State,* Norton, New York, 1971.

Schumpeter, J., *Capitalism, Socialism, and Democracy,* Harper and Row, New York, 1942.

Selznick, P., *Leadership in Administration,* Harper and Row, New York, 1957.

Simon, H. A., *Administrative Behavior,* Macmillan, New York, 1947.

Starbuck, W., "Organizations as Action Generators," *Amer. J. Sociology,* 48, 1 (1983), 91–115.

Terryberry, S., "The Evolution of Organizational Environments," *Admin. Sci. Quart.,* 12 (1968), 590–613.

Trist, E., "The Evolution of Sociotechnical Systems as a Conceptual Framework and as an Action Research Program," in A. Van de Ven and W. Joyce (Eds.), *Perspectives on Organization Design and Behavior,* Wiley, New York, 1981, 19–75.

Tushman, M. and E. Romanelli, "Organizational Evolution: A Metamorphosis Model of Convergence and Reorientation," in B. Staw and L. Cummings (Eds.), *Research in Organizational Behavior,* Vol. 7, JAI Press, Greenwich, Conn., 1985.

Tversky, A. and D. Kahneman, "Judgment under Uncertainty: Heuristics and Biases," *Science,* 185 (1974), 1124–1131.

Utterback, J., "The Process of Technological Innovation within the Firm," *Acad. Management J.,* 14 (1971), 75–88.

Van de Ven, A., "Problem Solving, Planning, and Innovation. Part 1. Test of the Program Planning Model," *Human Relations,* 33 (1980a), 711–740.

———, "Problem Solving, Planning, and Innovation. Part 2. Speculations for Theory and Practice," *Human Relations,* 33 (1980b), 757–779.

———, "Strategic Management Concerns among CEOs: A Preliminary Research Agenda," Presented at Strategic Management Colloquium, University of Minnesota, Minneapolis, October 1982.

——— and Associates, "The Minnesota Innovation Research Program," Strategic Management Research Center, Minneapolis, Discussion Paper #10, 1984.

——— and R. Hudson, "Managing Attention to Strategic Choices," in J. Pennings (Ed.), *Strategic Decision Making in Complex Organizations,* Jossey-Bass, San Francisco, 1984.

von Hippel, E., "Successful Industrial Products from Customer Ideas," *J. Marketing,* (January 1978), 39–40.

Weick, Karl, *The Social-Psychology of Organizing,* Addison-Wesley, Reading, Mass., 1979.

Wilson, J., "Innovation in Organizations: Notes toward a Theory," in J. Thompson (Ed.), *Approaches to Organizational Design,* University of Pittsburgh Press, Pittsburgh, 1966.

Zaltman, G., R. Duncan and J. Holbek, *Innovations and Organizations,* Wiley, New York, 1973.

Innovation and Corporate Strategy: Managed Chaos

James Brian Quinn

ABSTRACT. *Because major technological changes tend to progress in a highly probabilistic, tumultuous, interactive fashion—driven by a few champions—corporate innovation programs must be designed and managed accordingly. Based on a 2½-year study of large, innovative enterprises, this article suggests how some of the world's most innovative companies interlink careful strategic planning concepts with some novel organizational and motivational approaches to achieve their purposes. Neither structured formality nor unstructured chaos works well alone. Innovative companies seem to evolve a sophisticated approach to "managed chaos," which recognizes the realities of how major technological innovations evolve, and harnesses this process to corporate needs. While some general principles apply, the particular strengths, style, goals and competitive situation of each company may dictate very different specific solutions. But experience suggests certain common threads whose adoption increases the probability of succeeding.*

The competitive environment facing large US and European corporations has probably changed more in the past 25 years than in the preceding 100. In earlier years, most such corporations operated either in imperialistic situations or in islands of competition involving primarily their own domestic markets, a few large domestic competitors with similar cost structures, generally supportive governments, technologies which could be developed in depth and protected for substantial periods, and/or special access to low-cost capital which allowed

overwhelming scale economies or barriers to market entry. Now virtually all their markets and competitors are international in scale; domestic governments are often adversaries (on capital, tax, environmental or welfare issues); foreign governments are direct competitors; technologies and capital move rapidly across world borders[1]; and individual countries' materials costs, labor costs, monetary exchange rates, or intervention strategies create a bizarre array of cost patterns not present in the domestic market place.

How can a modern corporation produce domestically in a world where its capital costs are twice those of some competitors (Japan), its labor costs are 10 to 20 times higher than others (the LDCs),[2] its nation's safety and environ-

123

mental standards are among the world's highest, and its raw materials and energy bases have moved overseas. Strategically there are only a few options:

· Anticipating and servicing specific customers' needs better than anyone else in the world;

· Innovating faster and more continuously than all competitors; and

· Leveraging one's own capabilities by accessing and utilizing the world's best external science, technology and information sources.

To do this requires that companies integrate their research, technology, production, marketing and innovation management practices in a way few have achieved. Several streams of research suggest certain critical dimensions.

A Planning Perspective

Management research has defined well the formal methodologies for effectively aligning research and development activities with corporate goals.[3] The most successful enterprises seem to:

· Establish broad but challenging overall company objectives in light of predicted long-term economic, sociological and technological developments;

· Determine what specific unique strategy the overall company and each of its major divisions will use in effecting these objectives;

· Rank and balance research programs and projects into patterns which match these needs and deal with anticipated threats and opportunities in the external environment; and

· Develop clear mechanisms and a supporting motivational environment to transfer results from research to operations.

Continuous *minor changes in existing technologies* are often an important component in a technical strategy. These can and generally should be accomplished within such relatively planned frameworks. But such planning does not provide sufficient links for technology and strategy in today's complex world.

A Historical Perspective

Historians of technology have thoroughly documented that most *major technological advances* occur not in the linear mode suggested above, but in a relatively chaotic sequence of events typically involving an early vision; numerous fits, starts and lapses in progress; random interactions with the outside world; frequent intuitive insights and personal risk-taking; and even some lucky breaks which ultimately lead to success.[5] While in retrospect a technology may seem to progress (worldwide) doggedly up a Gaussian curve toward its ultimate theoretical limits,[6] more detailed observers see a series of "relatively small steps marked by an occasional obstacle and by constant random breakthroughs interacting across laboratories and borders."[7]

The unraveling of DNA's structure first followed a convoluted and somewhat disjointed route through various nations' biology, organic chemistry, x-ray crystallography and mathematics laboratories until its rather rapid, Nobel prize-winning conception as a spiral staircase of matched base pairs.[8] Then years of painstaking and highly individualistic research on gene structures and enzymes went through alternating periods of optimism and discouragement before the possibility of practical recombinant DNA suddenly emerged from Boyer and Cohen's 1972 discussions over sandwiches after a Hawaiian meeting. And now—almost 3 decades after Watson and Crick's insights and 12 years after Boyer and Cohen's first experiments—rDNA's first commercial applications are still moving forward in fits of insight and frustration as its expected major uses (like self-fertilizing plants, interferon, or commercial waste decomposition) constantly encounter new problems and create new surprises in their struggle toward the marketplace.

Companies seeking strategic advantage through significant technological innovation need to recognize the tumultuous, long-term realities of how this process operates, and design specific organizations and management practices to motivate and guide it. This paper is based on a two-and-one-half-year study of how some of the world's most innovative large companies stimulate such activities, yet maintain the practical controls so essential to efficient current operations.

TIME HORIZONS—VISION

Perhaps the most difficult issue for most enterprises to overcome is the time horizon needed for major innovations.[9] Most radical innovations—*e.g.,* jet aircraft engines, float glass, recombinant DNA, hovercraft, xerography, nuclear power, etc.—take over a decade from first discovery to net positive cash flows.[10] In my sample, delays ranged from three to 25 years.

For example, in the late 1930's Russell Marker was working on steroids called sapogenins, which make a soap-like foam in water. In the process he found a technique that would degrade one of these, diosgenin, into the female sex hormone progesterone. Marker found that diosgenin was abundant in certain Mexican yams. By processing about ten tons of these in rented and borrowed lab space, he extracted some 4 pounds of diosgenin, and in 1944 started a tiny company, Syntex, producing steroids for the laboratory market. But it was not until 1957 that Searle started marketing a derivative, norethynodril, as a menstrual regulator—with a specific warning against "possible contraceptive activity." In 1962, over 23 years

after Marker's early research, Syntex obtained FDA approval for its oral contraceptive.

Even smaller extensions of technology typically take two to five years from initiation to first positive rewards. Such delays are often anathema—to public companies, who gauge their success against quarterly reports of profitability, to managers who are rotated in two-year patterns, and to public laboratories dependent upon annual or biennial elections for their support.[11] Yet short time horizons need not be the norm. All the innovative enterprises in my sample had found ways to deal with this problem.

Long time horizons can occur in organizations for a variety of reasons. In some, the very nature of the product creates such views. Aircraft manufacturers, pharmaceutical companies, public utilities, electrical generation equipment producers, forest products companies, shipbuilders, etc., all have very long operating time horizons. Successful companies in these fields must develop internal management systems to plan and control on multiple-year horizons, or they fail. Their corporate cultures and financial sources accept such horizons as given. But such companies are not all innovative. Long time horizons alone do not lead to innovativeness.

Vision

Continuous innovation occurs largely because a few key executives have a broad vision of what their organizations can accomplish for the world and lead their enterprises toward it. They appreciate the role of innovation in achieving their goals and consciously manage their concerns' value systems and atmospheres to support it. Because familiarity fosters understanding and decreases one's fear of risk in a situation, technical people (or those who have personally participated in successful innovations) are often those who create such supportive atmospheres. But—as the non-technical top

125

managers of Genentech, IBM, Pilkington, Elf Aquitaine, AT&T, Merck and others in my sample amply illustrate—executive vision is more important than one's functional background.

Innovative managements—whether technical or not—project clear long-term visions for their organizations that go beyond simple economic performance measurements.[12] For example, Intel's chairman, Gordon Moore, says: "We intend [for Intel] to be the outstandingly successful company in our industry. We intend to be a leader in this revolutionary [semi-conductor] technology that is changing the way the world is run." Young people attracted by Intel's goals and style responded, "It's great to say you work at Intel. You know you're with the best. . . . There's a real pride in being first, in being on the frontier. You know you are part of something very big—very important." Genentech's original plan states: "We expect to be the first company to commercialize the [rDNA] technology, and we plan to build a major profitable corporation by manufacturing and marketing needed products that benefit mankind. . . . The future uses of genetic engineering are far reaching and many. Micro-organisms could be engineered to produce proteins to meet the world's food needs or to produce antibodies to fight viral infections. Any product produced by a living organism is eventually within the company's reach."[13]

Such visions, vigorously supported, have many practical applications. They attract quality people to the company, and give focus to their creative and entrepreneurial drives. People who share common values and goals can work independently, self-coordinating their activities without detailed controls. And they are more motivated to achieve goals they generally believe in. Overarching goals and values help power growth by concentrating on the actions which lead to profitability, rather than on profitability itself. Such visions establish a realistic

time-frame for innovation and, properly presented externally, attract the kind of investors who will support it. In essence, they provide the rationale for that ultimate expression of stockholder confidence and patience, high P/E ratios.

Market-Goal Orientation

Innovative companies then consciously anchor their visions to the practical realities of the marketplace. Each company develops specific techniques adapted to its particular style and strategy, but two elements are always present: (1) a strong market orientation at the very top of the company, and (2) explicit mechanisms to force market-technical interactions at lower levels. For example, Sony's Chairman Morita, who has personally started several of its most innovative ventures, purposely seeks informal settings where he can meet with and understand the lifestyles of trend-setting younger people. Soon after technical people are hired, Sony cycles them through weeks of retail selling. Sony engineers become sensitive to the ways retail sales practices, displays, and non-quantifiable customer preferences affect success.[14] Bell Laboratories has an Operating Company Assignment Program (OCAP) to rotate its researchers through AT&T-Western development and producing facilities. And until the AT&T break-up, Bell Laboratories maintained a rigorous Engineering Complaint System through which it collected technical problems and inquiries from its operating companies. These inquiries *had* to receive an "action response" within a month, *i.e.,* Bell Labs *had* to specify how it could solve or would attempt to resolve the problem and notify the operating company (in writing) of its action.

From top management to bench levels in my sample's "most innovative" companies, executives' primary focus seemed to be on opportunistically seeking and solving customers' emerging problems. For example, rather than follow many others companies' practices by cutting back or shutting down operations in the 1981–82 recession, Intel started its "20% solution." Key members of the professional staff agreed to work an extra day a week for several months to bring out new products earlier and to solve customer problems in ways that maintained Intel's own employment. Other examples follow.

A UNIQUE STRATEGIC FOCUS

How each enterprise can best satisfy identified needs depends on its own particular strengths and style—and the characteristics of the field it is in. Successful innovation managements carefully conceptualize the role they want innovation to play in their strategies, and organize for that purpose. This has led to quite different alignments in various companies based on their particular technologies, skills, resources, and desired risk patterns. Several distinctive strategies will make the point:

Extensions of Customers' R&D. Some suppliers of technical specialties to original equipment manufacturers (OEMs) regard their own technologists as extensions of their customers' R&D programs. Typically the supplier has knowledge depth in areas different from its customers—*i.e.,* a chemical specialties producer like Dewey & Almy or a semi-conductor specialist like Intel supplying a mechanical OEM. Such suppliers develop strong technical networks to discover and understand customer needs in depth and try to have their solutions designed into the customers' products. These companies maintain flexible applied technology groups working close to the marketplace—and even on customer premises. All technology is kept highly proprietary. They have quickly expandable production facilities and a cutting edge technology (not necessarily basic research) group that allows rapid selection of the best available technologies to solve problems.

Basic Research Enterprises. Some companies like Genentech, Merck, or Hoffman LaRoche have been built on basic research strategies. These companies offer researchers better equipped laboratories, higher pay, and often more freedom than universities in order to attract top flight people. To make sure their people are on the cutting edge of science, they allow researchers to publish extensively (after appropriate patent applications, if relevant), participate in meetings, and develop wide colleague networks externally. They further leverage their internal spending through research grants, clinical grants, and direct research relationships with outside groups throughout the world. These companies want a clear understanding of possible alternatives and a preemptive position in the sciences underlying the product before they invest $20 to $50 million to clear and produce a new drug. But their basic science activities must also be coupled with elaborate precautions to ensure that the drug is safe and effective in use and controllable in manufacture. Hence their structures are designed to place them at the frontiers of science, but to be quite conservative in animal testing, clinical evaluations and production control.

Entrepreneurial Start Ups. Some companies, like Hewlett Packard and 3M, have strategies to develop product lines from a series of small, discrete, freestanding new product concepts. These companies form units akin to entrepreneurial start ups. Each tends to be a small team, led by a product champion, in relatively low cost "incubator" facilities, where overheads are kept to a minimum. These companies allow many different proposals to come forward and test them—experimentally and in-

127

teractively—as early as possible in the market-place. Their control systems try to spot and prevent significant early losses on any entry. Like venture capitalists they look for their high gains on a few successful products. But, unlike venture capitalists, they also try to blend their less successful, smaller, individual entries into product lines their existing channels or facilities can support.

Major Systems Innovations. Other companies (like AT&T, the aircraft companies, or the oil majors) have to innovate large scale systems that only pay back if they last for decades. These concerns make very long term needs and competitive forecasts, with all possible precision. They often start several alternate designs in parallel to be as sure as possible of selecting the right technologies at the last moment before commercialization. They extensively test new system and subsystem technologies in actual use before making full-scale commitments. And they frequently sacrifice speed of entry for long-term low cost and reliability.

Many other examples, of course, exist. And even within a single company, individual divisions may have different strategic needs—hence different structures and practices.[15] No single approach to innovation suits all situations.

MANAGING CHAOS

If such formalities were all that is required for innovation, many more companies would be successful at it. The central problem is that innovation—first reduction of an idea to practice—is a process that rarely can be planned and controlled with the kind of analytical certainty managers associate with other operations. Instead, technology tends to advance in a bubbling, intuitive, tumultuous process—more akin to a fermentation vat than a production line.[16] Initial discoveries tend to be highly individualistic and serendipitous, advances chaotic and interactive, and specific outcomes unpredictable and chancy until the very last moment. How can one manage such a process in a large organization?

Multiple Approaches

All innovation is probabilistic. At first, one cannot know whether the desired result can be achieved or which of several possible approaches may dominate the field as it develops. Initial insights can come from almost anywhere—and the right insight can create a whole new industry or offer dominance to its discoverer for years.

Henry Perkin, trying to synthesize quinine from coal tar, discovered the synthetic purple dye that began the modern chemical industry. Leo Baekeland was looking for a synthetic shellac when he found Bakelite and started the modern plastics industry. A gust of wind, blowing mold on Fleming's cultures, created the antibiotic age. The microcomputer was born because Ted Hoff happened to work on a complex calculator for a Japanese client just when DEC's PDP8 architecture was fresh in his mind. Carl Djerassi's group at Syntex was not looking for an oral contraceptive when it created 19-nor-progesterone, the precursor to norethindrome which became the active ingredient in half of all contraceptive pills. Sir Alastair Pilkington was washing dishes when he got the critical insight that led to float glass. And so on.

To increase the probability of some program's succeeding and to multiply the kinds of interactions that lead to success, innovative companies (where possible) encourage several programs to proceed in parallel. Lucky "accidents" are involved in almost all major technological advances. When theory can predict everything, one is in production stages—not development. In fact, Murphy's Law works because engineers design for what they can foresee; hence what fails is what theory did not predict. And it is rare that the interactions of components and subsystems can be predicted over the complete performance envelope of a complex device's anticipated operations. For example, despite thorough theoretical design work, the first high-performance US jet engine literally tore itself to pieces on its test stand.[17] Later, well tested jet engines failed in an Iranian sand storm during the US attempt to rescue its Embassy's personnel held as hostages.

Recognizing these characteristics of development, all of the "most innovative" companies in my sample consciously started several prototype programs in parallel, thinning down the number of alternatives only as probabilities of success approached certainty or costs became prohibitive. Sony, for example, in its video-tape recorder (VTR) program pursued ten different major options, each with two to three subsystem alternatives. Such redundancy helps both in coping with some of the inherent randomness and unpredictability in development and in motivating people through constructive competition. The latter may easily be as important as the former.

Development Shoot-Outs

To gain maximum motivational and information benefits many of these companies consciously structure "shoot-outs" between competing approaches only after they reach advanced prototype stages. Managers find this practice: (1) provides the most objective possible data for decision purposes; (2) decreases risks by making choices as close as possible to the marketplace; and (3) ensures that whatever option wins will move ahead with a committed team behind it. Although anathema to many who worry about presumed *efficiencies* in R&D,

greater *effectiveness* in choosing a right solution can easily outweigh "duplication costs." This is likely to be the case when the market genuinely rewards higher performance or when large volumes justify increased technical or cost sophistication. Under these conditions competing approaches can both improve probabilities of success and decrease development times, making parallel development less costly in both the short and long run.

Properly rewarding and reintegrating losing teams is perhaps the most difficult and essential skill in managing competing projects. If the total company is expanding rapidly or if the successful project creates a substantial growth opportunity, competing team members can generally find another interesting program or sign on with the "winner" as it moves toward the marketplace. For the shoot-out system to work continuously, however, executives must create a climate that honors high performance, whether it wins or loses; reabsorbs people quickly into their technical specialities or onto other projects; and accepts and expects rotation among tasks and groups.

At Polaroid during its prolonged growth period it was an honor to work on one of Dr. Land's projects. Many of these grew into the major new product lines of the company with attendant management opportunities. Some specialized technical solutions (like high-performance batteries) became entire new divisions. Many engineers followed products into production, operating production prototype machines at first, then phasing back into development activities or forward into manufacturing management. The R&D organization was described as a "coat rack of talent to be assembled and reassembled on projects as it was needed, but with a base of attractive problems to attack whenever people were not involved on specific task forces." Those who were successful on these projects became the next generation of management at Polaroid.[18]

Small, Flexible Teams

To further enhance motivation, interactiveness, and rapid progress, innovative companies also emphasize small teams working in a relatively independent environment. The optimum number of key players sought per team varies from five to seven. This number seems to provide a critical mass of skills, foster maximum communications,[19] and allow sufficient latitude for individual creativeness and commitment. The epitome of this style is the "skunkworks"—named after "Kelly" Johnson's successful group at Lockheed—in which small teams of engineers, technicians and designers are placed together with no intervening organizational barriers to developing a new product or process from concept to commercial prototype stages.

Innovative European groups (like Pilkington Brothers or Elf Aquitaine) or Japanese companies (like Sony or Honda) also used this approach. Interestingly, Japan's much publicized *Ringii* decision-making was not evident in these latter situations. Sony's founder, Mr. Ibuka, or his technical successor, Mr. Kihara, participated directly with their design groups, and made rapid "on the spot" decisions at key junctures. Soichiro Honda was known for working with his engineering and design groups and emphasizing his own views by shouting at his engineers and occasionally even popping them with a wrench.[20]

Progress in technology is largely determined by the number of successful experiments made per unit time. Skunkworks help eliminate bureaucratic delays, allow fast unfettered communication, and permit the quick turnarounds and decisions that stimulate rapid advance. To further enhance fast, effective decision-making, many of the most innovative enterprises observed also tried to structure flat organizations to decrease the number of organizational layers between the bench worker and the top. By keeping total division sizes below 400, close communications can be maintained with only two intervening decision layers to the top. In units much larger than this, people quickly lose touch with the total concept of their product or process, bureaucracies grow, and projects must go through more and more formal screens to survive. Since it takes a chain of "yesses" to approve a project as it moves to the top and only one "no" to kill it, jeopardy multiplies as management layers increase.

Interactive Learning Processes

But the most innovative enterprises go even further in structuring interactiveness. They consciously tap into multiple outside sources of technological capability. The first of these is their own customers. Many studies have shown that demand and technology often drive one another. Von Hipple[21] suggests that over 50% of all innovation in some industries (like electronics) is performed by customers. In others (like textiles) materials or equipment suppliers provide most of the innovation. In still others (like biotechnology) universities are dominant, while foreign sources strongly contribute to industries such as pharmaceuticals, foods, or environmental controls.

No company or research group can spend more than a small portion of the world's (approximately $200 billion) R&D budget. Yet scientific knowledge and technological advances from outside sources and seemingly unrelated fields—like biology and medicine's increasing intersections with electronics—often interact to create totally new concepts or opportunities important to an enterprise. Recognizing this, many concerns have active strategies to develop information for trading with outside research or technology groups, and special teams to cultivate these sources.[22] Large Japanese companies, of course, have been notably effective at this. So have such diverse US companies as DuPont, AT&T, Apple Computer, Johnson & Johnson, and Genentech. Wholly new relationships in which large companies participate—as joint venturers, consortium members, limited partners, guarantors of first markets, venture capitalists, or spin-off equity holders—have begun to rival the imagination and variety of structures used in entrepreneurial start ups. These represent some of the most exciting new dimensions for corporate strategies in the current decade.

Champions, Experts and Rewards

Many studies have emphasized the essential role of a "determined champion" in overcoming the many soul-wrenching disappointments and setbacks major innovation always seems to encounter.[23] In large organizations, in particular, the process seems long term, ambiguous, uncertain, frustrating, and debilitating beyond belief. It seems to be in the nature of both the receiving organizations (production units, customers, or one's own peers) to actively adapt to resist change. Often only a fanatic can survive the strain. And even some of them ultimately give up.

These frustrations were so intense that Armstrong (creator of FM), Carothers (synthetic fibers), and Diesel (the diesel engine) committed suicide before their great inventions reached full commercial success. Goddard (champion of rocketry), Whittle (jet aircraft engines), Carlson (Xerox), Sikorski (helicopters), Ibuka (video tape recorders), and Ovshinski (amorphous semi-conductors) are typical of those who pour frustrating decades into their radical innovations before they become profitable.

Fortunately their fanaticism makes innovator-entrepreneurs underestimate the length of time and obstacles to success. Time horizons for radical innovations (averaging at least 12 years) make them essentially irrational from a calculated financial investment (present value) viewpoint. For these reasons individual champions (or fanatics) are the heart and soul of most small start-up entrepreneurial ventures, substituting sweat capital (or entre-

preneurial rent) for actual dollar investments. Similarly, a few such people almost always drive any innovation in a larger enterprise—in all cases in my study. But rarely can a single individual carry out an entire innovation alone. A champion desperately needs other expertise and support—both business and technical—at crucial times. Some of the most difficult problems for large organizations are: (1) how to tolerate and nurture the kinds of off-beat, driven personalities who tend to become champions; and (2) how to reward both champions and experts appropriately.

Professional Venture Teams?

Because so many large enterprises fail on both counts, cities like Boston, Minneapolis and San Francisco are surrounded by successful companies started by erstwhile champions not handled properly by their former employers. Recognizing the joint skills and motivation problems involved, many companies try to assign identified innovations to "new venture teams" with the professional (engineering, marketing and finance) skills to see them through. But these teams rarely succeed if the actual champion is not a member (preferably the head) of the team. Texas Instruments made a study of some 40 of its own innovations, and found no projects which were successful where management had chosen the team head. Innovations seem to be much like human babies. For a high probability of success an innovation needs a mother (champion) who loves it emotionally and will stay with it when others would give up, a father (authority figure with resources) who can support it, and pediatricians (experts) who can see it through technical difficulties. Unfortunately many companies assign the task only to their pediatricians.

Rewards pose a special complexity. Most large concerns do not feel they can offer internal innovators the millions of dollars they might make if they successfully build their own companies. Lack of personal investment by the innovator and concerns about "fair play to others" are the reasons most often cited. Yet the few concerns that have made innovative leaders very rich do not seem to have suffered inordinately. And, of course, those who do nothing are protected by the fact that losses caused by potential innovators leaving (or not performing) do not appear on income statements.

The most innovative companies do recognize that innovative champions tend to work for more than just monetary rewards.[24] They try to offer innovators significant personal recognition, independence in research, appropriate power or visibility within the company or in the division exploiting the innovation, or that most cherished of technical incentives—the right to a major role in the next big innovation.[25] Some examples will suggest the kinds of specific rewards offered. Sony gives "a small but significant" percentage of a new product's sales to its innovating teams. Pilkington, IBM and 3M's top executives have often been chosen from those who earlier headed successful new product entries. Intel let its Magnetic Memory Group operate like a small company with special performance rewards and simulated stock options. GE, Syntex and UTC have actively helped internal champions of "nonrelated" product innovations establish new companies and have taken equity positions in their enterprises. And so on.

Increasingly, however, companies are supplementing such recognition with much larger financial awards to keep their most productive people from jumping outside.

PLANNING VERSUS CHAOS

While understanding and permitting the essential chaos of innovation, effective managers also channel its main directions. As has been suggested, formal planning has severe limitations in this area, particularly in the early stages of the process. For example, formal market analyses may be very useful for product line extensions, but they are often quite misleading when applied to radical innovations.[26]

Such studies said: "Haloid would never sell more than 5000 xerographic machines"; Intel's microprocessor "would never sell more than 10% as many units as there were minicomputers"; Sony's transistor radios and miniature television sets "would fail in the marketplace." All were big winners. Yet more heralded concepts (like Hovercrafts, bio-engineered insulin, and Josephson-effect devices) have yet to fulfill their projections. Many of industry's largest failures were actually over-planned and -researched on paper (like Ford's Edsel, IBM's FS system, and FAA's SST) rather than interactively developed with genuine customer feedback.

Frequently, neither innovators, market researchers nor users can quite visualize a totally new product's real potential. Initial markets often depend on opportunistic adaptations to specific unsuspected needs.

Edison's lights first appeared on ships and in baseball parks. Astro Turf was intended to convert the flat roofs and asphalt playgrounds of city schools into more humane environments. Recording English language programs for Japanese schools was the unexpected first market for audio tape recorders. The hovercraft only became profitable when its original "amphibian" use concept was abandoned and rigid "side skirts" were extended into the water. Graphite and boron fibers unexpectedly found their major uses in sporting equipment. And so on.

Recognizing the limits of early-stage marketing forecasts, more innovative companies tend to move to interactive testing in lead customers' hands as soon as possible. They both benefit from some of these customers' own innovative ideas and learn what the product's actual use and desirability characteristics will be.

Hewlett Packard, 3M, Sony, and Raychem—among the most sophisticated diversifying innovators encountered—specifically acknowledge the complexity of assessing new markets and the benefits of highly interactive relationships with customers during development. They frequently introduce radical new products through small teams closely participating with lead customers, learning from them, and rapidly modifying designs and entry strategies around this information.

Chaos Within Milestones

Instead of relying on detailed plans and control systems, innovative top managements—like venture capitalists—tend to administer primarily by helping establish goals, selecting key people, and defining certain critical limits or decision points at which they will intervene. As technology leads or market needs emerge, they set a few—most crucial—performance targets or concept limits.

Pilkington's Board stipulated that "the float glass process must obsolete all existing plate processes in order to go ahead." Data General's management announced "no mode bit" for the Eagle computer. Mr. Ibuka, Sony's chairman, first set a target of "comparable performance with Ampex but at ¼ the cost" for Sony's first videotape recorder; then shortly thereafter targeted an "industrial video cassette recorder for ¹⁄₁₀ the price"; and finally envisioned the "first *home* video recorder at ⅓ the industrial cassette's price." Each of these was a clear target, but an incredible challenge at the time.

These managements then allow their technical units to decide how to achieve these, subject to defined constraints and program reviews at crucial junctures. Early bench-scale projects may at first pursue many options with little attempt to integrate them all into a total program. Only after key variables are understood—and perhaps measured and demonstrated in lab models—may more precise planning be meaningful. Even then many factors

may remain unknown, and competition can continue to thrive in pursuit of solutions. At agreed-on review points, however, only those options that can clear defined performance hurdles or milestones may continue (see Figure 1). But managers may still continue a few other options as small-scale "side bets" in case something goes wrong. In a surprising number of cases, these alternatives prove successful when the planned options fail.

As the program approaches ultimate exploitation, uncertainty decreases, and costs escalate. This is where formal planning, using the full array of program planning, economic evaluation, and progress monitoring techniques can pay high dividends. By then, however, incremental approaches[27] have built the information bases, options, certainty of outcomes, motivation, comfort levels in management, and political power bases necessary for the program to compete and succeed against options more familiar—and hence seemingly less risky—to the enterprise.

Complex Portfolio Planning

Such approaches increase the possibility of individual innovative programs developing successfully. But in the turmoil of a number of such endeavors proceeding throughout the organization, top managers also need some way to balance the requirements and benefits of current lines against those of intermediate-horizon and future prospects. While it is inappropriate to detail such a planning system here, this calls for a complex dual matrix planning approach (highlighted in Figure 2) in which the total corporate strategy is broken down into its key thrusts. To the extent possible, these are assigned to individual divisions (the vertical vectors in Figure 2), with differential investment returns defined for each division to allow them the time horizons to fulfill their roles. For example, divisions charged with developing the company's ten-year future markets cannot be expected to make the same returns (on a pre-

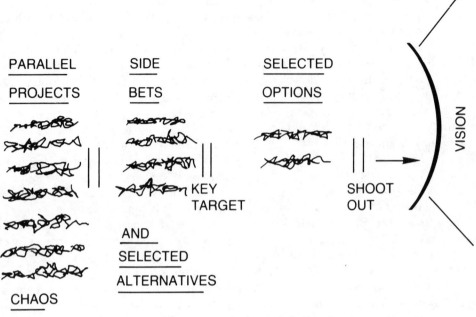

FIGURE 1.   *Somewhat Orderly Tumult*

sent value basis) as divisions charged with exploiting existing technologies. And so on.

But there will also be thrusts that cut across divisions (the horizontal vectors in Figure 2). These require that each appropriate division devote *enough* resources to fulfill its share of that thrust. If each division merely ranks and supports programs in terms of returns to that division, it will tend to eliminate or underfund needed corporate strategic programs, particularly longer-run or risky activities. It is essential that strategy override short term (tactical) or divisional (local) priorities. Corporate strategic controls must see that *enough* resources are devoted to each major thrust, even if this means divisions must perform some projects with lower seeming (divisional) returns than some projects which must be rejected. All this requires a much more complex concept of portfolio planning than the often discussed and popular Boston Consulting Group "four box" matrix. Figure 2 is a simplified version of the

practices many large innovative companies use for this purpose.

CONCLUSION

In successfully managing innovation, such formal techniques are important to help provide balance, adequate time horizons, and reasonable overall resource allocations to protect both the enterprise's present and future needs. They are essential to good strategy, but not sufficient for the purpose.

Because major technical changes tend to progress in a highly probabilistic, tumultuous, interactive fashion—driven by a few champions—innovation programs must be designed accordingly. This article suggests how some of the world's most innovative companies interlink careful strategic planning concepts with some novel organizational and motivational ap-

James Brian Quinn

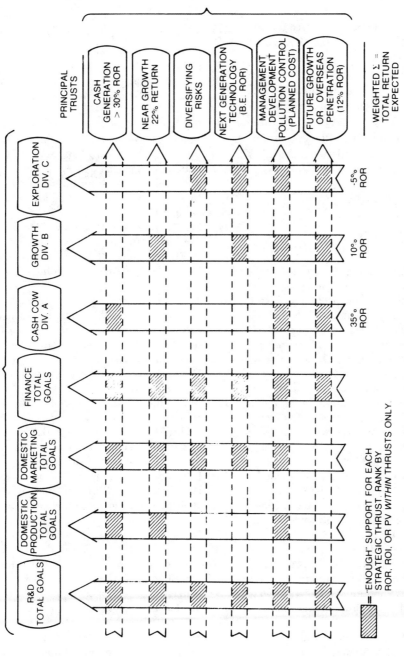

FIGURE 2.   *Division Goals Sum to Corporate Strategic Goals*

The total strategy is divided into 6-10 Principal Strategic Thrusts. Each of these has a different minimum hurdle rate (or allowed cost total) for its programs. Each division's or function's plans and budgets must show *sufficient* support for its share of that strategic thrust. When an entire thrust cannot be assigned to a specific division, efforts are coordinated across divisions to obtain the desired total corporate impact. Each division has its own expected rate of return within which some programs can achieve higher and others lower returns. The weighted sum of all programs and divisions returns must meet the overall corporation's targets.

proaches to achieve these dual purposes. Neither structured formality nor unstructured chaos will work well alone. Innovative companies seem to evolve a sophisticated approach to "managed chaos" which recognizes the realities of how major technological innovations evolve and harness this process to corporate needs. While some general principles apply, the particular strengths, style, goals and competitive situation of each company may dictate very different specific solutions. But experience suggests certain common threads whose adoption increases the probability of succeeding.

NOTES

1. R. Vernon and W. Davidson, *Foreign Production of Technology Intensive Products* (Washington, DC: National Science Foundation, 1979).

2. For specific data, see: J.B. Quinn, "Overview of the Current Status of US Manufacturing: Optimizing US Manufacturing" in *U.S. Leadership in Manufacturing* (Washington, DC: National Academy of Engineering, 1983).

3. M. Tushman and W. Moore, *Readings in the Management of Innovation* (Boston: Pitman Publishing, 1982) perhaps offers the best current single collection of these sources.

4. See J.B. Quinn, "Long Range Planning of Industrial Research," *Harvard Business Review,* July–August 1961, and "Transferring Research Results to Operations," *Harvard Business Review,* January–February 1964, for a more complete delineation of key issues and approaches.

5. D. DeSola Price, "Of Sealing Wax and String," *Natural History,* January 1984, offers an eminent technological historian's view of this process. J. Jewkes, D. Sawyers and R. Stillerman, *The Sources of Invention* (London: MacMillan & Co., 1958) is perhaps the first classical presentation of this theme.

6. J. Fisher and R. Pry, "A Simple Substitution Model of Technological Change," monograph. 1981. J. Birnbaum, "Computers: A Survey of Trends and Limitations," *Science* February 12, 1982. R. Ayers, "Envelope Curve Forecasting" in J. Bright, ed., *Technological Forecasting for Industry and Government* (New York: Prentice Hall, 1968).

7. T. Hughes, "The Inventive Continuum," *Science '84,* p. 83.

8. J. Watson, *The Double Helix* (New York: Athenaeum Publishing Co., 1968) offers an excellent narrative of many crucial events in this sequence.

9. R.C. Dean, "The Temporal Mismatch—Innovation's Pace vs. Management's Time Horizon," *Research Management,* May 1974.

10. Battelle Memorial Laboratories, "Science, Technology, and Innovation," Report to the National Science Foundation, Columbus, Ohio, 1973.

11. See R. Hayes and D. Garvin, "Managing As If Tomorrow Really Mattered," *Harvard Business Review,* May–June 1982; and S. Ramo, *America's Technology Slip* (New York: John Wiley & Sons, 1980).

12. T. Kono, *Strategy and Structure of Japanese Enterprises* (Armonk, NY: M. Sharpe, Inc., 1984).

13. From case studies: *Intel Corporation and Genentech Inc.,* copyright, J.B. Quinn, Amos Tuck School, Dartmouth College, 1981.

14. N. Lyons, *The Sony Vision* (New York: Crown Publishers, 1976).

15. A. Chandler, *Strategy and Structure: Chapters in the History of American Industrial Enterprise* (Cambridge, MA: MIT Press, 1962) provides the classic study suggesting that for effectiveness, structure must follow strategy.

16. D. Schon, *Technology and Change* (New York: Delacorte Press, 1967) was among the first to view the management implications of this reality.

17. In *Herman the German* (New York: W.C. Morrow Publishing, 1984), Gerhard Neumann, the longtime head of GE's aircraft engine division, describes this event and other aspects of managing development in his company.

18. M. Olshaker, *The Instant Image* (New York: Stein & Day, 1978).

19. My colleague, Victor McGee, has calculated the total number of one-way channels of communications in a group as $n[2^{(n-1)} - 1]$, which shows a large increase in channels from 75 to

1016 as numbers move from five to eight persons in a group.

20. T. Sakiya, *Honda Motors: The Men, The Management, The Machines* (Tokyo: Kodansha International Ltd., 1982).

21. E. Von Hipple, "Successful Products from Customer Ideas," *Journal of Marketing,* January 1978.

22. In "Information Flow in R&D Labs," *Administrative Science Quarterly* 14 (1969), T. Allen and S. Cohen analyze the very high leverage such "gatekeeper" activities can have in R&D organizations.

23. J. Schumpeter, *Capitalism, Socialism and Democracy* (New York: Harper & Row, 1975); and M. Maidique, "Entrepreneurs, Champions, and Technological Innovation," *Sloan Management Review,* Winter 1980, present the classic views and data on champions.

24. D. McClelland, *The Achieving Society* (Princeton: Van Nostrand Press, 1961); and G. Bylinski, *The Innovation Millionaires* (New York: C. Scribner's Sons, 1976) offer some specific and interesting insights on entrepreneurial motives.

25. T. Kidder, *The Soul of a New Machine* (Boston: Little Brown, 1981) refers to this as "playing pinball."

26. E. Tauber, "How Market Research Discourages Major Innovation," *Business Horizons,* June 1974.

27. J.B. Quinn, "Managing Strategies Incrementally," *Omega,* Vol. 10, no. 6 (1982).

The Mythology of Innovation, or a Skunkworks Tale, Part II

Thomas Peters

Yellow "Post-It Notepads" have quickly become a staple in the American office—and a $100 million product winner for 3M. The notion emerged from a 3M employee who sang in a choir. The slips of paper he used to mark the hymns kept slipping out, and he thought of an adhesive-backed paper. The technology was there, and a prototype was soon available. "Great story," you say. Not quite yet. Major office supply distributors thought it was silly. Market surveys were negative. But 3M secretaries got hooked—once they actually used it. The eventual breakthrough: a mailing of samples sent to *Fortune 500* CEO executive secretaries under the letterhead of the 3M chairman's secretary.

The above anecdote would merely amount to a charming story were it not repeated at Citicorp and Bell Labs, from jet engines to steel mills to computers to new financial instruments. The course of innovation—idea generation, prototype development, contact with initial user, breakthrough to final market—is highly uncertain, to say the least. Moreover, it always will be messy, sloppy, and unpredictable, and this is the important point. It's important, because we must learn to design organizations that take into account, explicitly, the irreducible sloppiness of the process and take advantage of it rather than attempt to fight it.

Unfortunately, most innovation management practice appears to be predicated on the implicit assumption that we can beat the sloppiness out of the process if only we'd get the plans tidier and the teams better organized. The role of experiments and skunkworks, the zeal of champions, the power gained from exploiting the innovative user as partner, is denigrated as an aid only fit for those who aren't smart enough to plan wisely.

In Part I of this article we looked at experiments and skunkworks, and examined the overpowering evidence that disorder is the norm, no matter how wise the planner or executive. We did so by contrasting reality with three predominant myths that lead us down a primrose path to bureaucratic inflexibility rather than responsive adaptation.

We will now take a look at seven additional prevailing myths that are also generally debilitating:

4. Time for reflection and thought built into the development process is essential to creative results.

"The Mythology of Innovation, or a Skunkwork's Tale, Part II" by Thomas Peters, from *The Stanford Magazine*, 1983. Copyright 1983 Stanford Alumni Association. Reprinted with permission.

Note that this is Part II of an article previously published by *The Stanford Magazine*.

5. Big projects are inherently different from small projects (or an airplane is not a calculator).

6. Strong functions (and, consequently, functional organizations) are imperative if the would-be innovators (R&D, engineering) are to get a fair hearing.

7. Product line/product family (and inter-product line) compatibility is the key to economic success.

8. Customers tell you only about yesterday's needs.

9. Technology push is the cornerstone of business (and presumably American economic) success.

10. Strive to optimize.

CREATIVITY (INNOVATION) FOLLOWS FROM REFLECTION

If big, functional, well-orchestrated teams *were* the heart of successful innovations, we would expect to find them, as a corollary, populated with powerful thinkers who regularly ascend to their mountaintop retreats to look out over the pines. While thus reflecting, they would accomplish necessary breakthroughs, presumably on schedule. If, on the other hand, rough and tumble skunkworks, hell-bent on out-producing some formal group, were the norm, we would expect to find bleary-eyed folks staring at CRTs or test tubes in dirty, forgotten corners of basements.

It does turn out that bleary eyes have quite a role to play. Perhaps it is tautological to state that when commitment, ownership, and year-long tasks routinely accomplished in five weeks are the apparent "norm" of the truly innovative enterprise, someone called a "champion" can be found at the heart of the operation. Tautological, perhaps, but nonetheless often ignored. But not always. Formal IBM in-

house studies of research project success always unearth a champion. Similar National Science Foundation studies suggest that the champion's role is crucial in pushing an idea to fruition. My own studies of consumer goods companies show that when the brand manager, even in a highly structured system, becomes a determined champion, the odds for success go up tenfold.

MIT's Don Schon argues convincingly that "the new idea either finds a champion or it dies. No ordinary involvement with a new idea provides the energy required to cope with the indifference and resistance that change provokes." . . . Brian Quinn of Dartmouth, whose studies of industrial innovation were mentioned in Part I, concludes that "fanaticism is crucial." . . . Peter Drucker, noted business consultant, educator, and writer, looking back 50 years in *Adventures of a Bystander*, remarks, "Whenever *anything* is accomplished, it is being done, I have learned, by a monomaniac with a mission." . . . A successful technology company manager argues that one of his prime criteria for promotion is that a candidate be "a thug," i.e., determined to relentlessly beat down barriers in his path (the prime trait for which Thomas Edison was renowned). . . . *The Nobel Duel*, Nicholas Wade's study of the discovery of a brain hormone release mechanism, presents persuasive arguments that the two researchers who were the subject of the book demonstrated "perseverance, not great leaps of imaginative insight." The author also cites others with greater expertise in the field who wrote off the Nobel-winning effort as "an intellectually barren exercise." Still another study of Nobel Prize winners quotes their peers, who time and again credit the Nobelists with having such traits as "peasant toughness," "a streak of brutality," and "killer instincts." They are "good finishers."

There are crucial implications here. In the halcyon days of Organizational Development, in the 1960s and 1970s, we seemed to place cooperation above all other desirable traits in any human being who works in a big corporation. Our champions, however, frequently are not the souls of sweetness and light. As John Jewkes, author of *The Sources of Invention,* concludes, "The most inventive spirits have confessed a constitutional aversion to cooperation." And UC-Berkeley Nobel Prize winner (physics) Luis Alvarez, commenting on Stanford's dominance (Paul Berg's ascendance in particular) and Berkeley's relative eclipse in physics, notes, "I'd say, if anything, there is *too much* teamwork, and there are no individual stars (at Berkeley)." In fact, the people who are going to be tenacious, committed stars are often royal pains in the neck (except to their immediate team members, who would often go over the hill for them). Further, these stars openly admit their ego.

A crucial corollary is that the corporation that would nourish inventors also *must* tolerate, even laud by a narrow definition, failures. General Georges Doriot, as responsible for industrial development on Route 128 in Massachusetts as Fred Terman of Stanford was for the burgeoning Silicon Valley, said recently in an address at Digital Equipment (one of the first companies in which he invested), "If failure can be explained, and it's not based on a lack of morality, then to me failure is acceptable." The best of the companies I've looked at—3M, HP, Apple, J&J, Emerson Electric, Mary Kay, Digital, PepsiCo—explicitly support failures in the sense that they admit that failure is normal. Their philosophies say it, and it is especially fostered through stories of their battles, usually concluding with success but marked by scores, if not hundreds, of setbacks along the way.

Going through 3M's senior officer roster with a 3M executive a couple of years ago, I discovered that virtually every 3M officer reached the top because of having introduced a couple of important new products. Moreover,

each story, as it was recounted in conventional form, focused on the tough road: the ten years of ups and downs when the product was too early for the marketplace, when it had to be reformulated, when the manufacturing scale-up didn't work. Setbacks are considered to be Standard Operating Procedure. The winners are seen, above all, to persist.

BIG IS DIFFERENT

Hold on a minute, you say. Maybe I buy your story for the likes of HP's hand-held calculators. And you did unearth the story of the locomotive. But what about the "real stuff"? The space program? The invention of the transistor at Bell Labs? Surely it's different for the big ones. No skunkworks, no wild-eyed champions, no 99 percent failure rate for the planning process.

My response to such queries is that I might disagree that big breakthroughs *are* the heart of it. When we write about even relatively "small" inventions, like Howard Head's metal sandwich ski, Head becomes the hero. But there must be 25 people who contributed advances in materials science that made it possible for Head to even conceive his solution. The world of computers on a chip is similarly indebted to hundreds of important breakthroughs from the hardest sciences—polymer chemistry, surface physics, optics, and the like. Evolutionary leaps do occur, but most advances are incremental and fairly cumulative.

But I don't want to dwell on that issue in this brief section. I want, in fact, to take the opposite tack, to assume that there *are* big projects. Then I want to suggest that they, too, can be treated to a substantial degree as collections of skunkworks. In particular, the principle I call "small within big" turns out in an important sense to be "optimal." The U-2 was developed in a skunkwork. So was the Polaris submarine.

So was the Basic Oxygen Furnace. The Bell digital switch and the IBM System 360 weren't, by some standards, built in skunkworks. But in many respects they were. Both were very sloppy processes. Most of the breakthroughs in each were the result of champions operating offline. Chairman Charles Brown of AT&T said recently, speaking exactly to this point, "The (long distance) network looks, *today,* like one big, perfectly conceived solution. The reality, that we often forget when we think about innovation planning, is that [the network] is a collection of thousands of small breakthroughs which occurred here and there, and certainly not according to schedule or by courtesy of a flawless master plan." Similarly, time and again, at IBM, a 25-person group would go off-site, usually into $19.95 motel rooms, for ten days, and break a logjam that had been holding up a multi-hundred-person unit for months.

The story of a recent Boeing missile development, the Air-Launched Cruise Missile (ALCM), is even more pertinent. The system was complex. It should have been developed "all at once," undoubtedly, with the aid of a 100,000-bubble PERT chart. (If you discern, in my attacks on 100,000-bubble charts, the zeal of a convert, you're reading correctly. My ancient engineering master's degree thesis was on combining probabilistic time distributions in multi-task PERT design.) Now let me talk about how the ALCM *actually* evolved. The missile was in fact broken down into seven major pieces. Modest-sized teams were assembled around the seven tasks. Each task was then accomplished in a remarkably short period of time, relative to the norm. Each had a champion. Each was in competition with all the others on several vital parameters.

Then what happened? You guessed it. Put the seven pieces together and they don't fit exactly right. So you have to spend some time, up to a few months, getting the interfaces just right, despite the prior effort that had gone into interface specification. (There were twice-a-week meetings of a "tie-breaker" group that sorted out many of the issues in progress.) Now here's the crux of the issue. Most people bridle against such retrofitting. And, in fact, once the interfaces have been "Rube Goldberged" together, the final design isn't as technically beautiful as theoretical perfection suggests is possible. But I observe the following: multiple passes usually take less time and result in the development of simpler, more practical systems than the single, all-at-once pass. (The ALCM was in fact delivered over a year ahead of schedule and well under budget.)

Thus we find that via the "multiple pass" route, we wind up with a handful of competitive "skunkworked," relatively simply designed hunks. Let's say it takes six months to get them done. Then we take a three-month pass to get the interfaces fixed. Then the five to ten teams go back to work again and repolish their apples, at the cost of another three months. Then another two months are needed to get the interfaces fixed one final time. We've thus made two complete passes through the system in, say, 14 months. We probably have a very reliable and practical, if not "optimal" and quite so beautiful, system. My experience is that the "single pass" (everything depends on everything else), "get it exactly right the first time" approach is likely to be a two-year effort to start with—and then the damn thing won't work anyway and will likely be too complex. And all this becomes clear only at the end of the two years. So the multiple passes take much less time and are more reliable—even if they result in a few less "oohs" and "aahs" by purists. (Tom West continually warned his Data General skunkwork team to ignore the "technology bigots.")

But back to the question of whether big *is* different. There is no question but that it is. The hand-held calculator and the MX missile system are *not* the same. The Boeing 767 and a french fry seasoning change at McDonald's are *not* the same. On the other hand, championing, small within big, the value of the turned-on

141

modest-sized group, the role of the overtight deadline, piece versus piece competition (e.g., the missile case or Data General's computer group isolation) . . . and commitment . . . are keys to big and small. Big system monomaniacs with missions: certainly Kelly Johnson, inventor of the U-2, qualifies, the Navy's Hyman Rickover qualifies.

ONLY STRONG FUNCTIONS SPAWN CONCERN FOR TECHNOLOGY

Engineers lose out to marketers and finance people in divisional organizations. The division is interested only in short-term profit. Consequently, only a strong functional monolith will keep the engineering (and innovation) view at the forefront. So runs the conventional argument. It's a nice argument on paper, but it doesn't hold much water in practice.

The functional monolith is, by definition, almost always bureaucratic, not commitment/small-team-action oriented. I remember an organization that went even a step further. It took its already monolithic engineering organization and decided that the key to technology advance was organization around technological disciplines or competency centers. Nice in theory. The difficulty was that the product or project took second fiddle. Many experts, basically demonstrating their narrow competence, were simultaneously working on five or six projects that spanned three or four divisions. My experience on this one is crystal clear. No person with "a seventh of a responsibility" for anything ever got committed to it. Drucker's essential "monomaniacs with missions" were pointedly *not* monomaniacs with *seven* missions.

Under some forms of management divisional organizations that are allowed to grow too big can, indeed, become hopelessly bureaucratic. On the other hand, "the division *is* the solution" (and the strategy) for Hewlett-Pac-

kard, 3M, Johnson & Johnson, Emerson Electric, and the like. (J&J constantly creates new divisions. Its corporate watchword is simply "growing big by staying small.")

These companies carefully manage division size. At HP, divisions are kept below a thousand people, so, in CEO John Young's words, "the general manager will know all his people by their first names." Bill Gore, of successful W.L. Gore, comments, "As the number approaches 200, the group somehow becomes a crowd in which individuals grow increasingly anonymous and significantly less cooperative." The low numbers, whether 200 or 1,000, are all aimed at enhancing our by-now old friends— ownership and commitment.

Another vital part of the small-team, small-division mentality is the ability to manage, with less muss and fuss, the pass-off interfaces that fatally delay so much development. Via its "triad" development team principle, HP ensures that the manufacturer and marketer (as well as the engineer) are full-scale partners in the development process starting *very* early in the design phase. The same principle is followed at 3M, where they put a full-time manufacturer on the team before such a move can be theoretically warranted. With an early start, the product can be manufactured more easily and the path to the marketplace is smoother.

A corollary to the theme is the degree to which these decentralized, collection-of-skunkwork companies are *niche-men*. The special magic at 3M: "We hit fast, price high (full economic value of the product to the user), and get the heck out when the 'me-too' products pour in." At 3M, new niches, generally ones of $10 million to $50 million, are dominated for a period that can be as short as five years. At 3M, they further insist that divisions spend most of their time worrying about the next generation of products rather than controlling the last two cents of margin on one that is dying.

Raychem Corporation of Menlo Park, a maker of sophisticated, irradiated plastic, heat-shrinkable connectors, and the like, is even more extreme. Chairman Paul Cook estimates that the highly profitable, half-billion-dollar company has fielded 200,000 products during the company's 25-year history. Raychem's idea of a great product is a 100 percent share, with a 90 percent operating margin, in a $10 million niche! Stories from some of Raychem's divisions suggest that finding and exploiting such niches—instantly—*is* Raychem. One case I stumbled across involved development of a substantially new product in five days. A leak occurred on an offshore oil platform in the North Sea. Researchers from Menlo Park, California, and Swindon, England, were sent to the site immediately. A little facility was set up and a new product (not a great leap, but genuinely new) was developed, prototyped, debugged, and installed under the sea . . . in just five days! As an executive participant in a workshop said upon hearing the tale, "In our company, at the end of 120 hours, we'd still be cutting travel orders to make the first trip from Menlo Park."

So Raychem, it turns out, *is* skunkworks. Raychem. Digital. Hewlett-Packard. 3M. Johnson & Johnson. Frito-Lay. P&G, in many senses. Strong functions are important, but champions and small-team zeal, and all-important commitment, are arguably even more important. As an old hand at the process said to me recently, "Tom, let's be clear about the magnitude. The charged up 10-50 person team isn't in the '10 percent productivity improvement' game. Its results are often 300, 400, 700 percent better!"

COMPATIBILITY DRIVES ALL

Product release is held up another three months. And then another 45 days. "We've got to make sure that the software is totally compatible with all the rest of the product family." So the logic goes. I buy it . . . to a point. Compatibility is important, particularly in higher technology, systems-related products. But some-

times the last 2 percent takes 12 months. In the meantime, ten competitors have approximated the solution, and gotten theirs into the marketplace more quickly. Digital products overlap: sometimes frustrated users find a piece here, a piece there that's incompatible with a product that is supposed to be compatible. Hewlett-Packard engineers, marketers, and salesmen, too, lament the incompatibility among some products. Another computer maker held up the introduction of a product by over six months; in the meantime, a major competitor came out with a product that was 70 percent as good— and stole the march on an important niche. A friend recently went to a major computer systems bazaar at Brooks Hall in San Francisco. In his words, "There were 1,500 people displaying: 750 of them were turning Apples into IBMs, and the other 750 were turning IBMs into Apples."

Likewise, customers use features in unintended ways. A Datapoint officer recalls, "The 2200 programmer terminal device, for instance, would emulate any number of terminals. The customers paid no attention to that and started using it in stand-alone computing operations. We brought this thing out without a compiler because it was made only to run terminal emulations. But from day one, users were writing software with only a primitive assembler in the machine."

I'm not against compatibility, certainly. But, particularly in extremely fast-paced markets associated with computers, data handling, and telecommunications, there are literally thousands of entrepreneurs who will (and do) fill in the spaces, do for you the last 2 percent. And they've got every reason in the world to try and work *with* the Digitals, the HPs, the Apples, the Tandems, and the IBMs. They'll try to make the stuff compatible, fit things together in ways that are helpful to the user. Those who wait, trying to get the last percent compatibility, may well go by the boards.

The same phenomenon, I might add, holds in many other markets, albeit with a little less intensity. That's the reason Procter & Gamble, 3M, Mars, and Johnson & Johnson are so insistent about spurring competition between their own divisions and brand managers. Bloomingdale's does the same thing among buyers and for floor space in their stores, and Macy's has done extremely well emulating them. In most markets; new stuff is happening all the time. The lion's share is virtually invisible. That is, you frequently don't see it until it's too late. To keep up with competition, you have to keep getting new items into the market.

Errors of premature early release can be (and often are) disastrous. So when I talk about early release relative to compatibility, I'm limiting "early" to that: I'm not talking about quality. Often a product hits the marketplace with the bugs not ironed out. Its technical superiority is blunted by rotten reliability or insufficient support. Things that get out there ought to work, and ought to have a handful of new features that give you position in the marketplace. But getting that last possible feature, that last degree of complexity (read over-complexity), that last degree of compatibility may cost you more of the market than you would have gained by perfection. It's an open issue and a tough one. The problem, as is the case with so much of what I'm describing, is that the pointed general tendency is to err in the direction of too much complexity, too many features, too much compatibility. It's not an "even" fight. The optimizers, the perfectionists, the last-quarter-percenters tend to win—because they always use the argument that "it will only take us another 30 days." But we all know that those last-30-day activities seem to always take 120 days . . . if you're lucky.

CUSTOMERS TELL YOU ONLY ABOUT YESTERDAY

I've been able to unearth about 80 studies of sources of new product ideas, covering perhaps 30 industries, from commodity chemicals to computers to shoes. In only one instance does the product development department seem to provide the lion's share of the ideas: bulk commodity chemicals. The great majority of the ideas for new products come from the users. This is true not only in high technology, but in the hamburger business as well. McDonald's: the greatest share of the new products come from franchisees, usually out in the boondocks.

But let us go on to higher tech. Studies at MIT by Eric Von Hippel of scientific instrument and component manufacturing equipment businesses are revealing. He reviewed 160 inventions, and found that over 70 percent of the product ideas came from users. There was a most intriguing and counterintuitive point: It wasn't just the bells and whistles or tweeks that came from users. Sixty percent of the minor modifications came from users, as did 75 percent of the major modifications . . . but, astonishingly, *100 percent* of the so-called "first of type" ideas (sophisticated devices such as the transmission electron microscope) were user generated. And Von Hippel's criteria were tough. "Idea from user" means that the user did a lot more than whisper into the producer's ear. The user got the idea, he prototyped it, he debugged it, and he had it working. Only then did he tap the producer for his experience in reliable production of multiple copies.

Example after example follows the same pattern. The General Electric "New Series" locomotive, terribly successful, came straight from user listening. GE knocked the socks off General Motors because they asked customers, for the first time in a long while, what they wanted. The task was accomplished by just one marketer with a couple of design engineers in tow. They went out to user premises and conducted lengthy seminars, and asked them what was needed.

An executive with an aircraft contractor provided an exceptional example in the fast moving CAD/CAM area (Computer Aided Design, Computer Aided Manufacturing). He's

144

working with McDonnell Douglas. Their CAD system was a leader only half a dozen years ago. But McDonnell Douglas decided that it was *so* good that they'd keep it proprietary and hold their advantage. My colleague is also working with the Lockheed CAD system. Lockheed arrived late to the marketplace and was positioned well behind McDonnell Douglas. But Lockheed took a different tack, deciding, in effect, to let users do the development for them. They didn't keep it proprietary, but sold it instead. Over a three-year period, they cornered 250 commercial customers. (I refer to them as 250 free development centers.) In just a couple of years, the late-to-the-market Lockheed system has leap-frogged the McDonnell Douglas system, thanks to the 250 user inputs.

The key word is *lead user.* After-the-fact analysis for every industry from blue jeans and hamburgers to mainframe computers and aircraft engines makes it possible to say that many products we won't see until 1990 were first invented and prototyped in 1980. The product of the next decade may have emerged as a test between a *lead producer* (more often than not a small company) and a *lead user,* someone who thought he could really take advantage of the new, largely untested, but highly useful technology.

The computer industry is rife with this. Much of the development money in computers has come from MIT, Carnegie-Mellon, Penn State, Michigan, and Bell Labs. They've all been the test sites for the computer industry. For instance, Carnegie-Mellon is now serving as the role model for a whole new concept of 1990 network architecture that IBM is testing. The ferocious fights between IBM and GE for sites in the early days of decentralized, on-line computing show the crucial importance of the lead user.

(There are also amusing stories I've unearthed. A classic "lead user" was a housewife involved with Corning. Her husband brought home a new glass container to be used for holding a certain acid in the laboratory. She put it in the oven, the story goes, accidentally. In fact, she heated something up in it, and it didn't break. Such are the origins of Pyrex! I've heard her referred to as "the Pyrex housewife.")

There's an interesting debate that surrounds this issue. Bob Hayes and Bill Abernathy *(Managing Our Way to Economic Decline)* and others have suggested that U.S. corporations are too market oriented, not driven enough by technology. I agree, if we're talking about after-the-fact polls about what users want; then, indeed, we hear only about yesterday's ideas from the *average* user. But I find, particularly in sophisticated industries, that users, like virtually everything else in life, can be placed in a normal probability distribution. At the "front" tip of the distribution are those who are often as much as 10 to 15 years ahead of their peers (e.g., GM or Boeing in CAD use). They're willing to take a risk in return for a new invention. Similarly, the lead producer (particularly if he's small with little to lose) in fact welcomes the lead users.

In summary, Eric Von Hippel, who has been studying this process intensely for years, goes so far as to suggest that "market research, now chartered to seek data and analyze it, should be reoriented to search out data on *user prototypes.*"

It is really all about being in touch with users. It's important from hamburgers to computers. Hewlett-Packard has the term MBWA—Management By Wandering Around. Wandering around *should* mean listening to the user in a direct, not in an abstract or shorthand, form. A general manager who designed a major new computer, as sophisticated a person as I know, did a neat trick: "I bought my uncle a computer store. I spent nights and weekends working there. My objective was to stay close to the ultimate user, observe his frustrations and needs firsthand, incognito." His learning was reflected in the eventual design in a thousand little ways and several big ones.

TECHNOLOGY PUSH IS THE SECRET

"More scientists in bigger labs" seems to be the watchword, with "better planning, better tools." The heck with skunkworks. But it's more than skunkworks. I've mentioned user listening. It's more than that, too.

Recently I was going to talk to a technology company president about commodities, more precisely the unfortunate tendency to call some high-technology products—chips, instruments, personal computers—commodities. My point (with which I find almost universal agreement): If you label "it" a commodity, you'll start to behave as if it *is*, in effect, a commodity, denigrating service and quality. I ran a little test, taking a full 20 minutes, in preparing for my visit with him. I went to my local cooperative grocery and priced generic toilet paper. I looked at a four-roll, 220-square-foot package of one-ply toilet paper: price—79 cents. I went only three blocks away to a 7-11. There I priced an apparently similar product (toilet paper: four-roll, one-ply, 220 square feet), but this time by Procter & Gamble–Charmin: price—$1.99. Two blocks, and the difference in distribution channels (7-11) and the quality difference (P&G) was obviously enough to add $1.20 to a $.79 product, or, more accurately, add $1.20 to a product that cost about a quarter to produce. Service and quality add as much value (or more) as does gee-whiz technology. This is a paper on innovation, so maybe it seems to be the wrong issue. I don't think it is. It's possible to innovate, which is really my point, on service and user friendliness. In fact, it is the big secret among those technology companies that are "winners." Good manuals, ready answers, and a commitment to be accessible could, I'd argue, revolutionize the personal computer business.

American Bell's Arch McGill, also an IBM alumnus, states wisely: "The customer perceives service in his own terms." I'd add only, in his own, unique, idiosyncratic, some- what crazy, realtime terms. He wants to "feel right" about the product, no matter what it is.

Technology push is crucial, but I would argue that it is not the principal reason America is having so many industry setbacks. User unfriendliness, the inability to realize that the customer *does* perceive it in his own terms, is at least as big a weakness. If you don't believe me, ask 'em in Detroit.

OPTIMIZE!

Oh, if it weren't for people! Ten-thousand-person groups would be the most efficient. Oh, if it weren't for people! One-hundred-thousand-bubble PERT charts and "all at once" execution would be most efficient. Oh, if it weren't for people! Huge amounts of money put into technical forecasting would anticipate competition, customer, and technological surprises. Oh, if it weren't for people (users)! De novo invention of every innard would be the best way to assure quality. Oh, yes, if it weren't for people.

Optimization. What's optimal? I find, simply, that the "suboptimal" is most often the truly optimal. Go back to the big-scale versus small-scale debate. Two "passes" through the system, the latter to patch up sloppy boundaries among the skunkwork interfaces (e.g., the seven pieces of the missile), is, it turns out, a heck of a lot faster *and* cheaper *and* a higher quality way to do the job than via the "optimization" route. Getting 90 percent compatibility, and letting the Darwinist marketplace do the rest, turns out to be optimal, not suboptimal. Getting the last 10 percent may cost you 60 percent of the market. A major National Science Foundation study says that scientific productivity goes down when groups surge above the number seven. Suboptimal? Optimal.

The real problem is the word itself, and our training, especially those of us who are en-

146

gineers. When we use the word suboptimal to describe a skunkwork, we are implicitly suggesting that the optimal *is* even possible. That's where I depart. The theoretical optimal, except as a very abstract proposition, is not really of much interest. *I* call optimal the product of a skunkwork—one that's simple, friendly, timely, fills a practical purpose—and works.

Optimality to me is the group that goes outside to buy, off the shelf, highly reliable parts; the group that gets an 18-month task done in 15 weeks. It's optimal: namely the practically obtainable optimal combination of technical and people resources applied over a finite (competitively plausible) time horizon. The technical optimal is another thing—and not very interesting in the world of commerce. The "people optimal" is another thing, too—also probably not very interesting as a pure proposition. But technical/people "suboptimization" *is* the optimal that I'm looking for.

Let's take a mundane example, far from weapons systems and supercomputers. Procter & Gamble is the master consumer goods advertiser. A former Unilever executive pointed out a surprising fact to me recently. He said, "You know, P&G basically uses just one measure of ad effectiveness: 24-hour recall (of ad content). The marketing journals serve up a new variable every week. At Unilever we used every one as it came along. We'd debate endlessly. The reality was that each new one *was* pretty good. But when the measures are shifting all the time and most of the debate and discussion is about the measures rather than the ad, then you're not getting anywhere. The P&G measure is not the world's best, in the old days or today, but they've used it for years. They have a feel, developed over the years, for what a good ad or a bad one is, based on that invariant measure." Twenty-four-hour recall explains only 60 per-

cent of the variance; maybe only 50 percent. A combination of 50 measures takes us up to 80 or so percent. But 50 measures is humanly unmanageable, virtually meaningless. Everyone loses "feel" with so much complexity. There's no time to worry about ad copy objectives because you're worried so much about post hoc measures of ad effectiveness. The suboptimal is optimal for P&G; the optimal is suboptimal for Unilever.

Remember, too, Tom West from Data General: He didn't care a whit about building a machine that the "technology bigots" would like. He was interested in people who "wanted to get a machine out the door with their name on it." West was a classic suboptimizer: "Not all jobs worth doing are worth doing well." The stories from Bell Labs, the U-2, the Polaris submarine, the missile broken into seven parts, the GE gas turbine and locomotive seem to be the same. Committed people, turned on people, people competing against the market and another corporation and another division tend to get the job done.

And the performance improvement is not a couple of percent here and there at the margin. Fred Hooven of Dartmouth talked about getting a model built in half an hour, instead of waiting for four months to get tech specs carefully drafted and codified "upstairs" in the Ford organization where he was a research director.

The tales that we've unearthed, hardly covering the entire universe, but covering a reasonable sample from all sorts of industries, suggest that half an hour versus four months *is* the right kind of improvement leverage (several hundred percent) that's possible, even regular.

That's innovation. Hail to the skunkworks!

A Model for Diagnosing Organizational Behavior

David A. Nadler

Michael L. Tushman

Managers must continually identify and find solutions to problems caused by mismatched components within the organization. A unique approach that will help managers perform this vital function is offered.

Management's primary job is to make organizations operate effectively. Society's work gets done through organizations and management's function is to get organizations to perform that work. Getting organizations to operate effectively is difficult, however. Understanding one individual's behavior is challenging in and of itself; understanding a group that's made up of different individuals and comprehending the many relationships among those individuals is even more complex. Imagine, then, the mind-boggling complexity of a large organization made up of thousands of individuals and hundreds of groups with myriad relationships among these individuals and groups.

But organizational behavior must be managed in spite of this overwhelming complexity; ultimately the organization's work gets done through people, individually or collec-

tively, on their own or in collaboration with technology. Therefore, the management of organizational behavior is central to the management task—a task that involves the capacity to *understand* the behavior patterns of individuals, groups, and organizations, to *predict* what behavioral responses will be elicited by various managerial actions, and finally to use this understanding and these predictions to achieve *control.*

How can one achieve understanding and learn how to predict and control organizational behavior? Given its inherent complexity and enigmatic nature, one needs tools to unravel the mysteries, paradoxes, and apparent contradictions that present themselves in the everyday life of organizations. One tool is the conceptual framework or model. A model is a theory that indicates which factors (in an organization, for example) are most critical or important. It also shows how these factors are related—that is, which factors or combination of factors cause other factors to change. In a sense then, a model is a roadmap that can be used to make sense of the terrain of organizational behavior.

The models we use are critical because they guide our analysis and action. In any organizational situation, problem solving involves the collection of information about the problem, the interpretation of that information to determine specific problem types and causes, and the development of action plans accordingly. The models that individuals use influence the kind of data they collect and the kind they ignore; models guide people's approach to analyzing or interpreting the data they have; finally, models help people choose their course of action.

Indeed, anyone who has been exposed to an organization already has some sort of implicit model. People develop these roadmaps over time, building on their own experiences. These implicit models (they usually are not explicitly written down or stated) guide behavior; they vary in quality, validity, and sophistication depending on the nature and extent of the experiences of the model builder, his or her perceptiveness, his or her ability to conceptualize and generalize from experiences, and so on.

We are not solely dependent, however, on the implicit and experience-based models that individuals develop. Since there has been extensive research and theory development on the subject of organizational behavior over the last four decades, it is possible to use scientifically developed explicit models for analyzing organizational behavior and solving organizational problems.

We plan to discuss one particular model, a general model of organizations. Instead of describing a specific phenomenon or aspect of organizational life (such as a model of motivation or a model of organizational design), the general model of organization attempts to provide a framework for thinking about the organization as a total system. The model's major premise is that for organizations to be effective, their subparts or components must be consistently structured and managed—they must approach a state of congruence.

In the first section of this article, we will discuss the basic view of organizations that underlies the model—that is, systems theory. In the second section, we will present and discuss the model itself. In the third section, we will present an approach to using the model for organizational problem analysis. Finally, we will discuss some of the model's implications for thinking about organizations.

A BASIC VIEW OF ORGANIZATIONS

There are many different ways of thinking about organizations. When a manager is asked to "draw a picture of an organization," he or she typically draws some version of a pyramidal organizational chart. This is a model that views the stable, formal relationships among the jobs and formal work units as the most critical factors of the organization. Although this clearly is one way to think about organizations, it is a very limited view. It excludes such factors as leadership behavior, the impact of the environment, informal relations, power distribution, and so on. Such a model can capture only a small part of what goes on in organizations. Its perspective is narrow and static.

The past two decades have seen a growing consensus that a viable alternative to the static classic models of organizations is to envision the organization as a social system. This approach stems from the observation that social phenomena display many of the characteristics of natural or mechanical systems. In particular, as Daniel Katz and Robert L. Kahn have argued, organizations can be better understood if they are considered as dynamic and open social systems.

What is a system? Most simply, a system is a set of interrelated elements—that is, a change in one element affects other elements. An *open system* is one that interacts with its environment; it is more than just a set of interrelated elements. Rather, these elements make up a mechanism that takes input from the environment, subjects it to some form of transformation process, and produces output. At the most general level, it should be easy to visualize organizations as systems. Let's consider a manufacturing plant, for example. It is made up of different related components (a number of departments, jobs technologies, and so on). It receives inputs from the environment—that is, labor, raw material, production orders, and so on—and transforms these inputs into products.

As systems, organizations display a number of basic systems characteristics. Some of the most critical are these:

- *Internal interdependence.* Changes in one component or subpart of an organization frequently have repercussions for other parts; the pieces are interconnected. Again, as in the manufacturing plant example, changes made in one element (for example, the skill levels of those hired to do jobs) will affect other elements (the productiveness of equipment used, the speed or quality of production activities, the nature of supervision needed, and so on).

- *Capacity for feedback*—that is, information about the output that can be used to control the system. Organizations can correct errors and even change themselves because of this characteristic. If in our plant example plant management receives information that the quality of its product is declining, it can use this information to identify factors in the system itself that contribute to this problem. However, it is important to note that, unlike mechanized systems, feed-

back information does not always lead to correction. Organizations have the potential to use feedback to become self-correcting systems, but they do not always realize this potential.

- *Equilibrium*—that is, a state of balance. When an event puts the system out of balance the system reacts and moves to bring itself back into balance. If one work group in our plant example were suddenly to increase its performance dramatically, it would throw the system out of balance. This group would be making increasing demands on the groups that supply it with the information or materials it needs; groups that work with the high-performing group's output would feel the pressure of work-in-process inventory piling up in front of them. If some type of incentive is in effect, other groups might perceive inequity as this one group begins to earn more. We would predict that some actions would be taken to put the system back into balance. Either the rest of the plant would be changed to increase production and thus be back in balance with the single group, or (more likely) there would be pressure to get this group to modify its behavior in line with the performance levels of the rest of the system (by removing workers, limiting supplies, and so on). The point is that somehow the system would develop energy to move back toward a state of equilibrium or balance.

- *Equifinality.* This characteristic of open systems means that different system configurations can lead to the same end or to the same type of input-output conversion. Thus there's no universal or "one best way" to organize.

- *Adaptation.* For a system to survive, it must maintain a favorable balance of input or output transactions with the environment or it will run down. If our plant produces a product for which there are fewer applications, it must adapt to new demands and develop new products; otherwise, the plant will ultimately have to close its doors. Any system, therefore, must adapt by changing as environmental conditions change. The consequences of not adapting are evident when once-prosperous organizations decay (for example, the eastern railroads) because they fail to respond to environmental changes.

Thus systems theory provides a way of thinking about the organization in more complex and dynamic terms. But although the theory provides a valuable basic perspective on organizations, it is limited as a problem-solving tool. This is because a model systems theory is too abstract for use in day-to-day analysis of organizational behavior problems. Because of the level of abstraction of systems theory, we need to develop a more specific and pragmatic model based on the concepts of the open systems paradigm.

A CONGRUENCE MODEL OF ORGANIZATIONAL BEHAVIOR

Given the level of abstraction of open theory, our job is to develop a model that reflects the basic systems concepts and characteristics, but that is more specific and thus more usable as an analytic tool. We will describe a model that specifies the critical inputs, the major outputs, and the transformation processes that characterize organizational functioning.

The model puts its greatest emphasis on the transformation process and specifically reflects the critical system property of interdependence. It views organizations as made up of components or parts that interact with each other. These components exist in states of relative balance, consistency, or "fit" with each other. The different parts of an organization

can fit well together and function effectively, or fit poorly and lead to problems, dysfunctions, or performance below potential. Our *congruence model of organizational behavior* is based on how well components fit together—that is, the congruence among the components; the effectiveness of this model is based on the quality of these "fits" or congruence.

The concept of congruence is not a new one. George Homans in his pioneering work on social processes in organizations emphasized the interaction and consistency among key elements of organizational behavior. Harold Leavitt, for example, identified four major components of organization as being people, tasks, technology, and structure. The model we will present here builds on these views and also draws from fit models developed and used by James Seiler, Paul Lawrence and Jay Lorsch, and Jay Lorsch and Alan Sheldon.

It is important to remember that we are concerned about creating a model for *behavioral* systems of the organization—the system of elements that ultimately produce behavior patterns and, in turn, organizational performance. Put simply, we need to deal with questions of the inputs the system has to work with, the outputs it must produce, the major components of the transformation process, and the ways in which these components interact.

Inputs

Inputs are factors that, at any one point in time, make up the "givens" facing the organization. They're the material that the organization has to work with. There are several different types of inputs, each of which presents a different set of "givens" to the organization (see Figure 1 for an overview of inputs).

The first input is the *environment*, or all factors outside the organization being examined. Every organization exists within the context of a larger environment that includes individuals, groups, other organizations, and

	Input Environment	Resources	History	Strategy
Definition	All factors, including institutions, groups, individuals, events, and so on, that are outside the organization being analyzed, but that have a potential impact on that organization.	Various assets to which the organization has access, including human resources, technology, capital, information, and so on, as well as less tangible resources (recognition in the market, and so forth).	The patterns of past behavior, activity, and effectiveness of the organization that may affect current organizational functioning.	The stream of decisions about how organizational resources will be configured to meet the demands, constraints, and opportunities within the context of the organization's history.
Critical Features for Analysis	1. What demands does the environment make on the organization? 2. How does the environment put constraints on organizational action?	1. What is the relative quality of the different resources to which the organization has access? 2. To what extent are resources fixed rather than flexible in their configuration(s)?	1. What have been the major stages or phases of the organization's development? 2. What is the current impact of such historical factors as strategic decisions, acts of key leaders, crises, and core values and norms?	1. How has the organization defined its core mission, including the markets it serves and the products/ services it provides to these markets? 2. On what basis does it compete? 3. What supporting strategies has the organization employed to achieve the core mission? 4. What specific objectives have been set for organizational output?

FIGURE 1. *Key Organizational Inputs*

even larger social forces—all of which have a potentially powerful impact on how the organization performs. Specifically, the environment includes markets (clients or customers), suppliers, governmental and regulatory bodies, labor unions, competitors, financial institutions, special interest groups, and so on. As research by Jeffrey Pfeffer and Gerald Salancik has suggested, the environment is critical to organizational functioning.

The environment has three critical features that affect organizational analysis. First, the environment makes demands on the organization. For example, it may require certain products or services at certain levels of quality or quantity. Market pressures are particularly important here. Second, the environment may place constraints on organizational action. It may limit the activities in which an organization may engage. These constraints range from limitations imposed by scarce capital to prohibitions set by government regulations. Third, the environment provides opportunities that the organization can explore. When we analyze an organization, we need to consider the factors in the organization's environment and determine how those factors, singly or collectively, create demands, constraints, or opportunities.

The second input is the organization's *resources*. Any organization has a range of different assets to which it has access. These include employees, technology, capital, information, and so on. Resources can also include less tangible assets, such as the perception of the organization in the marketplace or a positive organizational climate. A set of resources can be shaped, deployed, or configured in different ways by an organization. For analysis purposes, two features are of primary interest. One concerns the relative quality of those resources or their value in light of the environment. The second concerns the extent to which resources can be reshaped or how fixed or flexible different resources are.

The third input is the organization's *history*. There's growing evidence that the way organizations function today is greatly influenced by past events. It is particularly important to understand the major stages or phases of an organization's development over a period of time, as well as the current impact of past events—for example, key strategic decisions, the acts or behavior of key leaders, the nature of past crises and the organization's responses to them, and the evolution of core values and norms of the organization.

The final input is somewhat different from the others because in some ways it reflects some of the factors in the organization's environment, resources, and history. The fourth input is *strategy*. We use this term in its broadest context to describe the whole set of decisions that are made about how the organization will configure its resources against the demands, constraints, and opportunities of the environment within the context of its history. Strategy refers to the issue of matching the organization's resources to its environment, or making the fundamental decision of "What business are we in?" For analysis purposes, several aspects of strategy are important to identify. First, what is the core mission of the organization, or how has the organization defined its basic purpose or function within the larger system or environment? The core mission includes decisions about what markets the organization will serve, what products or services it will provide to those markets, and how it will compete in those markets. Second, strategy includes the specific supporting strategies (or tactics) the organization will employ or is employing to achieve its core mission. Third, it includes the specific performance or output objectives that have been established.

Strategy may be the most important single input for the organization. On one hand, strategic decisions implicitly determine the nature of the work the organization should be doing or the tasks it should perform. On the other hand, strategic decisions, and particularly decisions about objectives determine the system's desired outputs.

In summary, there are three basic inputs—environment, resources, and history—and a fourth derivative input, strategy, which determines how the organization responds to or deals with the basic inputs. Strategy is critical because it determines the work to be performed by the organization and it defines desired organizational outputs.

Outputs

Outputs are what the organization produces, how it performs, and how effective it is. There has been a lot of discussion about the components of an effective organization. For our purposes, however, it is possible to identify several key indicators of organizational output. First, we need to think about system output at different levels. In addition to the system's basic output—that is, the product—we need to think about other outputs that contribute to organizational performance, such as the functioning of groups or units within the organization or the functioning of individual organization members.

At the organizational level, three factors must be kept in mind when evaluating organizational performance: (1) goal attainment, or how well the organization meets its objectives (usually determined by strategy), (2) resource utilization, or how well the organization makes use of available resources (not just whether the organization meets its goals, but whether it realizes all of its potential performance and whether it achieves its goals by building resources or by "burning them up"), and (3) adaptability, or whether the organization continues to position itself in a favorable position vis-à-vis its environment—that is, whether it is capable of changing and adapting to environmental changes.

Obviously, the functioning of groups or units (departments, divisions, or other subunits within the organization) contribute to these organizational-level outputs. Organizational output is also influenced by individual behavior, and certain individual-level outputs (affective reactions such as satisfaction, stress, or experienced quality of working life) may be desired outputs in and of themselves.

The Organization as a Transformation Process

So far, we've defined the nature of inputs and outputs of the organizational system. This leads us to the transformation process. Given an environment, a set of resources, and history, "How do I take a strategy and implement it to produce effective performance in the organization, in the group/unit, and among individual employees?"

In our framework, the organization and its major component parts are the fundamental means for transforming energy and information from inputs into outputs. On this basis, we must determine the key components of the organization and the critical dynamic that shows how those components interact to perform the transformation function.

Organizational Components

There are many different ways of thinking about what makes up an organization. At this point in the development of a science of organizations, we probably do not know the one right or best way to describe the different components of an organization. The task is to find useful approaches for describing organizations, for simplifying complex phenomena, and for identifying patterns in what may at first blush seem to be random sets of activity. Our particular approach views organizations as composed of four major components: (1) the task, (2) the individuals, (3) the formal organizational arrangements, and (4) the informal organization. We will discuss each of these individually (see Figure 2 for overviews of these components).

The first component is the organization's *task*—that is, the basic or inherent work to be done by the organization and its subunits or the activity the organization is engaged in, particularly in light of its strategy. The emphasis is on the specific work activities or functions that need to be done and their inherent characteristics (as opposed to characteristics of the

Component	Task	Individual	Formal Organizational Arrangements	Informal Organization
Definition	The basic and inherent work to be done by the organization and its parts.	The characteristics of individuals in the organization.	The various structures, processes, methods, and so on that are formally created to get individuals to perform tasks.	The emerging arrangements, including structures, processes, relationships, and so forth.
Critical Features for Analysis	1. The types of skill and knowledge demands the work poses. 2. The types of rewards the work can provide. 3. The degree of uncertainty associated with the work, including such factors as interdependence, routineness, and so on. 4. The constraints on performance demands inherent in the work (given a strategy).	1. Knowledge and skills individuals have. 2. Individual needs and preferences. 3. Perceptions and expectancies. 4. Background factors.	1. Organization design, including grouping of functions, structure of subunits, and coordination and control mechanisms. 2. Job design. 3. Work environment. 4. Human resource management systems.	1. Leader behavior. 2. Intragroup relations. 3. Intergroup relations. 4. Informal working arrangements. 5. Communication and influence patterns.

FIGURE 2.   *Key Organizational Components*

work created by how the work is organized or structured in this particular organization at this particular time. Analysis of the task would include a description of the basic work flows and functions with attention to the characteristics of those work flows—for example, the knowledge or skills demanded by the work, the kinds of rewards provided by the work, the degree of uncertainty associated with the work, and the specific constraints inherent in the work (such as critical time demands, cost constraints, and so on). Since it's assumed that a primary (although not the only) reason for the organization's existence is to perform the task consistent with strategy, the task is the starting point for the analysis. As we will see, the assessment of the adequacy of other components depends to a large degree on an understanding of the nature of the tasks to be performed.

A second component of organizations involves the *individuals* who perform organizational tasks. The issue here is identifying the nature and characteristics of the organization's employees (or members). The most critical aspects to consider include the nature of individual knowledge and skills, the different needs or preferences that individuals have, the perceptions or expectancies that they develop, and other background factors (such as demographics) that may potentially influence individual behavior.

The third component is the *formal organizational arrangements.* These include the range of structures, processes, methods, procedures, and so forth that are explicitly and formally developed to get individuals to perform tasks consistent with organizational strategy. The broad term, organizational arrangements, encompasses a number of different factors. One factor is organization design—that is, the way jobs are grouped together into units, the internal structure of those units, and the coordination and control mechanisms used to link those units together. A second factor is the way jobs are designed within the context of organizational designs. A third factor is the work environment, which includes a number of factors that characterize the immediate environment in which work is done, such as the physical working environment, the available work resources, and so on. A final factor includes the organization's formal systems for attracting, placing, developing, and evaluating human resources.

Together, these factors create the set of formal organizational arrangements—that is, they are explicitly designed and specified, usually in writing.

The final component is the *informal organization.* Despite the set of formal organizational arrangements that exists in any organization, another set of arrangements tends to develop or emerge over a period of time. These arrangements are usually implicit and unwritten, but they influence a good deal of behavior. For lack of a better term, such arrangements are frequently referred to as the informal organization and they include the different structures, processes, and arrangements that emerge while the organization is operating. These arrangements sometimes complement formal organizational arrangements by providing structures to aid work where none exist. In other situations they may arise in reaction to the formal structure, to protect individuals from it. They may therefore either aid or hinder the organization's performance.

Because a number of aspects of the informal organization have a particularly critical effect on behavior, they need to be considered. The behavior of leaders (as opposed to the formal creation of leader positions) is an important feature of the informal organization, as are the patterns of relationships that develop both within and between groups. In addition, different types of informal working arrangements (including rules, procedures, methods, and so on) develop. Finally, there are the various communication and influence patterns that combine to create the informal organization design.

Organizations can therefore be thought of as a set of components—the task, the individuals, the organizational arrangements, and the informal organization. In any system, however, the critical question is not what the components are, but what the nature of their interaction is. This model raises the question: What are the dynamics of the relationships among the components? To deal with this issue, we must return to the concept of congruence or fit.

The Concept of Congruence

A relative degree of congruence, consistency, or "fit" exists between each pair of organizational inputs. The congruence between two components is defined as "the degree to which the needs, demands, goals, objectives, and/or structures of one component are consistent with the needs, demands, goals, objectives, and/or structures of another component."

Congruence, therefore, is a measure of how well pairs of components fit together. Consider, for example, two components—the task and the individual. At the simplest level, the task presents some demands on individuals who would perform it (that is, skill/knowledge demands). At the same time, the set of individuals available to do the tasks have certain characteristics (their levels of skill and knowledge). Obviously, if the individual's knowledge and skill match the knowledge and skill demanded by the task, performance will be more effective.

Obviously, too, the individual-task congruence relationship encompasses more factors than just knowledge and skill. Similarly, each congruence relationship in the model has its own specific characteristics. Research and theory can guide the assessment of fit in each relationship. For an overview of the critical elements of each congruence relationship, see Figure 3.

The Congruence Hypothesis

The aggregate model, or whole organization, displays a relatively high or low degree of system congruence in the same way that each pair of components has a high or low degree of congruence. The basic hypothesis of the model, which builds on this total state of congruence, is as follows: "Other things being equal, the greater the total degree of congruence or fit between the various components, the more effective will be the organization—effectiveness being defined as the degree to which actual organization outputs at individual, group, and organizational levels are similar to expected outputs, as specified by strategy."

The basic dynamic of congruence sees the organization as most effective when its pieces fit together. If we also consider strategy, this view expands to include the fit between the organization and its larger environment—that

Fit	Issues
Individual/Organization	How are individual needs met by the organizational arrangements? Do individuals hold clear or distorted perceptions of organizational structures? Is there a convergence of individual and organizational goals?
Individual/Task	How are individual needs met by the tasks? Do individuals have skills and abilities to meet task demands?
Individual/Informal organization	How are individual needs met by the informal organization? How does the informal organization make use of individual resources consistent with informal goals?
Task/Organization	Are organizational arrangements adequate to meet the demands of the task? Do organizational arrangements motivate behavior that's consistent with task demands?
Task/Informal organization	Does the informal organization structure facilitate task performance or not? Does it hinder or help meet the demands of the task.
Organization/Informal organization	Are the goals, rewards, and structures of the informal organization consistent with those of the formal organization?

FIGURE 3. Definitions of Fits

is, an organization is most effective when its strategy is consistent with its environment (in light of organizational resources and history) and when the organizational components are congruent with the tasks necessary to implement that strategy.

One important implication of the congruence hypothesis is that organizational problem analysis (or diagnosis) involves description of the system, identification of problems, and analysis of fits to determine the causes of problems. The model also implies that different configurations of the key components can be used to gain outputs (consistent with the systems characteristic of equifinality). Therefore the question is not how to find the "one best way" of managing, but how to find effective combinations of components that will lead to congruent fits among them.

The process of diagnosing fits and identifying combinations of components to produce congruence is not necessarily intuitive. A number of situations that lead to congruence have been defined in the research literature. Thus in many cases fit is something that can be defined, measured, and even quantified; there is, in other words, an empirical and theoretical basis for assessing fit. The theory provides considerable guidance about what leads to congruent relationships (although in some areas the research is more definitive and helpful than others). The implication is that the manager who wants to diagnose behavior must become familiar with critical aspects of relevant organizational behavior models or theories so that he or she can evaluate the nature of fits in a particular system.

The congruence model provides a general organizing framework. The organizational analyst will need other, more specific "submodels" to define high and low congruence. Examples of such submodels that might be used in the context of this general diagnostic model include the following: (1) the job characteristics model to assess and explain the fit between in-

dividuals and tasks as well as the fit between individuals and organizational arrangements (job design), (2) expectancy theory models of motivation to explain the fit between individuals and the other three components, (3) the information processing model of organizational design to explain the task-formal organization and task-informal organization fits, or (4) an organizational climate model to explain the fit between the informal organization and the other components. These models and theories are listed as illustrations of how more specific models can be used in the context of the general model. Obviously, those mentioned above are just a sampling of possible tools that could be used.

In summary, then, we have described a general model for the analysis of organizations (see Figure 4). The organization is seen as a system or transformation process that takes inputs and transforms them into outputs—a process that is composed of four basic components. The critical dynamic is the fit or congruence among the components. We now turn our attention to the pragmatic question of how to use this model for analyzing organizational problems.

A PROCESS FOR ORGANIZATIONAL PROBLEM ANALYSIS

The conditions that face organizations frequently change; consequently, managers are required to continually engage in problem-identification and problem-solving activities. Therefore, managers must gather data on organizational performance, compare the data with desired performance levels, identify the causes of problems, develop and choose action plans and, finally, implement and evaluate these action plans. These phases can be viewed as a generic problem-solving process. For long-

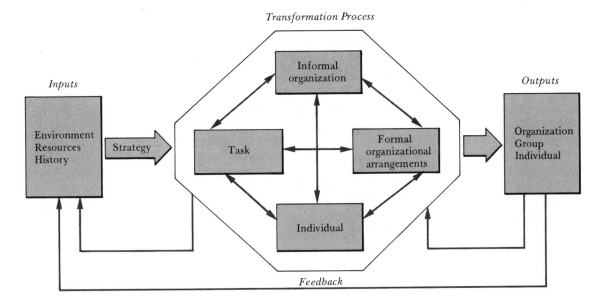

FIGURE 4. A Congruence Model for Organization Analysis

term organizational viability, some type of problem-solving process must operate—and operate continuously.

Experience with using the congruence model for organizations for problem analysis in actual organizational settings has led to the development of an approach to using the model that's based on these generic problem-solving processes (see Figure 5). In this section, we will "walk through" this process, describing each step in the process and discussing how the model can be used at each stage. Here are the steps in the problem-analysis process:

1. *Identify symptoms:* In any situation initial information (symptomatic data) may indicate that there are problems, but not what the problems are or what the causes are. Symptomatic data are important because the symptoms of problems may indicate where to look for more complete data.

2. *Specify inputs:* Once the symptoms are identified, the starting point for analysis is to identify the system and the environment in which it functions. This means collecting data about the nature of environment, the type of resources the organization has, and the critical aspects of its history. Input analysis also involves identifying the overall strategy of the organization—that is, its core mission, supporting strategies, and objectives.

3. *Identify outputs:* The third step is an analysis of the organization's outputs at the individual, group, and organizational levels. Output analysis actually involves two elements: (1) defining the desired or planned output through an analysis of strategy that explicitly or implicitly defines what the organization wants to achieve in terms of output or performance indicators, and (2) collecting data that indicate the type of output the organization is actually achieving.

4. *Identify problems:* Symptoms may indicate problems—in this case, significant difference between desired or planned output and actual output. Such problems might be discrepancies (actual vs. expected) in or-

Step	Explanation
1. Identify symptoms.	List data indicating possible existence of problems.
2. Specify inputs.	Identify the system. Determine nature of environment, resources, and history. Identify critical aspects of strategy.
3. Identify outputs.	Identify data that define the nature of outputs at various levels (individual, group/unit, organizational). This should include desired outputs (from strategy), and actual outputs being obtained.
4. Identify problems.	Identify areas where there are significant and meaningful differences between desired and actual outputs. To the extent possible, identify penalties; that is, specific costs (actual and opportunity costs) associated with each problem.
5. Describe components of the organization.	Describe basic nature of each of the four components with emphasis on their critical features.
6. Assess congruence (fits).	Conduct analysis to determine relative congruence among components (draw on submodels as needed).
7. Generate and identify causes.	Analyze to associate fit with specific problems.
8. Identify action steps.	Indicate the possible actions to deal with problem causes.

FIGURE 5.   Basic Problem Analysis Steps Using the Congruence Model

ganizational performance, group functioning, individual behavior, or affective reactions. These data tell us what problems exist, but they still don't tell us the causes. (Note: Where data are available, it's frequently also useful to identify the costs associated with the problems or the *penalties* the organization incurs by not fixing the problem. Penalties might be actual costs—increased expenses, and so on—or opportunity costs, such as revenue lost because of the problem.)

5. *Describe organizational components:* At this step the analysis to determine the causes of problems begins. Data are collected about the nature of each of the four major organizational components, including information about the component and its critical features in this organization.

6. *Assess congruence (fits):* Using the data collected in step 5 as well as applicable submodels or theories, an assessment is made of the positive or negative fit between each pair of components.

7. *Generate hypotheses about problem causes:* Once the components are described and their congruence assessed, the next step is to link together the congruence analysis with the problem identification (step 4). After analyzing to determine which are the poor fits that seem to be associated with, or account for, the output problems that have been identified, the patterns of congruence and incongruence that appear to cause the patterns of problems are determined.

8. *Identify action steps:* The final step in problem analysis is to identify possible action steps. These steps might range from spe-

cific changes to deal with relatively obvious problem causes to a more extensive data collection designed to test hypotheses about relatively more complex problems and causes.

In addition to these eight steps, some further steps need to be kept in mind. After possible actions are identified, problem solving involves predicting the consequence of various actions, choosing the course of action, and implementing and evaluating the impact of the chosen course of action. It is, of course, important to have a general diagnostic framework to monitor the effects of various courses of action.

The congruence model and this problem-analysis process outline are tools for structuring and dealing with the complex reality of organizations. Given the indeterminate nature of social systems, there is no one best way of handling a particular situation. The model and the process could, however, help the manager in making a number of decisions and in evaluating the consequences of those decisions. If these tools have merit, it is up to the manager to use them along with his or her intuitive sense (based on experience) to make the appropriate set of diagnostic, evaluative, and action decisions.

FUTURE DIRECTIONS

The model we've presented here reflects a particular way of thinking about organizations. If that perspective is significant, the model might be used as a tool for handling more complex problems or for structuring more complex situations. Some directions for further thought, research, and theory development could include these:

1. *Organizational change.* The issue of organizational change has received a good deal of attention from both managers and academics. The question is how to effectively implement organizational change. The problem seems to center on the lack of a general model of organizational change. It is hard to think about a general model of organizational change without a general model of organizations. The congruence perspective outlined here may provide some guidance and direction toward the development of a more integrated perspective on the processes of organizational change. Initial work in applying the congruence model to the change issue is encouraging.

2. *Organizational development over time.* There has been a growing realization that organizations grow and develop over time, and that they face different types of crises, evolve through different stages, and develop along some predictable lines. A model of organizations such as the one presented here might be a tool for developing typology of growth patterns by indicating the different configurations of tasks, individuals, organizational arrangements, and informal organizations that might be most appropriate for organizations in different environments and at different stages of development.

3. *Organizational pathology.* Organizational problem solving ultimately requires some sense of the types of problems that may be encountered and the kinds of patterns of causes one might expect. It is reasonable to assume that most problems encountered by organizations are not wholly unique, but are predictable. The often expressed view that "our problems are unique" reflects in part the lack of a framework of organizational pathology. The question is: Are there basic "illnesses" that organizations suffer? Can a framework of organizational pathology, similar to the physician's framework of medical pathology, be developed? The lack of a pathology framework, in turn,

reflects the lack of a basic functional model of organizations. Again, development of a congruence perspective might provide a common language to use for the identification of general pathological patterns of organizational functioning.

4. *Organizational solution types.* Closely linked to the problem of pathology is the problem of treatment, intervention, or solutions to organizational problems. Again, there's a lack of a general framework in which to consider the nature of organizational interventions. In this case, too, the congruence model might be a means for conceptualizing and ultimately describing the different intervention options available in response to problems.

SUMMARY

This article has presented a general approach for thinking about organizational functioning and a process for using a model to analyze organizational problems. This particular model is only one way of thinking about organizations; it's clearly not the only model, nor can we claim it's definitively the best model. It is one tool, however, that may be useful for structuring the complexity of organizational life and helping managers create, maintain, and develop effective organizations.

SELECTED BIBLIOGRAPHY

For a comprehensive review and synthesis of research in organizational behavior, see Marvin Dunnette's *Handbook of Industrial and Organizational Psychology* (Rand-McNally, 1976). Daniel Katz and Robert Kahn's seminal work on organizations as systems, *The Social Psychology of Organizations* (John Wiley & Sons, 1966), has been revised, updated, and ex-

tended in their 1978 edition. See their new book for an extensive discussion of organizations as open systems and for a unique synthesis of the literature in terms of systems ideas.

For a broad analysis of organizational behavior, see David Nadler, J. Richard Hackman, and Edward E. Lawler's *Managing Organizational Behavior* (Little, Brown, 1979) and see Charles Hofer and Daniel Schendel's *Strategy Formulation: Analytical Concepts* (West, 1978) for a discussion of strategy.

For an extensive discussion of output and effectiveness, see Paul Goodman and Johannes Pennings's *New Perspectives on Organizational Effectiveness* (Jossey-Bass, 1977) and Andrew Van de Ven and Diane Ferry's *Organizational Assessment* (Wiley Interscience, 1980).

For more detail on organizational arrangements, see Jay R. Galbraith's *Designing Complex Organizations* (Addison-Wesley, 1973); on job design and motivation, see J. Richard Hackman and Greg Oldham's *Work Redesign* (Addison-Wesley, 1979); and on informal organizations, see Michael Tushman's "A Political Approach to Organizations: A Review and Rationale" (*Academy of Management Review*, April 1977) and Jeffrey Pfeffer's new book, *Power and Politics in Organizations* (Pitman Publishing, Inc., 1980).

Submodels corresponding to the various components of our congruence model would include: J. Richard Hackman and Greg Oldham's job design model; Victor Vroom and Edward Lawler's work on expectancy theory of motivation and decision making—see Vroom's *Work and Motivation* (Wiley, 1964) and Lawler's *Motivation in Work Organizations* (Wadsworth Publishing Co., 1973); Jay R. Galbraith, Michael Tushman, and David Nadler's work on information processing models of organizational design; and George Litwin and Robert Stringer's work on organization climate—see Litwin and Stringer's *Motivation and Organizational Climate* (Harvard University Graduate School of Business Administration, 1968).

David Nadler's "An Integrative Theory of Organizational Change," to appear in the *Journal of Applied Behavioral Science* in 1981, uses the congruence model to think about the general problems of organizational change and dynamics. Several distinct levers for change are developed and discussed. Other pertinent books of interest include: Jay R. Galbraith's *Organization Design* (Addison-Wesley, 1979),

Jay R. Galbraith and Daniel A. Nathanson's *Strategy Implementation: The Role of Structure and Process* (West, 1978), George C. Homans's *The Human Group* (Harcourt Brace Jovanovich, Inc., 1950), Paul R. Lawrence and Jay W. Lorsch's *Developing Organizations: Diagnosis and Action* (Addison-Wesley, 1969), Harold J. Leavitt's "Applied Organization Change in Industry" in J. G. March's (ed.) *Handbook of Organizations* (Rand-McNally, 1965), Harry Levinson's *Organizational Diagnosis* (Harvard University Press, 1972), Harry Levinson's *Psychological Man* (Levinson Institute, 1976), Jay W. Lorsch and Alan Sheldon's "The Individual in the Organization: A Systems View" in J. W. Lorsch and P. R. Lawrence's (eds.) *Managing Group and Intergroup Relations* (Irwin-Dorsey, 1972), David A. Nadler and Noel M. Tichy's "The Limitations of Traditional Intervention Technology in Health Care Organizations" in N. Margulies and J. A. Adams's (eds.) *Organization Development in Health Care Organizations* (Addison-Wesley, 1980), Edgar H. Schein's *Organizational Psychology* (Prentice-Hall, 1970), and James A. Seiler's *Systems Analysis in Organizational Behavior* (Irwin-Dorsey, 1967).

3

BASIC ORGANIZATIONAL PROCESSES

Gunfire at Sea:
A Case Study of Innovation

Elting Morison

In the early days of the last war when armaments of all kinds were in short supply, the British, I am told, made use of a venerable field piece that had come down to them from previous generations. The honorable past of this light artillery stretched back, in fact, to the Boer War. In the days of uncertainty after the fall of France, these guns, hitched to trucks, served as useful mobile units in the coast defense. But it was felt that the rapidity of fire could be increased. A time-motion expert was, therefore, called in to suggest ways to simplify the firing procedures. He watched one of the gun crews of five men at practice in the field for some time. Puzzled by certain aspects of the procedures, he took some slow-motion pictures of the soldiers performing the loading, aiming, and firing routines.

When he ran these pictures over once or twice, he noticed something that appeared odd to him. A moment before the firing, two members of the gun crew ceased all activity and came to attention for a three-second interval extending throughout the discharge of the gun. He summoned an old colonel of artillery, showed him the pictures and pointed out this strange behavior. What, he asked the colonel, did it mean. The colonel, too, was puzzled. He asked to see the pictures again. "Ah," he said when the performance was over, "I have it. They are holding the horses."

This story, true or not, and I am told it is true, suggests nicely the pain with which the human being accommodates himself to changing conditions. The tendency is apparently involuntary and immediate to protect oneself against the shock of change by continuing in

Reprinted by permission from *Men, Machines and Modern Times* (Cambridge: MIT Press, 1966). This essay was delivered as one of three lectures at the California Institute of Technology in 1950. It has been reprinted in various truncated forms a good many times since.

the presence of altered situations the familiar habits, however incongruous, of the past.

Yet, if human beings are attached to the known, to the realm of things as they are, they also, regrettably for their peace of mind, are incessantly attracted to the unknown and things as they might be. As Ecclesiastes glumly pointed out, men persist in disordering their settled ways and beliefs by seeking out many inventions.

The point is obvious. Change has always been a constant in human affairs; today, indeed, it is one of the determining characteristics of our civilization. In our relatively shapeless social organization, the shifts from station to station are fast and easy. More important for our immediate purpose, America is fundamentally an industrial society in a time of tremendous technological development. We are thus constantly presented with new devices or new forms of power that in their refinement and extension continually bombard the fixed structure of our habits of mind and behavior. Under such conditions, our salvation, or at least our peace of mind, appears to depend upon how successfully we can in the future become what has been called in an excellent phrase a completely "adaptive society."

It is interesting, in view of all this, that so little investigation, relatively, has been made of the process of change and human responses to it. Recently, psychologists, sociologists, cultural anthropologists, and economists have addressed themselves to the subject with suggestive results. But we are still far from a full understanding of the process and still further from knowing how we can set about simplifying and assisting an individual's or a group's accommodation to new machines or new ideas.

With these things in mind, I thought it might be interesting and perhaps useful to examine historically a changing situation within a society; to see if from this examination we can discover how the new machines or ideas that introduced the changing situation developed;

to see who introduces them, who resists them, what points of friction or tension in the social structure are produced by the innovation, and perhaps why they are produced and what, if anything, may be done about it. For this case study the introduction of continuous-aim firing in the United States Navy has been selected. The system, first devised by an English officer in 1898, was introduced in our Navy in the years 1900 to 1902.

I have chosen to study this episode for two reasons. First, a navy is not unlike a society that has been placed under laboratory conditions. Its dimensions are severely limited; it is beautifully ordered and articulated; it is relatively isolated from random influences. For these reasons the impact of change can be clearly discerned, the resulting dislocations in the structure easily discovered and marked out. In the second place, the development of continuous-aim firing rests upon mechanical devices. It therefore presents for study a concrete, durable situation. It is not like many other innovating reagents—a Manichean heresy, or Marxism, or the views of Sigmund Freud—that can be shoved and hauled out of shape by contending forces or conflicting prejudices. At all times we know exactly what continuous-aim firing really is. It will be well now to describe, as briefly as possible, what it really is. This will involve a short investigation of certain technical matters. I will not apologize, as I have been told I ought to do, for this preoccupation with how a naval gun is fired. For one thing, all that follows is understandable only if one understands how the gun goes off. For another thing, a knowledge of the underlying physical considerations may give a kind of elegance to the succeeding investigation of social implications. And now to the gun and the gunfire.

The governing fact in gunfire at sea is that the gun is mounted on an unstable platform, a rolling ship. This constant motion obviously complicates the problem of holding a steady aim. Before 1898 this problem was solved in the following elementary fashion. A gun pointer estimated the range of the target, ordinarily in the nineties about 1600 yards. He then raised the gun barrel to give the gun the elevation to carry the shell to the target at the estimated range. This elevating process was accomplished by turning a small wheel on the gun mount that operated the elevating gears. With the gun thus fixed for range, the gun pointer peered through open sights, not unlike those on a small rifle, and waited until the roll of the ship brought the sights on the target. He then pressed the firing button that discharged the gun. There were by 1898, on some naval guns, telescope sights which naturally greatly enlarged the image of the target for the gun pointer. But these sights were rarely used by gun pointers. They were lashed securely to the gun barrel, and, recoiling with the barrel, jammed back against the unwary pointer's eye. Therefore, when used at all, they were used only to take an initial sight for purposes of estimating the range before the gun was fired.

Notice now two things about the process. First of all, the rapidity of fire was controlled by the rolling period of the ship. Pointers had to wait for the one moment in the roll when the sights were brought on the target. Notice also this: There is in every pointer what is called a "firing interval"—that is, the time lag between his impulse to fire the gun and the translation of this impulse into the act of pressing the firing button. A pointer, because of this reaction time, could not wait to fire the gun until the exact moment when the roll of the ship brought the sights onto the target; he had to will to fire a little before, while the sights were off the target. Since the firing interval was an individual matter, varying obviously from man to man, each pointer had to estimate from long practice his own interval and compensate for it accordingly.

These things, together with others we need not here investigate, conspired to make gunfire at sea relatively uncertain and ineffective. The pointer, on a moving platform, estimating range and firing interval, shooting while his sight was off the target, became in a sense an individual artist.

In 1898, many of the uncertainties were removed from the process and the position of the gun pointer radically altered by the introduction of continuous-aim firing. The major change was that which enabled the gun pointer to keep his sight and gun barrel on the target throughout the roll of the ship. This was accomplished by altering the gear ratio in the elevating gear to permit a pointer to compensate for the roll of the vessel by rapidly elevating and depressing the gun. From this change another followed. With the possibility of maintaining the gun always on the target, the desirability of improved sights became immediately apparent. The advantages of the telescope sight as opposed to the open sight were for the first time fully realized. But the existing telescope sight, it will be recalled, moved with the recoil of the gun and jammed back against the eye of the gunner. To correct this, the sight was mounted on a sleeve that permitted the gun barrel to recoil through it without moving the telescope.

These two improvements in elevating gear and sighting eliminated the major uncertainties in gunfire at sea and greatly increased the possibilities of both accurate and rapid fire.

You must take my word for it, since the time allowed is small, that this changed naval gunnery from an art to a science, and that gunnery accuracy in the British and our Navy increased, as one student said, 3000% in six years. This does not mean much except to suggest a great increase in accuracy. The following comparative figures may mean a little more. In 1899 five ships of the North Atlantic Squadron fired five minutes each at a lightship hulk at the conventional range of 1600 yards. After twenty-five minutes of banging away, two hits had been made on the sails of the elderly vessel. Six years later one naval gunner made fifteen hits in one minute at a target 75 by 25 feet at the same range—1600 yards; half of them hit in a bull's-eye 50 inches square.

Now with the instruments (the gun, elevating gear, and telescope), the method, and the results of continuous-aim firing in mind, let us turn to the subject of major interest: how was the idea, obviously so simple an idea, of continuous-aim firing developed, who introduced it into the United States Navy, and what was its reception?

The idea was the product of the fertile mind of the English officer Admiral Sir Percy Scott. He arrived at it in this way while, in 1898, he was the captain of H.M.S. *Scylla*. For the previous two or three years he had given much thought independently and almost alone in the British Navy to means of improving gunnery. One rough day, when the ship, at target practice, was pitching and rolling violently, he walked up and down the gun deck watching his gun crews. Because of the heavy weather, they were making very bad scores. Scott noticed, however, that one pointer was appreciably more accurate than the rest. He watched this man with care, and saw, after a time, that he was unconsciously working his elevating gear back and forth in a partially successful effort to compensate for the roll of the vessel. It flashed through Scott's mind at that moment that here was the sovereign remedy for the problem of inaccurate fire. What one man could do partially and unconsciously perhaps all men could be trained to do consciously and completely.

Acting on this assumption, he did three things. First, in all the guns of the *Scylla*, he changed the gear ratio in the elevating gear, previously used only to set the gun in fixed position for range, so that a gunner could easily elevate and depress the gun to follow a target throughout the roll. Second, he rerigged his telescopes so that they would not be influenced by the recoil of the gun. Third, he rigged a small target at the mouth of the gun, which was moved up and down by a crank to simulate a moving target. By following this target as it moved and firing at it with a subcaliber rifle rigged in the breech of the gun, the pointer could practice every day. Thus equipped, the ship became a training ground for gunners. Where before the good pointer was an individual artist, pointers now became trained technicians, fairly uniform in their capacity to shoot. The effect was immediately felt. Within a year the *Scylla* established records that were remarkable.

At this point I should like to stop a minute to notice several things directly related to, and involved in, the process of innovation. To begin with, the personality of the innovator. I wish there were time to say a good deal about Admiral Sir Percy Scott. He was a wonderful man. Three small bits of evidence must here suffice, however. First, he had a certain mechanical ingenuity. Second, his personal life was shot through with frustration and bitterness. There was a divorce and a quarrel with that ambitious officer Lord Charles Beresford, the sounds of which, Scott liked to recall, penetrated to the last outposts of empire. Finally, he possessed, like Swift, a savage indignation directed ordinarily at the inelastic intelligence of all constituted authority, especially the British Admiralty.

There are other points worth mention here. Notice first that Scott was not responsible for the invention of the basic instruments that made the reform in gunnery possible. This reform rested upon the gun itself, which as a rifle had been in existence on ships for at least forty years; the elevating gear, which had been, in the form Scott found it, a part of the rifled gun from the beginning; and the telescope sight, which had been on shipboard at least eight years. Scott's contribution was to bring these three elements appropriately modified into a combination that made continuous-aim firing possible for the first time. Notice also that he was allowed to bring these elements into combination by accident, by watching the unconscious action of a gun pointer endeavoring through the operation of his elevating gear to correct partially for the roll of his vessel. Scott, as we have seen, had been interested in gunnery; he had thought about ways to increase accuracy by practice and improvement of existing machinery; but able as he was, he had not been able to produce on his own initiative and by his own thinking the essential idea and modify instruments to fit his purpose. Notice here, finally, the intricate interaction of chance, the intellectual climate, and Scott's mind. Fortune (in this case, the unaware gun pointer) indeed favors the prepared mind, but even fortune and the prepared mind need a favorable environment before they can conspire to produce sudden change. No intelligence can proceed very far above the threshold of existing data or the binding combinations of existing data.

All these elements that enter into what may be called "original thinking" interest me as a teacher. Deeply rooted in the pedagogical mind often enough is a sterile infatuation with "inert ideas"; there is thus always present in the profession the tendency to be diverted from the *process* by which these ideas, or indeed any ideas, are really produced. I well remember with what contempt a class of mine which was reading Leonardo da Vinci's *Notebooks* dismissed the author because he appeared to know no more mechanics than, as one wit in the class observed, a Vermont Republican farmer of the present day. This is perhaps the expected result produced by a method of instruction that too frequently implies that the great generalizations were the result, on the one hand, of chance—an apple falling in an orchard or a teapot boiling on the hearth—or, on the other hand, of some towering intelligence proceeding in isolation inexorably toward some prefigured idea, like evolution, for example.

This process by which new concepts appear, the interaction of fortune, intellectual climate, and the prepared imaginative mind, is an interesting subject for examination offered by any case study of innovation. It was a subject as Dr. Walter Cannon pointed out, that momentarily engaged the attention of Horace Walpole, whose lissome intelligence glided over the surface of so many ideas. In reflecting upon the part played by chance in the development of new concepts, he recalled the story of the three princes of Serendip who set out to find some interesting object on a journey through their realm. They did not find the particular object of their search, but along the way they discovered many new things simply because they were looking for *something.* Walpole believed this intellectual method ought to be given a name, in honor of the founders, serendipity; and serendipity certainly exerts a considerable influence in what we call original thinking. There is an element of serendipity, for example, in Scott's chance discovery of continuous-aim firing in that he was, and had been, looking for some means to improve his target practice and stumbled upon a solution by observation that had never entered his head.

Serendipity, while recognizing the prepared mind, does tend to emphasize the role of chance in intellectual discovery. Its effect may be balanced by an anecdote that suggests the contribution of the adequately prepared mind. There has recently been much posthaste and romage in the land over the question of whether there really was a Renaissance. A scholar has recently argued in print that since the Middle Ages actually possessed many of the instruments and pieces of equipment associated with the Renaissance, the Renaissance could be said to exist as a defined period only in the mind of the historians such as Burckhardt. This view was entertainingly rebutted by the historian of art Panofsky, who pointed out that although Robert Grosseteste indeed did have a very rudimentary telescope, he used it to examine stalks of grain in a field down the street. Galileo, a Renaissance intelligence, pointed his telescope at the sky.

Here Panofsky is only saying in a provocative way that change and intellectual advance are the products of well-trained and well-stored inquisitive minds, minds that relieve us of "the terrible burden of inert ideas by throwing them into a new combination." Educators, nimble in the task of pouring the old wine of our heritage into the empty vessels that appear before them, might give thought to how to develop such independent, inquisitive minds.

But I have been off on a private venture of my own. Now to return to the story, the introduction of continuous-aim firing. In 1900 Percy Scott went out to the China Station as commanding officer of H.M.S. *Terrible.* In that ship he continued his training methods and his spectacular successes in naval gunnery. On the China Station he met up with an American junior officer, William S. Sims. Sims had little of the mechanical ingenuity of Percy Scott, but the two were drawn together by temperamental similarities that are worth noticing here. Sims had the same intolerance for what is called spit and polish and the same contempt for bureaucratic inertia as his British brother officer. He had for some years been concerned, as had Scott, with what he took to be the inefficiency of his own Navy. Just before he met Scott, for example, he had shipped out to China in the brand new pride of the fleet, the battleship *Kentucky.* After careful investigation and reflection he had informed his superiors in Washington that she was "not a battleship at all—but a crime against the white race." The spirit with which he pushed forward his efforts to reform the naval service can best be stated in his own words to a brother officer: "I am perfectly willing that those holding views differing from mine should continue to live, but with every fibre of my being I loathe indirection and shiftiness, and where it occurs in high place, and is used to save face at the expense of the vital interests of our great service (in which silly people place such a childlike trust), I want that man's blood and I will have it no matter what it costs me personally."

From Scott in 1900 Sims learned all there was to know about continuous-aim firing. He modified, with the Englishman's active assistance, the gear on his own ship and tried out the new system. After a few months' training, his experimental batteries began making remarkable records at target practice. Sure of the usefulness of his gunnery methods, Sims then turned to the task of educating the Navy at large. In thirteen great official reports he documented the case for continuous-aim firing, supporting his arguments at every turn with a mass of factual data. Over a period of two years, he reiterated three principal points: first, he continually cited the records established by Scott's ships, the *Scylla* and the *Terrible,* and supported these with the accumulating data from his own tests on an American ship; second, he described the mechanisms used and the training procedures instituted by Scott and himself to obtain these records; third, he explained that our own mechanisms were not generally adequate without modification to meet the demands placed on them by continuous-aim firing. Our elevating gear, useful to raise or lower a gun slowly to fix it in position for the proper range, did not always work easily and rapidly enough to enable a gunner to follow a target with his gun throughout the roll of the ship. Sims also explained that such few telescope sights as there were on board our ships were useless. Their cross wires were so thick or coarse they obscured the target, and the sights had been attached to the gun in such a way that the recoil system of the gun plunged the eyepiece against the eye of the gun pointer.

This was the substance not only of the first but of all the succeeding reports written on the subject of gunnery from the China Station. It will be interesting to see what response these met with in Washington. The response falls roughly into three easily identifiable stages.

First stage: At first, there was no response. Sims had directed his comments to the Bureau of Ordnance and the Bureau of Navigation; in both bureaus there was dead silence. The thing—claims and records of continuous-aim firing—was not credible. The reports were simply filed away and forgotten. Some indeed, it was later discovered to Sims's delight, were half-eaten-away by cockroaches.

Second stage: It is never pleasant for any man's best work to be left unnoticed by superiors, and it was an unpleasantness that Sims suffered extremely ill. In his later reports, beside the accumulating data he used to clinch his argument, he changed his tone. He used deliberately shocking language because, as he said, "They were furious at my first papers and stowed them away. I therefore made up my mind I would give these later papers such a form that they would be dangerous documents to leave neglected in the files." To another friend he added, "I want scalps or nothing and if I can't have 'em I won't play."

Besides altering his tone, he took another step to be sure his views would receive attention. He sent copies of his reports to other officers in the fleet. Aware as a result that Sims's gunnery claims were being circulated and talked about, the men in Washington were then stirred to action. They responded, notably through the Chief of the Bureau of Ordnance, who had general charge of the equipment used in gunnery practice, as follows: (1) our equipment was in general as good as the British; (2) since our equipment was as good, the trouble must be with the men, but the gun pointer and the training of gun pointers were the responsibility of the officers on the ships; and most significant (3) continuous-aim firing was impossible. Experiments had revealed that five men at work on the elevating gear of a six-inch gun could not produce the power necessary to compensate for a roll of five degrees in ten seconds. These experiments and calculations demonstrated beyond peradventure or doubt that Scott's system of gunfire was not possible.

This was the second stage—the attempt to meet Sims's claims by logical, rational rebuttal. Only one difficulty is discoverable in these arguments; they were wrong at important points. To begin with, while there was little difference between the standard British equipment and the standard American equipment, the instruments on Scott's two ships, the *Scylla* and the *Terrible,* were far better than the standard equipment on our ships. Second, all the men could not be trained in continuous-aim firing until equipment was improved throughout the fleet. Third, the experiments with the elevating gear had been ingeniously contrived at the Washington Navy Yard—on solid ground. It had, therefore, been possible to dispense in the Bureau of Ordnance calculation with Newton's first law of motion, which naturally operated at sea to assist the gunner in elevating or depressing a gun mounted on a moving ship. Another difficulty was of course that continuous-aim firing was in use on Scott's and some of our own ships at the time the Chief of the Bureau of Ordnance was writing that it was a mathematical impossibility. In every way I find this second stage, the apparent resort to reason, the most entertaining and instructive in our investigation of the responses to innovation.

Third stage: The rational period in the counterpoint between Sims and the Washington men was soon passed. It was followed by the third stage, that of name-calling—the *argumentum ad hominem*. Sims, of course, by the high temperature he was running and by his calculated over-statement, invited this. He was told in official endorsements on his reports that there were others quite as sincere and loyal as he and far less difficult; he was dismissed as a crackbrained egotist; he was called a deliberate falsifier of evidence.

The rising opposition and the character of the opposition were not calculated to discourage further efforts by Sims. It convinced him that he was being attacked by shifty, dishonest men who were the victims, as he said, of insufferable conceit and ignorance. He made

171

up his mind, therefore, that he was prepared to go to any extent to obtain the "scalps" and the "blood" he was after. Accordingly, he, a lieutenant, took the extraordinary step of writing the President of the United States, Theodore Roosevelt, to inform him of the remarkable records of Scott's ships, of the inadequacy of our own gunnery routines and records, and of the refusal of the Navy Department to act. Roosevelt, who always liked to respond to such appeals when he conveniently could, brought Sims back from China late in 1902 and installed him as Inspector of Target Practice, a post the naval officer held throughout the remaining six years of the Administration. And when he left, after many spirited encounters we cannot here investigate, he was universally acclaimed as "the man who taught us how to shoot."

With this sequence of events (the chronological account of the innovation of continuous-aim firing) in mind, it is possible now to examine the evidence to see what light it may throw on our present interest: the origins of and responses to change in a society.

First, the origins. We have already analyzed briefly the origins of the idea. We have seen how Scott arrived at his notion. We must now ask ourselves, I think, why Sims so actively sought, almost alone among his brother officers, to introduce the idea into his service. It is particularly interesting here to notice again that neither Scott nor Sims invented the instruments on which the innovation rested. They did not urge their proposal, as might be expected, because of pride in the instruments of their own design. The telescope sight had first been placed on shipboard in 1892 by Bradley Fiske, an officer of great inventive capacity. In that year Fiske had even sketched out on paper the vague possibility of continuous-aim firing, but his sight was condemned by his commanding officer, Robley D. Evans, as of no use. In 1892 no one but Fiske in the Navy knew what to do with a telescope sight any more than Grosseteste had known in his time what to do with

a telescope. And Fiske, instead of fighting for his telescope, turned his attention to a range finder. But six years later Sims, following the tracks of his brother officer, took over and became the engineer of the revolution. I would suggest, with some reservations, this explanation: Fiske, as an inventor, took his pleasure in great part from the design of the device. He lacked not so much the energy as the overriding sense of social necessity that would have enabled him to *force* revolutionary ideas on the service. Sims possessed this sense. In Fiske, who showed rare courage and integrity in other professional matters not intimately connected with the introduction of new weapons of his own design, we may here find the familiar plight of the engineer who often enough must watch the products of his ingenuity organized and promoted by other men. These other promotional men when they appear in the world of commerce are called entrepreneurs. In the world of ideas they are still entrepreneurs. Sims was one, a middle-aged man caught in the periphery (as a lieutenant) of the intricate webbing of a precisely organized society. Rank, the exact definition and limitation of a man's capacity at any given moment in his career, prevented Sims from discharging all his exploding energies into the purely routine channels of the peacetime Navy. At the height of his powers he was a junior officer standing watches on a ship cruising aimlessly in friendly foreign waters. The remarkable changes in systems of gunfire to which Scott introduced him gave him the opportunity to expend his energies quite legitimately against the encrusted hierarchy of his society. He was moved, it seems to me, in part by his genuine desire to improve his own profession but also in part by rebellion against tedium, against inefficiency from on high, and against the artificial limitations placed on his actions by the social structure, in his case, junior rank.

Now having briefly investigated the origins of the change, let us examine the reasons for what must be considered the weird response we have observed to this proposed change. Why this deeply rooted, aggressive, persistent hostility from Washington that was only broken up by the interference of Theodore Roosevelt? Here was a reform that greatly and demonstrably increased the fighting effectiveness of a service that maintains itself almost exclusively to fight. Why then this refusal to accept so carefully documented a case, a case proved incontestably by records and experience? Why should virtually all the rulers of a society so resolutely seek to reject a change that so markedly improved its chances for survival in any contest with competing societies? There are the obvious reasons that will occur to all of you—the source of the proposed reform was an obscure, junior officer 8000 miles away; he was, and this is a significant factor, criticizing gear and machinery designed by the very men in the bureaus to whom he was sending his criticisms. And furthermore, Sims was seeking to introduce what he claimed were improvements in a field where improvements appeared unnecessary. Superiority in war, as in other things, is a relative matter, and the Spanish-American War had been won by the old system of gunnery. Therefore, it was superior even though of the 9500 shots fired at various but close ranges, only 121 had found their mark.

These are the more obvious, and I think secondary or supporting, sources of opposition to Sims's proposed reforms. A less obvious cause appears by far the most important one. It has to do with the fact that the Navy is not only an armed force; it is a society. Men spend their whole lives in it and tend to find the definition of their whole being within it. In the forty years following the Civil War, this society had been forced to accommodate itself to a series of technological changes—the steam turbine, the electric motor, the rifled shell of great explosive power, case-hardened steel armor, and all the rest of it. These changes wrought extraordinary changes in ship design, and, therefore, in the concepts of how ships were to be used; that is, in fleet tactics, and even in naval strategy. The Navy of this period is a paradise for the historian or sociologist in search of evidence bearing on a society's responses to change.

To these numerous innovations, producing as they did a spreading disorder throughout a service with heavy commitments to formal organization, the Navy responded with grudging pain. For example, sails were continued on our first-line ships long after they ceased to serve a useful purpose mechanically, but like the holding of the horses that no longer hauled the British field pieces, they assisted officers over the imposing hurdles of change. To a man raised in sail, a sail on an armored cruiser propelled through the water at 14 knots by a steam turbine was a cheering sight to see.

This reluctance to change with changing conditions was not limited to the blunter minds and less resilient imaginations in the service. As clear and untrammeled an intelligence as Alfred Thayer Mahan, a prophetic spirit in the realm of strategy, where he was unfettered by personal attachments of any kind, was occasionally at the mercy of the past. In 1906 he opposed the construction of battleships with single-caliber main batteries—that is, the modern battleship—because, he argued, such vessels would fight only at great ranges. These ranges would create in the sailor what Mahan felicitously called "the indisposition to close." They would thus undermine the physical and moral courage of a commander. They would, in other words, destroy the doctrine and the spirit, formulated by Nelson a century before, that no captain could go very far wrong who laid his ship alongside an enemy. The fourteen-inch rifle, which could place a shell upon a possible target six miles away, had long ago annihilated the Nelsonian doctrine. Mahan, of course, knew and recognized this fact; he was, as a man raised in sail, reluctant only to accept its full meaning, which was not that men were no longer brave, but that 100 years after the

173

battle of the Nile they had to reveal their bravery in a different way.

Now the question still is, why this blind reaction to technological change, observed in the continuation of sail or in Mahan's contentions or in the opposition to continuous-aim firing? It is wrong to assume, as it is frequently assumed by civilians, that it springs exclusively from some causeless Bourbon distemper that invades the military mind. There is a sounder and more attractive base. The opposition, where it occurs, of the soldier and the sailor to such change springs from the normal human instinct to protect oneself, and more especially, one's way of life. Military organizations are societies built around and upon the prevailing weapons systems. Intuitively and quite correctly the military man feels that a change in weapon portends a change in the arrangements of his society. Think of it this way. Since the time that the memory of man runneth not to the contrary, the naval society has been built upon the surface vessel. Daily routines, habits of mind, social organization, physical accommodations, conventions, rituals, spiritual allegiances have been conditioned by the essential fact of the ship. What then happens to your society if the ship is displaced as the principal element by such a radically different weapon as the plane? The mores and structure of the society are immediately placed in jeopardy. They may, in fact, be wholly destroyed. It was the witty cliché of the twenties that those naval officers who persisted in defending the battleship against the apparently superior claims of the carrier did so because the battleship was a more comfortable home. What, from one point of view, is a better argument? There is, as everyone knows, no place like home. Who has ever wanted to see the old place brought under the hammer by hostile forces whether they hold a mortgage or inhabit a flying machine?

This sentiment would appear to account in large part for the opposition to Sims; it was the product of an instinctive protective feeling,

even if the reasons for this feeling were not overt or recognized. The years after 1902 proved how right, in their terms, the opposition was. From changes in gunnery flowed an extraordinary complex of changes: in shipboard routines, ship design, and fleet tactics. There was, too, a social change. In the days when gunnery was taken lightly, the gunnery officer was taken lightly. After 1903, he became one of the most significant and powerful members of a ship's company, and this shift of emphasis naturally was shortly reflected in promotion lists. Each one of these changes provoked a dislocation in the naval society, and with man's troubled foresight and natural indisposition to break up classic forms, the men in Washington withstood the Sims onslaught as long as they could. It is very significant that they withstood it until an agent from outside, outside and above, who was not clearly identified with the naval society, entered to force change.

This agent, the President of the United States, might reasonably and legitimately claim the credit for restoring our gunnery efficiency. But this restoration by *force majeure* was brought about at great cost to the service and men involved. Bitternesses, suspicions, wounds were made that it was impossible to conceal and were, in fact, never healed.

Now this entire episode may be summed up in five separate points:

1. The essential idea for change occurred in part by chance but in an environment that contained all the essential elements for change and to a mind prepared to recognize the possibility of change.

2. The basic elements, the gun, gear, and sight, were put in the environment by other men, men interested in designing machinery to serve different purposes or simply interested in the instruments themselves.

3. These elements were brought into successful combination by minds not interested in the instruments for themselves but in what

they could do with them. These minds were, to be sure, interested in good gunnery, overtly and consciously. They may also, not so consciously, have been interested in the implied revolt that is present in the support of all change. Their temperaments and careers indeed support this view. From gunnery, Sims went on to attack ship designs, existing fleet tactics, and methods of promotion. He lived and died, as the service said, a stormy petrel, a man always on the attack against higher authority, a rebellious spirit; a rebel, fighting in excellent causes, but a rebel still who seems increasingly to have identified himself with the act of revolt against constituted authority.

4. He and his colleagues were opposed on this occasion by men who were apparently moved by three considerations: honest disbelief in the dramatic but substantiated claims of the new process, protection of the existing devices and instruments with which they identified themselves, and maintenance of the existing society with which they were identified.

5. The deadlock between those who sought change and those who sought to retain things as they were was broken only by an appeal to superior force, a force removed from and unidentified with the mores, conventions, devices of the society. This seems to me a very important point. The naval society in 1900 broke down in its effort to accommodate itself to a new situation. The appeal to Roosevelt is documentation for Mahan's great generalization that no military service should or can undertake to reform itself. It must seek assistance from outside.

Now with these five summary points in mind, it may be possible to seek, as suggested at the outset, a few larger implications from this story. What, if anything, may it suggest about the general process by which any society attempts to meet changing conditions?

There is, to begin with, a disturbing inference half-concealed in Mahan's statement that no military organization can reform itself. Certainly civilians would agree with this. We all know now that war and the preparation for war are too important, as Clemenceau said, to be left to the generals. But as I have said before, military organizations are really societies, more rigidly structured, more highly integrated, than most communities, but still societies. What then if we make this phrase to read, "No society can reform itself"? Is the process of adaptation to change, for example, too important to be left to human beings? This is a discouraging thought, and historically there is some cause to be discouraged. Societies have not been very successful in reforming themselves, accommodating to change, without pain and conflict.

This is a subject to which we may well address ourselves. Our society especially is built, as I have said, just as surely upon a changing technology as the Navy of the nineties was built upon changing weapon systems. How then can we find the means to accept with less pain to ourselves and less damage to our social organization the dislocations in our society that are produced by innovation? I cannot, of course, give any satisfying answer to these difficult questions. But in thinking about the case study before us, an idea occurred to me that at least might warrant further investigation by men far more qualified than I.

A primary source of conflict and tension in our case study appears to lie in this great word I have used so often in the summary, the word "identification." It cannot have escaped notice that some men identified themselves with their creations—sights, gun, gear, and so forth—and thus obtained a presumed satisfaction from the thing itself, a satisfaction that prevented them from thinking too closely on either the use or the defects of the thing; that others identified themselves with a settled way of life

175

they had inherited or accepted with minor modification and thus found their satisfaction in attempting to maintain that way of life unchanged; and that still others identified themselves as rebellious spirits, men of the insurgent cast of mind, and thus obtained a satisfaction from the act of revolt itself.

This purely personal identification with a concept, a convention, or an attitude would appear to be a powerful barrier in the way of easily acceptable change. Here is an interesting primitive example. In the years from 1864 to 1871 ten steel companies in this country began making steel by the new Bessemer process. All but one of them at the outset imported from Great Britain English workmen familiar with the process. One, the Cambria Company, did not. In the first few years those companies with British labor established an initial superiority. But by the end of the seventies, Cambria had obtained a commanding lead over all competitors. The President of Cambria, R. W. Hunt, in seeking a cause for his company's success, assigned it almost exclusively to the labor policy. "We started the converter plant without a single man who had ever seen even the outside of a Bessemer plant. We thus had willing pupils with no prejudices and no reminiscences of what they had done in the old country." The Bessemer process, like any new technique, had been constantly improved and refined in this period from 1864 to 1871. The British laborers of Cambria's competitors, secure in the performance of their own original techniques, resisted and resented all change. The Pennsylvania farm boys, untrammeled by the rituals and traditions of their craft, happily and rapidly adapted themselves to the constantly changing process. They ended by creating an unassailable competitive position for their company.

How then can we modify the dangerous effects of this word "identification"? And how much can we tamper with this identifying process? Our security—much of it, after all—comes from giving our allegiance to something greater than ourselves. These are difficult questions to which only the most tentative and provisional answers may here be proposed for consideration.

If one looks closely at this little case history, one discovers that the men involved were the victims of *severely limited* identifications. They were presumably all part of a society dedicated to the process of national defense, yet they persisted in aligning themselves with separate parts of that process—with the existing instruments of defense, with the existing customs of the society, or with the act of rebellion against the customs of the society. Of them all the insurgents had the best of it. They could, and did, say that the process of defense was improved by a gun that shot straighter and faster, and since they wanted such guns, they were unique among their fellows, patriots who sought only the larger object of improved defense. But this beguiling statement, even when coupled with the recognition that these men were right and extremely valuable and deserving of respect and admiration—this statement cannot conceal the fact that they were interested too in scalps and blood, so interested that they made their case a militant one and thus created an atmosphere in which self-respecting men could not capitulate without appearing either weak or wrong or both. So these limited identifications brought men into conflict with each other, and the conflict prevented them from arriving at a common acceptance of a change that presumably, as men interested in our total national defense, they would all find desirable.

It appears, therefore, if I am correct in my assessment, that we might spend some time and thought on the possibility of enlarging the sphere of our identifications from the part to the whole. For example, those Pennsylvania farm boys at the Cambria Steel Company were, apparently, much more interested in the manufacture of steel than in the preservation of any particular way of making steel. So I would suggest that in studying innovation, we look further into this possibility: the possibility that any group that exists for any purpose—the family, the factory, the educational institution—might begin by defining for itself its grand object and see to it that that grand object is communicated to every member of the group. Thus defined and communicated, it might serve as a unifying agent against the disruptive local allegiances of the inevitable smaller elements that compose any group. It may also serve as a means to increase the acceptability of any change that would assist in the more efficient achievement of the grand object.

There appears also a second possible way to combat the untoward influence of limited identifications. We are, I may repeat, a society based on technology in a time of prodigious technological advance, and a civilization committed irrevocably to the theory of evolution. These things mean that we believe in change; they suggest that if we are to survive in good health we must, in the phrase that I have used before, become an "adaptive society." By the word "adaptive" is meant the ability to extract the fullest possible returns from the opportunities at hand: the ability of Sir Percy Scott to select judiciously from the ideas and material presented both by the past and present and to throw them into a new combination. "Adaptive," as here used, also means the kind of resilience that will enable us to accept fully and easily the best promises of changing circumstances without losing our sense of continuity or our essential integrity.

We are not yet emotionally an adaptive society, though we try systematically to develop forces that tend to make us one. We encourage the search for new inventions; we keep the mind stimulated, bright, and free to seek out fresh means of transport, communication, and energy; yet we remain, in part, appalled by the consequences of our ingenuity, and, too frequently, try to find security through the shoring up of ancient and irrelevant conventions, the

extension of purely physical safeguards, or the delivery of decisions we ourselves should make into the keeping of superior authority like the state. These solutions are not necessarily unnatural or wrong, but they historically have not been enough, and I suspect they never will be enough to give us the serenity and competence we seek.

If the preceding statements are correct, they suggest that we might give some attention to the construction of a new view of ourselves as a society which in time of great change identified with and obtained security and satisfaction from the wise and creative accommodation to change itself. Such a view rests, I think, upon a relatively greater reverence for the mere *process* of living in a society than we possess today, and a relatively smaller respect for and attachment to any special *product* of a society, a product either as finite as a bathroom fixture or as conceptual as a fixed and final definition of our Constitution or our democracy.

Historically such an identification with *process* as opposed to *product,* with adventurous selection and adaptation as opposed to simple retention and possessiveness, has been difficult to achieve collectively. The Roman of the early republic, the Italian of the late fifteenth and early sixteenth century, or the Englishman of Elizabeth's time appears to have been most successful in seizing the new opportunities while conserving as much of the heritage of the past as he found relevant and useful to his purpose.

We seem to have fallen on times similar to theirs, when many of the existing forms and schemes have lost meaning in the face of dramatically altering circumstances. Like them we may find at least part of our salvation in identifying ourselves with the adaptive process and thus share with them some of the joy, exuberance, satisfaction, and security with which they went out to meet their changing times.

I am painfully aware that in setting up my historical situation for examination I have, in a sense, artificially contrived it. I have been forced to cut away much, if not all, of the connecting tissue of historical evidence and to present you only with the bare bones and even with only a few of the bones. Thus, I am also aware, the episode has lost much of the subtlety, vitality, and attractive uncertainty of the real situation. There has, too, in the process, been inevitable distortion, but I hope the essential if exaggerated truth remains. I am also aware that I have erected elaborate hypotheses on the slender evidence provided by the single episode. My defense here is only that I have hoped to suggest possible approaches and methods of study and also possible fruitful areas of investigation in a subject that seems to me of critical importance in the life and welfare of our changing society.

Who Gets Power—And How They Hold on to It: A Strategic-Contingency Model of Power

Gerald R. Salancik

Jeffrey Pfeffer

Power adheres to those who can cope with the critical problems of the organization. As such, power is not a dirty secret, but the secret of success. And that's the path power follows, until it becomes institutionalized—which makes administration the most precarious of occupations.

Power is held by many people to be a dirty word or, as Warren Bennis has said, "It is the organization's last dirty secret."

This article will argue that traditional "political" power, far from being a dirty business, is, in its most naked form, one of the few mechanisms available for aligning an organization with its own reality. However, institutionalized forms of power—what we prefer to call the cleaner forms of power: authority, legitimization, centralized control, regulations, and the more modern "management information systems"—tend to buffer the organization from reality and obscure the demands of its environment. Most great states and institutions declined, not because they played politics, but because they failed to accommodate to the political realities they faced. Political processes, rather than being mechanisms for unfair and unjust allocations and appointments, tend toward the realistic resolution of conflicts among interests. And power, while it eludes definition, is easy enough to recognize by its consequences—the ability of those who possess power to bring about the outcomes they desire.

The model of power we advance is an elaboration of what has been called strategic-contingency theory, a view that sees power as something that accrues to organizational subunits (individuals, departments) that cope with critical organizational problems. Power is used by subunits, indeed, used by all who have it, to enhance their own survival through control of scarce critical resources, through the placement of allies in key positions, and through the definition of organizational problems and policies. Because of the processes by which power develops and is used, organizations become both more aligned and more misaligned with their environments. This contradiction is the most interesting aspect of organizational power, and one that makes administration one of the most precarious of occupations.

WHAT IS ORGANIZATIONAL POWER?

You can walk into most organizations and ask without fear of being misunderstood, "Which are the powerful groups or people in this organization?" Although many organizational informants may be *unwilling* to tell you, it is unlikely they will be *unable* to tell you. Most people do not require explicit definitions to know what power is.

Power is simply the ability to get things done the way one wants them to be done. For a manager who wants an increased budget to launch a project that he thinks is important, his power is measured by his ability to get that budget. For an executive vice-president who wants to be chairman, his power is evidenced by his advancement toward his goal.

People in organizations not only know what you are talking about when you ask who is influential but they are likely to agree with one another to an amazing extent. Recently, we had a chance to observe this in a regional office of an insurance company. The office had 21 department managers; we asked ten of these managers to rank all 21 according to the influence each one had in the organization. Despite the fact that ranking 21 things is a difficult task, the managers sat down and began arranging the names of their colleagues and themselves in a column. Only one person bothered to ask, "What do you mean by influence?" When told "power," he responded, "Oh," and went on. We compared the rankings of all ten managers and found virtually no disagreement among them in the managers ranked among the top

five or the bottom five. Differences in the rankings came from department heads claiming more influence for themselves than their colleagues attributed to them.

Such agreement on those who have influence, and those who do not, was not unique to this insurance company. So far we have studied over 20 very different organizations—universities, research firms, factories, banks, retailers, to name a few. In each one we found individuals able to rate themselves and their peers on a scale of influence or power. We have done this both for specific decisions and for general impact on organizational policies. Their agreement was unusually high, which suggests that distributions of influence exist well enough in everyone's mind to be referred to with ease—and we assume with accuracy.

WHERE DOES ORGANIZATIONAL POWER COME FROM?

Earlier we stated that power helps organizations become aligned with their realities. This hopeful prospect follows from what we have dubbed the strategic-contingencies theory of organizational power. Briefly, those subunits most able to cope with the organization's critical problems and uncertainties acquire power. In its simplest form, the strategic-contingencies theory implies that when an organization faces a number of lawsuits that threaten its existence, the legal department will gain power and influence over organizational decisions. Somehow other organizational interest groups will recognize its critical importance and confer upon it a status and power never before enjoyed. This influence may extend beyond handling legal matters and into decisions about product design, advertising production, and so on. Such extensions undoubtedly would be accompanied by appropriate, or acceptable, verbal justifications. In time, the head of the legal department may become the head of the corporation, just as in times past the vice-president for marketing had become the president when market shares were a worrisome problem and, before him, the chief engineer, who had made the production line run as smooth as silk.

Stated in this way, the strategic-contingencies theory of power paints an appealing picture of power. To the extent that power is determined by the critical uncertainties and problems facing the organization and, in turn, influences decisions in the organization, the organization is aligned with the realities it faces. In short, power facilitates the organization's adaptation to its environment—or its problems.

We can cite many illustrations of how influence derives from a subunits's ability to deal with critical contingencies. Michael Crozier described a French cigarette factory in which the maintenance engineers had a considerable say in the plantwide operation. After some probing he discovered that the group possessed the solution to one of the major problems faced by the company, that of troubleshooting the elaborate, expensive, and irrascible automated machines that kept breaking down and dumbfounding everyone else. It was the one problem that the plant manager could in no way control.

The production workers, while troublesome from time to time, created no insurmountable problems; the manager could reasonably predict their absenteeism or replace them when necessary. Production scheduling was something he could deal with since, by watching inventories and sales, the demand for cigarettes was known long in advance. Changes in demand could be accommodated by slowing down or speeding up the line. Supplies of tobacco and paper were also easily dealt with through stockpiles and advance orders.

The one thing that management could neither control nor accommodate to, however,

was the seemingly happenstance breakdowns. And the foremen couldn't instruct the workers what to do when emergencies developed since the maintenance department kept its records of problems and solutions locked up in a cabinet or in its members' heads. The breakdowns were, in truth, a critical source of uncertainty for the organization, and the maintenance engineers were the only ones who could cope with the problem.

The engineers' strategic role in coping with breakdowns afforded them a considerable say on plant decisions. Schedules and production quotas were set in consultation with them. And the plant manager, while formally their boss, accepted their decisions about personnel in their operation. His submission was to his credit, for without their cooperation he would have had an even more difficult time in running the plant.

Ignoring Critical Consequences

In this cigarette factory, sharing influence with the maintenance workers reflected the plant manager's awareness of the critical contingencies. However, when organizational members are not aware of the critical contingencies they face, and do not share influence accordingly, the failure to do so can create havoc. In one case, an insurance company's regional office was having problems with the performance of one of its departments, the coding department. From the outside, the department looked like a disaster area. The clerks who worked in it were somewhat dissatisfied; their supervisor paid little attention to them, and they resented the hard work. Several other departments were critical of this manager, claiming that she was inconsistent in meeting deadlines. The person most critical was the claims manager. He resented having to wait for work that was handled by her department, claiming that it held up his claims adjusters. Having heard the rumors about dissatisfaction among her subordinates, he attributed the situation to poor supervision. He was second in command in the office and therefore took up the issue with her immediate boss, the head of administrative services. They consulted with the personnel manager and the three of them concluded that the manager needed leadership training to improve her relations with her subordinates. The coding manager objected, saying it was a waste of time, but agreed to go along with the training and also agreed to give more priority to the claims department's work. Within a week after the training, the results showed that her workers were happier but that the performance of her department had decreased, save for the people serving the claims department.

About this time, we began, quite independently, a study of influence in this organization. We asked the administrative services director to draw up flow charts of how the work of one department moved onto the next department. In the course of the interview, we noticed that the coding department began or interceded in the work flow of most of the other departments and casually mentioned to him, "The coding manager must be very influential." He said "No, not really. Why would you think so?" Before we could reply he recounted the story of her leadership training and the fact that things were worse. We then told him that it seemed obvious that the coding department would be influential from the fact that all the other departments depended on it. It was also clear why productivity had fallen. The coding manager took the training seriously and began spending more time raising her workers' spirits than she did worrying about the problems of all the departments that depended on her. Giving priority to the claims area only exaggerated the problem, for their work was getting done at the expense of the work of the other departments. Eventually the company hired a few more clerks to relieve the pressure in the coding department and performance returned to a more satisfactory level.

Originally we got involved with this insurance company to examine how the influence of each manager evolved from his or her department's handling of critical organizational contingencies. We reasoned that one of the most important contingencies faced by all profit-making organizations was that of generating income. Thus we expected managers would be influential to the extent to which they contributed to this function. Such was the case. The underwriting managers, who wrote the policies that committed the premiums, were the most influential; the claims managers, who kept a lid on the funds flowing out, were a close second. Least influential were the managers of functions unrelated to revenue, such as mailroom and payroll managers. And contrary to what the administrative services manager believed, the third most powerful department head (out of 21) was the woman in charge of the coding function, which consisted of rating, recording, and keeping track of the codes of all policy applications and contracts. Her peers attributed more influence to her than could have been inferred from her place on the organization chart. And it was not surprising, since they all depended on her department. The coding department's records, their accuracy and the speed with which they could be retrieved, affected virtually every other operating department in the insurance office. The underwriters depended on them in getting the contracts straight; the typing department depended on them in preparing the formal contract document; the claims department depended on them in adjusting claims; and accounting depended on them for billing. Unfortunately, the "bosses" were not aware of these dependences, for unlike the cigarette factory, there were no massive breakdowns that made them obvious, while the coding manager, who was a hardworking but quiet person, did little to announce her importance.

The cases of this plant and office illustrate nicely a basic point about the source of power in organizations. The basis for power in an organization derives from the ability of a person or subunit to take or not take actions that are desired by others. The coding manager was seen as influential by those who depended on her department, but not by the people at the top. The engineers were influential because of their role in keeping the plant operating. The two cases differ in these respects: The coding supervisor's source of power was not as widely recognized as that of the maintenance engineers, and she did not use her source of power to influence decisions; the maintenance engineers did. Whether power is used to influence anything is a separate issue. We should not confuse this issue with the fact that power derives from a social situation in which one person has a capacity to do something and another person does not, but wants it done.

POWER SHARING IN ORGANIZATIONS

Power is shared in organizations; and it is shared out of necessity more than out of concern for principles of organizational development or participatory democracy. Power is shared because no one person controls all the desired activities in the organization. While the factory owner may hire people to operate his noisy machines, once hired they have some control over the use of the machinery. And thus they have power over him in the same way he has power over them. Who has more power over whom is a mooter point than that of recognizing the inherent nature of organizing as a sharing of power.

Let's expand on the concept that power derives from the activities desired in an organization. A major way of managing influence in organizations is through the designation of activities. In a bank we studied, we saw this principle in action. This bank was planning to install a computer system for routine credit evaluation. The bank, rather progressive-minded, was concerned that the change would have adverse effects on employees and therefore surveyed their attitudes.

The principal opposition to the new system came, interestingly, not from the employees who performed the routine credit checks, some of whom would be relocated because of the change, but from the manager of the credit department. His reason was quite simple. The manager's primary function was to give official approval to the applications, catch any employee mistakes before giving approval, and arbitrate any difficulties the clerks had in deciding what to do. As a consequence of his role, others in the organization, including his superiors, subordinates, and colleagues, attributed considerable importance to him. He, in turn, for example, could point to the low proportion of credit approvals, compared with other financial institutions, that resulted in bad debts. Now, to his mind, a wretched machine threatened to transfer his role to a computer programmer, a man who knew nothing of finance and who, in addition, had ten years less seniority. The credit manager eventually quit for a position at a smaller firm with lower pay, but one in which he would have more influence than his redefined job would have left him with.

Because power derives from activities rather than individuals, an individual's or subgroup's power is never absolute and derives ultimately from the context of the situation. The amount of power an individual has at any one time depends, not only on the activities he or she controls, but also on the existence of other persons or means by which the activities can be achieved and on those who determine what ends are desired and, hence, on what activities are desired and critical for the organization. One's own power always depends on other people for these two reasons. Other people, or groups or organizations, can determine the definition of what is a critical contingency for the organization and can also undercut the uniqueness of the individual's personal contribution to the critical contingencies of the organization.

Perhaps one can best appreciate how situationally dependent power is by examining how it is distributed. In most societies, power organizes around scarce and critical resources. Rarely does power organize around abundant resources. In the United States, a person doesn't become powerful because he or she can drive a car. There are simply too many others who can drive with equal facility. In certain villages in Mexico, on the other hand, a person with a car is accredited with enormous social status and plays a key role in the community. In addition to scarcity, power is also limited by the need for one's capacities in a social system. While a racer's ability to drive a car around a 90° turn at 80 mph may be sparsely distributed in a society, it is not likely to lend the driver much power in the society. The ability simply does not play a central role in the activities of the society.

The fact that power revolves around scarce and critical activities, of course, makes the control and organization of those activities a major battleground in struggles for power. Even relatively abundant or trivial resources can become the bases for power if one can organize and control their allocation and the definition of what is critical. Many occupational and professional groups attempt to do just this in modern economies. Lawyers organize themselves into associations, regulate the entrance requirements for novitiates, and then get laws passed specifying situations that require the services of an attorney. Workers had little power in the conduct of industrial affairs until they organized themselves into closed and controlled systems. In recent years, women and blacks have tried to define themselves as important and critical to the social system, using law to reify their status.

In organizations there are obviously op-portunities for defining certain activities as more critical than others. Indeed, the growth of managerial thinking to include defining organizational objectives and goals has done much to foster these opportunities. One sure way to liquidate the power of groups in the organization is to define the need for their services out of existence. David Halberstam presents a description of how just such a thing happened to the group of correspondents that evolved around Edward R. Murrow, the brilliant journalist, interviewer, and war correspondent of CBS News. A close friend of CBS chairman and controlling stockholder William S. Paley, Murrow, and the news department he directed, were endowed with freedom to do what they felt was right. He used it to create some of the best documentaries and commentaries ever seen on television. Unfortunately, television became too large, too powerful, and too suspect in the eyes of the federal government that licensed it. It thus became, or at least the top executives believed it had become, too dangerous to have in-depth, probing commentary on the news. Crisp, dry, uneditorializing headliners were considered safer. Murrow was out and Walter Cronkite was in.

The power to define what is critical in an organization is no small power. Moreover, it is the key to understanding why organizations are either aligned with their environments or misaligned. If an organization defines certain activities as critical when in fact they are not critical, given the flow of resources coming into the organization, it is not likely to survive, at least in its present form.

Most organizations manage to evolve a distribution of power and influence that is aligned with the critical realities they face in the environment. The environment, in turn, includes both the internal environment, the shifting situational contexts in which particular decisions get made, and the external environment that it can hope to influence but is unlikely to control.

THE CRITICAL CONTINGENCIES

The critical contingencies facing most organizations derive from the environmental context within which they operate. This determines the available needed resources and thus determines the problems to be dealt with. That power organizes around handling these problems suggests an important mechanism by which organizations keep in tune with their external environments. The strategic-contingencies model implies that subunits that contribute to the critical resources of the organization will gain influence in the organization. Their influence presumably is then used to bend the organization's activities to the contingencies that determine its resources. This idea may strike one as obvious. But its obviousness in no way diminishes its importance. Indeed, despite its obviousness, it escapes the notice of many organizational analysts and managers, who all too frequently think of the organization in terms of a descending pyramid, in which all the departments in one tier hold equal power and status. This presumption denies the reality that departments differ in the contributions they are believed to make to the overall organization's resources, as well as to the fact that some are more equal than others.

Because of the importance of this idea to organizational effectiveness, we decided to examine it carefully in a large midwestern university. A university offers an excellent site for studying power. It is composed of departments with nominally equal power and is administered by a central executive structure much like other bureaucracies. However, at the same time it is a situation in which the departments have clearly defined identities and face diverse external environments. Each department has its own bodies of knowledge, its own institutions, its own sources of prestige and resources. Because the departments operate in different external environments, they are likely to contribute differentially to the resources of the overall organization. Thus a physics department with close ties to NASA may contribute substantially to the funds of the university; and a history department with a renowned historian in residence may contribute to the intellectual credibility or prestige of the whole university. Such variations permit one to examine how these various contributions lead to obtaining power within the university.

We analyzed the influence of 29 university departments throughout an 18-month period in their history. Our chief interest was to determine whether departments that brought more critical resources to the university would be more powerful than departments that contributed fewer or less critical resources.

To identify the critical resources each department contributed, the heads of all departments were interviewed about the importance of seven different resources to the university's success. The seven included undergraduate students (the factor determining size of the state allocations by the university), national prestige, administrative expertise, and so on. The most critical resource was found to be contract and grant monies received by a department's faculty for research or consulting services. At this university, contract and grants contributed somewhat less than 50 percent of the overall budget, with the remainder primarily coming from state appropriations. The importance attributed to contract and grant monies, and the rather minor importance of undergraduate students, was not surprising for this particular university. The university was a major center for graduate education; many of its departments ranked in the top ten of their respective fields. Grant and contract monies were the primary source of discretionary funding available for maintaining these programs of graduate education, and hence for maintaining the university's prestige. The prestige of the university itself was critical both in recruiting able students and attracting top-notch faculty.

From university records it was deter-

185

mined what relative contributions each of the 29 departments made to the various needs of the university (national prestige, outside grants, teaching). Thus, for instance, one department may have contributed to the university by teaching 7 percent of the instructional units, bringing in 2 percent of the outside contracts and grants, and having a national ranking of 20. Another department, on the other hand, may have taught one percent of the instructional units, contributed 12 percent to the grants, and be ranked the third best department in its field within the country.

The question was: Do these different contributions determine the relative power of the departments within the university? Power was measured in several ways; but regardless of how measured, the answer was "Yes." Those three resources together accounted for about 70 percent of the variance in subunit power in the university.

But the most important predictor of departmental power was the department's contribution to the contracts and grants of the university. Sixty percent of the variance in power was due to this one factor, suggesting that the power of departments derived primarily from the dollars they provided for graduate education, the activity believed to be the most important for the organization.

THE IMPACT OF ORGANIZATIONAL POWER ON DECISION MAKING

The measure of power we used in studying this university was an analysis of the responses of the department heads we interviewed. While such perceptions of power might be of interest in their own right, they contribute little to our understanding of how the distribution of power might serve to align an organization with its critical realities. For this we must look to how power actually influences the decisions and policies of organizations.

While it is perhaps not absolutely valid, we can generally gauge the relative importance of a department of an organization by the size of the budget allocated to it relative to other departments. Clearly it is of importance to the administrators of those departments whether they get squeezed in a budget crunch or are given more funds to strike out after new opportunities. And it should also be clear that when those decisions are made and one department can go ahead and try new approaches while another must cut back on the old, then the deployment of the resources of the organization in meeting its problems is most directly affected.

Thus our study of the university led us to ask the following question: Does power lead to influence in the organization? To answer this question, we found it useful first to ask another one, namely: Why should department heads try to influence organizational decisions to favor their own departments to the exclusion of other departments? While this second question may seem a bit naive to anyone who has witnessed the political realities of organizations, we posed it in a context of research on organizations that sees power as an illegitimate threat to the neater rational authority of modern bureaucracies. In this context, decisions are not believed to be made because of the dirty business of politics but because of the overall goals and purposes of the organization. In a university, one reasonable basis for decision making is the teaching workload of departments and the demands that follow from that workload. We would expect, therefore, that departments with heavy student demands for courses would be able to obtain funds for teaching. Another reasonable basis for decision making is quality. We would expect, for that reason, that departments with esteemed reputations would be able to obtain funds both because their quality suggests they might use such funds effectively and because such funds would allow them to maintain their quality. A rational model of bureaucracy intimates, then, that the organizational decisions taken would favor those who perform the stated purposes of the organization—teaching undergraduates and training professional and scientific talent—well.

The problem with rational models of decision making, however, is that what is rational to one person may strike another as irrational. For most departments, resources are a question of survival. While teaching undergraduates may seem to be a major goal for some members of the university, developing knowledge may seem so to others; and to still others, advising governments and other institutions about policies may seem to be the crucial business. Everyone has his own idea of the proper priorities in a just world. Thus goals rather than being clearly defined and universally agreed upon are blurred and contested throughout the organization. If such is the case, then the decisions taken on behalf of the organization as a whole are likely to reflect the goals of those who prevail in political contests, namely, those with power in the organization.

Will organizational decisions always reflect the distribution of power in the organization? Probably not. Using power for influence requires a certain expenditure of effort, time, and resources. Prudent and judicious persons are not likely to use their power needlessly or wastefully. And it is likely that power will be used to influence organizational decisions primarily under circumstances that both require and favor its use. We have examined three conditions that are likely to affect the use of power in organizations: scarcity, criticality, and uncertainty. The first suggests that subunits will try to exert influence when the resources of the organization are scarce. If there is an abundance of resources, then a particular department or a particular individual has little need to attempt influence. With little effort, he can get all he wants anyway.

The second condition, criticality, suggests that a subunit will attempt to influence decisions to obtain resources that are critical to its own survival and activities. Criticality implies that one would not waste effort, or risk being labeled obstinate, by fighting over trivial decisions affecting one's operations.

An office manager would probably balk less about a threatened cutback in copying machine usage than about a reduction in typing staff. An advertising department head would probably worry less about losing his lettering artist than his illustrator. Criticality is difficult to define because what is critical depends on people's beliefs about what is critical. Such beliefs may or may not be based on experience and knowledge and may or may not be agreed upon by all. Scarcity, for instance, may itself affect conceptions of criticality. When slack resources drop off, cutbacks have to be made—those "hard decisions," as congressmen and resplendent administrators like to call them. Managers then find themselves scrapping projects they once held dear.

The third condition that we believe affects the use of power is uncertainty: When individuals do not agree about what the organization should do or how to do it, power and other social processes will affect decisions. The reason for this is simply that, if there are no clear-cut criteria available for resolving conflicts of interest, then the only means for resolution is some form of social process, including power, status, social ties, or some arbitrary process like flipping a coin or drawing straws. Under conditions of uncertainty, the powerful manager can argue his case on any grounds and usually win it. Since there is no real consensus, other contestants are not likely to develop counter arguments or amass sufficient opposition. Moreover, because of his power and their need for access to the resources he controls, they are more likely to defer to his arguments.

Although the evidence is slight, we have found that power will influence the allocations of scarce and critical resources. In the analysis of power in the university, for instance, one of the most critical resources needed by departments is the general budget. First granted by the state legislature, the general budget is later allocated to individual departments by the university administration in response to requests from the department heads. Our analysis of the factors that contribute to a department getting more or less of this budget indicated that subunit power was the major predictor, overriding such factors as student demand for courses, national reputations of departments, or even the size of a department's faculty. Moreover, other research has shown that when the general budget has been cut back or held below previous uninflated levels, leading to monies becoming more scarce, budget allocations mirror departmental powers even more closely.

Student enrollment and faculty size, of course, do themselves relate to budget allocations, as we would expect since they determine a department's need for resources, or at least offer visible testimony of needs. But departments are not always able to get what they need by the mere fact of needing them. In one analysis it was found that high-power departments were able to obtain budget without regard to their teaching loads and, in some cases, actually in inverse relation to their teaching loads. In contrast, low-power departments could get increases in budget only when they could justify the increases by a recent growth in teaching load, and then only when it was far in excess of norms for other departments.

General budget is only one form of resource that is allocated to departments. There are others such as special grants for student fellowships or faculty research. These are critical to departments because they affect the ability to attract other resources, such as outstanding faculty or students. We examined how power influenced the allocations of four resources department heads had described as critical and scarce.

When the four resources were arrayed from the most to the least critical and scarce, we found that departmental power best predicted the allocations of the most critical and scarce resources. In other words, the analysis of how power influences organizational allocations leads to this conclusion: Those subunits most likely to survive in times of strife are those that are more critical to the organization. Their importance to the organization gives them power to influence resource allocations that enhance their own survival.

HOW EXTERNAL ENVIRONMENT IMPACTS EXECUTIVE SELECTION

Power not only influences the survival of key groups in an organization, it also influences the selection of individuals to key leadership positions, and by such a process further aligns the organization with its environmental context.

We can illustrate this with a recent study of the selection and tenure of chief administrators in 57 hospitals in Illinois. We assumed that since the critical problems facing the organization would enhance the power of certain groups at the expense of others, then the leaders to emerge should be those most relevant to the context of the hospitals. To assess this we asked each chief administrator about his professional background and how long he had been in office. The replies were then related to the hospitals' funding, ownership, and competitive conditions for patients and staff.

One aspect of a hospital's context is the source of its budget. Some hospitals, for instance, are run much like other businesses. They sell bed space, patient care, and treatment services. They charge fees sufficient both to cover their costs and to provide capital for expansion. The main source of both their operating and capital funds is patient billings. Increasingly, patient billings are paid for, not by patients, but by private insurance companies. Insurers like Blue Cross dominate and represent a potent interest group outside a hospital's control but critical to its income. The insurance companies, in order to limit their own costs, attempt to hold down the fees allowable to hospitals, which they do effectively from their positions on state rate boards. The squeeze on hospitals that results from fees increasing slowly while costs climb rapidly more and more demands the talents of cost accountants or people trained in the technical expertise of hospital administration.

By contrast, other hospitals operate more like social service institutions, either as government healthcare units (Bellevue Hospital in New York City and Cook County Hospital in Chicago, for example) or as charitable institutions. These hospitals obtain a large proportion of their operating and capital funds, not from privately insured patients, but from government subsidies or private donations. Such institutions rather than requiring the talents of a technically efficient administrator are likely to require the savvy of someone who is well integrated into the social and political power structure of the community.

Not surprisingly, the characteristics of administrators predictably reflect the funding context of the hospitals with which they are associated. Those hospitals with larger proportions of their budget obtained from private insurance companies were most likely to have administrators with backgrounds in accounting and least likely to have administrators whose professions were business or medicine. In contrast, those hospitals with larger proportions of their budget derived from private donations and local governments were most likely to have administrators with business or professional backgrounds and least likely to have accountants. The same held for formal training in hospital management. Professional hospital administrators could easily be found in hospitals drawing their incomes from private insurance

and rarely in hospitals dependent on donations or legislative appropriations.

As with the selection of administrators, the context of organizations has also been found to affect the removal of executives. The environment, as a source of organizational problems, can make it more or less difficult for executives to demonstrate their value to the organization. In the hospitals we studied, long-term administrators came from hospitals with few problems. They enjoyed amicable and stable relations with their local business and social communities and suffered little competition for funding and staff. The small city hospital director who attended civic and Elks meetings while running the only hospital within a 100-mile radius, for example, had little difficulty holding on to his job. Turnover was highest in hospitals with the most problems, a phenomenon similar to that observed in a study of industrial organizations in which turnover was highest among executives in industries with competitive environments and unstable market conditions. The interesting thing is that instability characterized the industries rather than the individual firms in them. The troublesome conditions in the individual firms were attributed, or rather misattributed, to the executives themselves.

It takes more than problems, however, to terminate a manager's leadership. The problems themselves must be relevant and critical. This is clear from the way in which an administrator's tenure is affected by the status of the hospital's operating budget. Naively we might assume that all administrators would need to show a surplus. Not necessarily so. Again, we must distinguish between those hospitals that depend on private donations for funds and those that do not. Whether an endowed budget shows a surplus or deficit is less important than the hospital's relations with benefactors. On the other hand, with a budget dependent on patient billing, a surplus is almost essential; monies for new equipment or expansion must be drawn from it, and without them quality care becomes more difficult and patients scarcer. An administrator's tenure reflected just these considerations. For those hospitals dependent upon private donations, the length of an administrator's term depended not at all on the status of the operating budget but was fairly predictable from the hospital's relations with the business community. On the other hand, in hospitals dependent on the operating budget for capital financing, the greater the deficit the shorter was the tenure of the hospital's principal administrators.

CHANGING CONTINGENCIES AND ERODING POWER BASES

The critical contingencies facing the organization may change. When they do, it is reasonable to expect that the power of individuals and subgroups will change in turn. At times the shift can be swift and shattering, as it was recently for powerholders in New York City. A few years ago it was believed that David Rockefeller was one of the ten most powerful people in the city, as tallied by *New York* magazine, which annually sniffs out power for the delectation of its readers. But that was before it was revealed that the city was in financial trouble, before Rockefeller's Chase Manhattan Bank lost some of its own financial luster, and before brother Nelson lost some of his political influence in Washington. Obviously David Rockefeller was no longer as well positioned to help bail the city out. Another loser was an attorney with considerable personal connections to the political and religious leaders of the city. His talents were no longer in much demand. The persons with more influence were the bankers and union pension fund executors who fed money to the city; community leaders who represent blacks and Spanish-Americans, in contrast, witnessed the erosion of their power bases.

One implication of the idea that power

shifts with changes in organizational environments is that the dominant coalition will tend to be that group that is most appropriate for the organization's environment, as also will the leaders of an organization. One can observe this historically in the top executives of industrial firms in the United States. Up until the early 1950s, many top corporations were headed by former production line managers or engineers who gained prominence because of their abilities to cope with the problems of production. Their success, however, only spelled their demise. As production became routinized and mechanized, the problem of most firms became one of selling all those goods they so efficiently produced. Marketing executives were more frequently found in corporate boardrooms. Success outdid itself again, for keeping markets and production steady and stable requires the kind of control that can only come from acquiring competitors and suppliers or the invention of more and more appealing products—ventures that typically require enormous amounts of capital. During the 1960s, financial executives assumed the seats of power. And they, too, will give way to others. Edging over the horizon are legal experts, as regulation and antitrust suits are becoming more and more frequent in the 1970s, suits that had their beginnings in the success of the expansion generated by prior executives. The more distant future, which is likely to be dominated by multinational corporations, may see former secretaries of state and their minions increasingly serving as corporate figureheads.

THE NONADAPTIVE CONSEQUENCES OF ADAPTATION

From what we have said thus far about power aligning the organization with its own realities, an intelligent person might react with a resounding ho-hum, for it all seems too obvious: Those with the ability to get the job done are given the job to do.

However, there are two aspects of power that make it more useful for understanding organizations and their effectiveness. First, the "job" to be done has a way of expanding itself until it becomes less and less clear what the job is. Napoleon began by doing a job for France in the war with Austria and ended up Emperor, convincing many that only he could keep the peace. Hitler began by promising an end to Germany's troubling postwar depression and ended up convincing more people than is comfortable to remember that he was destined to be the savior of the world. In short, power is a capacity for influence that extends far beyond the original bases that created it. Second, power tends to take on institutionalized forms that enable it to endure well beyond its usefulness to an organization.

There is an important contradiction in what we have observed about organizational power. On the one hand we have said that power derives from the contingencies facing an organization and that when those contingencies change so do the bases for power. On the other hand we have asserted that subunits will tend to use their power to influence organizational decisions in their own favor, particularly when their own survival is threatened by the scarcity of critical resources. The first statement implies that an organization will tend to be aligned with its environment since power will tend to bring to key positions those with capabilities relevant to the context. The second implies that those in power will not give up their positions so easily; they will pursue policies that guarantee their continued domination. In short, change and stability operate through the same mechanism, and, as a result, the organization will never be completely in phase with its environment or its needs.

The study of hospital administrators illustrates how leadership can be out of phase with reality. We argued that privately funded hospitals needed trained technical administrators more so than did hospitals funded by donations. The need as we perceived it was matched in most hospitals, but by no means in all. Some organizations did not conform with our predictions. These deviations imply that some administrators were able to maintain their positions independent of their suitability for those positions. By dividing administrators into those with long and short terms of office, one finds that the characteristics of longer-termed administrators were virtually unrelated to the hospital's context. The shorter-termed chiefs on the other hand had characteristics more appropriate for the hospital's problems. For a hospital to have a recently appointed head implies that the previous administrator had been unable to endure by institutionalizing himself.

One obvious feature of hospitals that allowed some administrators to enjoy a long tenure was a hospital's ownership. Administrators were less entrenched when their hospitals were affiliated with and dependent upon larger organizations, such as governments or churches. Private hospitals offered more secure positions for administrators. Like private corporations, they tend to have more diffused ownership, leaving the administrator unopposed as he institutionalizes his reign. Thus he endures, sometimes at the expense of the performance of the organization. Other research has demonstrated that corporations with diffuse ownership have poorer earnings than those in which the control of the manager is checked by a dominant shareholder. Firms that overload their boardrooms with more insiders than are appropriate for their context have also been found to be less profitable.

A word of caution is required about our judgment of "appropriateness." When we argue some capabilities are more appropriate for one context than another, we do so from the perspective of an outsider and on the basis of reasonable assumptions as to the problems the organization will face and the capabilities they will need. The fact that we have been able to predict the distribution of influence and the characteristics of leaders suggests that our reasoning is not incorrect. However, we do not think that all organizations follow the same pattern. The fact that we have not been able to predict outcomes with 100 percent accuracy indicates they do not.

MISTAKING CRITICAL CONTINGENCIES

One thing that allows subunits to retain their power is their ability to name their functions as critical to the organization when they may not be. Consider again our discussion of power in the university. One might wonder why the most critical tasks were defined as graduate education and scholarly research, the effect of which was to lend power to those who brought in grants and contracts. Why not something else? The reason is that the more powerful departments argued for those criteria and won their case, partly because they were more powerful.

In another analysis of this university, we found that all departments advocate selfserving criteria for budget allocation. Thus a department with large undergraduate enrollments argued that enrollments should determine budget allocations, a department with a strong national reputation saw prestige as the most reasonable basis for distributing funds, and so on. We further found that advocating such self-serving criteria actually benefited a department's budget allotments but, also, it paid off more for departments that were already powerful.

Organizational needs are consistent with a current distribution of power also because of a human tendency to categorize problems in familiar ways. An accountant sees problems with organizational performance as cost ac-

countancy problems or inventory flow problems. A sales manager sees them as problems with markets, promotional strategies, or just unaggressive salespeople. But what is the truth? Since it does not automatically announce itself, it is likely that those with prior credibility, or those with power, will be favored as the enlightened. This bias, while not intentionally self-serving, further concentrates power among those who already possess it, independent of changes in the organization's context.

INSTITUTIONALIZING POWER

A third reason for expecting organizational contingencies to be defined in familiar ways is that the current holders of power can structure the organization in ways that institutionalize themselves. By institutionalization we mean the establishment of relatively permanent structures and policies that favor the influence of a particular subunit. While in power, a dominant coalition has the ability to institute constitutions, rules, procedures, and information systems that limit the potential power of others while continuing their own.

The key to institutionalizing power always is to create a device that legitimates one's own authority and diminishes the legitimacy of others. When the "Divine Right of Kings" was envisioned centuries ago it was to provide an unquestionable foundation for the supremacy of royal authority. There is generally a need to root the exercise of authority in some higher power. Modern leaders are no less affected by this need. Richard Nixon, with the aid of John Dean, reified the concept of executive privilege, which meant in effect that what the President wished not to be discussed need not be discussed.

In its simpler form, institutionalization is achieved by designating positions or roles for organizational activities. The creation of a new post legitimizes a function and forces organization members to orient to it. By designating how this new post relates to older, more established posts, moreover, one can structure an organization to enhance the importance of the function in the organization. Equally, one can diminish the importance of traditional functions. This is what happened in the end with the insurance company we mentioned that was having trouble with its coding department. As the situation unfolded, the claims director continued to feel dissatisfied about the dependency of his functions on the coding manager. Thus he instituted a reorganization that resulted in two coding departments. In so doing, of course, he placed activities that affected his department under his direct control, presumably to make the operation more effective. Similarly, consumer-product firms enhance the power of marketing by setting up a coordinating role to interface production and marketing functions and then appoint a marketing manager to fill the role.

The structures created by dominant powers sooner or later become fixed and unquestioned features of the organization. Eventually, this can be devastating. It is said that the battle of Jena in 1806 was lost by Frederick the Great, who died in 1786. Though the great Prussian leader had no direct hand in the disaster, his imprint on the army was so thorough, so embedded in its skeletal underpinnings, that the organization was inappropriate for others to lead in different times.

Another important source of institutionalized power lies in the ability to structure information systems. Setting up committees to investigate particular organizational issues and having them report only to particular individuals or groups, facilitates their awareness of problems by members of those groups while limiting the awareness of problems by the members of other groups. Obviously, those who have information are in a better position to interpret the problems of an organization, re-

gardless of how realistically they may, in fact, do so.

Still another way to institutionalize power is to distribute rewards and resources. The dominant group may quiet competing interest groups with small favors and rewards. The credit for this artful form of cooptation belongs to Louis XIV. To avoid usurpation of his power by the nobles of France and the Fronde that had so troubled his father's reign, he built the palace at Versailles to occupy them with hunting and gossip. Awed, the courtiers basked in the reflected glories of the "Sun King" and the overwhelming setting he had created for his court.

At this point, we have not systematically studied the institutionalization of power. But we suspect it is an important condition that mediates between the environment of the organization and the capabilities of the organization for dealing with that environment. The more institutionalized power is within an organization, the more likely an organization will be out of phase with the realities it faces. President Richard Nixon's structuring of his White House is one of the better documented illustrations. If we go back to newspaper and magazine descriptions of how he organized his office from the beginning in 1968, most of what occurred subsequently follows almost as an afterthought. Decisions flowed through virtually only the small White House staff; rewards, small presidential favors of recognition, and perquisites were distributed by this staff to the loyal; and information from the outside world—the press, Congress, the people on the streets—was filtered by the staff and passed along only if initialed "bh." Thus it was not surprising that when Nixon met war protestors in the early dawn, the only thing he could think to talk about was the latest football game, so insulated had he become from their grief and anger.

One of the more interesting implications of institutionalized power is that executive turnover among the executives who have structured the organization is likely to be a rare event that occurs only under the most pressing crisis. If a dominant coalition is able to structure the organization and interpret the meaning of ambiguous events like declining sales and profits or lawsuits, then the "real" problems to emerge will easily be incorporated into traditional molds of thinking and acting. If opposition is designed out of the organization, the interpretations will go unquestioned. Conditions will remain stable until a crisis develops, so overwhelming and visible that even the most adroit rhetorician would be silenced.

IMPLICATIONS FOR THE MANAGEMENT OF POWER IN ORGANIZATIONS

While we could derive numerous implications from this discussion of power, our selection would have to depend largely on whether one wanted to increase one's power, decrease the power of others, or merely maintain one's position. More important, the real implications depend on the particulars of an organizational situation. To understand power in an organization one must begin by looking outside it—into the environment—for those groups that mediate the organization's outcomes but are not themselves within its control.

Instead of ending with homilies, we will end with a reversal of where we began. Power, rather than being the dirty business it is often made out to be, is probably one of the few mechanisms for reality testing in organizations. And the cleaner forms of power, the institutional forms, rather than having the virtues they are often credited with, can lead the organization to become out of touch. The real trick to managing power in organizations is to ensure somehow that leaders cannot be unaware of the realities of their environments and cannot avoid changing to deal with those realities.

That, however, would be like designing the "self-liquidating organization," an unlikely event since anyone capable of designing such an instrument would be obviously in control of the liquidations.

Management would do well to devote more attention to determining the critical contingencies of their environments. For if you conclude, as we do, that the environment sets most of the structure influencing organizational outcomes and problems, and that power derives from the organization's activities that deal with those contingencies, then it is the environment that needs managing, not power. The first step is to construct an accurate model of the environment, a process that is quite difficult for most organizations. We have recently started a project to aid administrators in systematically understanding their environments. From this experience, we have learned that the most critical blockage to perceiving an organization's reality accurately is a failure to incorporate those with the relevant expertise into the process. Most organizations have the requisite experts on hand but they are positioned so that they can be comfortably ignored.

One conclusion you can, and probably should, derive from our discussion is that power—because of the way it develops and the way it is used—will always result in the organization suboptimizing its performance. However, to this grim absolute, we add a comforting caveat: If any criteria other than power were the basis for determining an organization's decisions, the results would be even worse.

SELECTED BIBLIOGRAPHY

The literature on power is at once both voluminous and frequently empty of content. Some is philosophical musing about the concept of power, while other writing contains popularized palliatives for acquiring and exercising influence. Machiavelli's *The Prince*, if read carefully, remains the single best prescriptive treatment of power and its use. Most social scientists have approached power descriptively, attempting to understand how it is acquired, how it is used, and what its effects are. Mayer Zald's edited collection *Power in Organizations* (Vanderbilt University Press, 1970) is one of the more useful sets of thoughts about power from a sociological perspective, while James Tedeschi's edited book, *The Social Influence Processes* (Aldine-Atherton, 1972) represents the social psychological approach to understanding power and influence. The strategic contingencies's approach, with its emphasis on the importance of uncertainty for understanding power in organizations, is described by David Hickson and his colleagues in "A Strategic Contingencies Theory of Intraorganizational Power" (*Administrative Science Quarterly,* December 1971, pp. 216-229).

Unfortunately, while many have written about power theoretically, there have been few empirical examinations of power and its use. Most of the work has taken the form of case studies. Michel Crozier's *The Bureaucratic Phenomenon* (University of Chicago Press, 1964) is important because it describes a group's source of power as control over critical activities and illustrates how power is not strictly derived from hierarchical position. J. Victor Baldridge's *Power and Conflict in the University* (John Wiley & Sons, 1971) and Andrew Pettigrew's study of computer purchase decisions in one English firm (*Politics of Organizational Decision-Making,* Tavistock, 1973) both present insights into the acquisition and use of power in specific instances. Our work has been more empirical and comparative, testing more explicitly the ideas presented in this article. The study of university decision making is reported in articles in the June 1974, pp. 135-151, and December 1974, pp. 453-473, issues of the *Administrative Science Quarterly,* the insurance firm study in J. G. Hunt and L. L. Larson's collection, *Leadership Frontiers* (Kent State University Press, 1975), and the study of hospital administrator succession will appear in 1977 in the *Academy of Management Journal.*

Managing Careers: The Influence of Job and Group Longevities

Ralph Katz

Any serious consideration of organizational careers must eventually explore the dynamics through which the concerns, abilities, and experiences of individual employees combine and mesh with the demands and requirements of their employing work environments. How do employees' needs for security, equitable rewards, and opportunities for advancement and self-development, for example, interact with the needs of organizations for ensured profitability, flexibility, and innovativeness? More important, how should they interact so that both prescription sets are filled satisfactorily?

Further complexity is added to this "matching" process with the realization that interactions between individuals and organizations are not temporarily invariant but can shift significantly throughout workers' jobs, careers, and life cycles. As employees pass from one phase in their work lives to the next, different concerns and issues are emphasized; and the particular perspectives that result produce different behavioral and attitudinal combinations within their job settings. Over time, therefore, employees are continuously revising and adjusting their perspectives toward their organizations and their roles in them. And it is the perspective that one has formulated at a partic-

ular point in time that gives meaning and direction to one's work and to one's career.

Because the effectiveness of a given organizational unit ultimately depends on the combined actions and performances of its membership, we must begin to examine more systematically the impact of such varying perspectives on the predilections of unit members for particular kinds of activities, interactions, and collective judgments. Clearly, a better understanding of the substantive nature of such dispositions and behavioral tendencies will help clarify accommodation processes between organizations and individuals so that eventual problems can be dealt with to their mutual benefits. To accomplish such objectives, however, we need to develop more process-oriented frameworks for analyzing the diverse kinds of concerns and associated behaviors that tend to preoccupy and characterize employees as they proceed through their respective jobs, project groups, and organizational careers.

A MODEL OF JOB LONGEVITY

Based on some recent findings in the areas of job satisfaction and task redesign, Katz (1980) has been working to develop a more general theory for describing how employees' perspectives unfold and change as they journey through their own discrete sequences of job

Adapted from "Managing Careers: The Influence of Job and Group Longevities," by Ralph Katz, in *Career Issues in Human Resource Management,* edited by Ralph Katz, pp. 154–181 © 1982. Reprinted by permission of Prentice–Hall Inc., Englewood Cliffs, NJ.

situations. In particular, a three-transitional stage model of job longevity has been proposed to illustrate how certain kinds of concerns might change in importance according to the actual length of time an employee has been working in a given job position. Generally speaking, each time an employee is assigned to a new job position within an organization, either as a recent recruit or through transfer or promotion, the individual enters a relatively brief but nevertheless important "socialization" period. With increasing familiarity about his or her new job environment, however, the employee soon passes from socialization into the "innovation" stage, which, in turn, slowly shifts into a "stabilization" state as the individual gradually adapts to extensive job longevity, (i.e., as the employee continues to work in the same overall job for an extended period of time). Table 1 summarizes the sequential nature of these three stages by comparing some of the different kinds of issues affecting employees as they cycle through their various job positions.[1]

Socialization

As outlined under the initial socialization stage, employees entering new job positions are concerned primarily with reality construction,

TABLE 1. *A Model of Job Longevity*

Job Longevity Stages	Primary Areas of Concern
Stage 1. SOCIALIZATION: Reality Construction[a]	a) To build one's situational identity b) To decipher situational norms and identify acceptable, rewarded behaviors c) To build social relationships and become accepted by others d) To learn supervisory, peer, and subordinate expectations e) To prove oneself as an important, contributing member
Stage 2. INNOVATION: Influence, Achievement, and Participation	a) To be assigned challenging work b) To enhance one's visibility and promotional potential c) To improve one's special skills and abilities d) To enlarge the scope of one's participation and contribution e) To influence one's organizational surroundings
Stage 3. STABILIZATION: Maintenance, Consolidation, and Protection	a) To routinize one's task activities b) To preserve and safeguard one's task procedures and resources c) To protect one's autonomy d) To minimize one's vulnerability e) To cultivate and solidify one's social environment

[a]The listed items are not meant to be exhaustive; rather they are intended to illustrate both the domain and the range of issues within each stage.

building more realistic understandings of their unfamiliar social and task environments. In formulating their new perspectives, they are busily absorbed with problems of establishing and clarifying their own situational roles and identities and with learning all the attitudes and behaviors that are appropriate and expected within their new job settings. Estranged from their previous work environments and supporting relationships, newcomers must construct situational definitions that allow them to understand and interpret the myriad of experiences associated with their new organizational memberships. They need, for example, to learn the customary norms of behavior, decipher how reward systems actually operate, discover supervisory expectations, and more generally learn how to function meaningfully within their multiple group contexts (Schein, 1978). Through information communicated by their new "significant others," newcomers learn to develop perceptions of their own roles and skills that are both supported within their new surroundings and which permit them to organize their activities and interactions in a meaningful fashion. As pointed out by Hughes (1958) in his discussion of "reality shock," when new employees suddenly discover that their somewhat "overglorified" work-related expectations are neither realistic nor mutually shared by their boss or co-workers, they are likely to feel disenchanted and will experience considerable pressure to either redefine more compatible expectations or terminate from their work settings.

The importance of such a "breaking-in" period has long been recognized in discussions of how social processes affect recent organizational hires trying to make sense out of their newfound work experiences. What is also important to recognize is that veteran employees must also relocate or "socialize" themselves following their displacements into new job positions within their same organizations (Wheeler, 1966). Just as organizational newcomers have to define and interpret their new territorial domains, veteran employees must also restructure and reformulate perceptions regarding their new social and task realities.[2] As they assume new organizational positions and enter important new relationships, veterans must learn to integrate their new perceptions and experiences with prior organizational knowledge in order to develop fresh situational perspectives, including perceptions about their own self-images and their images of other organizational members.

Such perceptual revisions are typically necessary simply because work groups and other organizational subunits are often highly differentiated with respect to their idiosyncratic sets of norms, beliefs, perceptions, time perspectives, shared language schemes, goal orientations, and so on (Lawrence and Lorsch, 1967). As communications and interactions within an organizational subunit continue to take place or intensify, it is likely that a more common set of understandings about the subunit and its environment will develop through informational social influence. Such shared meanings and awarenesses not only provide the subunit's members with a sense of belonging and identity but will also demarcate the subunit from other organizational entities (Pfeffer, 1981). Consequently, as one shifts job positions and moves within the organization, one is likely to encounter and become part of a new set of groups with their correspondingly different belief systems and perspectives about themselves, their operations, and their operating environments. It is in this initial socialization period, therefore, that organizational employees, and newcomers in particular, learn not only the technical requirements of their new job assignments but also the interpersonal behaviors and social attitudes that are acceptable and necessary for becoming a true contributing member.

Since employees in the midst of socialization are strongly motivated to reduce ambiguity by creating order out of their somewhat vague and unfamiliar surroundings, it becomes clear why a number of researchers have discovered organizational newcomers being especially concerned with psychological safety and security and with clarifying their new situational identities (Kahn et al., 1964; Hall and Nougaim, 1968). In a similar vein, Schein (1971) suggests that to become accepted and to prove one's competence represent two major problems that newcomers and veterans must face before they can function comfortably within their new job positions. It is these kinds of concerns that help to explain why Katz (1978a) discovered that during the initial months of their new job positions, employees are not completely ready to respond positively to all the challenging characteristics of their new task assignments. Instead, they appear most responsive to job features that provide a sense of personal acceptance and importance as well as a sense of proficiency through feedback and individual guidance.[3] Van Maanen's (1975) study of urban police socialization also demonstrated that for about the first three or four months of their initial job assignments, police recruits are busily absorbed in the process of changing and solidifying their own self- and job-related perceptions as they finally come to know the actual attitudes and behaviors of their veteran counterparts.

How long this initial socialization period lasts, therefore, probably depends on how long it takes employees to feel accepted and competent within their new work environments. Not only is the length of such a time period greatly influenced by the abilities, needs, and prior experiences of individual workers and influenced as well by the clarity and usefulness of the interpersonal interactions that take place, but it also probably differs significantly across occupations. Based on the retrospective answers of his hospital employee sample, for example, Feldman (1977) reports that on the average, accounting clerks, registered nurses, and engineering tradesmen reporting feeling accepted

after one, two, and four months, respectively although they did not feel completely competent until after three, six, and eight months, respectively. Generally speaking, one might posit that the length of one's initial socialization period varies positively with the level of complexity within one's job and occupational requirements, ranging perhaps from as little as a month or two on very routine, programmed-type jobs to as much as a year or more on very skilled, unprogrammed-type jobs, as in the engineering and scientific professions. With respect to engineering, for example, it is generally recognized that a substantial socialization period is often required before engineers can fully contribute within their new organizational settings, using their particular knowledge and technical specialties. Thus, even though one might have received an excellent education in mechanical engineering principles at a university or college, one must still figure out from working and interacting with others in the setting how to be an effective mechanical engineer at Westinghouse, DuPont, or Procter and Gamble.[4]

Innovation

With time, interaction, and increasing familiarity, employees soon discover how to function appropriately in their jobs and to feel sufficiently secure in their perceptions of their workplace. Individual energies can now be devoted more toward task performance and accomplishment instead of being expended on learning the previously unfamiliar social knowledge and skills necessary to makes sense out of one's work-related activities and interactions. As a result, employees become increasingly capable of acting in a more responsive, innovative, and undistracted manner.

The movement from socialization to the innovation stage of job longevity implies that employees no longer require much assistance in deciphering their new job and organizational surroundings. Having adequately constructed their own situational definitions during the socialization period, employees are now freer to participate within their own conceptions of organizational reality. They are now able to divert their attention from an initial emphasis on psychological safety and acceptance to concerns for achievement and influence. Thus, what becomes progressively more pertinent to employees as they proceed from socialization to the innovation stage are the opportunities to participate and grow within their job settings in a very meaningful and responsible manner.

The idea of having to achieve some reasonable level of psychological safety and security in order to be fully responsive to challenges in the work setting is very consistent with Kuhn's (1963) concept of "creative tensions." According to Kuhn, it is likely that only when conditions of *both* stability and challenge are present can the creative tensions between them generate considerable innovative behavior. Growth theorists such as Maslow (1962) and Rogers (1961) have similarly argued that the presence of psychological safety is one of the chief prerequisites for self-direction and individual responsiveness. For psychological safety to occur, however, individuals must be able to understand and attach sufficient meaning to the vast array of events, interactions, and information flows involving them throughout their workdays. Of particular importance to growth theorists is the idea that employees must be able to expect positive results to flow from their individual actions. Such a precondition implies that employees must have developed sufficient knowledge about their new job situations in order for there to be enough predictability for them to take appropriate kinds of actions.[5]

A similar point of view is taken by Staw (1977) when he argues that if employees truly expect to improve their overall job situations, they must first learn to predict their most relevant set of behavioral-outcome contingencies before they try to influence or increase their control over them. One must first construct a reasonably valid perspective about such contingencies before one can sensibly strive to manage them for increasingly more favorable outcomes. In short, there must be sufficient awareness of one's environment, sufficient acceptance and competence within one's setting, and sufficient openness to new ideas and experiences in order for employees to be fully responsive to the "richness" of their job demands.

Stabilization

As employees continue to work in their same overall job settings for a considerable length of time, without any serious disruption or displacement, they may gradually proceed from innovation to stabilization in the sense of shifting from being highly involved in and receptive to their job demands to becoming progressively unresponsive. For the most part, responsive individuals prefer to work at jobs they find stimulating and challenging and in which they can self-develop and grow. With such kinds of activities, they are likely to inject greater effort and involvement into their tasks which, in turn, will be reflected in their performances (Hackman and Oldham, 1975; Katz, 1978b). It seems reasonable to assume, however, that in time even the most challenging job assignments and responsibilities can appear less exciting and more habitual to jobholders who have successfully mastered and become increasingly accustomed to their everyday task requirements. With prolonged job longevity and stability, therefore, it is likely that employees' perceptions of their present conditions and of their future possibilities will become increasingly impoverished. They may begin essentially to question the value of what they are doing and where it may lead. If employees cannot maintain, redefine, or expand their jobs for continual challenge and growth, the substance and meaning of their work begins to deteriorate.

Enthusiasm wanes, for what was once challenging and exciting may no longer hold much interest at all.

At the same time, it is also important to mention that if an individual is able to increase or even maintain his or her own sense of task challenge and excitement on a given job for an extended period of time, then instead of moving toward stabilization, the process might be the reverse (i.e., continued growth and innovation). As before, the extent to which an individual can maintain his or her responsiveness on a particular job strongly depends on the complexity of the underlying tasks as well as on the individual's own capabilities, needs, and prior experiences. With respect to individual differences, for example, Katz's (1978b) findings suggest that employees with high growth needs are able to respond to the challenging aspects of their new jobs sooner than employees with low growth needs. At the same time, however, high-order-need employees might not retain their responsiveness for as long a job period as employees with low-growth-need strength.

It should also be emphasized that in addition to job longevity, many other contextual factors can affect a person's situational perspective strongly enough to influence the level of job interest as one continues to work in a given job position over a long period of time. New technological developments, rapid growth and expansion, the sudden appearance of external threats, or strong competitive pressures could all help sustain or even enhance an individual's involvement in his or her job-related activities. On the other hand, having to work closely with a group of unresponsive peers might shorten an individual's responsive period on that particular job rather dramatically. Clearly, the reactions of individuals are not only influenced by psychological predispositions and personality characteristics but also by individuals' definitions of and interactions with their overall situational settings (Homans, 1961; Salancik and Pfeffer, 1978).

Generally speaking, however, as tasks become progressively less stimulating to employees with extended job longevity, they can either leave the setting or remain and adapt to their present job situations (Argyris, 1957). In moving from innovation to stabilization, it is suggested that employees who continue to work in their same overall job situations for long periods of time gradually succeed in adapting to such steadfast employment by becoming increasingly indifferent and unresponsive to the challenging task features of their job assignments (Katz, 1978a). In the process of adaptation, they may also redefine what they consider to be important, most likely by placing relatively less value on intrinsic kinds of work issues. The findings of Kopelman (1977) and Hall and Schneider (1973) suggest, for example, that when individuals perceive their opportunities for intrinsic-type satisfactions and challenges to be diminishing, they begin to match such developments by placing less value on such types of expectations. And as employees come to care less about the intrinsic nature of the actual work they do, the greater their relative concern for certain contextual features, such as salary, benefits, vacations, friendly co-workers, and compatible supervision.

The passage from innovation to stabilization is not meant to suggest that job satisfaction necessarily declines with long-term job longevity. On the contrary, it is likely that in the process of adaptation, employees' expectations have become adequately satisfied as they continue to perform their familiar duties in their normally acceptable fashions. If aspirations are defined as a function of the disparity between desired and expected (Kiesler, 1978), then as long as what individuals desire is reasonably greater than what they can presently expect to attain, there will be energy for change and achievement. On the other hand, when employees arrive at a stage where their chances for future growth and challenges in their jobs are perceived to be remote, then as they adapt, it is likely that existing situations will become accepted as the desired and aspirations for growth and change will have been reduced. As a result, the more employees come to accept their present circumstances, the stronger the tendency to keep the existing work environment fairly stable. Career interests and aspirations may become markedly constricted, for in a sense, adapted employees may simply prefer to enjoy rather than try to add to their present job accomplishments.

Underpinning the descriptive changes represented by the stabilization stage is the basic idea that over time individuals try to organize their work lives in a manner that reduces the amount of stress they must face and which is also low in uncertainty (Pfeffer, 1980; Staw, 1977). Weick (1969) also relies on this perspective when he contends that employees seek to "enact" their environments by directing their activities toward the establishment of a workable level of certainty and clarity. In general, one might argue that employees strive to bring their work activities into a state of equilibrium where they are more capable of predicting events and of avoiding potential conflicts.[6]

Given such developmental trends, it seems reasonable that with considerable job longevity, most employees have been able to build a work pattern that is familiar and comfortable, a pattern in which routine and precedent play a relatively large part. According to Weick (1969), as employees establish certain structures of interlocked behaviors and relationships, these patterns will in time become relatively stable simply because they provide certainty and predictability to these interlinked employees. It is further argued here that as individuals adapt to their long-term job tenure and become progressively less responsive to their actual task demands, they will come to rely more on these established modes of conduct to complete their everyday job requirements. Most likely, adapted employees feel safe and

comfortable in such stability, for its keeps them feeling secure and confident in what they do, yet requires little additional vigilance or effort. In adapting to extended job longevity, therefore, employees become increasingly content and ensconced in their customary ways of doing things, in their comfortable routines and interactions, and in their familiar sets of task demands and responsibilities.

If change or uncertainty is seen by individuals in the stabilization period as particularly disruptive, then the preservation of familiar routines and patterns of behavior is likely to be of prime concern. Given such a disposition, adapted employees are probably less receptive toward any change or toward any information that might threaten to disturb their developing sense of complacency. Rather than striving to enlarge the scope of their job demands, they may be more concerned with maintaining their comfortable work environments by protecting themselves from sources of possible interference, from activities requiring new kinds of attention, or form situations that might reveal their shortcomings. Adapted employees, for example, might seek to reduce uncertainty in their day-to-day supervisory dealings perhaps by solidifying their attractiveness through ingratiating kinds of behavior (Wortman and Linsenmeier, 1977) or perhaps by isolating themselves from such supervisory contacts (Pelz and Andrews, 1966). Or they might seek to reduce uncertainty by trying to safeguard their personal allocations of resources and rewards through the use of standardized practices and policies. Whatever the specific behaviors that eventually emerge in a given setting, it is likely that employees who have become unresponsive to the challenging features of their assigned tasks will strongly resist events threatening to introduce uncertainty into their work environments.

One of the best examples of the effects of such long-term stability can still be found in Chinoy's (1955) classic interviews of automobile factory workers. Chinoy discovered that although almost 80% of the workers had wanted to leave their present jobs at one time or another, very few could actually bring themselves to leave. Most of the workers were simply unwilling to give up the predictability and comfortableness of their presently familiar routines and cultivated relationships for the uncertainties of a new job position.

SITUATIONAL VERSUS INDIVIDUAL CONTROL

In presenting this three-stage model of job longevity, I have tried to describe some of the major concerns affecting employees as they enter and adapt to their particular job positions. Of course, the extent to which any specific individual is affected by these issues depends on the particular perceptual outlook that has been developed over time through job-related activities and through role-making processes with other individuals, including supervisors, subordinates, and peers (Weick, 1969; Graen, 1976). Employees, as a result, learn to cope with their particular job and organizational environments through their interpretations of relevant work experiences as well as their expectations and hopes of the future. To varying degrees, then, situational perspectives are derivatives of both retrospective and prospective processes, in that they are built and shaped through knowledge of past events and future anticipations.

One of the more important aspects of the socialization process, however, is that the information and knowledge previously gathered by employees from their former settings are no longer sufficient or necessarily appropriate for interpreting or understanding their new organizational domains. Newcomers, for instance, have had only limited contact within their new institutional surroundings from which to construct their perceptual views. Similarly, the extent to which veterans who are assuming new job positions can rely on their past organizational experiences and perspectives to function effectively within their new work settings can also be rather limited, depending of course on their degrees of displacement.

Essentially, individuals in the midst of socialization are trying to navigate their way through new and unfamiliar territories without the aid of adequate or even accurate perceptual maps. During this initial period, therefore, they are typically more malleable and more susceptible to change (Schein, 1968). In a sense, they are working under conditions of high "situational control" in that they must depend on other individuals within their new situations to help them define and interpret the numerous activities taking place around them. The greater their unfamiliarity or displacement within their new organizational areas, the more they must rely on their situations to provide the necessary information and interactions by which they can eventually construct their own perspectives and reestablish new situational identities. And it is precisely this external need or "situational dependency" that enables these individuals to be more easily influenced during their socialization processes through social interactions (Salancik and Pfeffer, 1978; Katz, 1980).

As employees become increasingly cognizant of their overall job surroundings, however, they also become increasingly capable of relying on their own perceptions for interpreting events and executing their everyday task requirements. In moving from socialization into the innovation or stabilization stage, employees have succeeded in building a sufficiently robust situational perspective, thereby freeing themselves to operate more self-sufficiently within their familiar work settings. They are now working under conditions of less "situational" but more "individual" control, in the sense that they are now better equipped to determine for themselves the importance and meaning of the various events and information flows surrounding them. Having established their own social and task supports, their own perceptual outlooks, and their own situational identities, they become less easily changed and less easily manipulated. As pointed out by Schein (1973), when individuals no longer have to balance their situational perspectives against the views of significant others within their settings, they become less susceptible to change and situational influences. Thus, movement through the three stages of job longevity can also be characterized, as shown in Figure 1, by relative shifts to more individual and less situational control.

As the locus of "control" shifts with increasing job longevity and individuals continue to stabilize their situational definitions, other important behavioral tendencies could also materialize. In particular, strong biases could develop in the way individuals select and interpret information, in their cognitive abilities to generate new options and strategies creatively, and in their willingness to innovate or implement alternative courses of action. Table 2 outlines in more detail some of the specific possibilities within each of these three general areas. Fur-

thermore, it is the capacity either to prevent or overcome these kinds of tendencies that is so important to the long-term success of organizations; for, over time, each of these trends could lead to less effective performance and decision-making outcomes.

Problem-Solving Processes

It has been argued throughout this paper that as employees gradually adapt to prolonged periods of job longevity, they may become less receptive toward any change or innovation threatening to disrupt significantly their comfortable and predictable work practices and patterns of behavior. Individuals, instead, are more likely to develop reliable and effective routine responses (i.e., standard operating procedures) for dealing with their frequently encountered tasks in order to ensure predictability, coordination, and economical information processing. As a result, there may develop over time increasing rigidity in one's problem-solving activities—a kind of functional fixedness that reduces the individual's capacity for flexibility and openness to change. Responses and decisions are made in their fixed, normal patterns while novel situations requiring re-

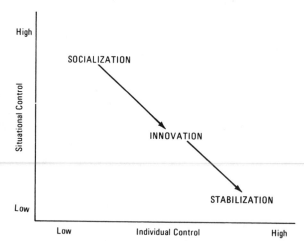

FIGURE 1.   *Situational versus Individual Control along the Job-Longevity Continuum*

TABLE 2. *Representative Trends Associated with Long-term Job Longevity*

Problem-solving processes
  Increased rigidity
  Increased commitment to established practices
    and procedures
  Increased mainlining of strategies

Information processes
  Increased insulation from critical areas
  Increased selective exposure
  Increased selective perception

Cognitive processes
  Increased reliance on own experiences and
    expertise
  Increased narrowing of cognitive abilities
  Increased homophyly

sponses that do not fit such established molds are either ignored or forced into these molds. New or changing situations either trigger responses of old situations or trigger no responses at all. It becomes, essentially, a work world characterized by the phrase "business as usual."

Furthermore, as individuals continue to work by their well-established problem-solving strategies and procedures, the more committed they may become to such existing methods. Commitment is a function of time, and the longer individuals are called upon to follow and justify their problem-solving approaches and decisions, the more ingrained they are likely to become. Drawing from his work on decision making, Allison (1971) strongly warns that increasing reliance on regularized practices and procedures can become highly resistant to change, since such functions become increasingly grounded in the norms and basic attitudes of the organizational unit and in the operating styles of its members. Bion (1961) and Argyris (1969) even suggest that it may be impossible for individuals to break out of fixed patterns of activity and interpersonal behavior without sufficiently strong outside interference or help.

With extended job tenure, then, prob-lem-solving activities can become increasingly guided by consideration of methods and programs that have worked in the past. Moreover, in accumulating this experience and knowledge, alternative ideas and approaches were probably considered and discarded. With such refutations, however, commitments to the present courses of action can become even stronger—often to the extent that these competing alternatives are never reconsidered.[7] In fact, individuals can become overly preoccupied with the survival of their particular approaches, protecting them against fresh approaches or negative evaluations. Much of their energy becomes directed toward "mainlining their strategies," that is, making sure their specific solution approaches are selected and followed. Research by Janis and Mann (1977) and Staw (1980) has demonstrated very convincingly just how strongly committed individuals can become to their problem-solving approaches and decisions, even in the face of adverse information, especially if they feel personally responsible for such strategies.

Information Processes
One of the potential consequences of developing this kind of "status-quo" perspective with respect to problem-solving activity is that employees may also become increasingly insulated from outside sources of relevant information and important new ideas. As individuals become more protective of and committed to their current work habits, the extent to which they are willing or even feel they need to expose themselves to new or alternative ideas, solution strategies, or constructive criticisms becomes progressively less and less. Rather than becoming more vigilant about events taking place outside their immediate work settings, they may become increasingly complacent about external environmental changes and new technological developments.

In addition to this possible decay in the amount of external contact and interaction, there may also be an increasing tendency for individuals to communicate only with those whose ideas are in accord with their current interests, needs, or existing attitudes. Such a tendency is referred to as selective exposure. Generally speaking, there is always the tendency for individuals to communicate with those who are most like themselves (Rogers and Shoemaker, 1971). With increasing adaptation to long-term job longevity and stability, however, this tendency is likely to become even stronger. Thus, selective exposure may increasingly enable these individuals to avoid information and messages that might be in conflict with their current practices and dispositions.

One should also recognize, of course, that under these kinds of circumstances, any outside contact or environmental information that does become processed by these long-tenured individuals might not be viewed in the most open and unbiased fashion. Janis and Mann (1977), for example, discuss at great length the many kinds of cognitive defenses and distortions commonly used by individuals in processing outside information in order to support, maintain, or protect certain decisional policies and strategies. Such defenses are often used to argue against any disquieting information and evidence in order to maintain self-esteem, commitment, and involvement. In particular, selective perception is the tendency to interpret information and communication messages in terms favorable to one's existing attitudes and beliefs. And it is this combination of increasing insulation, selective exposure, and selective perception that can be so powerful in keeping critical information and important new ideas and innovations from being registered.

Cognitive Processes

As individuals become more comfortable and secure in their long-tenured work environments, their desire to seek out and actively internalize new knowledge and new develop-ments may begin to deteriorate. Not only may they become increasingly isolated from outside sources of information, but their willingness to accept or pay adequate attention to the advice and ideas of fellow experts may become less and less. Unlike the socialization period in which individuals are usually very attentive to sources of expertise and influence within their new job settings, individuals in the stabilization stage have probably become significantly less receptive to such information sources. They may prefer, instead, to rely on their own accumulated experience and wisdom and consequently are more apt to dismiss the approaches, advice, or critical comments of others. As a result, adapted employees may be especially defensive with regard to critical evaluations and feedback messages, whether they stem from sources of outside expertise or from internal supervision.

It should also not be surprising that with increasing job stability one is more likely to become increasingly specialized, that is, moving from broadly defined capabilities and solution approaches to more narrowly defined interests and specialties. Without new challenges and opportunities, the diversity of skills and of ideas generated are likely to become narrower and narrower. And as individuals welcome information from fewer sources and are exposed to fewer alternative points of view, the more constricted their cognitive abilities can become. Essentially, there can be a narrowing of one's cognitive processes, resulting in a more restricted perspective of one's situation, coupled with a more limited set of coping responses. Such a restricted outlook, moreover can be very detrimental to the organization's overall effectiveness, for it could lead at times to the screening out of some vitally important environmental information cues.

Homophyly refers to the degree to which interacting individuals are similar with respect to certain attributes, such as beliefs, values, education, and social status (Rogers and

Shoemaker, 1971). Not only is there a strong tendency for individuals to communicate with those who are most like themselves, but it is also likely that continued interaction can lead to greater homophyly in knowledge, beliefs, and problem-solving behaviors and perceptions (Burke and Bennis, 1961; Pfeffer, 1980). The venerable proverb "birds of a feather flock together" makes a great deal of sense, but it may be just as sensible to say that "when birds flock together, they become more of a feather." Accordingly, as individuals stabilize their work settings and patterns of communication, a greater degree of homophyly is likely to have emerged between these individuals and those with whom they have been interacting over the long tenure period. And any increase in homophyly could lead in turn to further stability in the communications of the more homophilous pairs, thereby increasing their insulation from heterophilous others. Thus, it is possible for the various trends to feed on each other. Finally, it should be mentioned that although individuals may be able to coordinate and communicate with homophilous partners more effectively and economically, such interactions are also more likely to yield less creative and innovative outcomes (Pelz and Andrews, 1966).

Longevity and Performance

These problem-solving, informational, and cognitive tendencies, of course, can be very serious in their consequences, perhaps even fatal. Much depends, however, on the nature of the work being performed and on the extent to which such trends actually transpire. The performances of individuals working on fairly routine, simple tasks in a rather stable organizational environment, for example, may not suffer as a result of these trends, for their own knowledge, experiences, and abilities become sufficient. Maintaining or improving on one's routine behaviors is all that is required—at least for as long as there are no changes and no new

developments. However, as individuals function in a more rapidly changing environment and work on more complex tasks requiring greater levels of change, creativity, and informational vigilance, the effects of these long-term longevity trends are likely to become significantly more dysfunctional.

GROUP LONGEVITY

The degree to which any of these previously described trends actually materializes for any given individual depends, of course, on the overall situational context. Individuals' perceptions and responses do not take place in a social vacuum but develop over time as they continue to interact with various aspects of their job and organizational surroundings (Crozier, 1964; Katz and Van Maanen, 1977). And in any job setting one of the most powerful factors affecting individual perspectives is the nature of the particular group or project team in which one is a functioning member (Schein, 1978; Katz and Kahn, 1978).

Ever since the well-known Western Electric Studies (Cass and Zimmer, 1975), much of our research in the social sciences has been directed toward learning just how strong group associations can be in influencing individual member behaviors, motivations, and attitudes (Asch, 1956; Shaw, 1971; Katz, 1977). From the diffusion of new innovations (Robertson, 1971) to the changing of meat consumption patterns to less desirable but more plentiful cuts (Lewin, 1965) to the implementation of job enrichment (Hackman, 1978), group processes and effects have been extremely critical to more successful outcomes. The impact of groups on individual responses is substantial, if not pervasive, simply because groups mediate most of the stimuli to which their individual members are subjected while fulfilling their everyday task and organizational requirements. Accordingly, whether individuals experiencing long-term job longevity enter the stabilization period and become subjected to the tendencies previously described may strongly depend on the particular reinforcements, pressures, and behavioral norms encountered within their immediate project or work groups (Likert, 1967; Weick, 1969).

Generally speaking, as members of a project group continue to work together over an extended period of time and gain experience with one another, their patterns of activities are likely to become more stable, with individual role assignments becoming more well-defined and resistant to change (Bales, 1955; Porter et al., 1975). Emergence of the various problem-solving, informational, and cognitive trends, therefore, may be more a function of the average length of time the group members have worked together (i.e., group longevity) rather than varying according to the particular job longevity of any single individual. A project group, then, might either exacerbate or ameliorate the various trends (e.g., insulation from outside developments and expertise), just as previous studies have shown how groups can enforce or amplify certain standards and norms of individual behavior (e.g., Seashore, 1954; Stoner, 1968). Thus, it may be misleading to investigate the responses and reactions of organizational individuals as if they functioned as independent entities; rather, it may be more insightful to examine the distribution of responses as a function of different project teams, especially when project teams are characterized by relatively high levels of group longevity.

CONCLUSIONS

What is suggested by this discussion of job and group longevities is that employee perspectives and behaviors, and their subsequent effects on performance, might be significantly managed through staffing and career decisions. One could argue, for example, that the energizing and destabilizing function of new team members can be very important in preventing a project group from developing some of the tendencies previously described for long-tenured individuals, including insulation from key communication areas. The benefit of new team members is that they may have a relative advantage in generating fresh ideas and approaches. With their active participation, existing group members might consider more carefully ideas and alternatives they might have otherwise ignored or dismissed. In short, project newcomers can represent a novelty-enhancing condition, challenging and improving the scope of existing methods and accumulated knowledge.[8]

The longevity framework also seems to suggest that periodic job mobility or rotation might help prevent employees from moving into a stabilization stage. As long as the socialization period is positively negotiated, employees can simply cycle from one innovation period into another.[9] Put simply, movements into new positions may be necessary to keep individuals stimulated, flexible, and vigilant with respect to their work environments. Within a single job assignment, the person may eventually reach the limit to which new and exciting challenges are possible or even welcomed. At that point, a new job position may be necessary. To maintain adaptability and to keep employees responsive, what might be needed are career histories containing sequences of job positions involving new challenges and requiring new skills (Kaufman, 1974). As pointed out by Schein (1968), continued growth and development often come from adaptations to new or changing work environments requiring individuals to give up familiar and stable work patterns in favor of developing new ones.

As important as job mobility is, it is probably just as important to determine whether individuals and project groups can circumvent the effects of longevity without new assignments or rejuvenation from new project members. Rotations and promotions are not always possible, especially when there is little organizational growth. As a result, we need to learn considerably more about the effects of increasing job and group longevities. Just how deterministic are the trends? Can long-tenured individuals and project teams remain high-performing, and if so, how can it be accomplished?

In a general sense, then, we need to learn how to detect the many kinds of changes that either have taken place or are likely to take place within a group as its team membership ages. Furthermore, we need to learn if project groups can keep themselves energized and innovative over long periods of group longevity, or whether certain kinds of organizational structures and managerial practices are needed to keep a project team effective and high-performing as it ages.

In response to this issue, Tom Allen and I have undertaken an extensive study in 12 different organizations involving over 200 R&D project teams of which 50 or so have group longevity scores that exceed 5 years. More interesting, it turns out that a large number of these long-tenured project groups were judged to be high-performing teams. Although we are still processing the data, preliminary analyses suggest that the nature of the project's supervision may be the most important factor differentiating the more effective long-tenured teams from the less effective ones. In particular, engineers belonging to the high-performing, long-tenured project groups perceived their functional supervision to be significantly higher (1) in disseminating technical information, (2) in being well-informed professionally, and (3) in being concerned about their professional development.[10]

Such findings suggest that a strong functional competency dimension may be especially important in the effective management of long-tenured project groups. With respect to R&D settings, this may imply that the presence of certain technical specialists, labeled gatekeepers by Allen (1977), may be especially important to the success of long-term R&D project teams.[11] Such a role requirement may be necessary because with long-term group longevity, many project members have become increasingly overspecialized and more "locally" oriented (i.e., more organizationally oriented), thereby making it increasingly difficult for them to communicate effectively with outside sources of technology or with keeping themselves up to date within their technical specialities.

In a broader context, we need to learn how to manage workers, professionals, and project teams as they enter and proceed through different stages of longevity. Clearly, different kinds of managerial styles and behaviors may be more appropriate at different stages of longevity. Delegative or participative management, for example, may be very effective when individuals are vigilant and highly responsive to their work demands, but such supervisory activities may prove less successful when employees are unresponsive to their job environments, as in the stabilization stage. Furthermore, as perspectives and responsiveness shift over time, the actions required of the managerial role will also vary. Managers may be effective, then, to the extent they are able to recognize and cover such changing conditions. Thus, it may be the ability to manage change—the ability to diagnose and manage between socialization and stabilization—that we need to learn so much more if we truly hope to provide careers that keep employees responsive and also keep organizations effective.

NOTES

1. For a more extensive discussion of the job-longevity model, see Katz (1980). In the current presentation, the term "stabilization" is used in

place of "adaptation" since individuals are in effect adapting to their job situations in all three stages, albeit in systematically different ways.

2. The extent to which a veteran employee actually undergoes socialization depends on how displaced the veteran becomes in undertaking his or her new job assignment. Generally speaking, the more displaced veterans are from their previously familiar task requirements and interpersonal associations, the more intense the socialization experience.

3. After comparing the socialization reactions of veterans and newcomers, Katz (1978a) suggests that newcomers may be especially responsive to interactional issues involving personal acceptance and "getting on board," whereas veterans may be particularly concerned with reestablishing their sense of competency in their newly acquired task assignments.

4. One of the factors contributing to the importance of this socialization period lies in the realization that engineering strategies and solutions within organizations are often not defined in very generalizeable terms but are peculiar to their specific settings (Allen, 1977; Katz and Tushman, 1979). As a result, R&D project groups in different organizations may face similar problems yet may define their solution approaches and parameters very differently. And it is precisely because technical problems are typically expressed in such "localized" terms that engineers must learn how to contribute effectively within their new project groups.

5. It is also interesting to note that in discussing his career-anchor framework, Schein (1978) points out that career anchors seem to represent a stable concept around which individuals are able to organize experiences and direct activities. Furthermore, it appears from Schein's research that it is within this area of stability that individuals are most likely to self-develop and grow.

6. There are, of course, alternative arguments, such as in activation theory (Scott, 1966), suggesting that people do in fact seek uncertainty, novelty, or change. The argument here, however, is that as individuals adapt and become increasingly indifferent to the task challenges of their jobs, it is considerably more likely that

they will strive to reduce uncertainty and maintain predictability rather than the reverse.

7. As shown by Allen's (1966) research on parallel project efforts, such reevaluations can be very important in reaching more successful outcomes.

8. As discussed by Van Maanen (1975), the socialization process of individuals can greatly affect the extent to which newcomers may be willing to try to innovate on existing "wisdoms."

9. A discussion on effectively managing the socialization process is beyond the scope of this paper. The reader is referred to the descriptive theory presented in Schein (1968), Kotter (1973), Hall (1976), Katz (1980), and Wanous (1980).

10. For the 40 long-tenured project groups, the significant correlations between project performance ratings and project member perceptions of these three supervisory activities were .54, .58, and .44, respectively.

11. It is interesting to note that in the data presented from the large R&D facility, none of the long-tenured development project teams had a technical gatekeeper as part of their team membership.

REFERENCES

Allen, T.J. "Studies of the problem-solving processes in engineering designs." *IEEE Transactions in Engineering Management*, 1966, *13*, 72–83.

Allen, T.J. *Managing the Flow of Technology.* Cambridge, Mass.: M.I.T. Press, 1977.

Allison, G.T. *Essence of Decision: Explaining the Cuban Missile Crisis.* Boston: Little, Brown, 1971.

Argyris, C. *Personality and Organization.* New York: Harper Torch Books, 1957.

Argyris, C. "The incompleteness of social psychological theory: examples from small group, cognitive consistency and attribution research." *American Psychologist*, 1969, *24*, 893–908.

Asch, S.E. "Studies of independence and conformity: a minority of one against a unanimous majority." *Psychological Monographs*, 1956, *70*.

Bales, R.F. "Adaptive and integrative changes as sources of strain in social systems." In A.P.

Hare, E.F. Borgatta, and R.F. Bales, eds., *Small Groups: Studies in Social Interaction*. New York: Knopf, 1955, pp. 127–31.

Bion, W.R. *Experiences in Groups*. New York: Basic Books, 1961.

Burke, R.L., and Bennis, W.G. "Changes in perception of self and others during human relations training." *Human Relations*, 1961, *14*, 165–82.

Cass, E.L., and Zimmer, F.G. *Man and Work in Society*. New York: Van Nostrand Reinhold, 1975.

Chinoy, E. *Automobile Workers and the American Dream*. Garden City, N.Y.: Doubleday, 1955.

Crozier, M. *The Bureaucratic Phenomenon*. Chicago: University of Chicago Press, 1964.

Dewhirst, H., Arvey, R., and Brown, E. "Satisfaction and performance in research and development tasks as related to information accessibility." *IEEE Transactions on Engineering Management*, 1978, *25*, 58–63.

Dubin, S.S. *Professional Obsolescence*. Lexington, Mass.: Lexington Books, D.C. Heath, 1972.

Feldman, D. "The role of initiation activities in socialization." *Human Relations*, 1977, *30*, 977–90.

Graen, G. "Role-making processes within complex organizations." In M.D. Dunnette, ed., *Handbook of Industrial and Organizational Psychology*, Chicago: Rand McNally, 1976.

Hackman, J.R. "The design of self managing work groups." In B. King, S. Streufert, and F. Fielder, eds., *Managerial Control and Organizational Democracy*. New York: Wiley, 1978.

Hackman, J.R., and Oldham, G.R. "Development of the job diagnostic survey." *Journal of Applied Psychology*, 1975, *60*, 159–70.

Hall, D.T. *Careers in Organizations*. Pacific Palisades, Calif.: Goodyear, 1976.

Hall, D.T., and Nougaim, K.E. "An examination of Maslow's need hierarchy in an organizational setting." *Organizational Behavior and Human Performance*, 1968, *3*, 12–35.

Hall, D.T., and Schneider, B. *Organizational Climates and Careers*. New York: Seminar Press, 1973.

Homans, G.C. *Social Behavior: Its Elementary Forms*. New York: Harcourt, Brace and World, 1961.

Hughes, E.C. *Men and Their Work*. Glencoe, Ill.: Free Press, 1958.

Janis, I.L., and Mann, L. *Decision Making*. New York: Free Press, 1977.

Kahn, R.L., Wolfe, D.M., Quinn, R.P., Snoek, J.D., and Rosenthal, R.A. *Organizational Stress: Studies on Role Conflict and Ambiguity*. New York: Wiley, 1964.

Katz, D., and Kahn, R.L. *The Social Psychology of Organizations*. New York: Wiley, 1978.

Katz, R. "The influence of group conflict on leadership effectiveness." *Organizational Behavior and Human Performance*, 1977, *20*, 265–86.

Katz, R. "Job longevity as a situational factor in job satisfaction." *Administrative Science Quarterly*, 1978a, *10*, 204–23.

Katz, R., "The influence of job longevity on employee reactions to task characteristics." *Human Relations*, 1978b, *31*, 703–25.

Katz, R. "Time and work: toward an integrative perspective." In B. Staw and L.L. Cummings, eds., *Research in Organizational Behavior*, Vol. 2. Greenwich, Conn.: JAI Press, 1980, 81–127.

Katz, R., and Allen, T. "Investigating the not-invented-here syndrome." In A. Pearson, ed., *Industrial R&D Strategy and Management*, London: Basil Blackwell Press, 1981.

Katz, R., and Tushman, M. "Communication patterns, project performance and task characteristics: an empirical evaluation and integration in an R&D setting." *Organizational Behavior and Human Performance*, 1979, *23*, 139–62.

Katz, R., and Van Maanen, J. "The loci of work satisfaction: job, interaction, and policy." *Human Relations*, 1977, *30*, 469–86.

Kaufman, H.G. *Obsolescence of Professional Career Development*. New York: AMACOM, 1974.

Kiesler, S. *Interpersonal Processes in Groups and Organizations*. Arlington Heights, Ill.: AHM Publishers, 1978.

Kopelman, R.E. "Psychological stages of careers in engineering: an expectancy theory taxonomy." *Journal of Vocational Behavior*, 1977, *10*, 270–86.

Kotter, J. "The psychological contract: managing the joining-up process." *California Management Review*, 1973, *15*, 91–99.

Kuhn, T.S. *The Structure of Scientific Revolutions*. Chicago: University of Chicago Press, 1963.

Lawrence, P.R., and Lorsch, J.W. *Organizational and Environment*. Boston: Harvard Business School, 1967.

Lewin, K. "Group decision and social change." In H. Proshansky and B. Seidenberg, eds., *Basic Studies in Social Psychology*. New York: Holt, Rinehart, and Winston, 1965, pp. 423–36.

Likert, R. *The Human Organization.* New York: McGraw-Hill, 1967.

Maslow, A. *Toward a Psychology of Being.* Princeton, N.J.: D. Van Nostrand, 1962.

Menzel, H. "Information needs and uses in science and technology." In C. Cuadra, ed., *Annual Review of Information Science and Technology,* New York: Wiley, 1965.

Pelz, D.C., and Andrews, F.M. *Scientists in Organizations.* New York: Wiley, 1966.

Pfeffer, J. "Management as symbolic action: the creation and maintenance of organizational paradigms." In L.L. Cummings and B. Staw, eds., *Research in Organizational Behavior,* Vol. 3. Greenwich, Conn.: JAI Press, 1981.

Porter, L.W., Lawler, E.E., and Hackman, J.R. *Behavior in Organizations.* New York: McGraw-Hill, 1975.

Robertson, T.S. *Innovative Behavior and Communication.* New York: Holt, Rinehart and Winston, 1971.

Rogers, C.R. *On Becoming a Person.* Boston: Houghton Mifflin, 1961.

Rogers, E.M., and Shoemaker, F.F. *Communication of Innovations: A Crosscultural Approach.* New York: Free Press, 1971.

Salancik, G.R., and Pfeffer, J. "A social information processing approach to job attitudes and task design." *Administrative Science Quarterly,* 1978, *23,* 224–53.

Schein, E.H. "Organizational socialization and the profession of management." *Industrial Management Review,* 1968, *9,* 1–15.

Schein, E.H. "The individual, the organization, and the career: a conceptual scheme." *Journal of Applied Behavioral Science,* 1971, *7,* 401–26.

Schein, E.H. "Personal change though interpersonal relationships." In W.G. Bennis, D.E. Berlew, E.H. Schein, and F.I. Steele, eds., *Interpersonal Dynamics: Essays and Readings on Human Interaction.* Homewood, Ill.: Dorsey Press, 1973.

Schein, E.H. *Career Dynamics.* Reading, Mass.: Addison-Wesley, 1978.

Scott, W.E., "Activation theory and task design." *Organizational Behavior and Human Performance,* 1966, *1,* 3–30.

Seashore, S.F. "Group cohesiveness in the industrial work group." Ann Arbor, Mich.: Survey Research Center, University of Michigan, 1954.

Shaw, M.E. *Group Dynamics: The Psychology of Small Group Behavior.* New York: McGraw-Hill, 1971.

Staw, B. "Motivation in organizations: toward synthesis and redirection." In B. Staw and G.R. Salancik, eds., *New Directions in Organizational Behavior.* Chicago: St. Clair Press, 1977.

Staw, B. "Rationality and justification in organizational life." In B. Staw and L.L. Cummings, eds., *Research in Organizational Behavior,* Vol. 2. Greenwich, Conn.: JAI Press, 1980, pp. 45–80.

Stoner, J.A. "Risky and cautious shifts in group decisions: the influence of widely held values." *Journal of Experimental Social Psychology,* 1968, *4,* 442–59.

Tushman, M., and Katz, R. "External communication and project performance: an investigation into the role of gatekeepers." *Management Science,* 1980, *26,* 1071–1085.

Van Maanen, J. "Police socialization." *Administrative Science Quarterly,* 1975, *20,* 207–28.

Wanous, J. *Organizational Entry.* Reading, Mass.: Addison-Wesley, 1980.

Weick, K.E. *The Social Psychology of Organizing.* Reading, Mass.: Addison-Wesley, 1969.

Wheeler, S. "The structure of formerly organized socialization settings." In O.G. Brim and S. Wheeler, eds., *Socialization after Childhood: Two Essays.* New York: Wiley, 1966.

Wortman, C.B., and Linsenmeier, J. "Interpersonal attraction and techniques of ingratiation in organizational settings." In B. Staw and G.R. Salancik, eds., *New Directions in Organizational Behavior.* Chicago: St. Clair Press, 1977.

SECTION III

Technology and Business Strategy

Decisions regarding product, service, and process innovation need to be made with a knowledge of technological opportunities, market need, and competitor analysis. Chapter 5 focuses on technology and business strategy, a primary building block in the management of innovation. Foster's article discusses the importance of technological discontinuities in the evolution of product classes. Observing that major technological advance occurs more frequently and is a potential strategic opportunity and/or threat, Foster provides insight on how to manage technological progress more proactively. Pappas offers a complementary framework linking technology and business strategy. Maidique and Patch argue that technology policy should be anchored to business strategy and illustrate various types of corporate strategies and their technological and organizational correlates. Finally, Cooper and Schendel bring organization history back into the discussion of business-unit strategy. Building on the Smith and Morison articles, Cooper and Schendel trace response patterns to technological change. Their research finds that prior success often stifles the firm's response to major technological threat—success often sows the seeds for future failure. Cooper and Schendel discuss actions to reduce these dysfunctional behaviors.

4

TECHNOLOGY AND BUSINESS STRATEGY

Timing Technological Transitions

Richard N. Foster

ABSTRACT. Technological discontinuities are among the most disruptive challenges facing corporations, and they are on the increase. Empirically, it appears that few corporations are able to survive discontinuities. They either lose market share, money or their independence. Despite the importance of the discontinuity problem for top management, few have done much to better anticipate and deal with these increasingly frequent events. They act as if they assume that there is nothing that can be done. The problem appears to be one of misplaced faith in some conventional assumptions about what it takes to succeed. The author examines these assumptions, particularly those dealing with the supposed advantages of defenders, and pinpoints their errors. This knowledge then leads to positive solutions that hopefully, in turn, will lead to increased success for some of our most important companies.

When the transistor replaced the vacuum tube in the mid-1950s, every one of the top ten producers of vacuum-tube components was caught flat-footed. Six of them discounted the importance of the new development and held aloof from solid-state technology, while the other

four tried to change horses, but failed. Not one of the ten is a force in today's $10 billion semiconductor industry.

This familiar scenario is not confined to high-technology industries. Consider tire cords. Sixty years ago they were made of cotton. Then—in the mid-1940s—the lead shifted to rayon and the American Viscose Company became the leading supplier of cord. Early in the 1960s, however, Du Pont's nylon tire cords captured the top position. The lead shifted yet again in the late 1970s, when Celanese took control with its polyester fibers. Each time a new technology superseded the old, industry leadership changed hands.

Few companies, it seems, are able to make a successful transition to a profitable new technology. Typically, technology leaders wake up too late to the threat of a new technology.

How do technologies mature? Why is the process so often accompanied by management paralysis and competitive dislocation? And what steps can companies take to improve their ability to anticipate and manage transitions between technologies, reducing the chances of such dislocation in the face of future technological change? These are the questions that will be explored here.

TECHNOLOGICAL LIMITS AND POTENTIAL

Technological improvement in any field is ultimately limited by the laws of nature. The number of transistors that can be placed on a silicon chip is limited by the crystal structure of silicon, the ultimate strength of a fiber by the strength of its intermolecular bonds.

Most industries are far from these ultimate natural barriers, and much more likely to come up against practical technological limits. The efficiency of today's car engines, for instance, is limited by the ability of existing engine alloys to withstand heat; hence, researchers are investigating the possibility of using ceramics in their place.

The mission of research and development is to advance toward the limit—both practical and theoretical—of a particular technology. How much improvement is possible is determined by the difference between the current state of the art and the technical limit. This difference can be called the *technical potential.* As

the state of the art nears its technical limits, each increment of research effort produces less improvement since the most effective ideas presumably have already been used. Said another way, as the limits of technology are approached, the returns from further R&D diminish. Many companies don't understand where the limits of their technologies lie, and finding out can be a complicated business. But the payoff can be great.

In the late 1960s, for example, IBM investigated the limits of computer technology.[1] Thorough evaluation at the Watson Research Labs confirmed the prevalent assumption that these limits were ultimately set by the accuracy with which computer circuits could be printed by photolithography. It also disclosed, however, that the state of the art of component design, device design, and—most particularly—chip "package" design all posed more restrictive practical limits. To get around these limits, the scientists changed the existing design in what became known as the "multiple stacked array" concept on which IBM's 4300 Series computers were based—a major market success of the 1970s that would have been impossible without thorough understanding of technological limits.

The role of technology in corporate strategy is directly related to technological potential. If the potential is high—*i.e.,* if the state of the art is remote from the technical limit (a situation typical of "high-tech" industries)—technology is likely to be an important element in the strategy of the business. If the technological potential is low, then other business functions are likely to loom larger.

THE S-CURVE PHENOMENON

Of necessity, all R&D work is carried out in the context of technical limits. Early in an R&D program, knowledge bases need to be built, lines of inquiry must be drawn and tested, and technical problems surfaced. Researchers need to investigate and discard unworkable approaches. Thus, until this knowledge has been acquired, the pace of progress toward technological limits is usually slow. But then it picks up, typically reaching a maximum when something like half the technical potential has been realized. At this point, the technology begins to be constrained by its limits, and the rate of performance improvement begins to slow. Diminishing returns set in. This phenomenon can be depicted as an S-curve. Figure 1 shows the concept schematically, while Figure 2 displays a representative set of actual S-curves from the tire cord industry.

In the case of the tire cords for bias-ply passenger tires, technical performance is defined in terms of the combined parameters used by the B.F. Goodrich Company in evaluating suppliers' offerings. These include tensile strength, abrasion and corrosion resistance,

and adhesion to the rubber carcass of the tire—all factors related to desired end-product attributes, such as dimensional stability, longevity and cost. The cumulative R&D effort usually can be measured in man-years of R&D effort, or, as it is in this example, in dollars of R&D investment.

What is important to remember is that, no matter how vigorously a competitor backs its own technology, it cannot escape the realities of technical limits and the implications of the S-curve. Inevitably, the technology with the greatest potential takes control of the market. At the same time, competitive leadership is all too likely to change hands.

R&D productivity, like manufacturing productivity, may be defined as the ratio of the change in output to the change in input—*i.e.*, the improvement in the performance of the product or process divided by the incremental effort required. Accordingly, R&D productivity is determined by the relationship between

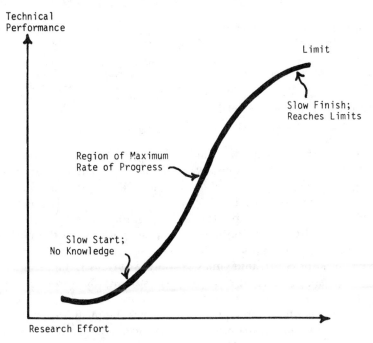

FIGURE 1. *The S-Curve Phenomenon*

217

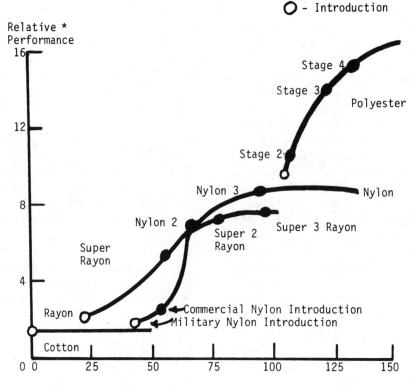

FIGURE 2.   *S-Curves in Tire Cord Technology*

technical progress and R&D investment at various stages of development. For example, Du Pont, pursuing a later stage advance in nylon, was getting 0.08 units of progress per $1 million of R&D investment. Its competitor, Celanese, then in the midstage of polyester development, was getting 0.33 units of improvement per $1 million of effort, four times as much as Du Pont.[2] Also, polyester tire cord was already superior to nylon, so Celanese was making faster progress on a better product. No company can withstand such competi-tion for long. As Celanese's victory over the redoubtable Du Pont demonstrated, in technological contests the attacker often has a decisive economic advantage.

Sometimes, of course, market factors can actually cancel the day of reckoning. The end objective, after all, is not R&D productivity, but R&D return: the profit earned on the R&D investment. R&D productivity is only one of the two variables determining R&D return; the other is the profit made from a given technical advance. This variable, which can be called the

R&D yield, is a function of the competitive structure of the industry: supply/demand conditions, relative power of customers and suppliers, company strategies, substitute products, external influences, and so on.

Decisions about which technology to work on and when to invest in it should be made on the basis of returns, and should involve both productivity and yield. Examining returns in terms of technological progress (productivity) and the economic value of that progress (yield) has organizational advantages, too. R&D productivity is determined largely by the R&D strategy and the efficiency of the R&D department in pursuing that strategy. Thus, controlling and improving R&D productivity is largely within the power of the R&D department. Yield, on the other hand, is the result of market conditions and competitor strategies, subjects that are the traditional responsibilities of the business units. R&D productivity and yield ultimately need to be coupled, of course, but this is best done after they have been examined independently.

Breaking down returns into productivity and yield also provides a more precise sense of the sources of low returns. For returns to be high, both productivity and yield need to be substantial. If either is low, their product will be low. If productivity is high and yield is low or negative, as it might be if there is a prospect for continued overcapacity in the industry, then returns will be low or negative. Accordingly, projects in this area should not be pursued, despite substantial technical potential and high productivity. Nor do projects make sense where yield is substantial, but the technology is mature. The low productivity of the mature technology indicates that little technical progress can be made. Investments in manufacturing, logistics, or marketing might be more productive. Companies that manage their technology well have recognized these pitfalls, and learned to avoid them.

THE IMPACT OF DISCONTINUITIES

By focusing R&D on improving the performance of current technology, companies limit themselves to evolutionary progress along a single S-curve until rapidly diminishing returns signal the imminent approach of technical limits. In failing to investigate the potential of new technologies, they often miss opportunities to exploit powerful competitive leverage.

Such opportunities, however, are not always easy to grasp. Almost by definition, the S-curves of different technologies are not linked. The gaps that separate them are often both substantial and strategically important, and to manage the transition is a difficult and delicate task.

The tire and cord technologies discussed previously, for example, are linked to bias-ply tires—a construction technology that has its own S-curve and its own technological potential, which happens to be significantly less than that of the newer radial construction. Goodyear clung to its bias-ply construction, which was beaten out by the Michelin-backed radials, just as Du Pont, sticking to nylon, was bested by Celanese. It is easier to detect a technological discontinuity from afar than it is when you are right in the middle of one.

One reason why it is so difficult to detect a technological discontinuity from the inside is that an existing competitor has more to lose with the introduction of a new technology, while a new entrant has little to lose and a great deal to gain. In tire cords, Du Pont felt it needed to protect its existing nylon plant assets. Celanese had no such need; it only felt a need to utilize its new polyester facilities. The need to protect its assets may have helped Du Pont believe that polyester had little market potential. History shows that this was wishful thinking.

This blindness to the advantages of new

technology has affected almost all industries at one time or another. In electronics, the dramatic discontinuity seen in the shift from vacuum tubes to transistors was followed by a succession of further discontinuities as transistors gave way to SSI, to LSI, and then to microprocessors. In metal packaging, the move from three-piece to two-piece cans represents an important discontinuity (the retort pouch may be another). In each case, a technological discontinuity led to a competitive dislocation—and technology leaders became losers.

The essence of managing technology well, it seems, is the ability to make smooth and timely transitions to new technologies with superior performance improvement potential— in other words, to cross discontinuities effectively. In many cases, this ability may be the only key to the problems of declining R&D productivity and deteriorating competitive position. Superior operational efficiency in R&D can rarely, if ever, compensate for an inferior technical strategy.

TECHNOLOGICAL MYOPIA

Strategic errors, however, are not always to blame when a technology leader is overtaken by a competitor. Incorrect perceptions of technical limits, inability to measure technological progress, faulty interpretation of market signals, and unrealistic faith in "understanding customer needs" tend to mask the deterioration that sets in when an existing technology matures and the superior potential of an alternative begins to affect the competitive balance.

Managements unaccustomed to thinking in terms of technological limits and systematically measuring changes in technical performance almost always grossly overestimate the improvement potential of existing technology. It is tempting to assume that the rapid progress that occurs midway through the S-curve will continue and, unless technical performance *per se* is carefully monitored, the gap between reality and expectation is typically perceived too late—on a 10- or 20-year S-curve, perhaps five to seven years too late.

Knowing when to investigate new technologies is certainly important. On a symmetrical S-curve, half the potential for improving the technology typically remains to be exploited when R&D productivity has reached its maximum. At the midpoint, the bugs are out of the system, cash flow has turned positive, and the business has begun to make a decent profit. It may seem economically perverse to begin investing in a new technology when there is still a lot of potential left in the old. But since it can easily take five or 10 years to develop a new technology, this is often the best time to begin the move by shifting the focus of R&D.

Managers intent on lowering break-even points and improving profits are normally inclined to build bigger plants and spread their overhead over more products. They segment markets to squeeze maximum profit out of customer groups. They work to reduce fixed and overhead costs. All these measures help to maximize profits. And increasing profits tend, in turn, to justify further investment—especially when decisions are based on accounting and financial criteria alone. Only when aggregate financial measures have been broken down into their components, is it possible to distinguish between technological progress and economic returns, and gain real insight into the underlying forces at work.

While R&D concentrates on extracting marginal improvements from old technology, windows of opportunity open up for competitors. And, when a competitor does move in with a new technology, the company's response is typically to increase its investment in the old. But, in choosing to defend an existing technology, rather than displace it with a new one, management only increases the inherent economic advantage of the attacker by enabling him to develop his product or process unhampered by entrenched competitors.

When the first steamships were launched in the mid-19th century, established sailing-ship builders set about improving their design technology, and were building vessels that carried more sail, had faster hulls, and required smaller crews. But the result was only a brief reprieve for a mature technology, not an escape from the S-curve. Thanks to the far greater technological potential of steamships, sails eventually disappeared from the world's shipping lanes.

Sylvania, for another example, was hamstrung by this sailing-ship phenomenon when transistors made their commercial debut. Despite announced plans to move into solid-state technology, it kept on pouring R&D effort into increasingly sophisticated vacuum-tube designs, bringing out complex multi-functional units as late as 1968. Some evidence suggests that, despite Sylvania's public comments to the contrary, the company still believed that the vacuum-tube had a future. For example, Sylvania was an active member of the Electron Tube Information Committee, which was established to explain the advantages of vacuum-tubes and the disadvantages of solid-state devices to the public. In its efforts to wrest incremental performance improvement from its outmoded vacuum-tube technology, Sylvania got aboard the solid-state S-curve too late with too little effort to become a force in the new market.

MISREADING MARKET SIGNALS

The tendency to discount signs of market penetration by new competitors is another costly common error. New products are typically introduced in market niches where they can be highly competitive—niches apparently too small or insufficiently related to the core market to provoke an immediate competitive reaction from the industry leader. Yet, even in a growing market, when the defending producer's sales begin to level off, management often misreads this as a sign of predictable market maturation. And, as the substitution process gathers momentum, the company's sales actually begin to sag. Finally, as the challenger penetrates deeply into its heartland markets, sales collapse and profits swiftly deteriorate—almost before the defender realizes that the battle has begun. It is already too late.

Consider the pattern of substitution cases already mentioned. Polyester tire cords went from a 20 to an 80% share of the bias-ply tire market in 10 years. Radials repeated that market performance against bias-ply tires in four years, and steel tire cords took three years to go from 20 to 80% of the radial tire market. And, in electronics, transistors took seven years to displace vacuum tubes. In each case competitive leadership changed hands.

Part of the problem of recognizing a technological discontinuity is the problem of correctly defining the market. The more mature the product, the more narrowly the market tends to be defined. In the mid-1960s, all tires were bias-ply tires. Hence, by definition, tire-cord makers measured their market performance in terms of their share of the bias-ply tire market. When radial tires entered, the leading tire cord manufacturers should have redefined the market to include tires of any kind, rather than dismissing the radials as "niche fillers." Too many businessmen fail to recognize that any new product—and particularly one based on new technology—potentially redefines the served market.

A final strategic pitfall in technology management is management's tendency to wait for market research cues before committing funds to a new technology. There are three dangers in this perversion of the marketing concept. First, customers are rarely the best judges of potential utility of products that might require them to do things differently; change is painful. Second, the same market cues are available to all qualified suppliers, and the customer, once given the whip hand, can often play suppliers off against one another to advantage, as Boeing has learned to do in the aerospace industry with its materials suppliers. Third, sustained competitive success demands the ability to meet a market need in a unique way. In many businesses it is easy to copy a competitor's market research, but far less easy to copy its technology and harder still to copy its methods of developing new technology. Thus, a company is far more likely to win a real competitive edge by superior technological prowess if the technical potential exists than through any unique knowledge of market needs.

THE CULTURE TRAP

The most serious and most elusive impediments to effective technology management, however, are the fundamental cultural weaknesses that underlie the strategic errors being discussed.

A company's product and production technology is an integral part of its corporate culture. Technology directs and conditions management's intuitive strategic responses to opportunities ("can we make a profit on this?"). It conditions the assumptions on which the strategy of the business is based. When the technology changes, the whole corporate culture frequently must change as well. And this is a difficult and painful process.

For electrical engineers designing vacuum tubes in the mid-1950s, the transition to solid-state technology was a real culture shock. To design solid-state devices, they needed to learn an entirely new technical discipline in-

volving a different mathematics. Expert vacuum-tube designers don't necessarily become good solid-state designers. But, because there were enough similarities between the two fields, some vacuum-tube producers assumed that making the transition would not be particularly difficult.

Sometimes a cultural preference for established technology is explicitly built into the management system; several major corporations still give top R&D priority to the defense of their existing product lines. The more doggedly they champion a policy of unyielding technological defense, the more vulnerable to new technology they become. Many R&D vice presidents, having earned their titles by successfully guiding their companies into new technologies, are not disposed to abandon their favorites easily. Indeed, they often inadvertently block the investigation of new and threatening technologies in the name of defending existing product lines.

Traditionally different industries have met technological challenges in different ways. Some observers see the emerging biochemical industry as a wave of the future. If they are correct, how well could established chemical manufacturers cope with the new technological environment? Consider just a few of the differences.

· The chemical business operates large plants whose capacities range from 100 million to one billion pounds a year. Biochemical plants might be small, with typical annual capacities of perhaps from one million to 100 million pounds.
· Whereas chemical plants usually produce a single product, biochemical products could be produced in multiproduct facilities.
· Chemical processes are typically high-temperature operations. Biochemical processes will probably take place at low temperatures, where the energy minimization

experience of the chemical industry becomes much less relevant. Biochemical processes will require far tighter processes controls and heavier instrumentation than most chemical processes; thus capital requirements may well differ.

- Biological sterility, generally unnecessary in the chemical industry, is an important requirement in some biochemical processes.

Clearly, to capitalize on the opportunities in biochemicals, established chemical companies would have to do many things differently. As centralized operations become decentralized, strategies, as well as organizational structures, might need to be altered. Management systems might require revision with more authority given to people close to the market and the competition. With shrinking plant size, the role of the plant manager would diminish. New skills and staff needs would develop. New employees, coming from different colleges, universities and disciplines, would move in different cultural circles, and might even speak a different social language. All this would lead to changes in culture, shared values, and management style. The company emerging from such a metamorphosis might resemble its former self in little but name. This transition might be closer than some think. Presently about $1 billion is being spent on biotechnology R&D with perhaps 20% of this going to chemical applications. If the productivity of this work is five to ten times that of conventional chemical R&D, not an uncommon ratio in other transitions, then the total technical programs expected from biochemical research could be on a par with the technical programs expected from all other chemical research.

Cultural barriers to successful technological transitions are particularly difficult to overcome because they are qualitative rather than quantitative. Many managers are accustomed to dealing with strategic issues by creating new research programs or developing new measurement systems; few have experience in changing the entire nature of their companies. But nothing less than that may be called for.

DETECTING DECAY

Rather than wait passively for a competitive threat to develop, it is necessary for farsighted management to head off technological decay by addressing the strategic and cultural issues as quickly as they can be detected and identified. A good way to begin is to assess the company's proximity to its technical limits. Typically, certain tell-tale symptoms tend to signal the onset of technological decay:

- *A feeling among top managers that the company's R&D productivity is declining.* Managements that don't measure technological progress have to rely on "gut feel" to know when the S-curve has flattened out. Unfortunately, by the time the flattening can be felt, the game may be lost.
- *A trend toward missed R&D deadlines.* Often interpreted as a sign that the R&D department is losing effectiveness, this may mean that the technology is approaching its limits, thus making performance improvement more difficult to achieve.
- *A shift in the company or industry from product- to process-oriented R&D.* As a technology approaches its limits, process improvements typically replace product improvements as the chief source of technological innovation. Many of our so-called "smokestack" industries are in this situation now. Paradoxically, the best corrective action is a move into unfamiliar product technologies or markets, a risky proposition at best.

223

- *An apparent loss of productivity in the R&D department.* There is no room for creativity in a technology that is bumping up against its limits.
- *Dissension among the R&D staff.* A particularly frustrating assignment—for example, closing a small gap between the "state of the art" and the technological limit—may turn scientists into pessimists. If the mood of the R&D people is bad, the limits of the technology are probably close.
- *A shift in the sources of sales growth toward narrower market segments.* As we have seen, the final stages of a typical substitution pattern are often dominated by product-line proliferation to meet the needs of small market segments. While this is usually a mandatory economic strategy, it does indicate increasing maturity, and should, therefore, be taken as a signal to look for new "S-curves" that could supersede the present technology.
- *A tendency for significant variations among competitors in R&D spending to produce ever less significant results.* When there is no market advantage to spending more, nor a market disadvantage to spending less, the technology has virtually reached its limits.
- *Dissatisfaction with the performance of a "new broom" R&D manager.* Sometimes a chief executive, frustrated at his company's declining R&D performance, will replace the senior executive with a new, more vibrant R&D manager who is expected to shake the department up and get it going again. He will fail too, if the technology is close to its limits.
- *A trend among smaller, weaker competitors in the industry to invest R&D effort in radical new approaches.* A small company, particularly if it is financially hard pressed, is more likely than an entrenched and profitable producer to invest in a really innovative solution to a customer's problem.

TOWARD EFFECTIVE ACTION

Using these criteria to form an impressionistic assessment of the company's technological health can be an effective prelude to a thorough R&D audit. The audit itself should be based on the principle that R&D effort and R&D investment in a given technology should be proportional to potential for productivity and yield improvement. This principle allows the definition appropriate for R&D strategies (in terms of, say, aggressiveness, risk profile, time for commercial introduction, or a balance of work between old and new markets) to be used in a consistent way for each product line (Figures 3 and 4). These are then compared with actions being taken, or implicit strategies, and the inconsistencies identified. These become the agenda for reshaping the company's R&D program.

Audits of this kind are particularly useful to CEOs of companies with many product lines, since it is typically quite difficult to remember the details of hundreds or thousands of R&D projects. Accurate summaries based on these principles can be extremely useful tools.

These audits freqently reveal that companies are overinvesting in mature technologies and underinvesting in newer ones. In terms of the matrix in Figure 3, this means that companies putting too much emphasis on product lines warranting only "limited defensive support" are not putting enough effort into "heavy emphasis" product lines. To avoid this trap, companies generally will have to take action in two areas in addition to resource reallocations: (1) knowledge building, and (2) measurement and planning systems.

Knowledge Building

Since a major source of mistakes in technology resource allocation is inadequate knowledge about the limits of present and potential future

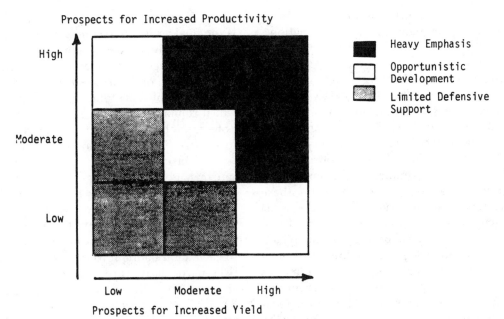

Prospects for Increased Productivity

FIGURE 3.   R&D Audit: Matrix for Sorting Priorities

	Heavy Emphasis	Opportunistic Development	Limited/Defensive Support
Appropriate Funding	High	Moderate	Low
Expected Payoff	Long-term	Medium-term	Short-term
Acceptable Technical Task	High	Moderate	Low
Primary Focus of Work	Balance New & Existing	Existing Products	Existing Processes
Appropriate Level of Basic Research	High	Low	Very Low

FIGURE 4.   R&D Audit: Typical Strategic Implications

technologies, efforts to correct this problem are obviously in order, but the utility of knowledge does not stop there. The new knowledge can also be useful in controlling the costs of development. A Rand Corporation study,[3] conducted for the US Department of Energy, for example, indicates that nearly 75% of cost overruns in the construction and operation of plants for new products and processes can be blamed on inadequate technical information.

Measurement and Planning

Effective knowledge building is only the first step in improving returns from technical investments. The second step is to use this knowledge to plan better future investments. Armed with an understanding of the appropriate technical performance parameters, the limits alternative technical approaches place on these parameters, and an estimate of future R&D productivity based on analyses of past S-curves, senior management will be in a position to conduct a reasoned dialogue about the broad alternatives facing the company. In the course of this dialogue, existing assumptions grounded in conventional economic performance measures and incremental thinking are bound to be severely tested.

The ability to measure S-curve and R&D productivity can be integrated into the technical planning process to permit new assumptions to be tested. Then management can use these new technical information systems to determine which technology to pursue, when to pursue it, and at what levels of investment— and measure the results of these strategic decisions. Managers can determine whether a given effort has actually yielded the anticipated degree of performance improvement—and if not, why not. The conclusions, fed back into the technological planning and development process, can further refine its effectiveness. Such measures can rejuvenate a company.

Tuning in to Change

Radical technological change cuts away the company's base of technical and perhaps strategic intuition. Leading through the resulting maze of shifting organizational priorities requires new management priorities as well.

In most US corporations today, the CEO is the final arbiter of technological choice. It is he who finally decides to reallocate resources from old to new technologies. It is he who controls the pace of S-curve transitions, determining the rate of R&D funds transfer, the rate at which capital spending is shifted to new technologies, the redirection of marketing efforts—even the reallocation of managers.

The chief executive, in fact, *is* the corporate manager of technology. Unfortunately, the job is difficult. Confronted by an R&D program composed of dozens or hundreds of individual projects, small and large, the CEO is likely to be baffled by the task of assessing the relative value of each and relating it to the company's objectives, particularly in areas of rapid technological change. Yet if he cannot assure himself that R&D will be effectively focused on the technical alternatives with the greatest potential, he may be compromising the company's future.

The executive best qualified to aid the CEO is obviously the vice president in charge of R&D. But in too many US corporations this executive busies himself with more functional tasks—within boundaries implicitly set by the chief executive—than with basic issues of technological choice. In a recent Conference Board survey,[4] only one CEO in four named the R&D VP as a member of his inner circle. Something will need to change here. Until the R&D vice president earns the CEO's full confidence, strategic changes in technology will be painful, and future competitiveness in doubt.

Had such an executive been at Sylvania in the mid-1950s, he might have recruited more solid-state engineers. Were he in charge of R&D at a major chemical company today, he might be busy hiring biochemists, setting them up in an independent facility, and holding down investments in incremental improvement of conventional product or process technologies.

The R&D VP will need to become well

tuned to overall competitive strategy and corporate options, and he may need to become more knowledgeable about corporate finance. He should establish a better working relationship with the top strategic planner and with the company's financial officers so that he can better empathize with their problems. In short, he will need to think like a CEO.

THE NEED TO ANTICIPATE

Since the shift to a new technology may take a decade or more to complete, most companies need to do a better job of anticipating technological development relatively far in advance. In short, they need a distant early warning capability.

In the mid-1930s, for example, Mervin Kelly, then research director of Bell Laboratories, projected telephone switching needs ahead to the early 1960s. He found that, with the current mechanical relay technology, an impossibly huge number of switches would be needed to support the future telephone network. This meant that a new technology would be needed. Kelly hypothesized that it might emerge from the new science of quantum mechanics. Accordingly, after World War II, he launched an R&D project to develop a quantum-mechanical amplifier. Bell Laboratories developed the first transistors in 1947; by 1955, transistors were an item of commerce. The whole process had taken 20 years (with 10 years out for the war).[5]

Development activities must be planned, funded, started, supported, and often completed before existing measurement systems can accurately predict their ultimate economic benefits for the company. But the rewards of anticipating future requirements are exceeded only by the costs of ignoring them.

One effective way a company can broaden its horizons is to push hard for the identification of technological alternatives. In the computer industry, for example, evidence is beginning to accumulate that solid-state components—designed and produced as they have been for the past 10 years—may be approaching the upper end of the S-curve. As the cost of creating ever more complex devices in the conventional way rises dramatically with each increment in performance, it is less and less economically feasible to persist with development at continually accelerating rates. Moreover, as the capacity of computer memories grows and the cost per bit of storage becomes infinitesimal, computer memories are becoming a commodity, and the market leverage of building bigger computers diminishes. This combination of increased expenditure needs and decreased leverage could create a scissors effect, constructing or even cutting off development. Already, in fact, the S-curve for random access memory chips (DRAMs) is exhibiting the classic phenomenon of diminishing returns (Figure 5), and many companies have already dropped out of the game.

As a result, it seems that the leverage in computer development is shifting from hardware to software, where no such diminishing returns are in sight. Monumental changes will accompany the transition. Manufacturing, always important in the computer hardware industry, may lose its place in the sun. Software manufacturers may already be beginning to embrace a strategy of creating only a few copies of each new design for distribution through telecommunications networks to the customers' own computers. With such an arrangement, the manufacturing process would take place, in a sense, on the customers' premises. The marketing manager would become a total business manager, controlling not only the sales of the product, but its manufacture and distribution as well.

Who would be most likely to survive

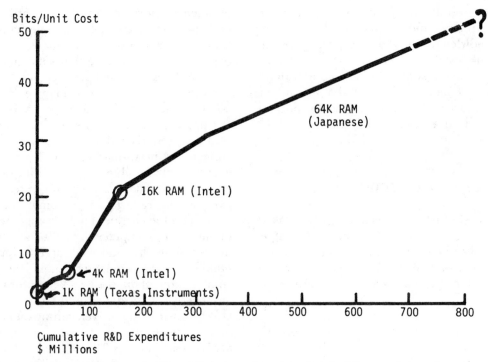

Bits/Unit Cost

Cumulative R&D Expenditures
$ Millions

Source: From McKinsey analysis based on industry interviews. Reprinted by permission of McKinsey and Company.

FIGURE 5.   Topping Out in Electronic Hardware

such fundamental changes—the large hardware manufacturers, like IBM, Siemens, and Fujitsu, where manufacturing is an entrenched and powerful corporate function, or the smaller companies that now dominate the software industry? It may be hard to imagine anyone challenging IBM's preeminent position in the computer industry. But it was just as hard to imagine in 1955 that a little company called Texas Instruments would be a dominant force in electronics 10 years later. In the history of technological change, however, such revolutions are not rare. Companies that anticipate them, by continually investigating new alternatives and responding boldly to new challenges, are handsomely rewarded.

NOTES

1. Robert W. Keyes, "Physical Limits in Semiconductor Electronics," *Science,* March 18, 1977, pp. 1230–1235.
2. McKinsey analysis based on industry interviews.
3. Merrow, E.W. *et al.,* "Understanding Cost Growth and Performance: Shortfalls in Pioneer Process Plants," RAND Corporation Report R-2569 (1981).
4. "Who Is Top Management?," The Conference Board, Report no. 821 (1982).
5. Based on conversations with Bell Laboratories executives. For more on the development of the transistor, see Jeremy Bernstein, *Three Degrees Above Zero* (New York: Charles Scribner's Sons, 1984), especially p. 119.

Strategic Management of Technology

Chris Pappas

The first step in making sure that R&D spending is productive is to be sure it is going in the same direction as the overall business strategy. Simple enough to say but not always so simple to do. Chris Pappas suggests that the corporate strategy process often focuses on financial factors and market share and neglects technology as a key resource to be planned. With competitive success as well as productivity and profitability becoming more directly tied to technology development, it is time to give technology a more important place in the corporate strategy process. Using the example of an actual firm, Pappas shows that the key to achieving a sustainable competitive advantage lies in formulating the right technology strategy and integrating it into the corporate planning process. His article includes a useful framework for analysis and planning.

INTRODUCTION

The need for strategically managing technology has never been greater. Today, there is overwhelming evidence that technology will be a prime source of economic growth in the United States and the world throughout the 1980s and 1990s. As a result, technology considerations will underlie virtually every major decision that management will make.

"Strategic Management of Technology" by Chris Pappas, from *Journal of Product Innovation Management*, Number 1, pp. 30–35 (1984). Copyright 1984 by Elsevier Science Publishing Co., Inc. Reprinted by permission.

This article and the methods it describes are the outgrowth of numerous client assignments performed by Booz, Allen's Technology Management Group and the firm's strategic and general management specialists. Among the key contributors to the ideas presented here are: Dr. Joseph Nemec, Jr., Dr. William Sommers, Mr. John Harris, and Mr. John Allen.

With the exception of a small handful of industry executives, top management in the United States is not yet prepared to deal with the strategic implementation of technology. While senior management in many industries today may publicly embrace the importance of technology, they are frequently uncomfortable with it. In most industries and businesses, technology is placed second, third, or even fourth in importance behind other functions—sales/marketing, finance, and operations. For the most part, many executives remain technological illiterates.

Thus, the field of business strategy in the United States has to be greatly enhanced in terms of technology management. This will be the arena in which we will be competing worldwide in the balance of this decade and the next. Clearly, we have taken a back seat in many areas, for example, Japanese reliability and productivity. History has proved that our technology shortfall cannot be overcome by other functional superiorities, real or perceived.

In both a 1979 and 1981 Booz, Allen survey of officers and senior managers in Fortune 1000 firms, changing technology was ranked as one of the more critical issues business will face in the 1980s. Technology was also viewed as complex and difficult to manage. In the 1981 survey, some 86% of the respondents indicated that their lack of an analytical approach to integrating technology and business planning was a significant barrier to managing technology and 79% saw the limited involvement of technology managers in the planning process as another major drawback.

INTEGRATING TECHNOLOGY AND STRATEGIC PLANNING

A key to closing the gap lies in repositioning technology as a strategic resource by elevating its importance in the business planning process. There appear to be several emerging principles of technology management which facilitate the technology integration process.

First, *the direction and timing of technology evolution can be anticipated.* There has been a traditional view that technology is unpredictable, extremely risky and unquantifiable. To a large degree, this is not so. Once innovation occurs, the steps in new product development leading to eventual market commercialization follow the same general pattern, and essentially the same timing, for literally dozens of inventions—from xerography to integrated circuits (see Figure 1). As a result, a systematic approach to tracking innovations can be applied. This systematic approach is a key ingredient in the effective management of new technology and products.

Next, *technology should be viewed as a capital asset.* Traditionally, technology has been treated as an isolated, project-based phenomenon, rather than as a strategic resource. Again, the traditional view is offbase. Technology is critical to competitiveness, and technology changes and external competitive position can be analyzed. Thus, technology can, and should, be treated as an allocable corporate asset.

Finally, *assuring the congruence of technology investment and business strategy is essential to successful technology management.* Technology priorities should dictate investment thrust. And, effective use of technology leverage should avoid mismatches among strategic objectives and technology investments.

On balance, the three principles outlined above will be essential in driving corporate technology planning in the 1980s. They should serve as the basis for the analytical framework underlying successful management of technology.

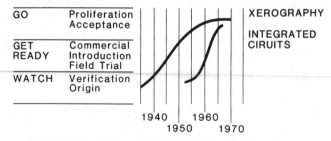

FIGURE 1.   Path and Timing of Technology Development

TECHNOLOGY MANAGEMENT: RECOMMENDED RECIPE

Technology strategy is the keystone of the recommended approach to successfully managing technology. Technology strategy is built on a penetrating analysis of a company's technology strengths and weaknesses, relative to the importance of these technologies to its businesses. Together with the corporate business strategy, the technology strategy defines how a company can most effectively invest its technology resources to achieve sustainable competitive advantage.

Planning a technology strategy is a complex and challenging four-step process involving:

· Technology situation assessment
· Technology portfolio development
· Technology and corporate strategy integration
· Setting technology investment priorities.

Each step in the process is described below.

Technology Situation Assessment

This assessment requires an internal and external scan of the technology environment beyond the scope of the traditional business portfolio. As a first step, the specific technologies employed in each of the firm's businesses, products, and processes should be analyzed. The importance of each technology to specific products and/or business should also be determined. Next, the priorities dictating past and current technology investments should be reviewed. This will normally involve some rethinking about the rationale for setting priorities and basis for determining level of investment.

Finally, the external competitive environment must be scanned to pinpoint the investment patterns of competitors on both the product and process side for each of the firm's vital technologies. The importance of evolving technologies must also be evaluated: Which companies developed them? What is the likely path of future innovation?

Technology Portfolio Development

The technology portfolio is a tool that can be used to identify and systematically analyze key corporate technology alternatives and to set technology priorities. There are two dimensions of the technology portfolio:

· *Technology importance,* that is, the relative importance that a specific technology plays in a given business segment. Technology importance would be based on criteria that include value added; rate of changes; and potential markets and their attractiveness.
· *Relative technology position* is basically a measure of the firm's investment in a given technology. It would be determined by assessing current and future position in that technology and its expected future development. Some quantitative criteria used to determine relative technology position would be number of patents; human resource strengths; product history, and cost; and technology expenditures, current and projected.

In developing the technology portfolio, it is convenient to think in terms of a two-by-two matrix as shown in Figure 2. The upper left quadrant of the matrix, for example, would conceptually be a good place for a firm to be. In this case, the firm would be in an excellent position technologically in a business segment where that technology is important. This would represent the business where commitment to the newest equipment, to the risky R&D project, or new product experimentation should be made.

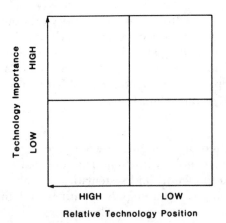

FIGURE 2. *Technology Profile*

a rapidly changing industry, such as electronics or engineered materials, where existing technology is continually being supplanted by new techniques.

If a firm finds itself in the lower right-hand quadrant it may or may not have a problem. The firm is weak technologically but in a field where technology is unimportant. Other resources of the firm might make up for the technology weakness. If the firm has already invested heavily, it would probably need to view the investment as a sunk cost.

The upper right-hand quadrant represents a potentially serious problem and is inherently unstable. One of two decisions needs to be made:

· Bet against the competition and invest to attain a leadership position, or
· Develop a plan to disengage from, or even abandon, that business and then invest in more lucrative areas.

In the lower left quadrant, a firm would be in a strong position technologically, but the technology is not really important in marketplace terms. This situation occurs most often in

Technology and Corporate Strategy Integration

Generally, a business portfolio is product oriented: it measures a firm's product lines in terms of market position and importance. The Boston Consulting Group portfolio model is one example of a business portfolio framework. By contrast, the technology portfolio is technology based and defines the firm's relative product and process technologies in terms of marketplace position and their importance to the basic business of the corporation (see Figure 3).

Though the business and technology portfolios provide fundamentally different perspectives, they must be compatible in order to gain advantage in a technology-related busi-

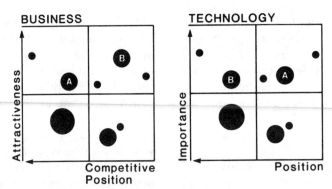

FIGURE 3. *Matching Business and Technology Portfolios*

ness. In other words, a technology portfolio will be effective only if it is consistent with the strategy implied by a company's business portfolio.

The reasons for linking the business and technology portfolios and for ensuring their consistency are compelling. On the one hand, if they are incompatible, a company runs the risk of developing a potentially attractive strategy based on financial data and other business portfolio information only to discover that it lacks the technology strengths needed to achieve its objectives. On the other hand, when analyzed in isolation, a technology portfolio can provide an unrealistic or distorted picture of market attractiveness and competitive position.

On balance, a technology portfolio, when viewed correctly, that is, in conjunction with the business portfolio, serves a number of purposes:

- It establishes a common planning base for all priority technologies.
- It provides an overview of a corporation's technological position, and a method for timing corporate technological investments in sync with its business plan.
- It identifies positions of strength to be leveraged and technology requirements to be strengthened or acquired to achieve corporate objectives.

- It provides a basis for focusing on high-potential, new business opportunities that could be built on current technological strengths.

SETTING TECHNOLOGY INVESTMENT PRIORITIES

As a final step, a specific technology strategy, critical to the survival and success of each business, should be developed. A number of key questions arise in this process: What resources are required to achieve corporate strategic objectives? What should the level and rate of technology investment be? What additional investments are needed to achieve corporate goals?

Technology investment options and relative technology expenditures can be developed from the portfolio analysis previously described. As illustrated in Figure 4, it is clear that for a situation represented by the upper left quadrant, spending should be high relative to the industry leader. In fact, a company should concentrate on becoming the leader. In the upper right-hand category, with a weak position in a technology of high importance, expenditures should be increased in order to move to the left, or perhaps play a wait and see game until increased expenditures can be better justi-

TECHNOLOGY PORTFOLIO QUADRANT	TECHNOLOGY IMPORTANCE	RELATIVE TECHNOLOGY POSITION	TECHNOLOGY INVESTMENT PRIORITY
UPPER LEFT	HIGH	HIGH	
UPPER RIGHT	HIGH	LOW	
LOWER LEFT	LOW	HIGH	
LOWER RIGHT	LOW	LOW	

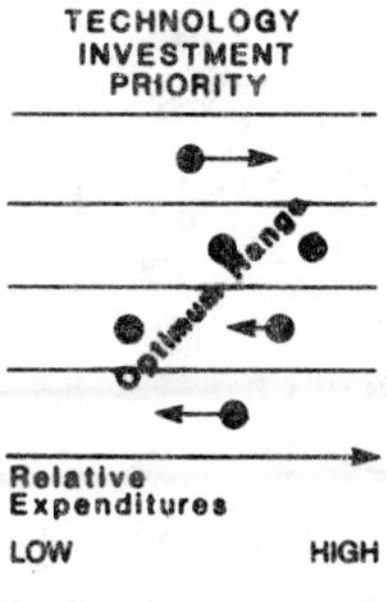

Relative Expenditures
LOW   HIGH

FIGURE 4. Technology Investment Matrix. If technology is outside the optimum range, the investment strategy should move it toward the optimum.

233

fied. Finally, with situations in the lower left or right quadrants, investments generally need either to be reduced or terminated, and concentration placed on recapturing resources on technologies with higher leverage.

TECHNOLOGY STRATEGY IN ACTION: A CASE EXAMPLE

The situation profiled here involved a major multinational firm (referred to here as the Engmat Company, a fictitious name) in a seemingly ideal situation: it had a strong competitive position with proprietary products in the engineered materials field. The market was growing rapidly in response to 15 years of applications development work. Engmat's synthetic materials were increasingly replacing glass and metals in transportation applications and a variety of consumer products, such as appliances, where high strength-to-weight ratio was important. The business was both technology dependent and capital intensive, and becoming more so. Growth and profitability of their engineered materials business had been strong.

Engmat, however, was concerned about maintaining a leadership position in light of some disconcerting trends. First, some of its original patents were expiring. Second, new competition appeared to be coming from world-class chemical and petrochemical companies based in the United States, Japan, and Europe. These firms were developing materials that competed well with Engmat's major products. Third, adding capacity and developing a really strong manufacturing cost position would require a massive capital investment in world-scale plants—over a billion dollars. Finally, because of rapid growth, Engmat had been forced to allocate production, giving emerging competition an opportunity to "qualify" inferior products as second sources for some customer applications.

The business portfolio (see Figure 5) showed that Engmat was in a strong overall position with high market share in commodity products A and B, and with leadership positions in the downstream fabricated products businesses C, D, and E. Only product F, the past generation product that had largely been replaced by A and B, appeared in a bad position.

The technology portfolio (see Figure

ORIGINAL POSITION

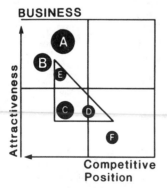 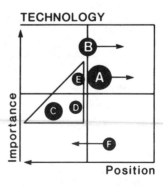

FIGURE 5. Case Example: Engineering Materials Business and Technology Portfolios

234

5), developed after considerable competitive analysis, highlighted areas where Engmat's technology program was out of balance with its business strategy. That is, recent emphasis had centered on the downstream products C, D, and E, and, as a result, Engmat's position in the basic business—products A and B—had been allowed to erode. Management was surprised to learn that, relative to competitive activity, its position in the cornerstone technologies was being liquidated as it poured resources into developing markets for the fabricated products. Management felt their spending levels had been adequate on products A and B. The C, D, and E teams did not appreciate their ultimate dependence on maintaining technology leadership in A and B. Second, product F was suffering from the "young tiger syndrome." The product had been properly identified by the planning staff as being in the last stage of a distinguished career—clearly a "harvest" strategy was appropriate. However, several senior managers had favorable memories of the product and a succession of young tigers had been given a year or two opportunity to reposition F for growth. Consequently, a number of development projects had been initially directed toward improving Engmat's competitive position in this first-generation material.

The technology investment matrix (see Figure 6) was used to highlight these issues. Subsequently, Engmat took the following actions which proved effective:

- Technology budgets for A and B were increased and plant expansions were accelerated.
- Programs for C, D, and E were modestly scaled back and some of their best people transferred to support the A and B programs.

RECOMMENDED STRATEGY

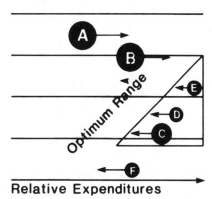

FIGURE 6. *Case Example: Engineered Materials Investment Strategy*

- Product F was handed over to a hard-nosed development veteran and managed for cash flow—not growth.

SUMMARY

As this article suggests, technology can be planned and managed using formal techniques similar to those used in business investment planning. An effective technology strategy is built on a penetrating analysis of technology strengths and weaknesses, and an assessment of the relative importance of these technologies to overall corporate strategy. Together with business strategy, the technology strategy defines how resources can be used most effectively to achieve sustainable competitive advantage.

The process is complex, but not mysterious; it can be planned, implemented, and exploited. The alternative to managing technology is to be mastered, and finally, overwhelmed by it. Surely this is too high a price to pay in today's global marketplace.

Corporate Strategy and Technological Policy

Modesto A. Maidique

Peter Patch

Technology is a vital force in the competitive environment of the modern firm. This is especially so in technology intensive industries such as aerospace, computers, chemicals, electronics and pharmaceuticals. Yet, even in these industries, technology is rarely an explicit element of corporate strategy. It is, along with manufacturing, a "missing link" in corporate strategy.[1] Babcock and Wilcox's widely publicized nuclear plant fiasco is often cited as a failure of top management to grasp the role of manufacturing. The failure was, however, just as much the result of a cavalier attitude towards technological change.[2]

Managements wishing to explicitly include technology in their corporate strategy will find little help in the traditional literature. Technology is generally viewed simply as either "high" or "low" or often not at all. Andrew's, *Concept of Corporate Strategy,* a classic in its field, argues that technological choices are of singular importance and should be carefully weighed, but, paradoxically, provides little assistance to the reader in making these choices.

American business education has in general given short rift to technological decision making. More often than not these choices are delegated to a corporate version of Galbraith's "technostructure" which, in principle, has the technical expertise to make the proper choices. Recently one of the authors interviewed the executive responsible for R & D in a technology-based Fortune 500 firm concerning his role in technology intensive decisions. When asked to describe how he participated in technological decisions the executive explained, "That's simple. I don't."

What many executives fail to recognize is that all technological choices, when made in a corporate context, are also business decisions. Design choices, for instance, involve tradeoffs between cost, performance, reliability and ease of use that may have long lasting competitive impact. The engineers, systems analysts, and programmers that developed the architecture of the IBM Systems 360 operating system delineated basic rules of technological and market competition in the computer industry for decades to come. Subsequent IBM products and associated peripherals, and most competitive offerings are compatible with that original technological choice. A decade later a senior IBM executive explained to one of the authors, "our customers don't want any operating system changes that would obsolete their equipment, so we now introduce technological changes very gradually."

On the other hand, it would be foolish to argue that senior executives should involve themselves in the minutia of technological decisions. What are needed are ways to cull the fundamental technological choices from the rest. To do this two things are needed: 1) a theory of technological progress, and 2) a framework that highlights and brings together those policies and decisions that impact technological progress in the firm. It is this set of policies and decisions that we characterize as the technological policy of the firm. The purpose of this note is to develop such a framework. In another paper one of the authors has presented his views on how technology progresses.[3] Such a framework can aid firms to develop a well-defined, coherent posture toward technology and to facilitate and encourage executive decision-making by integrating technology with business planning.

Technological policy and manufacturing policy should be differentiated. While the two are closely intertwined elements of business strategy, they, nevertheless, address distinct sets of decisions. Manufacturing policy principally involves decisions regarding the location, scale and organization of productive resources. As such it is formulated within the bounds of a given technology.

Technological policy, on the other hand, involves choices between alternative new technologies, the criteria by which they are embodied into new products and processes and the deployment of resources that will allow their successful implementation.

An example will illustrate this distinction. A firm choosing to enter the integrated circuit memory business may face a choice between three major technologies: MOS, bipolar and magnetic bubble, and a number of subdivisions thereof. For instance, MOS technology may be further segmented into n-MOS, p-MOS and CMOS. The choices the firm makes, say it opted for MOS technology in its CMOS variety, are major elements of the firm's technological policy. Subsequent choices to embody this technology into, say, memories that emphasize performance, particularly speed, rather than memory capacity (e.g., 64K bit fast, instead of 64K bit slow) are also central elements of technological policy.

Given these technological choices the firm then has to choose a manufacturing policy to implement these technological choices. For instance, plant location, scale and loading decisions have yet to be made.

The firm may choose, for instance, to build a 100,000 sq. foot plant in Phoenix, Arizona, and to staff it on a two shift basis, but to run it only at 80% of capacity. Beyond these decisions there remain manufacturing process and equipment decisions.[4] Levels of automation remain to be set. One choice might be to opt for a highly automated wafer manufacturing process and a labor intensive off-shore assembly process.

This example is meant as an illustrative rather than comprehensive description of the dimensions of technological policy. Technological policy properly defined cuts a wide swath across such functional policies as manufacturing, marketing, finance, R & D, as well as corporate-wide policies with respect to product-market focus, personnel resource allocation and control. As such, technological policy, as defined here is a corporate wide concept.

The remainder of this note examines in detail technological policy in its various dimensions, as well as its relationship to some of the functional components of corporate strategy. A set of four strategies are defined to illustrate the relationship between technological policy and the requirements in the functional areas of manufacturing and marketing. The emphasis throughout this note is on technological policy in the single business firm.

TECHNOLOGICAL POLICY: THE DIMENSIONS OF CHOICE

Technological policy consists of the portfolio of choices and plans that enables the firm to respond effectively to technological threats and opportunities. In formulating its technological policy, the firm must make choices in at least the following six areas: (See Exhibit 1)

- *Selection, specialization and embodiment:* What technologies to invest in? What technologies are promising from the perspective of the existing product line, or for new or related products? What technologies provide opportunities for improved product performance or lower product cost? How should these technologies be embodied into new products? What performance parameters should dominate? How should

proposals for new technologies/products be evaluated?

- *Level of competence:* How proficient to become in understanding and applying the technology? How close to the state of the art should the firm be in this technology to achieve its objectives in its products and markets, given the competitive environment? How much emphasis should be placed on advancing knowledge of the technology through basic or applied research, as opposed to straightforward applications of the technology through product development engineering?
- *Sources of Technology:* To what extent should external sources be relied upon, including contract research and licensing from individual inventors, research and engineering firms, or competitors? To what extent should we rely on internal development?

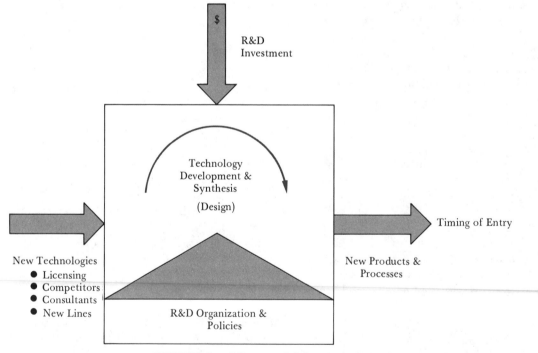

EXHIBIT 1.   A Framework for Technological Policy

- *R & D investment level:* How much to invest in these technologies? What level of internal staffing or external expenditure is appropriate? Do we let R & D investment or profit oscillate?
- *Competitive timing:* Should we lead or lag competitors in new product introduction? Does the benefit from leading competitors outweigh the risk of uncertain market acceptance of a new product? Are there benefits in allowing a competitor to go first, evaluating market acceptance of that product, and developing an improved product if market conditions warrant? What response is appropriate to a competitive product introduction?
- *R & D organization and policies:* Should there be a central R & D lab? How should it be structured? Should there be a separate career track for scientists? Should we use project teams? Or a matrix organization arrangement to allow sharing of scarce technical resources? Should we reward scientists and engineers with a level that is compatible with our industry? Or should we be leaders in compensation? How closely should top management be involved in technological decisions? What decision rules will we use to allocate funds to R & D projects? How should we protect our technological know-how? What should be our patent policy? Our publication policy?

FOUR SAMPLE STRATEGIES

To highlight these choices within the context of alternative corporate strategies, we will define four broad strategies commonly found in high-technology industries. Note that these are not complete definitions of the respective strategies, since they do not fully determine the basis on which all key decisions should be made. Nor are they "mutually exclusive" or "collectively exhaustive." A wide spectrum of other strategies are logically possible and potentially effective. They are simply intended to crystallize the concept of strategy in a way that is relevant in a technology-intensive environment. These strategies, which will be used as illustrative examples throughout this piece, are defined below:[5]

The *early, first-to-market* or leader strategy aims to get the product to the market before the competition. It provides the advantages of a *temporary monopoly* in exploiting a new technology during the period preceding the adoption of the new technology by competitors. Such a strategy normally requires a strong commitment to applied research and development, in order to achieve a position of technological leadership. The benefits of this strategy can be exploited by either of two principal approaches: 1) skimming by pricing high to achieve an immediate profit, or 2) through a penetration approach, by pricing low to achieve a higher market share and higher long-term profitability. The appropriate form in which to extract these "monopoly" benefits (short-term profit, market share, tying up technical resources and distribution channels) is a major issue in formulating a first-to-market strategy. An excellent example of a firm following this strategy is the Intel Corporation.

The second-to-market or fast follower strategy involves entry early in the growth stage of the life cycle and quick imitation of innovations pioneered by a competitor. This strategy generally requires a strong and nimble development and engineering capability, with little attention to applied or basic research. Marketing emphasis will generally be more on winning customers away from the technological innovator, with less emphasis on primary demand generation, as compared to the first-to-market strategy. All of the "follow-the-leader" strategies also try to learn from the innovator's mistakes, so as to develop an improved, more reliable product that may include "advanced features" while avoiding entirely those product innovations which prove to be market failures. An example of a firm following this strategy is the Zenith Corporation, RCA's perennial follower.[6]

The *cost minimization or late-to-market strategy* achieves a relative cost advantage over competitors through economies-of-scale and/or jointness across product lines in manufacturing and distribution, through process and product design modifications to reduce costs, and through overhead minimization and operating cost control. It requires product and process engineering skills to achieve a low-cost position. Entry into the market is generally in the growth stage or later, to allow market volume to grow to the point where significant economies of scale can be achieved, and to avoid investment in capital-intensive plant before product designs have become reasonably standardized. An example of a firm following this strategy is General Motors, especially with respect to compact cars.

The *market-segmentation or specialist* strategy focuses on serving small pockets of demand with special applications of the basic technology. Entry typically occurs in the early or growth stages of the product life cycle but may also occur at later stages as the market is segmented further. This strategy requires a strong capability in applied engineering, as well as flexibility in the manufacturing area. Large size or mass production competence is not required for this strategy, and may even be a handicap, since the scheduling and control requirements for a large number of special applications can be exceedingly complex. An example of a firm following this strategy is Silicon Valley Specialists, a manufacturer of high performance integrated circuits.

The essence of the late-to-market strategies is a reduced emphasis on basic and applied research, and a resulting reduction in the risk associated with the R & D investment of the firm. Of course, these defensive strategies do not absolve the firm of technological risk—risk of technological obsolescence in particular. These strategies are also characterized by increased investment (at least in relative terms) in various dimensions of marketing and production capacity, investments which carry their own risk.

TAILORING TECHNOLOGICAL POLICY TO BUSINESS STRATEGY

The technological policy appropriate for an individual firm would clearly depend on whether the firm was adopting a first-to-market, second-to-market, late-to-market or a market-segmentation strategy. For example, for firms selecting a given technological area for specialization, the early-to-market strategies require a higher level of competence in the technology, reflected in proximity to the state of the art achieved and increased emphasis on applied research, than the late-to-market segmentation strategies. Among the early-to-market strategies, the first-to-market strategy would tend to imply more emphasis on basic research, while the second-to-market strategy would imply relatively more emphasis on development engineering.

In terms of overall resource commitments, the first-to-market approach would naturally imply earlier commitments, but the second- and late-to-market strategies could involve as much or more than the leader in total dollar expenditures. The first-to-market strategy might imply more internal research capability, while the late-to-market approach could rely more heavily on external sources of technological information through licensing agreements or hiring of experienced personnel.

The relation between strategy and technological policy can also feed back in the other direction, from technological policy to strategy. A high level of competence achieved through the commitment of substantial resources to basic and applied research can lead to discoveries of new products or processes, which can provide an opportunity to lead competitors in introducing a product, although the firm might view itself as pursuing a second-to-market strategy in general. However, such a firm should be viewed as following its traditional second-to-market strategy, despite the occasional instances of leading the market with a new product introduction.

For each of the strategies discussed, there are natural implications for the capabilities required of the different functional areas within the business. It is useful to spell out in some detail what these might do for a firm applying each strategy. Typical functional requirements associated with each strategy are presented in Exhibit 2, along with typical organizational characteristics and the timing of inaugurating each strategy within the product life cycle.

As suggested above, the appropriate technological policies for an individual firm depend largely on the strategy adopted by the firm. The appropriate strategy depends, in turn, on the capabilities and resources available to the firm, on the competitive opportunities and threats faced by the firm, and on the objectives of the firm. In this section of the note, we look at how individual dimensions of technological policy may vary with the strategy adopted by the firm. Building on the six areas of technological policy introduced earlier, we will examine the following dimensions of policy:

1. Technology selection, specialization or embodiment
2. With respect to the level of competence: proximity to technological state of the art; emphasis on basic research, applied research, and development engineering;

3. Sources of technological capability: internal vs. external.
4. R & D investment: levels of investment and staffing
5. Competitive timing: initiative vs. responsiveness
6. R & D organization and policies: flexible or structured

1. Technology Selection or Specialization

Obviously, the selection of the technology or technologies in which the firm will specialize is of paramount importance for the technology-intensive firm. For instance, a fundamental element of Texas Instruments' (TI) technological policy is its choice of semiconductors as its core technology. For the small consumer electronics firm such as Advent, technological specialization may be simply determined by the skills of the founders, while for the larger firm such as Pilkington, existing technological resources will heavily influence the firm's choice with respect to technological specialization. TI will search far and wide within the confines of its core technology for products before committing to a completely new technological area. However, there is always an element of choice: what relative emphasis to place on the mix of existing technologies, and what *new* technologies to add to the firm's core set of technological skills?

The appropriate choice of technologies is an issue of particular importance for firms adopting the first-to-market strategy such as Hewlett Packard, and Intel since they must adapt the technology to their product needs earlier in the development of the technology. But the choices of a second-to-market firm trying to identify which of the technologies adopted by first-to-market firms will prove successful, or of a late-to-market firm trying to determine when to adopt a new technology for large-scale development are not necessarily easier.

241

Typical Functional Requirements of Alternative Technological Strategies

	R & D	Manufacturing	Marketing	Finance	Organization	Timing
First-to-Market	Requires state of the art R & D	Emphasis on pilot and medium-scale manufacturing	Emphasis on stimulating primary demand	Requires access to risk capital	Emphasis on flexibility over efficiency; encourage risk taking	Early-entry inaugurates the product life cycle
Second-to-Market	Requires flexible, responsive and advanced R & D capability	Requires agility in setting up manufacturing medium scale	Must differentiate the product; stimulate secondary demand	Requires rapid commitment of medium to large quantities of capital	Combine elements of flexibility and efficiency	Entry early in growth stage
Late-to-Market or Cost Minimization	Requires skill in process development and cost effective product	Requires efficiency and automation for large-scale production	Must minimize selling and distribution costs	Requires access to capital in large amounts	Emphasis on efficiency and hierarchical control; procedures rigidly enforced	Entry during late growth or early maturity
Market-Segmentation	Requires ability in applications, custom engineering, and advanced product design	Requires flexibility on short- to medium runs	Must identify and reach favorable segments	Requires access to capital in medium or large amounts	Flexibility and control required in serving different customers' requirements	Entry during growth stage

EXHIBIT 2. Corporate Strategy and Technological Policy

2. Level of Competence

a. *Proximity to Technological State of the Art.*
Proximity to the state of the art is an important dimension of the nature of the technological commitment of a business, and is closely related to the timing-to-market dimension of its strategy. Proximity to the state-of-the art, in turn, has important implications for the planning and control environment of the firm, and for the research vs. development mix within the R & D function.

The first-to-market strategy implies high proximity to, or development of, the state of the art. This often implies a significant emphasis on applied research. The second-to-market strategy also requires proximity to the state of the art, although the emphasis shifts from original development to a monitoring function followed by imitative, developmental engineering. In some cases the firm may have the resources and basic research capability to be first to market, but because of its own dominant market position will be slow in introducing new technology; IBM in the 70s is an example. The market segmentation and late-to-market strategies are typically a substantial distance from the state of the art, but are characterized by nimbleness in adopting new developments.

Proximity to the state of the art results in lower stability in the relevant technology for the firm, and reduced predictability as to the direction in which technology will change. Thus strategies relying heavily on high proximity to the state of the art become vulnerable to rapid change in technology.

Strategies relying on mass production of products incorporating proven technologies are generally less susceptible to such rapid, unpredictable shifts. However, such competitors may find it more difficult to shift to new technology when it becomes economic to do so, since they are heavily committed to their technology through plant and equipment investment. Known competitors, employing the same basic technology in parallel, mass production

strategies, are particularly unlikely to inaugurate such rapid changes in technology. For these reasons radical change usually comes from outside the industry or from smaller, more fluid, competitors in the industry.

The lack of stability and predictability associated with state of the art strategies implies greater difficulty in controlling, and greater need for freedom and flexibility, in the R & D function. Proximity to the state of the art requires a greater emphasis on research. As Ansoff and Stewart[7] have indicated, research-intensive organizations are generally characterized by indefinite design specifications and nondirective work assignments. In such an environment, there is an emphasis on innovation over efficiency, resulting in increased emphasis on *direction* control for projects ("Are we heading in the right direction?) and decreased emphasis on *cost* control ("How much will it cost?").

By way of contrast, the second-to-market and market-segmentation strategies are more development intensive. Development intensive organizations are characterized by well-defined design specifications, highly directive supervision, and structured sequencing of tasks and responsibilities.

b. *Relative Emphasis on Basic Research, Applied Research and Development Engineering.*
Basic research, applied research and development engineering differ with respect to their predictability (risk), the required creativity, expenditure level, and financial leverage. They also differ with respect to the sophistication of technical personnel required, the appropriate organizational environment or climate, and the kinds of control procedures which can be adopted. Thus, the choice with respect to R & D emphasis affects the risk profile and organizational climate of the firm. Since increased emphasis on being near the state of the art, and on fast response to technological opportunities has implications for the balance between fundamental research, applied research and development engineering, these strategic choices also

effect the risk profile of the firm and its appropriate organizational structure.

As suggested above the strategic choice with respect to proximity to the state of the art has direct implications for R & D emphasis. The commitment to developing the state of the art generally implies a commitment to applied research.[8] A commitment to meeting the state of the art through a second-to-market strategy implies, at a minimum, a commitment to a strong development engineering capability, and perhaps a limited capability to applied research so as to monitor and capitalize on new technology quickly, once it has been developed and brought successfully to market.

The R & D effort can also be characterized by its relative emphasis on product development (including both new product development and product improvements), and process development. Over the product life cycle, the relative emphasis generally moves from product development toward process development. Early in the product's life, a product satisfying the basic design objectives is acceptable. During the growth phase as more competing products enter the market, product improvements are often emphasized in order to effectively differentiate the product. At the same time, process improvements designed to reduce the manufacturing costs of the product become increasingly important. This is particularly true to the extent that products become standardized, and buyers become more price sensitive. In the mature phase, large production volumes make cost reduction through process improvements critical. Meanwhile, opportunities for major product improvements often become exhausted.

3. Sources of Technological Capability: Internal vs. External

Although an internal R & D unit is a common source of technological capability for the firm, it is not the only source. Licensing—buying the technological output of another firm's R & D unit—can be an important alternative source of technological know-how, particularly for firms adopting follower strategies, although some first-to-market firms use it to good advantage also (see following paragraph on DuPont). The individual inventor or the small innovative firm may not have the resources to take full commercial advantage of an invention, and may therefore be often willing to license the technology to a larger firm which can exploit it more effectively. Henry Kloss, for instance, was initially willing to license his large screen TV design for $300,000.

An analysis of DuPont's twenty-five major product and process innovations over the period 1920-1950 indicates that licensing technology made an important contribution to DuPont's success. Of the twenty-five innovations, only ten were invented by DuPont's own R & D staff. Of the eighteen new products invented, DuPont was responsible for the discovery of six. Out of seven product and process improvements, DuPont bore responsibility for five.[9] Products introduced into the U.S. market by DuPont but invented by others include tetraethyl lead, cellophane, freon, polyethylene, and titanium.

An additional source of technological capability is through acquisition—either of individuals, groups or entire firms. Hiring scientists and technical managers from competitors is a common approach to expanding a firm's technological capability, and is often used as a source of information regarding a competitor's R & D investment, although the ethics and legality of such a maneuver may be subject to question. Fairchild, for instance, hired away a complete top management group from Motorola in the early 1970s that came to be called Hogan's Heroes (after William Hogan, a member of the group, who was named president of Fairchild).

Hiring a complete research group or acquiring an entire firm is a complex maneuver.

Although this may be the quickest way to develop or expand a firm's R & D capability, the payoff can be uncertain. Of particular concern is whether the acquired unit will fit in with the organizational "culture" of the acquiring firm, whether its research efforts can be redirected by the acquiring firm.

4. R & D Investment Level

Setting the aggregate level of investment in equipment and staff that is committed to understanding, development and application of the selected technologies is a fundamental decision for the technology intensive firm. Investment in research, development and engineering can range from few percent of sales for firms in the automotive or electrical equipment industries to as high as 10 percent or more for leading computer or electronics firms. Thus, the firm that makes a major commitment to R & Ds investment may be sacrificing up to 5 percent of its short-term profits to gain long-term competitive advantage.

The firm that attempts to be consistently first to market needs to make a substantial R & D investment. This investment should be balanced between applied research and development engineering. The second-to-market generally makes a smaller investment and emphasizes product development and technology monitoring. The investment made by the late-to-market firm or cost minimizing firm is typically smaller yet, though these firms make substantial investment in process engineering. Firms following a market segmentation strategy, even if they adhere to early-to-market policies, may require a lower R & D investment for the fundamental technology may be developed by firms following broader market policies.

A key decision that these firms have to make is whether to allow R & D investment to diminish in the face of short-term pressures on profits. While most firms will look more closely at R & D investment in difficult times, the firm that purports to be first-to-market and yet sacrifices its long-term R & D investment may be implicitly, in the long run, changing its policy to a second-to-market policy.

5. Competitive Timing: Initiative vs. Responsiveness

The issue of competitive timing—of taking the initiative vs. responding to the initiatives of competitors—has been a central theme of this discussion. A firm may wish to lead its competitors to market, taking risks with respect to technological development and market acceptance, to gain a competitive advantage such as a larger market share. Silicon Valley Specialists, for instance, calculated that 10-20% larger market shares could result from an early entry. Alternatively, the firm may prefer to allow other competitors to take those risks, and rely on superior marketing or manufacturing capabilities to respond with a more attractive or less expensive product later in the life cycle of the product. In so doing, the firm takes the marketing and financial risks associated with these strategies.

The issue of timing is particularly important for the second-to-market strategy. Given that customers normally experience some costs and inhibitions in switching suppliers, earlier entry leaves a larger portion of the potential market available for penetration without having to overcome these switching costs. And of course, the competition for this market intensifies as more competitors bring their entries to the market.

6. R & D Organization and Policies

There is a variety of options that technology intensive firms exercise to deal with the unique needs and characteristics of their technical personnel. Among these are the decision to establish an R & D laboratory, the establishment of a special career track for scientists and the level and flexibility of compensation for technical and scientific personnel. There are, however, many other options. These are given as illustrative examples.

Firms that enter early into a new technology generally depend more on technological experts than firms that enter late. There may be in the early stages of a technological development only a handful of men and women that have a good understanding of the new devices and processes. For these reasons these firms are likely to consider extraordinarily high levels of compensation for their scientists, to exhibit a high level of flexibility in their compensation patterns, to treat scientists and engineers in a special way, and to establish, if their scale of operations permits, well-equipped central R & D labs.

On the other hand firms that emphasize competition later on in the product life cycle often find that:

1. There is ample availability of technical personnel with the requisite skills;
2. That much of the requisite technology is embodied, not in people, but in equipment.

For these reasons such firms are much less likely to adjust corporate-wide policies of the firm to satisfy the needs of the technical staff.

TECHNOLOGICAL POLICY AND MANUFACTURING

As pointed out earlier, technological policy involves choices among technologies, or to develop technologies, which form the backdrop for manufacturing decisions. Manufacturing policy involves choices with respect to the scale, location interrelationship of productive resources, given a basic set of technological choices. From one perspective, technological policy represents a dynamic counterpart to manufacturing policy, in that it determines what new technological capabilities are to be developed by the firm, which can then be incorporated into the product design and manufacturing decisions of the firm in a future period.

The technological staff must also interface with the manufacturing organization on an ongoing basis in the design and pilot manufacture of new products. Manufacturing is generally concerned with producing goods at the lowest cost, and with the most efficient use of productive resources (through long production runs, standard and easy-to-manufacture designs, and large manufacturing tolerances), while engineering must preserve the functional standards required for the product to perform its intended function for the end-user. The technological policies and the manufacturing policies of the firm must be made explicit and integrated, so that these natural conflicts can be resolved.

The technological policy of the firm must also recognize the demands placed on manufacturing by the corporate strategy, and provide the technical support necessary for manufacturing to perform its function. The first-to-market strategy requires rapid expansion in productive capacity for a successful product, so that manufacturing may require assistance in scaling up the manufacturing process to handle the increased volume. The second-to-market strategy similarly places heavy demands on manufacturing to rapidly start up volume production; there may not even be time to evaluate production processes on a pilot basis. Late-to-market strategies require strong process engineering capabilities to reduce manufacturing cost. Market segmentation requires the design of a flexible manufacturing process to handle short-to-intermediate volume production runs.

TECHNOLOGICAL POLICY AND MARKETING

In technology-intensive industries, the evaluation of demand for products in development is a particularly important function for the marketing organization. The product development process represents the crucial interface between the applied research, development engineering and marketing groups. The technological staff bears responsibility for determining and communicating what products are feasible and at what cost, while the marketing group bears responsibility for determining what products are desired by the market, and at what price. The nature of the marketing problem varies to some extent with the strategy adopted by the firm:

- First-to-market: Does demand exist for this product? How should it be sold? What price will people pay for such a set of performance attributes?
- Second-to-market: What product attributes are desired by the largest share of the market? What are the weaknesses of the market innovator's product? How can we effectively differentiate our product entry?
- Late-to-market: How sensitive is demand to price? What level of demand can be expected if costs and price can be cut by X%?
- Market segmentation: What pockets of demand exist that are not being effectively served by existing products? What product attributes are required to serve these market segments? How large are these segments, and what price can they each afford to pay?

Product differentiation is an important marketing issue in a broad range of technology intensive industries.[10] For example, being first to market with new design features has been an important element of differentiation in the home appliance industry.

The offensive, first-to-market strategy provides the strongest basis for differentiating a product. However, if early product introductions are technical or commercial failures, that differentiation may be unfavorable rather than favorable. A second-to-market strategy can also provide favorable differentiation for a product, particularly if the product represents an improvement over earlier introduction by competitors. The market segmentation strategy can provide a significant degree of differentiation, depending on the extent to which the product is modified for specific market segments. For example, in the computer industry, the development of applications software for certain markets can represent a very significant basis for differentiating one machine over another. The cost minimization strategy tries to deemphasize differentiation, by competing on price through mass production, operating cost controls, and reduced overhead levels (R & D is less important and administrative expense is relatively low).

In general, marketing and technological policies should be defined so that they are mutually consistent and in support of corporate strategy. One area where this consistency is of importance is in the planned performance of the product line relative to competitors, especially where performance is a function of proximity to the technological state of the art. Similarly, the frequency of new product introductions envisioned by marketing must be supported by adequate staffing of the product engineering group.

SUMMARY

Technological policy involves choices within six major areas: technology selection, specialization or embodiment, level of technological competence (including proximity to the state of the art; relative emphasis on basic and applied research and development engineering); internal vs. external sources of technological capability; R & D investment level; competitive timing and R & D organization and policies. These choices must be made on a basis that is consistent and mutually reinforcing with the other dimensions of corporate strategy, including manufacturing policy and marketing policy. To highlight the nature of these relationships between the dimensions of technological policy and corporate strategy, four sample strategies were defined: first-to-market, second-to-market, late-to-market (or cost-minimization) and market segmentation. These strategies each have different implications for the level of technological competence, the relative emphasis on research and development, the use of external sources of technological information, the timing and level of technological investment and staffing for the firm, and the R & D management policies and organization.

For the technology-intensive firm the development of such a framework is an essential step in defining its technological posture. In the 1950s and 1960s we saw the first explicit use of the concept of corporate strategy. In the 1970s manufacturing became part of corporate strategy. In the 1980s, technology must become a full partner in corporate strategy.

NOTES

1. Wick Skinner, "Manufacturing—Missing Link in Corporate Strategy," *Harvard Business Review,* May-June 1969.
2. Harold B. Meyers, "The Nuclear Fizzle at Old B&W," *Fortune,* November 1969.
3. M.A. Maidique, *The Performance Curve,* unpublished working paper.
4. This decision is often called the manufacturing technology selection. Wick Skinner calls it the EPT (equipment and Process Technology) decision.
5. These strategies roughly parallel the four strategies defined by Ansoff and Stewart, "Strategies for a Technology-Based Business, *Harvard Business Review,* Nov.-Dec., 1967, p. 81.
6. In the 70s Zenith fast follower strategy ran into trouble. After capitalizing on following RCA's entry into black and white and color TVs, it waited in vain for another major new product to follow.
7. Ansoff and Stewart, ibid.
8. Very few industrial firms do any significant amount of basic research. Basic research is usually done in government and university labs and independent research contractors such as the Mitre Corporation, Jet Propulsion Lab and Los Alamos National Laboratories.
9. Cited in Twiss, *Managing Technological Innovation,* Chapter 2, Strategies for Research and Development, pp. 58–59.
10. A product is differentiable if a buyer perceives differences between like products of competing firms.

Strategic Responses to Technological Threats

Arnold C. Cooper

Dan Schendel

Several industries threatened by major technological innovations are studied, and conclusions are drawn which have implications for other firms confronted with new technologies.

Technological innovation can create new industries and transform or destroy existing ones. At any time, many businesses are confronted with a host of external technological threats. Managements of threatened firms realize that many threats may not materialize, at least in the short run. However, one or more of those potential threats may develop in ways that will have devastating impact. Providers of kerosene lamps, buggy whips, railroad passenger service, steam radiators, hardwood flooring, passenger liner service and motion pictures all have had to contend with such threats. Few environmental changes can have such important strategic implications.

A typical sequence of events involving the traditional firm's responses to a technological threat begins with the origination of a tech-

"Strategic Responses to Technological Threats" by Arnold C. Cooper and Dan Schendel, from *Business Horizons*, February 1976, pp. 61–69. Copyright 1976 by the Foundation for the School of Business at Indiana University. Reprinted with permission.

The authors wish to acknowledge the contributions of Forrest S. Carter, Edward E. Demuzzio, Kenneth J. Hatten, Elija J. Hicks, David A. Robbins and Donald G. Tock.

nological innovation outside the industry, often pioneered by a new firm. Initially crude and expensive, it expands through successive submarkets, with overall growth following an S-shaped curve. Sales of the old technology may continue to expand for a few years, but then usually decline, the new technology passing the old in sales within five to fourteen years of its introduction.

The traditional firms fight back in two ways. The old technology is improved and major commitments are made to develop products utilizing the new technology. Although competitive positions are usually maintained in the old technology, the new field proves to be difficult. In addition to the major traditional competitors (who are also fighting for market share in the new field), a host of new competitors must be confronted. Despite substantial commitments, the traditional firm is usually not successful in building a long-run competitive position in the new technology. Unless other divisions or successful diversifications take up the slack, the firm may never again enjoy its former success.

Most previous research on technological

249

innovation has been concerned with the practices and problems of innovators. This research is concerned with major technological innovations from the viewpoint of firms in established industries threatened by innovation.

THREATENED INDUSTRIES

The industries and technologies selected for study were the following:

- steam locomotives vs. diesel-electric
- vacuum tubes vs. the transistor
- fountain pens vs. ball-point pens
- boilers for fossil fuel power plants vs. nuclear power plants
- safety razors vs. electric razors
- aircraft propellers vs. jet engines
- leather vs. polyvinyl chloride and poromeric plastics.

Within these traditional industries, twenty-two separate firms were studied, using data available in the secondary literature where over 200 separate sources were examined. The accompanying table lists these firms. Two broad questions are of concern to the study:

- What was the nature of the substitution of the new technology for the old?
- What response strategies were used to counter the technological threats?

The findings must be regarded as tentative. The data are incomplete in some areas, as should be expected from secondary data. For example, data on the performance and strategies of smaller firms in the threatened industries are not readily available. The relatively small number of industries studied and the complexity of the processes prevent definitive conclusions. However, the experiences of these industries and firms suggest a number of conclusions with implications for managers of threatened firms.

The first section which follows deals with substitution patterns of new technology for the old. The second deals with response strategies by the threatened firms, a subject which has received very little previous attention in the literature. Finally, a section is devoted to implications and conclusions.

PATTERNS OF SUBSTITUTION

The nature of the substitution of one technology for another is not well known. The product life cycle concept suggests that products move through a classical S-shaped curve: this implies that the sales of a new product grow, slowly at first and then rapidly, and finally mature to a plateau from which they decline. Presumably, this would apply to the sales of a new technology, with the sales of the old technology declining accordingly.

Empirical studies of ethical drug products and of nondurable goods found that new products did not always follow the S-shaped sales curve.[1] J. C. Fisher and R. H. Pry postulated that the substitution of one technology for another follows a hyperbolic tangent or S-shaped curve.[2] On a semilog scale, the market share of the new technology divided by the market share of the old technology plots as a straight line. They reported on some seventeen substitutions, most of which closely followed this pattern. Kenneth Hatten and Mary Louise Piccoli tested the Fisher-Pry model on over forty substitutions and reported that the model was useful, so long as care was exercised in the selection of units and in the application of the results.[3] Generally, then, an S-shaped curve of growth for the new technology and a similar S-shaped curve of decline for the old technology would be expected.

The data required to plot the product

TABLE 1. Traditional Industries Studied

	Locomotives	Vacuum (Receiving) Tubes	Fountain Pens	Safety Razors	Fossil Fuel Boilers	Propellers	Leather Industry
Sales decline immediately after new technology introduced?	No	No	*	No	No	No	Yes[1]
Sales eventually begin longterm decline?	Yes	Yes	Yes	No	No	Yes	Yes
Time from introduction of new technology until sales of new technology exceeded old.	Fourteen years[2]	Eleven years	Nine years	Twenty-five years[3]	Not during the twenty years since first sale	Five years[4]	†
New markets created by new technology?	No	Yes	Yes	No	No	No	Yes
New technology limited in application or crude at first?	Yes	Yes	Yes	Yes	Yes	Yes	Yes
New technology applied sequentially to submarkets?	Yes	Yes	Yes[5]	No	No	Yes	Yes
First commercial introduction by a firm in traditional industry?	‡	Yes	No	No	§	Yes[6]	No
First commercial introduction by a new firm?	No	No	Yes	Yes	No	No[6]	Yes
Old firms participate in new technology?	Yes	Yes	Yes (4 of 5)	Yes (briefly)	Yes	Yes (2 of 3)	No (1 of 4)[7]
Acquisitions a means of participating in new technology?	No	Yes Raytheon	Yes Parker	Yes Gillette[8]	No	No	Yes Allied Kid
Old technology improved after new technology was introduced?	Yes	Yes	Yes	Yes	Yes	Yes	Yes

	Locomotives	Vacuum (Receiving) Tubes	Fountain Pens	Safety Razors	Fossil Fuel Boilers	Propellers	Leather Industry
Traditional firms involved in improving old technology and in entering new technology?	Yes	Yes	Yes (4 of 5)	Yes (participation in electric razors short-lived)	Yes	Yes	Yes
Attempt to establish barriers to new technology?	No	No	No	No	No	No	No

Firms Studied: Locomotives—American Locomotive Co., Baldwin Locomotive Works; **Vacuum (receiving) tubes**—Columbia Broadcasting System (CBS), Radio Corp. of America (RCA), Raytheon Mfg. Co., Sylvania Electric Products, Inc.; **Fountain pens**—Esterbrook Pen Co., Eversharp, Inc., Parker Pen Co., Sheaffer Pen Co., Waterman Pen Co.; **Safety razors**—American Safety Razor Corp., Gillette Safety Razor Co.; **Fossil fuel boilers**—Babcock & Wilcox Co., Combustion Engineering Inc.; **Propellers**—Koppers Corp., Curtiss-Wright Corp., United Aircraft Corp.; **Leather industry**—A. C. Lawrence Leather Co., Armour Leather Co., Allied Kid Co. (Cudahy), Seton Leather Co.

1. Production of three of the four types of leather declined in the year after vinyl was first used as a leather substitute.
2. Available sales data relate to units sold rather than sales dollars, but it appears that diesel-electric sales exceeded steam locomotive sales by 1938, fourteen years after the first diesel-electric switcher was introduced. Subsequently, steam locomotive unit sales exceeded diesel-electric unit sales during World War II, but steam locomotive sales then dropped sharply after the War.
3. During 1956–1958, electric razor sales exceeded sales of razor blades. Subsequently, however, razor blades regained a sales lead and have maintained it to the time of the study.
4. Unit production of jet engines exceeded unit production of piston engines during a three-year period in the early 1950s. It appears that the dollar value of jet engines produced exceeded the value of the smaller, less powerful piston engines within about five years of their introduction in the United States.
5. The pen market is segmented by price. Initially, the ball-point pen was relatively expensive.
6. Power Jets, a new British firm, developed the first jet engine. General Electric developed and introduced the first American jet engine, relying upon Power Jets' designs.
7. Allied Kid bought Corfam from DuPont in 1965. Also, all the firms began coating hides with synthetic materials to improve their qualities.
8. Gillette acquired Braun, A. G., and thereby entered the overseas market for electric razors. Gillette has not reentered the U. S. market since 1938, when its internally developed electric razor was introduced and subsequently withdrawn.
   *Data were not found to indicate whether sales of fountain pens declined the year the ball-point pen was introduced.
   †Results are mixed by type of application. By 1950, synthetics had captured 50% of the shoe sole market.
   ‡The first mainline diesel-electric was introduced by General Motors, a firm which never made steam locomotives. However, American Locomotive had earlier introduced an experimental diesel-electric switcher.
   §The first nuclear power plant was developed by Westinghouse, a firm with a strong position in turbines. However, for the producers of boilers, it was not a traditional competitor which introduced the new technology.

life cycles for the new and old technologies are not always available in the form desired. For instance, the unit sales of piston engines and of jet engines over time do not accurately reflect the much greater horsepower of jet engines. Comparison of leather and vinyl is made difficult by the wide range of uses of both materials, often in applications where they do not compete with each other. Nevertheless, a number of questions of managerial interest can be considered with the data available.

An examination of the sales over time for both the new and old technologies showed variable patterns which do not always duplicate the classical S-shaped pattern. Analysis of this sales data, coupled with extensive examination of other information, leads to a number of conclusions concerning the substitution pattern of new for old technologies.

1. After the introduction of the new technology, the sales of the old technology did not always decline immediately; in four out of seven cases, sales of the old technology continued to expand.
2. In two cases, sales of the old technology continued to expand for the entire period studied, despite growth in sales of the new technology.
3. When sales of the old technology did decline, the time period from first commercial introduction to the time when dollar sales of the new technology exceeded dollar sales of the old ranged from about five to fourteen years.
4. The first commercial introduction of the new technology was, in four out of seven cases, made by a firm outside the traditional industry. It might have been expected that the traditional competitors would have been the logical sources of industry innovation because of their strong customer relationships, well-developed channels of distribution and organizations oriented toward serving those industries.
5. In three of the four industries in which capital requirements were not excessive, new firms were the first to introduce the new technology.
6. The new technology often created new markets which were not available to the old technology. Although the initial ball-point pens were expensive, low-priced pens were later developed which opened up a new market—the "throw-away" pen. It was also estimated that 50% of the applications for the transistor were in equipment made possible by the invention of the transistor. Vinyls were used in floor coverings and build-

ing materials, applications not open to leather.

7. The new technology was expensive and relatively crude at first. Often, its initial shortcomings led observers to believe it would find only limited applications. Although the first ball-point pens wrote under water, they blotted, skipped and stopped writing on paper and even leaked into pockets; after an initial fad phase, public disenchantment set in and sales dropped dramatically. The first transistors were expensive and had sharply limited frequencies, power capabilities and temperature tolerance; some observers thought they would never find more than limited application. The jet-powered airplane was initially thought to be suitable only for the military market.
8. The new technology often invaded the traditional industry by capturing sequentially a series of submarkets. Although the new technology was crude it often had performance advantages for certain applications. Some submarkets were insulated from competition for extended periods. General Motors' diesel-electric locomotive first invaded the submarket for passenger locomotives, subsequently the submarket for switcher locomotives, and then freight locomotives—the major submarket—accounting for about 75% of industry sales. The transistor found early application in hearing aids and pocket radios, but not in radar systems and television.
9. The new technology did not necessarily follow the standard S-shaped growth curve. Erratic patterns were caused by abnormal economic and social conditions (World War II in the case of the electric razor, propellers and steam locomotives), by faddish phases of sales (ball-point pens), and by a newer technology replacing the original new technology (transistors and integrated circuits).

Some Pitfalls of Appraisals

Many factors affect the rate of penetration of a new technology: it does not capture markets overnight. Substantial sales opportunities may exist in the old technology for extended periods. It may be difficult for management in the traditional firms to judge the eventual impact of a developing threat, but at least there is usually time to develop a new strategy.

However, response presumes the ability to recognize and assess the threatening innovation. Intelligence activities focusing only upon traditional competitors are not enough, inasmuch as nontraditional competitors and new firms may be the originators of the threatening technology. It may be necessary to monitor a variety of innovations, many of which may never have significant impact.

Surviving past technological threats does not confer future immunity. In 1934, when General Motors introduced the first mainline diesel-electric locomotive, the producers of steam locomotives could look back upon two earlier threats which they had survived: the electric locomotive, and, in the 1920s, passenger cars with individual gasoline-powered engines. Both of these prior threats captured only small segments of the American locomotive market. There was no indication that the next threat, the diesel-electric, would destroy the traditional industry within fifteen years.

It would be a mistake to wait until decline in sales of the old technology triggered the need for appraisal of the threat. By then, much of the lead time would have passed. However, this means that the new technology must be appraised when it is still relatively crude. In an earlier article, James C. Utterback and James W. Brown emphasized that hypotheses about directions for change aid in selection of parameters which can be observed and evaluated.[4] For instance, early diesel engines had such a high weight-to-horsepower ratio that a diesel-electric locomotive would have been impossibly large. Managements of steam locomotive firms might have hypothesized that any changes leading to improvements in this weight-to-power ratio were of critical importance and deserved continuous monitoring.

It is not enough to judge that someday a new technology will replace an old one. Rates of penetration must be determined. When the Baldwin Locomotive Works was founded in 1831, it would have been of little value to tell founders that someday their principal product would be obsolete. However, when Sylvania introduced a new line of vacuum tubes for computers in 1957, the rate of improvement of transistors then taking place was extremely relevant.

The forecaster needs to understand differences in needs of market segments and relate these to probable improvements in the new technology. Some market segments in a traditional industry are threatened earlier and to a greater extent than others. Firms should consider strategies involving emphasis on the less threatened segments.

RESPONSE STRATEGIES

Once the threat has been recognized, what kind of response is made by the traditional firm? If it decides not to participate in the new technology, management might elect one or a combination of the following specific actions.

Do nothing.

Monitor new developments in the competing technology through vigorous environmental scanning and forecasting activity.

Seek to hold back the new threat by fighting it through public relations and legal action.

Increase flexibility so as to be able to

respond to subsequent developments in the new technology.

Avoid the threat through decreasing dependence on the most threatened submarkets.

Expand work on the improvement of the existing technology.

Attempt to maintain sales through actions not related to technology, such as promotion or price-cutting.

A firm might, however, choose to participate in the new technology. The degree of commitment could vary widely, ranging from a token involvement, such as defensive research and development, to seeking leadership in the new technology through major and immediate commitments. Important dimensions of a strategy for participation in the technology include decisions about the level of acceptable risk, the magnitude of commitments to the new technology, the timing of those commitments and the extent of reliance on internal development versus acquisition. Against this background of possible responses, the seven industries were studied to determine the response strategies actually used by the threatened firms. Their strategies are shown in the accompanying table.

Participation in the New Technology

Of the twenty-two firms studied, all but five made at least some effort to participate in the new technology. Fifteen of the firms made major efforts to establish positions in the new technology. Firms with small market shares in the old technology were not the focus of this study. However, it does appear that they either did not attempt to establish positions in the new technology, or they achieved no visible success. For instance, the hundreds of small razor blade firms never had successful electric razors, and the five smallest locomotive producers never made the transition to diesel-electrics.

Nature of Participation

The timing of traditional firms' entries in the new technology varied widely. Raytheon and RCA vacuum tube producers were among the first to enter the transistor market. By contrast, Parker Pen brought out its first ball-point pen nine years after its first commercial introduction. Of the nine firms which had traditionally emphasized research and development in their various divisions, six were early entrants in the new technology. By contrast, only two of the firms with a low research and development emphasis were early entrants.

Acquisition was not a widely used means of entry into the new technology. Only four of twenty-two traditional firms used acquisition, and two of these used acquisitions to supplement their internal development. Parker acquired the Writing Division of Eversharp as a means of successfully entering the low-priced ball-point pen market after having first developed a high-priced ball-point pen. Raytheon, having previously made major commitments to germanium transistors, acquired Rheem Semiconductor as a means of entering the silicon transistor field.

Emphasis on Old Technology

In every industry studied, the old technology continued to be improved and reached its highest stage of technical development *after* the new technology was introduced. For instance, the smallest and most reliable vacuum tubes ever produced were developed after the introduction of the transistor. No threatened firm adopted a strategy of early withdrawal from the old technology in order to concentrate on the new. Moreover, all but one of the twenty-two companies continued to make heavy commitments to the improvement of the old technology.

Most of the firms followed a strategy of dividing their resources, so as to participate in a major way in both the old and new technologies. Baldwin Locomotive developed both advanced turbine-powered electric locomotives and diesel-electric locomotives. CBS and Raytheon developed new lines of vacuum tubes and also made major investments in research and development and production facilities for transistors. This dual strategy was not usually successful, particularly in relation to building a strong competitive position in the new technology. There were no apparent actions taken by the traditional firms to create or strengthen the barriers to adoption and diffusion of the innovations.

Firms that pioneered the new technology generally did not enter the old technology. The only exception was BIC, a successful French producer of low-priced ball-point pens, which acquired Waterman, an American fountain pen manufacturer. The acquisition was apparently for Waterman's U. S. distribution system rather than its product line, inasmuch as the fountain pen line was discontinued four years later.

Overall Performance

The new technical innovations did not always lead to immediate financial returns and, in fact, sometimes presented all participants with severe competitive challenges. The nuclear power field involved very heavy investments for many years by all participants before the first profits were earned. The precipitous sales decline, which occurred after the first cycle of ball-point pen sales, drove more than 200 new firms, as well as several established firms, from the market. DuPont's poromeric leather substitute, Corfam, reportedly resulted in losses of $100 million: Goodrich and Armstrong were also entrants who later withdrew from the leather substitute field.

The new technology often evolved rapidly. Transistors, nuclear power plants and jet engines all confronted participants with a succession of decisions about commitments to evolving technologies. Early leaders, such as Raytheon in transistors and Curtiss-Wright in jet engines, lost their competitive positions as the technology changed.

Where the old technology continued to grow, traditional firms were able to maintain their competitive positions and enjoy financial success. But many of the most successful firms in the new technology had never participated in the old technology. In industries in which capital barriers were not great, new firms were among the most successful. Examples of successful new firms were Papermate in ball-point pens, Fairchild Semiconductor in transistors and Schick in electric razors.

Over the long run most of the traditional firms that tried to participate in the new technology were not successful. Of the fifteen firms making major commitments, only two, Parker in ball-point pens and United Aircraft in jet engines, enjoyed long-term success as independent firms participating in the new technology.

Patterns of Commitment

Managers of threatened firms must decide how to allocate resources in choosing between improving the old technology and attempting to establish a competitive position in the new. If sales of the old continue to grow, as in safety razors or fossil fuel power plants, then the strengthening of the firm's position in the business it knows so well can be rewarding. However, if sales of the old technology are declining, heavy, across-the-board commitments seem questionable. Management should carefully segment its markets and identify those which appear protected from the threat. Strategies based upon maintaining strong competitive positions in these segments seem justified.

It is interesting that the traditional firms studied here continued to make substantial commitments to the old technologies, even when their sales had already begun to decline because of the competitive pressures of the new technologies. Perhaps this demonstrates the difficulty of changing the patterns of resource allocation in an established organization. Decisions about allocating resources to old and new technologies within the organization are loaded with implications for the decision makers; not only are old product lines threatened, but also old skills and positions of influence.

It was common for spokesmen for the traditional firms to emphasize the shortcomings of the new technology with comments such as "It is no wonder if the public feels that the steam locomotive is about to lay down and play dead," and "It is certain that substantially all airplanes which operate at speeds of 550 mph or less will use propeller propulsion." The executives who made these statements, conditioned by life-long involvements with the old technology, may have been slower than others to recognize the declining opportunities for their traditional products.

Commitment to the new technology, with its expanding opportunities and lack of entrenched competitors, may seem attractive. Certainly most of the firms studied here made such strategic investments. Yet such decisions are fraught with risk, as evidenced by the traditional firms being relatively unsuccessful in the new fields.

For these companies, the patterns of commitment seem to be related to the firm's characteristics. One group of firms was relatively undiversified and did not have strong research and development orientations. The producers of locomotives, fountain pens, safety razors and two of the leather producers might be so classified. Except for several of the pen companies, these firms usually were *not* early entrants, and furthermore, never captured substantial market shares in the new technologies. It is tempting to conclude that an innovative technical and managerial organization is required to make a successful transition from the old to the new technology.

Another group of firms had relatively strong research and development traditions and were accustomed to managing multibusiness organizations. Most of this group, which included the producers of vacuum tubes, boilers and propellers, made major commitments to the new technologies and in several instances achieved substantial early success. However, these technologies continued to evolve rapidly, so that it was necessary to generate successive generations of successful new products. Here, companies such as Curtiss-Wright in jet engines and RCA and Raytheon in transistors were unable to continue their early successes.

The reasons for these firms' inability to build and maintain strong competitive positions are not obvious. Resource limitations apparently were not a major factor in the transistor industry, inasmuch as a number of new companies were relatively successful. The traditional firms not only had to develop new products based upon different technologies, but also had to adapt to changing methods of marketing, servicing and manufacturing. Their lack of long-term success may be an indication of the relative difficulty of changing organizational strategy successfully. The skills, attitudes and assumptions which undergird successful strategy in a traditional technology may require modification in ways both major and subtle to bring about equivalent success in the new technology. Apparently, many organizations found this difficult to do.

Managers of threatened firms should consider carefully commitments to the new technology. Where such commitments are made, it is desirable to recognize explicitly the different strategic requirements for success in the new field. Acquisition, although not widely used by the firms studied here, merits particular consideration. This may be a way to acquire not only technical capabilities, but also organizations attuned to competition in the new field. There are no easy paths to success when faced with major technological threats. However, the experiences of these firms illustrate some of the approaches and pitfalls which management should consider.

NOTES

1. William E. Cox, "Product Life Cycles in Marketing Models," *Journal of Business* (October 1967), pp. 375–384. Rolando Polli and Victor Cook, "Validity of the Product Life Cycle," *Journal of Business* (October 1969), pp. 385–400.

2. J. C. Fisher and R. H. Pry, "A Simple Substitution Model of Technological Change," *Technological Forecasting and Social Change* (1971), pp. 75–88.

3. Kenneth Hatten and Mary Louise Piccoli, "An Evaluation of a Technological Forecasting Method by Computer Based Simulation" (Proceedings of the Division of Business Policy and Planning, Academy of Management, August 1973).

4. James C. Utterback and James W. Brown, "Monitoring for Technological Opportunities," *Business Horizons* (October 1972), pp. 5–15.

SECTION
IV

Managing Functional Areas

Successful innovation involves linking technological opportunities with market needs with operational capabilities. Section IV takes a closer look at building functional competence. Chapter 5 explores managing RD&E and managing professionals. Tushman discusses the importance of informal networks in the innovation process. Then, Roth's article reviews the literature on careers of scientists and engineers, with particular reference to dual ladders. Katz and Allen build on Tushman's article and, in turn, investigate communication patterns over time, exploring the generic "not-invented-here" syndrome that so frequently plagues the innovation process. Next, Roberts and Fusfeld summarize the research on informal roles so critical in the management of R&D and so critical to the innovation process. Finally, Westney and Sakakibara contrast US and Japanese R&D facilities and discuss the implications of these differences for firms desiring to establish R&D facilities in Japan.

Marketing plays a crucial role in the reduction of new product failures by searching for ideas that satisfy customer needs or solve customer problems. Marketing may also be involved in developing and testing product or service offerings. Chapter 6 includes a set of articles that reflect different marketing issues in the innovation process. Howard and Moore provide insight on different consumer behavior patterns over a product's life cycle. Knowledge of consumer behavior can, in turn, assist in concept/market testing as well as in advertising strategy. The von Hippel and Moore articles explore the special issues involved in getting market input for major new product, service, or process innovation. Von Hippel finds that lead users are a potential source of new product information as well as the product, service, or process itself, while Moore reports field research on concept testing. Page and Rosenbaum discuss conjoint analysis as a powerful new product-development tool. Using a detailed example of food processors, they describe how conjoint analysis assists in managing product line development within the Sunbeam Appliance Company.

Manufacturing and production strength is crucial to effective innovation over time. If an organization loses its ability to make its product or service, it

259

will not have the competence to be innovative. Chapter 7 concentrates on manufacturing's role in the innovation process. Jelinek and Goldhar explore the linkages between business strategy and manufacturing strategy. This article also discusses the consequences of CAD-CAM technology on manufacturing capabilities. Hayes and Wheelwright discuss the evolving linkages between manufacturing and other functional areas. This article builds on Abernathy and Utterback's work as they discuss the linkages between dominant design and process innovation, and between technological discontinuities and revolutionary shifts in manufacturing. Wheelwright's following article discusses the linkages between manufacturing and engineering, with a focus on reducing the time between design engineering and product introduction. Finally, no discussion of manufacturing could be complete without some discussion of organizing for total quality. Deming's article presents his fourteen managerial points toward achieving total quality.

5

RESEARCH AND DEVELOPMENT

Managing Communication Network in R&D Laboratories

Michael L. Tushman

This article is concerned with the problem of actively managing communication networks in R&D settings. A series of studies comparing and contrasting the communication networks of high- and low-performing projects are summarized. The results are linked with the larger literature of managing R & D and lead to a contingent approach to managing communication. Finally, the author presents a communication design model and process as a framework for managing these networks. Ed.

The process of developing and introducing technological innovation is central to industrial firms. Innovation can lead to competitive and sales advantages in a growing industry, or to diversification and new applications for existing products in more mature companies. In short, managers must consider innovation with respect to both the growth and survival of new products and markets.[1]

While innovation is of vital importance, research indicates that the innovation process is relatively costly and inefficient. Indeed, research conducted over the past decade indicates that organizational factors (i.e., nontechnical factors) are often the most critical barriers to effective innovation.[2] While several organizational factors affect the innovation process (e.g., organization design or control systems), mechanisms by which projects in research and development (R & D) laboratories acquire and process information are of particular importance.[3]

Reprinted from "Managing Communication Networks in R&D Laboratories" by Michael L. Tushman, *Sloan Management Review*, Winter 1979, pp. 37–49, by permission of the publisher. Copyright 1979 by the Sloan Management Review Association. All rights reserved.

*The author thanks Professors David Nadler, George Farris, Michael Ginzberg, and Noel Tichy for their constructive comments and encouragement.

This article focuses on managing communication networks in R & D laboratories. It will synthesize, and link with a larger literature on managing R & D, a series of studies which have investigated the association between communication networks and technical performance in research and development. This synthesis will, in turn, be used to generate a model for managing communication networks to enhance innovation.*

To anticipate the conclusion, this article argues that communication networks are an important process in R & D settings, that these processes can be actively managed, and that there is no one best communication pattern. Rather, communication networks for more innovative projects are contingent on the nature of the work involved.

COMMUNICATION NETWORKS IN R&D

Field studies indicate that engineers and applied scientists spend between 50 and 75 percent of their time communicating with others.[4] Why should so much time be devoted to verbal interaction? One reason is that for problem-solving purposes, verbal communication is a more efficient information medium than written or more formal media (e.g., management information systems). Verbal communication permits timely information exchange, rapid feedback, and critical evaluation, as well as the opportunity for real-time recoding and synthesis of information.

The importance of verbal communication is accentuated in R & D settings given the importance of information acquisition and exchange and given the complexity of technological problem solving. Supporting the importance of verbal communication, research has consistently demonstrated the link between widespread and extensive patterns of verbal

communication and both individual and project performance.[5] Verbal communication is, then, pervasive and important in R & D settings.

Given the utility of verbal communication, more specific questions can be raised. Does a project's *pattern* of verbal communication affect technical performance, or is communication equally effective throughout the laboratory and larger organization? More specifically:

- Does the amount and structure of communication within projects affect technical performance?
- Does the pattern of project communication with external areas affect technical performance?
- Finally, does the mechanism for linking the project to external areas affect technical performance? Should widespread communication be encouraged or should more specialized linking mechanisms be developed?

To address these questions, an extensive field study was conducted in the corporate R & D facility of a large American corporation. The laboratory is physically isolated from the rest of the corporation and is organized into seven divisional laboratories. Each divisional laboratory is, in turn, organized by project area (such as fiber forming development, yarn technology, etc.). The work in these projects (sixty-one in total) ranges from basic/exploratory research (research projects) to more routine cost/performance work on existing products (technical service projects).

All technical personnel were asked to report all work-related, verbal communication at the end of selected days; these data were collected for fifteen days over a four-month period (each of the weekdays equally represented). Information on technical performance was gathered by asking the laboratory's senior management team ($N = 9$) to rate the overall technical performance of projects with which they were familiar. As their individual ratings were highly correlated, individual ratings were pooled to produce project performance scores. Information on organizational conditions was obtained by interview and questionnaire.

With these data, several analyses were conducted to address the questions raised previously. Analyses were based on the assumption that, due to information requirements and demands, high-performing projects facing different tasks would have systematically different communication networks. Therefore, the research questions were addressed by contrasting the communication networks of different types of projects within the laboratory.[6]

COMMUNICATION PATTERNS, PROJECT WORK CHARACTERISTICS, AND TECHNICAL PERFORMANCE

Impact of Task Characteristics on Communication Networks

Research projects centered on work oriented towards developing new knowledge and concepts in several technologies important to the firm. Tasks in these projects were complex and universally defined.[7] All technical staff members in research projects had advanced degrees (e.g., medical doctors as well as doctorates in different disciplines) and were professionally oriented (in terms of papers presented and attendance at professional meetings).

High-performing research projects evidenced extensive intra-project communication. The amount of both problem solving and administrative communication was positively associated with project performance. Further, decentralized patterns of communication within the project were characteristic of high-performing research projects. With such projects, reli-

ance was less on supervisory direction and more on peer decision making and problem solving to deal with their substantial information processing requirements.

Communication outside the project but within the corporation was highly specialized for more effective research projects. The more innovative research projects were strongly connected only to other areas within the laboratory that could provide effective technical feedback and evaluation. High-performing research projects were weakly connected to areas in the larger organization.

Communication outside the firm was also specialized. Research projects were relatively strongly connected to universities and professional societies. Special boundary individuals[8] evolved to span the laboratory-professional communication boundary, and a strong positive association between professionally oriented communication and project performance was also evident. Given the importance of external information and the staff's ability to communicate across this organizational boundary, high-performing research projects were linked to professional domains through both direct staff contact and technological gatekeepers. On the other hand, research projects were relatively weakly connected to customers and suppliers. Further, there was an inverse relationship between operational communication and performance for research projects.[9]

In all, consistent with the distribution of expertise and the nature of research tasks, high-performing research projects showed extensive and decentralized communication within the project, strong linkages only to professionally oriented areas within the firm, and a reliance on both gatekeepers and direct contact to acquire information from professional areas outside the firm. The more effective research projects included diverse and extensive communication only with those areas that could provide technical input and critical evaluation.

These consistent patterns (see Table 1) were not found for low-performing research projects.

Technical service projects considered problems specific to this firm in which core technologies were relatively stable and well known (e.g., minor modifications on existing products). Tasks in these projects were, then, less complex and more closely tied to the firm than research projects. Technical service project members did not have advanced degrees and were significantly less professionally oriented than members of research projects. Consistent with these task and professional differences, high-performing technical service projects evi-

denced systematically different communication patterns than high-performing research projects.

One distinguishing characteristic of high-performing technical service projects was the relatively little problem solving or administrative communication within the project (less than half that of research projects). Indeed, increased administrative communication within the project was associated with lower performance. Moreover, high-performing technical service projects typically involved centralized patterns of intra-project communication. These results indicate that the more effective technical service projects required less peer decision

TABLE 1.   *Communication Patterns for High-Performing Research, Development, and Technical Service Projects*

	Research		Development		Technical Service	
	Amount	Mechanism	Amount	Mechanism	Amount	Mechanism
Intra-Project	Intense	Peer decision making and problem solving	Moderate	Combination of peer and supervisory-oriented decision making	Weak	Supervisory-dominated decision making and problem solving
Structure	Decentralized		Moderate		Centralized	
Intra-Firm						
Laboratory	Intense	Direct contact	Weak	Not investigated	Weak	Not investigated
Organization	Weak	Not investigated	Intense	Direct contact	Intense	Communication mediated by supervisors
Extra-Firm						
Operational	Weak	Not investigated	Moderate	Communication mediated by boundary-spanning individuals	Moderate	Communication mediated by supervisors
Professional	Moderate	Gatekeepers and direct contact	Weak	Not investigated	Weak	Not investigated

making and consultation and more supervisory direction and involvement.

As with research projects, clearly specialized communication patterns outside high-performing technical service projects were evident. Consistent with the need for marketing and manufacturing inputs, more effective technical service projects were strongly connected to organizational areas outside the laboratory (more than three times the amount of organizational communication seen in high-performing research projects). However, while communication was specialized to organizational areas, there was an *inverse* relationship between organizational communication and project performance. Rather than wide-spread communication to organizational areas, more effective technical service projects were dependent on supervisors for the bulk of this organizationally oriented communication. Thus, intra-organizational communication *and* the mechanism by which this linkage was accomplished were specialized for technical service projects.

Communication outside the firm was also specialized. Consistent with the information needs of technical service projects, extra-organizational communication was focused on suppliers, vendors, and customers (again, more than three times the amount of operational communication observed in research projects). However, in contrast to research projects, there was no association between wide-spread communication with suppliers and vendors and the performance of technical service projects. Rather, results indicate that supervisors, again, served as boundary-spanning individuals linking their project to external information areas.

Technical service tasks were organizationally defined and were based on relatively mature technologies. As such, the locus of task expertise was at the supervisory level, while operationally oriented (as opposed to professionally oriented) external areas could provide technical input and feedback. Consistent with these task requirements and the distribution of information, high-performing technical service projects showed supervisory-oriented communication within the project, organizational communication focused on marketing and manufacturing areas, and communication outside the project focused on suppliers, vendors, and customers. Moreover, extra-project communication was not accomplished by a majority of the technical staff. Rather, supervisors of technical service areas served as liaisons between their project and external information areas. These patterns (see Table 1) were not found for the less effective technical service projects.

Development projects were centered on tasks that were relatively less complex than research tasks but more complex than technical service projects. Development projects were locally defined yet involved technologies that were not well understood (for instance, utilizing new technologies to deal with a particular product problem).

Development projects required intra-project communication patterns between the decentralized nature of research projects and the centralized nature of technical service projects. Communication outside the project was also specialized for development projects. Consistent with the nature of their tasks, development projects were strongly linked to marketing and manufacturing areas in the larger organization. However, in contrast to technical service projects, the amount of organizational communication was strongly associated with technical performance. Thus, where technical service projects required that supervisors provide the link with organizational areas, high-performing development projects were characterized by widespread direct verbal contact with marketing and manufacturing areas.

Finally, extra-organizational communication in development projects was focused on operationally oriented areas (consultants, customers, suppliers). As with technical service projects, the bulk of this extra-organizational communication was mediated by a few boundary-spanning individuals. However, these technological gatekeepers were not supervisors; rather, they were more professionally oriented project members.

In all, consistent with the distribution of expertise and the nature and specification of development tasks, high-performing development projects showed communication patterns inclined towards operationally oriented areas both within and outside the firm. Due to the locus of task expertise within the project and the similarity in time frames and goal orientations between development projects and organizational areas, communication with organizational areas was most effective when widespread and direct. However, due to the local nature of the task and the attendant communication barriers, communication to operational areas outside the firm was mediated by the more professionally oriented project members (see Table 1).

Impacts of Environmental Conditions and Task Interdependence on Communication Networks

Communication networks are one mechanism to attend to work-related uncertainty. Task interdependence (i.e., the need to work with other areas to accomplish the task) and the nature of the project's task environment (the extent to which the project faces changing informational requirements) are two additional sources of project uncertainty. What are their effects on project communication patterns? Are there differential impacts for more vs. less effective projects?

Task interdependence is associated with problem solving and coordination requirements. As might be expected, projects facing substantial laboratory or organizational interdependence evidenced more than twice the amount of laboratory or organizational communication than projects with only a small amount of interdependence with these areas. These differences were only found for high-

performing projects. There were, however, differences in the nature of the linking mechanisms. For interdependence between areas with different time frames and goal orientations (for instance, the interdependence of research projects with marketing areas or technical service projects with research projects), communication was most effective if mediated by boundary-spanning individuals. These key linking people were peers or supervisors, depending on the locus of expertise within the project. If, however, the interdependent areas had similar time frames and goal orientations (for instance, development projects and marketing areas), then direct communication between the two areas was most effective. In short, the mechanism for linking the project with external areas was contingent on both the degree of differentiation among the interdependent areas and the distribution of expertise within the project (see Table 2).

Task environment: As the rate of change of the project task environment increases, the project must attend to increased environmental-based uncertainty. Environmental turbulence had systematically different effects on high-performing research and technical service projects. Environmental variability was associated with even more intense and decentralized communication for research projects, but with less intense and more centralized communication for technical service projects. In terms of leadership style, research supervisors were more participative, while technical service supervisors were more directive, under turbulent environmental conditions.

Environmental turbulence was also associated with decreased amounts of external communication for each task category; these decreases were accentuated for more effective projects. If communication to external areas decreased as environmental turbulence increased, and if external areas are important sources of information, then how were projects linked to external information domains? Evidence suggests that high-performing projects

TABLE 2. *Impacts of Environmental Conditions and Task Interdependence on Communication Networks*

	Environmental Turbulence Associated with:	Increased Task Interdependence Associated with:
Research Projects:		
Intra-Project	More intense and more decentralized communication patterns	Not investigated
Extra-Project	Decreased contact with laboratory and professional areas; increased reliance on boundary-spanning individuals	Increased communication between two interdependent areas; direct contact with laboratory; communication with organization mediated by boundary-spanning individuals
Development Projects:		
Intra-Project	More intense yet more centralized communication patterns	Not investigated
Extra-Project	Decreased contact with organizational and operational areas; increased reliance on boundary-spanning individuals	Increased communication between two interdependent areas; direct contact with both laboratory and organization
Technical Service Projects:		
Intra-Project	Less intense and more centralized communication patterns	Not investigated
Extra-Project	Decreased contact with organizational and operational areas; increased reliance on supervisors to span boundaries	Increased communication between two interdependent areas; communication mediated by supervisors

which faced turbulent environmental conditions relied on more boundary-spanning individuals to attend to the project's external information needs. Thus, while the overall amount of external communication decreased under turbulent environmental conditions, the responsibility for external communication shifted to a small number of boundary-spanning individuals. These patterns (see Table 2) were not found in low-performing projects.

In summary, higher-performing projects facing different work requirements had systematically different communication patterns: differences in the amount and structure of communication within the project, in the linkages to external areas (both within and outside the firm), and in the mechanisms used to link the project with external areas. The

less effective projects did not have these specialized networks. Managers of the more effective projects, then, facilitated communication patterns consistent with the needs of their project's work.

A CONTINGENCY MODEL FOR MANAGING COMMUNICATION IN R & D

These interrelated results, along with existing research, can be synthesized into a communication design model. Basic assumptions of this model are: (1) that a primary function of a project's communication network is the transmission and processing of information, and

(2) that to be effective, projects with different work characteristics will require systematically different communication networks in order to deal with the problem-solving requirements of their work. There will, then, be no one best communication pattern; rather, the communication networks of more effective projects will be contingent on the nature of their work.

To specify this communication model, four basic propositions can be presented:

1. PROJECT WORK REQUIREMENTS VARY IN THEIR DEGREE OF UNCERTAINTY: Projects within R & D laboratories face different work requirements. One way to characterize these differences is in terms of information processing requirements. Research indicates three basic sources of uncertainty and, therefore, of information processing requirements.

Nature of the Project Task.   Tasks differ in the degree to which the work is routine or nonroutine. For example, routine tasks (technical service) may be preplanned or solved using a set of standard operating procedures, whereas more complex tasks (research) are better handled on a case-by-case basis. As the nature of the project task becomes more complex, the degree of work-related uncertainty increases, and the amount of information processing required to complete the task increases.[10]

Nature of the Project Task Environment.   The external environment which the project faces is another source of uncertainty (e.g., suppliers, vendors, customers, new knowledge). As the rate of change in these information areas increases, the project staff must attend to and deal with increased external information requirements and, therefore, increased work-related uncertainty.[11]

Nature of the Project Task Interdependence. Task interdependence requires coordination and joint problem solving with other organiza-

tional areas. The more substantial the interdependence, the greater the requirements for coordination, and the less the project can work independently. Task interdependence is, then, another source of work-related uncertainty and increased information processing requirements.[12]

In summary, the nature of the uncertainty facing a project can be evaluated by examining the "routineness" of the work involved, the rate of change of the task environment, and the degree of interdependence with other areas. As the project task becomes less routine, as the environment becomes more turbulent, and as the interdependence becomes more complex, the project must cope with increased work-related uncertainty and, therefore, increased information processing requirements (see Figure 1).

2. COMMUNICATION NETWORKS HAVE VARYING CAPABILITIES TO DEAL EFFECTIVELY WITH UNCERTAINTY: Patterns of communication affect the ability of project staff members to deal with uncertainty. Three interrelated components of a project's communication network can be isolated.

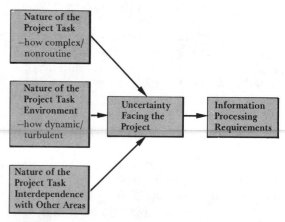

FIGURE 1.   *Sources of Uncertainty and Information Processing Requirements*

Intra-Project Communication Patterns. The amount and structure of communication patterns are particularly relevant in this case. Intense and decentralized intra-project communication patterns increase the opportunity for feedback, error correction, and the synthesis of different points of view. Further, decentralized communication patterns are associated with less formality, less attention to rules and regulations, and greater peer consultation. Each of these factors enhances effective attention to uncertainty. Projects with intense and decentralized intra-project communication will, then, be more effective in dealing with work-related uncertainty than projects with more centralized patterns of intra-project communication.[13]

While fully connected communication networks are better able to deal with substantial uncertainty than more hierarchical networks, there are, of course, costs associated with this information processing capacity. Fully connected networks consume more time, effort, and energy, and are less amenable to managerial control. Thus, the benefits of increased information processing capacity must be weighed against the costs of less control and potentially increased response time.

Extra-Project Communication. Typically, all the necessary information or insight required to complete the tasks will not be contained solely within the project. Information, ideas, and feedback must be obtained from other areas both within and outside the firm. However, not all external areas are equally effective in providing problem-solving assistance. Given the costs and difficulties of communicating across organizational boundaries,[14] communication specialized to areas with similar technical orientations, problem focus, or professional backgrounds will be most able to provide technical feedback, information, and stimulation. Thus, extra-project communication with selected areas will be better able to deal with uncertainty than more fully connected patterns of external communication.[15]

Boundary-Spanning Individuals. In order to more effectively accomplish the work required for any project, distinctive norms, values, and language schemes typically evolve. These local languages make communication across boundaries difficult and prone to bias and distortion. The more differentiated the subunits, the greater the difficulty of communicating across these boundaries.[16]

Since communication across boundaries is difficult, and given that external information is vital to the project, one technique for attending to information from highly differentiated areas is through the use of boundary-spanning individuals. These individuals straddle several communication boundaries and serve as liaisons to external areas. Information from external areas enters the laboratory through a set of key individuals who, in turn, channel this information into their project. These boundary-spanning individuals exist to mediate the laboratory to organization and extra-organization boundaries. If, however, project work requirements are not associated with differences in language and technical orientation between the project and some external area (e.g., between research projects and researchers in universities), this communication boundary will not exist and boundary-spanning individuals will not be necessary.[17]

3. PROJECTS WILL BE MOST EFFECTIVE IF THE COMMUNICATION NETWORKS FIT THE INFORMATION PROCESSING DEMANDS OF THE WORK: Different types of projects face different amounts of uncertainty and, therefore, distinct information processing requirements. Similarly, different communication patterns, both within and outside the project, result in varying capabilities to deal with uncertainty. What, however, is the relationship between work-related uncertainty and communication patterns? Based on several studies, it can be suggested that effectiveness will be contingent on matching communication patterns (one source of information processing capacity) to the na-

ture of the work (information processing requirements). A mismatch between work requirements and communication patterns will, then, be associated with lower performance.[18] These ideas are diagrammed in Figure 2.

In cells A and D, the information processing requirements facing the project are matched by the information processing capabilities of the particular communication network. Projects achieving this consistency will be most effective. However, in cell B, the extensive uncertainty posed by the task is not matched with appropriate communication patterns (for example, the extensive use of formal rules in highly uncertain conditions). Decisions will, therefore, be made with less than optimal amounts of information. On the other hand, it is possible to have too much information processing capacity for the requirements of the task. In cell C, the extra information processing capacity is redundant and costly (for example, the extensive use of peer communication where tasks are routine and weakly interdependent). In all, maximum effectiveness (other things being equal) will occur when the communication network fits the information demands of the required work.

4. If Projects Face Different Work Conditions Over Time, the Communication Networks of More Effective Projects Will Be Adapted To Meet the New Information Processing Demands: What are the implications of changing work requirements (for example, moving from the problem-solving to the implementation phase of a project)? This proposition suggests that not only may different projects require different communication networks, but as the nature of the work changes, the same project may exhibit different networks over time.[19] This process perspective, then, directs attention away from a static approach to communications, toward a more dynamic approach to managing communication over time.

These four propositions form the basis of a communication design model (see Figure 3). The basic idea is that projects within R & D laboratories face different amounts of uncertainty and that, to be successful, projects must match information processing capacity to processing requirements. Since different communication patterns imply different information processing capabilities, there will be no one best communication network. Rather, communication networks for high-performing projects will be contingent on the nature of the work. Finally, since information processing requirements change over time, the issue of managing communication will never be fully accomplished.

MANAGING COMMUNICATION IN R & D

This synthesis of the research on communication in R & D laboratories indicates that communication networks can be utilized as an important managerial lever; communication networks are an important factor in the innovation process *and* are amenable to managerial influence. The communication design model provides tools to examine systematically communication networks and to evaluate the extent to which the networks are consistent with the demands of the laboratory's work. More specifically, the model can be used to derive a set of steps for analyzing or designing communication networks.

Information Processing Requirements	Information Processing Capacity	
	High	Low
Extensive	A (Match)	B (Mismatch)
Minimal	C (Mismatch)	D (Match)

FIGURE 2.   *Relationships between Information Processing Capacity and Information Processing Requirements*

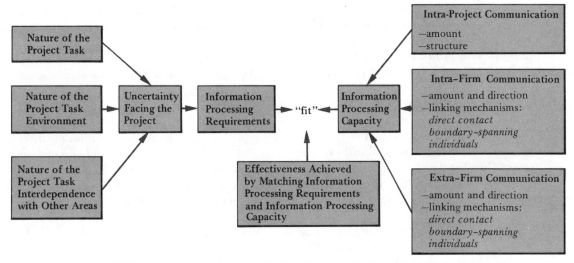

FIGURE 3. *An Information Processing Model for Managing Communication*

Before presenting these steps, two cautions should be noted. First, the determinants of communication patterns are not yet entirely clear. What is clear, however, is that a range of factors affect communication, such as leadership styles, control and reward systems, individual differences, and organizational structure.[20] Managing communication, then, really means managing a set of formal and informal organizational factors. A second caution relates to the quality of communication. Communication can be noisy, distorted, inaccurate, and late. While this article has assumed high-quality communication, much work remains to be done to better understand the determinants of efficient and effective communications.[21]

Steps in Using the Model

This communications model suggests a three-stage process for designing or managing communication networks. First, the amount and pattern of communication within the project must be able to attend to the information processing requirements of the particular task and task environment. Second, the project must be

linked to interdependent areas within the firm and to those areas that can provide technical feedback and support. Third, the project must be linked to external sources of information through direct contact and/or boundary-spanning individuals. This three-step approach to designing communication networks can be made more specific.

1. *Analyze information processing requirements of each project based on the uncertainty posed by the task characteristics and the task environments.* This step involves examining information processing requirements within the project based primarily on the nature of the project task and task environment.

2. *Based on the information processing requirements of each project, determine appropriate intra-project communication patterns.* Projects requiring substantial internal information processing requirements must develop more intense and decentralized communication patterns, while projects facing less uncertainty should be structured with less intense and more centralized communication patterns.

271

3. *Analyze information processing requirements between the project and other areas of the firm.* How should projects be linked to other areas in the firm? The primary issues here are the nature of the project task interdependence and the degree of differentiation between the project and external areas. The greater the interdependence, the greater the information processing requirements between the two respective areas. Further, the greater the degree of differentiation (that is, the more dissimilar the respective areas), the greater the communication difficulties.

4. *Based on information processing requirements between the project and other areas of the firm, choose appropriate linking mechanisms.* This step links the interdependent areas with appropriate communication mechanisms. If the interdependent areas are not highly differentiated (that is, they have similar time frames and goal orientations), then direct communication is most effective. If, however, the interdependent areas are highly differentiated, then direct contact will be inefficient. Under these conditions, boundary-spanning individuals provide an effective linking mechanism. These boundary-spanning individuals should be either supervisors or peers depending on the locus of expertise within the project.

5. *Analyze the need for information from areas outside the firm.* The innovation process is very sensitive to information from professional (universities, professional societies) and operational (suppliers, vendors, customers) areas outside the firm. Research projects require state-of-the-art information from professional areas, while technical service tasks require current information from operational areas.

6. *Based on the project's external information requirements, choose appropriate linking mechanisms.* Research projects should be linked to professional areas through direct contact and through gatekeepers. Develop-

ment and technical service projects, however, should be linked with operational areas by way of boundary-spanning individuals.

7. *If information processing requirements facing the project change over time, then the project's communication network must be adapted to those changes.* This final step suggests that the problem of managing communication is never fully completed. Managers must reevaluate the appropriateness of their project's communication network as the work changes over time.

A simplified representation of the communication design process is presented in Table 3. This table describes the basic steps of analysis and choice involved in managing communication.

CONCLUSION

In summary, substantial research indicates the utility of actively thinking about communication networks in R & D settings. What is the current state of the laboratory's communication network? What organizational factors affect who talks with whom? What is the quality of the communications? To what extent do these networks meet the informational needs of work areas within the laboratory? As communication patterns have an important impact on performance in research and development, and since managers can influence communication networks, communication networks can be used as a managerial design lever. The diagnostic questions suggested above are important, then, since there will be no one best communication network for innovation. Rather, managers of more effective projects will develop communication patterns to meet the information processing demands of the work involved in their particular projects.

TABLE 3. *The Communication Design Process*

	Intra-Project	Intra-Firm	Extra-Firm
Analysis	1. Analyze intra-project information processing requirements by evaluating nature of project task and task environment.	3. Analyze information processing requirements between project and other areas of firm by evaluating both task interdependence and degree of differentiation.	5. Analyze need for information from external areas.
Choice	2. Based on information processing requirements facing project, determine appropriate amount and pattern of communication within project.	4. Based on information processing requirements, determine appropriate linking mechanisms.	6. Based on extra-firm information needs, determine appropriate linking mechanisms.

Time Dimension

7. As information processing demands change, project's communication network must evolve to attend to new requirements.

REFERENCES

1. E. Mansfield, J. Rapoport, A. Romeo, E. Villani, S. Wagner, and F. Husic, *The Production and Application of New Industrial Technology* (New York: Norton Co., 1977).

2. T. J. Allen, *Managing the Flow of Technology* (Cambridge, MA: MIT Press, 1977); P. Kelly and M. Kranzberg, *Technological Innovation: A Critical Review of Current Knowledge* (Atlanta, GA: Advanced Technology and Science Studies Group, Georgia Institute of Technology, 1975); A. Gerstenfeld, *Effective Management of Research and Development* (Reading, MA: Addison-Wesley, 1970); J. Goldhar, L. Bragaw, and J. Schwartz, "Information Flows, Management Styles, and Technological Innovation," *IEEE Transactions on Engineering Management*, 1976, pp. 51–62.

3. M.L. Tushman and D. Nadler, "Communication and Technical Roles in R & D Laboratories: An Information Processing Approach," in *Management of Research and Innovation*, ed. by B. Dean and J. Goldhar. TIMS Studies in Management Sciences. (New York: North-Holland, 1980); E.A. von Hippel, "The Dominant Role of Users in the Scientific Instrument Innovation Process," *Research Policy* 5(1976): 212–239.

4. R.S. Rosenbloom and F.W. Wolek, *Technology, Information, and Organization: Information Transfer in Industrial R&D* (Boston, MA: Graduate School of Business Administration, Harvard University, 1967); L. Sayles, *Managerial Behavior* (New York: McGraw-Hill, 1964).

5. D. Pelz and F. Andrews, *Scientists in Organizations* (New York: John Wiley & Sons, 1966); T. J. Allen, "Communication Networks in R&D Labs," *R&D Management*, 1970, pp. 14–21.

6. The results of this research have been reported in the following papers: M.L. Tushman, "Communications across Organizational Boundaries: Special Boundary Roles in the Innovative Process," *Administrative Science Quarterly* 22(1977): 587–605; M.L. Tushman, "Task Characteristics and Technical Communication in Research and Development," *Academy of*

Management Journal 21 (1978): 624–645; M.L. Tushman, "Impact of Perceived Environmental Variability on Patterns of Communication," *Academy of Management Journal* 22 (1979): 482–500; M.L. Tushman, "Determinant of Subunit Communication Structure: A Contingency Analysis," *Administrative Science Quarterly* 23 (1979): 82–98; R. Katz and M.L. Tushman, "Effectiveness of Communication Patterns in R & D: A Focus on Problem-Solving and Administrative Communication," *Organizational Behavior and Human Performance* 23 (1979): 139–162; M.L. Tushman and T. Scanlan, "Characteristics and External Orientations of Boundary-Spanning Individuals." *Academy Management Journal*, 1981, Vol. 24, 83–98.

7. Universally defined tasks are those in which parameters and specifications are not defined by the firm (e.g., research on fiber optics). Locally defined tasks, however, are defined and constrained by the given organization's norms, values, and history (e.g., applied development work integrating a given technology with an existing product).

8. Boundary-spanning individuals, sometimes termed gatekeepers, are those key people in the laboratory's communication network who are strongly connected to external information areas and strongly connected within the laboratory's communication network. These individuals link more locally oriented laboratory members to external sources of information.

9. Operational communication (communication with suppliers, vendors, customers, or consultants) reflects an interest in gathering information contributing to the accomplishment of particular organizational objectives. Professional communication (communication within universities or professional societies) reflects an interest in gathering information which defines principles or natural laws underlying a particular technology.

10. J. Galbraith, *Designing Complex Organizations* (Reading, MA: Addison-Wesley, 1973).

11. T. Connolly, "Communication Nets and Uncertainty in R & D Planning," *IEEE Transactions on Engineering Management*, 1975, pp. 50–54.

12. P. Gerstberger, "The Preservation and Transfer of Technology in R & D Organizations" (Ph.D. diss., Massachusetts Institute of Technology, 1971).

13. G. Farris, "The Effect of Individual Roles on Performance in Innovative Groups," *R & D Management*, 1972, pp. 23–28; M. Shaw, "Communication Networks," in *Advances in Experimental Social Psychology*, ed. L. Berkowitz (New York: Academic Press, 1964).

14. T.J. Allen and S. Cohen, "Information Flow in R & D Labs," *Administrative Science Quarterly* 14 (1969): 12–19.

15. M.L. Tushman and D. Nadler, "An Information Processing Approach to Organizational Design," *Academy of Management Review* 3 (1978): 613–624. M.L. Tushman and D. Nadler, "Communication and Technical Roles in R & D Laboratories: An Information Processing Approach," in *Management of Research and Innovation*, ed. by B. Dean and J. Goldhar. TIMS Studies in Management Sciences. (New York: North-Holland, 1980); R. Whitley and P. Frost, "Task Type Information Transfer in a Government Research Lab," *Human Relations* 25 (1973): 537–550.

16. J. March and H. Simon, *Organizations* (New York: John Wiley & Sons, 1966); Tushman (1977); Allen (1977).

17. T. Allen, M.L. Tushman, and D. Lee, "Technology Transfer as a Function of Position on Research, Development, and Technical Service Continuum (Sloan School of Management Working Paper, 1978).

18. M.L. Tushman and D. Nadler, "An Information Processing Approach to Organizational Design," *Academy of Management Review* 3 (1978): 613–624.

19. Taylor and J.M. Utterback, "A Longitudinal Study of Communication and Research," *IEEE Transactions on Engineering Management* EM22, 1975, pp. 80–87; G. Zaltman, R. Duncan, and J. Holbeck, *Innovation and Organizations* (New York: John Wiley & Sons, 1973).

20. G. Farris, "Managing Informal Dynamics in R & D" (Paper presented at TIMS meetings, New York, 1978); F. Andrews and G. Farris, "Supervision Practices and Innovation in Scientific Teams," *Personnel Psychology* 20 (1967): 497–515.

21. Kelly and Kranzberg (1975).

A Critical Examination of the Dual Ladder Approach to Career Advancement

Laurie Michael Roth

In recent years the fastest growing segment of the American labor force has been that of the "knowledge worker." This emergent class of organizational professionals is comprised of a variety of occupational groups—salaried managers, technical professionals, business specialists, and computer specialists. They are employed in business, industry, and government to seek, create, and disseminate new knowledge, which has become the organization's "primary instrument of progress and competition (Zand 1981, 11)." Not surprisingly, the increased professionalization of the American workforce has necessitated changes in personnel management and organization. Among the most critical tasks facing organizations is to motivate, reward, and provide career paths for their professionals and staff specialists, whose values concerning status and success may differ from those held by the work establishment.

Career advancement in organizations has traditionally assumed the pyramidal model: authority, status, and pay increase as the individual is promoted by virtue of his or her ability to manage. The pyramidal model has explicitly defined success as advancement on the organizational ladder. Yet for many professionals, the reward of promotion into management conflicts with the goals of professional work and leads them away from the practice of their chosen specialty. Organizations dependent upon individual creativity may also suffer when scarce technical talent is diverted to the managerial ranks simply because alternative advancement strategies for specialist groups are lacking.

Organizations have adopted various strategies to accommodate the problems arising from their increased employment of specialist groups. One strategy designed specifically to provide career paths and a reward system for specialists and individual contributors is known as the "dual ladder":

> The dual ladder approach is generally formalized into parallel hierarchies; one provides a managerial career path, and the other advancement as a professional or staff member. Ostensibly, the dual ladder promises equal status and rewards to equivalent levels in both hierarchies. (Kaufman 1974, 125)

Dual ladder systems of career advancement have been utilized widely in scientific and engineering firms during the last twenty-five years. Although few observers report that dual ladder programs have successfully achieved their intended goals, dual ladders remain popular in high technology settings and variations of the concept have been adopted in other types of organizations engaged in creative professional work.

This paper reports on a study of dual ladder systems conducted from February to October 1981. The primary objective of the study was to gather state-of-the-art information on the current utilization of the dual ladder approach to career advancement including:

- the kinds of organizations currently using dual career ladders and the employees for whom they do so;
- the reasons why some organizations have recently implemented and others continue to use dual ladder systems;
- the problems organizations face when designing and implementing dual ladder systems;
- the characteristics of more effective dual ladder approaches; and
- employees' evaluations of the dual ladder approach as a vehicle for career management.

The research project included (1) a review of the literature on dual ladder systems, (2) both formal and informal surveys of dual ladder systems used in over twenty organizations with differing professional populations and industry affiliations, and (3) a focused study of the design and implementation phases of a technical ladder program at a research facility of a large consumer products company. Since the objective for the formal interview study was to gather data from organizations with varying experience with dual ladders, that sample of fourteen organizations was selected as follows: Nine organizations identified by other researchers, human resource executives, or consultants in this area as having "well-established" dual ladders or as currently involved in implementing or evaluating their dual ladder programs participated in the study at the au-

thor's request. Human resource personnel from five other organizations in a variety of industries learned about the ongoing research, contacted the author, and volunteered to participate. Representatives from most of these latter organizations revealed that they "had problems" with their dual ladder programs and wanted to discuss their experiences. Participants generally consisted of human resource executives responsible for implementing and/ or overseeing a dual ladder program. Each participant in the formal study was interviewed for one and a half to two hours in accordance with a structured interview instrument focusing on the issues outlined earlier. While the names of the participating organizations and respondents will not be disclosed, a description of the companies and the titles of the respondents appears in Appendix I.

Additional data were collected from a consulting firm and several other organizations not formally participating in the interview study. The author learned about company programs through informal discussions with human resource managers and from brochures and documents obtained through various sources. See Appendix II for a description of the companies constituting the informal sample.

One organization that had participated in the formal interview study volunteered to serve as a case study. The author was allowed to monitor the design and implementation phases of the technical ladder program in the company's research facility.

Outline of Topics
This paper has several objectives. Its central concern is to describe, evaluate, and recommend modifications in the dual ladder approach to career advancement based on findings from the study just outlined. Its second concern is to provide human resource practitioners with a diagnostic framework they can

use to evaluate the structure and functioning of their organization's dual ladder programs. Its third concern is to review and elaborate upon theoretical issues concerning the career development of professionals in organizations.

The introductory section places the dual ladder concept within a theoretical context. It includes a brief discussion of the role orientations of organizational professionals, a presentation of the theory of the dual ladder, and a summary of the weaknesses of the dual ladder approach outlined in the existing literature.

The second section summarizes the research findings. State-of-the-art information is presented identifying: the kinds of organizations currently using dual ladders and their reasons for use; the problems encountered when designing, implementing, and administering systems; and the characteristics of "effective" dual ladder approaches.

The final section outlines a diagnostic framework developed by the author to aid practitioners in evaluating their organization's dual ladder programs.

PART I: CAREER ISSUES CONCERNING PROFESSIONALS EMPLOYED IN ORGANIZATIONS

Theoreticians contend that professionals possess a distinct set of attitudes toward work which is the product of their socialization into a profession. For example, Hall (1968) identifies five attitudinal attributes of professionals that distinguish them from other occupational groups:

1. The use of the professional organization as a major reference—i.e., both the formal organization and informal colleague groupings serve as the major source of ideas for professionals in their work.
2. A belief in service to the public.

3. A belief in self-regulation—i.e., a belief that the person best qualified to judge the work of a professional is a fellow professional.
4. A sense of calling to the field—i.e., dedication of the professional to his or her work and the feeling that he or she would probably want to do the work even if fewer extrinsic rewards were available.
5. Autonomy—i.e., the feeling that practitioners ought to be able to make their own decisions without external pressures from employers, clients, or individuals who are not members of the professional community.

Professional work incentives, definitions of status, and criteria for success clearly differ from those prevailing in bureaucratic organizations. As Kornhauser (1962) and others note, there are built-in strains between work establishments and professional institutions with respect to the goals, controls, and incentives for professional work as well as the influence of professional work. For instance, in a bureaucratic organization, control is exercised according to the hierarchical principle: authority is delegated through a series of positions that are hierarchically ordered. In a profession, control is exercised by a company of equals, and the professional assumes the final responsibility for exercising judgments in his or her area of competence.

Professional incentives also clash with organization incentives. In theory, professionals are motivated by a desire to contribute to a field of knowledge and to establish a distinguished reputation within their professional community. Bureaucrats, on the other hand, desire upward mobility in the organizational hierarchy. Thus, while professionals acquire status from their colleagues in the profession, bureaucrats acquire it from their superiors in the organization.

Many researchers have studied the career orientations of specialists employed in business and industry. Most utilize reference group theory to distinguish types of occupational orientation. In his classic theory, Gouldner (1957) conceptualizes occupational orientation as a unidimensional continuum labeled "local" and "cosmopolitan" at the poles. Three criteria measure the local-cosmopolitan dimension: (1) loyalty to the employing organization, (2) commitment to specialized role skills, and (3) reference group orientation. Basically, "locals" are oriented toward their employing organization, while "cosmopolitans" are oriented toward their profession. Cosmopolitans have low loyalty to their organizations, high commitment to specialized role skills, and a strong outer reference group orientation. On the other hand, locals are committed to organizational goals, seek organizational approval, and focus upon developing their organizational careers.

Additional typologies of career orientations have been formulated by Reissman (1949), Wilensky (1956) and others. Reissman, for example, distinguishes between functional, job, specialist, and service bureaucrats, while Wilensky describes the professional service expert, the careerist, the missionary, and the politico. In his well-known study *Scientists in industry,* Kornhauser (1962) has shown that many of these typologies identify comparable career orientations that simply are labeled differently. Treating Gouldner's local-cosmopolitan characterization as two independent dimensions, Kornhauser distinguishes four types of career orientations that vary in their support for or rejection of a "professional orientation" and an "organizational orientation." Then he uses his own classification scheme to summarize the types of career orientations described by other researchers. In a similar vein, Glaser (1963) and Delbecq and Elfner (1970) conceptualize an individual's career strategy as a product of two occupational orientations—one emphasizing the employing organization; the other emphasizing the speciality. Like Kornhauser, Delbecq

and Elfner identify four career strategies which are associated with high or low levels of loyalty to one's organization and one's specialty.

Rewarding the Professional

Organizations employing large numbers of specialists are faced with the dilemma of establishing an incentive system that both motivates the professional and is efficient for the organization. If the work establishment encourages too much integration of its professionals into the organization by stressing organizational incentives, the professional performance of its specialists is likely to decline. On the other hand, if the organization grants its specialists too much autonomy with their work, then specialist groups may not be sufficiently motivated to achieve the goals of the organization.

Few would dispute Kornhauser's (1962) contention that "the balance of incentives in industry . . . is . . . heavily weighted on the organizational rather than the professional side" (p. 128). Of course, organizations do vary, and those that rely strongly upon research functions tend to provide more professional incentives. Kornhauser offers two reasons for industry's heavy reliance upon organizational incentives. First, the long and satisfactory use of those incentives for other job categories encourages the organization to resist altering the structure of incentives to accommodate professional employees. Second, since the organization's incentive system embodies organizational values, alternative systems tend to be viewed as competing sources of loyalty.

Traditionally, the highest rewards in business and industry have been conferred on those who assume administrative responsibility. Movement up the managerial ladder secures increases in status, recognition, and salary. For many specialists employed in business and industry, movement into management has been the only available advancement strategy. Given the limited number of managerial positions, even those professionals with a strong organizational orientation may suffer from the lack of alternative advancement opportunities. As Kornhauser (1962) points out, specialists with a strong professional orientation are subject to a different kind of frustration because the opportunities to gain high status and rewards without undertaking managerial responsibilities are even more limited.

Organizations dependent upon creative professional work may also be negatively affected by their own failure to provide alternative reward structures and advancement opportunities for specialist groups. The movement of highly competent individual contributors into management seriously depletes the pool of professional talent. In addition, retention of competent specialists is often difficult when career opportunities and professional rewards are limited in the organization but available elsewhere.

One Solution: The Dual Ladder

One organizational arrangement that has been developed to provide meaningful rewards and alternative career paths for organizational professionals is the "dual ladder system of career advancement." The concept of the dual ladder is quite simple:

> The dual ladder refers to the side-by-side existence of the usual ladder of hierarchical positions leading to authority over greater and greater numbers of employees and another ladder consisting of titles carrying successively higher salaries, higher status, and sometimes greater autonomy or more responsible assignments. (Goldner and Ritti 1967, 489)

Also known as the "professional," "technical," or "individual contributor" ladder,[1] the dual ladder was developed about twenty-five years ago to reward professionals (specifically, scientists and engineers) for good scientific or technical performance without removing them

from their professional work (Shepard 1958). By providing cosmopolitan or professionally oriented individuals with an opportunity and incentive to remain in their fields and stay up to date, the dual ladder aims to secure for the high technology organization an adequate pool of technical talent.

Figure 1 illustrates a typical dual ladder system of career advancement that might be used in a research and development laboratory of an engineering firm. The "Y" concept of career progression is popular and provides for a parallel "branching" of career paths along either the managerial or professional ladder often at the third or fourth level of the hierarchy. Although the present illustration provides for an equal number of advancement levels on both ladders, this is not necessarily the case. While the height of the managerial ladder is contingent upon the organizational design, the "height" of the professional ladder is determined by the level of work engaged in by the organization and the labor market value of individual contributors as compared to managers. In some dual ladder programs, recipients of the highest technical ladder positions are desig-

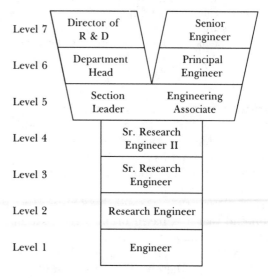

Level 7	Director of R & D	Senior Engineer
Level 6	Department Head	Principal Engineer
Level 5	Section Leader	Engineering Associate
Level 4		Sr. Research Engineer II
Level 3		Sr. Research Engineer
Level 2		Research Engineer
Level 1		Engineer

FIGURE 1. Dual Ladder in an R & D Laboratory

279

nated as company "fellows." Special status, rewards, privileges, and recognition are conferred on this very select group of outstanding performers, who generally have national or international reputations for their professional achievements. Most dual ladder systems allow for lateral movement between the managerial and professional ladders but tend to discourage "crossovers" at the extreme upper levels of the hierarchy.

While the dual ladder promises equal status and financial rewards to those at equivalent levels in the managerial and professional hierarchies, the set of incentives associated with each ladder differs sharply. Movement up the managerial ladder leads to positions of power and participation in the affairs of the organization; by contrast, advancement up the professional ladder secures autonomy in the practice of one's specialty. As Goldner and Ritti (1967) explain:

> The professional ladder makes no pretense in the direction of increasing the authority of titleholders. It is based on the same assumptions that support an ideology of professionalism, assumptions stating that upward mobility in the power hierarchy is of no importance compared to autonomy in the practice of one's special competence and that success for the professional is independent of such mobility. (p. 491)

Criticism of the Dual Ladder

A fair amount of the literature on dual ladder systems is highly critical of the approach. Kaufman (1974) describes it as "a dubious reward system" that suffers from "flaws in its logic and application that have made it unworkable in practice" (p. 125). Goldner and Ritti (1967) contend that it is based on incorrect assumptions concerning the career orientations of most professional groups employed in business and industry. Dalton, Thompson, and Price

(1977) assert that the dual ladder typically is poorly operationalized and thereby fails to accomplish its intended goals of providing increased status and compensation as well as more freedom for individual contributors.

Kaufman's as well as Goldner and Ritti's chief complaint is that the dual ladder approach is based on the incorrect assumption that most professionals employed in organizations are generally cosmopolitan (i.e., professional) and not local (i.e., organizational) in orientation. They claim that engineers, business professionals, and computer scientists in fact enter industry with non-professional goals, soon strongly identify with the organization and its objectives, and want to participate in decisions affecting their work. Advancement along the professional ladder defeats these goals because it removes the professional from the organizational mainstream.[2]

Goldner and Ritti further contend that the professional ladder fails to satisfy a basic need of the professional: influence and responsibility in the practice of one's special competence. Professional ladder positions are not "equivalent" in status to managerial positions because they do not "provide the power to allocate limited resources or to pursue alternative goals, and this power is intrinsic to traditional prestigious and successful professional performance" (p. 502).

Another problem with the technical ladder stems from its reliance upon invention or technical creativity as the criterion for promotion. Since technical creativity is sometimes more prolific early in one's career, long-term advancement possibilities for individual contributors remaining on the technical ladder may be rather limited (confidential company document).

The following additional shortcomings of the dual ladder approach have been identified by other researchers (e.g., Shepard 1958; Kaufman 1974; Kornhauser 1962):

1. The professional ladder may be used by management as a way of shelving senior staff members who are regarded as lacking in managerial ability. Alternatively, it may be used as a dumping ground for incompetent or excess managers.
2. There tends to be a shortage of rungs on the professional ladder and a ceiling for advancement well below the top of the supervisory ladder, thus making the two ladders non-equivalent.
3. Assignment to the professional ladder may isolate the individual contributor from the organization. The reward of freedom may not be desired by some organizational professionals who associate it with "loneliness, rejection, [and] the feeling of not belonging." (Shepard 1958, 182)
4. Professional ladder positions may be viewed by recipients and others as an ambiguous status symbol. Titles awarded on the professional ladder may not correspond to those within one's profession. Moreover, professional ladder positions do not entail supervisory responsibility, a widely recognized criterion of status.
5. Promotion to the professional ladder often serves as a reward for past performance rather than an opportunity to develop one's future potential.
6. Assignment to the professional ladder may be regarded as proof of inadequacy. Such positions are typically reserved for competent professionals lacking in leadership skills who therefore are judged as unfit to exercise power.
7. Supervision of individual contributors on the professional ladder may be difficult since they are granted a good deal of independence with their work.
8. Professional ladder positions tend to be less secure than managerial positions because the productivity of the individual contributor is more easily assessed.
9. Dual ladders are often poorly publicized within organizations. Employees may not be aware of the options available to them or the advancement patterns associated with each ladder.
10. Because the rewards associated with the professional ladder may be ambiguous or negatively valued, organizational professionals may be reluctant to use the ladder for career planning.

In short, a critical tone pervades the literature on dual ladder systems. Researchers and observers have identified a host of problems in their design, implementation, and utilization. Nevertheless, some organizations which have implemented dual ladder programs claim that the approach has been "successful" and is responsible for the high morale and satisfaction of their individual contributors. Even with their shortcomings, professional ladders are widely utilized because they offer one means of providing for the career development of an increasingly professionalized workforce during a period of slow economic growth when large organizations must exhibit restraint in their hiring and advancement practices. In Part II we review the findings of a recent study of dual ladders. We examine the difficulties organizations are having with their dual ladders and the characteristics of more effective programs.

PART II: FINDINGS—WHAT KINDS OF ORGANIZATIONS CURRENTLY USE DUAL LADDERS, FOR WHICH EMPLOYEES, AND WHY?

Dual ladders (in some shape or form) are currently popular in a wide variety of organizations engaged in creative professional work. While they predominate in scientific laboratories and new product engineering departments, dual ladders can also be found in legal, marketing, manufacturing, employee relations, accounting, and sales departments of large companies. They also are extensively utilized throughout the Federal government.

Appendices I and II list the industry affiliations of organizations participating in both our formal and informal surveys. While the participants do not constitute a random sample of organizations using dual ladders, they do form a representative sample. Traditionally, research and development settings have been a natural spawning ground for professional ladders, for the success of such facilities depends upon motivating and retaining scarce scientific and technical talent. Thus, as we would expect, the dual ladder approach is popular in the pharmaceutical, chemical, computer, and electronics industries. More recently, professional ladders have been implemented in specialized departments of organizations in the manufacturing and financial industries to provide non-managerial career opportunities for increasingly large numbers of professional staff members.

During the formal interviews, respondents discussed the reasons why their organizations had recently implemented or continue to use dual ladders. Several explanations were given repeatedly. One of the typical responses was that the dual ladder provides an incentive system tailor-made for the professional who is either not interested in pursuing a managerial career or not qualified for one. Organizations realize the importance of recognition, status, and challenge to the scientist, engineer, or other specialist even though these professionals are known as "self-motivators." They are aware that professional performance is likely to suffer if professionals are not rewarded for good work with increases in title, salary, and perquisites. Some respondents explained that their organizations are turning to dual ladders because other advancement strategies routinely used in prior years, such as corporate transfers, have become prohibitively expensive. Organizations are interested in "keeping the head count down," and in making sure that the managerial pyramid does not become top-heavy. The dual ladder becomes a substitute means of promoting specialists who might have made good managers if sufficient managerial positions were available. At the same time, organizations dependent upon creative and innovative ability desperately want to retain their most competent professionals. Many respondents in the engineering, computer, and chemical industries claimed that one of their most pressing concerns in the 1980's will be the attraction and retention of scarce technical talent. Since competition for the best available talent is keen, organizations are anxious to create an environment that will induce competent professionals to stay. Dual ladder programs are appealing because they promise advancement opportunities and special rewards for the high-performing individual contributor.

In short, it appears that organizations are currently using dual ladders to cope with the problems of attracting, retaining, and motivating their technical and specialist human resources in an environment characterized by high competition, scarce technical talent, and economic instability. Many of the organizations participating in the study reported a sense of urgency in responding to these problems, while others anticipated an intensification of the problems during the next decade.

Dual Ladders: Major Problem Areas

As expected, the dual ladder systems of the participating organizations varied extensively along a number of dimensions. In addition to differences in design, implementation, and utilization, there was a good deal of variation in the extent to which organizations formalized and documented their systems, publicized them to employees, monitored their operation, and modified them to better accomplish their objectives.

One of the objectives of our study was to learn about the problems organizations face when they design, implement, and utilize dual ladder systems. Our research shows that despite variations in their dual ladder approaches, our respondents tend to share a rather consistent set of problems. This section identifies the typical problems organizations are having with their dual ladder systems of career advancement. Some are the direct result of poor designs and mismanagement, while others are a function of the professional environment in which the dual ladder must operate.

Interestingly enough, all but one of the respondents in the formal interview study disclosed that their organization "had problems" with their dual ladder system. Several respondents had recently participated in internal studies of dual ladder issues and were in the process of implementing new programs or reconstructing old ones. Throughout the study it was generally observed that organizational confusion about dual ladder systems is not at all uncommon. In fact, there is often considerable disagreement within an organization about whether a dual ladder system really exists, what it looks like, how it operates, and with what effect.

Of course, the kinds of problems our respondents identified reflected not only their organization's experience with their dual ladder arrangements but also their particular diagnostic and problem-solving approaches to human resource issues. While most were primarily concerned about technical or design issues, a few ventured beyond these more self-evident problems and discussed cultural and political issues within the organization that affected the functioning of their dual ladder systems. The problem areas to be discussed include those typically identified by our respondents as well as those commonly overlooked by them but perceived by the author during the course of

the study as recurrent problems facing organizations with dual ladder systems.

Design Problems

One of the basic problems of several of the dual ladder programs examined in our study, and one which ironically escapes the attention of too many human resource executives, is the inexplicit design of their dual ladder systems. Our research shows that it is not uncommon for organizations to fail to clearly define and document their dual ladder programs in terms of the performance standards, qualification criteria, and accountability standards for the various ladder positions. While these data are typically prepared for the managerial hierarchy, they often are scant or nonexistent for the professional hierarchy. Obviously, these definitions are fundamental to a dual ladder system for they determine the structure of the formal career paths as well as the criteria for advancement. Without such documentation, advancement decisions may appear to employees as management subjectivity. In addition, the dual ladder program is likely to be viewed by professional employees as an arbitrary system that is neither administered fairly nor communicated to employees nor utilized for career planning purposes.

Despite the pervasiveness of imprecise designs, only a few of the organizations participating in our study had attempted to formalize and/or redesign existing dual ladder structures. One strategy used to attack this problem was to convene a task force comprised of managers, professional employees, and employee relations personnel who attempted first to distinguish the steps in the professional ladder by analyzing the job levels that existed within departments, and second, to define titles, job responsibilities, qualification criteria, and accountability standards for all steps in the dual ladder structure.

While dual ladder programs can generally be criticized for having inexplicit designs, many also suffer from faulty design. The purpose of implementing a dual ladder structure is to provide meaningful career opportunities and a special reward system for professional employees who choose to develop their careers in the organization as individual contributors. In theory, at least, the professional ladder is supposed to serve as an alternative career path that enjoys equivalent prestige and comparable rewards to that of the traditional managerial ladder. Nevertheless, our respondents commonly complained that their dual ladder designs prevented any semblance of equivalence between the different career ladders. They explained that the professional ladder failed to provide the same degree of recognition, status, or rewards because it had fewer levels, its ceiling was lower, and/or it offered fewer rewards and perquisites at comparable rungs.

Additional design problems were mentioned by some respondents. A few were concerned that their organization's technical ladder had gaps in their promotional chains which resulted in untimely career plateauing of professional employees. In one company, such gaps were responsible for the additional burden of a shortage of mentors for employees at lower levels of the technical ladder. Others described the structure of their professional ladders as unwieldy: It contained too many levels which did not correspond to "real" and distinguishable levels of work within a given department.

Establishing and Safeguarding the Credibility of the Professional Ladder

One of the biggest challenges facing all organizations with dual ladders is to establish and safeguard the professional ladder as a means of providing motivation, recognition, and compensation for superior technical or professional performance. All too often our respondents reported that their technical ladders were used not to reward excellence in technical performance but to pacify an individual who had been "passed over" for a managerial position, to prevent a "quit," or to shelve an incompetent or unneeded manager. Rather than housing the best technical talent, the technical ladder often became a major "dumping ground" for managerial rejects.

One unfortunate result of such misuse is that technical ladders often suffer from severe "image" problems. When undeserving individuals secure technical ladder positions, the integrity of the ladder is necessarily undermined. The morale of competent professionals on the ladder is likely to decline if they view their own promotions as tarnished. In addition, even the most professionally oriented specialist may be reluctant to choose a technical career because it is not considered a respectable route to career advancement.

Several of our respondents were working on solutions to the problem of "policing" the technical ladder to maintain its technical purity. In many dual ladder systems special technical review committees act as judicial bodies that evaluate candidates' qualifications for promotions to higher-level ladder positions. Needless to say, the credibility of dual ladder systems is highly dependent upon these peer review bodies' commitment to standards of professional excellence.

Cultural Commitment to the Professional Ladder

While many of our respondents realized the negative consequences of misusing the technical ladder to solve managerial problems, few recognized the extent to which the cultural values within the organization threatened the viability of their technical ladders. Clearly, the success of any dual ladder system rests largely upon corporate commitment to both its managers and its specialists. Such commitment implies recognition of the invaluable contribution that both groups make to the organization. It also implies a willingness to provide a challenging professional environment that will enhance the attractiveness of a specialist career.

One of the most difficult tasks in designing and implementing an equitable dual ladder system is to overcome the traditional bias in favor of managerial skill as the principal criterion of success within the organization. Many of our respondents noted the predominance of those traditional values in their organization, but only a few realized that such a cultural orientation is bound to sabotage the design and administration of the professional ladder. Our research shows that dual ladder systems implemented in management-dominated cultures tend to suffer from most of the problems already outlined. In addition to having unbalanced designs which designate the lion's share of rewards and opportunities to those in management, these systems are typically misused by management to take care of problem cases. It is no surprise that those on the technical ladder are commonly dissatisfied with their lower status as individual contributors and that few technical specialists seek technical careers in the organization.

The task of changing the value system of an organization's culture is both challenging and difficult. Cultural commitment to the professional ladder must begin with management's respect for the talents of its individual contributors and its willingness to relinquish its perception of technical specialists as merely "individuals lacking in managerial skills." Rather, specialists must be valued for their distinct set of skills which are essential in meeting the organization's objectives. It involves management's commitment to developing a technical ladder that uses the needs of the technical specialist and not those of the manager as its major point of reference. It also involves designing a professional environment that allows for not only the professional development of its specialists but also the involvement of top-level individual contributors in corporate planning and decision-making.

Other Organizational Issues

Part of the challenge in designing, implementing, and administering a "successful" dual ladder system consists of addressing and resolving other organizational problems which affect the professional environment in which the dual ladder structure must operate. During our interviews we learned that some of the "dual ladder problems" our respondents identified were the by-products of other organizational problems. In addition, such problems were not generally specific to one organization, but rather were evident in similar kinds of organizations. This section examines three additional problem areas impacting upon the functioning of a dual ladder program which illustrate that dual ladder problems, like any category of organizational problems, tend not to occur in a vacuum.

The Organization's Commitment to Technology. Our previous discussion briefly alluded to the importance of providing a challenging professional work environment for recipients of technical ladder positions. In order for the technical ladder to be recognized by specialists and managers alike as a legitimate, valued, and prestigious avenue of career advancement, it is essential that specialists view their work on the technical ladder as challenging, that they witness that "real" work is to be done at every level of the hierarchy, and that they sense management's commitment to their specialties as vital for the accomplishment of organizational objectives.

It is not uncommon for dual ladder programs to fare poorly, either when management is uncertain about the future direction of its research and development functions or when it undergoes a reevaluation of those objectives.

One of the participating organizations was especially plagued by the first problem. In response to growing technical service needs and the possibility that it might engage in new product development in the near future, the company had recently developed a research and technical service facility. It attracted high-level technical professionals by offering excellent salaries and promising future growth. Yet, three years after its inception, few staff members had been promoted and since many questioned the company's commitment to "knowledge building," most believed that their career opportunities in that facility were limited. Because there was no demonstration of organizational commitment to advanced technical work, employees did not perceive the technical ladder as a viable means of professional growth. In their opinion, the level of work "needed" to accomplish organizational objectives rarely exceeded the fourth level of a proposed seven-level hierarchy.

Similar problems occurred in other organizations participating in the study which also were in the process of developing new research functions or which had recently altered their short- or long-range business objectives such that work in certain disciplines was no longer needed. This often occurred when a production process was mastered or when economic slowdown forced cutbacks in research activities. Obviously, those organizations relatively immune from this category of problems were those with proven records of commitment to competitive high technology work.

Non-cash Forms of Recognition.   Many organizations with dual ladder programs fail to realize that the task of recognizing good professional performance should not end with the creation of a formal dual ladder structure. Once the professional ladder is established, there remains the challenge of creating a system of non-cash rewards which will attract, retain, and motivate superior professional talent. The failure to provide alternative forms of recognition is the undoing of many dual ladder systems.

While technical ladder promotions serve as formal recognition of good technical performance, professionals are rarely satisfied with new titles and increased compensation alone. Professionals seek recognition from management of their contribution to corporate decision-making. They also seek recognition of their technical accomplishments from professional colleagues outside their companies.

A few of the organizations participating in our study have recently formulated programs designed to provide non-cash recognition of good professional performance. Such programs typically offer perquisites (e.g., vacation accrual, more luxurious offices, special laboratory equipment), special awards and publicity for outstanding performance, and increased opportunities to participate in outside professional activities.[3]

Role of Top-level Specialists in Corporate Decision-making.   While the non-cash forms of recognition just identified serve to boost the morale of technical contributors, they do not redress one of the major grievances of high-level specialists on the technical ladder: their isolation from the top management decision-making group. Top-level specialists in many companies with dual ladders are discontent with their role as the silent partners in corporate decision-making. For example, they complain that they act as problem-solvers behind the scene, but rarely enjoy the prestige and recognition gained in presenting their findings to top management. They are frustrated by management's failure to engage them early enough in the planning of a project and to consult with them regularly before making decisions of a technical nature. In short, they are dissatisfied with the lack of collaboration between managers and technical contributors which results from management's monopolization of decision-making power.

Unless they are permitted to assume an active role in making corporate decisions requiring technical expertise, top-level individual contributors in a dual ladder system are likely to feel like second-class citizens in relation to their managerial counterparts. But giving professionals this measure of authority requires a change in the traditional power bases of an organization, a process which is not likely to be undertaken in many management-dominated organizations.

This is not to say that organizations with dual ladder structures necessarily isolate high-level technical staff members from planning and decision-making activities. In fact, in some of the organizations participating in our study, primarily those engaged in high technology work, the primary role of top-level individual contributors is that of troubleshooter throughout the organization and consultant to corporate staff. Even in the research and technical service facility with only limited interest in advanced technical work, efforts were made to increase the participation of higher-level technical contributors in decision-making activities by including them in "staff" meetings which had previously been reserved for managers. Based on technical staff recommendations, management also considered assigning a "knowledge base development responsibility" to a group of higher-level individuals on the technical ladder who would periodically review the research facility's knowledge resources, identify resource needs, and recommend to management research activities designed to meet the identified needs.

Characteristics of Effective Professional Ladders

In the previous section's enumeration of the major problems organizations tend to have with their dual ladder programs, we also alluded to the characteristics of more effective approaches. The purpose of this section is to briefly summarize those characteristics, many of which may be self-evident by this point in our discussion.

Clearly, the sine qua non of a successful dual ladder program is a well-articulated dual ladder structure, one which precisely defines each job position. The jobs should represent "distinguishable" levels of "real" work in the organization. Employees should be able to define the differences between adjacent job positions, and new positions should not be created merely to provide higher pay levels or to create a facade of higher-level job opportunities. To satisfy these requirements, most effective dual ladders distinguish between five and eight levels of work. Finally, each job within the dual ladder structure should be formally described, and the title, qualification criteria, job skills and responsibilities, performance standards, and accountabilities associated with each level clearly defined. These definitions are essential because they determine the structure of the formal career paths as well as the advancement criteria for each of the levels.

While these are the basic requirements of a well-articulated dual ladder structure, additional factors must be taken into consideration:

1. The career paths defined by the dual ladder structure should be achievable and not just available on paper. If the rewards associated with top-level positions are to motivate good performance, then individuals on the professional ladder must witness that their colleagues actually advance to the highest positions.
2. Positions on the professional ladder should extend high enough to give professionals a truly challenging job.
3. The rewards that are associated with the various ladders should be meaningful and appealing to the groups they aim to motivate. In addition, they should be distributed equitably on the various ladders. To insure some measure of equivalence

287

between the different ladders, compensation packages for the parallel ladders should be identical at each of the levels.

4. Organizational recognition of outstanding performance should go beyond well-publicized promotions. Organizations should work at enhancing their professional environments with a variety of non-cash forms of recognition.

5. The organization's performance appraisal system should be tailored to evaluate both individual contributor and managerial positions. Unfortunately, most performance appraisal systems focus primarily upon management skills and are ill-designed to assess the skills of individual contributors.

6. To insure that technical ladder promotions are based upon technical performance, a formal peer review committee should oversee the promotion consideration process for technical ladder positions.

7. High-level specialists on the technical ladder should not be isolated from the top management decision-making group. They should be actively involved in planning and decision-making activities of a technical nature. Such a change in the traditional power bases of an organization will not only raise the status of technical contributors and improve their morale but will also lead to more balanced decisions.

8. It is essential that both management and specialists view the technical ladder as a legitimate and prestigious avenue of career advancement. Cultural commitment to the professional ladder implies recognition of the critical contributions of technical specialists to the achievement of organizational objectives.

9. The dual ladder structure should be clearly communicated to employees early in their careers so that they will be aware

of formal career paths and can use this information for their own career planning. Employees should also be aware of long- and short-range job opportunities, as well as job requirements, so that they can take an active role in their career development.

10. Finally, the dual ladder program should be evaluated regularly. A close monitoring of the system will help to identify general problems that need to be addressed.

PART III: A DIAGNOSTIC FRAMEWORK FOR EVALUATING A DUAL LADDER PROGRAM

This final section outlines a diagnostic framework which can be used by human resource managers to evaluate the structure and functioning of their organization's dual ladder system. The proposed framework is not intended to be comprehensive in its design; rather, it suggests a series of tasks to guide the evaluation process.

I. Stating the Objectives of the Dual Ladder Program

The evaluation should begin with a clarification of the objectives of the organization's dual ladder system of career advancement. In reviewing the programs' objectives, human resource managers should identify the needs or problems of the organization and those of different employee groups which the dual ladder approach aims to address.

II. Identifying the Dual Ladder Structure

Before human resource managers can diagnose the problem areas in a dual ladder program, they must have a clear understanding of what the system looks like. The following exercise is designed to help the human resource manager identify the basic structure of the organization's dual ladder:

A. Sketch the design of the dual ladder structure. Include the following data:
    a. Levels in the hierarchies
    b. Job titles
    c. Major responsibilities and skills, qualification criteria and accountability standards for each job position
    d. Typical career paths, including selection points and crossovers
    e. Salary ranges and perquisites for each level in the hierarchies
B. Note which of the data specified in (a) through (d) above are missing. These characteristics of the dual ladder structure should be formally defined.
C. In order to evaluate patterns of career advancement:
    a. Record the number of employees in each position,
    b. Compute the percentages of employees at each level of the respective ladders, and
    c. Record the average number of years in each position

III. Examining the Design and Operation of the Dual Ladder Program

Examine the dual ladder structure, as diagrammed, in light of the following issues:

A. Job Definitions
    a. Can jobs be distinguished? How?
    b. Are different skills and responsibilities required at each level?
    c. Does "real" work exist at each level of the hierarchy?

B. Career Paths
    a. Are there pressure points, plateaus, gaps in career paths? Where and what, if any, are the consequences?
    b. Is advancement faster on one ladder than the other?
    c. Are higher-level positions within career paths achievable?
    d. When and how do employees learn about formal career paths?
C. Promotion Considerations
    a. Does the performance appraisal system evaluate managerial *and* technical skills?
    b. Who makes promotion decisions at each level of the hierarchy? What procedure has been set up? Is it consistently followed?
D. Rewards and Recognition
    a. Are technical ladder promotions based on technical excellence? Are they ever given for other reasons?
    b. Are rewards (e.g., money, information, power, perquisites) equivalent on the various ladders?
    c. Which non-cash forms of recognition are provided for those on the professional ladder?

IV. Surveying Employees' Evaluations of the Dual Ladder Program

A survey of both professional and managerial employees' views of the dual ladder program is critical for the evaluation process. The survey should focus on the following kinds of issues:

a. How do managers and professional employees perceive the professional ladder? Do they consider it a prestigious route to career advancement within the organization?

b. How do the two groups feel technical ladder positions compare with managerial positions on the following dimensions:
- status
- rewards and perquisites
- management's recognition of their contributions
- influence within the organization?
c. Are the rewards associated with the different ladders meaningful and valued by the groups they aim to motivate?
d. Are employees satisfied with their ladder assignments?
e. What do managers and professional employees consider to be the major problems of the dual ladder program?
f. What are their suggestions for improving it?
g. Overall, do managers and professional employees judge the dual ladder approach as useful in recruiting, motivating, and retaining competent professionals?

CONCLUSION

The primary objective of this paper was to describe, evaluate, and recommend modifications in the dual ladder approach to career advancement based on findings from a recent study of dual ladder programs in over twenty organizations with different professional populations and industry affiliations. State-of-the-art information has been presented identifying the kinds of organizations utilizing the dual ladder approach; the major problems organizations face when designing, implementing, and administering dual ladders; and the characteristics of more effective dual ladder programs. In addition, a diagnostic framework has been outlined to guide human resource practitioners in evaluating their organization's dual ladder.

Throughout this report we have shown

that human resource managers must attend to a variety of technical, political, and cultural tasks when developing and administering a dual ladder program of career advancement. The primary technical task is to design a dual ladder structure that clearly defines challenging career paths for managerial and professional employees and distributes rewards equitably on the parallel ladders. The critical political task is to alter the traditional power bases within the organization to allow for the sharing of decision-making powers by top-level managers and technical specialists. Finally, the predominant cultural task is to establish and maintain organizational commitment to the professional ladder by cultivating an awareness of and appreciation for the professional's contributions to the organization.

If the organization is not prepared to tackle these difficult tasks and thus provide the necessary environmental conditions for a well-functioning dual ladder program, then it is doubtful that a dual ladder approach to career advancement would be appropriate. Before choosing the dual ladder approach, the organization employing a large number of professionals must determine whether it can live with the unwanted consequences of its traditional pyramidal model of career advancement—high turnover, low morale, and insufficient motivation among professional employees. Then it must determine whether the resources and commitment are available to undertake the challenge of designing, implementing, and administering a dual ladder or whether an alternative strategy for the career planning of professionals is more suitable.

NOTES

1. These alternative labels for dual ladder systems will be used interchangeably throughout this text.

2. Research by Badawy (1975) shows that the value systems and role expectations of scientists differ from those held by engineers. In contrast to engineers, scientists tend to be oriented toward their profession rather than the goals of the organizations for which they work. In addition, scientists define recognition in their profession rather than advancement within the organization as the major criterion of success. These findings suggest that the "cosmopolitan" assumption of the dual ladder is more in line with the career orientations of scientists than with those of engineers.

3. In addition to non-cash awards, several of the participating organizations considered implementing cash incentive awards to top-level professionals to encourage and reward innovation and risk-taking. In one company, a special incentive plan was designed for senior professionals who made technological or scientific advancements judged as important to company objectives. The awards were to range in size from 10 to 50 percent of the recipient's base salary. In addition, the recipient's base salary was to be increased by 10 to 50 percent of the value of the incentive award.

REFERENCES

Badawy, M. K. Organizational designs for scientists and engineers: Some research findings and their implications for managers. *IEEE Transactions on Engineering Management* EM-22, no. 4 (Nov. 1975).

Dalton, G. W.; P. H. Thompson; and R. L. Price. The four stages of professional careers—A new look at performance by professionals. *Organizational Dynamics* 6 (Summer 1977): 19–42.

Delbecq, A. L., and E. S. Elfner. Local-cosmopolitan orientations and career strategies for specialists. *Academy of Management Journal* 13 (1970): 373–87.

Glaser, B. G. The local-cosmopolitan scientist. *American Journal of Sociology* 69 (1963): 249–59.

Goldner, F. H., and R. R. Ritti. Professionalization as career immobility. *American Journal of Sociology* 72 (1967): 489–502.

Kaufman, H. G. *Obsolescence and professional career development.* New York: AMACOM, 1974.

Kornhauser, W. *Scientists in industry: Conflict and accommodation.* Berkeley: University of California Press, 1962.

Reissman, L. A study of role conceptions in bureaucracy. *Social Forces* 27 (1949): 305–10.

Shepard, H. A. The dual hierarchy in research. *Research Management* 1 (1958): 177–87.

Wilensky, H. L. *Intellectuals in labor unions.* Glencoe, Illinois: The Free Press, 1956.

Zand, D. E. *Information, organization, and power: Effective management in the knowledge society.* New York: McGraw-Hill, 1981.

APPENDIX I

Titles of Respondents and Industry Affiliations of Organizations Participating in the Formal Interview Study

Program Manager, Management Development
Large international computer company

Vice President of Personnel; Director of Corporate Personnel; and Manager, Employee Relations, small research and technical service facility
Large consumer products company

Organization and Personnel Advisor, Employee Relations Research and engineering division
Large international petroleum company

Senior Employee Relations Consultant
Large international chemical company

Director of Education of the Data Processing Division
Large insurance company

291

Director, Career and Executive Development, Corporate Personnel, and Manager, Executive Development, Corporate Personnel
Large insurance company

Personnel Manager of Technical Research Facility
Large consumer products company

Director of Technical Human Resources
Large petroleum company

Assistant Vice President, Personnel Resources
Large computer company

Management Supervisory Trainer/Career Counselor
Medium-sized electronics company

President
Small science research facility

Director of Personnel, Manager of Compensation; and Manager, Management Development and Training
Large international pharmaceutical company

Director, Employee Relations
Large international pharmaceutical company

Staff Development Officer and Compensation Manager
Public sector financial institution

APPENDIX II

Industry Affiliations of Organizations Participating in the Informal Study
Large consumer products company (2)

Large computer company

Large petroleum company

High technology company (2)

Large consumer and industrial products company

Management consulting firm

Branch of the armed services

Investigating the Not Invented Here (NIH) Syndrome: A Look at the Performance, Tenure, and Communication Patterns of 50 R&D Project Groups

Ralph Katz

Thomas J. Allen

Abstract: The Not-Invented-Here (NIH) syndrome is defined as the tendency of a project group of stable composition to believe it possesses a monopoly of knowledge of its field, which leads it to reject new ideas from outsiders to the likely detriment of its performance. The authors have carried out an empirical test of the extent to which the rate of communication between a project group and the outside world decreases with mean project tenure and how far performance decreases with project tenure.

The study, carried out in a large laboratory, shows that performance increases up to 1.5 years tenure, stays steady for a time but by five years has declined noticeably. This tendency is best accounted for by the marked decline in communication rate among group members and between them and critical external sources of information. The authors analyse the significance of this finding and suggest means of maintaining the vitality of long-standing project teams.

Engineers have long recognized the problems facing a technical group should its membership

remain constant too long. General folklore among R & D professionals holds that a group of engineers whose membership has been relatively stable for several years may begin to believe that it possesses a monopoly on knowledge in its area of specialization. Such a group therefore does not consider very seriously the possibility that outsiders might produce important new ideas or information relevant to the group. This has come to be known in the R & D community as the "Not Invented Here" or "NIH" syndrome.

NIH and Increasing Project Insulation

According to the NIH syndrome, then, stable project teams become increasingly cohesive over time and begin to separate themselves from external sources of technical information and influence by communicating less frequently with professional colleagues outside their teams. Rather than striving to enlarge the scope of their information processing activities, long-tenured groups become increasingly complacent about outside events and new technological developments. The extent to which they may be willing or even feel they need to expose themselves to new or alternative ideas, suggestions, and solutions lessens with time.

In spite of this belief, the fact remains that groups, including those of long standing membership, must still collect and process information from outside sources in order to keep current technically. Project members rarely have all the required knowledge and expertise to complete their tasks successfully; information and assistance must be drawn from many sources outside the project. Moreover, research findings by Allen (1977), Menzel (1965), and many others have consistently shown that personal contacts, rather than written technical reports or publications, are the primary means by which engineering professionals acquire important technical ideas and information from outside sources. In testing for the NIH syndrome, therefore, it is hypothesized that project groups whose members have been working together over a long period of time (i.e., project teams whose members are averaging high levels of group or project tenure) will have significantly less communication with other professionals both within and outside the organization.

NIH and Decreasing Project Performance

If project communication with internal and external colleagues diminishes significantly as mean group tenure increases, and if such communications are essential to technical performance, then one should also expect a strong inverse relationship between group tenure and project performance. Several studies, in fact, have presented evidence to support such an association. Shepard (1956) was the first to relate performance to mean group tenure as measured by averaging the lengths of time individual members had been working within the group. He found that performance increased up to about 16 months average tenure, but that performance gradually decayed with increasingly higher levels of group stability. Pelz and Andrews (1976) uncovered a similar curvilinear connection between mean tenure and performance. In their study, however, the "optimum" group tenure seemed to fall at about the four year mark. Smith (1970) also showed R & D group performance peaking at a mean tenure of about three to four years.

The present study investigates once again the association between mean tenure and technical performance of R & D groups. This time, however, the research will focus on clearly defined project teams. It is not clear in the previous research studies whether "groups" are project teams or whether they are functional, disciplinary, or specialty-based groups. It is presumed that there was a mix of both types in each of the three previous studies. The reason for our project focus is a practical one. It is expected that results could differ considerably for project as opposed to functional or disciplinary groups. The project team with its more intense focus on a specific product or problem could be expected to obsolesce more rapidly than a functional group (Marquis and Straight, 1965). In the latter case, the fact that members are normally all working within a single discipline or technical specialty can help group members keep in closer touch with developments within that particular specialty. Contrastingly, members of project teams tend to become over time more narrow and more highly specialized in the technical problem areas associated with their specific project assignments. In this process, they are drawn away from and begin to lose touch with the more recent developments within their technical specialties.

In addition to this distinction, our study will also examine the explanatory part of the NIH syndrome. Communication by project teams is expected to decline significantly with mean group tenure. As a result, one should be able to account for the decline in performance, with increasing mean tenure, through the intervening variable, communication.

Accordingly, the following hypotheses will be tested:

1. The relationship between the mean tenure of project members and the project's overall technical performance will be curvilinear with performance reaching its highest levels between a mean project tenure of two to four years and decaying thereafter.

1A. As a corollary to the above hypothesis, it is expected that project performance will also be related to regular and gradual turnover of project personnel. To test this, the variance across the project tenures of individual group members should also vary curvilinearly with project performance.

2. Technical communication to information sources outside the project team will follow a pattern similar to that of project performance, peaking between two and four years of mean group tenure and decreasing thereafter. In particular:

a. Technical communication by project members with professional colleagues *inside* the organization (i.e., internal communications) will be significantly lower for teams with high levels of mean group tenure.

b. Technical communication by project members with professional colleagues *outside* the organization (i.e., external communications) will be significantly lower for teams with high levels of mean group tenure.

RESEARCH SETTING AND METHOD

This study was conducted among all the R & D professionals (N = 345) of a large corporate R & D facility. The laboratory's professionals were organized into seven separate departments which, in turn, were divided into a total of 61 distinct projects. These project groupings remained stable over the course of the study, and each professional was a member of only one project team. Complete project data were successfully obtained on a total of 50 groups, representing 82% of all projects within this facility. Other findings from this same data base have recently been reported in Katz and Tushman (1981).

Technical Communication

To measure communication data, professionals were asked to report (on specially provided lists) those individuals with whom they had technical communication on a randomly chosen day each week for 15 weeks. Sampling was constrained to provide equal representation of all weekdays. Respondents reported all personal contacts both within and outside the organization, including the name of the individual with whom they talked and how many times they talked to that person during the day. They did not report communications which were strictly social, nor did they report written communications. During the 15 weeks, the overall response rate was 93 per cent. Although the overall response rate was extremely high, the raw communications data for incomplete respondents were proportionately adjusted by the number of missing weeks. Moreover, 68 percent of all communication interactions within the R & D organization were reciprocally reported by both parties. These research procedures are similar to those used in other sociometric communication studies such as Allen and Cohen (1969), Whitley and Frost (1973), and Tomlin (1981) and provide a clear, accurate picture of the professionals' communication patterns.

As discussed by Katz and Tushman (1979), six mutually exclusive communication measures were operationalized for each project group as follows:

Internal Communications

1. Intraproject: The amount of communication reported among all project team members.
2. Departmental: The amount of communication reported between the project's members and other R & D professionals within the same functional department.
3. Laboratory: The amount of communication reported between the project's members and R & D professionals outside their functional department but within the R & D organization.
4. Organizational: The amount of communication reported by the project's members with other individuals outside R & D but within other corporate functions such as marketing and manufacturing.

External Communications

5. Professional: The amount of communication reported by project members with professionals outside the parent organization including universities, consulting firms, and professional societies.
6. Operational: The amount of communication reported by project members with external operational areas including vendors and suppliers.

The amount of communication with these four internal and two external categories of people was calculated by summing the number of interactions reported during the 15 weeks and normalizing for the number of project members (see Katz and Tushman, 1979 for details). Except for the correlation between Department and Laboratory Communication ($r = 0.31$; $p < .05$), none of the six measures of project communication were significantly intercorrelated (.10-level of significance or less).

Project Performance

Since the laboratory's management could not develop objective performance measures which would be comparable across the laboratory, a subjective measure similar to that used by Lawrence and Lorsch (1967) was employed. All Department Managers ($N = 7$) and Laboratory Directors ($N = 2$) were interviewed individually and asked to evaluate the overall technical performance of all projects with which they were technically familiar. Each was asked to make an informed judgment based on knowledge of and experience with the projects. If they could not make an informed judgment for a particular project, they were asked not to rate that project. Criteria the managers considered (but were not limited to) included: schedule, budget, and cost performance; innovativeness; adaptability; and cooperation with other organizational areas. Each project was independently rated by about five managers using a seven-point scale that ranged from very low to very high. Since there was a very strong consensus across the performance ratings of the nine judges (Spearman-Brown reliability = 0.81), individual ratings were averaged to yield overall project performance scores.

Project Characteristics and Individual Demographics

Each professional was asked to specify the degree to which his project assignment involved research, development, or technical service activities (see Katz and Tushman, 1979). By pooling individual members' responses to obtain project scores, we could easily identify a project as being predominantly either research, development, or technical service. As discussed in Tushman (1977), analysis of variance was used to ensure that there was sufficient agreement among project members with respect to the classification of their project.

Finally, each professional provided information on age, education, and an estimate of the number of years and months of association with a particular project group and with the overall R & D organization. These individual age, project, and organizational tenures were averaged across project members to obtain measures of mean age, mean project tenure, and mean organizational tenure. It is important to recognize that mean group tenure is not the length of time the project has been in existence, but rather it measures the average length of time project members have interacted with each other. Thus, the measure of group tenure is not tied to project phase nor necessarily related to how long R & D professionals per se have been working in that particular problem area within the company. The mean, then, is used to obtain a representative picture of how long project members have worked together and shared mutual experiences. The mean, however, only represents the central tendency

of the project tenure distribution among project members. As a result, we will also examine the distribution about the mean using variance measures.

RESULTS

Project Performance

The 50 projects have mean group tenures ranging from several months to almost 13 years with an overall mean of 3.41 years and a standard deviation of 2.67 years. The mean rating of project performance, as provided by the evaluators, ranged from a low of 3.0 to a high of 6.4. Mean performance for the 50 projects was 4.59. When project performance is plotted as a function of mean project tenure of team members (Figure 1), there is some indication that performance is highest in the two to four year interval, with lower performance scores both before and after.

To get a clearer picture of the underlying relationship between mean group tenure and project performance, these original data were subjected to a smoothing technique using a moving median procedure (Tukey, 1977). The specific smoothing technique used is what Tukey calls a 3RSSH smooth (Tukey, 1977, p. 231ff). The results, plotted in Figure 2, illustrate very clearly that performance is highest for projects in the medium range of mean tenure. More interestingly, the smoothed data show a marked decline in performance for projects whose members have worked together for more than five years. Given this exploratory analysis, more confirmatory statistical procedures will now be used to investigate the relationship among project performance, mean group tenure, and project communication using only the *original* raw data. Smoothed data are presented only to better illustrate relationships. All analyses are performed on the unsmoothed data.

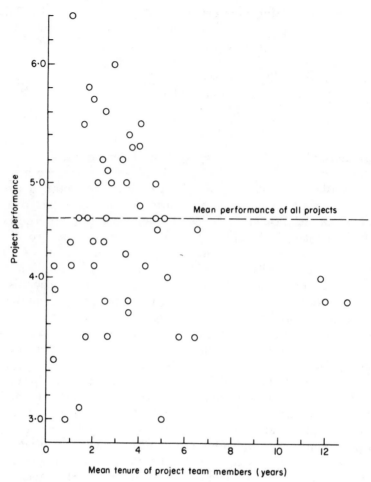

FIGURE 1.   *Project Performance as a Function of the Mean Tenure of Project Team Members (Raw Data)*

To test for significant differences in the distribution of project performance as a function of mean project tenure, the 50 groups were divided into five separate mean group tenure categories. Based on the curvilinear patterns in Figures 1 and 2, there appear to be at least three different tenure periods represented within the original data: (1) 0 to 1.5 years; (2) 1.5 to 4.9 years; and (3) 5 or more years. For additional exploratory purposes, the 30 project groups falling within the middle tenure range

were equally subdivided into three separate categories, as shown in Table 1. The first 0 to 1.5 year interval corresponds to the initial learning or building phase previously discussed by Shepard (1956), Pelz and Andrews (1966) and Smith (1970). In a similar fashion, the last category of project groups, representing teams whose members have worked together for more than five years, corresponds to the low performance interval shown in these same studies as well as to the time period commonly used

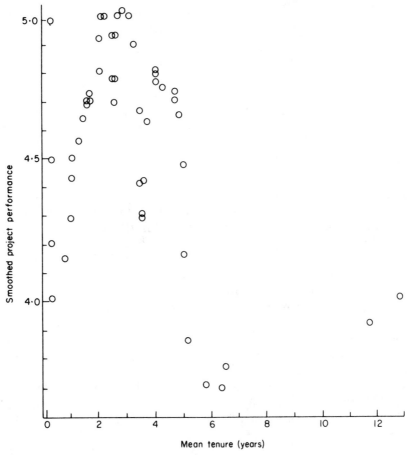

FIGURE 2.   *Smoothed Project Performance as a Function of the Mean Tenure of Project Team Members*

to estimate the half-life of technical information (Dubin, 1972).

An examination of the actual mean performance scores of projects within each of the five tenure categories of Table I strongly supports the curvilinear association hypothesized between project performance and mean project tenure. Within this organization, performance was, on the average, significantly lower for project teams whose mean tenures were either less than 1.5 years or were more than 5 years. In contrast, performance was significantly higher across all three middle tenure categories (i.e., between 1.5 and 4.9 years).

Group Tenure or Age of Individuals?

Almost by definition, projects with higher mean tenures are also staffed by older engineers. This raises, of course, the possibility that the performance decay associated with longer tenure had little to do with the team per se. It may have resulted, instead, from the increasing obsolescence of individuals as they aged. For the

299

TABLE 1.   *Project Performance as a Function of the Mean Tenure of Project Team Members*

	Mean Project Tenure (in years)				
	0–1.4 (N = 10)	1.5–2.4 (N = 10)	2.5–3.4 (N = 10)	3.5–4.9 (N = 10)	5 or greater (N = 10)
Mean Project Performance*	4.29	4.89	4.87	4.82	4.07
Standard Deviations	0.99	0.67	0.70	0.59	0.52

*Using a 1-way ANOVA test, the mean project performance scores are significantly different across the five tenure categories.
$(F(4,45) = 2.89; p < .05)$.

entire population, the correlation between project performance and the mean age of project team members is slightly negative $(r = -0.18)$ but far from significant statistically. In the downward sloping interval of project performance, that is beyond a mean project tenure of 2.5 years, there is a slightly stronger negative relation, although still not significant. For those 30 projects with a mean group tenure of at least 2.5 years, the correlation between performance and the mean age of project members was -0.28; the corresponding relationship between performance and mean group tenure of project members is both negative and significant $(r = -0.39; p < 0.05)$. A third variable,

mean organizational tenure, is also correlated with the other two aging variables and is included in the next analysis.

The partial correlations of Table 2 demonstrate convincingly that it is tenure with the project team and not age or organizational tenure that influences project performance. Neither individual age nor organizational tenure shows any negative association with performance when project tenure is controlled. In fact, organizational tenure correlates positively, although not significantly, with performance when project tenure is held constant. It may be that projects staffed by longer term employees perform slightly better, provided these veteran

TABLE 2.   *Partial Correlations Between Project Performance and Various Aging Variables for Projects with Average Member Tenure of at Least 2.5 Years*

Aging Variables	Correlations with Project Performance	Partial Correlations	Variables Controlled
(a) Mean project tenure of project members	−0.39**	−0.28*	(Mean age)
		−0.33**	(Mean organizational tenure)
(b) Mean organizational tenure of project members	−0.23	0.20	(Mean project tenure)
		−0.05	(Mean age)
(c) Mean age of project members	−0.28	−0.08	(Mean project tenure)
		−0.19	(Mean organizational tenure)

N = 30; *p < 0.10; **p < 0.05

employees are not retained on any single project team for too long a time.

Clearly, there are any number of strategies for assigning or rotating individual engineers among project groups. All or nearly all of the team members could be replaced every few years, or members could be changed individually at more frequent intervals. Different strategies such as these can obviously result in markedly different distributions of project tenure among team members even though mean tenures may be similar. In the organization under study, it is evident that many such strategies were pursued, resulting in a wide variety of distributions of project tenure.

Using the standard deviation of project tenure across team members as one measure of these distributions, we explored the relationship between project performance and these variance measures. As before, the smoothing

procedures developed by Tukey (1977) were used to determine the general form of any possible association. The smoothed pattern shows a distinct curvilinear relation between project performance and the standard deviation of project tenure (Figure 3). Project performance is highest when the standard deviation in group tenures is about three years. This is true for all 50 projects as well as for the relatively long-tenured project teams. In other words, it appears that project teams perform best when their team memberships have not been completely stable, but where there has been some turnover of team personnel. On the other hand, when tenures are too widely dispersed, performance is also low. Such findings suggest that project groups must balance their needs for gradual turnover with reasonable amounts of team stability. Periodic turnover of personnel may help to keep a team alert and vigilant,

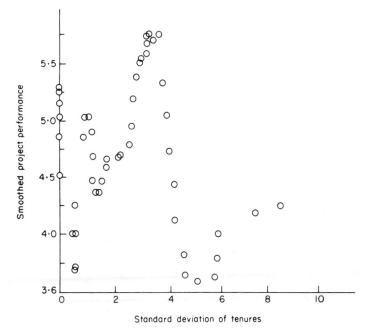

FIGURE 3. *Smoothed Project Performance as a Function of Standard Deviation in Individual Project Tenures*

but constantly changing membership will create confusion and detract from performance.

Project Communication

Having demonstrated a strong connection between group tenure and technical performance for R & D project teams, we can now turn to factors that might be inhibiting group effectiveness as team membership ages. As part of the NIH syndrome, it was hypothesized that if performance was found to vary with mean project tenure, then technical communication to sources outside the project team would follow a similar pattern. More specifically, one of the contributing reasons for the decline in project performance with increasing mean tenure might be decreased outside contact and interaction. Members of such project groups would be paying less attention to outside sources of ideas and information, relying more on their own expertise and wisdom.

To examine this issue empirically, we tested for significant differences in the *actual* communication behavior of research, development, and technical service projects as a function of mean tenure. Communication varies significantly with tenure in three cases: Intraproject, Organizational and External Professional. Communications to the other three areas, i.e., Department, Laboratory, and External Operational, did not diminish significantly with increasing tenure for any type of project, i.e., research, development, or technical service.

Communication, as performance, follows a pattern of decline with mean group tenure after 2.5 years (Table 3). It is a different type of communication, however, that shows the greatest decay with each of the three types of R & D activity. For technical service projects, communication among project members themselves shows the greatest decline. For development projects, it is communication with the rest of the organization; and for research projects, it is external professional communication.

The remarkable thing about this set of results is that it appears to be the most important type of communication for each activity that has the greatest decline with tenure. Previous research (e.g., Allen, 1964; 1977; Allen, et al., 1980; Baker, et al., 1967; Pelz and Andrews, 1976) has shown very clearly that for development projects the best sources of technical information lie within the organization. Communication with other members of the R & D staff, of marketing and manufacturing, and of other functions shows consistently positive correla-

TABLE 3.   *Correlations between Group Tenure and Performance and Communications for Projects with Mean Tenure of at Least 2.5 Years*

Variables Correlated With Mean Group Tenure	Project Type			All Projects (N = 30)
	Research (N = 6)	Development (N = 12)	Technical Service (N = 12)	
CORRELATIONS:				
a) Project Performance	−0.62*	−0.39*	−0.44*	−0.43***
Internal Communications				
b) Intraproject Communication	−0.26	−0.14	−0.72***	−0.39***
c) Organizational Communication	0.27	−0.53**	−0.12	−0.20
External Communications				
d) External Professional Communication	−0.51	−0.23	−0.38	−0.37**

*p < 0.10; **p < 0.05; ***p < 0.01

tions with performance for development projects. In the present study, it is just this type of communication that decays most with increasing team tenure. In the case of research projects, the findings are just as clear. Communication with colleagues outside the organization is most important for members of research teams (Allen, et al., 1979; Katz and Tushman, 1979; Hagstrom, 1965).

For technical service projects, the evidence is not as strong. However, there is some indication that communication among project members is most important for the performance of these projects. Allen, et al. (1980) have shown, for example, that communication activity between project members and the project manager correlates significantly with performance for technical service projects.

So it is almost as though some demon were at work, selecting the most useful form of communication in each instance and causing it to decay most with increasing mean tenure of project members. Development project teams isolate themselves most from organizational colleagues; research teams from external colleagues, and technical service team members from each other.

It is impossible, at this point, to determine why this should be so. We can, however, look more deeply into the relationships to determine whether the decreases of different forms of communication with tenure are in fact related to project performance. This is indeed the case (Table 4). When organizational communications are held constant, the relationship between tenure and performance for development projects changes from a significant negative value to nearly zero. Similarly, when intraproject communication is held constant for technical service projects, the negative performance-tenure correlation approaches zero. Unfortunately, there were too few research groups to permit such an analysis for research projects. Nevertheless, it appears that increasing team tenure operates on the particular form of communication that most affects performance to create the effect of decreasing performance with increasing stability in project membership.

DISCUSSION

The findings presented here emphasize the important influence of mean group tenure on the communication activities and performances of project teams. In examining the technical per-

TABLE 4. *Relations Between Project Performance and Mean Tenure of Project Members Controlling for Communication (Projects with Mean Tenure of 2.5 Years or More)*

| Type of Project | Relation Between Tenure and Performance | | Type of Communication Held Constant |
	r	r partial	
Research (N = 6)	−0.62*	*	
Development (N = 12)	−0.39*	−0.46*	Intraproject
		−0.19	Organizational
		−0.42*	External Professional
Technical Service (N = 12)	−0.44*	−0.17	Intraproject
		−0.45*	Organizational
		−0.36	External Professional

*p < 0.10;
* = Too few research projects for partial analyses

formance of project groups, a curvilinear relationship is found to exist between performance and mean project tenure. As in several previous studies, performance is found to increase steadily until mean tenure reaches about 1.5 years after which performance remains high but then gradually declines. The decline sets in clearly by the fourth or fifth year. This decay in project performance operates independently of the actual age of project team members and is independent of the type of R & D being performed. Similar decays in performance were found for all categories of projects whether research, development, or technical service.

By itself, the observation that R & D project performance declines significantly with high levels of group tenure raises more questions than it answers. Why were the performances of longer-tenured project teams significantly lower on the average? Were they staffed by proportionately less able or less motivated engineers, or were there important differences in how project members actually conducted their day-to-day activities that could help to account for these significant differences in technical performance?

Information gathered during a recent follow-up visit to the organization shows that the same proportion of professionals from both the long and medium tenured project teams were promoted to higher laboratory positions during the five year interval since the collection of the original data. Fifteen percent of the engineers who had been working on projects in the medium range of group tenure (i.e., between 1.5 and 5.0 years) were promoted to higher level managerial positions. The comparable percentage for engineers working in the ten long-tenured projects was 13 percent. In addition to managerial promotions, 12 percent of the engineers from medium-tenured project teams were promoted to positions on the technical side of the organization's dual ladder system. The comparable percentage for the long-tenured teams was slightly higher, roughly 19

percent. Such promotional histories strongly suggest that neither individual competence nor perhaps the importance or visibility of the project can account for the significant difference in technical performance between medium and long-tenured project groups.

As hypothesized, it is the reduction in communications of project members to key information sources that accounts for the performance differences. For projects whose group memberships had remained relatively stable over time, team members were communicating less often amongst themselves, less with individuals in other parts of the organization, and less with external professionals from the larger R & D community. Since the discussion and transfer of technical information and new ideas is important to effective project performance in R & D, it seems reasonable to attribute, at least in part, the lower technical performances of long-tenured groups to these differences in communication.

It is important to emphasize that it is not a reduction in project communication per se that leads to the deterioration in performance. Indeed, some of the measures of project communication did not diminish with increasing mean tenure. Rather a decline in performance is more likely to stem from a project team's tendency to ignore and become increasingly isolated from sources that provide more critical kinds of evaluation, information, and feedback. Since research, development, and technical service projects differ significantly in the kinds of communication important to them, projects in each of these categories are more likely to suffer when members are isolated from their most critical information sources. Thus, overall performance will suffer when research teams fail to pay sufficient attention to new advances and information within their relevant external R & D community, when technical service groups fail to interact among themselves, or when development project members fail to communicate with individuals from other parts of the organization, particularly R & D, marketing, and manufacturing.

This is not to say that external developments in technology are unimportant for development projects. On the contrary, they are exceedingly important! What is implied by our findings is simply that the performances of development projects are not affected adversely by having all of their members communicate less often with external professionals. This occurs because development projects, unlike research groups, are most effectively linked with their external technical environments through specialized boundary-spanning individuals labelled gatekeepers (Allen, 1977, Katz and Tushman, 1981) rather than through direct external interactions by all project members. Gatekeepers are defined as those key R & D professionals who are both high internal and external communicators and who are also able to effectively transfer external ideas and information into their project groups. As a result, the impact of project tenure on development project performance may be more sensitive to the emergence and use of technical gatekeepers rather than being affected by the amount of external contact conducted by all project members. Although such an analysis cannot be done with the present data base, it is interesting to note that of the five development projects with an average tenure of at least five years, none had a technical gatekeeper within their membership. Thus, while lower levels of external contact may not directly affect the performance of development projects, reduced project communication in general might affect the extent to which gatekeepers are able to emerge in long-tenured project groups.

Implications

To gain additional insight into the curvilinear relationship portrayed in Figure 2, a regression curve was fitted to the smoothed data. By observation, the relationship appears to be of the form $Y = aX^b e^{-cX}$ where Y and X represent project performance and average project tenure respectively. Fitting the smoothed data to this type of nonlinear equation, the regression analysis yielded the following functional model:

$$Y = 4.77X^{0.08}e^{-0.04X}$$
where Y = Project Performance
X = Mean Tenure of Project Members

This equation, moreover, seems to be a reasonably good fit as it accounts for over 49 percent of the variance in the smoothed performance data ($R = 0.71$).

Based on this regression model, one can think of project performance as a function of the product (or interaction) of two distinct factors. The first factor influencing performance is a positive component of the form $Y = aX^b$, resulting from team members developing better and more effective working relationships; e.g., a kind of team-building component. In contrast, the second factor is inversely associated with performance, stemming perhaps from the development of the NIH syndrome. As team membership remains stable, communication with the rest of the technological world is reduced leading to an exponential type decay in performance of the form $Y = e^{-cX}$. Using the previously derived regression parameters, each of the component factors and their intersection are drawn in Figure 4.

While the regression relationship between project performance and mean team tenure is an inverted U-shaped curve, its two major component shapes are very different. The first component term rises rapidly with mean tenure showing the positive effects of "team-building." Team members develop better understanding of one another's capabilities, better understanding of the relevant technologies, better working relationships, etc., and such improvements are reflected in rapidly increasing performances. The team-building effect gradually tapers off, and as a result, its gradient with

305

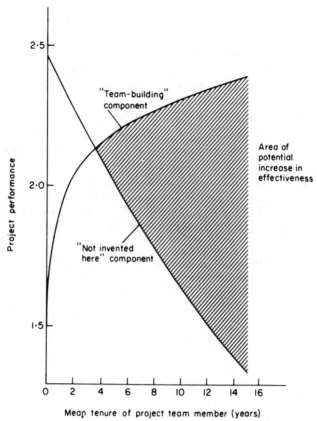

FIGURE 4.   *The Relationship between Mean Group Tenure and Project Performance Analyzed into Its Components*

performance diminishes. At the same time, the exponential decay term has set in, resulting in part from reduced communication. Between these two component curves lies the area for potentially influencing project performance. As we gain additional understanding of the reasons underlying this exponential decay, policies can be implemented to counter such effects in order to have the relationship between mean project tenure and performance approximate more closely to the team-building curve.

In particular, the data suggest that the communication of project team members and its subsequent effects on technical performance might be strongly influenced and managed through staffing decisions. Specifically, it would seem that the energizing and destabilizing function of new members can prevent a project group from developing interactions and behaviours characteristic of the NIH syndrome. Whether or not project groups can circumvent the NIH syndrome without some rejuvenation from newcomers is the next question that needs to be investigated. In the present organization, none of the ten project groups with a mean team tenure of five or more years was among the higher performing projects. We cannot, as a result, determine from the present sample the extent to which effective long-tenured project groups might have been able to maintain their effectiveness through appropriate high levels of communication and interaction with their more critical areas. Clearly, additional research is needed to ascertain just how deterministic the current findings are with respect to project performance, mean tenure, and project communication.

A Temporal Perspective

Underlying these kinds of change is the basic idea that over time individuals try to organize their work environments in a manner that reduces the amount of stress and uncertainty they must face (Katz, 1980; Pfeffer, 1981). According to this argument, employees strive to direct their activities toward a more workable and predictable level of certainty and clarity. The lower levels of intraproject communication, for example, strongly suggest that as project members work together and gain experience with one another, their project activities are likely to become more stable with individual role assignments and contributions becoming better defined and more resistant to change.

Based on such a perspective, we must begin to develop more comprehensive frameworks for analyzing how individuals and groups adapt to their situations over long tenure periods (Katz, 1981). Working in a given position for a considerable period of time, an employee establishes work patterns that are familiar and comfortable, patterns in which routine and precedent play a relatively large part. Recent findings by Katz (1978a, 1978b) suggest, for example, that with increasing project tenure, members may gradually become less responsive to the challenging aspects of their technical assignments. Instead, they become increasingly committed to their current problem-solving strategies, their customary ways of doing things, and their traditional modes of conduct. The longer individuals have actively participated in and become responsible for a given set of policies or strategy decisions, the

stronger their attachment to such policies and strategies even though they may eventually become outdated and inappropriate (Staw and Ross, 1978). Furthermore, in the process of solidifying their contributions and commitments, individuals may come to rely more heavily on their own knowledge, views, experiences and capabilities, thereby reducing their attentiveness to outside sources of information and expertise. It is possibly trends like these that cause project communication to deteriorate with prolonged group tenure. In short, as employees adapt to increasing amounts of job stability, they may become less open and receptive to new and innovative approaches and procedures, preferring instead the predictability of their secure and familiar environments and the confidence which it brings.

The degree to which these tendencies actually materialize for any given individual depends, of course, on the extent to which the overall situation either reinforces or extinguishes such tendencies. Ever since the Hawthorne experiments, it has been generally acknowledged that the pressures and interactions within a given work group can significantly influence the behaviours, motivations, and attitudes of its individual members. In essence, the group controls the stimuli to which the individual is subjected.

How individuals eventually adapt to prolonged tenure on a given project, therefore, is probably influenced to a great extent by their project colleagues. In particular, the greater the mean tenure of project team members, the more these previously described tendencies are likely to occur and be reinforced. In the current organizational sample, for example, it is important to point out that there was no clear trend in any of the communication patterns of individual engineers when plotted as a function of job tenure. Only when the engineers were grouped according to their projects was there a clear and obvious decrease in certain communication measures as a function of mean project tenure. Insulation, then, may be more a function of the average length of time group members have worked together rather than varying according to the particular project tenure of any single individual. Furthermore, our findings suggest that it is not just the mean that is important, it is also the distribution of project tenures among team members that must be considered.

In a general sense, we need to consider the many kinds of changes that are likely to take place within a group as its team membership ages, and more importantly, we need to uncover the kinds of managerial pressures, policies, and practices that can be used to keep a project effective and high performing under such tendencies. In addition to such managerial interventions, it would be even more important to determine if and how a project group can keep itself highly energized and innovative. The challenge to industry in general, and to organizations in particular, is to learn to effectively organize and manage projects in a world characterized by a more rapidly changing and complex technology coupled with a more maturing and stable population.

REFERENCES

Allen, T. J. (1964). *The use of information channels in R & D proposal preparation.* M.I.T. Sloan School of Management Working Paper.

Allen, T. J. (1977). *Managing the Flow of Technology.* Cambridge, MA: M.I.T. Press.

Allen, T. J., and Cohen, S. (1969). Information flow in R & D laboratories. *Administrative Science Quarterly, 14,* 12–19.

Allen, T. J., Lee, D., and Tushman, M. (1980). R & D performance as a function of internal communication, project management, and the nature of work. *IEEE Transactions on Engineering Management, 27,* 2–12.

Allen, T. J., Tushman, M., and Lee, D. (1979). Technology transfer as a function of position on research, development, and technical service continuum. *Academy of Management Journal, 22,* 694–708.

Baker, N. R., Siegmann, J., and Rubenstein, A. H. (1967). The effects of perceived needs and means on the generation of ideas for industrial research and development projects. *IEEE Transactions on Engineering Management, 14,* 156–162.

Dewhirst, H. D., Arvey, R. D., and Brown, E. M. (1978). Satisfaction and performance in research and development tasks as related to information accessibility. *IEEE Transactions on Engineering Management, 25,* 53–63.

Dubin, S. S. (1972). *Professional Obsolescence.* Lexington, Mass: Lexington Books, D. C. Heath.

Hagstrom, W. (1965). *The Scientific Community,* New York: Basic Books.

Katz, R. (1978a). Job longevity as a situational factor in job satisfaction. *Administrative Science Quarterly, 23,* 204–223.

Katz, R. (1978b). The influence of job longevity on employee reactions to task characteristics. *Human Relations, 31,* 703–725.

Katz, R. (1980). Time and Work: Towards an integrative perspective. In B. Staw and L. L. Cummings (eds.), *Research in Organizational Behavior, 2,* JAI Press, 81–127.

Katz, R. (1981). Managing careers: The influence of job and group longevities. In R. Katz (ed.), *Career Issues in Human Resource Management.* New York: Prentice-Hall.

Katz, R., and Tushman, M. L. (1979). Communication patterns, project performance and task characteristics: An empirical evaluation and integration in an R & D setting. *Organizational Behavior and Human Performance, 23,* 139–162.

Katz, R., and Tushman, M. L. (1981). An investigation into the managerial roles and career paths of gatekeepers and project supervisors in a major R & D facility. *R & D Management, 11: 3,* 103–110.

Lawrence, P. R., and Lorsch, J. W. (1967). *Organization and Environment.* Boston: Harvard Business School.

Marquis, D., and Straight, D. (1965). *Organizational Factors in Project Performance.* M.I.T. Sloan School of Management Working Paper No. 133–65.

Menzel, H. (1965). Information needs and uses in science and technology. In C. Cuarda (ed.), *An-*

nual Review of Information Science and Technology, New York: Wiley.

Pelz, D.C., and Andrews, F. M. (1976). *Scientists in Organizations.* Revised Edition, Ann Arbor: University of Michigan Press.

Pfeffer, J. (1981). Management as symbolic action: The creation and maintenance of organizational paradigms. In L. L. Cummings and B. Staw (eds.), *Research in Organizational Behavior, 3,* JAI Press.

Shepard, H. A. (1956). Creativity in R & D teams. *Research and Engineering,* 10–13.

Smith, C. G. (1970). Age of R and D groups: A reconsideration. *Human Relations, 23,* 81–92.

Staw, B., and Ross, J. (1978). Commitment to a policy decision: A multi-theoretical perspective. *Administrative Science Quarterly, 23,* 140–146.

Tomlin, B. (1981). Inter-location technical communication in a geographically dispersed research organization. *R & D Management, 11:* 1, 19–23.

Tukey, J. W. (1977). *Exploratory Data Analysis.* Reading, Mass: Addison-Wesley.

Tushman, M. L. (1977). Communication across organizational boundaries: Special boundary roles in the innovation process. *Administrative Science Quarterly, 22,* 587–605.

Tushman, M. L., and Katz, R. (1980). External communication and project performance: An investigation into the role of gatekeepers. *Management Science, 26,* 1071–1085.

Weick, K. E. (1969). *The Social Psychology of Organizing,* Reading, Mass: Addison-Wesley.

Whitley, R., and Frost, P. (1973). Task type and information transfer in a government research lab. *Human Relations, 25,* 537–550.

ACKNOWLEDGMENT

This research was sponsored by a grant from the Department of the Army DASG-60-77-C-0147.

Staffing the Innovative Technology-Based Organization

Edward B. Roberts

Alan R. Fusfeld

In this article, the authors identify and describe five informal but critical behavioral functions, which are needed for effective execution of technology-based innovative projects. They observe that some individuals are capable of performing concurrently more than one of these critical functions. They also find that a person's role in the innovation process may change over the course of his or her career. The authors discuss the managerial implications of their findings, particularly with respect to manpower planning, objective setting, and performance measurement and rewards. They then illustrate how an organizational assessment can be carried out. Finally, they discuss how critical functions concepts may be appropriate in other types of organizations.

This article examines the technology-based innovation process in terms of certain behavioral functions. These functions are usually informal, but they are critical. They can be the key to an effective organizational base for innovation. This approach to the innovation process is similar to that taken by early industrial theorists, such as Frederick W. Taylor who focused on organizing efficiently the production process. However, examination of how industry has organized its innovation tasks—those tasks needed for product or process development and for responses to nonroutine demands—indicates an absence of comparable theory. Many corporations' attempts to innovate consequently suffer from ineffective management and inadequately staffed organizations. Yet, through studies conducted largely in the last fifteen years, we now know much about the activities that are requisite to innovation. We also know about the characteristics of the people who perform these activities most effectively.

THE INNOVATION PROCESS

The major steps involved in the technology-based innovation process are shown in Figure 1. Although the project activities do not neces-

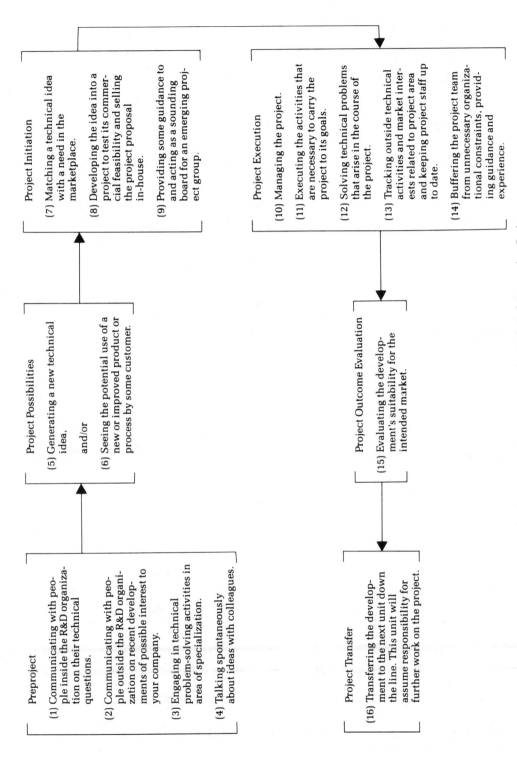

Preproject

(1) Communicating with people inside the R&D organization on their technical questions.

(2) Communicating with people outside the R&D organization on recent developments of possible interest to your company.

(3) Engaging in technical problem-solving activities in area of specialization.

(4) Talking spontaneously about ideas with colleagues.

Project Possibilities

(5) Generating a new technical idea,

and/or

(6) Seeing the potential use of a new or improved product or process by some customer.

Project Initiation

(7) Matching a technical idea with a need in the marketplace.

(8) Developing the idea into a project to test its commercial feasibility and selling the project proposal in-house.

(9) Providing some guidance to and acting as a sounding board for an emerging project group.

Project Execution

(10) Managing the project.

(11) Executing the activities that are necessary to carry the project to its goals.

(12) Solving technical problems that arise in the course of the project.

(13) Tracking outside technical activities and market interests related to project area and keeping project staff up to date.

(14) Buffering the project team from unnecessary organizational constraints, providing guidance and experience.

Project Outcome Evaluation

(15) Evaluating the development's suitability for the intended market.

Project Transfer

(16) Transferring the development to the next unit down the line. This unit will assume responsibility for further work on the project.

FIGURE 1.   A Multi-stage View of a Technical Innovation Project

311

sarily follow each other in a linear fashion, there is more or less clear demarcation between them. Each stage and its activities, moreover, require a different mix of "people" skills and behaviors to be carried out effectively.

The figure portrays six stages as occurring in the typical technical innovation project. It also shows sixteen representative activities that are associated with innovative efforts. The six stages are identified as:

1. Preproject;
2. Project possibilities;
3. Project initiation;
4. Project execution;
5. Project outcome evaluation; and
6. Project transfer.

These stages often overlap and frequently recycle.[1] For example, problems or findings that are generated during project execution may cause a return to project initiation activities. Outcome evaluation can restart additional project execution efforts. And, of course, project cancellation can occur during any of these stages, thus redirecting technical endeavors back into the preproject phase.

A variety of different activities are undertaken during each of the six stages. Some of the activities, such as generating new technical ideas, arise in all innovation project stages, from preproject to project transfer. Our research studies and consulting efforts in dozens of companies and government labs, however, have shown other activities to be concentrated in specific stages, as discussed below.

Preproject.   Before formal project activities are undertaken in a technical organization, considerable technical work is done, which provides a basis for later innovation efforts. Scientists, engineers, and marketing people find themselves involved in discussions that are internal and external to the organization. Ideas are discussed in rough-cut ways and broad parameters of innovative interest are established. Techni-

cal personnel work on problem-solving efforts to advance their own areas of specialization. Our discussions with numerous industrial firms in the U.S. and Europe suggest that from 30 to 60 percent of all technical effort is devoted to work outside of or prior to formal project initiation.

Project Possibilities.   Specific ideas for possible projects arise from the preproject activities. They may be technical concepts for developments that are assumed to be feasible. They may also be perceptions of possible customer interests in product or process changes. Customer-oriented perspectives may originate with technical, marketing, or managerial personnel, who develop these ideas out of their own imaginations or from direct contact with customers or competitors. Recent evidence indicates that many of these ideas enter as "proven" possibilities inasmuch as these ideas have already been developed by the customers themselves.[2]

Project Initiation.   As ideas evolve through technical and marketing discussions and exploratory technical efforts, the innovation process moves into a more formal project initiation stage. Activities occurring during this phase include attempts to match the directions of technical work with perceived customer needs. (Of course, such customer needs may exist either in the production organization or in the product marketplace.) Inevitably, a specific project proposal is written, proposed budgets and schedules are produced, and informal pushing as well as formal presentations are undertaken in order to sell the project. A key input during this stage is the counseling and encouragement that senior technical professionals or lab and marketing management may provide to the emerging project team.

Project Execution.   When the project is approved formally, activities increase in intensity and focus. Usually, someone undertakes planning, leadership, and coordinating efforts.

These efforts are related to the many continuing activities of the engineers and scientists assigned to the project. These activities include problem solving and the generation of technical ideas. Technical people make special attempts to monitor (and transfer in) the results of previous activity as well as relevant external information. Management or marketing people take a closer look at competitors and customers to be sure the project is appropriately targeted.[3] Senior people try to protect the project from being controlled too tightly or from being cut off prematurely. The project manager and other enthusiasts fight to defend their project's virtues (and budget). Unless cancelled, the project continues toward completion of its objectives.

Project Outcome Evaluation. When the technical effort seems complete, most projects undergo another intense evaluation to see how the results compare with prior expectations and current market perceptions. If successful innovation is to occur, some further implementation must take place. The interim results are either transferred to manufacturing (where they are either embodied in the manufacturing process or produced in volume) or transferred to further stages of development. All such later stages involve heavier expenditures. The project outcome evaluation can then be viewed as a way to screen projects prior to their possible transfer into these later stages.

Project Transfer. If the project results survive this evaluation, transfer efforts take place (e.g., from central research to product department R&D, or from development to manufacturing engineering).[4] The project's details may require further technical documentation to facilitate the transfer. Key technical people may be shifted to the downstream unit to transfer their expertise and enthusiasm, since downstream staff members in the technical or marketing areas often need instruction to assure effective continuity. Within the downstream organizational unit, the cycle of stages may begin again, perhaps bypassing the earliest two stages and starting with project initiation or even project execution. This "pass down" continues until successful innovation is achieved, unless project termination occurs first.

NEEDED ROLES

Assessment of activities involved in the several-stage innovation process, as just described, points out that the repeated direct inputs of five different work roles are critical to innovation. The five roles arise in differing degrees in each of the several steps. Furthermore, different innovation projects obviously call for variations in the required role mix at each stage. Nevertheless, all five work roles must be carried out by one or more individuals if the innovation is to pass effectively through all six steps. The five critical work functions are:

- *Idea Generating:* Analyzing or synthesizing information about markets, technologies, approaches, or procedures, from which is generated an idea for a new or improved product or service, a new technical approach or procedure, or a solution to a challenging technical problem.[5] The analysis or synthesis may be implicit or explicit; the information may be formal or informal.
- *Entrepreneuring or Championing:* Recognizing, proposing, pushing, and demonstrating a new technical idea, approach, or procedure for formal management approval.[6]
- *Project Leading:* Planning and coordinating the diverse sets of activities and people involved in moving a demonstrated idea into practice.[7]
- *Gatekeeping:* Collecting and channeling information about important changes in the internal and external environments. Information gatekeeping can be focused on developments in the market, in manufacturing, or in the world of technology.[8]

313

- *Sponsoring or Coaching:* Guiding and developing less experienced personnel in their critical roles; behind-the-scenes support, protection, advocacy, and sometimes "bootlegging" of funds.[9]

Lest the reader confuse these roles as mapping one-for-one with different people, three points need emphasis: (1) some roles, e.g., idea generating, frequently need to be fulfilled by more than one person in a project team in order for the project to be successful; (2) some individuals occasionally fulfill more than one of the critical functions; (3) the roles that people play periodically change over a person's career with an organization. The latter two points will be discussed in more depth later in this article.

Critical Functions

These five critical functions represent the various roles that must be carried out for successful innovation to occur. They are critical from two points of view. First, each role is unique and demands unique skills. A deficiency in any one of the roles contributes to serious problems in the innovation effort, as we shall illustrate below. Second, each role tends to be carried out primarily by relatively few individuals, thereby making the critical role players even more unique. If any one of these individuals leaves, the problem of recruiting a replacement is very difficult. The specific qualities needed in the replacement usually depend on unstated role requirements. Most critical functions cannot be fulfilled by new recruits to an organization.

We must add at this point that another role clearly exists in all innovative organizations, but it is not an *innovative* role. "Routine" technical problem solving must be carried out in order to advance innovative efforts. Indeed, the vast bulk of technical work is probably routine. It requires professional training and competence, to be sure, but it is nonetheless routine in character for an appropriately prepared individual. A large number of people in innovative organizations do very little critical functions work; others who are important performers of the critical functions also spend a good part of their time in routine problem-solving activity. Our estimate, supported now by data from numerous organizations, is that 70 to 80 percent of technical effort falls into this routine problem-solving category. But, the 20 to 30 percent that is unique and critical is the part we emphasize here.

Generally, the critical functions are not specified within job descriptions, since they tend to fit neither administrative nor technical hierarchies. But they represent necessary activities for R&D, such as problem definition, idea nurturing, information transfer, information integration, and program pushing. Consequently, these role behaviors are the underlying informal functions that an organization carries out as part of the innovation process. Beyond the five roles described earlier, different business environments may also demand that additional roles be performed in order to assure innovation.[10]

It is desirable for every organization to have a balanced set of abilities for carrying out these roles as needed. Unfortunately, few organizations have such a balanced set. Some organizations overemphasize one role (e.g., idea generating) and underplay another role (e.g., entrepreneuring). Technical organizations tend to assume that the necessary set of activities will somehow be performed. As a consequence, R&D labs often lack sensitivity to the existence and importance of these roles, which, for the most part, are not defined within the formal job structure. Yet, the way in which critical functions are encouraged and made a conscious part of technology management is probably an organization's single most important area of leverage for maintaining and improving effective innovation.

Impact of Role Deficiencies

Such an analytic approach to developing an innovative team has been lacking in the past. Consequently, many organizations suffer because one or more of the critical functions are not being performed adequately. Certain characteristic signs can provide evidence that a critical function is missing.

Idea generating is deficient if the organization is not thinking of new and different ways of doing things. However, when a manager complains of insufficient ideas, we commonly find the real deficiency to be that people are not aggressively entrepreneuring or championing ideas—either their own or others'. Pools of unexploited ideas that seldom come to managers' attention are evidence of an entrepreneuring deficiency.[11]

Project leading is suspect if schedules are not met, activities "fall through cracks" (e.g., coordinating with a supplier), people do not have a sense of the overall goal of their work, or units that are needed to support the work back out of their commitments. Project leading is most commonly recognized by the formal appointment of a project manager. In research, as distinct from development, this formal role is often omitted.

Gatekeeping is inadequate if news of changes in the market, technology, or government legislation comes without warning. It is also inadequate if needed information is not passed along to people within the organization. If, six months after the project is completed, you suddenly realize that you have just succeeded in reinventing a competitor's wheel, your organization is deficient in needed gatekeeping! Gatekeeping is further lacking when the wheel is invented just as a regulatory agency outlaws its use.

Inadequate or inappropriate sponsoring or coaching often explains how projects are pushed into application too soon, why project managers have to spend too much time defending their work. It also explains why personnel complain that they do not know how to "navigate the bureaucracy" of their organizations.

The importance of each critical function varies with the development stage of the project. Initially, idea generation is crucial. Later, entrepreneurial skill and commitment are needed to develop the concept into a viable activity. Once the project is established, good project leadership is needed to guide its progress. Of course, the need for a critical function does not abruptly appear and disappear. Instead, the need grows and diminishes. Each function is the focus at some points, but it is of lesser importance at others. Thus, the absence of a function at a time when it is potentially important is a serious weakness, regardless of whether or not the role had been filled at an earlier, less crucial time. As a corollary, assignment of an individual to a project, at a time when the critical role that he or she provides is not needed, leads to frustration for the individual and to a less effective project team.

Frequently, we have observed that personnel changes that occur because of career development programs often remove critical functions from a project at a crucial time. Although these roles are usually performed informally, job descriptions are made in terms of technical specialties. Thus, personnel replacements are chosen on the basis of their technical qualifications rather than on their ability to fill the needs of the vacated critical roles. Consequently, the project team's innovative effectiveness is reduced, sometimes to the point of affecting the project's success.

CHARACTERISTICS OF THE ROLE PLAYERS

Compilation of several thousand individual profiles of staff in R&D and engineering organizations has demonstrated patterns in the characteristics of the people who perform each innovation function.[12] These patterns are shown in Table 1. The table indicates which persons are predisposed to be more interested in one type of activity than another and to perform certain types of activities well. For example, a person who is comfortable with abstractions and theory might feel more suited to the idea-generating function than would someone who is very practical. In any unit of an organization, people with different characteristics can work to complement each other. For instance, a person who is effective at generating ideas can be teamed with a colleague who is good at gate-keeping and with another colleague who has good entrepreneurial abilities. Of course, each person must understand his or her own expected role in a project and must appreciate the roles of others in order for the teaming process to be successful. As will be discussed later, some people have sufficient breadth to perform well in multiple roles.

Table 1 underlies our conclusion that each of the several roles required for effective technical innovation presents unique challenges and must be filled with different types of people. Each type must be recruited, managed, and supported differently; offered different sets of incentives; and supervised with different types of measures and controls. However, most technical organizations seem not to have grasped this concept. The result is that all technical people tend to be recruited, hired, supervised, monitored, evaluated, and encouraged as if their principal roles were those of creative scientists, or, worse yet, of routine technical problem solvers. In fact, only a few of these people have the personal and technical qualifications for scientific inventiveness and prolific idea generating. A creative, idea-generating scientist or engineer is a special kind of professional. This person needs to be singled out, cultivated, and managed in a special way. He or she is probably innovative, technically well educated, and enjoys working on advanced problems, often as a "loner."

The technical champion or entrepreneur is a special person, too. He or she shows creativity, but it is an aggressive form of creativity that is appropriate for selling an idea or product. The entrepreneur's drives may be less rational and more emotional than those of the creative scientist; he or she is committed to achieving but is less concerned about how to do so. This person is as likely to pick up and successfully champion someone else's original idea as to push something of his or her own creation. Such an entrepreneur may well have a broad range of interests and activities. He or she must be recruited, hired, managed, and stimulated very differently from the way an idea-generating scientist is treated in the organization.

The person who effectively performs project leading or project managing activities is yet a different kind of person. He or she is an organized individual, is sensitive to the needs of the several different people who are being coordinated, and is an effective planner. The ability to plan is especially important if long lead time, expensive materials, and major support are involved in the project development.

The information gatekeeper is a communicative individual and is the exception to the truism that engineers do not read (especially that they do not read technical journals). Gatekeepers provide links to sources of the technical information which flows into and within a research and development organization and which can enhance new product development or process improvement. But those who do research and development need market information as well as technical information: What do customers seem to want? What are competitors providing? How might regulatory shifts affect the firm's present or contemplated products or processes? For answers to questions such as these, research and development organizations need people we call the "market gatekeepers." These people are engineers, scientists, or possibly marketing people with technical backgrounds who focus on market-related information sources and communicate effectively to

TABLE 1. Critical Functions in the Innovation Process

Critical Function	Personal Characteristics	Organizational Activities
Idea Generating	Expert in one or two fields. Enjoys conceptualization; comfortable with abstractions. Enjoys doing innovative work. Usually is an individual contributor. Often will work alone.	Generates new ideas and tests their feasibility. Good at problem solving. Sees new and different ways of doing things. Searches for the breakthroughs.
Entrepreneuring or Championing	Strong application interests. Possesses a wide range of interests. Less propensity to contribute to the basic knowledge of a field. Energetic and determined; puts self on the line.	Sells new ideas to others in the organization. Gets resources. Aggressive in championing his or her "cause." Takes risks.
Project Leading	Focus for decision making, information, and questions. Sensitive to the needs of others. Recognizes how to use the organizational structure to get things done. Interested in a broad range of disciplines and in how they fit together (e.g., marketing, finance).	Provides the team leadership and motivation. Plans and organizes the project. Insures that administrative requirements are met. Provides necessary coordination among team members. Sees that the project moves forward effectively. Balances the project goals with organizational needs.
Gatekeeping	Possesses a high level of technical competence. Is approachable and personable. Enjoys the face-to-face contact of helping others.	Keeps informed of related developments that occur outside the organization through journals, conferences, colleagues, other companies. Passes information on to others; finds it easy to talk to colleagues. Serves as an information resource for others in the organization (i.e., authority on who to see or on what has been done). Provides informal coordination among personnel.
Sponsoring or Coaching	Possesses experience in developing new ideas. Is a good listener and helper. Can be relatively objective. Often is a more senior person who knows the organizational ropes.	Helps develop people's talents. Provides encouragement, guidance, and acts as a sounding board for the project leader and others. Provides access to a power base within the organization—a senior person. Buffers the project team from unnecessary organizational constraints. Helps the project team to get what it needs from the other parts of the organization. Provides legitimacy and organizational confidence in the project.

317

their technical colleagues. Such individuals read trade journals, talk to vendors, go to trade shows, and are sensitive to competitive information. Without them, many research and development projects and laboratories become misdirected with respect to market trends and needs.

Finally, the sponsor or coach is, in general, a more experienced, older project leader or former entrepreneur, who has a "softer touch" than when he or she was first in the organization. As a senior person, he or she can coach and help subordinates in the organization and can speak on their behalf to top management. This activity makes it possible for ideas or programs to move forward in an effective, organized fashion. Many organizations totally ignore the sponsor role, yet our studies of industrial research and development suggest that many projects would not have been successful were it not for the subtle and often unrecognized assistance of such senior people acting in the role of sponsors. Indeed, organizations are most successful when chief engineers or laboratory directors naturally behave in a manner consistent with this sponsor role.

The significant point here is that the staffing needed for effective innovation in a technology-based organization is far broader than the typical research and development director usually has assumed. Our studies indicate that many ineffective technical organizations have failed to be innovative solely because one or more of these five quite different critical functions has been absent.

All of these roles can be fulfilled by people from multiple disciplines and departments. Obviously, technical people—scientists and engineers—might carry out any of the roles. But, marketing people also generate ideas for new and improved products, act as gatekeepers for information of key importance to a project (especially about use, competition, and regulatory activities), champion the idea, sometimes spon-sor projects, and sometimes even manage new projects, especially for new product development. Manufacturing people periodically fill similar critical roles, as do general management personnel.

Multiple Roles

As indicated earlier, some individuals have the skills, breadth, inclination, and job opportunity to fulfill more than one critical function in an organization. Our data collection efforts with R&D staffs show that a few clusters explain most of these cases of multiple role playing. One common combination of roles is the pairing of gatekeeping and idea generating. Idea-generating activity correlates, in general, with the frequency of person-to-person communication, especially with that which is external to the organization.[13] Moreover, the gatekeeper, who is in contact with many sources of information, can often connect synergistically these bits into a new idea. This ability seems especially true of market gatekeepers who can relate market relevance to technical opportunities.

Another role couplet is between entrepreneuring and idea generating. In studies of formation of new technical companies, the entrepreneur who pushed company formation and growth was found in half the cases also to have been the source of the new technical idea underlying the company.[14] Furthermore, in studies of M.I.T. faculty, 38 percent of those who had ideas that they perceived to be of commercial value also took strong entrepreneurial steps to exploit their ideas.[15] The idea generating-entrepreneuring pair accounts for slightly less than one-half the entrepreneurs.

Entrepreneuring individuals often become project leaders. This progression is thought to be a logical organizational extension of the effort of effectively "selling" the idea for the project. Some people who are strong at entrepreneuring also have the interpersonal and plan-oriented qualities needed for project leading. The responsibility for managing a project, though, is often mistakenly seen as a necessary reward for successful idea championing. This mistake arises from a lack of attention to the functional differences between the two roles. One should not necessarily assume that a good salesman will be a good manager. If the entrepreneur can be rewarded appropriately and more directly for his or her own function, many project failures caused by ineffective project managers might be avoided. Perhaps giving the entrepreneur a prominent project role, while clearly designating a different project manager, might be an acceptable compromise.

Finally, sponsoring occasionally evolves into a takeover of any or all of the other roles, even though it should be a unique role. Senior coaching can degenerate into idea domination, project ownership, and direction from the top. This confusion of roles can become extremely harmful to the entire organization: Who will bring another idea to the boss once he steals some junior's earlier concept? Even worse, who can intervene to stop the project once the boss is running amok with his new pet?

The performance of multiple roles can affect the minimum size group needed for attaining "critical mass" in an innovative effort. To achieve continuity of a project, from initial idea all the way through to successful commercialization, a project group must effectively fill all five critical roles. It must also satisfy the specific technical skill requirements for project problem solving. In a new, high-technology company, this critical mass may sometimes be ensured by as few as one or two cofounders. Similarly, an elite team—such as Cray's famed Control Data computer design group, Kelly Johnson's "skunk works" at Lockheed, or McLean's Sidewinder missile organization in the Navy's China Lake R&D center—may concentrate in a small number of select multiple-role players the staff needed to accomplish major objectives. But, the more typical medium-to-large company had better not plan on finding Renaissance persons or superstars to fill its job requirements. Staffing assumptions should more likely rest on estimates that 70 percent of scientists and engineers will turn out to be routine problem solvers only, and that even most critical role players will be single dimensional in their unique contributions.

Career-Spanning Role Changes

We showed above how some individuals fulfill multiple critical roles concurrently or in different stages of the same project. Even more people are likely to contribute critically but differently at different stages of their careers. This difference over time does not reflect change of personality, although such changes do seem partly due to the dynamics of personal growth and development. The phenomenon also clearly reflects individual responses to differing organizational needs, constraints, and incentives.

For example, let's consider the hypothetical case of a bright, aggressive, young engineer who has just joined a company upon graduation. What roles can he play? Certainly, he can quickly become effective at solving routine technical problems and, hopefully, at generating novel ideas. But, even though he may know many university contacts and be familiar with the outside literature, he can't be an effective information gatekeeper, for he doesn't yet know the people inside the company with whom he might communicate. Nor can he lead project activities, since no one would trust him in that role. He can't effectively serve as entrepreneur, as he has no credibility as champion for change. And, of course, sponsoring is out of the question. During this stage of his career, the limited legitimate role options may channel the young engineer's productive energies and reinforce his tendencies toward the output of creative ideas.

Alternatively, if he wants to offer and do more than the organization will allow, this high-

319

potential young performer may feel rebuffed and frustrated. His perception of what he can expect from the job and, perhaps more importantly, what the job will expect from him, may become set in these first few months on the job. Though he may remain with the company, he will "turn off" in disappointment from his previously enthusiastic desire to make multidimensional contributions. More likely, he will leave the company in search of a more rewarding job. He will perhaps be destined to find continuing frustration in his next one or two encounters. For many young professionals, the job environment moves too slowly from the stage of encouragement of idea generating to a time when entrepreneuring is even permitted.

The engineer's role options may broaden after two or three years on the job, however. Though routine problem solving and idea generating are still appropriate, some information gatekeeping may now also be possible as communication ties increase within the organization. Project leading may start to be seen as legitimate behavior, particularly on small efforts.[16] The young engineer's work behavior may begin to reflect these new possibilities. Nevertheless, his attempts at entrepreneurial behavior might still be seen as premature and sponsoring as still an irrelevant consideration.

After another few years at work, the role options are still wider. Routine problem solving, continued idea generating, broad-based gatekeeping (even bridging to the market or to manufacturing), responsible project managing, and project championing may become reasonable alternatives. Even coaching a new employee becomes a possibility. Though most people tend usually to focus on one of these roles (or on a specific multiple-role combination) during this midcareer period, the next several years can strengthen all these role options.

Losing touch with a rapidly changing technology may later narrow the available role alternatives as the person continues in his or her job. Technical problem-solving effectiveness may diminish in some cases, idea generating may slow down or stop, and technical information gatekeeping may be reduced. Market or manufacturing gatekeeping, however, may continue to improve with increased experience and outside contacts. Project managing capabilities may continue to grow as he or she tucks more projects under his or her belt. Entrepreneuring may be more important and for higher stakes. Sponsoring of juniors in the company may be more generally sought and practiced. This career phase is too often seen to be characterized by the problem of technical obsolescence, especially if the organization has a fixation on assessing engineer performance in terms of the narrow but traditional stereotypes of technical problem solving and idea generating. Channeling the engineer into a role that is more appropriate for an earlier stage in his or her career can be a source of mutual grief to both the organization and the individual. Such a role will be of little current interest and satisfaction to the more mature, broader, and now differently directed professional. An aware organization, thinking in terms of critical role differences, can instead recognize the self-selected branching in career paths that has occurred for the individual. Productive, technically trained people can carry out critical functions for their employers up to retirement if employers encourage the full diversity of vital roles.

At each stage of his or her evolving career, the individual can encounter severe conflicts between the organization's expectations and his or her personal work preferences. This conflict is especially likely if the organization is inflexible in its perception of appropriate technical roles. In contrast, if both the organization and the individual are adaptable in seeking mutually satisfying job roles, the engineer can contribute continuously and significantly to innovation. As suggested in this illustrative case, in the course of a productive career in industry, the technical professional may begin as a technical problem solver, spend several years primarily as a creative idea generator, and add technical gatekeeping to his or her repertoire while maintaining his or her earlier roles. He or she may then begin to serve as a project entrepreneur and lead projects forward. Gradually, he or she will develop greater market linking and project managing skills and eventually will assume senior sponsoring role, maintaining a position of project, program, or organizational leadership until retirement. This fully productive career would not be possible if the engineer were pushed to the side early as a technically obsolete contributor. The perspective taken here can lead to a very different approach to career development for professionals than is usually taken by industry or government.

MANAGING THE CRITICAL FUNCTIONS FOR ENHANCED INNOVATION

To increase organizational innovation, a number of steps can be taken to facilitate a balance of time and energy among the critical functions. These steps must be addressed explicitly or organizational focus will remain on the traditionally visible functions, such as problem solving, which produce primarily near-term incremental results. Indeed, the results-oriented reward systems of most organizations reinforce this short-run focus, causing the other, more significant activities to go unrecognized and unrewarded.[17]

Implementation of the results, language, and concepts of a critical functions perspective is outlined below for the selected organizational tasks of manpower planning, job design, and selection of measurement and rewards. If managers thought in critical functions terms, other tasks, not dealt with here, would also be

carried out differently. These tasks include R&D strategy, organizational development, and program management.

Manpower Planning

The critical functions concept can be applied usefully to the recruiting, job assignment, and development or training activities within an organization. In recruiting, for example, an organization needs to identify not only the specific technical or managerial requirements of a job, but also the critical function activities that the job infers, e.g., the organization needs to ask whether the job requires that less experienced personnel be coached and developed in order to insure the longer-run productivity of that area. If the job requires entrepreneuring, then the applicant who is more aggressive and has shown evidence of championing new ideas in the past should be preferred over the less aggressive applicant who has shown more narrowly technically oriented interests in the past.

Industry, at best, has taken a narrow view of manpower development alternatives for technical professionals. The "dual ladder" concept envisions an individual rising along either scientific or managerial steps. Attempted by many but with only limited success ever attained, the dual ladder reflects an oversimplification and distortion of the key roles needed in an R&D organization.[18] As a minimum, the critical function concept presents "multiladders" of possible organizational contribution; individuals can grow in any or all of the critical roles, while benefiting the organization. Depending on an organization's strategy and manpower needs, manpower development along each of the paths can and should be encouraged. Furthermore, there is room for individual growth and development from one function to another, as people are exposed to different managers, different environments, and jobs that require different activities.

Job Design and Objective Setting

Most job descriptions and statements of objectives emphasize problem solving and sometimes project leading. Rarely do job descriptions and objectives take into account the dimensions of a job that are essential for the performance of the other critical functions. Yet, the availability of unstructured time in a job, for example, can influence the performance of several of the innovation functions, and it needs to be designed into corresponding jobs. To stimulate idea generating, some slack time is necessary so that employees can pursue their own ideas and explore new and interesting ways of doing things. For gatekeeping to occur, slack time needs to be available for employees to communicate with colleagues and pass along information learned, both internal to and external to the organization. The coaching role also requires slack time, during which the "coach" can guide less experienced personnel.[19]

Essential activities for filling alternative roles also need to be included explicitly in a job's objectives. An important goal for a gatekeeper, for example, should be to provide useful information to colleagues. A person who has the attitudes and skills to be an effective champion or entrepreneur could be given responsibility for recognizing good new ideas. This person might have the character to roam around the organization, talk with people about their ideas, and encourage their pursuit of these ideas. He could even pursue these ideas himself.[20]

Performance Measures and Rewards

We all tend to do those activities that will be rewarded. If personnel perceive that idea generating will not be recognized but that idea exploitation will, they may withhold their ideas from those who can exploit them. They may try to exploit ideas themselves, no matter how unequipped or uninterested they are in carrying out the exploitation activity.

For this reason, it is important to recognize the distinct contributions of each of the separate critical functions. Table 2 identifies some measures relevant for each function, indicating both quantity and quality dimensions. For example, an objective for a person who has the skills and information to be effective at gatekeeping could be to help a number of people during the next twelve months. At the end of that time, his or her manager could survey the people who the gatekeeper felt he or she had helped to assess the gatekeeper's effec-

TABLE 2.   *Measuring and Rewarding Critical Function Performance*

Dimension of Management	Critical Function Idea Generating	Entrepreneuring or Championing	Project Leading	Gatekeeping	Sponsoring or Coaching
Primary contribution of each function for appraisal of performance	Quantity and quality of ideas generated.	Ideas picked up; percent carried through.	Project technical milestones accomplished; cost/schedule constraints met.	People helped; degree of help.	Success in developing staff; extent of assistance provided.
Appropriate rewards	Opportunities to publish; recognition from professional peers through symposia, etc.	Visibility; publicity; further resources for project.	Bigger or more significant projects; material signs of organization status.	Travel budget; key "assists" acknowledged; increased autonomy and use for advice.	Increased autonomy; discretionary resources for support of others.

322

tiveness in communicating key information. In each organization, the specific measures chosen will necessarily be different.

Rewarding an individual for the performance of a critical function makes the function more manageable and open to discussion. However, what is perceived as rewarding for one function may be seen as less rewarding, neutral, or even negative for another function because of the different personalities and needs of those filling the roles. Table 2 presents some rewards that seem appropriate for each function. Again, organizational and individual differences will generate variations in the rewards selected. Of course, the informal positive feedback of managers in their day-to-day contacts is a major source of motivation and recognition for any individual performing a critical innovation function, or any job for that matter.

Salary and bonus compensation are not included here, but not because they are unimportant to any of these people. Financial rewards should be employed as appropriate, but they do not seem to be linked explicitly to any one innovative function more than to another.

PERFORMING A CRITICAL FUNCTIONS ASSESSMENT

The preceding sections demonstrate that the critical functions concept provides an important way to describe an organization's resources for effective innovation activity. To translate this concept into an applied tool, one needs to be able to assess the status of an R&D unit in terms of critical functions. Such an assessment potentially provides two important types of information: (1) inputs for management evaluations of the organization's ability to achieve goals and strategy; and (2) assistance to R&D managers and professionals in performance evaluation, career development, and more effective project performance.

Method of Approach

The methodology chosen for a critical functions assessment is contingent on the situation. From experience gained with a dozen companies and government agencies in North America, the authors have found the most flexible approach to be a series of common questionnaires, which are developed from replicated academic research techniques on innovative contributors and modified as needed for the situation. Questionnaires are supplemented by a number of structured interviews or workshops. Data are collected and organized in a framework that represents: (a) the critical functions; (b) special characteristics of the organization's situation; (c) additional critical functions required in the specific organization; and (d) the climate for innovation provided by management. The results include a measure of an organization's current and potential strengths in each critical function; an evaluation of the compatibility of the organization's R&D strategy with these strengths; and a set of personnel development plans for both management and staff that support the organization's goals. This information is valuable for both the organization and the individual.[21]

Some Actions Taken in One Firm

As a result of a critical functions analysis in a company, multiple actions are usually taken. In order to consider some of the typical steps, we draw here from the outcomes implemented in one medium-sized R&D organization. The first action was that every first line supervisor and above, after some training, discussed with each employee the results of the employee's critical functions survey. (In other companies, employee anonymity has been preserved; data were returned only to the individual. In these companies, employees frequently have used the results to initiate discussions with their immediate supervisors regarding job fit and career development.) The purpose of the discus-

323

sion was twofold: to look for differences in how the employee and his or her boss each perceived the employee's job skills; and to engage in developmental career planning. The vocabulary of the critical functions plus the tangible feedback gave the manager and the employee a meaningful, commonly shared basis for the discussion.

Several significant changes resulted from these discussions. A handful of the staff recognized the mismatch between their present jobs and skills. With the support of their managers, job modifications were made. Another type of mismatch that this process revealed was between the manager's perception of the employee's skills and the employee's own perception. Most of the time the manager was underutilizing his or her human resources.

In this particular firm, the data also prompted action to improve the performance of the project leading function. An insufficient number of people saw themselves performing this function. Moreover, they saw themselves as lacking skills in this area. As a result of these deficiencies, upper management conducted several "coaching" sessions, worked to further clarify roles, and showed increased support for project leadership efforts.

Important changes also were made in how the technical organization recruited. The characteristic strengths behind each critical function were explicitly employed in identifying the skills necessary to do a particular job. This analysis led to a useful framework for interviewing candidates. It helped determine how the candidates might fit into and grow within the present organization. Upper management also became conscious of the unintended bias in the recruiting procedure. This bias was introduced both by the universities at which the company recruited and by the recruiters themselves. (In this case, the senior researchers, who conducted most of the interviews, were primarily interested in idea generating.) As a result of the analyses, upper management was careful to

have a mix of the critical functions represented by the people who interviewed job candidates.

The analyses led to other results that were less tangible than the above but equally important. Jobs were no longer defined solely in technical terms, i.e., in terms of required educational background or work experience. For example, if a job involved idea generation, the necessary skills and the typical activities for that critical function were included in the description of the job. Furthermore, the need for a new kind of teamwork developed since it was rare that any single person could perform effectively all five of these essential functions. Finally, the critical functions concept provided the framework for the selection of people and division of labor on the innovation team that became the nucleus for all new R&D programs.

CONCLUSION

We have examined the technology-based innovation process in terms of a set of informal but critical behavioral functions. Five critical roles have been identified within the life cycle of activities in an R&D project. These roles are idea generating, entrepreneuring or championing, project leading, gatekeeping, and sponsoring or coaching. In our surveys of numerous North American R&D and engineering organizations, we have made two key observations: some unique individuals are able to perform concurrently more than one of the critical roles; and patterns of roles for an individual often change over the course of his or her productive work career.

These critical functions concepts have managerial implications in such areas as manpower planning, job design, objective setting, and performance measurement and rewards. They provide a conceptual basis for design of a more effective multiladder system to replace many R&D organizations' ineffectual dual ladder systems.

Several years of development, testing, and discussion of this critical functions perspective have also led to applications outside of R&D organizations. We have seen the perspective extended to such areas as computer software development and architectural firms. Recent discussions with colleagues suggest an obvious appropriateness for marketing organizations. A more difficult translation is expected in the areas of finance and manufacturing. To the extent that innovative outcome rather than routine production is the output sought, we have confidence that the critical functions approach will afford useful insights for organizational analysis and management.

REFERENCES

1. For a different and more intensive quantitative view of project life cycles, see E. B. Roberts, *The Dynamics of Research and Development* (New York: Harper & Row, 1964).
2. See E. von Hippel, "Users as Innovators," *Technology Review,* January 1978, pp. 30–39.
3. For issues that need to be highlighted in a competitive technical review, see A. R. Fusfeld, "How to Put Technology into Corporate Planning," *Technology Review* 80.
4. For further perspectives on project transfer, see E. B. Roberts, "Stimulating Technological Innovation: Organizational Approaches," *Research Management* (November 1979) pp. 26–30.
5. See D. C. Pelz and F. M. Andrews, *Scientists in Organizations* (New York: John Wiley & Sons, 1966).
6. See E. B. Roberts, "Entrepreneurship and Technology," *Research Management* (July 1968): 249–266.
7. See D. G. Marquis and I. M. Rubin, "Management Factors in Project Performance" (Cambridge, MA: M.I.T. Sloan School of Management, Working Paper, 1966).

8. See: T. J. Allen, *Managing the Flow of Technology* (Cambridge, MA: The MIT Press, 1977); R. G. Rhoades et al., "A Correlation of R&D Laboratory Performance with Critical Functions Analysis," *R&D Management,* October 1978, pp. 13–17. Our empirical studies have pointed out three different types of gatekeepers: (1) technical—relates well to the advancing world of science and technology; (2) marketing—senses and communicates information relating to customers, competitors, and environmental and regulatory changes affecting the marketplace; and (3) manufacturing—bridges the technical work with the special needs and conditions of the production organization. See Rhoades et al. (October 1978).
9. See Roberts (July 1968): 252.
10. One role we have observed frequently is the "quality controller" who stresses high work standards in projects. Other critical roles relate more to organizational growth than to innovation, e.g., the "effective trainer" who could absorb new engineers productively into the company, seen as critical to one firm that was growing 30 percent per year.
11. One study that demonstrated this phenomenon is N. R. Baker et al., "The Effects of Perceived Needs and Means on the Generation of Ideas for Industrial Research and Development Projects," *IEEE Transactions on Engineering Management,* EM-14 (1967): 156–165.
12. Section VI describes a methodology for collecting these data.
13. See Allen (1977).
14. See Roberts (July 1968).
15. See E. B. Roberts and D. H. Peters, "Commercial Innovations from University Faculty," *Research Policy,* in press.
16. One study showed that engineers who eventually became managers of large projects began supervisory experiences within an average of 4.5 years after receiving their B.S. degrees. See I. M. Rubin and W. Seelig, "Experience as a Factor in the Selection and Performance of Project Managers," *IEEE Transactions on Engineering Management,* EM-14 (September 1967): 131–135.
17. For further perspectives on the consequences of this short-run view by U.S. managers, see R.

H. Hayes and W. J. Abernathy, "Managing Our Way to Economic Decline," *Harvard Business Review,* July-August 1980, pp. 67–77.

18. For a variety of industrial approaches to the dual ladder, see the special July 1977 issue of *Research Management* or, more recently, *Research Management,* November 1979, pp. 8–11.

19. In a more macroscopic way, March and Simon observed years ago that innovation could only occur in the presence of organizational slack. See J. G. March and H. A. Simon, *Organizations* (New York: John Wiley & Sons, 1958).

20. For more details on various job design dimensions appropriate to the critical functions, see E. B. Roberts and A. R. Fusfeld, "Critical Functions: Needed Roles in the Innovation Process," in *Career Issues in Human Resource Management,* ed. R. Katz (Englewood Cliffs, NJ: Prentice-Hall, forthcoming).

21. For samples of questionnaire items, more details on diagnostic uses of the resulting data, and numerical outputs from one company's assessment, see Roberts and Fusfeld (forthcoming).

The Role of Japan-Based R & D in Global Technology Strategy

D. Eleanor Westney

Kiyonori Sakakibara

ABSTRACT. For US technology-intensive corporations striving to develop effective global strategies, the establishment of R&D facilities in Japan is increasingly seen as a necessary means of tailoring products to the large local market, of effective technology scanning of Japanese competitors, and of tapping into Japan's growing capabilities in science and technology. The establishment and management of such facilities, however, is no easy task. A comparative study of US and Japanese computer firms identified major differences on five dimensions: the corporate research structure, the linkage between R&D and manufacturing, mechanisms of recruitment of R&D personnel, career patterns and the locus of responsibility for careers, and reward and incentive systems. These differences have important implications for US firms trying to set up research facilities in Japan, both in terms of managing those facilities effectively and in terms of integrating them into the firm's overall technology strategy.

Global technology strategy has at least two elements: the effective exploitation of the firm's technology in all of the world's major markets, and the utilization and coordination of the technological resources of all the world's "cradles of innovation", not simply those of the home country.[1] For many US firms, accustomed to decades of US dominance of scientific and technical innovation, the transition to a

genuinely global technology strategy is a painful process, and nowhere are the difficulties greater than those posed by Japan.

Japan is one of the world's three major markets (the second largest single market after the United States). Yet with some notable exceptions US high technology firms, which are encountering Japanese competitors in the US market as well as in third country markets, have a relatively low level of market penetration there. Both competitor and technology scanning in Japan are therefore comparatively difficult for US firms. Given Japanese technological strengths and the prospects for future enhancement of those strengths, this is a serious barrier to the development of genuinely global technology strategies.

A growing number of US high technology firms are turning to the establishment of R and D facilities in Japan as a key element of their global competitive strategies. Corning Glass, Digital Equipment, and Teradyne have set up their own research and development facilities in Japan in the past four years; Data-General and Hewlett-Packard have acquired majority ownership of Japanese firms with R&D capabilities. Other firms are considering following suit.

One powerful motive behind the establishment of such research facilities is a growing commitment on the part of major US firms to developing a significant presence in the Japanese market, both because of the sheer size of that market and because of a perceived need for a strategic response to the increasing presence of Japanese competitors in the US domestic market. A research facility in Japan can improve significantly the firm's ability to tailor products to the demands of Japanese customers.

A second and perhaps equally powerful motive is the difficulty of effectively monitoring developments in Japanese science and technology and the technology strategy of Japanese firms from a vantage point in the United States.

Finally, there is growing recognition that Japan is developing comparative advantages in certain areas of technology that can make an important contribution to the technology portfolio of the firm. The perceived success of IBM's subsidiary in Japan, whose research labs have made major contributions to the corporation's worldwide product lines, may also be an incentive for US firms, especially those in the computer industry.[2]

Managing R & D facilities in Japan, and integrating them effectively with the overall technology strategy of the firm is no easy task for a US company. The differences between Japanese and US research organizations are often imperfectly understood, and the management problems can be exacerbated by a lack of strategic coherence in the expectations generated within the firm by the establishment of a Japanese research operation. The article examines the factors behind the growing interest of US high technology firms in developing a research presence in Japan, and draws on the findings of a comparative study on the organization of R&D in Japanese and US computer firms to indicate some of the major problems in effectively pursuing this internationalization of R&D as part of a global technology strategy.

JAPAN'S GROWING TECHNOLOGY RESOURCES

In general, large Japanese firms have two major areas of competitive advantage in technology over US firms. First, they have developed ways of embodying technology in products and moving from development through manufacturing to the marketplace quickly, with high quality, and with a relatively low purchase price.[3] Second, Japanese firms have extremely effective systems of global technology scanning, developed over decades of playing "catch-up" in science and technology. These include systems of global patent scanning, in-house international scientific and technical information systems, and high-level technology scanning systems in their foreign subsidiaries. They have been aided in their efforts by the Japanese government, which funded the establishment of a global scientific and technical information system through JICST (the Japan Information Center for Science and Technology), 60% of which covers non-Japanese material.[4] The fact that the Japanese school system provides six years of English language instruction means that most Japanese scientists and engineers have an effective reading knowledge of English and can easily keep abreast of English-language technical publications.[5]

Until very recently, however, the success of Japanese firms at process innovations and the rapid application of technology, and their ability to identify and acquire technical innovations developed in other countries have been widely interpreted abroad as indicative of Japan's inability to innovate.[6] The resulting sense of superiority has meant that US and European firms have been relatively slow to respond to the growing capacity of the Japanese research community with effective scanning systems of their own.

The long-standing image of the Japanese as clever technology imitators but not technology generators, however, is giving way to a recognition of the growing strength of Japanese scientific and technological research capacity. There is both an increased international appreciation of the importance of Japanese process and engineering innovations, and general agreement among technical experts that, in fields such as VLSI, robotics, computers, materials science and engineering, and biotechnology, Japanese researchers are working at the frontiers of existing knowledge.[7] Japan now ranks third in the world, after the USSR and the United States, in the scale of its research capacity, measured either by total expenditures on research or by the total number of active researchers. Scientific and technological research has absorbed a steadily increasing proportion of Japan's expanding national income. The results can be seen in the fact that although Japan still has a deficit in its technology balance of payments, owing to the payment of royalties on long-standing licensing agreements, on new technology agreements Japan has—since the mid-1970s—earned more by selling its technology abroad than it has paid to import technology.[8]

In the coming years, therefore, the significance of the contribution of Japanese researchers is likely to continue to increase, and to move more and more into "state of the art" research. Both Japanese firms and the Japanese government have identified the expansion of the country's basic research capacity as a major priority for the coming decade.[9] The government, with its already high deficits, is somewhat limited in its contribution to this effort. Its rigid restrictions on the acceptance by national universities of outside research funding have been relaxed, and it remains committed to funding research projects in its government labs. However, even when defense expenditures are eliminated from the calculations, direct government contributions to R&D in Japan remain considerably below the levels of the United States and Europe, and are likely to remain so.[10] The major research push is coming from the leading Japanese firms.

THE "LIMITS OF FOLLOWERSHIP"

They are motivated by a belief that Japan is approaching the "limits of followership." Growing resentment of Japanese "technology copying" is producing an increasing reluctance on the part of US and European firms and individuals to license technology to Japanese firms, and greater aggressiveness in pursuing legal redress for perceived infringements on proprietary technology. Even more important is that Japan is—in many areas of science and technology—approaching the limits of existing research, and is itself on the frontiers of knowledge. Large Japanese firms already maintain large central research labs and they are moving in significant numbers to enhance their research capacity. As *Business Week* recently reported, Japanese manufacturers have established more than 25 new industrial research labs during the past two years.[11]

In other words, Japanese firms are moving to add enhanced research capacity to their already impressive capacity for engineering, process and product innovations, and their highly developed systems of global technology scanning. The image of absolute US superiority in science and technology is one which is difficult for many Americans to shed, but there is growing recognition that to keep abreast of important developments in science and technology it is increasingly necessary to monitor Japanese research.[12]

This is not an easy task for US researchers and US firms. Estimates put the amount of Japanese scientific and technical literature that is translated into English at about 25% of the total,[13] and less in some fields. However, even if 25% of published articles find their way into English, this does not mean that 25% of the total body of research is available in English, nor that it is available in timely fashion. Technical translations often leave much to be desired: a translation of a technical article, however well-intentioned the translator, is usually more difficult to follow and less clear in details than the original. There is also a substantial bias in what gets routinely translated. Basic research in scientific journals is more likely to be translated than technical articles in company journals, for example. Translation is expensive, and more important, time-consuming. An additional problem is that getting access to Japanese-language publications within the United States is not easy. The routine acquisition of the flood of Japanese publications in scientific and technical fields is well beyond the resources and perhaps policies of most of the best research libraries.

In addition, experts in some fields have recently noted a new and potentially significant trend: As Japan's own research establishment grows and Japanese researchers have a larger and more sophisticated domestic audience for their research, they seem to be less likely to undergo the difficulties of writing in English and submitting to the laborious review processes necessary to publish in English-language journals. Instead, they can increasingly get useful and speedy feedback by publishing only in Japanese journals.[14]

Moreover, being abreast of published research is only part of the task of keeping up with scientific and technical developments. Few US researchers would rely solely on publications to keep them informed of developments in their field: The informal networks that link researchers are an important part of the scientific and technical information system. Given the explosion in research publication over the last decade, these information networks provide an invaluable guide to what publications are significant, as well as to what research is being carried out prior to the publication stage. Such networks are perhaps even more important in Japan than they are in the United States: Less of the scientific and technical literature is indexed in Japan, and there is a greater tendency to publish in "house organs" of the firm or the university, rather than in nationally refereed journals. The filtering and alerting functions of informal professional networks are therefore extremely important in the Japanese context. Yet comparatively few US researchers have access to such networks within Japan.[15]

TWO DIMENSIONS

US firms therefore have strong strategic reasons to develop at least technology scanning facilities in Japan. Such scanning has two dimensions: One is keeping abreast of Japanese developments so as not to have the firm's researchers covering problems already dealt with by their Japanese counterparts; the other is monitoring the technology strategy of major Japanese competitors, a critically important part of effective competitor scanning. The first scanning dimension is most effectively carried out by people who are actively involved in research within the firm: They alone have the expertise to identify what information is of critical importance. In other words, scanning to identify key scientific and technical resources is most effectively carried out by a subunit that itself possesses a research capability.

Still another argument for setting up research facilities in Japan is that in a number of areas the Japanese are developing comparative research advantages that are unlikely to be matched within the United States in the near future. One of these is in materials science and engineering, where the Japanese—driven by their poverty of natural resources and dependence on imported materials—have greater incentives to develop new materials (such as ceramics) than US firms in their comparatively resource-rich environment.[16] Another critical area is the computer and communications field, where Japanese-language processing (with its two 50-character syllabaries and 1780 "everyday use" Chinese characters) makes technical demands that simpler English-language processing does not. The technical solutions to the problems, however, (greatly increased memory, distinctive storage architecture, greater screen definition for monitors, more versatile printers, voice recognition systems for input) have broad applicability to general products. Tapping into the distinctive competence of Japanese research in these areas provides yet another incentive for developing a research presence in Japan in such industries.

In summary, there are four major motivations behind the establishment of R&D facilities in Japan by US high-technology firms: technology scanning to identify scientific and technical research that would be a significant input into the research process of the firm as a whole; technology scanning to analyze the technology strategy of Japanese competitors; research to tailor products to the demands of the Japanese market (particularly important in fields such as computers that demand Japanese-language capability); and research that taps into the comparative advantages of Japanese science and technology to contribute to worldwide product development and the firm's overall technology strategy.

CONTRASTS IN R&D ORGANIZATION IN JAPAN AND THE UNITED STATES

The interest in Japanese management in recent years has produced a considerable body of research on the differences between large US and Japanese firms in terms of the organization and careers of managers and of blue-collar workers.[17] There has been very little comparative study, however, of the organization and careers of research and technical personnel. In 1984, therefore, we carried out a comparative study of the organization and careers of engineers in R&D in three Japanese firms in the computer industry (Fujitsu, NEC, and Toshiba) and three US firms (Data-General, the Digital Equipment Corporation, and Honeywell Information Systems).[18] We were concerned with two critical questions. First, are there aspects of Japanese R&D organization from which US firms would benefit by applying in their own operations? Second, what are the implications for US firms trying to internationalize their R&D strategy by setting up research facilities in Tokyo?

The study identified major differences between the US and Japanese firms on five major dimensions: the corporate research structure and the differentiation of R&D units; the linkage between R&D and manufacturing, especially in the hand-off of research; mechanisms and criteria of recruitment of R&D personnel; career patterns and the locus of responsibility for careers; and reward and incentive systems.

The three Japanese and three US firms

are not strictly comparable as firms, but their differences are representative not only of the different structure of the computer industries of the two countries but of the differences in the structure of high technology industries in general.

Each of the Japanese firms is a large, integrated electrical equipment manufacturer, with a broad range of products and a structure that is highly standardized across those product lines. All three Japanese firms were founded before World War II, and the emerging divisions in computers and information systems adopted and adapted existing corporate patterns of R&D. All three Japanese firms have two levels of R&D facilities; corporate-level central research laboratories, and divisional-level R&D units attached to the business divisions, which are centered on manufacturing.

The central research labs conduct basic research and new product development (usually to the stage of the development of the first prototype); the divisional-level research labs carry new product development through manufacturing and testing, carry out research on product enhancement and improvement, and work on improvements in production technology. The firms' desire to maintain the standardization of structures and careers across product lines within these facilities has severely limited the firms' flexibility in developing distinctive career and reward structures in high technology areas. It has led several firms in this industry to set up separate companies in fast-growing and competitive research areas, such as software, to escape from the standardized patterns of the parent firm.

Among the US firms, two are much younger and more specialized than the Japanese firms, and have developed their R&D organization in response to the emerging demands of the industry, and the needs and predilections of the engineers who have dominated it. Even in the case of the third firm, Honeywell, which is part of a huge and highly

diversified corporation, the Office Management Systems Division, on which our research was focused, seems to be a fairly autonomous facility that has developed along the same lines as those of its competitors in the industry rather than adapting and integrating with other Honeywell structures.

All three US firms also have two levels of R&D organization, which may at first glance seem to be analogous to the Japanese structures. All three US firms have Advanced Technology Groups which are working, as the name implies, on new technologies and applications. There are clear differences across the two societies, however. The American Advanced Technology Groups are much smaller than the Japanese central labs, they specialize more in the initial stages of R&D, and they play a less crucial role in new product development.

A STRIKING DIFFERENCE

One of the most striking differences between the US and the Japanese firms was that US development and design groups were much less closely linked to manufacturing than their Japanese counterparts, at all levels: the corporate level, in terms of R&D funding and the physical location of research activities; the level of the project group and internal technology transfer; and the level of the individual engineer, in terms of career patterns and significant reference groups.

In the Japanese firms, half the research budget is allocated directly to the central R&D lab, and half is allocated through the business divisions, which can use their funding either to carry out development activities within their own facilities or to commission specific research projects from the central labs. Because of this quasi-market relationship, the business divisions, which in Japan are centered on manufacturing, are, in effect, important internal clients of the central labs. Their power and status within the company as a whole and especially in their interactions with the central R&D lab is therefore very high, especially in comparison to the status of manufacturing in the US firms.

Because the divisions have their own development labs, the handoff of research projects from the central lab to manufacturing takes place at a fairly early stage, often while the formal specifications are still flexible. Engineers in the divisional labs therefore play an important role in the R&D process, and ease of manufacture is a major consideration in the design process. Moreover, the research hand-off is usually accomplished by dispatching one of the engineers from the central lab's project team to the divisional lab. This transfer is permanent, not temporary; it is a step in the engineer's career path, and takes place after he has been with the central lab for six or seven years. It is the first step in a career ladder that leads into line management in the divisions. This contrast with the US firms in terms of the dual career ladder should be noted. In the US firms, the managerial rungs of the dual career ladder led into R&D management; in Japan, they led into line management in the operating divisions.

Internal technology transfer in the Japanese firms seems to follow the maxim that, to move information, you move people. Perhaps because this transfer is standard and expected by the individual engineers, people in the manufacturing are a much more important reference group for central lab engineers than they are for their US counterparts, as Table 1 indicates.

The recruitment of R&D personnel is another aspect of R&D organization on which US and Japanese firms differ dramatically. Recruitment is much more centralized and standardized in the Japanese firms. In our US firms, recruitment was the responsibility of the individual research groups, whose members were heavily involved in the recruitment process. In Japan, recruitment is the responsibility of the corporate personnel department; it takes place once a year, and virtually all recruits to the central research labs are new graduates.

These graduates are brought into the firm by highly routinized paths. Key professors in the major Japanese universities routinely al-

TABLE 1. Engineers' Reference Group

Respondents were asked to select three of the following groups in whose eyes they wished to do well, and to rank order them from 1 (most important) to 3 (third most important). The following table shows the mean ranking for the samples as a whole.

	RANKING Japan (n = 206)	(Mean Score) U.S. (n = 101)	
Research administrator of your group	1 (2.389)	5 (.660)	***
Professional colleagues within your own group	2 (1.320)	1 (2.155)	***
People in Manufacturing divisions	3 (.827)	7 (.112)	***
Top executives in your company	4 (.716)	4 (.730)	***
Professional colleagues elsewhere in your company	5 (.385)	2 (1.005)	***
People in sales/marketing	6 (.216)	8 (.102)	
Respected friends outside your company	7 (.058)	6 (.282)	**
Your family	8 (.053)	3 (.870)	***

***$p < 0.001$
**$p < 0.01$

locate their students to the major companies. Although student desires are consulted to some extent, the professor in charge of placement will usually write only one letter of recommendation for each student, and that will be to a major company that will accept without question that recommendation. For a company to disregard the recommendation would be to forfeit the opportunity to obtain future graduates from that professor. For the student to refuse to go to the company for which the professor has written the letter would mean that he would forfeit all chance of working for a major Japanese firm, since the companies do not recruit students for which such letters are not written.

As a result of this system, the leading Japanese companies tend to hire a fixed number of students from the major universities each year. They also make ongoing efforts to establish and maintain close relationships with leading universities through the supply of equipment, research grants to faculty, and personal contacts (for example, personnel staffers frequently visit professors and university placement officers). The lower-ranking universities have a less secure place in the technical job market, and professorial recommendations do not carry nearly the same weight with companies. For graduates of such universities, companies are more inclined to make hiring decisions on the basis of an assessment of the engineer's quality (an assessment assisted by a rigorous written technical exam).

The transfer of engineers from the central labs to the divisional labs is an example of the third major contrast between the Japanese and US firms: the standardization of career patterns. The US engineers and personnel managers we interviewed scoffed at the idea of describing a "typical" career in R&D in their firm. The frequent mobility across firms, the extent of individual options for pursuing new specialities through outside study, the level of individual choice in moving across projects—all these factors make it difficult to describe a "typical" career.

Japanese engineers and managers, on the other hand, had no hesitation in describing the "typical" career for an engineer who joined the central research labs. He (and it was invariably a "he"; to date, no women have been hired as engineers in the central labs of the three Japanese firms) would be recruited directly from university, usually with a master's degree. None of the three companies had any established mechanism for recruiting mid-career engineers, and they said it was very rare for them to do so.

The fact that recruitment takes place once a year allows the companies to put each year's hires through an introductory training program that includes both technical and managerial recruits. This program introduces the recruits to the company, exposes them to the range of functions, and provides them with an intense, shared experience that establishes the basis for horizontal communication after they have dispersed to their new positions.

The engineer whose new position is in the central research lab would spend the first two years in the central labs on a succession of projects, largely in an apprenticeship role. In all three companies, research managers estimated that it took two years before an engineer was capable of making an independent contribution to the research process. He would become more and more active on a succession of projects over the next four to five years, and then would be transferred to the divisional labs, usually as the principal carrier of a research project on which he had taken the major role. Very few engineers remain for their entire careers in the central labs.

As one might gather from the recruitment patterns and from the earlier discussion of career paths within the firm, in the US firms, the locus of responsibility for the engineer's career lies unquestionably with the individual;

in the Japanese firms, it lies with the firm. To give one example, the questionnaire respondents in the US firms cited their own expressed wishes as the most important factor in the assignment to their last research project; for Japanese engineers, it was the supervisor of their previous project (see Table 2).

The same pattern could be seen in responses to a question about the motives for taking technical courses since graduation (see Table 3).

Japanese engineers were much more likely to have been assigned to the course by the company; US engineers were much more likely to take the course to improve their career opportunities.

The standardization of careers in Japan extends even to the area of rewards and incentives. In contrast to US firms, where outstanding performance is quickly rewarded with salary increases and promotions, personnel managers in all three Japanese firms insisted that neither

TABLE 2. *Factors in Assignment to Last Project*

Question: "In your assignment to your last project, how much influence did each of the following have in your getting that assignment?
NOTE: Mean score of 5-point, Likert scale from 1 = very little influence to 5 = very great influence.

	Mean		
	Japan (n = 202)	U.S. (n = 95)	t
Your previous supervisor	4.44	2.77	***
Manager of your department/job	3.44	2.62	***
Head of your project	3.92	3.07	***
Personnel staff	1.47	1.35	
Your own expressed wishes	2.94	3.90	***

***p < 0.001

TABLE 3. *Comparison of Motives for Taking Courses*

Question: "How important was each of the following motives for taking these courses?

Motives	Mean		
	Japan (n = 123)	U.S. (n = 55)	t
To update existing skills	4.10	4.07	
To add new skills	3.98	4.58	***
To improve chances of promotion	2.02	2.70	***
To improve chances of assignment to more interesting activities	2.35	3.08	***
Assigned to course by company	3.17	1.30	***

(1) Mean score of 5-point, Likert scale from 1 = unimportant to 5 = very important.
(2) *** indicates a t-test with probability p < 0.01

salary nor rapid promotion were used to reward exceptional performance. Even the most brilliant engineer proceeded up the salary ladder at the same pace as his peers. The principal rewards for outstanding performance were intrinsic (the respect of superiors and peers) and long-term (the opportunity to go abroad for advanced study, for example, and the prospect of staying in the central lab rather than transferring to the divisions).

The standardization of Japanese career paths is accentuated by the fact that the organizational structure of the R&D groups is the same as that of manufacturing or sales: the hierarchy of sections *(ka)* and departments *(bu)* is identical, and the titles of section chief and department head carry the same status in every function. They also carry much the same salary across functions.

This standardization of careers and rewards is part of the difference between the two sets of firms in the factors that influence promotion to project leader, although other factors are also involved. Our US respondents rated "technical expertise" as the most important factor in being promoted to project leader;

it ranked fourth of five factors for the Japanese respondents. This reflects an important difference in the role of the project leader. In the US firms, the project leader is expected to be a technical leader, who has won his position and commands respect by virtue of his mastery of engineering and his ability to find solutions to problems. In the Japanese firms, his seniority has much more importance, and his role is more that of the effective chairman, encouraging and guiding his research team. In both sets of firms, as we can see from Table 4, a track record of participation in successful projects is seen as very important.

But our interviews confirm the fact that such a track record indicates different things in the two contexts. In the United States, it is an indicator of technical mastery; in Japan, it is an indicator of the ability to manage a team successfully.

Four of these five areas of difference—linkage with manufacturing, recruitment, career patterns and staffing, and rewards and incentives—indicate clearly that developing effective research operations in Japan will be a demanding task for a US firm. Each presents a

TABLE 4.   *Important Factors in the Promotion to Project Leader*

Question: "What factors do you think are the most important in being promoted to project leader in your company?"

NOTE: Mean score of importance (3 points for the most important factor, 2 points for the second, 1 point for the third, and 0 point for others).

	Japan (n = 206)		U.S. (n = 103)	
	Mean	(Ranking)	Mean	(Ranking)
Seniority***	1.36	(2)	0.66	(4)
Track record of participation in successful projects	1.42	(1)	1.67	(2)
Administrative ability***	1.27	(3)	0.52	(5)
Technical expertise***	1.17	(4)	1.81	(1)
Ability to work well with others***	0.74	(5)	1.48	(3)

***$p < 0.001$

significant challenge both to a US firm's organizational structures for its subunit and to its general technology strategy *vis-a-vis* Japan.

THE STRATEGIC IMPLICATIONS FOR US FIRMS

The linkage with manufacturing is perhaps the most wide-reaching in its implications. The close linkages between design and manufacturing explain why Japanese firms in this industry are perceived by engineering managers on both sides of the Pacific as more successful than US firms at moving products quickly from the design stage through development and manufacturing to the market with high quality and reliability. Indeed, the desire to tap into this kind of engineering expertise is often an important motivation behind the US firm's establishment of R&D facilities in Japan. But our study suggests strongly that this close linkage is not inherent in the formal training of Japanese engineers or in Japanese "engineering culture," but is due to the multiple linkages between design and manufacturing within the firm. If a US firm establishes stand-alone R&D facilities in Japan without a substantial manufacturing presence, it is unlikely to realize this key competitive advantage of Japanese R&D organization. Moreover, because the movement into line management in manufacturing is a standard part of the design engineer's career path in the Japanese context, the absence of a major manufacturing presence in Japan may well deter many Japanese engineers from joining the US firm, because of doubts about promotion and career possibilities.

The linkage between design and manufacturing is also an area of difference with significant implications for US firms in general, not just in terms of any operations they may have or plan to have in Japan. The quasi-market relationship that exists between corporate level R&D facilities and the operating divisions is one that US firms might do well to consider emulating. Neither unit (the R&D facility or the operating division) is completely bound to the other in terms of resource allocation, but neither are they mutually independent. The negotiations involved between the two in R&D budgeting and the enhanced status of the operating division as an "internal client" of the R&D lab increase the linkages between the two functions. The steady flow of people across the R&D subunits into the operating divisions is another element of the linkage that might have significant benefits for the technology interface within firms in general, not just within Japan.

The structure of the technical labor market and the difference in patterns of recruitment pose problems for the US firm (and indeed for smaller Japanese firms as well). A major foreign company can decide to emulate the methods used by the large Japanese firms: that is, to cultivate close ties with key professors at the top universities, so as to enter the allocation system. IBM-Japan has done this with great success, but it is a process that takes time. US firms may therefore be tempted to rely instead on recruiting engineers from the lower-ranking universities.

Such a strategy may have costs, however: the prestige of a firm is closely associated with its attractiveness to graduates of the top universities, and a firm that is seen as recruiting only from second-rate universities may itself be seen as second-rate. The prestige factor is sometimes one that is unappreciated by US firms, the more so because it is by no means clear that the education offered at the most prestigious universities is of higher quality than that at lower-ranking schools.

On the other hand, there are more immediate opportunities for recruiting mid-career engineers out of the major Japanese firms. Not all Japanese engineers wholeheartedly ap-

337

prove of the standardization of career paths. Some would prefer to remain in research roles in the central lab, rather than transferring to the divisions, but their only alternative to the transfer is to leave the company. This provides a "window of opportunity" for US firms attempting to set up or expand research facilities in Japan: Engineers from the central labs of major Japanese companies who are at the stage of their careers when they are facing a transfer to the divisions may be targets for recruitment into a research facility which will allow them to continue active research. There is also considerable dissatisfaction with the complete absence of immediate rewards for outstanding performance. In our questionnaire survey, Japanese engineers were significantly more likely than their American counterparts to express dissatisfaction with the level of financial rewards they received. This too widens the "window of opportunity" for US research facilities in Japan.

This should not, however, be taken to mean that US firms can simply apply their own patterns and expectations to the personnel administration of a Japanese facility. One implication of the general contrast in the locus of responsibility for careers in the two countries is that Japanese engineers expect greater career planning and company responsibility for their careers than their US counterparts. Even the "mavericks" who find the tight control of the large Japanese firm claustrophobic, and who are the most likely prospects for recruitment to a US firm will expect the company to assume more responsibility for their careers than would US engineers. This means that developing appropriate personnel strategies and structures will be a difficult and time-consuming process for firms starting up wholly owned research facilities. To completely adopt Japanese patterns of recruitment, career structure, and personnel administration would mean sacrificing one of the key competitive advantages of the firms in recruiting mid-career engineers.

However, to transplant completely US patterns would likely produce great dissatisfaction and uncertainty among the Japanese staff. It is necessary to develop "hybrid" patterns, and this will take time and commitment. But the development of a coherent and long-term personnel strategy is a key ingredient for success in establishing effective research subsidiaries in Japan.

A SOURCE OF DANGER

The differences in career patterns and incentive structures can provide yet another source of danger for US firms setting up research facilities in Japan. They may well look for research managers who have the qualities that make for successful project leaders in the United States. As a result, they may hire researchers who are technically brilliant, but who lack the skills that are essential for successful project management in working with Japanese researchers. They may also encounter problems when it is necessary to develop a cooperative working arrangement between a project group working in the United States and one working in the Japanese facility. The US researchers may quickly develop a contempt for someone who is a very effective project leader in Japan, but who lacks the technical superiority they value; the Japanese engineers may find it hard to work with an American project leader who commands the respect of his own subordinates by his technical expertise, rather than his management skills. An awareness of the different criteria of excellence in the two settings may go far in anticipating and dealing with such problems.

Finally, there is a set of strategy management issues confronting both the parent firm and the Japan-based R&D subsidiary. In the first section, we outlined the four major motivations behind the establishment of R&D facilities in Japan by US high-technology firms. They were: technology scanning to identify scientific and technical research that would be a significant input into the research process of the firm as a whole; technology scanning to analyze the technology strategy of Japanese competitors; research to tailor products to the demands of the Japanese market; and research that taps into the comparative advantages of Japanese science and technology to contribute to worldwide product development and the firm's overall technology strategy.

Each of these mandates has a different constituency within the firm. Central R&D headquarters is the primary internal client for the first, and they are likely to want technology scanning for research purposes to have a high priority in the activities of the Japanese operation. Competitor scanning is likely to have a different, but equally urgent client: those groups at corporate headquarters who are developing global strategy. The client for research that tailors products to the local market (the classic role of overseas R&D facilities) will be the local marketing organization, one whose demands may have especial force for the research facility because of its physical proximity. Finally, in terms of corporate budgeting and long-term survival of the R&D operation, making a major contribution in research terms to the development of new products for worldwide markets is likely to be the most visible and therefore highest priority activity of researchers within the Japanese facility.

Given the growing Japanese capacity in science and technology, and the nature of the technical job market in Japan, it is the fourth mandate—developing products on a worldwide basis—that would seem to be both the most effective use of the resources necessary to build an effective operation in Japan and the one which would be most attractive to potential Japanese recruits. But it is also the mandate with the least specific internal corporate constituency. Without a strong high-level recognition of the forces that make such a strategic mandate desirable, and a long-term commitment at the top levels of the US firm, it is unlikely that the time and resources necessary to meet organizational challenges of developing effective management and liaison systems will be available.

There is an obvious danger of overloading the Japanese research operation, especially in its early stages, when it is also struggling with the organizational problems of recruitment and of developing structures that combine features of US and Japanese research organizations. The choices made in these early stages are crucial for the long-term viability of the operation. For example, if the technology scanning activities take precedence, it will become increasingly difficult for the facility to attract first-class researchers. After all, the key pool of potential recruits to a new operation consists of the mid-career engineers from the large firms who are willing to leave the security of their current jobs for the opportunity to do challenging research instead of moving into line management. They are unlikely to be attracted by the prospect of technology scanning. Such recruits are, in turn, essential in beginning to build linkages with the major universities from which they were graduated.

In the long term, without people who are seen by the Japanese research establishment as "first-rate"—by virtue of their university background and their research activities—the Japanese facility will be unlikely to be capable of doing an effective job even of technology scanning: Its people will lack the access to the informal information networks of the research establishment that are so critical to obtaining and assessing timely information.

The first section of this article presented the factors that were pushing US firms into establishing or considering establishing research facilities in Japan. This section has laid out—on the basis of comparative research into the organization of R&D in the Japanese and US computer industries—some of the major difficulties that they are likely to encounter in doing so.

339

Clearly, this list of potential problems indicates that building effective research facilities in Japan will be a slow and difficult process for US firms, and will involve major commitments from them.

The importance of Japan as a market, as the home base of major internationally competitive firms, and as a source of new technology, however, requires that US technology-intensive corporations take the time and effort to seriously consider development of such facilities and the organizational systems to support them, if these firms wish to remain genuinely "global players" in technology.

NOTES

1. Robert Ronstadt and Robert J. Kramer, "Internationalizing Industrial Innovation," *Journal of Business Strategy*, Vol. 3, no. 3 (Winter 1983), pp. 3–15.

2. See "A Global Reach in the R&D Realm," *Datamation* 30 (April 1, 1984), pp. 153–156.

3. See for example the observation of Jules J. Duga, principal research scientist at Battelle Memorial Institute: "The Japanese have obviously been much, much better than us when it comes to taking new technology and doing something with it." (*Business Week*, July 8, 1985, p. 87). This viewpoint was repeatedly enunciated by informants on both sides of the Pacific in the computer industry study cited below.

4. Robert Gibson, "Japanese Scientific and Technical Information", pp. 14–26 in *Japanese scientific and technical information in the United States* (*Proceedings of a Workshop at MIT*), published by the National Technical Information Service, Department of Commerce, (PB83-179903); also Kagaku gijutsu-cho (Science and Technology Agency), *Kagaku gijutsu hakusho (Showa 58)* (Science and Technology White Paper, 1983) published in Tokyo by the Okura-sho.

5. This is not a new development. A British engineer, Henry Dyer, writing in 1904, noted that "Japanese engineers and scientific men are often found better informed regarding the contents of British journals than are many in this country." From *Dai Nippon: A Study in National Evolution* (London: Blackie and Son, Ltd., 1904), p. 176.

6. Leonard Lynn, "Technology Transfer to Japan: What We Know, What We Need to Know, and What We Know That May Not Be So," paper given at the SSRC Conference on International Technology Transfer: Concepts, Measures, and Comparisons, June 2–3, 1983, New York City. We should make it clear that while Lynn cites examples of this viewpoint, his paper is largely a critique of the lack of solid evidence for it.

7. See the transcript of the testimony at these hearings published under the title of "The Availability of Japanese Scientific and Technical Information in the United States," hearings before the Subcommittee on Science, Research, and Technology of the Committee on Science and Technology of the House of Representatives, March 6, 7, 1984 (Washington: US Government Printing Office, 1984).

8. See data from the Prime Minister's Office cited in Lynn, *op. cit.*, p. 8.

9. See the series of annual White Papers on Science and Technology (*Kagaku Gijutsu Hakusho*) published by the Science and Technology Agency, especially those for 1983 and 1984.

10. In 1981, the US government share of the nation's total R&D expenditures (non-defence-related) was 30.3%; in West Germany it is 40.9%; in France, 46.7%; and in Japan, 24.5%. OECD data, published in the Kagaku Gijutsucho, *Kagaku gijutsu yoran* (Indicators of Science and Technology), Tokyo, 1983.

11. *Business Week*, Feb. 25, 1985, p. 96.

12. See, for example, the Battelle Memorial Institute report for the US Department of Energy on ten energy-related areas of science and technology in Japan: G.J. Hane, P.M. Lewis, R.A. Hutchinson, B. Rubinger, and A. Willis, "Assessment of Technical Strengths and Information Flow of Energy Conservation Research in Japan", Vol. 1, (Richland, WA: Pacific Northwest Laboratory [PLN-5244 Vol. 1], September 1984).

13. Gibson, *op. cit.*, p. 17.

14. Congressional Hearings testimony, March 1984, cited above; pp. 324, 342.

15. The suggestion that immediate steps should be taken by the US government to begin to remedy this lack of cultivation of networks with Japan's scientific community is one of the recommendations of the Battelle Memorial Institute study cited above (Hane *et al.*, pp. 17–19).

16. See the survey of Japanese efforts in this area written by George B. Kenney and H. Kent Bowen, "High Tech Ceramics in Japan: Current and Future Markets", *American Ceramics Society Bulletin*, Vol. 62, no. 5 (May 1983), pp. 590–596.

17. Among the major empirical research studies have been: Ronald Dore, *British Factory Japanese Factory* (Berkeley: University of California Press, 1973); Rodney C. Clark, *The Japanese Company* (New Haven: Yale University Press, 1979); Robert E. Cole, *Work, Mobility, and Participation* (Berkeley: University of California Press, 1980); Satoshi Kamata, *Japan in the Pass-ing Lane: An Insider's Account of Life in a Japanese Auto Factory* (New York: Pantheon Books, 1982).

18. The study involved interviews in each company with research managers, personnel managers, and individual engineers. More aggregate data were collected through questionnaires administered to a stratified sample of engineers working in development projects in the companies. There were a total of 206 questionnaire respondents in the Japanese central research labs and 98 in the divisional labs; and 103 US respondents. The data presented in this paper use the responses from the central R&D labs, because the emphasis in US firms setting up research facilities in Japan, which are more analogous to the central labs than to the divisional labs. The research was carried out under the auspices of the MIT-Japan Science and Technology Program and the Harvard Program on US-Japan Relations, and funded by a research grant from the Digital Equipment Corporation.

6

MARKETING

Changes in Consumer Behavior over the Product Life Cycle

John A. Howard

William L. Moore

INTRODUCTION

One of the most frequently cited reasons for the failure of new products is the inability to judge consumer reaction to them. The purpose of this paper is to provide some general observations about the way consumers react to innovations at different points in the product life cycle (PLC). (For a discussion of some of these concepts in an industrial setting see Howard 1980, Webster 1969 and Zaltman and Wallen-

dorf 1979, Chapter 19). This will be accomplished through an integration of Howard's theory of buyer behavior (Howard 1963, 1977, Howard and Sheth 1969) with a number of concepts drawn from the literature on adoption and diffusion of innovations.

In this theory, Howard classified consumer choice processes into three categories. Extensive Problem Solving (EPS) occurs when a consumer confronts a new brand in a new product class. Limited Problem Solving (LPS) is associated with a new brand in a familiar product class and Routinized Response Behavior (RRB) a familiar brand in a familiar product class.

Thus when an innovation creates a new product class, all customers are initially in EPS. However, when other companies enter the market and consumers encounter a second or third brand, they move to LPS. Finally if the product does not change substantially be-

---

Parts of this chapter are an adaptation from John A. Howard, "The Empirical Theory for Managing the Market," in B. Enis (ed.) *Review of Marketing,* Chicago: American Marketing Association (1981). In this much longer paper, Howard relates his theory of buyer behavior to a number of concepts such as: a prescriptive cash flow allocation model, changes in seller behavior over the product life cycle by drawing heavily upon Abernathy and Utterback's discussion in reading 2 of this book, competitive structure, and product hierarchy, to construct a market process model.

tween purchases, a number of consumers will become familiar with their alternatives and will be in RRB. Thus, the proportion of customers in each problem solving stage will shift over the PLC, as seen in Figure 1. For various reasons not all customers are alike at any point on the cycle. However, there is still a strongly dominant type of problem-solving behavior in each stage. In the introduction stage of the PLC, it is extensive problem solving (EPS) that is dominant; in the growth stage it is limited problem solving (LPS); and in the stable stage it is routinized response behavior (RRB). Because of this correspondence between problem solving behavior and stages in the PLC, these types of behavior will form the framework for the paper.

Because this theory treats new brands in new product classes different from new brands in established categories, one is able to compare the reactions to products with varying degrees of newness. This gives some theoretical backing to a number of observations in the diffusion of innovations. The discussion of changes in problem solving behavior over the PLC is preceded by a short review of adoption and diffusion of innovations.

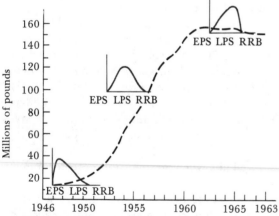

FIGURE 1.   PLC and Changes in Customer Behavior

Source: J. A. Howard, *Consumer Behavior: Application of Theory*, McGraw-Hill, 1977, p. 13

ADOPTION AND DIFFUSION OF INNOVATIONS

Adoption

Adoption is the acceptance and continued use of a product, service or idea. The adoption process refers to a series of mental and behavioral states that a person passes through leading to the adoption or rejection of an innovation. Rogers and Shoemaker (1971) have classified the process into the following stages.

1. *Knowledge:* This stage begins when the consumer receives some stimulus regarding the innovation. During this time, the consumer becomes aware of the product, but does not form an attitude toward it.

2. *Persuasion:* This period involves the formation of a positive or negative attitude toward the innovation. Additional information about the product is gathered and the consumer may imagine what it would be like to adopt the innovation (vicarious trial) or the innovation may be tried on a limited basis (e.g. a test drive).

3. *Decision:* Here a decision is made to accept or reject the innovation. This can be reached through additional trial and/or discussions with other people. Of course, this decision may be reversed at some future time.

4. *Confirmation:* After making a decision a consumer may seek additional information for reinforcement. Contrary evidence may cause a reversal of the adoption decision. Discontinuance is most likely when the innovation is not well integrated into the consumer's life style.

In summary, Rogers and Shoemaker propose a model that is similar to the hierarchy of effects model by Lavidge and Steiner (1961). Both of these models assume that consumers move sequentially through distinct cognitive, affective, and behavioral stages. While there is a lack of research with a process orientation on this model, it is generally believed that mass communication is more effective in the Knowledge stage but in the latter points in the adoption process, personal communication is more important.

In spite of their wide-spread use, these models have been criticized in a number of ways. For example the stages are not as distinct as suggested here and people may not move through them sequentially, but may "back track" at certain times. Furthermore, a number of researchers (Krugman 1965, Ray 1973) have criticized the basic cognitive-affective-behavioral flow when there are minimal product differences. They suggest that in low involvement situations, consumers may buy a product before forming a positive attitude toward it. However, this basic model does appear to be a reasonable (possibly simplified) way of representing the adoption process for products perceived to be new and different. Moreover, it has considerable value if it does no more than get the marketer to think carefully about new product acceptance.

Diffusion

Diffusion is the spread of an innovation throughout a social system. It is widely recognized that not all consumers try, or adopt, an innovation at the same time and that some innovations diffuse much more quickly than others. Therefore the diffusion of innovations has been studied from the perspectives of both the types of people that adopt it first and the characteristics of the innovation that speed or retard its diffusion.

Some consumers are usually the first to adopt new fashion innovations, some doctors are the first to prescribe new drugs and some farmers are the first to adopt new agricultural techniques. Rogers (1962) developed a classification of adopter categories based on their relative time of adoption and related each of these categories to certain values. These categories and values are depicted in Figure 2. Additionally, a number of studies have attempted to relate innovativeness to a number of demographic and attitudinal variables. In general, characteristics such as income, socio-economic status, education, product category knowledge, and achievement motivation have been positively associated with early adoption of innovations. In spite of these general relationships to innovativeness, there does not appear to be a generalized innovator. That is, a person who is an innovator with respect to clothing fashion may not be so with respect to books. Instead, Robertson (1971) suggests that while innovativeness cannot be expected across categories it can be within a category and, sometimes, within related categories.

Finally, the characteristics of the innovation itself influence the speed of diffusion. Characteristics positively related to adoption are:

1. Relative advantage—The degree to which it is superior to competitive products.
2. Compatibility—How consistent the innovation is with the values and experiences of potential adopters.
3. Divisibility—The degree to which it can be tried on a limited basis.
4. Communicability—The relative ease of observing and describing the benefits to others.

Additionally, complexity and perceived risk (economic, social and/or physical) have been found to be negatively related to the speed of diffusion.

These concepts have proven to be useful to marketers. Adoption theory has caused them to look at new products from the perspective of the consumer and has suggested the use of different promotional tools at various stages in the process. Diffusion theory has pointed out who the early adopters of a new product are likely to be. Also it has shown ways to speed up

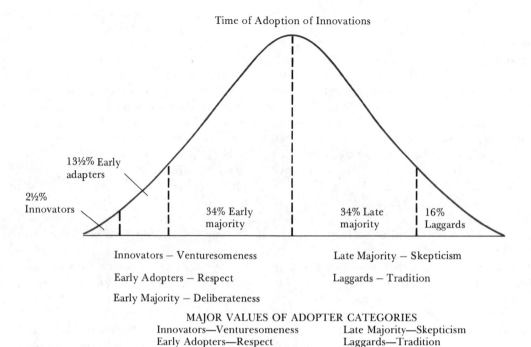

Time of Adoption of Innovations

13½% Early adapters

2½% Innovators

34% Early majority

34% Late majority

16% Laggards

Innovators – Venturesomeness

Early Adopters – Respect

Early Majority – Deliberateness

Late Majority – Skepticism

Laggards – Tradition

MAJOR VALUES OF ADOPTER CATEGORIES

Innovators—Venturesomeness	Late Majority—Skepticism
Early Adopters—Respect	Laggards—Tradition
Early Majority—Deliberateness	

FIGURE 2.

the process of diffusion. However, we feel that an integration of these concepts with Howard's problem solving categories will be even more useful.

EXTENSIVE PROBLEM SOLVING

Extensive Problem Solving (EPS) occurs when a consumer confronts a new brand in a new product class. This process corresponds to what psychologists call concept formation, the process of developing criteria for the identification and evaluation of a new product. While consumers start this process at different times and take differing amounts of time to complete it, most consumers will be in EPS during the introductory and growth stages of the PLC. However, some people will be in EPS in the maturity stage. A few may be just entering the

market at that time—called laggards by Rogers (1962). Others may be in the stage because it has been so long since their last purchase and the brands have changed enough that they do not know what criteria to use in evaluation of the current alternatives.

Considerable psychological evidence indicates that product concepts are formed by first grouping the new product with products it is similar to, then distinguishing it from them. For example, when consumers first became aware of instant coffee they might have associated it with regular coffee, then distinguished it as being more convenient, but worse tasting. The result of these two processes is the placement of the concept in the consumers' semantic structure, or product hierarchy.

The concept of a product hierarchy can be illustrated with a vegetable bacon introduced into test market by General Foods in 1974. The vertical dimension in Figure 3 de-

346

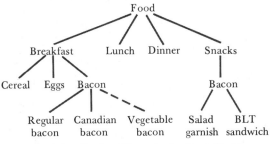

FIGURE 3. *Product Hierarchy for Vegetable Bacon*

picts the level of inclusiveness of a category. For example, breakfast (at the top of the Figure) includes foods other than bacon, and the three subcategories of bacon are found at a lower point in the hierarchy. The dotted line in Figure 3 indicates the addition of vegetable bacon to this person's product hierarchy. The consumer recognized the product as bacon, then differentiated it from other types of bacon.

In this case, the formation of a concept of vegetable bacon was probably fairly easy as it was one level subordinate to the *basic level—* bacon. In product hierarchies, a level that is more basic than the others has always been found. It is basic in the sense that it is the level of greatest cognitive economy—at this level concepts are differentiated enough to give the consumer the information he/she needs but still keep the number of categories as small as possible. This is the level at which people think and form images. For example a basic level might be chair rather than the subordinate kitchen chair, or stuffed chair, or the superordinate furniture. The basic level is the highest level at which it is possible to form a mental image that is isomorphic to the average member of the class. New products are first recognized to be members of their basic classes, then only with additional processing are they identified as members of superordinate or subordinate classes. Therefore products that form new basic classes may be the hardest for consumers to form concepts of, and products that are sev-

eral levels removed from the basic level probably present more problems in concept formation than those that are close to, but not on, the basic level.

Along with the identification of the concept the consumer must also learn how to evaluate it. This process starts with the consumer's instrumental values acquired in the home, community, church, schools, etc. These values interact with variables such as where the product was placed in the semantic structure, communication with friends and marketer dominated communication, to arrive at choice criteria, or the *benefits* the consumer wants from the product. Additionally if more than one benefit is desired, the relative importance, or salience of each benefit must also be determined. Furthermore, the consumer must find how much of each benefit the product is able to provide. Finally, he must have some confidence in the information. At this point the consumer can form an attitude toward, or a customer value for, a new product. This portion of EPS is illustrated with the vegetable bacon in Table 1.

During this stage, the customer will be relatively insensitive to the price. It can be changed substantially within wide limits without causing him to shift. Specifically, his price elasticity is less than one (Simon 1978).

Relation to Adoption and Diffusion

EPS takes place when a person is adopting a new brand in a new category. Adoption or rejection occurs when a consumer has enough confidence to form a value for the new product. While the adoption model outlines a general process, EPS provides considerably more detail about a person's reaction to a product that forms a new product class. Specifically, EPS gives a more complete understanding of how the Knowledge and Evaluation phases of the adoption process take place and it provides marketers with additional insights into ways of speeding up the process. For example, because

TABLE 1.   *Customer Value in Extensive Problem Solving for Lean Strips*

(1) Instrumental Values	(2) Customer Benefits	(3) Importance of Benefits	× 	(4) Amount of Benefits	= 	(5) Contribution of Benefits
Sense of accomplishment		?		?		?
Pleasure +	(Taste)	?		?		?
Equality +						
Cheerful	(Nutrition)	?		?		?
	(Convenience)	?		?		?
	(Expensiveness)	?		?		?
			Customer Value =			SUM

concepts are formed by grouping with similar objects, then distinguishing from them, one is able to influence the process by first suggesting what the new concept is like. If a home computer is grouped with calculators in a person's product hierarchy, most people would say calculators are too complex as it is, and they don't want a more complex calculator. On the other hand if a home computer is compared to a record keeping service, an energy monitoring system, or a personal accountant, the reaction may be much different. Moreover, the marketer may be able to help the consumer distinguish the new product in ways that are most favorable. For example, Lite Beer is like regular beer except it is less filling (not more watery, less alcoholic, or more sissified). Thus, while adoption theory gives considerable guidance about the use of mass media and personal communication and shows what stages a person must be moved through, EPS gives more guidance about the message that will help get him/her through each stage.

EPS also provides additional insight into which individuals will adopt an innovation earlier and which innovations will diffuse faster than others. The people who can fit the innovation into their product hierarchies most easily are going to be the first to adopt or reject the innovations. These will be the people with better developed product hierarchies in that area

or ones who are best able to build new structures rapidly. Typically, these people tend to be interested in that general product category, and/or better educated and of higher socioeconomic status. Furthermore, those people that adopt the innovation are the ones who form positive attitudes towards or perceive a large value for the product. As shown in Table 1, this is a function of the desired benefits and the relative importance of each. Both of these factors will vary across the population.

A number of the characteristics that have been shown to affect the rate of diffusion are related to the ease of grouping the innovation with other products or distinguishing it from them. For example, innovations that are compatible with the experiences of society will be easier to group with other products. Innovations with large relative benefits that are easily observed and communicated can be distinguished from other products more easily. Complex innovations will tend to be harder both to group and to distinguish from competing products. In addition to explaining why these characteristics affect the rate of diffusion, EPS suggests ways of overcoming some of the problems. For example, while the inherent complexity of an innovation cannot be changed, it is possible to make it easier for the consumer to form a concept of it by grouping and distinguishing it appropriately.

LIMITED PROBLEM SOLVING

From the experience of being in EPS, consumers have formed a concept for the product class (attributes used in identification and evaluation have been discovered). They are able to fill in all five columns of Table 1 for the first brand in the product class they encountered.

During the growth stage of the market, new brands will enter the product class either as me too products or as products with different amounts of the various benefits and/or different prices. EPS occurs on the first brand encountered. New brands that the consumer comes in contact with at this stage will be evaluated under limited problem solving (LPS). If these brands are considered to be new by the customers they are also classified as innovations. However, the adoption process for these innovations is going to be much different than the one observed for the first brand in the product class.

While the consumer must still become aware of the new brand, it will be much easier to gain knowledge and make an evaluation of it because the product hierarchy is more fully developed in this area. In Table 1, the first three columns are already filled in. One just needs to determine how much of each desired benefit the new brand possesses and to obtain confidence that he/she can judge it adequately so that the information can be further processed to form an overall attitude. Even these steps should be taken more quickly due to the consumer's experience curve. In summary, the decision to adopt or reject the innovation can be made much more quickly.

Price may become much more of a factor at this time. Partially this is because the consumer has learned enough about the various brands to consider price. It is also due to the fact that some marketers are beginning to use price as a marketing variable. As a dominant design emerges and products become more standardized inter-brand price elasticity will increase. At the same time market segments and specialized products to serve them will be formed. Some may want additional benefits and will be willing to pay for them. Other customers may want a stripped-down version at a lower price.

Relation to Adoption and Diffusion

LPS gives a more complete understanding of the adoption process and suggests that the process is going to differ depending upon the relative newness of the innovation. Furthermore, it gives more guidance to the message that should be presented to the consumer. If many of the people have passed through EPS, a marketer may gain more by focusing the communication program on the specific brand rather than the product class. Boyd, Ray, and Strong, (1973) point out this can be done by (1) focusing on how much of a given benefit the new brand possesses, (2) trying to change the relative importance of the benefits, or (3) trying to change the desired level of each benefit. However, if a seller emphasizes substantially different benefits, he would be moving the buyer back into EPS.

The structure of LPS also explains why second brands may be adopted or rejected more quickly than the first brand. As the product class concept has already been formed, the new brand will not be perceived to be as complex as the first one was, there will generally be less perceived risk, it will be higher in compatibility. Moreover, the seller can focus on the relative advantages of the new brand without needing to spend too much time explaining what the product is.

ROUTINIZED RESPONSE BEHAVIOR

In the mature stage of the PLC, new brands are no longer gaining access to the market. In the extreme case, all buyers know about the available brands. Consequently, the only thing for buyers to consider at this point is whether the benefits offered by each brand have changed since it was last purchased. The benefits that are most likely to have changed, are price and availability. Product quality tends to be stable and product benefits have already been learned from LPS.

Hence in terms of Table 1, they will have to form new beliefs only about those changing benefits of price and availability. Consequently, very little information is required and because it is easily understood, the decision to buy or not buy will be quick.

The content of the information should be mainly to remind him to buy the product class if its consumption can be varied and it is not subject to rigid habits which render its consumption inflexible. However, many customers, even if the market is stable, will not remain loyal to the brand. After a period they begin to look around for something different (Howard, 1981). These need some brand image information for a brand they are not highly familiar with. Consequently, advertising and promotion can play a role for them. The form of the information should stress attention-getting.

Finally, in RRB, price is becoming a more commonly used competitive tool by companies because the competitive brands are becoming much more standardized. Although price can be used to attract customers, the size of the price change does not much affect how much of the product they use. Consequently, the price is again inelastic (elasticity < 1) for the product class, but is more elastic for individual brands.

Relation to Adoption and Diffusion

This type of problem solving behavior does not have an analog in the theory of adoption and diffusion of innovations, because these products are not considered to be new.

SUMMARY

This paper has attempted to integrate two different streams of theory and research relating to new product acceptance. Instead of finding them to be competing approaches they appear to be quite complementary—each broadens the understanding of the other. Howard's model provides an explanation for many of the findings in studies of adoption and diffusion of innovations. Similarly the study of new product adoption increases understanding of consumer behavior generally and integrating theories like Howard's in particular. Moreover, Howard's model provides a way of overcoming the lack of a process orientation in adoption and diffusion research cited by Rogers (1976) in a recent review. It is hoped that by using both of these perspectives, one can better understand consumer behavior and the product life cycle.

REFERENCES

Boyd, Harper W., Jr., Michael L. Ray, and Edward S. Strong (1973), "An Attitudinal Framework for Advertising Strategy," in *Marketing Management and Administrative Action*, Stewart H. Britt and Harper W. Boyd (eds), New York: McGraw-Hill Book Co.

Howard, John A. (1963), *Marketing Management: Analysis and Planning*, Homewood, Ill.: Richard D. Irwin, Inc.

——— (1977), *Consumer Behavior: Application of Theory*, New York: McGraw-Hill Book Co.

——— (1978), "Progress in Modeling Extensive Problem Solving" paper presented at 1978 American Psychological Association Convention.

——— (1980), "Concept of Product Hierarchy" in John R. Rossiter (ed) *Marketing and Competition*, mimeo.

——— (1981a), "Promotion in a Stable Market" *Proceedings, Association for Consumer Research*, forthcoming.

——— (1981b), "The Empirical Theory for Managing the Market," in B. Enis (ed) *Review of Marketing,* American Marketing Association.

——— and Jagdish M. Sheth (1969) *The Theory of Buyer Behavior,* New York: John Wiley & Sons, Inc.

Krugman, Herbert E. (1965) "The Impact of Television Advertising: Learning Without Involvement" *Public Opinion. Quarterly,* 349–56.

Lavidge, Robert J. and Gary A. Steiner, (1961), "A Model for Predictive Measurements of Advertising Effectiveness," *Journal of Marketing.*

Ray, Michael L. (1973), "Marketing Communication and the Hierarchy-of-Effects, Cambridge, Mass.: Marketing Science Institute.

Robertson, Thomas S. (1971), *Innovative Behavior and Communication,* New York: Holt, Rinehart and Winston.

Rogers, Everett M. (1962), *Diffusion of Innovations,* New York: Free Press.

——— (1970), "New Product Adoption and Diffusion," *Journal of Consumer Research,* 2, 290–301.

——— and F. Floyd Shoemaker (1971), *Communication in Innovation,* New York: Free Press.

Simon, Herman (1978), "Dynamics of Price Elasticity and the Product Life Cycle: An Empirical Study," Working Paper, Sloan School of Management, M.I.T.

Webster, Fredrick E., Jr. (1969), "New Product Adoption in Industrial Markets: A Framework for Analysis," *Journal of Marketing,* 33, 35-9.

Zaltman, Gerald and Melanie Wallendorf (1979), *Consumer Behavior: Basic Findings and Managerial Implications,* New York: J. Wiley & Sons, Inc.

Lead Users:
A Source of Novel Product Concepts

Eric von Hippel

Accurate marketing research depends on accurate user judgments regarding their needs. However, for very novel products or in product categories characterized by rapid change—such as "high technology" products—most potential users will not have the real-world experience needed to problem solve and provide accurate data to inquiring market researchers. In this paper I explore the problem and propose a solution: Marketing research analyses which focus on what I term the "lead users" of a product or process.

Lead users are users whose present strong needs will become general in a marketplace months or years in the future. Since lead users are familiar with conditions which lie in the future for most others, they can serve as a need-forecasting laboratory for marketing research. Moreover, since lead users often attempt to fill the need they experience, they can provide new product concept and design data as well.

In this paper I explore how lead users can be systematically identified, and how lead user perceptions and preferences can be incorporated into industrial and consumer marketing research analyses of emerging needs for new products, processes and services. (MARKETING—NEW PRODUCTS; RESEARCH AND DEVELOPMENT; INNOVATION MANAGEMENT)

1. INTRODUCTION

Accurate understanding of user need has been shown near-essential to the development of

Reprinted by permission of the publisher from "Lead Users: A Source of Novel Product Concepts" by Eric von Hippel, from *Management Science*, July 1986, Volume 32, Number 7. Copyright 1986 by The Institute of Management Sciences.

I wish to gratefully acknowledge the helpful comments and suggestions provided by Professor Glen L. Urban of MIT's Sloan School of Management and Professor John H. Roberts of the Australian Graduate School of Management.

commercially successful new products (Rothwell et al. 1974, Achilladelis et al. 1971). Unfortunately, current market research analyses are typically not reliable in the instance of very novel products or in product categories characterized by rapid change, such as "high technology" products. In this paper I explore the problem and propose a solution: marketing research analyses which focus on what I term the "lead users" of a product or process.

Lead users are users whose present strong needs will become general in a marketplace months or years in the future. Since lead users are familiar with conditions which lie in the future for most others, they can serve as a need-forecasting laboratory for marketing research. Moreover, since lead users often attempt to fill the need they experience, they can provide new product concept and design data as well. How lead users can be systematically identified, and how their perceptions and preferences incorporated into industrial and consumer marketing research analyses of emerging needs for new products, processes and services is examined below.

2. MARKETING RESEARCH CONSTRAINED BY USER EXPERIENCE

Users selected to provide input data to consumer and industrial market analyses have an important limitation: Their insights into new product (and process and service) needs and potential solutions are constrained by their own real-world experience. Users steeped in the present are thus unlikely to generate novel product concepts which conflict with the familiar.

The notion that familiarity with existing product attributes and uses interferes with an individual's ability to conceive of novel attributes and uses is strongly supported by research into problem solving (Table 1). We see that experimental subjects familiar with a complicated problem-solving strategy are unlikely to devise a simpler one when this is appropriate (Luchins 1942). Also, and germane to our present discussion, we see that subjects who use an object or see it used in a familiar way are strongly blocked from using that object in a novel way (Duncker 1945, Birch and Rabinowitz 1951, Adamson 1952). Furthermore, the more re-

TABLE 1.   *The Effect of Prior Experience on Users' Ability to Generate or Evaluate Novel Product Possibilities*

Study	Nature of Research	Impact of Prior Experience on Ability to Solve Problems
Luchins (1942)	Two groups of subjects ($n =$) were given a series of problems involving water jars, e.g.: 'If you have jars of capacity A, B and C how can you pour water from one to the other so as to arrive at amount D?' Subject group 1 was given 5 problems solvable by formula, $B - A - 2C = D$. Next, both groups were given problems solvable by that formula *or* by a simpler one (e.g. $B - C = D$).	81% of experimental subjects who had previously learned a complex solution to a problem type applied it to cases where a simple solution would do. No control group subjects did so ($p =$ NA[a]).
Duncker (1945)	The ability to use familiar objects in an unfamiliar way was tested by creating 5 problems which could only be solved by that means. (For example, one problem could be solved only if subjects bent a paper clip provided them and used it as a hook.) Subjects were divided into two groups. One group of problem solvers saw the crucial objects being used in a familiar way (e.g. the paper clip holding papers), the other did not (e.g. the paper clip was simply lying on a table unused).	Subjects were much more likely to solve problems requiring the use of familiar objects (e.g. paper clips) in unfamiliar ways (e.g. beat into hooks) if they had not been shown the familiar use just prior to their problem-solving attempt. Duncker called this effect "functional fixedness" ($n = 14$; $p =$ NA[a]).

353

Study	Nature of Research	Impact of Prior Experience on Ability to Solve Problems
Birch and Rabinowitz (1951)	Replication of Duncker, above	Duncker's findings confirmed ($n = 25$; $p < 0.05$)
Adamson (1952)	Replication of Duncker, above	Duncker's findings confirmed ($n = 57$; $p < 0.01$)
Adamson and Taylor (1954)	The variation of "functional fixedness" with time was observed by the following procedure. First, subjects were allowed to use a familiar object in a familiar way. Next, varying amounts of time were allowed to elapse before subjects were invited to solve a problem by using the object in an unfamiliar way.	If a subject uses an object in a familiar way, he is partially blocked from using it in a novel way. ($n = 32$; $p < 0.02$) This blocking effect decreases over time (see graph).

Functional fixedness as a function of log time

80
70
60
50 — Chance
40

1 min. / ½ hr. / 1 hr. / 1 day / 1 wk.
Log time |
| Allen and Marquis (1964) | Government agencies often buy R & D services via a "Request for Proposal" (RFP) which states the problem to be solved. Interested bidders respond with Proposals which outline their planned solutions to the problem and its component tasks. In this research, relative success of eight bidders' approaches to the component tasks contained in 2 RFPs was judged by the Agency buying the research ($n = 26$). Success was then compared to prior research experience of bidding laboratories. | Bidders were significantly more likely to propose a successful task approach if they had prior experience with that approach only, rather than prior experience with inappropriate approaches only. |

[a]This relatively early study showed a strong effect but did not provide a significance calculation—or present data in a form which would allow one to be determined without ambiguity.

354

cently objects or problem-solving strategies have been used in a familiar way, the more difficult subjects find it to employ them in a novel way (Adamson and Taylor 1954). Finally, we see that the same effect is displayed in the real world, where the success of a research group in solving a new problem is shown to depend on whether solutions it has used in the past will fit that new problem (Allen and Marquis 1964). These studies thus suggest that typical users of existing products—the type of user-evaluators customarily chosen in market research—are poorly situated with regard to the difficult problem-solving tasks associated with assessing unfamiliar product and process needs.

As illustration, consider the difficult problem-solving steps which potential users must go through when asked to evaluate their need for a proposed new product. Since individual industrial and consumer products are only components in larger usage patterns which may involve many products, and since a change in one component can change perceptions of and needs for some or all other products in that pattern, users must first identify their existing multiproduct usage patterns in which the new product might play a role. Then they must evaluate the new product's potential contribution to these. (For example, a change in the operating characteristics of a computer may allow a user to solve new problem types if he makes related changes in software and perhaps in other, related products and practices. Similarly, a consumer's switch to microwave cooking may well induce related changes in food recipes, kitchen practices, and kitchen utensils.) Next, users must invent or select the new (to them) usage patterns which the proposed new product makes possible for the first time, and evaluate the utility of the product in these. Finally, since substitutes exist for many multiproduct usage patterns (e.g., many forms of problem analysis are available in addition to the novel ones made possible by a new com-

puter) the user must estimate how the new possibilities presented by the proposed new product will compete (or fail to compete) with existing options. This problem-solving task is clearly a very difficult one, particularly for typical users of existing products whose familiarity with existing products and uses interferes with their ability to conceive of novel products and uses when invited to do so.

The constraint of users to the familiar pertains even in the instance of sophisticated consumer marketing research techniques such as multiattribute mapping of product perceptions and preferences (Silk and Urban 1978, Shocker and Srinivasan 1979, Roberts and Urban 1985). "Multiattribute" (multidimensional) marketing research methods, for example, describe a consumer's perception of new and existing products in terms of a number of attributes (dimensions). If and as a complete list of attributes is available for a given product category, a consumer's perception of any particular product in the category can be expressed in terms of the amount of each attribute the consumer perceives it to contain, and the difference between any two products in the category can be expressed as the difference in their attribute profiles. Similarly, consumer preferences for existing and proposed products in a category can in principle be built up from consumer perceptions of the importance and desirability of each of the component product attributes.

Although these methods frame user perceptions and preferences in terms of attributes, they do not offer a means of going beyond the experience of the users interviewed. First, for reasons discussed above, user subjects are not well positioned to accurately evaluate novel product attributes or "amounts" of familiar product attributes which lie outside the range of their real-world experience. Second, and more specific to these techniques, there is no mechanism to induce users to identify all product attributes potentially relevant to a product category, especially attributes which are currently not present in any extant category member. To illustrate this point, consider two types of such methods, similarity-dissimilarity ranking and focus groups.

In similarity-dissimilarity ranking, data regarding the perceptual dimensions by which consumers characterize a product category are generated by inviting a sample of consumers to compare products in that category and assess their similarity-dissimilarity. In some variants of the method, the consumer specifies the ways in which the products are similar or different. In others, the consumer simply provides similarity and difference rankings, and the market analyst determines—via his personal knowledge of the product type in question, its function, the marketplace, the consumer, etc.—the important perceptual dimensions which "must" be motivating the consumer rankings obtained.

The similarity-dissimilarity method clearly depends heavily on an analyst's qualitative ability to interpret the data and correctly identify all the critical dimensions. Moreover, by its nature, this method can only explore perceptions derived from attributes which exist in or are associated with the products being compared. Thus, if a group of consumer evaluators is invited to compare a set of cameras and none has a particular feature—say, instant developing—then the possible utility of this feature would not be incorporated in the perceptual dimensions generated. That is, the method would have been blind to the possible value of instant developing prior to Edwin Land's invention of the Polaroid camera.

In focus group methods, market analysts assemble a group of consumers familiar with a product category for a qualitative discussion of perhaps two hours' duration. The topic for the focus group, which is set by an analyst, may be relatively narrow (e.g., "35 mm amateur cameras") or somewhat broader (e.g., "the photo-graphic experience as you see it"). The ensuing discussion is recorded, transcribed, and later reviewed by the analyst whose task it is to identify the important product attributes which have implicitly or explicitly surfaced during the conversation. Clearly, as with similarity-dissimilarity ranking, the utility of information derived from focus group methods depends heavily on the analyst's ability to accurately and completely abstract from the interview data the attributes which consumers feel important in products.

In principle, however, the focus group technique need not be limited to only identifying attributes already present in existing products, even if the discussion is nominally focused on these. For example, a topic which extends the boundaries of discussion beyond a given product to a larger framework could identify attributes not present in any extant product in a category under study. If discussion of the broad topic mentioned earlier, "the photographic experience as you see it," brought out consumer dissatisfaction with the time lag between picture taking and receipt of the finished photograph, the analyst would be in possession of information which could induce him to identify an attribute not present in any camera prior to Land's invention, instant film development, as a novel and potentially important attribute.

But how likely is it that an analyst will take this creative step? And, more generally, how likely is it that either method discussed above, similarity-dissimilarity ranking or focus groups, will be used to identify attributes not present in extant products of the type being studied, much less a complete list of all relevant attributes? Neither method contains an effective mechanism to encourage this outcome, and discussions with practitioners indicate that in present-day practice, identification of any novel attribute is unlikely.

Finally, both of these methods conventionally focus on familiar product categories.

This restriction, necessary to limit the number of attributes which "completely describe" a product type to a manageable number, also tends to limit consumer perceptions to attributes which fit products within the frame of existing product categories. Modes of transportation, for example, logically shade off into communication products as partial substitutes ("I can drive over to talk to him—or I can phone"), into housing and entertainment products ("We can buy a summer house—or go camping in my recreational vehicle"), indeed, into many other of life's activities. But since a complete description of life cannot be compressed into 25 attribute scales, the analysis is constrained to a narrower—usually conventional and familiar—product category or topic. This has the effect of rendering any promising and novel cross-category new product attributes less visible to the methods I have discussed.

In sum, then, we see that marketing researchers face serious difficulties if they attempt to determine new product needs falling outside of the real-world experience of the users they analyze.

3. LEAD USERS' EXPERIENCE IS NEEDED FOR MARKETING RESEARCH IN FAST-MOVING FIELDS

In many product categories, the constraint of users to the familiar does not lessen the ability of marketing research to evaluate needs for new products by analyzing typical users. In the relatively slow-moving world of many consumer products, new cereals and new car models do not often differ radically from their immediate predecessors. Therefore, even the "new" is reasonably familiar, and the typical user can thus play a valuable role in the development of new products.

In contrast, in high technology industries, the world moves so rapidly that the related real-world experience of ordinary users is often rendered obsolete by the time a product is developed or during the time of its projected commercial lifetime. For such industries I propose that "lead users" who *do* have real-life experience with novel product or process concepts of interest are essential to accurate marketing research. Although the insights of lead users are as constrained to the familiar as those of other users, lead users are familiar with conditions which lie in the future for most—and so are in a position to provide accurate data on needs related to such future conditions.

I define "lead users" of a novel or enhanced product, process or service as those displaying two characteristics with respect to it:

· Lead users face needs that will be general in a marketplace—but face them months or years before the bulk of that marketplace encounters them, *and*
· Lead users are positioned to benefit significantly by obtaining a solution to those needs.

These two lead user characteristics are shown schematically in Figure 1. Two specific examples of lead users: Firms who today need and could obtain significant benefit from a type of office automation which the general market will need tomorrow are lead users of office automation; a semiconductor producer with a current strong need for a process innovation which many semiconductor producers will need in two-years' time is a lead user with respect to that process.

Users whose present needs foreshadow general demand exist because important new technologies, products, tastes, and other factors related to new product opportunities typically diffuse through a society, often over many years, rather than impact all members simultaneously (Rogers and Shoemaker 1971). Thus, when

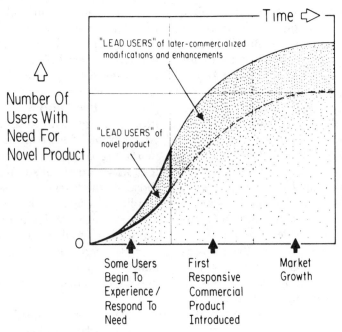

FIGURE 1.   A Schematic of Lead Users' Position in the Life Cycle of a Novel Product. Process or Service. Lead Users (1) Encounter the Need Early and (2) Expect High Benefit from a Responsive Solution (Higher expected benefit indicated by deeper shading)

Mansfield (1968) explored the rate of diffusion of twelve very important industrial goods innovating into major firms in the bituminous coal, iron and steel, brewing, and railroad industries, he found that in 75 percent of the cases it took over 20 years for complete diffusion of these innovations to major firms. Accordingly, some users of these innovations could be found far in advance of the general market.

Users of new products and processes have been shown to differ on the level of benefit they can obtain from these (Mansfield 1968). The greater the benefit a given user can obtain from a needed novel product or process, the greater his effort to obtain a solution will be. (This link between innovation activity and expectation of economic benefit was first empirically established by Jacob Schmookler (1966), who conducted a careful study of the correlation between changes in sales volumes of some

capital goods and appropriately lagged changes in rates of patent applications in categories related to those goods.) I therefore reason that users able to obtain the highest net benefit from the solution to a given new product (or process or service) need will be the ones who have devoted the most resources to understanding it. And it follows that this subset of users should have the richest real-world understanding of the need to share with inquiring market researchers.

4. UTILIZING LEAD USERS IN MARKETING RESEARCH

How then can lead users be incorporated into marketing research? I suggest a four-step process:

1. Identify an important market or technical trend;
2. Identify lead users who lead that trend in terms of (a) experience and (b) intensity of need;
3. Analyze lead user need data;
4. Project lead user data onto the general market of interest.

I consider how each of these steps might be approached below with regard to industrial and consumer products.

4.1. Identifying an Important Trend

Lead users are defined as being in advance of the market with respect to a given important dimension which is changing over time. Therefore, before one can identify lead users in a given product category of interest, one must identify the underlying trend on which these users have a leading position.

Identification of important trends affecting promising markets is already commonly performed by many firms as a necessary component of their corporate strategy. Methods used range from the intuitive judgments of experts, perhaps formalized in a technique such as the "Delphi" method, to simple trend extrapolation to more complex correlational or econometric models. (See Chambers, Mullick, and Smith (1971), for a useful practitioner's overview. See Martino (1972) for the special case of forecasting trends in technology.) Despite the existence of formal trend assessment methods, however, trend identification and assessment remains something of an art. Thus, analysts typically must judge which of many important trends in a market they will focus on, or must combine several into a suitable index variable.

In the case of industrial goods, trend identification and assessment can often be both informal and accurate. Since potential buyers typically measure the value of proposed new industrial products in economic terms, impor-tant underlying trends related to product value are often inescapably clear to those in the industry. For example, it is clear to those in the semiconductor and computer fields that computer memory and microprocessor chips are getting more capable and less expensive for a given capability every year. It is also clear that, for technical reasons, this trend is also likely to continue for a number of years. Finally, it is clear that this trend has very important cost/performance implications for firms which incorporate these semiconductors in computers or myriad other increasingly "intelligent" products.

In the case of consumer goods, accurate trend identification is often more difficult because there is often no underlying stable basis for comparison such as that played by economic value for industrial goods. Therefore, while consumer perceptions of trends and their subjective assessment of the importance of these can be determined straightforwardly by survey at any given point in time, these perceptions may not be consistent over time. (For example, we cannot expect to predict the trend in consumer interest in auto fuel economy as a function of fuel cost as accurately as we could predict industrial buyer interest in fuel economy on that basis.)

In sum, reliable methods for formal prediction of trends over time which will have an important effect on a given product area are not yet well developed. In some product areas, however, notably in industrial goods, the needed data on important trends are clear to those with expertise, and in these instances the poor state of formal methods is not an impediment to incorporating analyses of lead users into marketing research.

4.2 Identifying Lead Users

Once a firm has identified one or more significant trends which appear associated with promising new product opportunities, the market researcher can begin to search for lead users, users (1) who are at the leading edge of each identified trend in terms of related new product and process needs and (2) who expect to obtain a relatively high net benefit from solutions to those needs. Let us consider practical means for identifying lead users in the instance of industrial goods and consumer goods in turn.

The first task, identifying users at the leading edge of a given trend, is usually straightforward in the case of industrial goods because a given firm's position on a range of trends is usually well known to industry experts. Thus, in many instances, industrial good manufacturers have only a few or a few score major potential customers for a given product type and often know the characteristics of each user quite well. As illustration, recall the important trend toward cheaper, more capable computer memory and microprocessor semiconductor chips. Manufacturers of semiconductor process equipment would recognize that cheaper, more capable semiconductors are achieved in major part by the packing of circuit elements ever more densely on chip surfaces by semiconductor makers (equipment users), a trend involving significant new user needs for semiconductor process equipment. They would also know that many users at the leading edge of the need for density trend are makers of VLSI memories, with makers of other types of semiconductors such as linear ICs lying further back on the curve. Therefore, VLSI memory manufacturers can be flagged as potential lead users of process equipment with regard to this trend.

The second task is identifying the subset of those user firms positioned at the forefront of the trend under study who are also able to obtain relatively high "net benefit" from adopting a solution to trend-related needs. In the case of industrial goods, net benefit is typically measured in economic terms. And when this is so, the net benefit (B) which a user firm expects to obtain from a solution to a given need can be

stated as: $B = (V)(R) - C - D$ where (V) is the dollar "volume" of product sales or processing activity to which the user plans to apply his solution; (R) is the increased rate of profit per dollar of this volume resulting from application of that solution; (C) is the user's anticipated costs in developing and/or adopting the solution; and (D) is the net benefit which the user would have obtained from old practices; equipment, etc., displaced by the novel solution. (Industrial firms typically make the calculation described above when they assess the return on investment they may anticipate from investing in a new product, process or service. I will not describe specific methods of making such calculations here. General methods can be found in accounting texts. A discussion of net benefit calculations useful in the specific instance of innovations will be found in von Hippel (1982).)

An additional, very practical method for identifying lead users involves identifying those users who are actively innovating to solve problems present at the leading edge of a trend. Thus, the semiconductor process equipment makers mentioned in our earlier example could seek those few VLSI memory manufacturers (equipment users) who are actively developing processes for the manufacture of denser chips. A user conducting such R & D is probably a lead user because innovation is expensive, and the user engaged in it surely expects to reap high net benefit from a problem solution. Identifying lead users by seeking innovating users can be very economical, because the identity of users conducting R & D on a given problem area is often common knowledge to industry participants.

In the case of consumer goods, lead users with respect to specified trends can readily be identified by appropriately designed surveys. For example, if the trend toward increasing consumption of "health foods" is selected, a survey of consumer food preferences could identify those on the leading edge of that trend.

The lead users among this group could then be identified by additional questions concerning the value respondents place on improvements in the healthfulness of food. (Such a screening question might be: "How much extra would you be willing to pay for X food free of Y additive?") Those found to place a significantly higher value than most on such improvements, e.g., x standard deviations above the mean, are the users who anticipate obtaining the highest net benefit from a solution to the need. They are therefore lead users with respect to this trend.

Finally, three important complexities with regard to identifying lead users should be noted. First, key lead users should not necessarily be sought within the usual customer base of the manufacturer performing the market research. They may be customers of a competitor—or totally outside of the industry he serves. For example, if a manufacturer of composite materials used in autos identifies an important trend toward lighter, higher strength materials, he may find the lead users at the front of this trend are aerospace firms rather than auto firms, because aerospace firms may be willing to pay more than auto firms for improvements on these attributes. Often, consumer products manufacturers will find valuable lead users among users of analogous industrial goods, because the benefit which an industrial user can expect from a given advance often far outstrips that which an individual consumer could expect. Thus, a manufacturer seeking to develop centralized controllers for home heating, lighting and security systems might well seek lead users among firms seeking to use controllers of similar function in commercial buildings. (Note that one must always be aware of both similarities and differences between the lead users one is assessing and the user population one intends to serve. This point will be developed in §4.4.)

The second complexity with respect to identifying lead users is that one need not be restricted to identifying lead users who can illumine the *entire* novel product, process or service which one wishes to develop. One may also seek out those who are lead users with respect to only a few of its attributes—or indeed of a single attribute, defined as narrowly as one likes. Thus, to elaborate on the example begun above, a manufacturer of centralized controllers of home systems might seek lead users with respect to the energy management aspects of such a controller among firms using controllers of analogous function in commercial or industrial applications where a great deal of energy is used and/or energy costs are high. At the same time, he might seek lead users with respect to the security system attributes of such a home controller system among a totally different set of users—perhaps individuals or firms who feel at high risk for burglary and have very valuable goods to protect.

The third and final complexity regarding the identification of lead users which I will mention has the following source: Users driven by expectations of high net benefit to develop a solution to a need might well have solved *their* problem and no longer feel that need. Therefore a survey seeking to identify lead users on the basis of high unmet need might not identify these particular users. This can be a significant loss, since lead users who perceive that they have successfully developed a responsive solution to the need at issue clearly have valuable data for market researchers. In practice, however, I find that in the instance of industrial products, users at the leading edge of an important (moving) trend have to innovate again and again to maintain a level of satisfaction with their current practice, and so will seldom express expectations of low net benefit from additional improvements over their present practice. In the instance of consumer products, the problem can be addressed via additional survey questions specifically inquiring about possible consumer-developed solutions to the needs under study.

4.3. Analyzing Lead User Data

Data derived from lead users and their real-life experience with novel attributes and/or product concepts of commercial interest can be incorporated in market research analyses using standard market research methods. However, the analyst might wish to be on the lookout for somewhat more user-developed product solutions and more substantive need statements in lead user data than he is used to finding in analyses of other user populations. Recall that, since problem-solving activity has been shown to be motivated by expectations of economic benefit, and since lead users have been defined in part as users positioned to obtain high net benefit from a solution to their needs, it is reasonable that lead users may have made some investment in solving the need at issue. Sometimes lead user problem-solving activity takes the form of applying existing commercial products or components in ways not anticipated by their manufacturers. Sometimes lead users may have developed complete new products responsive to their need.

Product development by users receiving relatively high returns from such activity has been empirically documented (von Hippel 1982). In some product areas (e.g., semiconductor process machinery (von Hippel 1977) and scientific instruments (von Hippel 1976)), moreover, users have developed *most* of the commercially successful product innovations. To illustrate, the results of several studies of the functional locus of innovation are summarized in Table 2. (The absence of actual product development by users in a given area (e.g., engineering plastics (Berger 1975) and conductor attachment equipment (VanderWerf 1982)) does not mean that lead users with their rich insights are absent here; it simply means that the distribution of economic benefits flowing from product development in that area makes product development by manufacturers or other nonuser groups so attractive that nonusers preempt user product development activity (von Hippel 1982).)

Users develop both industrial and consumer products. An example of each will help convey the flavor:

> IBM designed and built the first printed circuit card component insertion machine of the X-Y Table type to be used in commercial production. (IBM needed the machine to insert components into printed circuit cards which were in turn incorporated into computers.) After building and testing the design in-house, IBM sent engineering drawings of their design to a local machine builder along with an order for eight units. The machine builder completed this and subsequent orders satisfactorily and, two years later, applied to IBM for permission to build essentially the same machine for sale on the open market. IBM agreed and the machine builder became the first commercial manufacturer of X-Y Table component insertion machines. (The above episode marked that firm's first entry into the component insertion equipment business. They are a major factor in the business today.) (von Hippel 1977)

> In the early 1970's, store owners and salesmen in southern California began to notice that youngsters were fixing up their bikes to look like motocycles, complete with imitation tailpipes and "chopper-type" handlebars. Sporting crash helmets and Honda motorcycle T-shirts, the youngsters raced their fancy 20-inchers on dirt tracks.

> Obviously on to a good thing, the manufacturers came out with a whole new line of "motocross" models. By 1974 the motorcycle-style units accounted for 8 percent of all 20-inch bicycles shipped. Two years later half of the 3.7 million new juvenile bikes sold were of the motocross model. . . . (*New York Times* 1978)

Of course, completely developed new products are not the only useful "solution data" available from lead users. All need statements implicitly or explicitly contain more or less information about possible solutions to the need at issue. Consider the following sequence of need statements which deal with a consumer product:

TABLE 2. *Data Regarding the Role of Users in Product Development*

Study	Nature of Innovations and Sample Selection Criteria	R	Innovative Product Developed by:[a]		
			User	Mfg.	Other
Knight (1963)	Computer innovations 1944–1962:				
	—systems reaching new performance high	143	25%	75%	
	—systems with radical structural innovations	18	33%	67%	
Enos (1962)	Major petroleum processing innovations	7	43%	14%	43%[b]
Freeman (1968)	Chemical processes and process equipment available for license, 1967	810	70%	30%	
Berger (1975)	All engineering polymers developed in U.S. after 1955 with > 10^6 pounds produced in 1975	6	0%	100%	
Boyden (1976)	Chemical additives for plastics—all plasticizers and UV stabilizers developed post World War II for use with 4 major polymers	16	0%	100%	
Lionetta (1977)	All pultrusion processing machinery innovations first introduced commercially 1940–1976 which offered users a major increment in functional utility[c]	13	85%	15%	
von Hippel (1976)	Scientific instrument innovations:				
	—first of type (e.g. first NMR)	4	100%	0%	
	—major functional improvements	44	82%	18%	
	—minor functional improvements	63	70%	30%	
von Hippel (1977)	Semiconductor and electronic subassembly manufacturing equipment:				
	—first of type used in commercial production	7	100%	0%	
	—major functional improvements	22	63%	21%	16%[d]
	—minor functional improvements	20	59%	29%	12%[d]
VanderWerf (1982)	Wirestripping and connector attachment equipment	20	11%	33%	56%[e]

[a]NA data excluded from percentage computations.
[b]Attributed to independent inventors/invention development companies.
[c]Figures shown are based on reanalysis of Lionetta's (1977) data.
[d]Attributed to joint user-manufacturer innovation projects.
[e]Attributed to connector suppliers.

- I am unhappy . . .
- about my children's clothes . . .
- which are often not fully clean even when just washed.
- I find that X type stains on Y type clothes are especially hard to remove.

- If I mix my powdered detergent into a paste and apply it to the stain before washing, I find it helps get things clean.

Each succeeding statement clearly provides a valuable increment of data useful for defining a new product need and devising a responsive solution. On the basis of the last statement we see that liquid detergent could be invented. We also are able to learn that the user is approaching the problem as a "stain removal" problem rather than "keep the kids away from X staining agent" problem. And, probably, the user is ranking this choice after having experimented with both approaches. In essence, such *experience* with the need/problem is what makes lead user's data so valuable.

4.4. Projecting Lead User Data onto the General Market of Interest

The needs of today's lead users are typically not precisely the same as the needs of the users who will make up a major share of tomorrow's predicted market. Indeed, the literature on diffusion suggests that, in general, the early adopters of a novel product or practice differ in significant ways from the bulk of the users who follow them (Rogers and Shoemaker 1971). Thus, analysts will need to assess how lead user data apply to the more typical user in a target market rather than simply assume such data straightforwardly transferable.

In the instance of industrial goods, the translation problem is typically not serious. As we pointed out earlier, industrial products are typically evaluated on economic grounds whereby users calculate the relative costs and benefits of the proposed product. When an objective economic analysis is possible, all users—not just lead users—will make similar calculations and thus provide a common basis for market projections.

In the instance of consumer goods, and in the instance of industrial goods for which the costs and benefits of the proposed product for the user do not form the basis for product preferences, a test of the applicability of lead user needs and concepts to the future general market is not so simple. One approach involves prototyping the novel product and asking a sample of typical users to use it. Such users would then be in a position to provide accurate product evaluation data to market research (a) if presenting the user with the product created conditions for him similar to the conditions a future user would face, and (b) if the user were given enough time to fully explore the new product and fully adapt his usage patterns to it. If a new product were being tested in this manner in a field where little else was expected to change by the time of product introduction—say, a new detergent for the home laundry—conditions (a) and (b) could probably be effectively met. However, in rapidly moving fields in which the proposed new product will interact with many other not-yet-developed products in unforeseen ways, new approaches may be needed.

5. SUMMARY

In this paper I have defined lead users, and explored the valuable insights they can offer regarding needs—and, often, prototype solutions—for novel products, processes and services. I have also presented four general steps by which one may identify and analyze lead users in any given instance. Practitioners may wish to use the method in its present early form, while researchers may wish to explore, test and refine it. Both are possible, and we make suggestions regarding each below.

I suggest that interested practitioners have no hesitation about experimenting with the general methodology described here. Lead users are often accessible enough to allow successful identification of interesting data regarding desirable new products and/or product modifications with little effort—given that the practitioner has a good knowledge of the customers and application area he is analyzing. As evidence, during the past two years Professor Glen Urban and I have helped approximately

100 MIT Master's students to undertake short projects involving the identification of lead users in areas they were familiar with. With very little coaching, almost all have succeeded. (Examples: lead users of sports equipment have been identified and studied in sports ranging from rock climbing to trail biking to street hockey. Other projects have dealt with lead users of various types of industrial process equipment and various types of computer hardware and software.) Therefore, I urge practitioners to "learn by doing," and conduct a rough initial test for their own interest and satisfaction. If the results are positive, I hope they will be motivated to do still more.

Researchers who wish to systematically explore the value of lead user methods will find many possible approaches. I propose that initial empirical studies of the value of lead user data be focused on industrial goods rather than consumer goods. (As was noted earlier, lead users of industrial goods can typically be identified more reliably than lead users of most consumer goods given today's state of the art.) The value of lead user data under real-world conditions can be assessed via a longitudinal study design which tests the predictive accuracy of data collected earlier from lead users against the actual future general market as it evolves. A less ambitious effort could focus on industrial products and compare the economic performance of novel product concepts proposed by lead versus typical users.

Researchers who wish to improve lead user methods will find much needs to be done in both industrial and consumer goods arenas. For example, the means for identifying lead users and analyzing their needs obviously must be improved and extended. Also, organizational schemes for routinely acquiring lead user data (special interface groups, special incentives which will induce lead users to interact on a continuing basis, etc.) need to be developed and tested. Valuable research on all these topics appears to be exciting and well within reach.

REFERENCES

Rothwell, R., C. Freeman, et al., "SAPPHO Updated—Project SAPPHO Phase II," *Research Policy*, 3 (1974), 258–291. See also Achilladelis, B., A. B. Robertson and P. Jervis, *Project SAPPHO: A Study of Success and Failure in Industrial Innovation*, 2 vols., Center for the Study of Industrial Innovation, London, 1971.

Luchins, Abraham S., "Mechanization in Problem Solving: The Effect of Einstellung," *Psychological Monographs*, 54 (6) (1942) (Whole No. 248).

Duncker, K., "On Problem-Solving," trans. Lynne S. Lees, *Psychological Monographs*, 58 (5) (1945) (Whole No. 270).

Birch, Herbert G. and Herbert J. Rabinowitz, "The Negative Effect of Previous Experience on Productive Thinking," *J. Experimental Psychology*, 41 (2) (February 1951), 121–125.

Adamson, Robert E., "Functional Fixedness as Related to Problem Solving: A Repetition of Three Experiments," *J. Experimental Psychology*, 44 (4) (October 1952), 288–291.

Adamson, Robert E. and Donald W. Taylor, "Functional Fixedness As Related to Elapsed Time and to Set," *J. Experimental Psychology*, 47 (2) (February 1954), 122–126.

Allen, T. J. and D. G. Marquis, "Positive and Negative Biasing Sets: The Effects of Prior Experience on Research Performance," *IEEE Trans. Engineering Management*, EM-11 (4) (December 1964), 158–161.

Silk, Alvin J. and Glen L. Urban, "Pre-Test-Market Evaluation of New Packaged Goods: A Model and Measurement Methodology," *J. Marketing Res.*, 15 (May 1978), 189; Also, Shocker, Allan D. and V. Srintvasan, "Multiattribute Approaches for Product Concept Evaluation and Generation: A Critical Review," *J. Marketing Res.*, 16 (May 1979), 159–180.

Roberts, John H. and Glen L. Urban, "New Consumer Durable Brand Choice: Modeling Multiattribute Utility, Risk, and Dynamics," Working Paper WP 1636–85, MIT Sloan School of Management, Cambridge, MA 1985.

Rogers, Everett M. with F. Floyd Shoemaker, *Communication of Innovations: A Cross-Cultural Approach*, 2nd ed., The Free Press, New York, 1971.

Mansfield, Edwin, *Industrial Research and Technical Innovation: An Econometric Analysis,* W. W. Norton, New York, 1968, 134–235.

Schmookler, Jacob, *Invention and Economic Growth,* Harvard University Press, Cambridge, MA. 1966.

Chambers, John C., Satinder K. Mullick and Donald D. Smith, "How to Choose the Right Forecasting Technique," *Harvard Business Rev.,* (July–August 1971), 45–74, provides a useful practitioner's overview.

Martino, Joseph P., *Technological Forecasting for Decision-Making,* American Elsevier, New York, 1972.

von Hippel, Eric, *The Sources of Innovation,* Oxford University Press, New York, forthcoming. Also, Eric von Hippel, "Appropriability of Innovation Benefit as a Predictor of the Source of Innovation," *Research Policy,* 11 (1982), 95–115.

Knight, K. E., "A Study of Technological Innovation: The Evolution of Digital Computers," Ph.D. dissertation, Carnegie Institute of Technology, 1963. Data shown in Table 2 of this paper obtained from Knight's Appendix B, parts 2 and 3.

Enos, John Lawrence, *Petroleum Progress and Profits: A History of Process Innovation,* MIT Press, Cambridge, 1962.

Freeman, C., "Chemical Process Plant: Innovation and the World Market," *National Institute Economic Rev.,* 45 (August 1968), 29–57.

Berger, Alan J., "Factors Influencing the Locus of Innovation Activity Leading to Scientific Instrument and Plastics Innovations," S.M. thesis, MIT Sloan School of Management, Cambridge, MA. 1975.

Boyden, Julian W., "A Study of the Innovative Process in the Plastics Additives Industry," S.M. thesis, MIT Sloan School of Management, Cambridge, MA, 1976.

Lionetta, William G., Jr., "Sources of Innovation Within the Pultrusion Industry," S.M. thesis, MIT Sloan School of Management, Cambridge, MA. 1977.

von Hippel, Eric, "The Dominant Role of Users in the Scientific Instrument Innovation Process," *Research Policy,* 5 (1976), 212–239.

von Hippel, Eric, "The Dominant Role of Users in Semiconductor and Electronic Subassembly Process Innovation," *IEEE Trans. Engineering Management,* EM-24 (2) (May 1977), 60–71.

VanderWerf, Pieter, "Parts Suppliers and Innovators in Wire Termination Equipment," Working Paper WP 1289-82, MIT Sloan School of Management, Cambridge, MA, 1982.

New York Times, 29 January 1978, F3.

Concept Testing

William L. Moore

This paper is a review of concept testing based on the published literature and a series of personal interviews with leading practitioners. While there is considerable agreement on the usefulness of concept testing, practitioners disagree on the best way to perform them. In addition to highlighting these disagreements, the paper covers general suggestions for improving concept testing.

Even a cursory review of marketing literature [8, 12, 36, 37] reveals that a major problem facing companies today is reducing new product failures and improving the return on funds invested in new product activities. While many factors contribute to the high failure rate, a major factor is the inability to predict consumer response to new products and services [9, 10, 11, 12, 33].

One way to decrease new product failures is through systematic testing which includes concept testing, product testing (including extended use testing), and market testing. The purpose of this paper is to examine one part of this process, concept testing, with regard to current practices and possible future directions. It is based on a review of the pub-

Reprinted by permission of the publisher from "Concept Testing" by William L. Moore, *Journal of Business Research,* Volume 10, pp. 279–294. Copyright 1982 by Elsevier Science Publishing Co., Inc.

The author thanks Newton Frank for suggesting a number of the participants in the study, and Noel Capon, Stephen J. Cook, Newton Frank, Neil B. Holbert, Morris B. Holbrook, Ian M. Lewis, John L. McMennamin, Tibor Weiss, Pamela J. Welker, and Stephen K. Zrike for comments on an earlier draft.

lished literature and a number of personal interviews with leading practioners. (See the table below for a list of participants and a discussion of methodology.)

CONCEPT TESTING PROCESS

The primary purpose of concept testing is to estimate consumer reactions to a product idea before committing substantial funds to it. Additionally, concept tests are used to determine the potential target market and how the concept might be improved. There can be some confusion about concept tests because several different tests are called by this name. Also, concept tests are built into a number of sophisticated prediction systems (e.g., Burke's BASES and NPD's ESP®) that use a concept score along with company estimates of spending and distribution in a regression equation to estimate trial. Concepts can be presented in forms that very from simple statements to finished advertisements, so the line between concept tests and copy tests becomes hazy. Further-

PERSONAL INTERVIEWS

The following individuals (listed in alphabetical order of company) participated in open-ended interviews that lasted from 45 to 60 min.

Tibor Weiss, AHF Marketing Research
Melvin Harbinger, Bristol Myers Co.
Pamela J. Welker, Burke Marketing Research
Lynn S. Whitton and Stephen K. Zrike, Colgate-Palmolive Co.
Edward M. Tauber, Dancer Fitzgerald Sample, Inc.
Newton Frank, Data Development Corp.
Bernard Ruderman and Steve Roth, Decisions Center, Inc.
Jay Friedland, Guideline Research Corp.
Whitney J. Coombs and Ian M. Lewis, Lever Brothers Co.
Stephen J. Cook, Market Facts—New York, Inc.
Lawrence Newman, Newman-Stein, Inc.
John L. McMennamin, Norton Simon Communications
Neil B. Holbert, Philip Morris, Inc.

Four general questions were asked in each interview: "How do you perform concept tests?" "What are the primary problems you see with concept tests?" "Under what circumstances do concept tests work best and when are they most likely to fail?" and "What changes over the next few years do you see taking place in the way concept tests will be conducted?" Each of these general questions was followed with a series of more specific questions based on the initial answers. The views expressed in the paper are the author's perceptions of generally held feelings. On any particular point, several of the above-mentioned participates may disagree with the paper.

more, concept tests never test concepts. Concepts are ideas; they are in someone's mind. Concept tests measure consumers' reactions to concept statements. Making the distinction may seem like academic squabbling, but it is of great practical interest as it raises the question: "what is the consumer responding to—concept, positioning, or execution?" Obviously, the consumer is reacting to all three.

In order to sort out the differences between types of concept tests, a general concept-testing sequence is described. Each of the practitioners pointed out that the specific test or combination of tests used by a firm depends on a number of factors, including the objectives of the research, what information is already known, the availability of time and money, the number of concepts to be tested, and the type of product or service the concept describes. However, the general description provides a common terminology and a background for discussion.

Concept Screening Tests

While a test presupposes the existence of an idea, the method of idea generation may have an effect on the number of ideas uncovered and, therefore, on the type of concept tests that are performed initially. Some approaches, such as joint space analysis [44] and conjoint analysis [15, 16, 21, 55], tend to generate a single or at most a small number of concepts per segment. However, a number of other techniques tend to generate a larger number of concepts for a given product class. Examples of these methods include benefit structure analysis [35] and problem detection analysis [13]. Lanatis [29] suggests several additional ways of generating concepts. Other ideas may come from a variety of sources [32].

When a large number of ideas (over ten) is generated, it is typically reduced to a more manageable set through screening concept testing [14]. In this type of test, usually 10 to 30 (but sometimes up to 50) bland concept statements are presented to each respondent in sentence or short paragraph form. Typically these concept statements represent only the core idea. They are rated on one primary scale, such as intention to purchase, interest, or liking and possibly a few secondary scales such as uniqueness or believability. They are then compared based on the number of people rating each concept in the "top box" (most favorable category). These top box scores can be weighted by expected usage if the concepts will go into different categories but are usually unweighted. The concepts are also compared on secondary criteria to chose between concepts that score similarly on the primary criterion or to determine potential trouble.

Because this process tends to be reasonably complex, these data are usually collected through personal interviews using central locations or in-home sampling. Sample sizes for this stage can vary from 40–50, when few concepts are involved and the analysis is univariate, to 300–500, when each respondent rates only a subset of the concepts and/or a multivariate technique such as Q-sort is employed.

The biggest question with regard to concept screening tests is not how but *whether* they should be done. A significant number of practitioners felt this was a highly unrealistic task and that managers should determine which four or five ideas were worthy of further study. However, at least one firm has done some validation research indicating that concepts scoring well at this stage also score well in subsequent concept tests, and concepts scoring poorly here also score poorly at a later phase.

Concept Generation Tests

The next step is usually the concept generation test, the qualitative phase of concept testing [14] (called Diagnosis by Holbert [24]). According to Holbert (pp. 11–12), the purpose of this phase is "to end up with a statement that tells (as clearly and meaningfully as we know how to present it) all about the product—its

physical characteristics and sensory associations—and its benefits to the consumer." This does not necessarily mean a long concept statement, or one with lots of execution and "fluff," but one that is clear and concise.

Devising a statement is accomplished through one or more focus groups or a series of individual personal interviews. During interviews, respondents are shown a preliminary concept statement and asked to respond to it. The researcher is trying to find the answers to questions such as, Is the concept statement clear and forceful? Is there a better way of stating it? Is it unique and believable? What are its advantages and disadvantages? This interative process can be though of as concept optimizing, with consumer reactions on one iteration used to improve the concept statement for the next iteration. Pietrzak [41] refers to this phase not as market analysis but as creative work. This work should be done by the concept statement writers themselves [24, 41]. In practice, time pressure tends to reduce the number of iterations and push the burden of testing onto someone other than the concept statement writer.

Virtually everyone agrees that concept generation testing is an important step when used to express the concept in a clear and forceful manner. However, Holbert [24], Pietrzak [41], and Linsky [30], as well as most of the people interviewed in this survey, cautioned strongly against the use of this qualitative research in place of the more quantitative research that should follow.

Concept Evaluation, Positioning, and Concept/Product Tests

The next step is to measure a larger number of consumer responses to the concept statement in a more quantitative manner. If the concept statement does not include a positioning, it is an evaluation test; if it does, it is a positioning test. If the consumers try the product after the concept test, and the reaction to the product and concept are compared, it is called a concept/product test. If a company performs all three tests, they are done in this order. Possibly the most typical sequence is to perform either an evaluation or a positioning test first, and then a concept/product test once the product has been developed.

While these three tests are different, they can be discussed as a group. The primary difference between the first two is the form of the concept statement(s). In an evaluation test, the concept is typically a fairly bland statement typed on a 4″ × 6″ card. such a statement might read: "A powdered product that adds considerable nutrition when mixed with milk." This concept could be positioned as an instant breakfast, a diet or health food, or a snack food. It could be presented in any form from a 4″ × 6″ card to a finished print or television ad. (The concept portion of a concept/product test can also be presented in any of these forms.) Unless different positionings of the same concept are being tested, the distinction between evaluation and positioning tests can become fairly arbitrary. The more important question is usually how much "sell" or embellishment should be given to the concept statement.

Usually 200–300 people are sampled per concept or per positioning. Concept/product tests may use sample sizes of 300–400. These tests are typically personal interviews at shopping centers or in-home personal interviews, but household mail panels are employed occasionally. The procedure chosen depends on a number of factors, such as the desired amount of dispersion on certain consumer characteristics, the number of open-ended questions, cost, time urgency, and whether a focus group discussion is desired after the initial interview.

While the concepts can be presented in a number of designs, the two most popular designs by far are monadic and competitive tests. Competing concepts are not directly compared in either monadic or competitive tests, but comparisons are made using matched samples.

Monadic tests are conducted by giving one concept to each respondent. In the competitive design, commonly called competitive environment tests (e.g., AHF Research's CET®), the concept being tested is presented along with concepts written for leading brands in the product class. The CET® is a pre-post design. First, concept statements for the leading brands, which may include price and brand name, are presented to the respondent, who distributes ten chips across the brands in proportion to expected purchases, preferences, or affect. Then the new concept is presented and the respondent is asked to redistribute the chips. A variation on this basic design is Eric Marder's STEP® test where the old and the new concepts are presented at the same time (i.e., the "post" only design).

Next, a number of questions are asked. These questions may be divided into four categories: primary criterion, comprehension, diagnostic, and classification. One of two questions is usually used as the primary criterion to decide whether to proceed with or to kill the concept. The most popular predictive question with monadic tests is an intention to purchase (ITP) and usually employs a five-point scale. The constant sum scale is used with CET® type designs. Another, less popular prediction method, (which is more widely used in concept/product tests and simulated tests markets), is actual choice. Typically, at least one question is asked to determine whether the respondent understood the concept statement. Diagnostic questions are asked to determine why the consumer responded as he or she did, how to improve the concept, and how to give additional insight into the possible success of the concept. Finally, some demographic or psychographic questions are included to determine potential market segments.

The responses to the primary criterion question—the percentage of respondents marking the top box or top two boxes on a ITP scale, or the percentage of chips allocated to the new concept on a constant sum scale—are used to evaluate concept tests. These scores can be used three ways. First, they can be compared with what is usually a category-specific norm (because concepts in some categories [e.g., desserts] tend to score higher than others). The norm may be an average or a lower limit of the scores of successful concepts in the past, or it may be the score for a concept written for a brand currently in the category.

The second method is to use the score to predict trial. Although there are some fairly sophisticated systems (e.g., Burke's BASES I) that incorporate seasonality, distribution build, category and brand development indexes and so forth, most companies use a fairly simple procedure to estimate trial. For example, estimated percent trial is equal to the percentage that marked top box. This result is adjusted for expected awareness, distribution, and the percentage receiving samples or coupons. Frank [14] cites several other examples of trial estimates derived from top box scores.

The third method is to judge the concept in financial terms. That is, managerial judgment and secondary data are used to make estimates of trial and repeat rates prior to any testing. These estimates are refined through concept and product testing. Therefore a concept should be passed on to the next phase of development if the predicted trial level and best estimate of repeat sales give a prediction of financial success. The estimate of repeat sales can come from managerial judgment, category averages, or responses to an expected usage question on the concept test. The expected usage responses are usually reduced to account for probable overstating (e.g., estimated usage is forecast to be one-half of stated usage). This number may be further reduced by making an estimate of the percentage of triers expected to actually adopt the product.

While the decision to go with a product is based mainly on the answers to these expected trial questions, the answers to other questions such as uniqueness, believability, or need fulfillment are also considered. Typically they are not employed in a systematic fashion but are used as tie-breakers or are compared with some lower limit as an indication of potential problems with the product.

User Satisfaction with Concept Tests

One of the most frequently mentioned limitations of concept tests is that they do not always predict market success. This lack of predictive power may be the result of (a) the product's not living up to the benefits promised in the concept, (b) changes in the concept, positioning, or physical product between the concept tests and introduction of the product, or (c) changes in the legal or social environment. No amount of improvement in current concept testing practices can remedy these problems. Similarly, Tauber [50] points out that the success of a frequently purchased product hinges on a number of people adopting it, i.e., using it regularly. Concept testing cannot determine whether this will occur, but, this should not be its objective. Concept testing should be used as an early screening device to obtain some consumer reaction to an idea and to predict the trial rate.

The empirical evidence on the ability of concept testing to perform this more limited role is fragmentary but nonetheless encouraging. Relatively few systematic attempts to validate the predictive ability of concept testing exist, either at the individual or aggregate level. This small number of validation attempts is due partly to the many changes that can occur between concept testing and test marketing or national introduction. However, a few companies have conducted concept tests just prior to market introduction, and the general feeling is that concept tests using an ITP scale reaction to a concept on a 4" × 6" card can predict trial rate within 20% (i.e., if the test predicts a trial rate of 50% the actual rate will fall between 40% and 60%) about 80% of the time. Frank

[14] mentioned similar figures for three companies using slightly different projection techniques. One respondent, using print ads and a constant sum scale prior to market introduction, indicated that about 90% of the time the predictions were very good. Another respondent mentioned an explained variance in trial rate of 80% on a series of predictive tests. Several respondents talked about instances in which the results were so good as to be "uncanny" or "scary." These predictions are based almost exclusively on concepts that scored well on concept tests (otherwise they would not be developed further).

However, the question of how many good concepts are screened out through concept testing remains unanswered. Tauber [47] suggests that this loss may be much greater than that resulting when products are not killed quickly enough. Furthermore, this danger is probably greater with radically new products [48]. However, several respondents mentioned that most any time someone's "gut feel" overrode the negative results of a concept test, the product did, in fact, turn out to be a failure.

In summary, the evidence to date indicates that properly executed concept tests can do a good job of predicting trial for concepts that are not radically different from products on the market.

PRIMARY DECISIONS TO BE MADE IN CONCEPT EVALUATION TESTS

This section focuses on four specific decisions associated with evaluation or positioning tests: concept statement, test design, questions asked, and prediction of success. Whereas the discussion has been separated into four categories, the division is somewhat artificial because decisions made in one area have an impact on other areas. For example, if a tester decides to do a monadic test, he or she can use either a bland or a promotional concept statement, but an ITP scale would be used virtually every time. Similarly, a competitive environment test is usually paired with a constant sum scale and more promotional concept statements.

Concept Statement

Virtually all concept statements give a description and list the principal benefits, information that is sometimes called the pure idea. A positioning statement is usually added to compare the product with possible competitors. The primary questions are how much "sell" or embellishment the concept is given and in what form it is presented. Conventional wisdom suggests that relatively bland concept statements should be used when a large number of concepts is evaluated, as it is quicker and cheaper, and when the concept is radically new, as positioning may limit its appeal. Similarly, promotional concept statements should be used when a concept is going into an existing product class and should be presented in the typical medium for that class.

Many of the people interviewed in this study tended to use almost exclusively either bland or promotional statements. Use of these types of statements enables a company to develop norms and is reinforced by past success with the method chosen. People who usually use only bland statements do so because a) they want a reaction to the pure concept. b) personal experience has found relatively small differences in consumer reactions to bland and promotional concept statements, or c) it is impossible to tell whether the consumer is reacting to the concept, positioning, or embellishment in more promotional statements. Others use promotional statements because of the added realism and because they are ultimately looking for the best combination of concept, positioning, and embellishment and the best pure concept, does not necessarily lead to the optimal combination.

Five pieces of research shed some light

on this question. Haley and Gatty [22] performed an experiment in which three copywriters provided statements for eight positionings of a concept. Their findings indicate that consumer reaction is based not only on the concept, the positioning, and the copywriter but also on the interaction between positioning and copywriter.

Tauber [46] held the idea, concept, and execution constant to test the effect of communication form on respondent ratings. Three concepts were presented in both print ad and paragraph form. He found that scores were much higher for the concept boards than the paragraphs but that the relative scores did not change. Thus the one big idea came through in either form of communication.

Armstrong and Overton [2] looked at the effect of brief vs comprehensive descriptions of a service when testing the intention to purchase a minicar leasing system. They found no significant differences between the two methods of presentation with regard to the respondents' intention to purchase or the price they would pay for the service. In contrast, Pessemier and Wilton [39, 40] varied the amount and importance of the information given to consumers about electric automobiles and competing products. The type of treatment had a significant effect on perceptions and preferences for the cars.

One related proprietary study compared consumers' responses with rough and finished advertisements. The correlation of the responses was reasonably high, but, as the authors pointed out, this was partly because in a few pairs both scored very high or very low. The finished ads tended to score equal to, or higher than, the rough ones, but there was no systematic relationship between the two scores. In a few cases the finished ads scored lower.

In summary, practitioners agree that great care should be taken in the writing of concept statements. This does not mean writing more promotional concept statements but better ones. We might hypothesize that the amount of positioning and sell is a function of how great the benefit is, how well it is understood, how socially acceptable it is to admit a certain need, and how emotional the benefit is. As a general rule, concept writers should use the minimum amount of sell required. A company would be advised to try to present all concepts for a given category similarly to build up a forecasting track record.

Design of Test

No published evidence indicates whether a concept should be presented monadically or in a competitive environment. The primary reason given for using monadic tests is past success with this method. As mentioned earlier, many firms are quite satisfied with the ability of monadic tests to predict trial rate. People using monadic tests argued that the consumer is responding to a new concept in a comparative nature anyway, so a direct comparison is not needed. Most of these people also favored relatively bland concept statements and did not know how realistic consumer reaction to bland, disguised statements for existing brands would be. Reactions to the concept being tested are influenced by competitive concepts, and in some categories (e.g., wine) it is hard to determine the proper competitive products. Some of these people also thought that both the pre-post design in CET®s and the equal prominance given to all competing concepts resulted in decreased realism. The people who favored comparative testing did so because they felt it was more realistic. They tended to use print ads for their concepts and used their competitor's ads for the comparison concepts. So the issue of comparative testing is confounded with that of the amount of "sell" in the concept statement.

In summary, people have reported considerable success in predicting trial with each method. Apparently, the opportunity exists to conduct research comparing the predictive and diagnostic powers of these two designs.

Specifically, it would be valuable to determine under what circumstances each method should be used. In the absence of such research, we might hypothesize that competitive environment tests would be preferable when knowledge of the alternatives is relatively low and there is considerable search prior to purchase. This is the case for many durable goods such as major appliances. Monadic tests would be preferable when it is hard to identify direct competitors or when there is little external search prior to purchase.

Questions Asked

While practitioners agree on what questions should be asked, they often disagree on two issues: price and intended usage.

Price. A widely used method of incorporating price into concept tests is to state the price in the concept statements; which may require checking the price of competing products in area stores. Another method used frequently is to ask respondents what they thought the product would cost. A number of practitioners in this survey said that consumers could give the cost accurately when the concept was in a very frequently purchased category, but that accuracy dropped off as purchases became less frequent. However, there seems to be a problem with the interpretation of this question. For example, if a consumer said that he or she would definitely purchase a product but understated its price, should the consumer be counted as a trier or a nontrier? A method similar to Pessemier's dollar metric procedure was occasionally used to determine price elasticity. A person who would purchase at a given price was asked whether he or she would do so at a higher price. Conversely, consumers who would not purchase at a given price, were asked whether they would purchase at a lower price.

Making price part of the concept statement appears to be the best way of incorporating price into the test. The use of multiple questions to estimate price elasticities has considerable face validity, but systematic research is needed to determine its predictive power.

Volume. Although the primary purpose of concept tests is to predict trial, the decision on whether a concept should be advanced to the next stage of development should be based on an estimate of financial success. The trial estimate must therefore be converted into an estimate of cash flow, either implicitly by using category-specific norms or explicitly by estimating sales volume.

A number of companies ask respondents what their expected usage would be or when they might use it (assuming the product was satisfactory). However, several studies, referenced in [56], found a systematic underprediction of the purchases of some brands and overprediction of others, and the total amount in a category tended to be overstated. Whether this percentage of overstating differs substantially across product categories is unknown. How highly correlated are stated and actual usage is also unknown. These two questions need to be answered before respondent estimates of usage are incorporated into concept tests, except as a check on preliminary estimates.

Prediction of Success

Aggregate Level. Even though the most popular method of evaluating concept tests involves comparison with a norm, several practitioners recommended using an estimate of financial success. This method involves estimating volume from trial based on the concept test and the best available measure of repeat sales.

A number of systems (e.g., BASES and ESP®) use concept/product tests to predict trial and first repeat, which are combined with an estimate of the decay rate to estimate volume. Simulated test markets (e.g., ASSESSOR,

LTM, and COMP) could also be viewed as fairly similar to concept/product tests in which the respondents purchases a product rather than rates a concept on an ITP scale. Finally, sales waves studies (e.g., Data Development Corporation) can be combined with concept tests to estimate ongoing volume. Examining, these tests, which involve consumer contact with the product, are beyond the scope of this paper.

Little empirical evidence exists on the question of using norms or sales estimates to evaluate concepts. In one study [50], Tauber shows that the concept with the highest top box score is not necessarily the most successful. Therefore decisions should be based on financial estimates instead of comparisons with norms. In the absence of any track record, testers should use the percentage marking top box adjusted for awareness, distribution, and sampling as the estimate of the number of category users that will try the product.

Individual Level. Favorable ratings on an attitude or intention-to-purchase scale are positively correlated with trial or usage [31, 45, 51]. However, this procedure obscures some predictive problems at the individual level. A frequently cited problem was of "yea saying." Two studies shed some light on this problem [8, 49, 52].

DISCONTINUOUS INNOVATIONS

Tauber [48] has suggested that current new product research techniques, including concept testing, may discourage major innovations because consumer attitudes, upon first exposure to discontinuous innovations, are not good predictors of what their actions will be after a prolonged exposure. This results partly from a lack of knowledge about the product and partly from the social system's influence on adoption decisions. Increased education prior

to measuring a reaction helps to reduce part of this problem. Pessemier and Wilton [39, 40] found that increased levels of information do have an effect on the predicted market share of an electric car. However, there was no way to determine whether the increased information resulted in a better prediction of the level of adoption of the electric car.

Wolpert [57] has found that the automobile market can be segmented by preferences for style. One segment prefers a conservative functional styling with little change. Another segment prefers nonfunctional, show-type styling. He feels that the second segment may be able to give valid predictions of the acceptance of revolutionary changes in styling. Innovators and early adopters in other categories could also be located. Their reactions to concepts may serve as leading indicators of the acceptance of major innovations.

Behavioral simulations have played a part in the study of various types of decisions. It is also possible that some type of simulation could be devised that would model the word-of-mouth communication that occurs in the diffusion process.

Moving further from concept testing, techniques such as conjoint analysis [15, 16] can estimate acceptable tradeoffs between desired characteristics for a product or service without describing the concept except as a bundle of characteristics. With this technique, practitioners may be able to measure the need for some benefit without confounding it with reactions to a product that is very different from ones currently on the market. Howard [25] is using several concepts from human categorization in an attempt to model extensive problem solving. Some of this work may enable us to predict the response to more discontinuous innovations.

There is lots of room for the creative development of ways to measure consumer reactions to discontinuous innovations. We will probably have to be content with the face validity of these techniques because no norms for products that create their own product classes will be available.

SUMMARY

The purpose of this paper has been to examine current concept testing practices and controversies as reported by the published literature and as seen by some of the leading practitioners. It appears that the basic knowledge of how to conduct good concept tests is available.

A company may gain the most not by developing new techniques but by improving the execution of the different phases of concept testing. Specific suggestions include

1. spending sufficient money on idea generation or strategic research to finding concepts that have true benefits,
2. forcing the concept writer to conduct individual or group interviews to insure that the ideas are communicated clearly and forcefully,
3. setting the action standards for the quantitative part of the test based on trial needed to meet financial criteria prior to the test,
4. spending the amount of money required to sample a sufficient number from the proper market segment, and
5. choosing test methodology to suit the nature of the concept (i.e., not testing discontinuous innovations and line extensions in the same manner).

In addition to these suggestions, the paper has highlighted a number of areas for research. An area of prime interest is a comparative study of the predictive and diagnostic ability of the two major methods: 1) monadic design, paragraph form of concept, and top box estimate of trial, and 2) multiple concept com-

parison, print ad form of concept, and constant sum estimate of trial rate. There are questions regarding each of these general methods, but the greatest difficulty for a practitioner is determining which method is appropriate for the circumstances. A second area of research is improvement of the predictive ability of concept tests at the individual level, primarily to get better estimates of market segments. A final, very large area of interest is the prediction of the success of discontinuous innovations. This issue is the most challenging but ultimately may yield the greatest payoff.

REFERENCES

1. Abrams, Jack, Reducing the Risk of New Product Marketing Strategies Testing, *J. Marketing Res.* 6 (May 1969): 216–220.
2. Armstrong, J. Scott, and Overton, Terry, Brief vs. Comprehensive Descriptions in Measuring Intention to Purchase, *J. Marketing Res.* 8 (February 1971): 114–117.
3. Axelrod, Joel N., Reducing Advertising Failures by Concept Testing, *J. Marketing* 28 (October 1964): 41–44.
4. Batsell, Richard S., and Wind, Yoram, Product Testing: Current Methods and Needed Developments, Working paper, The Wharton School, University of Pennsylvania, May 1979.
5. Belkin, Marvin, and Lieberman, Seymour, Effect of Question Wording on Response Distribution, *J. Marketing Res.* 4 (August 1967): 312–313.
6. Blankenship, A. B., Let's Bury Paired Comparisons, *J. Advertising Res.* 6 (March 1966): 13–17.
7. Booz, Allen & Hamilton, *Management of New Products*, Chicago, 1968.
8. Clancy, Kevin J., and Garsen, Robert, Why Some Scales Predict Better, *J. Advertising Res.* 10 (October 1970) 33–38.
9. Cochran, Betty, and Thompson, G., Why New Products Fail, *The National Industrial Conference Board Record*, October 1964, pp. 11–18.

10. Conference Board, *Market Testing of Consumer Products,* New York, 1967.

11. Cooper, Robert G., The Dimensions of Industrial New Product Success and Failure, *J. Marketing* 43 (Summer 1979): 93–103.

12. Davidson, J. Hugh, Why Most New Consumer Brands Fail, *Harvard Bus. Rev.* (March-April 1976): 117–122.

13. Dillion, Tom, Forecasting Wants and Needs of the Consumer, Paper given at the New York Chapter of AMA New Product Conference, 1978.

14. Frank, Newton, Can We Predict New Product Success from Concept Testing, Paper given at AMA New York Chapter 1972 New Products Conference.

15. Green, Paul E., and Rao, Vithala R., Conjoint Measurement for Quantifying Judgmental Data, *J. Marketing Res.* 8 (August 1971): 355–363.

16. Green, Paul E., and Wind, Yoram, New Way to Measure Consumers' Judgments, *Harvard Bus. Rev.* 53 (July–August 1975): 107–115.

17. Greenberg, Allan. Paired Comparisons Versus Monadic Tests, *J. Advertising Res.* 3 (August 1963): 44–47.

18. Greenhalgh, Colin, Research for New Product Development, in Consumer Market Research Handbook. R. M. Worster, ed., McGraw-Hill, London, 1971, pp. 378–410.

19. Gold, Bertram, and Salking, William, What Do Top Box Scores Measure? *J. Advertising Res.* 14 (March 1974): 19–23.

20. Golden, Hal, Concept Tests—Often Used, But How Well, *Marketing Rev.* 27 (September 1973): 20–24.

21. Gruber, Alin, Purchase Intent and Purchase Probability, *J. Advertising Res.* 10 (February 1970): 23–27.

22. Haley, Russell I., and Gatty, Ronald, The Trouble with Concept Testing, *J. Marketing Res.* 8 (May 1971): 230–232.

23. Haller, Terry P., Let's Not Bury Paired Comparisons, *J. Advertising Res.* 6 (September 1966): 29–30.

24. Holbert, Neil, *Research in the Twilight Zone,* American Marketing Association Monograph, Series #7, AMA, Chicago, 1977.

25. Howard, John A., Progress in Modeling Extensive Problem Solving: Consumer Acceptance of Innovation, Working paper, Columbia University, New York, 1978.

26. Johnson, Richard M., Trade-Off Analysis of Consumer Values, *J. Marketing Res.* 11 (May 1974): 121–128.

27. Juster, F. Thomas, *Consumer Buying Intentions and Purchase Probability: An Experiment In Survey Design,* National Bureau of Economic Research, Occasional Paper 99, distributed by Columbia University Press, New York, 1966.

28. Kassarjian, Harold, H., and Nakanishi, Masao, A Study of Selected Opinion Measurement Techniques, *J. Marketing Res.* 4 (May 1967): 148–153.

29. Lanatis, Tony, How to Generate New Product Ideas, *J. Advertising Res.* 10 (June 1970): 31–35.

30. Linsky, Barry R., Eliminate "Bombs" With A Systematic Approach to Concept Evaluation, Paper presented at 1975 ANA New Product Marketing Workshop, New York.

31. Longman, Kenneth A., "Promises, Promises," in *Attitude Research on the Rocks.* L. Adler and I. Crespi, eds., American Marketing Association, Chicago, 1968, pp. 28–37.

32. McGuire, E. Patrick, *Generating New Product Ideas,* The Conference Board, New York, 1972.

33. McGuire, E. Patrick, *Evaluating New-Product Proposals,* The Conference Board, New York, 1973.

34. Moriarity, Mark, and Venkatsen, M., Concept Evaluation and Market Segmentation, *J. Marketing* 42 (July 1978): 82–86.

35. Myers, James H., Benefit Structure analysis: A New Tool for Product Planning, *J. Marketing* 40 (October 1976): 23–32.

36. Neilsen, A. C. Co., New Product Success Ratio, *The Nielsen Researcher* No. 5, 1971, pp. 1–10.

37. O'Connor, J. J., R. J. R. Monitors 105 New Brands, Classified 13 as Successful, *Advertising Age,* July 12, 1976, p. 3.

38. Pessemier, Edgar A., An Experimental Method for Estimating Demand, *J. Bus.* 3 (1960): 373–383.

39. Pessemier, Edgar A., and Wilton, Peter, The Effects of Information on Perceptions and Pref-

erences for New Choice Objects, Institute Paper #683, Krannert Graduate School, Purdue University, February 1979.

40. Pessemier, Edgar A., and Wilton, Peter, Pretesting the Acceptance of Innovations, Institute Paper #696, Krannert Graduate School, Purdue University, April 1979.

41. Pietrzak, Robert J., Screening and Developing Concept After Idea Generation, Paper presented at 1975 ANA New Product Marketing Workshop.

42. Reibstein, David J., The Prediction of Individual Probability of Brand Choice, *J. Consumer Res.* 5 (December 1978): 163–169.

43. Seaton, Richard, Why Ratings are Better Than Comparisons, *J. Advertising Res.* 14 (February 1974): 45–48.

44. Shocker, Allan D., and Srinivasan, V., Multiattribute Approaches for Product Concept Evaluation: A Critical Review, *J. Marketing Res.* 16 (May 1979): 159–180.

45. Stapel, Jan, Predictive Attitudes, in *Attitude Research on the Rocks.* L. Adler and I. Crespi, eds., American Marketing Association, Chicago, 1968, pp. 96–115.

46. Tauber, Edward M., What is Measured by Concept Testing, *J. Advertising Res.* 12 (December 1972): 35–37.

47. Tauber, Edward M., Reduce New Product Failures: Measure Needs as Well as Purchase Interest, *J. Marketing* 37 (July 1973): 61–64.

48. Tauber, Edward M., How Marketing Research Discourages Major Innovations, *Bus. Horizons* (June 1974): 24–27.

49. Tauber, Edward M., Predictive Validity in Consumer Research, *J. Advertising Res.* 15 (October 1975): 59–64.

50. Tauber, Edward M., Why Concept and Product Tests Fail to Predict New Product Results, *J. Marketing* 39 (October 1975): 69–71.

51. Tauber, Edward M., Forecasting Sales Prior to Test Market, *J. Marketing* 41 (January 1977): 80–84.

52. Tauber, Edward M., The Decision Risks with New Product Concept Testing, Dancer, Fitzgerald, Sample Inc., New York, 1979.

53. Taylor, James W., Houlahan, John R., and Gabriel, Alan C., The Purchase Intention Question in New Product Development: A Field Test, *J. Marketing* 39 (January 1975): 90–92.

54. Wilkie, William L., and Pessemier, Edgar A., Issues in Marketing's Use of Multiattribute Models, *J. Marketing Res.* 10 (1973): 428–441.

55. Wind, Yoram, A New Procedure for Concept Evaluation, *J. Marketing* 37 (October 1973): 2–11.

56. Wind, Yoram, and Lerner, David, On the Measurement of Purchase Data: Surveys Versus Purchase Diaries, *J. Marketing Res.* 16 (February 1979): 39–47.

57. Wolpert, Henry W., Why Conventional Automobile Styling Research May Become Obsolete, in *Advances in Consumer Research Vol. VII.* Jerry C. Olson, ed., Association for Consumer Research, Ann Arbor, 1980.

58. Yuspeh, Sonia, Diagnosis—The Handmaiden of Prediction, *J. Marketing* 39 (January 1975): 87–89.

Redesigning Product Lines with Conjoint Analysis: How Sunbeam Does It

Albert L. Page

Harold F. Rosenbaum

Sunbeam Appliance Company, faced with the task of redesigning its lines of small appliances, developed a procedure that couples consumers' evaluations of possible new designs with a marketing research technique—conjoint analysis. In this article, Albert Page and Harold Rosenbaum describe Sunbeam's redesign of its food processor product line. They report detailed information, including key design attributes, alternative product designs, market segments, and competitive positions, and describe the use of a simulation model that predicts the market share of alternative product line configurations before they are even developed or introduced into the market. The procedure provides product line managers with a powerful tool to assist them in redesigning their lines.

PRODUCT VS. PRODUCT LINE REDESIGN DECISIONS

In 1981, Sunbeam Corp. was acquired by Allegheny International and new management was introduced into Sunbeam Corp. and its Sunbeam Appliance Co. (SAC) division. The new management team at SAC determined that it was necessary to redesign its many mature lines of small kitchen appliances. In addition to

Reprinted by permission of the publisher from "Redesigning Product Lines with Conjoint Analysis: How Sunbeam Does It" by Albert L. Page and Harold F. Rosenbaum, *Journal of Product Innovation Management*, Volume 4, pp. 120–137. Copyright 1987 by Elsevier Science Publishing Co., Inc.

all the creative and managerial issues normally involved in the redesign of an individual product, such product line redesign decisions also have an additional unique complexity that sets them apart from individual product redesign decisions since a product line is composed of a group of products that are "closely related" [7]. The individual models of the product in the line are interrelated in their demand and/or cost so that decisions made about one model in a line also affect the sales and/or costs of the line's other items as well. In this article we will be primarily concerned with the implications of interrelated demand for product line redesign.

Usually a product line is composed of a number of models of the same basic product where the models reflect different combinations of sizes, features, and price points, etc. such as a line of engine lathes, tractors, console television sets, cookies, or lipsticks. Managerial decisions to stretch a product line, to fill it out or to redesign it, lead to concern about new entries in the line stealing sales from the rest of the line, or cannibalizing it. The approach to product line redesign which SAC developed explicitly recognizes this demand interaction by simulating the effects of product line changes on the entire line and the different models in it.

The purpose of this article is to describe how Sunbeam Appliance Co. has successfully confronted the problem of redesigning its product lines by developing a sophisticated redesign procedure that is built around market research, conjoint analysis, and product line simulations to optimize the configuration of the redesigned line. The succeeding sections of this article explain SAC's specific situation, outline the procedure they developed, and present a detailed application of the procedure to the redesign of SAC's food processor line. An appendix presents a nontechnical overview of the conjoint analysis technique and the data collection it requires.

SUNBEAM'S SITUATION

In 1983, SAC embarked on its program to redesign its appliance product lines over a period of years and to become more responsive to the needs and dictates of its markets. Management's goal for the redesign of its product lines was to optimize their market share of each product category. During 1982, SAC's marketing research and product planning groups had worked together to develop a sound procedure for product line redesign that would enable them to achieve management's goal and could be applied in turn to each of its lines. The three marketing issues they identified as central to the problem of redesign were:

1. What models should be in the line?
2. What should their physical appearance be?
3. What should their performance characteristics be?

The product line redesign procedure which SAC's personnel developed is based upon the research technique of conjoint analysis and provides management with answers to these three questions. The complete procedure consists of four parts. In sequential order they are:

1. *A Consumer Usage and Attitude Survey.* The purpose of this part of the procedure is to determine how and for what purpose products in the product category are used, frequency of use, brand ownership, brand awareness, and attitudes towards the appliance. The results of this part of the process are useful for product and advertising planning.

2. *A Consumer Attribute and Benefit Survey.* Importance ratings of product attributes and benefits are collected along with their perceived presence or absence in each of the products of the key competitive brands including SAC. This data provides the basis for identifying "need gaps" where the perceived benefits are not being supplied by the existing brands.

3. *A Conjoint Analysis Study.* Insights from exploratory group discussions lead to a consumer survey that produces the data from which the structure of consumers' preferences for different product features is determined via the conjoint measurement technique. This structure of preferences is then used as an input into the fourth step of the procedure.

4. *Product Line Simulations.* Product management prepares various possible versions of the SAC product line, and their shares of the market are predicted from the structure of consumers' preferences. The market share generated by each of the simulations of the product line is used to determine the best configuration for SAC's redesigned product line in the category.

This four part procedure provides for maximum consumer input into the redesign process. Most importantly, it allows consumers who purchase appliances to provide SAC with their design and feature preferences before the appliances are redesigned. This in turn permits product line managers to test alternative product line configurations against the buyers' preferences in the simulation developed from the results of the conjoint analysis research. Since this procedure was developed it has been successively applied to SAC's food processor, stand mixer, iron, can opener, coffeemaker, and hand mixer product lines with good commercial results.

The most interesting and sophisticated parts of SAC's redesign procedure are the conjoint analysis study and the product line simulations that are based upon it. Because of their wide applicability for redesigning other types of product lines, the rest of this article will focus on their use in redesigning SAC's product lines. An Appendix presents an overview of conjoint analysis and associated data collection issues for readers who desire a description of these procedures.

SUNBEAM'S FOOD PROCESSOR REDESIGN STUDY

Because of their strategic position in SAC's product portfolio, food processors were chosen as the first product line to which the new conjoint based redesign procedure would be applied. In the spring of 1983 when the study commenced, SAC's food processor line was comprised of three basic models with two slightly different versions of each for a total of six items in the line. A second special set of the three models were derivative of SAC's regular set and were in the line in order to give department stores their own set of Sunbeam food processors to which they could apply their higher margins. The regular set of three mod-els was sold to and through SAC's normal channels of distribution for small kitchen appliances.

The food processor market contains three price points: less than $60.00, between $60.00 and $125.00, and over $125.00. The three models in SAC's regular line were located at two of these three price points. There were two midrange models priced between $60.00 and $125.00 and a promotional model priced under $60.00. SAC had no high-end model in its line to compete with Cuisinart, Kitchen Aid, and others. In 1982, the six models in the old line had less than a 10% unit share of the entire market, giving SAC the fifth place share rank in the food processor market behind Cuisinart, G.E., Hamilton Beach, and Moulinex. The shares of each of its six models in their respective price ranges varied widely, with SAC's greatest share coming from the midpriced segment of the market. SAC's overall objective for the product line redesign study was to substantially increase its share of the food processor category.

The food processor redesign project began with a series of group interviews. The purpose of these was to learn consumers' thoughts and feelings about food processors, their attributes and benefits, the uses they made of food processors, and their preferences regarding features. Also, the interviews provided an opportunity for SAC's research people to learn the vocabulary consumers use when they talk about and think about food processors. This was intended to allow the questionnaires and interviewers used in the conjoint data collection to use words and phrases regarding food processors that the responding consumers could understand.

Using the group interview results along with the experience and judgement of SAC's food processor product manager and market research staff the attributes and levels to be used in the conjoint study were determined. The 12 attributes of food processors and their levels actually manipulated in the study are shown in Table 1. These attributes and levels produce 69,984 possible food processor configurations ($3^7 \times 2^5$).

SAC decided to use main effects only conjoint measurement so a set of 27 orthogonal arrays of the 12 attributes and their levels were selected to be preference ranked by the appliance buyers in the conjoint data collection. SAC's industrial designers produced sketches of the 27 food processor designs implied by each of the specific combinations of the levels of the 12 attributes contained in each orthogonal array. In addition, printed labels about the features of each configuration were then added to the sketches. Four of the more interesting and varied sketches of the orthogonal array food processor configurations used in the study are shown in Figure 1. The respondents were shown, and had to preference order, all 27 of these sketches.

THE DATA COLLECTION

The actual data collected included a great deal of other information along with the preference ranking for the conjoint analysis. Besides a set of screening questions, the complete survey involved collecting the following kinds of data:

1. Food processor brand awareness.
2. Food processor usage behavior.
3. Food processor ownership and gift giving.
4. Conjoint analysis preference ranking.
5. Interest in new feature concepts for food processors like an automatic shut-off when food is processed to the desired consistency.
6. Interest in attachments that could be purchased for a food processor like a juice extractor.
7. Preference for different designs and shapes for a food processor.
8. A set of demographic questions.

TABLE 1.   Food Processor Attributes and Levels

   I. Price
     1. $49.99
     2. $99.99
     3. $199.99
  II. Motor Power
     1. Regular
     2. Heavy duty
     3. Professional power
 III. Number of Processing Blades
     1. Three
     2. Five
     3. Seven
 IV. Bowl Size
     1. 1½ quarts
     2. 2½ quarts
     3. 4 quarts
  V. Number of Speeds
     1. One
     2. Two (high/low)
     3. Seven
 VI. Other Uses
     1. Food processor only
     2. As a blender also
     3. As a blender and mixer also
 VII. Configuration
     1. Motor compartment and bowl situated side by side
     2. Bowl is situated on top of motor
     3. Processor is installed under the cabinet or counter top
VIII. Bowl Type
     1. Regular
     2. Side discharge
 IX. Type of Feed Tube Pusher
     1. Regular solid pusher
     2. A pusher made up of three interchangeable components to push smaller pieces of food
  X. Size of Feed Tube
     1. Regular
     2. Large
 XI. Bowl Shape
     1. Cylindrical
     2. Spherical
 XII. Pouring Spout
     1. Present
     2. Absent

The data collection took place in June 1983. Over 500 women were interviewed in four geographically dispersed central location facilities in high traffic shopping malls with an approximately equal number from each mall. The women had to be the female head of their household and own and use, on a regular basis, three or more portable cooking or heating appliances. Furthermore, the sampling plan specified certain respondent age and food processor ownership quotas as well.

Each completed interview took between 35 and 40 minutes. Most of this time was taken up by the preference ranking of the 27 food processor configuration sketches. The actual ranking was done in two stages to further reduce the difficulty of the respondents' task. In the first stage, the respondents were asked to sort the 27 food processor sketches into three groups of approximately equal size. The three groups were: "those food processors you would be *most* likely to buy, those food processors you feel neutral about, and those food processors you would be *least* likely to buy." In the second stage of the sorting each of the three groups from the first stage were separately sorted out on a continuum from "you would be *most* likely to buy that food processor" to "you would be *least* likely to buy that food processor." This sorting procedure is similar to the procedure of selecting an item from a mail order catalog—a task familiar to most consumers. It produced a complete preference ranking of the 27 sketches. The complete data collection cost $30,000.

THE CONJOINT ANALYSIS RESULTS

The utility functions for the food processor attributes were estimated from the preference data using the monotonic analysis of variance (MONANOVA) procedure [8].[1] The conjoint analysis and simulation results shown in this section of the article as well as the next have been disguised to preserve SAC's proprietary interest in them: they are *not* the actual findings. But the disguised results can still convey an excellent idea of the kind of insights into the product line redesign problem that SAC's conjoint based procedure can provide.

Overall Results

The initial results from the study were those for the conjoint analysis that showed the respondents' overall preference for the food processor features. The set of 12 utility functions was estimated from the composite rankings of the attributes by the entire sample. They revealed that appliance buyers preferred the following levels of each of the twelve attributes:

> A standard cylindrical bowl shape.
> Two speeds rather than one or seven.
> A medium-sized bowl capacity of 2½ quarts.
> The ability to also use the processor as a mixer but not a blender.
> The most amount of blades (7).
> A low profile rear motor design.
> A bowl with regular rather than side discharge.
> A heavy duty motor rather than a regular or professional power motor.
> A solid rather than a three-piece pusher.
> A pouring spout.
> A larger rather than a regular size feed tube.
> The lowest possible price ($49.99).

Thus, if SAC were to create a new food processor to match the preferences of the average food processor buyer in the study, it would have the product attribute profile described above. However, while all these attribute levels are preferred, they are not all equally important to the buyer when she makes her food processor purchase decision.

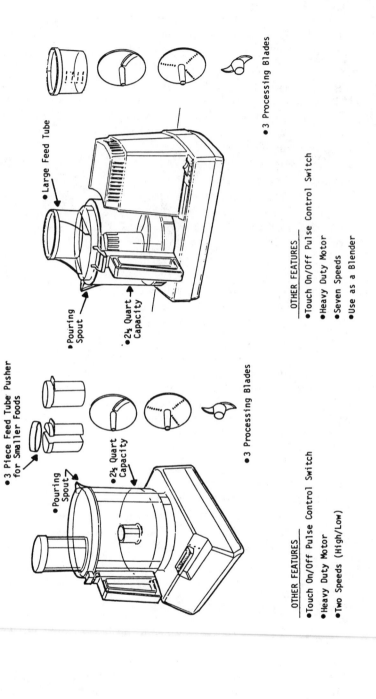

$199.99

• Large Feed Tube

• Pouring Spout

• 2½ Quart Capacity

• 3 Processing Blades

OTHER FEATURES
• Touch On/Off Pulse Control Switch
• Heavy Duty Motor
• Seven Speeds
• Use as a Blender

$199.99

• 3 Piece Feed Tube Pusher for Smaller Foods

• Pouring Spout

• 2½ Quart Capacity

• 3 Processing Blades

OTHER FEATURES
• Touch On/Off Pulse Control Switch
• Heavy Duty Motor
• Two Speeds (High/Low)

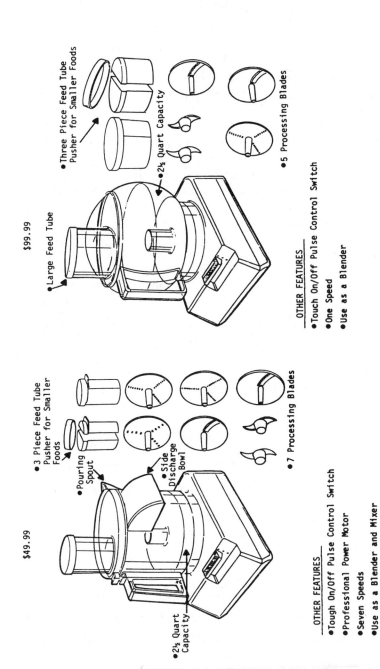

$49.99

- 3 Piece Feed Tube Pusher for Smaller Foods
- Pouring Spout
- Side Discharge Bowl
- 7 Processing Blades
- 2½ Quart Capacity

OTHER FEATURES

- Tough On/Off Pulse Control Switch
- Professional Power Motor
- Seven Speeds
- Use as a Blender and Mixer

$99.99

- Three Piece Feed Tube Pusher for Smaller Foods
- Large Feed Tube
- 2½ Quart Capacity
- 5 Processing Blades

OTHER FEATURES

- Touch On/Off Pulse Control Switch
- One Speed
- Use as a Blender

FIGURE 1. Four Food Processor Sketches

The bar chart in Figure 2 indicates the relative importance of each attribute in the purchase decision with importance being in direct proportion to the size of the bars and measured by the percentages on the chart. The larger bars for price, number of blades, and motor power indicated that these attributes are the most important to the homemaker in making her decision to buy a food processor. Furthermore, as we have seen, a large feed tube is preferred to a regular size one, but the bar chart tells us that this factor is of very little importance to these food processor buyers as compared to others that might enter their decision like the price or number of blades. In fact, the utility functions indicate that the average buyer would trade-off a larger feed tube for better levels of more important attributes like the price or the number of blades. This is because the additional utility or value to her that is derived from moving up to more blades is greater than what is lost from trading down to a regular size feed tube from a large one. Thus, if SAC could not design to cost a new food processor intended to be sold for $49.99 with all the attribute levels of the above profile, SAC would be better off to trade-off the size of the feed tube in favor of keeping seven processor blades if the cost of the design had to be reduced.

Segmentation Results

The next step in the analysis was conjoint utility segmentation. At this point, the 12 utility functions for each one of the more than 500 re-

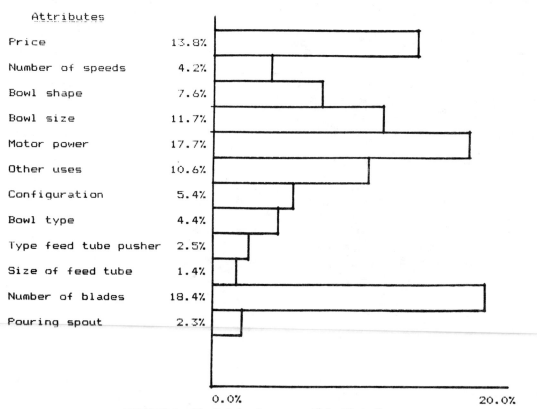

Attributes

Price	13.8%
Number of speeds	4.2%
Bowl shape	7.6%
Bowl size	11.7%
Motor power	17.7%
Other uses	10.6%
Configuration	5.4%
Bowl type	4.4%
Type feed tube pusher	2.5%
Size of feed tube	1.4%
Number of blades	18.4%
Pouring spout	2.3%

0.0%   20.0%

FIGURE 2.   *The Relative Importance of the 12 Attributes*

spondents were computed and entered into a hierarchical-cluster analysis program, the clustering groups respondents who have similar patterns of utility functions.

A four-cluster solution to the analysis was selected because it produced a set of clusters that SAC thought were a good representation of the market (face validity) as well as useful for marketing planning. The clustering results revealed four natural market segments of food processor buyers who vary according to the importance they attach to specific features and who desire different combinations of features and benefits when purchasing a food

processor. Separate utility functions and bar graphs, like that in Figure 2, were computed for each of the four segments of homemakers. Comparing the bar graphs for the four segments reveals market opportunities in the form of food processor designs to suit each segment's combination of product attributes that are very important, moderately important, and of minor importance to them. A comparison of these results for two of the segments, which are shown in Table 2, will illustrate this clearly.

One of these segments was labeled the *Cheap and Large* segment because of the particular preferences for food processor features of

TABLE 2. *A Comparison of Two Food Processor Market Segments*[a]

	The Cheap and Large Segment	The Multispeeds and Uses Segment
Very important features	$49.99 price 4-quart bowl	seven speeds can be used as a blender and a mixer
Moderately important features	two speeds seven processing blades heavy duty or professional power motor cylindrical bowl pouring spout	$99.99 price 2-quart bowl cylindrical bowl regular discharge bowl
Features of minor importance	side discharge bowl three-part feed tube pusher machine that is only a food processor large feed tube undercabinent design	three-part feed tube pusher regular size feed tube heavy duty or professional power motor seven processing blades bowl over the motor design
Other demographic and psychographic features of the segment	least likely segment to already own a food processor have higher than average ownership of Oster and Sears brands most likely segment to give a food processor as a gift older in age have midrange incomes comprise 22% of the food processor market	most likely to own a GE brand food processor younger in age have lower incomes comprise 28% of the food processor market

a. Within the very moderately and minor importance groups, the attributes are listed in declining order of importance.

the buyers who comprise the segment. The two attributes that are most important to them are price and bowl capacity. They will select a cheap food processor with the largest possible bowl (four quarts). As a group, these food processor buyers are extremely price sensitive. As Table 2 shows, they are the least likely of the four segments to already own a food processor and have higher than normal ownership rates for Oster and Sears brand processors. This segment's members are older, have midrange incomes, are the most likely to give a food processor as a gift, and comprise 22% of the consumer market.

The second segment was termed the *Multiple Speeds and Uses* segment because of its members expressed preferences for food processor features. Their preference for a food processor increases as the number of speeds increases and as the uses of the appliance increase to include, both mixing and blending. Thus they will strongly prefer a machine with many speeds and uses. The members of this segment of the market seem to distrust a low price of $49.99 for a food processor but are very sensitive to prices above $99.99. They are also indifferent to the presence or absence of a pouring spout, are more likely to own a G.E. brand processor, and are younger and have lower incomes. The group of buyers with these preferences and demographic characteristics comprises 28% of the food processor market.

The attribute profiles of the two market segments shown in Table 2 differ substantially from each other as well as from the attribute profile for the average buyer in the sample. They differ in their relative preferences for eight of the 12 attributes in the study including all four of the attributes that are most important to either of the two segments. Specifically, the Cheap and Large segment most strongly wants a $49.99 price and a 4-quart bowl while the Multispeeds and Uses segment prefers a $99.99 price and a 2½ quart bowl size. On the other hand, the Multispeeds and Uses segment

strongly prefers many speeds and the ability to also use the processor as a mixer and blender while the Cheap and Large segment wants only two speeds and a machine that is only a food processor.

The two profiles are so different that it is unlikely that members of both segments would buy the same food processor. The Cheap and Large segment calls for a basic low end machine while the Multispeeds and Uses segment calls for a more sophisticated and expensive machine. If no food processor on the market at the time of the study were offering either of these two combinations of features SAC could use these profiles to design new machines specifically targeted at one or both of these segments. If they did, then SAC could expect the new design to receive a very favorable reception among the members of the target segment because it was configured to match their set of expressed preferences more closely than any other machine in the marketplace.

The overall conjoint results along with those from the segmentation demonstrated to SAC which features should be stressed and which could be de-emphasized in the product line redesign to follow. Perhaps of even more importance was the fact that these results led SAC's product line managers to believe that they could greatly increase demand for the food processor line by changing the configuration and number of the models in the line.

PRODUCT LINE REDESIGN VIA SIMULATION

The total utility of any attribute profile can be computed for any of the individual respondents using the sets of 12 utility functions for each of the respondents. The profile with the highest utility would be the one she would most prefer and purchase if the opportunity were available. The computerized simulations use each buyer's set of utility functions to find the one food processor out of an exhaustive set of profiles of competitive food processor models which she would prefer the most. When the choices, or the single most preferred model, for each buyer are aggregated for all the buyers in the sample the result is a share estimate for each model and each brand in total which is the output of the simulations.

As Cattin and Wittink [1] point out, however, it is not advisable to use the conjoint analysis estimates directly as share predictions. This is because there are usually other marketing factors that also influence a product's market share that are not included in the conjoint analysis research. In the case of SAC's small appliance studies, two important factors influencing share which are not included are advertising weight and product availability. Therefore, SAC adjusts the estimates from the conjoint analysis to take into account differences in brand awareness and distribution across the competing brands. The weights for these adjustments to the conjoint based share estimates come from other marketing research studies that SAC conducts regularly. SAC bases its product line redesign decisions on these adjusted market share estimates.

The Base Case Simulation

The first use of this share computation procedure was to predict market shares for what is termed the "base case." The base case consists of all the food processor models currently sold by the major competitors, including SAC, at the time of the study. In order to do this, SAC identified all the major competitive brands and models and specified them by their levels of the 12 attributes used in the study.

Table 3 lists the brands and models of food processors that SAC identified and the profiles of those models in terms of the attributes measured in the conjoint analysis research. The table shows nine brands and 32 models of

food processors, including SAC's, that it considered to make up the vast majority of the food processor market. The numbers in parentheses in the 12 columns in Table 3 show the codes for each level of each attribute for each model as rated by SAC's product line managers while the numbers in the first five columns that are not in parentheses are the actual numerical values of that attribute for each model. Thus, the 12 by 32 matrix in Table 3 represents a comprehensive portrait of the food processor market in the summer of 1983 and was the input for the base case simulation.

Market share estimates were computed for each model from the base case cross-sectional portrait of the food processor market and aggregated together for each brand. To do this, the 32 base case attribute profiles were inserted into each of the respondent buyer's sets of 12 utility functions and each one "chose" the one food processor model that she would prefer most strongly. The predicted market shares (which have been weighted for brand awareness and distribution) are the results of the base case simulation. They are shown in the next to last column of the base case table, while the brands' actual market shares as determined from a separate national tracking study conducted by SAC are shown in the last column. The correlation between the predicted brand shares and the reported shares was very high (r = .96), which builds confidence in the quality of the market share predictions produced from the conjoint based simulations.

Table 3 shows that SAC's old six-model food processor line had a predicted market share of 10.2% while Cuisinart led the market with a 29.8% predicted share and the strongest position in the high end of the market. The base case market portrait and market share predictions provides a background and reference point for SAC's product line managers to plan to improve the performance of the line. To do that, the performance of alternative SAC food processor product line configurations is simu-

TABLE 3. *Food Processor Market Base Case Situation and Predicted Market Shares—Summer 1983[a]*

Products		Food Processor Attributes												Adjusted Predicted Share	Reported Share
	I	II	III	IV	V	VI	VII	VIII	IX	X	XI	XII			
Sunbeam															
14011 Le Chef	(1.58) $79	(2)	(1) 3	(2) 2½	(1) 1	(1)	(2)	(1)	(1)	(1)	(1)	(2)	.6%	—
14021 Le Chef (Side Chute)	(1.72) $86	(2)	(1) 3	(2) 2½	(1) 1	(1)	(2)	(2)	(2)	(1)	(1)	(2)	1.9	—
14056 Food Processor	(.90) $45	(1)	(1) 3	(2) 2½	(1) 1	(1)	(1)	(1)	(1)	(1)	(1)	(2)	1.9	—
84071 Vista Professional	(2.99) $199	(3)	(1) 3	(3) 4	(1) 1	(1)	(2)	(1)	(2)	(2)	(1)	(1)	3.6	—
84048 Vista	(1.72) $86	(2)	(1) 3	(2) 2½	(1) 1	(1)	(2)	(2)	(2)	(1)	(1)	(2)	1.9	—
84068	(1.64) $82	(2)	(1) 3	(2) 2½	(1) 1	(1)	(2)	(1)	(2)	(1)	(1)	(2)	.3	—
Brand total	—	—	—	—	—	—	—	—	—	—	—	—	10.2%	7%
Cuisinart															
DLC-7 Pro	(3.00) $199	(3)	(1.5) 4	(2.33) 3	(1) 1	(1)	(2)	(2)	(2)	(2)	(1)	(2)	1.4%	—
DLC-8F	(2.55) $155	(2)	(1.5) 4	(2) 2½	(1) 1	(1)	(2)	(1)	(2)	(2)	(1)	(2)	1.3	—
DLC-10E	(1.90) $95	(2)	(1.5) 4	(1.5) 2	(1) 1	(1)	(2)	(1)	(2)	(2)	(1)	(2)	5.3	—
DLCX	(3.99) $299	(3)	(1.5) 4	(3) 4	(1) 1	(1)	(2)	(1)	(2)	(2)	(1)	(2)	21.8	—
Brand total	—	—	—	—	—	—	—	—	—	—	—	—	29.8%	29%
General Electric															
FP1	(1.0) $50	(1)	(.5) 2	(.75) 1¼	(1) 1	(1)	(1)	(1)	(1)	(1)	(1)	(2)	4.6%	—
FP3	(1.36) $68	(1)	(.5) 2	(.75) 1¼	(2.8) 6	(2)	(1)	(1)	(1)	(1)	(1)	(2)	15.3	—
FP6	(1.80) $90	(2)	(1.5) 4	(1.5) 2	(1) 1	(1)	(2)	(2)	(1)	(1)	(1)	(2)	3.2	—
Brand total	—	—	—	—	—	—	—	—	—	—	—	—	23.1%	24%
Hamilton Beach															
702	(.94) $47	(1)	(1) 3	(1.5) 2	(2) 2	(1)	(1)	(1)	(1)	(1)	(1)	(2)	2.2%	—
707	(1.08) $54	(1)	(1.5) 4	(1.5) 2	(1) 1	(1)	(1)	(1)	(1)	(1)	(1)	(2)	.0	—
710	(1.30) $65	(1)	(1.5) 4	(1.5) 2	(2.8) 6	(1)	(1)	(1)	(1)	(1)	(1)	(2)	.1	—
737	(1.44) $72	(1)	(2) 5	(1.5) 2	(1) 1	(1)	(1)	(1)	(1)	(1)	(1)	(2)	.8	—
736	(1.36) $68	(1)	(2) 5	(1.5) 2	(2) 2	(1)	(1)	(1)	(1)	(1)	(1)	(2)	2.3	—
738	(1.14) $57	(1)	(.5) 2	(1.5) 2	(2.8) 6	(1)	(1)	(2)	(1)	(1)	(1)	(2)	4.9	—
2002	(1.80) $90	(1)	(2) 5	(1.5) 2	(4.8) 16	(1)	(1)	(1)	(1)	(1)	(1)	(2)	.5	—
Brand total	—	—	—	—	—	—	—	—	—	—	—	—	10.8%	8%
Kitchen Aid															
KFP 700	(2.9) $190	(3)	(1.5) 4	(3) 4	(1) 1	(1)	(2)	(1)	(1)	(1)	(1)	(2)	.1%	1.0%

390

Food Processor Attributes

Products	I	II	III	IV	V	VI	VII	VIII	IX	X	XI	XII	Adjusted Predicted Share	Reported Share
Moulinex														
LM 2	(.84) $42	(1)	(1.0) 3	(1.75) 2¼	(1) 1	(1)	(2)	(1)	(1)	(1)	(1)	(2)	1.1 %	—
LM 5	(1.60) $80	(1)	(2.0) 5	(1.75) 2¼	(3.8) 11	(1)	(2)	(1)	(1)	(1)	(1)	(2)	20.6	—
Brand total	—	—	—	—	—	—	—	—	—	—	—	—	21.7%	19%
Robot Coupe														
RC 2000	(1.66) $83	(2)	(1.5) 4	(1.5) 2	(1) 1	(1)	(2)	(1)	(1)	(1)	(1)	(2)	.1%	—
RC 2100W	(2.40) $140	(2)	(1.5) 4	(1.5) 2	(1) 1	(1)	(2)	(1)	(1)	(1)	(1)	(2)	.1	—
RC 3500	(3.2) $220	(3)	(1.5) 4	(3) 4	(1) 1	(1)	(2)	(1)	(1)	(1)	(1)	(2)	.1	—
RC 3600	(3.3) $230	(3)	(1.5) 4	(3) 4	(1) 1	(1)	(2)	(1)	(1)	(2)	(1)	(2)	.2	—
RC 2800	(2.85) $185	(2)	(1.5) 4	(2) 2½	(1) 1	(1)	(2)	(1)	(1)	(2)	(1)	(2)	.2	—
Brand total	—	—	—	—	—	—	—	—	—	—	—	—	.7%	1.0%
J.C. Penney														
R784 3246	(1.16) $58	(1)	(.5) 2	(1.75) 2¼	(1) 1	(1)	(2)	(1)	(1)	(1)	(1)	(2)	.2%	—
R784 3337	(1.60) $80	(1)	(2) 5	(1.75) 2¼	(1) 1	(1)	(2)	(1)	(1)	(1)	(1)	(2)	.2	—
Brand total	—	—	—	—	—	—	—	—	—	—	—	—	.4%	1.0%
Sears														
34A 82288	(1.0) $50	(1)	(1) 3	(1.5) 2	(1) 1	(1)	(1)	(1)	(1)	(1)	(1)	(2)	0 %	—
348 82368	(1.8) $90	(1)	(2.5) 6	(1.5) 2	(3) 7	(1)	(1)	(1)	(1)	(1)	(1)	(2)	2.1%	—
Brand total	—	—	—	—	—	—	—	—	—	—	—	—	2.1%	9.0%

a. Numbers in parentheses indicate level in research design. See Table 1 for full explanation of these numbers as well as the roman numerals in the heading of this table.

lated and their predicted share results are compared to those for the actual six-model line in the base case to determine whether or not the alternative configuration will improve on the present line's market share. In the subsequent product line simulations the features and pricing of the other competitors' models were assumed to remain constant while the configuration of SAC's line was varied to observe the resulting changes in shares. In this way the market share yield of SAC's line of food processors could be optimized.

Simulating "What If" Scenarios

The purpose of the "what if" simulations is to guide the redesign of the product line regarding the number of models, features, and pricing through the impact that changes in these factors are predicted to have on the line's market share. Management can ask the question "what if we redesign our line so that it is reconfigured as . . . ?" and answer it with the results from a simulation of the redesigned line's perform-

ance in the market place. Used in this way, the simulations provide product line managers with a powerful analytical tool to assist them in redesigning product lines.

The actual redesign began with SAC's product line managers studying the conjoint analysis segmentation results to develop hypothetical product line configurations that matched the needs of the market segments as expressed through the sample of homemakers. Brainstorming sessions identified a great many alternative configurations, and almost 50 of the more logical and realistic ones were fleshed out for testing as "what if" scenarios. Table 4 describes the first five alternative product line scenarios that were tested via the simulations.

Each of the scenario descriptions was converted into a set of product attribute profiles that described each of the models in that product line configuration. Figure 3 illustrates the set of profiles for one of the hypothetical five model line scenarios examined by SAC. The set of profiles for a scenario were the input data for the "what if" simulation of the market

TABLE 4.   Descriptions of Five Food Processor Product Line Simulations

Scenario 1: A basic four model line with the following configuration:

Keep the professional processor (high end) and the 14056 (low end) with an added pour spout and drop the other four current models.

Add two multispeed models with larger 4-quart bowls, side discharge, and large feed tube and priced at $59.00 and $79.00 to strengthen the middle of the line.

Scenario 2: A basic five model line with the following configuration:

Same as scenario 1 plus an additional low end model with two speeds, a 3-quart bowl, a regular size feed tube, no side discharge, and priced at $49.00.

Scenario 3: A four model line with the following configuration:

Same as scenario 1 but upgrade the low end 14056 model with a larger 3-quart bowl and increase its price by $5.00.

Scenario 4: A four model line with the following configuration:

Same as scenario 1 but upgrade model 14056 even more than in scenario 3 with a larger 3-quart bowl, a large feed tube, and increase its price by $9.00.

Scenario 5: A five model line with the following configuration:

Same as scenario 1 but with an upgraded low end provided by the low end model of scenario 3 and the two speed model from scenario 2.

	Model 1	Model 2	Model 3	Model 4	Model 5	Model 6	Model 7
I. Price	#199	#45	#85	#109	#59		
II. Motor Power (1, 2 or 3)	3	1	2	2	2		
III. Number of Blades	4	3	4	4	4		
IV. Bowl Size	4	2½	3	3	2		
V. Number of Speeds	1	1	7	7	5		
VI. Other Uses	1	1	1	1	1		
VII. Configuration	2	1	2	2	2		
IIX. Bowl Type	1	1	1	2	1		
IX. Type of Feed Tube Pusher	2	1	2	2	1		
X. Size of Feed Tube	2	1	2	2	2		
XI. Bowl Shape	1	1	1	1	1		
XII. Pouring Spout	1	0	1	1	1		

FIGURE 3. Input for Market Share Simulations: Scenario #7

share performance of that possible product line configuration. Although the product lines of all the other competitors were assumed to remain constant at the configurations shown in Table 3, while SAC's line was varied, the procedure also permits changing the competitors' product line(s) at the same time to simulate their hypothetical response to SAC's new line.

The output of the simulations allows SAC's managers to see the overall performance of a line configuration and the effect of the line on those of the competitors as well as the per-formance of each model in the line and the possible cannibalization effects created by additions and changes to the line. Table 5 shows the predicted market share results from simulations of the performance of scenarios 1 and 2 in Table 4. Comparison of the results for scenario 1 with those for the base case situation in Table 3 allowed SAC's managers to see that substantial improvement in line performance was possible with a product line configured to more closely meet the needs of the segments in the marketplace. In this scenario the addition of

TABLE 5. *Predicted Market Shares for Two "What If"*
Simulations

Brands and Models	Predicted Market Shares (%)	
	Scenario 1	Scenario 2
Sunbeam		
Model 1—Professional ($179)	7.1	7.0
Model 2—14056 ($30)	3.3	3.2
Model 3—Five speeds ($59)	5.8	4.0
Model 4—Seven speeds ($79)	8.7	8.7
Model 5—Two speeds ($49)	—	3.9
Brand total	24.9	26.8
Cuisinart		
DLC-7 Pro	1.1	1.1
DLC-8F	5.0	5.0
DLC-10E	0.0	0.0
DLCX	19.7	18.9
Brand total	25.8	25.0
General Electric		
FP 1	5.1	4.7
FP 3	13.0	13.0
FP 6	3.4	3.4
Brand total	21.5	21.1
Hamilton Beach		
702	1.8	1.6
736	3.4	3.0
738	3.8	3.6
2002	0.7	0.7
Brand total	9.5	8.9
Kitchen Aid		
KFP 700	0.2	0.2
Moulinex		
LM 2	1.0	0.9
LM 5	15.0	15.0
Brand total	16.0	15.9
Robot Coupe		
RC 2000	0.1	0.0
RC 2100W	0.1	0.1
RC 3500	0.1	0.1
RC 3600	0.2	0.2
RC 2800	0.2	0.2
Brand total	0.7	0.6

Brands and Models	Predicted Market Shares (%)	
	Scenario 1	Scenario 2
J. C. Penney		
R 784 3246	0.3	0.3
R 784 3337	0.1	0.1
Brand total	0.4	0.4
Sears		
34A 82288	0.0	0.0
34B 82368	1.3	1.3
Brand total	1.3	1.3

two new appropriately featured midrange models would provide the line with much more share than did the four old midrange models listed in Table 2 they would be intended to replace.

Looking at the share results for the other competitive brands reveals the shifts in market share that would take place and the sources of SAC's predicted share gain. The Sears and Moulinex lines show the biggest proportional declines in share with 38.1 and 26.3%, respectively, while Cuisinart and Hamilton Beach show smaller declines of 18 and 17.4%. On the other hand, General Electric, with the third rank share position in the base case situation, and the other three brands with minor share positions, would be unaffected by this particular configuration of a resigned line of food processors.

Comparison of the share results for the two hypothetical product line configurations in Table 5 also allowed SAC's managers to explicitly consider the demand interrelationships between the individual models in their line. This is illustrated by moving from a four model line in scenario 1 to a five model line in scenario 2. The only difference between the two configurations is the addition of a second low-end two-speed model in scenario 2. This addition to the line improves the line's overall share performance by 1.9% but the share for the added two-speed model is predicted to be 3.9%. Thus, some of the new model's share will come at the expense of the line's other models and the figures reveal that it cannibalizes the five-speed model above it for 1.8 share points, or 31% of what its share would be in the four-model line of scenario 1. At the same time, the high-end and low-end models are predicted to be only slightly affected to the tune of the loss of one tenth of a share point each while the seven-speed model should not be affected at all.

THE REDESIGNED PRODUCT LINE

Using the share results from the simulations, SAC's managers selected a number of the more attractive hypothetical product lines for further analysis. This involved an examination of their anticipated profitability. The expected number of units to be sold of each model in a candidate line configuration was computed from its predicted market share and the expected market volume.

These unit sales estimates were used in conjunction with the line's levels of the price attribute, prospective profit margins, and estimates of marketing costs as well as tooling and any other investments to compute estimates of the profitability and return on investment of that line configuration. Based on these considerations, one of the configurations was chosen as the redesigned product line.

The new, redesigned food processor product line ended up with three basic models that SAC felt would cover the market segments they wanted to target and provide it with a substantially increased market share. The actual line contained three all new models, a full featured high end professional model intended to sell for about $200.00, a two-speed promotionally priced low-end model intended to retail at about $35.00–$45.00, and a seven-speed, heavy duty, midrange machine intended to retail at

about $99.00. In addition, a derivative version of the midrange model was also added to the line for department stores.

The redesigned four model line appeared during 1984 and by the end of that year the new line's off-the-shelf market share had exceeded 10% and moved SAC into the fourth ranked share position in the market. Late in 1984 SAC also introduced a new second-generation food processor model that radically changed the market and invalidated the direct comparison of the redesigned line's 1985 share with that of the old line. In the subsequent redesign studies conducted for its other product lines where the share results have been clearer, SAC has been pleased by how close the actual share performance of the redesigned lines have come to those predicted by the conjoint based simulations.

CONCLUSIONS

In the 2½ years that SAC has been using its product line redesign procedure it has been very pleased with the results the procedure has produced. This success can be attributed not only to an effective application of the conjoint analysis technique but also to the continuing involvement of product management in the research studies as exemplified by the description of the food processor study.

The involvement of product managers and planners with all phases of the redesign process produces several benefits. Their involvement strengthens the quality of the research and makes it more relevant to their needs. The high level of involvement is also very motivating to them and eliminates the mistrust that frequently is engendered when marketing research is done without the participation of product management. Instead, they believed the results of the research and made use of them because they were deeply involved in its design and execution. The product managers were actually responsible for creating the alternative product line scenarios and made full use of the simulations to help them redesign the food processor line.

Thus, the redesign procedure and its research studies are truly a marketing department effort involving both the product and research staffs rather than simply a research study given to the product managers which they are told to use. They perceive it as "our study" and welcome the assistance it can provide them.

The combination of advanced marketing research methods with a more rational, market-driven approach to product development within the company and product and research people working together has resulted in improvements in market share and profits for all the product lines to which SAC has applied the procedure. Furthermore, with each succeeding study the sophistication of its use of the conjoint analysis technique has grown as the company's experience with it has grown. Conjoint analysis based product line redesign has proved to be a very valuable new tool for Sunbeam Appliance Company and it has wide applicability for other companies that have product lines which need to be redesigned.

REFERENCES

1. Cattin, Philippe and Wittink, Dick R. Commercial use of conjoint analysis: a survey. *Journal of Marketing* 46:44–53 (Summer 1982).
2. Green, Paul E. On the design of choice experiments involving multifactor alternatives. *Journal of Consumer Research* I:61–68 (September 1974).
3. Green, Paul E., and Rao, Vithala R. Conjoint measurement for quantifying judgmental data. *Journal of Marketing Research* VIII(3):355–363 (August 1971).
4. Green, Paul E. and Srinivasan, V. Conjoint analysis in consumer research: issues and outlook. *Journal of Consumer Research* 5:103–123 (September 1978).
5. Green, Paul E. and Wind, Yoram. *Multiattribute Decisions in Marketing: A Measurement Approach.* Hinsdale, IL: The Dryden Press, 1973.
6. Green, Paul E. and Wind, Yoram. New way to measure consumers' judgments. *Harvard Business Review* 53:107–117 (July–August 1975).
7. Kotler, Philip. *Marketing Management: Analysis, Planning and Control.* Englewood Cliffs, NJ: Prentice Hall, 1983.
8. Kruskal, Joseph B. Analysis of factorial experiments by estimating monotone transformations of the data. *Journal of the Royal Statistical Society, Series B* 27:251–63 (March 1965).

APPENDIX

Technical Description of Conjoint Analysis Procedures

At the heart of SAC's product line redesign procedure is an empirical marketing research study using conjoint analysis. Conjoint analysis is a measurement technique developed in mathematical psychology in the early 1960s and introduced into marketing research in the early 1970s [4]. Since then it has been widely adopted and in 1982, Cattin and Wittink [1] estimated that approximately 1000 conjoint analysis studies had been conducted in the United States. SAC's use of conjoint analysis to simultaneously optimize the number and design of the models of a product to appear in a product line is believed to be the first such application to appear in the literature.

The purpose of conjoint analysis is to enable a researcher to take a respondent's overall evaluations of a set of objects or concepts and decompose those evaluations into separate scores, called utility values, for the various attributes of the objects that led to the respondent evaluating them the way he did. A simple example will provide the basis for a nontechnical explanation of conjoint analysis and how it works.

Consider an appliance purchaser who

goes shopping for a new electric hand mixer. Her shopping identifies four different models of hand mixers from which she will make a purchase selection. Their major attributes, or features, and only points of difference are:

1. A $10.00 model with a 90 watt motor, three mixing speeds, a plastic body, and a retractable electric cord.
2. A $20.00 model with a 160 watt motor, five mixing speeds, a plastic body, and is cordless/rechargeable.
3. A $45.00 model with a 200 watt motor, 12 mixing speeds, a chrome body, and a removable electric cord.
4. A $20.00 model with a 90 watt motor, five mixing speeds, a chrome body, and a nonremovable electric cord.[2]

These alternative purchase choices represent four different combinations of five product attributes of hand mixers that are motor power, number of mixing speeds, body material, electric cord arrangement, and price. In this instance, there are three different levels of the motor power, mixing speeds, and price attributes, two levels of the body material attribute, and four levels of the electric cord arrangement attribute. Thus, any product like the hand mixer or a television set, personal computer, lipstick, or cash management account can be described as a multiattribute object.

The hand mixer descriptions above are attribute profiles of four different configurations of a mixer. In this situation, the appliance buyer is faced with making a choice decision from among a number of multiattribute alternatives. He or she will make an overall judgement about the relative value, or utility, of the combinations of attribute levels represented by each of the available product alternatives in order to reach a purchase decision. During this evaluation process, the buyer will probably make complex trade-offs since it is likely that no alternative is clearly better than another on

every attribute. Thus, the decision maker may have to trade-off economy for either more speeds or power or trade-off a chrome body for the cordless/rechargeable feature. This sort of complex choice among multiattribute alternatives is the kind of real choice problem that decision makers, both consumer and industrial, commonly face and which conjoint analysis can measure.

When conjoint analysis is used it is assumed that the decision maker has an implicit set of utility functions, one for each product attribute, which is used when the alternatives are evaluated. While he or she cannot articulate these functions, and may not even be aware of them, the overall evaluations of each of the multiattribute alternatives in terms of preference for purchase reflects these utility functions. In fact, for the most commonly used form of conjoint analysis, the overall preference for purchase of a given alternative is an additive combination of the utility given by the appropriate utility function for each discrete level of each attribute of the alternative.

The objective of the conjoint analysis technique is to uncover the decision maker's utility functions that determined his/her preferences for the different alternatives among a set of multiattribute alternatives. This can be done if the attributes and levels of the products and the decision maker's preference ranking of the alternatives are known.

Lets assume that, in a carefully designed experiment, a group of small appliance buyers was asked to rank the four hand mixer attribute profiles in order of their preference for purchase. From their preference rankings conjoint analysis can uncover a separate utility function for each of the five hand mixer attributes for each one of the respondents. The hypothetical utility functions for two of the five mixer attributes for one of the respondent buyers are shown in Figure 4. The vertical axes in the graphs of these utility functions measure utility in common, interval-scaled units which repre-

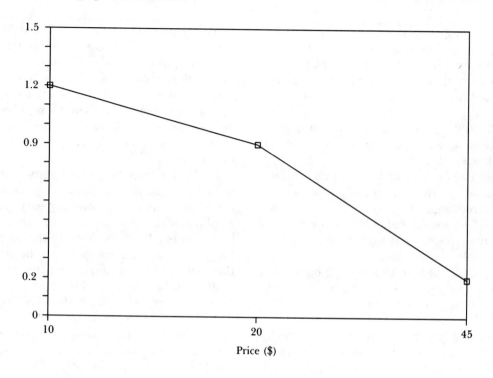

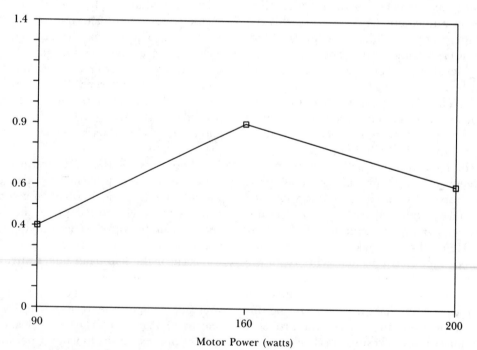

FIGURE 4.   *Illustrative Utility Functions for Two Hand Mixer Attributes*

sent this respondent's degree of preference for each level of each attribute. Because the utilities are measured in common units, preferences for different features can be compared and added together as shown in the following paragraphs.

These two utility functions for the price and motor power of a hand mixer allow demonstration of the type of results that conjoint analysis can provide through the illustration of the preferences of the hypothetical buyer for mixer prices and power. First, it shows that this respondent's preferences for mixer prices is highest at the $10.00 price and declines with increases in price, as we would expect, and that her preference for mixer power is strongest for 160 watts. Second, it shows that price has a stronger influence on this buyer's overall preferences for mixers than does power.

Third, the utility functions allow us to compute the degree of preference the buyer has for each of the four price/power combinations shown in the four attribute profiles. In this example the $45.00 and 200 watt combination of profile three has a combined utility of 0.8 (.2 + .6), whereas the $20.00 and 160 watt combination of profile two has a combined utility score of 1.8 (.9 + .9). Clearly, in this instance, the buyer exhibits a stronger preference for the price/power combination of profile two. It is also the highest combined two attribute utility score of the four profiles. Thus, the $20.00 and 160 watt attribute combination, were it made available in the marketplace, should be more favorably received by this appliance buyer than a mixer with any of the other three combinations of the two attributes would be.

The overall utility scores for these two price/power profiles also illustrate the kinds of tradeoffs consumers make in coming to decisions in multiattribute choice situations. Here, while profile one has the most attractive price, the buyer is willing to trade-off some of the utility of the low price for the added utility of more motor power as highlighted by the fact

that profile two with a higher price but more power has a higher two attribute utility than does profile one.

The utility functions also allow us to compute this buyer's preference for other price/power attribute combinations which he or she was not originally asked to rank. For instance, the utility of the $45.00 and 160 watt price/power combination is 1.1 (.2 + .9). Finally, the functions also allow us to identify the respondent's most strongly preferred price/power combination, $10.00 and 160 watts. These attribute levels have the highest combined utility score of 2.1 (1.2 + .9) and thus, this respondent should prefer this combination over any of the nine possible combinations of price and power which might appear in the marketplace. Of course, it should be pointed out here, that this combination of features and price, while most preferred by this shopper, may not turn out to be feasible from a design or price cost perspective, in which case other highly preferred combinations of attributes would be considered by the company.

Thus conjoint analysis procedures employ data that describe the respondents' overall preference ranking of a set of multiattribute alternatives. These rankings are decomposed into their underlying utility functions for each attribute. The preference data is ordinal scale level, while the derived utility functions are at an interval scale level.[3]

With the utility function output the original preference rankings can be reconstituted and other new configurations of attribute levels can also be ranked in terms of their relative total utility scores as illustrated above. This is the reason that SAC uses conjoint analysis as the basis for its product line redesign procedure.

Conjoint Analysis Data Collection

In the simple hand mixer example above there are 216 possible combinations of the three levels each of price, power, and speeds, the two levels of body material and the four levels of the cord arrangement $(3 \times 3 \times 3 \times 2 \times 4 = 216)$. In conjoint analysis problems of practical interest the numbers of attributes and/or levels can be much larger resulting in a very large number of possible attribute combinations. In the case of SAC's actual hand mixer study, 16 attributes were studied and each had two, three, or four levels. This complex study produced over 1.9 million possible combinations of attribute levels.

It would be a formidable task for any person to rank all 216 attribute combinations. Ranking thousands or millions of combinations is clearly impossible. Thus, when using conjoint analysis, it is necessary to reduce the number of actual combinations that are ranked to a more manageable number. This reduced number needs to be small enough to minimize the effect of respondent confusion, boredom, and/or fatigue on the integrity of the data collection process while at the same time also bringing the cost of the data collection down to a more reasonable level.

Fortunately, an experimental design called orthogonal arrays can be used to accomplish the reduction in the number of attribute combinations that must be preference ordered in a conjoint analysis study. By giving up the chance to measure the joint, or interaction, effects between the attributes in the study, the research can still clearly measure the separate main effect of each attribute on the ranking of the alternatives while vastly reducing the number of attribute combinations which the respondents must rank. By use of a main effects orthogonal array design in SAC's hand mixer study the number of attribute combinations the respondents had to rank order was drastically reduced to only 27.[4]

NOTES

1. Other procedures for estimating utility functions from various types of preference data include PREFMAP, LINMAP, LOGIT, PROBIT, and OLS regression. For more information about these estimation methods and when they should be used, see [4].
2. The four hand mixer models described here are only illustrative and are not meant to accurately portray the actual attributes of real offerings in the market.
3. For further information about conjoint analysis see the progressively more detailed explanations in [6], [3], and [5]. For a complete description of the steps in conducting a conjoint analysis study and the issues envolved in their execution, see [4].
4. For more information on main effects, interactions, and orthogonal arrays see [2].

7

MANUFACTURING/OPERATIONS

The Interface between Strategy and Manufacturing Technology

Mariann Jelinek

Joel D. Goldhar

Manufacturing systems typically represent the lion's share of a firm's human and financial assets, yet all too often, operations fail to enter into strategic planning except as a stepchild to financial, marketing or research options. Most often, the product idea is the focus, and manufacturing is simply expected to deliver the goods at the lowest possible cost. What is now needed is a framework for understanding manufacturing technology as a driving force in corporate strategy.[1]

In the past, manufacturing decisions have often been delegated downward as "routine," though we know that manufacturing excellence is one important source of competitive advantages in worldwide competition.[2,3] In the past, too, the rationale for conventional specialized equipment and automation has centered

on economies of scale and intensive capital investment.[4] Production processes were seen as moving from fluid and highly flexible processes for new products, to systematic and far less flexible, more capital intensive processes as products matured.[5]

All these assumptions reflect the manufacturing experience of the past, when "know-how" was captured by building it into special-purpose machines. Chart 1 lists some of the principal differences between the past and present manufacturing environment. The tradeoff involved was *increased efficiency* (lower per unit cost, greater precision and higher production volume) against *flexibility*. In manufacturing, these trends have pushed toward even greater specialization at the cost of increasing rigidity.[6]

Technological developments in the miniaturization of electronic components, in integrated computerization and in communications system design are radically transforming the interface between manufacturing and strategy. Some examples include:

Old Style Technology:	CAD/CAM Environment:
Economy of Scale	Economy of Scope
Experience Curve	Truncated (or expanded) Product Life Cycle
Task Specialization	Multi-Mission Firms
Work as a Social Activity	Unmanned Systems
Separable Variable Costs	Joint Costs
Standardization	Variety
Flexibility and Variety are Expensive	Flexibility and Variety Create Profits
Desirable Operating System Characteristics	
Centralization	Decentralization
Large Plants	Disaggregated Capacity
Balanced Lines	Flexibility
Smooth Flows Ability	Surge & Turn-around
Standard Product Design	Many Custom Products
Low Rate of Change, Responsiveness High Stability	Innovation & .
Inventory as a Decoupler Demand	Production Tied to .
Focus as an Organizing Concept Repeated	Functional Range for Reorganization .
Job Enrichment and Enlargement Rewards	Responsibility Tied to . .
Batch Systems	Flow Systems

CHART 1.   *Generally Accepted Fundamental Assumptions*

- National Semiconductor, like many components makers, found itself making increasingly complex components that constituted a growing portion of the computer systems sold by Itel, a National customer. When Itel ran into difficulties, acquisition seemed a logical, short step—and National, as National Advanced Systems, was in the computer business as well as components manufacture.

- Similarly, minicomputer makers like Digital are manufacturing an increasing portion of their own components, and finding themselves moving, along with technology, into new markets for personal computers on the low side, and supermini "almost mainframes" on the high side.

"Manufacturing" decisions in each case redirected the business in significant ways, yet the potential for strategic impact far exceeds these examples. Thus, if a range of manufacturing operations is possible in, say, metalworking, a range of products is possible. Exploitation of these options may entail new customers, new channels of distribution, new methods of selling, and new industries. Above all, exploiting the new options entails generating a new vision of the firm and its strategic mission.

The proper return on sizeable manufacturing technology investments required will be realized only when management is prepared to reorganize and to undertake new strategic activities. Informed choices about "manufacturing" decisions, then, involve new perspectives on the implications of technology choices. Conversely, if decisions such as these are treated as purely "manufacturing" decisions, either their high cost will preclude them or the firm will fail to realize the substantial additional benefits implicit in the new technology. Competitors, making similar decisions in a more informed fashion, may gain significant competitive advantage at lower risk.

NEW DEVELOPMENTS IN MANUFACTURING TECHNOLOGY

Four principal technologically based systems are being implemented worldwide in manufacturing operations. They are discussed below.

Robots

A new generation of small and relatively inexpensive robots is in use today. The biggest users are major mass producers like automobile companies or machine tool companies. In the United States, only 4,000 robots are in use today, compared to Japan, where some 14,000 are in use. The range of tasks the robots perform is wide—and applicable across the spectrum of manufacturing. A single robot can typically be fitted with a variety of heads to enable it to pick up, reposition, lubricate, drill, weld or paint. Such multi-purpose robots illustrate a key characteristic of the present trend of automation, which separates it distinctly from earlier attempts: flexibility. In place of the very expensive, rigidly specific machines of the past, manufacturers can now purchase a range of options. Thanks primarily to microprocessors, automation payback is now no longer a question of single-product production volume as it is of *function throughput*—and with multiple-function capacities, the opportunities for using such machines are markedly increased. It is no longer necessary to justify machines in terms of a single product or even a single process. With decreasing microprocessor costs, such economies, accuracy and volume as were once the domain of large scale manufacturers become accessible to the smallest batch producer and custom shop.

Integrated Flexible Systems

Many robots today are working side by side with human laborers, eliminating human participation in the hazardous, unpleasant, highly routinized or particularly difficult jobs. Such integration is a first step, but it limits production to human capacities. The sequence of manufacturing processes is limited by human abilities, especially speed. Clearly major productivity increases require going beyond human speeds, accuracies and staminas. Programmable controllers can oversee several machines, providing substantial benefits. Texas Instruments cites a 33% increase in productivity, dollar sav-

ings in production, and improved accuracy in thermostat calibrations, for instance. The next step beyond individual robots is the integrated manufacturing system, combining automated manufacturing processes with automated materials-handling and process monitoring. Here too, as with robots, the key to the future is flexibility. To be feasible for most of US industry, such systems must deal with frequent shifts from one product to another.

In both robots and integrated systems, the key to efficiency is careful coordination—of materials, machine utilization, sequencing, inventory removal, and the like—to insure maximum machine utilization.

Automated, computer-monitored job-shop systems are already in place today. They typically consist of a group of general machine tools, integrated with materials handling systems to perform a sequence of operations on a number of different products under computer direction.

- Ingersoll Rand installed the first computerized integrated system in 1970—over a decade ago. Palletized parts are automatically shuttled to the correct tool, loaded, processed and returned to the materials-handling conveyor after processing. The system manufactures some 150 different parts in batches of two to ten pieces. The system reduces parts costs by an average of 45% compared with conventional machine shop operations.
- Caterpillar Tractor purchased its system in 1971. The system machines housings for two families of automatic transmissions, producing the finished, assembled case and cover from rough castings. After the rough castings are manually loaded, the system recognizes the part, delivers it to the machine, and handles the remaining machining, inspection, transportation and other operations. In addition, the computer performs an elaborate array of scheduling, prioritizing, reporting and auditing functions.

403

- Deere & Company ordered a robot-tended flexible machining system in 1980. This system, designed to produce a variety of parts—many of which have not yet been designed—will cost over $2,000,000 as a turnkey project.
- The first General Motors flexible machining system was installed at the Chevrolet Gear and Axle plant in Hamtramck, Michigan, early in 1982. This line will machine a variety of short cycle parts in small batches (but totaling over 100,000 pieces per year in demand). The flexible machining center offers the potential for substantial cost savings in retooling to meet product design changes, and constitutes a very visible move away from traditional "hard-tooled" transfer lines.

CAD/CAM

Computer-Assisted Design with Computer-Aided Manufacture (CAD/CAM) is another step beyond the integrated manufacturing system. Coupling the computer into the design process offers substantial increases in productivity at each step in the process. For instance, a designer can draw a design on a monitor screen with a light pen. The computer will automatically generate different views, rotate the image or reverse it on command. Lettering can be added, the drawing scale changed, or other modifications easily made at the keyboard. Once a satisfactory monitor image has been produced, "hard copy" can be automatically generated by the computer. In addition, the computer can be instructed to produce a tape that will guide numerically-controlled production machinery.

Computer-assisted design can be used for revisions of existing work, too. It is estimated that some 80% of drafting work consists of modifying existing designs. CAD data bases can store large numbers of standard designs to be retrieved and modified. Computer-generated drawings reduce repetitious drafting, as well as increasing drafting productivity.

Beyond such economies, CAD/CAM can also integrate other sorts of data—materials constraints, stress and load factors, engineering formulae and the like—to be evaluated automatically in the course of a design. Thus routine constraints can be automatically attended to, flaws or inadequacies automatically detected, and adjustments made at early design stages with substantial consequent savings. In addition, such systems permit human designers to concentrate on creativity and aesthetics, confident that the underlying parameters have been taken into account. All of these capacities are in use today.

- St. Gobain designs elaborate perfume and liquor bottles by means of CAD. Precise volume capacity is attained, along with elegant design. Design productivity has increased sevenfold, while the needed turnaround time from design to manufacture has decreased.
- GM used CAD to downsize the Cadillac Seville, producing a shorter, lighter, more fuel-efficient car. In 1978, CAD cut the X-body car design process by a year, according to GM estimates.
- Boeing used CAD to design two new planes, the 757 and the 767, simultaneously—a feat made possible by productivity increases due to the new systems. Moreover, by enabling the company to incorporate engineering data-more readily into the design process, costly design flaws could be identified and eliminated before manufacturing began.

Computer-Integrated Manufacturing

Computer-integrated systems sequence and optimize a number of production processes, achieving order-of-magnitude improvements in equipment utilization and capital productivity by cutting down on queuing time, waiting time, machine down-time (through more predictable maintenance and operation) and elimination of in-process inventory. The significance of computer integration for process improvement can be estimated by the typical traditional machine-usage times: of time in the shop, a part spends typically only 5% on a machine, and less than 30% of this (1.5% of the total time in shop) is incut time. Elimination of substantial portions of the 95% moving and waiting time promises fuller utilization of machine capacities as well as substantial reductions of in-shop time and thus inventory in process. In one major integrated facility, operating now at Messerschmitt-Bolkow-Blohm in Augsburg, West Germany, incut time has increased to 75% or better. Production lead time for the Tornado fighter plane is 18 months, in comparison with 30 months for planes produced by more conventional means. The system, which cost more than $50 million, reduced the number of required NC machines by 44%, required personnel by 44%, required floor space by 30%, part flow time by 25% and investment costs by 9%.

Unmanned machines at Niigata Converter's Kamo Works, Niigata, Japan, have proven, on average, nine times more productive than the conventional machines they replaced, with incut time ranging around 50% and often going as high as 75%.

Integrated systems at John Deere were cited as the sources for improvements in manufacturing efficiency that boosted Deere's net 28% despite foreign currency losses and a sales decline of 0.8% for the quarter ending October 31, 1981. Deere's efforts include computerizing engineering, planning and analytical methods for tooling, part and process sequencing and automated machining and assembly.

In each system, computer controls and programmable, "smart" machines offer the advantages of specialized, automated processes and the flexibility of easily changeable specifications. The most important characteristics of these sophisticated, computer-based manufacturing systems include:

- Extreme flexibility of product design and product mix. The new machines perform a variety of tasks equally well, so the traditional logic of batch size versus inventory costs, for instance, will no longer be meaningful. New systems will process an almost unlimited variety of specific product designs within a reasonable family of design options, including alternative materials.
- Rapid responsiveness to changes in demands, in product mix and design, output rates and equipment scheduling.
- Greater control, accuracy and repeatability of process operations, leading to better quality products and much more reliable manufacturing operations.
- Reduced waste, lower training and changeover costs, and lower, more predictable maintenance costs.
- Greater predictability in all phases of manufacturing operations and vastly increased amounts of information. This will lead to more intensive management and control of the system.
- Faster throughput due to better utilization of all machines, less in-process inventory, fewer stoppages for missing parts or materials or machine breakdowns, and the use of higher speeds and a variety of exotic new processing techniques made possible (and economically feasible) by the sensory and control capabilities of the smart machines and the information management abilities of the new CAM software.
- Distributed processing capability made possible and economical by the encoding of process information in easily replicable software instead of hardware.

ECONOMY OF SCALE AND ECONOMIES OF SCOPE: THE NEW PRODUCT LINE

As we have noted, most current assumptions about manufacturing technology and strategy rest upon the notion of *Economy of Scale*—greater production volume is more economical in unit cost than lesser production, because of special-purpose equipment, which in turn is justified by large-scale operations. Typically, such machines are both faster and more accurate than their human counterparts. They were also more expensive and less flexible, necessitating vastly increased production of identical parts or products to amortize their cost. In contrast, the new technology is based upon *Economies of Scope*. A broader range of capabilities in a single machine is computer controlled. Because of computer controls, programmed production sequences and electronic memory, advanced techniques can now be applied to the production of many small runs, and automated production capabilities formerly economical only in very large plants, and for production of many identical units, are now possible in much smaller operations. Bigger is no longer better. Smaller factories and shorter production runs of any given design, as well as easy shifts from one design to another, are logical concomitants of computer-assisted manufacturing. Moreover, a total cost, maximum profits approach to manufacturing system design may lead to manufacturing organizations which are very different from the product-oriented types of today.

Economies of Scope

Economies of scope exist where multiple products can be more cheaply produced in combination than separately. Where the same equipment can produce multiple products, the potential for economies of scope exists. A computer-controlled machine tool does not "care" whether it works on a dozen units in succession of the same design or a dozen different product designs in random sequence (within a reasonable family of design limits—but that range is getting broader with each new generation of technology). The changeover time (and therefore cost) is almost negligible, since it involves simply reading a computer program. The machine "sets up" for each new design with electronic speed.

Economies of scope speak directly to product line and to inventory. It may be cheaper to make replacement parts to order than to warehouse them, for instance. A far broader product line is feasible with a library of programmed designs. The effect is to move the set-up costs back into the design process, for the programs that contain and reliably reproduce the required production sequences are part of the design. Moreover, such programming and computer controls at once demand and provide greater knowledge about manufacturing processes. Products can be designed for better manufacturability, and knowledge about manufacturing can stimulate new product designs. As well, tighter control of in-process inventory, a sort of a super "just-in-time" system, also becomes feasible.

The impact of the new technology is thus more than an increase in precision or speed or accuracy. Older systems trade off primarily behavioral art, craft or human skill against knowledge "built into" a special-purpose machine. The machine incorporated a portion of that knowledge as control hardware, but this advantage bore a cost: reduced flexibility. Newer systems offer both replicable accuracy and speed, and flexibility—through software controls.

The Importance of Software

The old trend toward specialized hardware is being reversed by a new emphasis on specialized *software*. Broad, general machine capability is matched with special-purpose programming to provide the necessary custom details. Whole sequences can be programmed and integrated.

Increasingly, as important information is programmed, the trend is for all types of production to move gradually toward operating systems: a series of operations, all computer controlled, flowing together in a smooth sequence. Such production looks more like chemical process operations than like the individual or batch operations of the past—it becomes ever more continuous. Electronics is the key, along with electronic memory. With the advent of low-cost, readily reprogrammable microprocessors, economy of scope becomes the new basis for operation. Instead of *identical* throughput (identical processes resulting in large numbers of identical products) to amortize equipment cost, variety is possible because the same equipment does many things equally well. And it is now easier and cheaper to make multiple products through multiple processes on the same machinery. This potential for relatively low-cost change has significant strategic consequences.

The Strategic Consequences: A New Dimension of Complexity

A manufacturer is no longer limited to only one production sequence; numerous sequences become possible. How many? And for what products? These questions were to some extent answered formerly by the limitations and constraints of the old equipment. Now they come up for discussion again. Imagination and strategic vision become the new constraining factors. A range of markets is possible, as well as a range of customers and channels of distribution. Because these options are possible, they represent a potential competitive advantage available to the firm with sufficient strategic sophistication to use them effectively.

The implications of these changes are far reaching. For instance, the Economic Order Quantity (EOQ) is one, and trade-off between costs of changing production versus costs of holding inventory becomes a trivial problem as the advantages of holding inventory decrease.

It is essentially as economical to manufacture one unit as to manufacture many. Learning curve cost changes also do not occur—at least not in the same way. The machine tool is as "smart" on the first unit as it ever will be—unless the program is changed. In essence, we are looking at fully fixed cost manufacturing systems. The responsibility for learning curve improvements, typically the domain of operations management in the past, will shift to manufacturing engineering. Rather than changing worker practice, learning curve improvements will require reprogramming and process change. Because the computerized machines make more information available, process changes will be based more on actual results. Because information is both available and crucial, manufacturing input into design decisions will be increased as well.

One key impact affects marketing and corporate strategy. The EOQ$=1$ phenomenon is viable only if the firm's marketing strategy emphasizes customized products and frequent product changes, its sales force is set up to handle such products, and its R&D/engineering can provide a constant stream of product modifications and process improvements. All of this means a different way of competing—a change in strategy. Ability to customize to satisfy unique requirements, rather than simply lower cost for standard products, may be the basis for sales, for instance. Ability to master diverse markets in order to more fully utilize broad-scope manufacturing capabilities may also be advantageous, providing a new rationale for diversifying product mix.

Organizational Effects

The concept of the strategic business unit or SBU is affected, too. An SBU is much less easy to define where quite different businesses share a common manufacturing core. Allocating manufacturing costs and overheads, and thus evaluating business performance, depends more upon transfer pricing, suggesting that conventional accounting methods like ROI for assessing business results may be of limited usefulness. The structure of the organization, and its organization chart, may look the same, but what lies behind the chart is more of a "joint venture in manufacturing." Production planning and access decisions logically become more centralized, while product design, research, marketing and sales diverge in the pursuit of different product-customer targets. Philips N.V. in Europe is already moving in this direction, with a shared manufacturing core.

With new electronic controls and programmable machines, economies of scope will more likely be the rule than the exception. Through economies of scope, manufacturing technology decisions affect future strategic options. For example, the most important contribution of CAM technology is its ability to do things that could not be done before in "traditional technology factories." For example, the Boeing Company has a parts control system that manages 1½ billion parts. Without computers, it is doubtful whether there would be enough tub files, index cards, clerks and floor space to manage that many parts in the traditional, non-automated fashion. CAD permitted effective development of several complex aircraft simultaneously, and the on-going parts support for them. Without the ability to manage this large a number of parts in design, during manufacture and afterwards, it is unlikely that Boeing could have offered the variety of aircraft designs it currently sells. Thus its whole strategy would have been different.

In light of such pervasive consequences, a rethinking of old strategic assumptions about manufacturing is essential, for manufacturing decisions now carry major long-term strategic implications. With new capabilities come new strategic options. They must be thoroughly integrated into the strategic thinking of the firm. It is important to emphasize here that computer-aided manufacturing technology is not a quick technological "fix" for all manufacturing

problems. Indeed, the new technology's capabilities underline the need for careful attention to the details of extracting maximum performance from manufacturing operations, and for thoroughgoing integration of manufacturing policies with corporate strategy. Nowhere is this more clear than in the effects of new controls.

With the advent of low-cost microprocessors, virtually all machines and techniques can be "smarter," incorporating electronic and electromechanical capacities to see, feel and compare that were formerly provided by people. By automating these processes, more rapid processing of both materials and information becomes possible. Computer driven or monitored equipment also makes possible automatic movement of materials from one sub-process to another. In addition, overall information monitoring becomes possible. As we have argued, their joint impact has been to make possible a range and, most importantly, a flexibility of automated processes far greater than ever before. With the new systems, information becomes the key integrating mechanism.

Information as an Integrator

The new technology is increasingly integrating manufacturing systems from design to implementation around sophisticated information systems. A substantial amount of very specific information must be gathered to make automation possible—thus manufacturing comes to rely increasingly on "science," rather than "art." New technologies of adaptive controls and feedback, electro-mechanical sensors and pattern recognition require substantial amounts of "tool point" information—and also help to acquire it. It is a relatively short step to unite the various subsystems of the manufacturing process. Once integrated, information usage accelerates, as does potential for improved control. Since more is known, it becomes more feasible to collect data, forecast and control, schedule and optimize. Bottlenecks are highlighted and well specified, so it is more possible to solve them. Moreover, this is real time, real world data, not textbook formulae or engineering guesstimates. With actual information in hand, routine practices can be more precisely specified, programmed and monitored.

All of this tends to subdue the routine and to highlight the potential for human creativity in both old and new processes. New exotic technologies are being applied to everyday processes, in part because of more information about the exotics, but in part also because of more information about the everyday processes. As a John Deere executive commented, "The software and the analytical routines are giving us insights into manufacturing that we never had before."[7] Since more is known, the marriage of new methods and standard manufacturing needs becomes possible. Thus chemical milling, the use of lasers for precise marking and penetration, enzymes for chemical processes, and the substitution of exotic materials for scarce or expensive traditional materials are all proliferating. These advances affect strategic decisions. New methods remove existing constraints by changing critical materials requirements, by making possible new processes and thus new products, and by opening new markets or radically altering old ones.

Process Choices

The wider process options, like the increased flexibility of computer-controlled manufacturing systems, act to radically shift the strategic possibilities. These shifts are not limited to new products. To understand the potentials for process choice in terms of the marketplace, we must look at the interaction between product life cycle and product design specialization, as shown in Chart 2.

Product Design/Degree of Tailoring

		Standard Low	Custom High
Product Life Cycle Stage	Commodity— Mature	Ethylene Plastic pipe Grain 8080 Chip	Automobiles Large airplanes Plastic resins Made-to-measure shirts Single family house
	Specialty— New Entry	Nuclear submarine Toys Pacemakers New food items	Nuclear power plants Engineer products Job shop orders

CHART 2.

Many products initially come into being as custom, specialty items. At first, they are understood by perhaps only a single producer and a few customers. The characteristics of the product can become increasingly standardized, however, as others imitate the original producer over time, and as the product's useful characteristics are more widely understood. Under these conditions, the basis of competition moves from product characteristics (or even product availability) to price. Especially under the older, more traditional modes of manufacture, with their limited options for flexibility, there comes increasingly to be a single, widely accepted "one best way" to compete in making a product. As the product becomes more widely understood and accepted, it tends to move toward commodity status—as stainless steel did after World War II. Moreover, the older, rigid technology can seriously inhibit innovation and productivity. (This is one explanation offered for the difficulties of the US automakers.)[8]

The new technology works against this older trend to homogenization. Instead, it moves toward increasing market segmentation, competition on perceived "special options," and custom products, making it possible to avoid commodity-oriented, price-based competition. Items that are generally available in a standard, basic conformation are differentiated to meet customers' preferences—often long after the product and its technology of manufacture are well understood. Even a generally well-understood technology can offer the basis for renewed competition, especially when new processes or materials emerge.[9] Similarly, tailoring products to individual need regenerates competition on product characteristics. It can also create massive coordination and inventory control problems in traditional production processes like automobile assembly. Such differentiation may occur in mature products, but it can also occur in early stages, as a strategy to

create a protected market niche within which to exploit specialized technical knowledge.

Awareness of these possibilities adds breadth to strategic decisions about manufacturing technology. Particularly under older constraining assumptions, where custom products meant necessarily higher costs, there were decided limits to the custom strategy. The trends in manufacturing technology we have been discussing remove many of the old limits, however. Custom production is no longer prohibitively expensive, nor does it reside solely in highly skilled production workers. Instead, design and programming are the keys, along with the ability to integrate a new, deeper understanding of production process possibilities into product design.

Product Design and Process Configuration

We can look at the other side of product positioning if we characterize the manufacturing technology in terms of product design and process configuration. Again, the degree of flexibility is central, for it helps us to highlight the strategic options available to management—here, in terms of product customizing. Process technologies can vary through the following range:

- A process technology that is independent of the design of the product and could be turned to many different products and designs. Examples include simple manual tools, stand-alone numerical-controlled tools and job shops.
- A programmable process technology that can accommodate a range of different configurations, each reflecting a different product design (computer-controlled machining centers, for instance).
- A flexible process technology that can accommodate a range of product designs

410

with a single configuration (mass production lines for automobiles, making different models, papermaking machines, etc.).

- A production technology that is dedicated to a single product design (transfer line, chemical plant).

As we have been arguing, under old assumptions flexibility was traded off against control, economy and precision: cost per unit may have been lower, and cost for initial capital investments higher, for more specialized equipment. Under new assumptions of economies of scope, new choices apply, reflecting new criteria for decisions about manufacturing technology. Flexibility is still the issue—but its dimensions are significantly different. The new economies of choice are displayed in Chart 3, which depicts these choices in terms of expected volume and variety mix.

Process technology has always been a major commitment. It tends to endure longer than either product design or market characteristics. The product life cycle typically moves through its phases more rapidly than manufacturing system hardware can be changed. This gives programmable technology special importance, for it permits continued easy, accurate response to market changes via stockpiled programs while existing hardware is "recycled" by reprogramming it for new products. Both Programmable and (to a lesser extent) Flexible Systems share this capacity. What the chart highlights is the interaction of product and process characteristics and the impact of newly emergent control technologies for strategic choice. This analysis can be applied at all levels of process and technology choice. Any organization can be broken down into combinations of these four basic process-product/market situations, and strategic choices made accordingly. The choice of process technology can be made on the basis of potential variety, potential change, and cost and capacity trade-offs. Following this line of analysis, the manager can generate a more complete strategic plan that takes into account both the constraints and the opportunities inherent in the new manufacturing technology for a particular product/market situation (or a particular portfolio of such situations).

Implications for Management

The new integrated manufacturing technologies, although less expensive in cost per unit, often require substantial initial outlays. Particularly for integrated manufacturing systems, the cost is typically an order of magnitude greater than for comparable non-integrated or non-controlled equipment. (The cost differences for integrated systems are comparable to those of the early numerically-controlled machine tools in contrast with prior manual systems.) With increased cost, the ante is up for management: the risk is higher, the gamble is greater. Indeed, the cost is often of the "you bet your company" magnitude. Perhaps even more important than the cost, however, are the strategic implications—often unrealized by management until after the fact. "You bet your company" here, too. Let's take a closer look.

The new technologies permit orders-of-magnitude improvements in manufacturing efficiency. They also offer advantages that transcend the manufacturing function simply because they both make possible and encourage a significant proliferation of product lines in "tailor made" products that require only reprogramming—not retooling. The same precision, almost identical costs, and approximately equal time-to-process are as accessible on short production runs as on long. Indeed, with computer-assisted design, it is possible to economically reprogram and thus offer not only tailored products but guaranteed replacement without costly physical inventory: programs can be stockpiled instead. The appropriate time-horizon, then, becomes not a product life cycle but

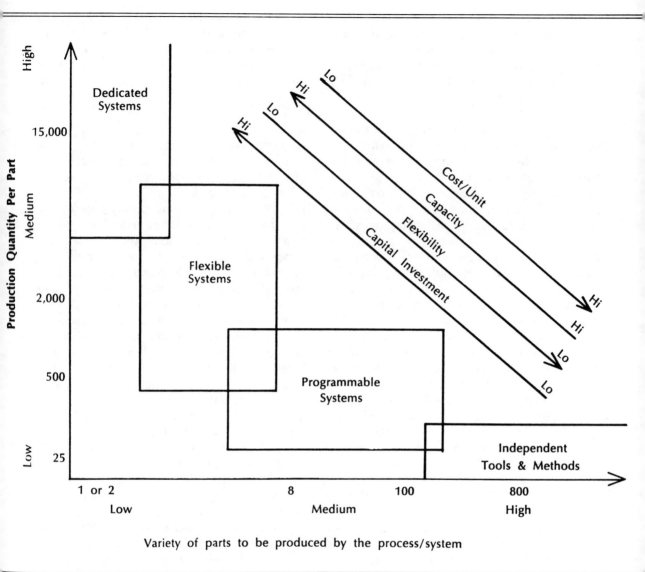

Production Quantity Per Part

High

Dedicated
Systems

15,000

Medium

Flexible
Systems

2,000

Programmable
Systems

500

Independent
Tools & Methods

Low   25

1 or 2   8   100   800
Low   Medium   High

Variety of parts to be produced by the process/system

Cost/Unit
Capacity
Flexibility
Capital Investment

Hi   Lo   Lo   Hi   Hi   Lo   Lo

CHART 3.

a manufacturing system life cycle, especially as this interacts with long range forecasts of market developments, product design possibilities and corporate strategy. Chart 4 highlights manufacturing system characteristics.

While initial cost is significant for some of the new technology, the major barriers to its effective use are ignorance and lack of understanding about its impact in broadly strategic areas. As we have suggested, managers face a new order of complexity in such decisions, for manufacturing, marketing and design factors

Manufacturing Systems Characteristics Based Upon Technology Choices

	TYPES OF TECHNOLOGY			
	Dedicated	**Flexible**	**Programmable**	**Independent**
Relationship to Product Design:	Single, unchangeable process configuration—produces a single standard product design.	Single, unchangeable process configuration—produces a limited variety of product design.	A combination of equipment and technology which can be reconfigured into a number of processes, each of which produces a single product design.	A combination of equipment and technology which can produce a wide variety of product designs without reconfiguration.
Capital Investment:	Highest			Lowest
Cost per unit/output:	Lowest			Highest
Flexibility:	Lowest			Highest
Labor:	Mostly overhead			Mostly direct
"Best Suited" to:	Long runs, low variety	Long runs, medium variety	Short runs, medium variety	Short runs, high variety
Machine Control:	Mechanical or electronic.	Electronic or computer	Computer	Manual
Material Handling:	Fully automated and built into the process	Automated movement between stations	Automated load/unload	Manual
Instrumentation:	Learning capability	Closed feedback	Learning capability	Direct read, manual
Management Tasks:	Sell	Plan	Design	Coordination
In-Process Inventory:	Least			Most
Competitive Options:	Few	Many	Many	Few
Technical Skills:	Process	Product	Product	Process
Flow:	Continuous			Independent
Innovation Thrust:	Process			Product
Product Change Frequency:	None			Often
Market Strategy	Dominant Design			Custom

CHART 4. *Manufacturing Systems Characteristics Based upon Technology Choices*   413

(among others) interact around new manufacturing capabilities. Set in a larger strategic context, "cost" must be reinterpreted in light of expanded options, while the strategic limitations of existing commitments assume their proper place in management decision-making. Of particular importance are the following:

- Organization Strategy: Especially for smaller firms, new machines mean levels of production, accuracy and product line scope to expand their concept of organization strategy in light of the new technologies. Programmable controls mean enhanced potential for innovation and productivity for all users. The consequences are huge and potentially enormously profitable. To harvest the potential, the vision of the firm must expand to take into account the new potentials made possible by the new technology. "Doing things right" may mean eliminating inventory or increasing productive tool utilization time by orders of magnitude from present levels through new approaches. "Doing the right things" or "What business we are in?" must also be reexamined and perhaps changed radically. Of course, such re-examination has never been easy.
- Organization Structure: The new integration of marketing and design, of design and manufacture, of manufacture and strategic positioning suggest the pending demise of older, functionally specialized structures— or, perhaps, their radical revision. Since structure typically represents power arrangements, the new technologies and their strategic implications suggest potential major shifts in power and thus potential resistances. Administrative arrangements, communications systems and organization structure can now respond to these new needs, for the new technologies also have impact there. But while people have always informally worked around inconvenient

structures, new needs for data access and input imply new flows of authority and a far broader legitimate participation in decisions formerly sequestered at high departmental levels, and thus new formal structures. Old notions of the division of work that the older structures embody are already breaking down in high technology industries, where the temporary task forces, responsibility without direct authority, distributed computer facilities and information access, and matrix management have become commonplace.

- Marketing: Old thinking about "custom" markets versus "mass" markets may well be obsolete. Big firms will be able to provide "custom" service, and small firms will be able to contemplate "mass" markets hitherto beyond their reach. A more complex approach to the marketplace and a more sophisticated appreciation of the marketing possibilities generated by the new technology will be required. Closer integration of manufacturing with marketing will require a greater appreciation of technological possibilities. Indeed, closer integration of marketing and design appear both essential and newly feasible.

CONCLUSION

Computers and microprocessors are making all machines "smart." Lasers, other exotic processing techniques and computer integration are increasing production rates dramatically. New materials are broadening both process and product design options and also demanding new process capabilities. Sophisticated software for manufacturing management and control leads to integrated systems, lower in-process inventories, higher equipment utilization and reduced lead or turn-around times. Instead of the old assumptions and rules of thumb based on economies of scale, the new technology calls for thinking based on understanding economies of scope. The implications for strategic management require a rethinking of many old truths and point to a new order of complexity in manufacturing decisions.

The picture now is one of a much closer relationship between the design of manufacturing systems, the selection of equipment and process technology, the management of the manufacturing operation, the design of individual products, the marketing mix choices, and segmentation and positioning decisions. In turn, this implies a reorganization of research and engineering activities; redefined linkages between marketing, production and technical activities; new communications patterns; new task allocation and reward systems; and redesigned cost accounting, marketing reporting and production control systems. But the strategic decisions required are most important of all.

"Smart machines" and the beginnings of computer controls—microprocessor driven pieces—are accessible today at relatively low cost, but the strategic thinking required to exploit fully options for repositioning and market alignment constitute major strategic shifts. These changes clearly highlight the strategic nature of the manufacturing decision to acquire new capabilities.

The decision process is still more complex for integrated systems, which promise the greatest returns from the new technology. At the front end, the cost for integrated CAD/CAM systems and the minimum increments for introducing them into existing operations is much greater than manufacturing technology in the past. The implications of the fully-developed CAD/CAM system underline the pervasive nature of the changes demanded in strategic thinking. The covert costs (and opportunities) attached to the new technologies reside in these changes.

What we have described is a substantial, pervasive and discontinuous change in manufacturing technology with multiple ramifications for all aspects of strategic management. All levels and functions of corporate management must be able to understand and participate in these changes. Corporate strategy must define distinctive competence also in terms of manufacturing systems capability, rather than solely in terms of product design or market sector dominance. Moreover, both the time horizons and the criteria for judging investment in such manufacturing systems must change to integrate the firm's manufacturing philosophy into its strategy. The appropriate question is not "What will the ROI be on this equipment?" but "What will the ROI be for the firm as a whole in five to ten years if we do not invest in the new technology?"

The impact of the new technologies are far-reaching, and widespread changes in organization design, human factors, manufacturing systems management, product designs, marketing and corporate strategy will be required if firms are to realize the full benefits of investment.

NOTES

1. See, for example, "Rediscovering the Factory," in *Fortune*, July 13, 1981, and "The Speedup in Automation" in *Business Week*, Aug. 3, 1981, on the importance of manufacturing.
2. Wickham Skinner, "Manufacturing—Missing Link in Corporate Strategy," *Harvard Business Review* (May-June 1969).
3. On the strategic implications of manufacturing policies, see Robert H. Hayes, "Why Japanese Factories Work," and Steven C. Wheelwright, "Japan—Where Operations Really Are Strategic," both in *Harvard Business Review* (July-August, 1981), and William J. Abernathy, Kim B. Clark and Alan M. Kantrow, "The New Industrial Competition," *Harvard Business Review* (Sept.-Oct. 1981).

4. Wickham Skinner, "The Focused Factory," *Harvard Business Review* (May-June 1974).

5. Steven C. Wheelwright, "Reflecting Corporate Strategy in Manufacturing Decisions," *Business Horizons* (February 1978), and Robert H. Hayes and Steven C. Wheelwright, "Link Manufacturing Process and Product Life Cycles," *Harvard Business Review* (January-February 1979).

6. Alfred D. Chandler, Jr., *The Visible Hand: The Managerial Revolution in American Business* (1977). Chandler discusses the impact of special-purpose machinery, volume throughput and interchangeable parts in mass production. William J. Abernathy, *The Productivity Dilemma: Roadblock to Innovation in the Automobile Industry* (1978). Abernathy thoroughly documents manufacturing experience and the constraints imposed by old-style automation in the automobile industry.

7. See "Paperless Factory," by Jack Thornton, *Metalworking News,* May 25, 1981.

8. Abernathy, op. cit.

9. Abernathy, Clark and Kantrow, op. cit.

Matching Process Technology with Product/Market Requirements

Robert H. Hayes

Steven C. Wheelwright

1 INTRODUCTION AND OVERVIEW

In the preceding chapter [of *Restoring Our Competitive Advantage*], issues related to manufacturing process technology were examined from three organizational levels: the technical manufacturing specialist (such as a process or industrial engineer), the operations manager, and the general manager. This chapter expands upon the discussion of the general manager's concerns relating to process technology. However, rather than starting with a perspective from within the manufacturing function, we adopt that of the senior nonmanufacturing manager who wishes to assess the potential contribution that process technology can make to the overall business.

A basic schematic of a business's various functions, and their relationship to manufacturing, is presented in Figure 1. We deal in this chapter with the blocks labeled "Processes," "Products," and "Customers." That is, we want to concentrate on how the marketing and sales functions interact with the manufacturing func-

tion in the selection and development of process technology. We examine the relationship between the blocks labeled "Suppliers" and "Processes" in Chapter 9.

The primary interface between the manufacturing function and the customer was once the *sales force,* which placed orders directly with the factory and often was in personal contact with plant managers and others in the manufacturing organization. This interface was supervised by a general manager, who resolved major differences when they arose. With the increasing size and complexity of today's business organizations, such a structure often has proved inadequate for integrating these two group's differing objectives. Consequently, in many firms the *marketing function* has grown to become the main link between its sales force and manufacturing organization. It is responsible for defining the firm's product strategy, selecting and positioning the products in the product line, and, frequently, transferring information from sales to manufacturing.

To assist in managing these interfaces, marketing managers have often found it useful to define positions such as "market managers" and "product managers." As suggested in Figure 1, these managers perform a coordinating function, representing not only the special interests of the two functions with which they

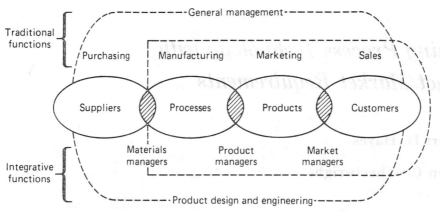

FIGURE 1.   *Manufacturing Interfaces with Other Functions*

work most closely, but also the concerns of general managers, product designers, and plant engineers. Such interface managers frequently alternate their attention between operating level issues (such as expediting a customer order through production) and general management level issues (such as assessing the effect of modifying a product on the process used to manufacture it).

But such organizational artifices, no matter how elaborate, are by themselves seldom capable of dealing adequately with all the possible linkages between manufacturing and the rest of the firm. An "informal organization"—a network of well-informed people who have developed good working relationships over a long period of time—is needed to handle all the contingencies that the formal organization inevitably neglects to provide for. And both the formal and informal organizations need a frame of reference, a common understanding of where the company is going.

The major purpose of this chapter is to describe some of the concepts and approaches that leading firms have found useful for integrating marketing/sales with manufacturing, and developing an effective set of manufacturing capabilities. Our premise is that for a company to develop a sustainable competitive edge, senior managers outside the manufacturing function need to have a broad understanding of

the firm's process technology and how it interacts with other parts of the business. As we pointed out in Chapter 2, the extent to which manufacturing is able to make a contribution to a firm's competitive success depends importantly on the fit between the process technology chosen and the firm's overall competitive strategy.

The next section begins with an overview of a group of ideas deriving from the notion of life cycle analysis. Then we describe the complementary concepts of the product life cycle (which places changing demands on the manufacturing function) and the manufacturing process life cycle (which provides changing capabilities that can be exploited in the marketplace).

Next we discuss the interaction of process technology capabilities with various product/market needs, using a two-dimensional representation known as the "product/process matrix." After applying it to some practical situations, we examine a few of its limits. Two other forms of product/process interaction also are discussed. One deals with issues of innovation, based on some of the findings of Abernathy and Utterback (1975). The other describes an approach used by Abell (1980) to define customers and markets in a way that has implications for process technology.

Chapters 8 and 9 explore some of the

specific process technology decisions that require the guidance of senior managers, and Chapter 10 discusses the overall task of managing major changes in process technology.

2 THE INTERFACE OF MANUFACTURING TECHNOLOGY WITH SALES AND MARKETING

Many managers, when asked about the relationship between the marketing and manufacturing functions in their companies, are likely to describe it as troubled and strained—or, at best, ambivalent. In one well-known article, Shapiro (1977) identified eight major areas in which problems tend to arise between these two functions. Shapiro's list of the typical views that marketing and manufacturing personnel have of each other is reproduced in Table 1. While this list focuses on a few, fairly specific kinds of interaction, it captures nicely the distinction between market needs and process technology capabilities. While not all of these eight problem areas relate directly to manufacturing technology, each has links with the definition of technology that we provide in Chapter 6.

TABLE 1. Functional Level Interactions of Marketing and Manufacturing

Problem Area	Typical Marketing Comment	Typical Manufacturing Comment
Capacity planning and long-range sales forecasting	Why don't we have enough capacity?	Why didn't we have accurate sales forecasts?
Production scheduling and short-range sales forecasting	We need faster response. Our lead times are ridiculous	We need realistic customer commitments and sales forecasts that don't change like wind direction
Delivery and physical distribution	Why don't we ever have the right merchandise in inventory?	We can't keep everything in inventory
Quality assurance	Why can't we have reasonable quality at reasonable cost?	Why must we always offer options that are too hard to manufacture and that offer little customer utility?
Breadth of product line	Our customers demand variety	The product line is too broad—all we get are short uneconomical runs
Cost control	Our costs are so high that we are not competitive in the marketplace	We can't provide fast delivery, broad variety, rapid response to change, and high quality at low cost
New product introduction	New products are our life blood	Unnecessary design changes are prohibitively expensive
Adjunct services such as spare parts inventory support, installation, and repair	Field service costs are too high	Products are being used in ways for which they weren't designed

In analyzing these eight problem areas, Shapiro focuses on such issues as evaluation and reward systems, inherent complexity, and differences in manager orientation/experience/"culture" as the basic causes of friction between marketing and manufacturing. While we tend to agree with much of his analysis, it is worth noting that similar differences tend to exist between any two of the functions in a business. However, our experience suggests that the marketing/manufacturing interface is the focal point of much more frequent and heated disagreement than occurs between other pairs of functions.

Dealing one-by-one with each of the problem areas in Table 1 is not by itself likely to lead to a substantial increase in harmony between marketing and manufacturing. Instead, one needs to understand, in managerial terms, why that interface can so easily become a fault line in the firm—where the requirements that marketing places on manufacturing and the capabilities that the manufacturing process technology provides to marketing grind against each other, in opposite directions. One approach to developing that kind of understanding is based on an analysis of how product and process life cycles interact.

The regularity of the growth cycles of living organisms has long fascinated thoughtful observers, and invited a variety of attempts to apply the same principles—of a predictable sequence of rapid growth followed by maturation and decline—to companies and industries. The "product life cycle," for example, has been studied in a wide range of organizational settings, although there are sufficient questions (see, for example, Dhalla and Yuspeh, 1976) to raise doubts as to the universal application of the concept.

Irrespective of whether the product life cycle pattern is a general rule or holds only for isolated cases, it does provide general managers with a useful and provocative framework for thinking about the growth and development of

a new product, a company, or an entire industry. However, one major shortcoming of this approach is that it focuses primarily on the marketing implications of the life cycle pattern, often to the exclusion of its manufacturing implications. In so doing it implies either that other aspects of the business and industry environment move in concert with the product life cycle, or that they are inconsequential. While such a view may help one reflect upon the kinds of changes that have occurred in different industries, an individual company or product line manager may find it too simplistic to be useful as a planning tool. In fact, the concept may even be misleading if it is used as the primary basis for strategic planning.

In attempting to relate the life cycle that governs products and markets to manufacturing technology, we begin by reviewing the way the life cycle is typically used in marketing analyses. Then we discuss an analogous concept called the manufacturing *process* life cycle. While the product life cycle describes how the growth and maturation of products and markets place changing demands on manufacturing, the process life cycle describes how the nature of and the capabilities provided by a process technology evolve through different stages.

2.1 Market Requirements and the Product Life Cycle

When marketers use this concept, they usually attempt to model a product's evolution over time by identifying a series of distinguishable stages that it passes through: introduction, rapid growth, competitive turbulence, maturation, and decline. Figure 2 describes the changes in strategic objectives, competition, product design, pricing, promoting, distribution, and informational requirements that are associated with each stage.

While the product life cycle is useful primarily in planning a firm's marketing strategy,

420

	Market development (introductory period for high learning products only)	Rapid growth (normal introductory pattern for a very low learning product)	Competitive turbulence	Saturation (maturity)	Decline
Strategy objective	Minimize learning requirements, locate and remedy offering defects quickly, develop widespread awareness of benefits, and gain trial by early adopters	To establish a strong brand market and distribution niche as quickly as possible	To maintain and strengthen the market niche achieved through dealer and consumer loyalty	To defend brand position against competing brands and product category against other potential products, through constant attention to product improvement opportunities and fresh promotional and distribution approaches	To milk the offering dry of all possible profit
Outlook for competition	None is likely to be attracted in the early, unprofitable stages	Early entrance of numerous aggressive emulators	Price and distribution squeezes on the industry, shaking out the weaker entrants	Competition stabilized. Few or no new entrants. Market shares relatively stable except when a brand gains substantial added perceived value through product improvement or price repositioning	Similar competition declining and dropping out because of decrease in consumer interest
Product design objective	Limited number of models with physical product and offering designs both focussed on minimizing learning requirements. Designs cost and use engineered to appeal to most receptive segment. Utmost attention to quality control and quick elimination of market revealed defects in design	Modular design to facilitate flexible addition of variants to appeal to every new segment and new use system as fast as discovered	Intensified attention to product improvement, tightening up of line to eliminate unnecessary specialties with little market appeal	A constant alert for market pyramiding opportunities through either bold cost and price penetration of new markets or major product changes. Introduction of flanker products. Constant attention to possibilities for product improvement and cost cutting. Reexamination of necessity of design compromises	Constant pruning of line to eliminate any items not returning a direct profit
Pricing objective	To impose the minimum of value perception learning and to match the value reference perception of the most receptive segments. High trade discounts and sampling advisable	A price line for every taste, from low end to premium models. Customary trade discounts. Aggressive promotional pricing, with prices cut as fast as costs decline due to accumulated production experience. Intensification of sampling	Increased attention to market broadening and promotional pricing opportunities	Price repositioning wherever demand pattern and competitors' strategies permit. Defensive pricing to preserve product category franchise. Search for incremental pricing opportunities, including private label contracts, to boost volume and gain an experience advantage	Maintenance of profit level pricing with complete disregard of any effect on market share
Promotional guidelines — Communications objectives	a) Create widespread awareness and understanding of offering benefits b) Gain trial by early adopters	Create and strengthen brand preference among trade and final users. Stimulate general trial	Maintain consumer franchise and strengthen dealer ties	Maintain consumer and trade loyalty, with strong emphasis on dealers and distributors. Promotion of greater use frequency	Phase out, keeping just enough to maintain profitable distribution
Most valuable media mix	In order of value Publicity Personal sales Mass communications	Mass media Personal sales Sales promotions, including sampling Publicity	Mass media Dealer promotions Personal selling to dealers Sales promotions Publicity	Mass media Dealer oriented promotions	Cut down all media to the bone - use no sales promotions of any kind
Distribution policy	Exclusive or selective, with distributor margins high enough to justify heavy promotional spending	Intensive and extensive, with dealer margins just high enough to keep them interested. Close attention to rapid resupply of distributor stocks and heavy inventories at all levels	Intensive and extensive, and a strong emphasis on keeping dealer well supplied, but with minimum inventory costs	Intensive and extensive, with strong emphasis on keeping dealer well supplied, but at minimum inventory cost to him	Phase out outlets as they become marginal
Intelligence focus	To identify actual developing use systems and to uncover any product weaknesses	Detailed attention to brand position, to gaps in model and market coverage, and to opportunities for market segmentation	Close attention to product improvement needs, to market broadening chances, and to possible fresh promotion themes	Close analysis of competitors' strategies. Regular monitoring of trends in use patterns and possible product improvements. Sharp alert for potential new technological and new interproduct competition or other signs of beginning product decline	Information helping to identify the point at which the product should be phased out

Source: Chester R. Wasson, *Dynamic Competitive Strategy and Product Life Cycles.* Austin, Texas: Austin Press, 1978.

FIGURE 2. Dimensions of the Product Life Cycle Concept Important to Marketing

one can also relate it indirectly to the firm's manufacturing strategy. Figure 3 suggests the implications that different product life cycle stages have for four issues that are directly linked to manufacturing production volume, product variety, industry structure, and the dominant form of competition. For example, manufacturing has a major stake in decisions that may affect such variables as product customization (versus standardization), volume per model, and the average time before obsolescence or replacement. Given this perspective, the product life cycle can be used to summarize the customer and product requirements that must be satisfied by the manufacturing function and its product technology.

A second aspect of the product life cycle that has a direct impact on manufacturing has to do with the nature of industry competition and the firm's major competitors. As suggested

421

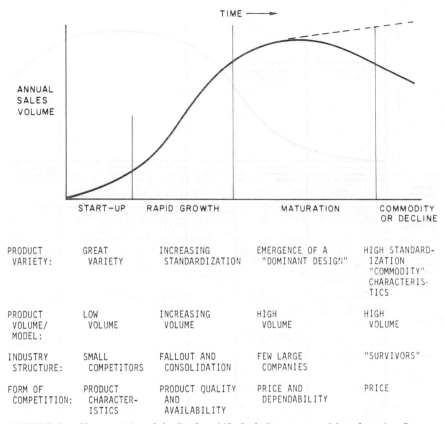

	START-UP	RAPID GROWTH	MATURATION	COMMODITY OR DECLINE
PRODUCT VARIETY:	GREAT VARIETY	INCREASING STANDARDIZATION	EMERGENCE OF A "DOMINANT DESIGN"	HIGH STANDARD-IZATION "COMMODITY" CHARACTERIS-TICS
PRODUCT VOLUME/MODEL:	LOW VOLUME	INCREASING VOLUME	HIGH VOLUME	HIGH VOLUME
INDUSTRY STRUCTURE:	SMALL COMPETITORS	FALLOUT AND CONSOLIDATION	FEW LARGE COMPANIES	"SURVIVORS"
FORM OF COMPETITION:	PRODUCT CHARACTER-ISTICS	PRODUCT QUALITY AND AVAILABILITY	PRICE AND DEPENDABILITY	PRICE

FIGURE 3.   Characteristics of the Product Life Cycle Important to Manufacturing Process Technology

in Figure 3, the maturation of a market generally leads to fewer competitors, increasing industry concentration, and competition based more on price and delivery than on unique product features. As the competitive focus shifts during the different stages of the product life cycle, the requirements placed on manufacturing (in terms of cost, quality, flexibility, and delivery dependability) also shift.

A third aspect has to do with the nature of the product itself. The stage of the product life cycle affects the product's design stability, the length of the product development cycle, the frequency of engineering change orders, and the commonality of components—all of which have implications for the manufacturing process technology.

In short, the product life cycle concept provides a framework for thinking about both a product's evolution through time and the kind of market segments that are likely to develop at various points in time. It also highlights the need to change the priorities that govern manufacturing behavior as products and markets evolve. These priorities, in turn, have important repercussions for the process technology employed. As we see in Section 3, it is not always easy to determine when shifts take place in a product's position along its life cycle; trying to ascertain implications of such shifts for the manufacturing function (for example, adjusting the process technology so that its capabilities better meet market requirements) can be even more difficult.

2.2 Manufacturing Capabilities and the Process Life Cycle

While the life cycle concept has not been applied to manufacturing processes nearly as extensively as it has to product and market development, a number of authors have suggested that such cycles exist. As summarized in Table 2, Abernathy and Townsend (1975) have broken down the evolution of a production process into three major stages—early, middle, and mature. For each stage they describe six important characteristics that are of concern to manufacturing managers. Using technology that is analogous to that used in conjunction with the product life cycle, Table 2 suggests that a process life cycle begins with a "fluid" production process: one that is highly flexible but not very cost efficient. Then it proceeds toward increasing standardization, mechanization, and automation until it eventually becomes "systemic": very efficient, but much more capital-intensive, interrelated, and hence less flexible than the original "fluid" process (see Section 6-3).

The description of the process life cycle in Table 2 can be very useful in manufacturing planning and decision making, but it also can be used at a general management level to relate specific manufacturing capabilities to various stages of the process life cycle. For example, it can be used to predict how the product's manufacturing cost per unit is likely to change over time (Figure 4). While the stages through which the product technology passes do not necessarily match exactly those of the product life cycle, we attach the same names to them.

The first stage in the development of a process technology has the characteristics of a job shop. It is flexible (and therefore able to cope with low volumes) because it has few rigid interconnections; it is characterized by little automation or vertical integration. As the process matures, it passes through intermediate stages that may involve decoupled line flows (batch processes) and/or connected line flows (assembly lines). Eventually, the process technology may evolve into a continuous flow operation with high throughput volumes, low rates of process innovation, and less flexibility due to high levels of automation and vertical integration. A number of dimensions can be used to characterize these various stages. Four of these are of particular importance to manufacturing: process organization, throughput volume, rate of process innovation, and the levels of automation and vertical integration. These are summarized in Figure 4.

The concept of a process life cycle can be of great usefulness to general managers for two reasons. First, it conveys a sense of the capabilities provided by different stages of the process cycle, and second, it identifies the key tasks that must be carried out if those capabilities are to be provided. The stage of the process life cycle in which a specific business finds itself at a given point in time is determined by decisions made both outside manufacturing (such as product line breadth, sales volume, and product design) and inside (equipment, materials flow, and information systems). Thus the evolution of a process technology can be viewed as a natural complement to the evolution of product technology; general managers should understand and work with both.

3 INTEGRATING PROCESS TECHNOLOGY CAPABILITIES AND PRODUCT/MARKET REQUIREMENTS THROUGH THE PRODUCT-PROCESS MATRIX

Figure 5 suggests one way in which the interaction of the product life cycle and the process life cycle can be represented. The rows in this matrix represent the major stages through which a production process tends to pass in going from the fluid form (top row) to the systemic form (bottom row). The columns represent product life cycle phases that progress from the great variety associated with the product's initial introduction (on the left) to the standardi-

TABLE 2. *Dimensions of Process Technology Evolution Important to Manufacturing Management*

Stage in the Productive Unit Life Cycle	Process Characteristics					Modes of Process Change (in Transition from One Stage to the Next)
	Material and Parts—Inputs	Technology	Labor	Scale	Product	
I. Early	Raw materials and parts used as available from supplier Types and quality vary widely Limited influence over supplier	General-purpose equipment and tools used as available from industry Special adaptations to general-purpose machines are made by user (jigs, fixtures, etc.). Flow-through process needs careful management control	Most workers have a broad range of performance skills Considerable flexibility exists in type of tasks each worker can and must perform Labor organization (if any) is along craft or skill (trade unionism)	Capacity ill defined Greater volume achieved by paralleling existing processes Short-run economies of scale achieved through learning curve improvement of manual operations Few scale barriers to entry into industry segment	Great variety of products with different features and quality Frequent design change Market relatively insensitive to price and quality (imperfect market that is price inelastic)	Process rationalization Standardize tasks Develop even flow through all process steps Automate easy tasks Introduce systematic or mechanized materials handling Redesign product and process to automate difficult tasks
II. Middle	Suppliers are strongly dependent Tailored material specifications imposed on supplier	Process automation is evident for some process tasks and systematized work flow Level of automation varies widely within process; islands of highly automated equipment are linked by manual operations Unique process equipment is designed for some tasks (often by outside firms)	Manual tasks are highly structured and standardized Labor is specialized with technical skills becoming more important Overhead labor functions such as maintenance scheduling and control are a significant cost	Capacity increased by equipment addition and advances to debottleneck particular operations Minimum size process necessary to compete in industry segment	Some-segments of market sensitive to price and quality (encouraging standard products and scale economies) Significant volume achieved in some product lines	Systemic development Separate difficult-to-automate tasks from process or eliminate them Design products to have maximum common process elements Arrange administrative organization for congruence of control over process flow

| Stage in the Productive Unit Life Cycle | Process Characteristics | | | Scale | Product | Modes of Process Change (in Transition from One Stage to the Next) |
	Material and Parts—Inputs	Technology	Labor			
III. Mature	Input's characteristics are optimized to process needs Supplier process integrated into overall process design Tasks that cannot be automated are segregated from process and are often subcontracted or performed by suppliers	Single units of equipment perform multiple process tasks and are integrated into automatic material handling equipment Formal systems engineering is required for process change Process equipment is designed as an integrated system, often by separate engineering groups or engineering companies Licensed technology may dominate, depending on the industry	Direct labor does monitoring and maintenance tasks Most important skills concern technical process equipment operation Labor classifications are rigid and are of primary concern to labor organization	Complete new facilities designed to achieve economies through spread costs Market growth and technological evolution pace scale increase Antitrust laws, logistics, or external factors eventually limit scale growth	Product variability is low and volume is high Standard products if price competition is prevalent or standard groups of products if product differentiation is prevalent Co- and by-products play greater role	Product and process realignment to meet changing markets and technological advances (may reset to earlier stage, 1 or 2 or stagnate during maturation)

Source. Reprinted by permission of the publisher from "Technology, Productivity, and Process Change" by William J. Abernathy and Phillip L. Townsend, *Technological Forecasting and Social Change*, volume 7, number 4, pp. 379–396. Copyright 1975 by Elsevier Science Publishing Co., Inc.

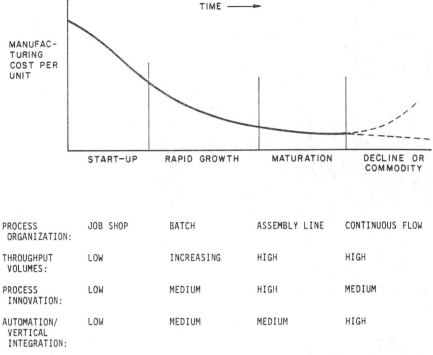

	START—UP	RAPID GROWTH	MATURATION	DECLINE OR COMMODITY
PROCESS ORGANIZATION:	JOB SHOP	BATCH	ASSEMBLY LINE	CONTINUOUS FLOW
THROUGHPUT VOLUMES:	LOW	INCREASING	HIGH	HIGH
PROCESS INNOVATION:	LOW	MEDIUM	HIGH	MEDIUM
AUTOMATION/ VERTICAL INTEGRATION:	LOW	MEDIUM	MEDIUM	HIGH

FIGURE 4.   *Characteristics of the Process Life Cycle*

zation associated with commodity products (on the right).

A company, a business unit within a diversified firm, or a product line can be characterized as occupying a particular region in this matrix, as determined by its stage in the product life cycle and its stage in the process life cycle. Some simple examples will illustrate this. Typical of a company positioned in the upper left-hand corner is a commercial printer. In such a firm each job is unique, and a jumbled flow or job-shop process (having the characteristics described in Chapter 6) is most effective in meeting product/market requirements. The market requires variety and relatively low volume per order, and competition consists of many firms offering a variety of product characteristics. A job shop process permits the economic production of relatively small lots, and requires great flexibility in workers and equipment.

Farther down the diagonal of this matrix, heavy equipment manufacturers usually choose a production technology characterized as a disconnected line flow (batch) process. This provides them with the capability to produce somewhat higher volumes, at somewhat lower unit costs, than would be possible with a job shop but retains considerable flexibility to produce a wide variety of products and customized features. The market, on the other hand, requires less variety than that served by the commercial printer, say, and may accept a basic catalog of models having a variety of options. Competition is likely to be among fewer competitors and is typically based on product quality, features, and availability.

Even farther down the diagonal are found producers of such products as automobiles or major home appliances. These companies generally choose to use a relatively mechanized and connected production process, such

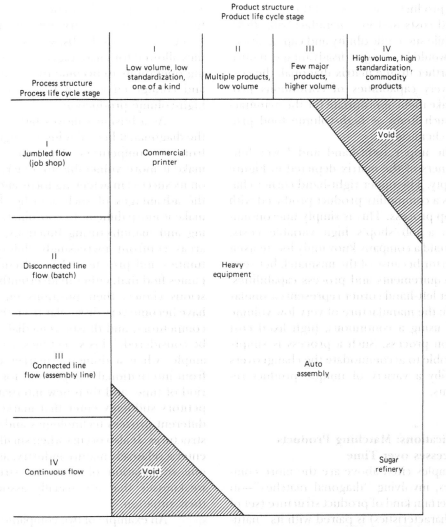

Product structure
Product life cycle stage

Process structure Process life cycle stage	I Low volume, low standardization, one of a kind	II Multiple products, low volume	III Few major products, higher volume	IV High volume, high standardization, commodity products
I Jumbled flow (job shop)	Commercial printer			Void
II Disconnected line flow (batch)		Heavy equipment		
III Connected line flow (assembly line)			Auto assembly	
IV Continuous flow		Void		Sugar refinery

FIGURE 5. Matching Major Stages of Product and Process Life Cycles—The Product-Process Matrix

as a moving assembly line, that offers still lower unit costs but is much less flexible. The capabilities and constraints of this process fit with the market's requirements and competitive behavior.

Finally, down in the lower right-hand corner of the matrix are found such businesses as sugar or oil refining, where the product is essentially a commodity (in that price and delivery terms may vary slightly from competitor to competitor, but other product characteristics are essentially standard). The production process used is based on a continuous flow technology that makes possible low variable costs and high product consistency, at the expense of high fixed costs and low manufacturing flexibility. While such inflexibility and capital intensiveness would be a major disadvantage in businesses further up the matrix diagonal, the cost and delivery capabilities of continuous processes make them appropriate for the manufacture of such items as high-volume food products or chemicals.

The upper right-hand and lower left-hand corners of the matrix depicted in Figure 5 are empty. The upper right-hand corner characterizes a commodity product produced with a job shop process. This is simply uneconomical given a job shop's high variable costs. Rarely would a company knowingly locate itself in that sector because of the mismatch between market requirements and process capabilities. The lower left-hand corner represents a similar mismatch: the manufacture of very low volume products using a continuous, high fixed cost production process. Such a process is simply too inflexible to accommodate the changeovers required by a variety of unique product requirements.

3.1 Applications: Matching Products and Processes over Time

The examples cited above are the more common ones, involving "diagonal matches"—in which a certain kind of product structure (set of market characteristics) is paired with its "natural" process structure (set of manufacturing characteristics). However, a business may consciously seek a position away from the diagonal in order to differentiate itself from its competitors. Rolls-Royce Ltd., for example, still makes a very narrow line of motor cars using a process that is more like a job shop than an assembly line. On the other hand, a company that allows itself to drift away from the diagonal without understanding the likely implications of such a shift may end up with an unintended mismatch. This can spell significant trouble for the organi-

zation, as apparently occurred in the manufactured housing industry during the housing boom of the early 1970s, when several companies allowed (or encouraged) their manufacturing operations to become too capital intensive and configured around the needs of stable, high-volume production.

As a business moves farther away from the diagonal, it becomes increasingly dissimilar from its competitors. This may or may not make it more vulnerable to attack, depending on its success in achieving focus and exploiting the advantages of such a niche. It may also make it more difficult to coordinate its marketing and manufacturing functions, as the two areas confront increasingly different opportunities and pressures. Not infrequently companies find that, either inadvertently or by conscious choice, their positions on the matrix have become very dissimilar from those of their competitors, and drastic remedial action must be considered. This sometimes occurs, for example, when a domestic market is insulated from international competition for a long period of time, and then new international competitors suddenly enter that market with very different process technologies and/or product structures. It also occurs when small companies enter a relatively mature industry, and provides one explanation of both the strengths and weaknesses that are usually associated with their situation.

An example of two companies that chose very different approaches to matching their movements along these two dimensions with industry changes involves Zenith Radio Corporation and RCA in the mid-1960s, the high growth stage of the color TV industry's product life cycle. Zenith had traditionally followed a strategy of maintaining a high degree of flexibility in its manufacturing facilities. This would be characterized on the matrix as being somewhat above the diagonal. As described in Chapter 3, when planning additional capacity for color TV manufacturing in 1966 (on the basis of forecasts that industry sales would double over the next three years). Zenith chose to expand in a way that represented a clear move down the process dimension towards the matrix diagonal. It consolidated color TV assembly into two large plants, one of which was in a relatively low-cost labor area in the United States. While Zenith continued to have manufacturing facilities that were more flexible than those of other companies in its industry, this decision reflected corporate management's assessment of the need to stay within range of the rest of its competitors on the process dimension so that its excellent marketing strategy would not be constrained by significant manufacturing inefficiencies.

During this same period, RCA (which had traditionally chosen to lead the industry in adopting newer, more mechanized manufacturing technologies) was introducing highly automated and specialized equipment, such as transfer lines which automatically inserted electronic components into printed circuit boards. As the market evolved toward higher volumes and more standardized products, this represented a move down the process dimension to a position below the diagonal. This strategy backfired when the introduction of integrated circuits and totally solid-state designs obsoleted much of this automated equipment.

Six years after Zenith's 1966 realignment of its position on the process life cycle, it made another decision to keep all of its assembly of color TV sets in the United States rather than lose the flexibility and incur the cost of moving production to the Far East. This decision, in conjunction with others made during the mid-1970s, was called into question toward the end of that decade. Zenith again found itself too far above the diagonal in comparison with its large, primarily Japanese, competitors, most of whom had mechanized their production processes, positioned them in low-wage countries, and embarked on other cost-reduction programs typical of a position farther down and

below the matrix diagonal. Zenith then decided to move most of its subassembly production to low wage locations outside the United States. These alternative competitive approaches can be depicted on the product–process matrix as shown in Figure 6.

Separating the major stages of the process life cycle from the stages of the product life cycle can be extremely helpful when general managers are grappling with the need to match process technology capabilities with product market requirements. Using this kind of two-dimensional representation encourages more creative thinking about organizational competences and the competitive role that various parts of the business can play. It can also lead to more informed predictions about the changes that are likely to occur in a particular industry, and to the development of a richer set of functional strategies for responding to such changes. Finally, it provides a natural way to bring manufacturing managers and marketing managers together so that they can relate their opportunities and decisions more effectively to overall business objectives. While this chapter deals primarily with the impact of process technology shifts within the product–process matrix, it also has implications for other aspects of manufacturing strategy, as we see in the next section.

3.2 Implications of Different Product–Process Choices

In this section we consider three issues that arise naturally when considering the interaction of product and process life cycles: (1) the concept of an organization's distinctive competence; (2) the management implications of selecting a particular product and process combination in light of competitors' selections; and (3) organizing different operating units so that they can specialize (focus) on separate portions of the total manufacturing task while still maintaining effective overall coordination.

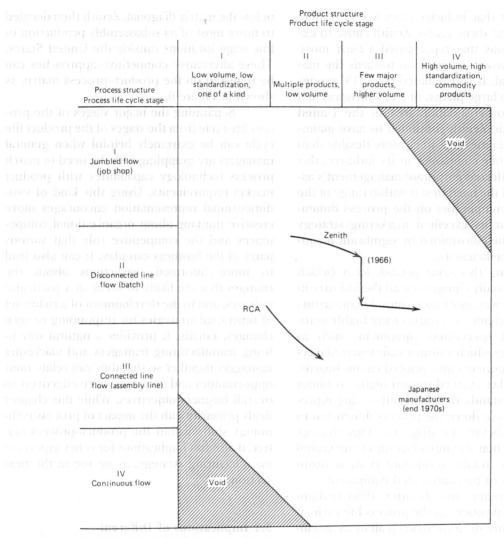

Product structure
Product life cycle stage

| | I
Low volume, low
standardization,
one of a kind | II
Multiple products,
low volume | III
Few major
products,
higher volume | IV
High volume, high
standardization,
commodity
products |

Process structure
Process life cycle stage

I
Jumbled flow
(job shop)

II
Disconnected line
flow (batch)

III
Connected line
flow (assembly line)

IV
Continuous flow

Zenith

(1966)

RCA

Japanese
manufacturers
(end 1970s)

Void

Void

FIGURE 6.   Product–Process Matrix for U.S. Color Television (1960s to mid-1970s)

The Concept of Distinctive Competence. Most companies like to think of themselves as being particularly good, relative to their competitors, in certain areas whereas they try to avoid head-to-head competition in other areas. Their objective is to guard this distinctive competence against outside attack or internal aimlessness and to exploit it wherever possible. Unfortunately, companies sometimes become preoccupied with the marketing aspects of their distinctive competence and lose sight of the

nature of their competence in manufacturing. When this happens, the company's strategic thinking tends to be dominated by product, market, and product life cycle considerations. In effect, management concentrates its resources and planning efforts on a relatively narrow column of the matrix shown in Figure 5.

One of the advantages of a two-dimensional point of view is that it encourages a business unit to be more precise about what its distinctive competence really is, and to concen-

430

trate its attention on a restricted set of process technology alternatives as well as a restricted set of marketing and product alternatives. Real focus is achieved only when one's attention is concentrated on a "patch" in the matrix—implying a process focus as well as a product focus. As we argue in our discussion of facilities strategy in Chapter 4, narrowing the focus of a business unit's activities, and particularly its manufacturing activities, often can lead to substantial improvement in organizational performance. In fact, the notion of facilities focus discussed there relates both to the specific focus of a given physical facility and to the general focus of a firm's overall process technology capabilities.

Thinking about both process and product dimensions can even affect the way a company defines its "product." For example, the management of one specialized manufacturer of printed circuit boards initially assessed its position on the product–process matrix as being in the lower left-hand corner: producing a low-volume, one-of-a-kind product using a highly connected assembly line process. On further reflection, however, management decided that while the company did specialize in small production batches, the "product" it really was offering was a design capability for special-purpose circuit boards. In a sense it was mass producing designs rather than boards. Hence the company was not far off the diagonal after all. This reconceptualization of the company's distinctive competence was helpful when management began to consider a set of proposed investments which varied greatly in terms of their fit with the company's *actual* position on the matrix.

Not only can the use of a product–process matrix help make explicit a firm's (or business unit's) distinctive competence, it can also help it avoid the dangers of product or process proliferation. Introducing a new product or entering a new market, either in an attempt to increase the utilization of existing facilities or simply to take advantage of the apparent profitability of a customer request for a modified product, can lead to a continually expanding product line—in effect causing the business unit to move horizontally to the left on the matrix. Unfortunately this often sets in motion a scenario that is difficult for most firms to deal with, because both the production and marketing sides of the business tend to encounter problems (different but complementary) at the same time. Marketing is trying to adapt itself to new products (and, possibly, markets) for which its procedures are not adequately suited, while production is trying to adapt its processes to new products which put analogous strains on its operation.

This can lead to what has been described as the "creeping breakeven" phenomenon: in an effort to stimulate demand a company enters a new market or introduces a new product. While this move may be successful, the existing process technology is incapable of meeting this added scale and complexity without additional investment (more capacity, different equipment, or other changes). Success breeds failure. The increased investment causes the company's breakeven point to rise, offsetting the expected gains from the increased sales volume. This motivates the company to pursue additional markets and products so as to "break out of the box" in which it finds itself. Within the context of the product–process matrix, the business finds itself trying to move along one dimension while not adequately adjusting its position on the other. Eventually it is forced to move along that other dimension as well. If this represents an *expansion* of its process (for example, adding a job shop to what is essentially an assembly line process) rather than a *repositioning*, however, it tends to dilute the company's manufacturing focus, making it more and more difficult to match the success that other firms—who continue to focus on their distinctive competence—are able to achieve.

The packaging division of one major consumer products company provides an illustration of this syndrome. The sole reason for the division's existence in the corporation was to offer a low cost source of supply for a highly specialized packaging product. Being a profit center, this division realized that if it could pick up some less specialized, high-priced business from outside customers, it would be able to increase substantially its profitability. (At least that is what its cost accounting system indicated!) As the division moved in that direction, however, it encountered pressure from its new customers to change its process technology so that it could better meet their needs. As the division began to dilute the focus it had previously maintained, it experienced increasing friction with its original in-house customers.

This scenario is also observed when an industry leader finds its standardized product line being challenged by smaller firms who attempt to segment the mass market and target specialized forms of the product for different segments. Over time such competition may slowly erode the leading firm's market share to the point where its relatively high-volume, standardized process is no longer economical. In an attempt to counterattack, it may introduce specialized products of its own (in effect, moving to the left on the product dimension), only to find that its process technology cannot compete effectively with competitors who have focused their process technologies around the specific volume and product characteristics best suited to each segment of the market.

Management Implications of Different Product–Process Positioning Strategies.   As firms alter their positions on the matrix by making different product and process choices, their competitive priorities and management tasks are profoundly affected. Looking at the process technology dimension, for example, we observe that the chief competitive advantage of a job shop process is its flexibility to both product and volume changes. As a firm moves toward more standardized process technologies,

its distinguishing capabilities shift from flexibility and customization (product specialization) to product reliability, delivery predictability, and cost. A similar shift in competitive emphasis occurs as the firm moves along the product structure dimension. These movements and their associated priorities are illustrated in Figure 7 and described in more detail in Chapter 10. In general, a company that chooses a given process structure can reinforce the characteristics of that structure by adopting the corresponding product structure.

For a given product structure, a company whose competitive strategy is based on offering customized products or features and rapid response to market shifts would tend to choose a much more flexible production technology than would a competitor that has the same product structure but follows a low-cost strategy. The former approach positions the company above the matrix diagonal; the latter positions it somewhere along or below the diagonal.

A company's location on the matrix also reflects what we referred to as its "dominant orientation" in Chapter 2. Most firms tend to be relatively aggressive along the dimension—product or process—where they feel most competent, and assume that the other dimension is a "given," in that it is determined by competitors and the general state of technological process in the industry. For example, a marketing-oriented company that is seeking to be responsive to the needs of a given market is more likely to emphasize flexibility and rapid delivery than is a more manufacturing-oriented company that seeks to mold the market to its own cost position or product specifications.

These two contrasting competitive approaches are illustrated in the electric motor industry by Reliance Electric and Emerson Electric. Reliance typically has chosen production processes that placed it above the diagonal for a given product and market, and it emphasizes product customization and performance.

Process Structure—Process Life Cycle Stage	Product Structure—Product Life Cycle Stage				Priorities	Key Management Tasks
	I Low-volume/ low-stand- ardization, one of a kind	II Low- volume, multiple products	III Higher volume few major products	IV High-volume/ high-stand- ardization, commodity products		
I Jumbled flow (job shop)					Flexibility quality	Fast reaction Loading plant, estimating ca- pacity
					Product customization	Estimating costs and delivery times
					Performance	Breaking bottle- necks
						Order tracing and expediting
II Disconnected line flow (batch)						Systematizing diverse elements
						Developing standards and methods, im- provement
						Balancing pro- cess stages
III Connected line flow (assembly line)						Managing large, specialized, and complex opera- tions
						Meeting material requirements
						Running equip- ment at peak ef- ficiency
						Timing expan- sion and tech- nological change
IV Continuous flow						Raising required capital
					Dependability- cost	

Priorities: Flexibility-quality → Dependability-cost

Dominant Competitive Mode			
Custom de- sign General pur- pose High margins	Custom design Quality control Service High margins	Standar- dized design Volume manufac- turing Finished goods inventory Distribu- tion Backup suppliers	Vertical inte- gration Long runs Specialized equipment and processes Economies of scale Standardized material

FIGURE 7. *Competitive Priorities and Key Tasks on the Product–Process Matrix*

433

Emerson, on the other hand, has tended to position itself below the diagonal, emphasizing low-cost production. The majority of Reliance's products are in the upper left quadrant, while Emerson's products tend to be in the lower right quadrant. Even where the two companies' product lines overlap, Reliance is likely to look for the more customized applications, and to use a more fluid process to produce that product, while Emerson is likely to use a more standardized production process. Each company is continually seeking to develop a set of competitive skills in manufacturing and marketing that will make it more effective within its selected quadrant.

The decision to concentrate on the upper left versus the lower right quadrant has many additional implications for a business. A company that chooses to compete primarily in the upper left, for example, has to decide when to drop a product or abandon a market that appears to be progressing inexorably along its product life cycle toward maturity, while a company that chooses to compete in the lower right must decide when to enter that market. The latter company does not need to be as flexible as the company that positions itself in the upper left. Moreover, since product and market changes typically occur less frequently during the latter phases of the product life cycle, it has more room for error.

Organizing (Focusing) Operating Units to Encourage Specialization and Coordination. A company that takes into consideration the process dimension when formulating its competitive strategy can usually focus its operating units much more effectively on their individual tasks. For example, many companies face the problem of how to organize the production of spare parts for their primary products. As the sales volume of its primary products increases, the company tends to move down the matrix diagonal. The follow-on demand for spare parts for these products, however, may imply a combina-tion of product and process structures much more toward the upper left-hand corner of the matrix. There are many more items to be manufactured, each in smaller volume, and the appropriate process tends to be more flexible than is the case for the primary products.

To accommodate the specific requirements of spare parts production, some companies develop a special manufacturing facility for them; others simply separate their production within the same facility. The least appropriate—and, unfortunately, most common—approach is to combine the production of spare parts with that of the basic product, which forces the manufacturing process to span a broad range of both product and process structures, reducing its effectiveness for both product categories.

The combined choice of product and process also determines the kind of manufacturing problems that are likely to be experienced. Some of the key tasks related to a particular process technology are indicated on the right-hand side of Figure 7. Recognizing the impact that the company's position on the matrix has on these important tasks often suggests changes in the policies and procedures used in managing the company's manufacturing function, particularly its manufacturing control system. The measurements used to monitor and evaluate manufacturing performance should also reflect the matrix position selected. Unfortunately, as Richardson and Gordon's survey (1980) of 15 manufacturing companies illustrates, most companies tend to use standardized measurement and control systems, no matter what their position on the matrix. Not only can designing more customized systems help a company avoid the loss of control over manufacturing that often results when its position on the matrix changes, it also suggests the changes in management skills, attitudes, and mindsets that may be needed.

While a fairly narrow focus may be required to succeed in any single product market, large companies generally produce multiple products for multiple markets. These products are often in different stages of their life cycles. Such companies can often benefit by separating their manufacturing facilities, and organizing each to meet the specific needs of different products.

In Chapter 4 we described one company that chose to separate its total manufacturing capabilities into a group of carefully specialized units: the Lynchburg Foundry Corporation. As outlined there, Lynchburg Foundry operates several different facilities, each representing a different position on the matrix. While each plant uses a somewhat different basic technology, they have many similarities. However, other elements of the manufacturing process used in each plant, such as the production layout, equipment, workforce organization, and control system, are very different. Lynchburg has chosen to design its facilities so that each meets the needs of a specific segment of the market.

This example of positioning individual manufacturing facilities (and their process technologies) to meet certain market segment requirements suggests another use for the matrix: identifying the suppliers who are most capable of meeting a company's needs. For example, high-volume automotive facilities need suppliers who can handle high-volume standardized products, whereas manufacturers of custom equipment who position themselves in the upper left quadrant are likely to be served much more effectively by suppliers who are also positioned in that quadrant.

On the other hand, companies that specialize their manufacturing units according to the needs of narrowly defined patches on the matrix may encounter problems integrating those units organizationally into a coordinated whole. Companies seem to be most successful when they organize their manufacturing function around either a product/market focus or a process focus, but not both. That is, individual operating units should either manage themselves relatively autonomously, responding directly to the needs of the particular markets they serve, or else they should be divided according to process stages (for example, fabrication, subassembly, and final assembly) and coordinated by a central staff (see Hayes and Schmenner, 1978).

Companies in the major materials industries—steel, oil, and paper, for example—provide classic examples of process-organized manufacturing operations. Most companies that broaden the span of their process through vertical integration tend to adopt such an organization, at least initially. By contrast, companies with a strong product/market orientation are usually unwilling to accept the organizational rigidity and lengthened response time that often accompany centralized coordination. For example, most companies in the packaging industry adopt such product/market-focused manufacturing organizations. They set up regional plants to serve geographical market areas in an attempt to reduce transportation costs and provide better response to customer needs.

Sometimes major competitive opportunities and entirely new market segments can be identified with the assistance of a product—process matrix. The restaurant industry, for example, has recently experienced major changes because it recognized such opportunities. As illustrated in Figure 8, the traditional short-order cafe uses a job shop process to produce low volumes of a wide variety of standard food items. The competitive emphasis of such local restaurants tends to be quick service and reasonable prices.

On the other hand, first-class restaurants almost invariably build their reputation by offering high-quality meals at high prices: service is slower but more elegant. Such restaurants are located in the extreme upper-left corner of the matrix. (In fact, it appears that some country restaurants in France are more properly

Product Structure—Product Life Cycle Stage

Process Structure Process Life Cycle Stage	I Low volume/low-standardization, one of a kind	II Low volume multiple products	III Higher volume few major products	IV High-volume high-standardization commodity products
I Jumbled flow (job shop)	Classic French restaurant Traditional restaurant			
II Disconnected line flow (batch)		Short-order cafe		
III Connected line flow (assembly line)		Steak house	Burger King/ McDonald's	
IV Continuous flow				

FIGURE 8. Restaurant Examples of Product and Process Matching

located off the matrix—one sometimes gets the impression that the chef goes out and orders the raw materials for the meal after the customer places the order.)

In recent years, two new types of restaurants have made widespread gains in the marketplace by positioning themselves differently on the matrix. One of these is the narrow-menu steak house, which concentrates on a single major type of food, offers a limited variety of side dishes, and employs a line flow production process. Several companies who were among the first to recognize the need for this type of service, and who tailored their skills and processes to meet the specific requirements of this market segment, have grown into substantial chains.

The so-called fast-food restaurants, like McDonald's, have positioned themselves even farther down the diagonal. They offer standardized products with few options and produce them in high volume with automated (or tightly controlled) processes. To ensure process standardization in such a service setting, McDonald's has designed its restaurants and chosen equipment in such a way that its site managers and workers are compelled to follow the intended production process.

But not all firms that attacked the fast-food market segment have followed the same philosophy and systems as McDonald's. Burger King, for example, has chosen to position itself in a slightly different location on the matrix, at least in the customers' eyes. As illustrated in Figure 8, Burger King has allowed customers a little more flexibility (they can select their own pickles, onions, catsup, and mustard), hoping to steal away from McDonald's those who prefer such "customization." While it may be hard to argue that offering customized condiments really represents a major difference in strategy, these two firms have also selected somewhat different production processes: Burger King "produces to order" (cooking hamburgers in response to individual orders), while McDonald's "produces to inventory."

Even in their advertising these two firms have sought to differentiate themselves. For example, McDonald's has used the phrase "We do it all for you." This really means "We run the production process the way it was set up to be run, and you don't have to worry about it (in fact, you can't interfere with it)." Burger King counters with "Have it *your* way." This suggests they will be responsive to individual customer requests, although they are constrained significantly by the narrow menu and standardized process they have adopted.

Like the U.S. auto companies, Burger King has adopted essentially the same position on the matrix as its major competitors, but it has sought to convince customers that they are actually farther to the left in their product structure. It has done this by offering options on items that have little effect on the production process but may affect customers' perceptions of the product and service being delivered (and perhaps the price they are willing to pay).

Offering customized products necessitates a close and effective interface between manufacturing and marketing. From a marketing point of view, almost all companies would prefer to offer broader product lines in order to match competitors' products. However, some manage such breadth much more effectively than do others. A firm that makes marketing decisions without taking manufacturing considerations into account may inadvertently cripple itself. For example, adding products whose market impact is marginal may seriously impair manufacturing effectiveness. In firms where marketing and manufacturing are closely coordinated, it is much more likely that the product options selected—like those at Burger King—will not detract from the firm's basic philosophy and matrix position.

As important as the marketing and manufacturing interface is in maintaining a desired position on the matrix, it is often no more critical than the interface between manufacturing and product engineering. When either engineering or manufacturing alters its basic strategy, the other must respond or the same types of mismatches that occur between manufacturing and marketing will occur. If communication across this interface is poor, engineering may design products that make it difficult, if not impossible, for manufacturing to employ the process technology it has chosen. In companies where the two functions cooperate closely and effectively, product design characteristics that are consistent with manufacturing capabilities are more likely.

3.3 Limits of the Product–Process Matrix Framework

Using the product–process matrix as a means for matching process technology and product line decisions has limitations, as does any theoretical construct. While these do not necessarily detract from the usefulness of the concept, it is important to keep in mind the fact that no single framework can ever handle all situations equally well. To suggest the nature of some of these limitations, we provide two illustrations in this section.

In this chapter we assume that the evolution of process technology along a process life cycle takes place through an ordered series of steps: standardization (of products, components, and equipment), rationalization (of flows, bottlenecks, and inventories), mechanization (of material conversion or handling steps, to replace labor), and, finally, automation (introduction of integrated systems that handle both material conversion and movement). This sequence is accompanied by increasing capital intensity, reduced process flexibility, and increased specificity to a narrow set of product and task requirements. Not all "progress" in process technology follows the same pattern, however.

For example, the recent development of flexible machining centers appears to offer firms both low cost and far greater flexibility for product changeovers than do older, less automated, and less capital-intensive processes. Similarly, some of the production practices adopted in Japan as part of "just-in-time" production and materials management systems require higher levels of equipment investment (together with lower machine utilization) but provide significantly increased production flexibility. Such improvements in production flexibility, in the absence of movement along the diagonal, might be thought of as a third dimension to the matrix. This dimension would represent increased overall effectiveness without a major change in the basic match between product life cycle and process life cycle. There are other approaches for handling changes in process technology that lead to improved performance without requiring compensating shifts in product/market offerings. The product-process matrix appears to be able to capture some of these, but not all; those other issues must be considered outside of the matrix.

A second example of the concept's limitations is when there is a breakdown in the assumption that a product's life cycle is equivalent to a market life cycle. While the two generally move in the same direction, they do not necessarily move at the same rate or to the same extent. The U.S. market for color television sets, for example, has clearly matured, yet numerous recent developments in product technology have caused the product life cycle to reverse direction—toward more variety and more options.

Another source of divergence between the product life cycle and the market life cycle occurs when the same product is sold into multiple markets. Again the color TV industry provides a good example. The personal computer and word processing markets, both of which were in the rapid growth stage of their life cycles in the early 1980s, utilize display devices that are similar in many respects to those found in TV receivers. Since they are often manufactured in the same facility, with the same production processes as TV displays, such products appear to be in the final stage of the product life cycle. Yet some of the products they go into are in the rapid growth stage. In a sense this limitation is simply the complement of one described earlier. In the case of flexible machining centers, a unidirectional process life cycle oversimplifies the realities of the situation, whereas in the case of display devices, a unidirectional product life cycle oversimplifies reality.

This latter difficulty also occurs when a market splits into price categories, and the products and customers in each major price segment follow separate product life cycles. In such a situation the low-end price segment may move very quickly to the final stage of the product life cycle, whereas higher price segments may never move beyond the middle stages. Many businesses appear to have these characteristics. For example, oscilloscopes, one of the most basic of electronic measuring devices, are positioned all along the matrix diagonal if the entire industry is lumped together. If split into two or three major price (and feature) segments, however, each can be represented by its own product–process matrix.

3.4 Integrating Product and Process Innovations

Shifts in position on the product-process matrix are often triggered by product or process innovations. Abernathy and Utterback (1975) have explored such innovations in some detail, and Figure 9 summarizes their research relating the rate of innovation along each dimension as the product life cycle evolves. Early in a product's life, great effort is expended on product design, and product innovation is rapid as competitors try to find a design that best fits the needs of potential users. Abernathy and Utterback refer to this early phase of product innova-

tion as the search for a "dominant design"—a standardized product that can form the basis for rapid growth and market development. Ford's Model T car, the DC-3 airplane, the Xerox 914 copier, and the Kodak Instamatic camera are examples of such dominant designs.

According to Abernathy and Utterback, as the dominant design catches hold in the marketplace, cost reduction and process innovation geared primarily to lowering production costs, increasing yields, and building production volume begin to replace product innovation as the major focus of management attention. Product changes become less frequent and less radical, and process innovation begins to get more of the R&D budget. However, as investment in such activities moves the production technology closer to the continuous flow end of the process life cycle, both product and process become increasingly vulnerable to the introduction of a radically different new product (usually produced with a different process technology) that provides the same functions. Examples of this would be the replacement of mechanical calculators with electronic calculators, and the replacement of mechanical watches with electronic watches.

Achieving the appropriate balance between process innovation and product innovation is a critical management task. Some organizations simply focus their attention on a certain section of the process–product matrix, and concentrate their innovation efforts on one or the other of these two types of change. A classic example is Hewlett-Packard's instrument business, which has chosen to position itself in the upper left-hand quadrant of the product–process matrix. As most of its competitors have moved down the diagonal—where, according to Abernathy and Utterback, product innovation becomes less important and process innovation more important—HP has countered by introducing new product generations, and thereby moving back up to the left-hand corner. Thus it avoids the necessity to develop the

439

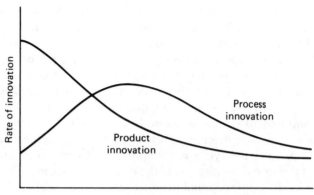

	Fluid Pattern	**Transitional Pattern**	**Specific Pattern**
Competitive emphases on	Functional product performance	Product variation	Cost reduction
Innovation stimulated by	Information on users' needs and users' technical inputs	Opportunities created by expanding internal technical capability	Pressure to reduce cost and improve quality
Predominant type of innovation	Frequent major changes in products	Major process changes required by rising volume	Incremental for product and process, with cumulative improvement in productivity and quality
Product line	Diverse, often including custom designs	Includes at least one product design stable enough to have significant production volume	Mostly undifferentiated standard products
Production processes	Flexible and inefficient major changes easily accommodated	Becoming more rigid with changes occurring in major steps	Efficient, capital-intensive, and rigid; cost of change is high
Equipment	General purpose, requiring highly skilled labor	Some subprocesses automated, creating "islands of automation"	Special purpose, mostly automatic with labor tasks mainly monitoring and control
Materials	Inputs are limited to generally available materials	Specialized materials may be demanded from some suppliers	Specialized materials will be demanded if they are not available, vertical integration will be extensive
Plants	Small-scale, located near user or source of technology	General purpose with specialized sections	Large-scale, highly specific to particular products
Organizational control	Informal and entrepreneurial	Through liaison relationships, project and task groups	Through emphasis on structure, goals, and rules

Source Reprinted with permission from *Omega*, Volume 3, Number 6, William J. Abernathy and James Utterback, "Dynamic Model of Process and Product Innovation," Copyright 1975, Pergamon Journals, Ltd.

FIGURE 9. Patterns of Product and Process Innovation

organizational capabilities required for process innovation. In fact, many of HP's instrument businesses consider "innovation" to be synonymous with product innovation, and the manufacturing processes used are not changed unless required to do so by the next generation of product. When this happens, new processes are likely to be adapted, on an as-needed basis, from other industries where they are already well developed. Process and product transfer, not new process development or radical process innovation, becomes management's chief concern when contemplating changes in the manufacturing process.

A firm like Texas Instruments, on the other hand, has tended to concentrate its attention in the lower right-hand quadrant where process innovation tends to be more important than product innovation. TI often waits while others do much of the early product innovation (in the upper left-hand quadrant) until it identifies an appropriate entry point. After it enters, it uses its skills in process innovation to push rapidly down the matrix diagonal, displacing some of the product's original developers who neglected, either intentionally or unintentionally, to develop similar skills.

3.5 Product Technology Options and Manufacturing Process Technology

Up to this point we have assumed that there was a single dominant product technology, even when discussing product innovation. In some cases, however, companies have the choice of pursuing quite different product technologies, each of which may require a different process technology. One author who has explored some of the managerial issues associated with choosing among alternative product technologies is Abell (1980). In developing concepts for "defining one's business," he highlights three primary dimensions of concern to marketing managers: customer groups, customer functions, and alternative technologies. These dimensions and the way they might be used in developing alternative business definitions are illustrated in Figure 10.

In Abell's terminology, firms (particularly their marketing functions) make decisions that define the range of their activities along each of three dimensions. (In a sense, he is splitting what we have called the product market dimension into three parts.) He uses the

example of computerized tomography (CT) to illustrate the importance of looking at these dimensions separately. In the late 1970s four very different product technologies were available for diagnostic imagining: X-rays, computerized tomography (linking an X-ray machine to a computer), ultrasound, and nuclear. Each had been adopted as the primary product technology by at least one of the major competitors in the industry. Pfizer was pursuing computerized tomography, EMI and Ohio Nuclear were pursuing ultrasound, and GE was pursuing nuclear. Several other old-line equipment manufacturers were still heavily involved in traditional X-ray technology. In each firm, the selection of a particular product technology established different requirements for its process technology capabilities.

Refining the product market dimension, as suggested by Abell, can aid management's direction of process technology in three ways. First, by separating out the impact of product technology, its interaction with manufacturing technology can be addressed explicitly and systematically. Second, it highlights the need to coordinate and manage not only the manufacturing–marketing linkage, but also the manufacturing–product design and product design–marketing linkages. Third, the concept of focus is enriched by suggesting additional dimensions for specifying the degree of differentiation that a firm might pursue.

4 SUMMARY AND CONCLUSIONS

In this chapter we described and illustrated the use of a framework, the product–process matrix, that can help general managers link their company's process technology capabilities with product/market requirements. Other supporting concepts and techniques were also presented and outlined. Whatever techniques are used in a given situation, three major conclusions seem to emerge:

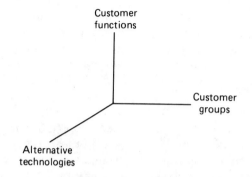

(a) Three dimensions for defining a business

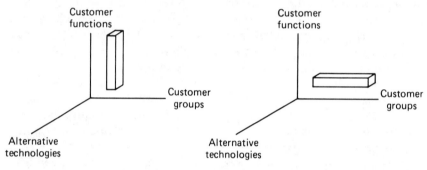

(b1) A business serving multiple
customer functions

(b2) A business serving multiple
customer groups

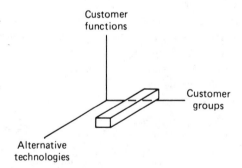

(b3) A business with products based on
several alternative technologies

Source: Abell, 1980. Used by permission.

FIGURE 10.   A Marketing Definition of a Business Involving Three Dimensions

1. The development of a process technology strategy requires substantial general management inputs, not just functional expertise.
2. The integration of marketing and manufacturing is an iterative process. The firm must continually monitor market, product, manufacturing, and technological developments to insure that the desired match is pursued throughout the organization.
3. Designing a manufacturing process technology should not be an afterthought, a hurried response to market selection or product design. It must be configured around the needs of a particular product design and competitive strategy, while exploiting the availability of potentially applicable manufacturing technologies.

We pick up on these ideas in Chapter 10, where we discuss the management of changes in process technology.

SELECTED REFERENCES

Abell, Derek F. *Defining the Business: The Starting Point of Strategic Planning.* Englewood Cliffs, NJ: Prentice-Hall, 1980.

Abernathy, William J., and Phillip L. Townsend, "Technology, Productivity, and Process Change." *Technological Forecasting and Social Change,* 1975. Vol. 7. No. 4, pp. 379–396.

Abernathy, William J., and James Utterback. "Dynamic Model of Process and Product Innovation." *Omega,* Vol. 3, No. 6, 1975, pp. 639–657.

Blois, K. J. "Market Concentration—Challenge to Corporate Planning." *Long-Range Planning.* Vol. 13, August 1980, pp. 56–62.

Dhalla, N. K., and S. Yuspeh. "Forget the Product Life Cycle Concept." *Harvard Business Review,* January–February 1976, pp. 102–112.

Hayes, Robert H., and Roger W. Schmenner. "How Should You Organize Manufacturing?" *Harvard Business Review,* January–February 1978, pp. 105–118.

Hayes, Robert H., and Steven C. Wheelwright. "Link Manufacturing Process and Product Life Cycles." *Harvard Business Review,* January–February 1979, pp. 133–140.

Richardson, P. R., and Gordon, J. R. M. "Measuring Total Manufacturing Performance" *Sloan Management Review,* Winter 1980, pp. 47–58.

Shapiro, Benson P. "Can Marketing and Manufacturing Coexist?" *Harvard Business Review.* September–October 1977, pp. 104–114.

Wasson, Chester R. *Dynamic Competitive Strategy and Product Life Cycles.* Austin. TX: Austin Press, 1978.

Wells, Louis T., Jr. (ed.) *The Product Life Cycle in International Trade.* Cambridge, MA: Harvard University Press, 1972.

Product Development and Manufacturing Start-Up

Steven C. Wheelwright

What makes some companies' manufacture of new products successful—meeting time, cost and performance objectives—and others' fraught with bugs, delays and overruns?

Current research into the new product development and manufacturing start-up experience of high-technology firms, both in the United States and in Japan, may lead to some answers. Although still in its early stages, the research is uncovering interesting facts about the manufacturing savvy of various companies, the length of their "cycle time" from research and development to product introduction, their organization of product development "projects," and the nature of the relationship between design engineering and manufacturing operations.

MANUFACTURING'S STRATEGIC ROLE

High-tech firms often seem oblivious to the role of manufacturing, seeing product innovation as the way out of any problem. When pressed for an assessment of manufacturing, they typically view it as secondary and downstream: "R&D designs the product, marketing

sells it and *then* manufacturing builds it." Perhaps more descriptive of manufacturing's role is, "Manufacturing is supposed to make up engineering delays and meet whatever promises are made by sales and marketing." As described by one electronics firm with tongue in cheek, manufacturing's job is seen as beat-to-fit, paint-to-match. These views are not universal, but surprisingly widespread.

In our recent book[1] Bob Hayes and I have defined a framework to describe the various stages of manufacturing's role within an organization (Figure 1), ranging from the most defensive to the most progressive. Most businesses in the United States start out at stage 1—they begin with marketing and/or technical expertise as their primary strength, not operations. That may last for some time, until the reality of costs and margins and competition sets in. Then they quickly move toward stage 2, where most of traditional U.S. smokestack industry resides—steel, auto, rubber companies—as well as electronic instruments. In this mode, manufacturing is doing okay as long as "industry practice" is followed and no one breaks out of the pack. But once the rules are broken, by a competitor like Toyota, Nissan or Honda in autos or Nucor steel in minimills—manufacturing competition takes on a different dimension. This leads firms to move toward the third stage, where manufacturing actively seeks to support the business strategy. Most of what is written in the business press about manufac-

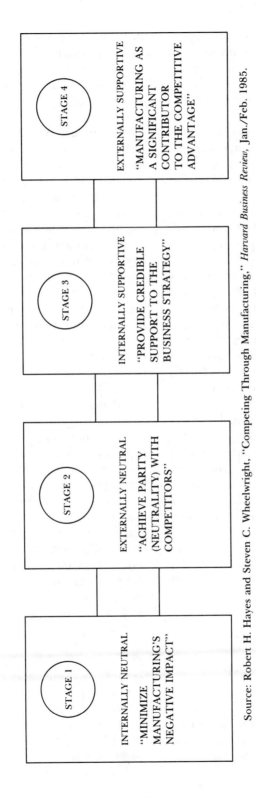

Source: Robert H. Hayes and Steven C. Wheelwright, "Competing Through Manufacturing," *Harvard Business Review*, Jan./Feb. 1985.

FIGURE 1. Four Stages in Manufacturing's Strategic Role

turing coming of age is focused on this stage—manufacturing going beyond the minimum and getting smarter about how to improve the product as it comes down the line. However, even in stage 3, manufacturing's role is still a derived one.

All three of the phases described so far are characterized by "manufacturing fixing itself." It is getting more progressive, but is still an isolated entity. The organization's view is that to fix something in manufacturing, you do not have to change the rest of the organization. All you have to change is manufacturing. And, certainly there is tremendous opportunity to do that.

Stage 4, however, is entirely different and sets up the link that ideally should exist between design engineering and manufacturing. In stage 4 companies, manufacturing makes a difference to the customer. Organizationally, manufacturing is more of a team sport. The vertical, divisible nature of activities (R&D, marketing, manufacturing) in the first three stages yields to more horizontal kinds of activities at stage 4 that cut across multiple functions at multiple levels. This is the ideal setting for new product development/manufacturing start-up activities in all types of companies. Each function is no longer doing its own thing, but each is integrated with the others in performing tasks that cut across boundaries. There are other features that characterize stage 4 companies, one of the most quantifiable being shorter cycle time, the time it takes from the start of product development to the shipment of manufactured products to customers.

EXPERIENCE WITH NEW PRODUCT DEVELOPMENT AND MANUFACTURING START-UP

The reality of manufacturing new products, even in renowned high-technology companies, is usually at odds with the dream. Time and

time again, products fail to meet expectations—in timing, cost, quality, specifications, customer satisfaction, market success and organizational learning. The causes, and cures, are the focus of my current research.

The example of a major Silicon Valley electronics manufacturer that utilizes the traditional product development/manufacturing interface is a useful starting point for understanding the reality of new product development. Figure 2a shows resource requirements over time, at least according to theory, in this leading firm. Marketing is involved up front to help define and focus the concept; design engineering (R&D) has a substantial role early on; and manufacturing starts slowly, peaks and then recedes as experience builds. Design engineering gets out of the game partway through, and manufacturing picks up the ball entirely. Such a scenario was expected to generate the production cost curve shown in the lower portion of the figure.

When this theoretical model was compared against the reality of recent projects, the actual curves turned out to be quite different than planned (Figure 2b). Invariably, the design engineering staff did not get out at the point expected, its job having been completed. Instead, a series of major redesigns of the product were required, until finally the product was abandoned and resources shifted to the next-generation product. The actual cost curve reflected blips that corresponded with each of the redesign efforts. Although the production cost curve slopes down to some degree, the cost reduction is nowhere near that expected. Such projects on average cost about twice the resources approved at the outset of the effort.

In trying to discover why this happens—why cost and development curves look so different from those anticipated—we have identified characteristics common to companies that follow the traditional, segmented, new product development approach. As a starting point, the traditional model establishes reference stakes (expectations or goals) covering completion

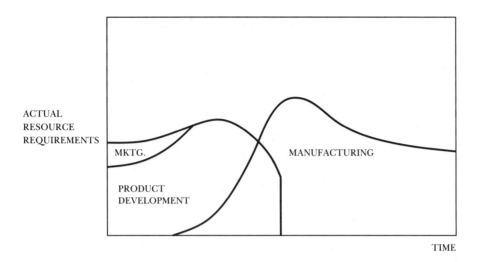

ACTUAL
RESOURCE
REQUIREMENTS

MKTG.

PRODUCT
DEVELOPMENT

MANUFACTURING

TIME

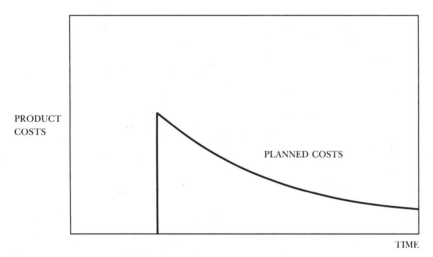

PRODUCT
COSTS

PLANNED COSTS

TIME

FIGURE 2A.   Expected Results of Traditional Approach to Product Development/Manufacturing Startup

schedules, product cost and product perform-ance specifications very early in the project. As soon as these stakes are set, the project team faces uncertainties—details about how two widgets are put together and which materials to use.

The second type of uncertainty (which we have found common to almost all projects that failed to meet their early stakes) is that to varying degrees the technologies needed for

the product are still in the invention stage. The project team must either wait for the invention to occur or try to push it, but inevitably they cannot program the time it will be available. Finally, uncertainties relating to organizational policies arise, which the project team does not have the power to control: Who is the target customer? What are the relationships of vari-ous product generations? What level of vertical integration is desired?

447

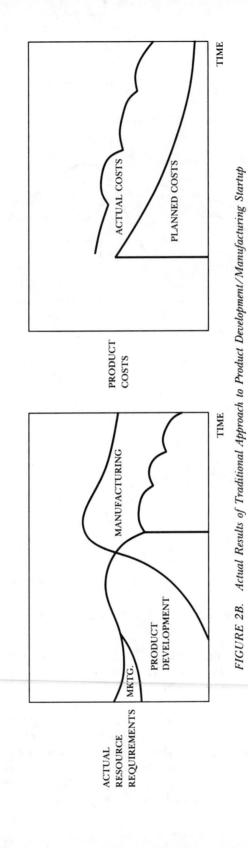

FIGURE 2B. *Actual Results of Traditional Approach to Product Development/Manufacturing Startup*

448

Many of these policy uncertainties are beyond the scope of the single project, but the project becomes the focal point or driving force for addressing such policies. If these strategic policy decisions have not been made, the project team must put the product on hold and use its time to push them through the organization. Soon the project misses its target with regard to the reference stakes. At some point the company either must change the stakes, which rarely happens, particularly if the project is very far along, or can skip steps so it misses the stakes by an "acceptable" margin. Many of these steps must be put back in after the project is handed to the operating (manufacturing) organization. Thus, operations completes the tasks that should have been part of the project.

How do you stem this trend and its associated problems of schedule and cost overruns and compromised product performance specifications? An important key to improving the design engineering/manufacturing interface that is essential to effective product development/manufacturing start-up is reducing "cycle time." In fact, cycle time for new product efforts is a good measure of the degree to which these functions have achieved stage 4 status. In data collected by one of my colleagues on both a U.S. and a Japanese electronics manufacturer, we found that the U.S. firm spent 30 months from design to manufacturing build-up (Figure 3). The Japanese competitor, in contrast, took only 24 months. While the lengths of the activity arrows were different, the essential difference was the overlap of Japanese functions. In the U.S. firm, activities were disjointed—there was a clear hand-off between functions. The Japanese firm had a smooth, evolving, back-and-forth passing of things: manufacturing involvement began before the design activities were finished. And the Japanese firm conducted more thorough up-front investigations.

Another example of this preparatory thoroughness recently became apparent at

GE's Lighting Products group recently. They are buying some high-speed equipment from a Japanese manufacturer and also developing some equipment themselves. Early into the project, while the Japanese were still investigating, GE kept reminding them of "The Schedule," which had a firm completion date. The Japanese continually responded that they recognized the schedule, but were not ready to start producing equipment yet. It now appears that some of the GE-made equipment that was started early will be somewhat over budget, late, and take longer to get up and running, while the Japanese equipment appears to be coming in on budget, on time and is up and running very quickly.

The bottom line is that the Japanese have more activity earlier on, yet shorter overall cycle times. Cycle time appears to be an excellent surrogate for how well the functions (R&D, manufacturing and marketing) are integrated, that is stage 4 in nature. (This presumes that shorter cycles were achieved through systematic improvement, not by the ad hoc cutting of important steps.) Also, shorter cycle times have a number of benefits. They materially reduce risk and uncertainty and provide the option of making a preemptive competitive move based on timing. Even more important, shorter cycles result in reduced design costs and lower production costs (fewer repeats, quicker start-up, more efficient-to-produce designs).

APPROACHES TO SHORTENING DEVELOPMENT CYCLES

If research observations support the conclusion that shorter product development cycles are characteristic of companies with the most advanced design and manufacturing functions (stage 4) and result in numerous cost, time and product performance benefits, how can a company shorten its development cycles and what

449

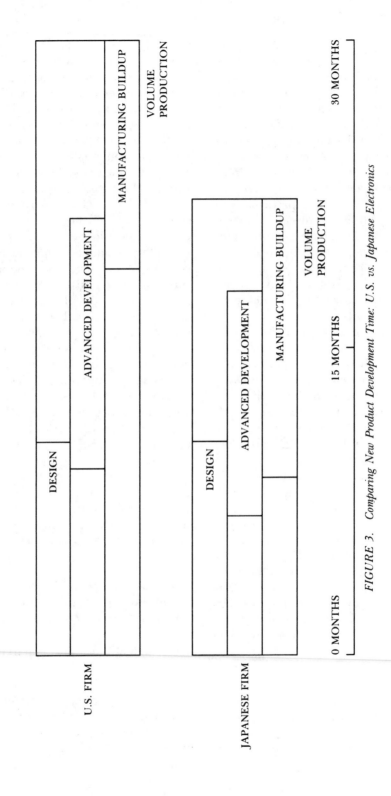

FIGURE 3. Comparing New Product Development Time: U.S. vs. Japanese Electronics

organizational framework is most conducive to shorter cycle times?

There are several alternative approaches to shortening cycle times: skipping (combining) steps, pursuing steps in parallel, putting more steps into the functions and tackling smaller but more numerous projects. When used wisely by stage 4 organizations, all of these have a role in continuing to shorten development cycles.

MAKING IT HAPPEN

Some of the keys to shorter, more effective, new product/new process development cycles are: better strategic management of the functions, better project management and continued learning across a series of projects. These three things need to be done well if developing and commercializing new products are to become significant competitive advantages.

In the first key area—improved strategic management of the functions—much of the project team's strategic policies work, ranging from technology innovation to pricing strategy and vertical integration policy, should not be necessary. These decisions and background work should be developed by the functions, independent of specific projects and products. They should serve as "maps" for product and process development efforts.

Figure 4A describes how product development is handled traditionally. As indicated by the bell curve, there is some manufacturing input, some marketing input and a substantial amount of design engineering input, along with the normal integration and coordination activities. Because the center of gravity is in design engineering, that function is in charge of the project. This means that the project team, which is primarily middle-level engineering, must resolve issues of marketing strategy and manufacturing strategy that it is not in a position to decide directly and thus must get the rest of the organization to act.

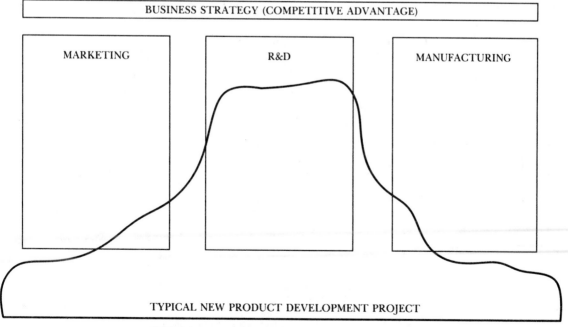

FIGURE 4A. Traditional Product Development Projects

Figure 4B describes how some more forward-looking high-technology firms have chosen to tackle product development. Their approach is a more horizontal cut with *less* R&D in each new product development project and a better balance in the team from all functions. Importantly, more R&D is focused on critical technologies and inventions, outside of individual product projects. The results of these efforts become a technical resource.

In this model, the new product project is narrowed and it becomes more fast-moving and free-flowing. This is facilitated when each function outlines as part of its strategy a series of maps that provide a sense of direction to itself, other functions, and product project teams. For example, R&D should develop product technology maps and product generation maps that should be agreed upon by the organization before specific projects are started. Marketing should develop basic customer maps, market maps and competitor maps. Likewise, essential manufacturing maps would include process technology maps, facility maps and work-force skill maps.

The second key area is effective project management. This deals with issues rather than functions—staffing issues, organizational issues, structure and control issues. It addresses how individual projects and groups of projects are to be run so that, assuming the up-front functional work is done, the project goes as planned.

The final key area relates to collective experience—learning and improving new product/manufacturing start-up management across a series of projects. Most companies do not appear to learn much about new product development management across projects. In fact, they often atrophy; they forget what they have learned to the point that they must go back to the drawing board and produce a new procedure for new product development every three or four years.

The bottom line is that companies that do not pursue effective practices in these three areas never realize the full competitive potential of their new product development efforts. An enlightened company with a stage 4 set of functions says, "Let's learn from experience,

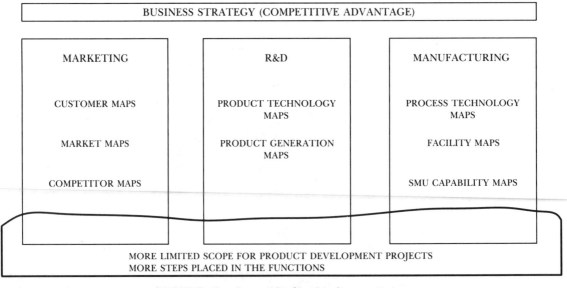

FIGURE 4B.   Stage 4 Product Development Projects

let's systematically get better. Let's shorten cycle times. Let's integrate functions horizontally and do the up-front R&D, marketing, manufacturing and project management that facilitate successful product development.''

NOTE

1. (Robert H. Hayes and Steven C. Wheelwright, *Restoring Our Competitive Edge: Competing Through Manufacturing,* John Wiley, New York, 1984).

Improvement of Quality and Productivity through Action by Management

W. Edwards Deming

Management can increase productivity by improving quality. Dr. Deming's fourteen points show the way.

WHY PRODUCTIVITY INCREASES WITH IMPROVEMENT OF QUALITY

Some simple examples will show how productivity increases with improvement of quality. Other benefits are lower costs, better competitive position, and happier people on the job. It is important to note that a gain in productivity is also a gain in capacity of a production-line.

Best Efforts Are Not Sufficient

"By everyone doing his best."

This is the answer that someone in a meeting volunteered in response to my question, "And how do you go about it to improve quality and productivity?"

It is interesting to note that this answer was wrong—wrong in the sense that best efforts are not sufficient. Best efforts are essential, but unfortunately, best efforts alone will not ac-

complish the purpose. Everyone is already doing his best. Efforts, to be effective, must move in the right direction. Without guidance, best efforts result in a random walk.

Some Folklore. Folklore has it in America that quality and production are incompatible: that you can not have both. It is either or. Insist on quality, and you will fall behind in production. Push production and you will find that your quality has suffered.

The fact is that quality is achieved by improvement of the process. Improvement of the process increases uniformity of output of product, reduces mistakes, and reduces waste of manpower, machine-time, and materials.

Reduction of waste transfers man-hours and machine-hours from the manufacture of defectives into the manufacture of additional good product. In effect, the capacity of a production-line is increased. The benefits of better quality through improvement of the process are thus not just better quality, and the long-range improvement of market-position that goes along with it, but greater productivity and much better profit as well. Improved morale of this work force is another gain: they now see that the management is making some effort themselves, and not blaming all faults on the production-workers.

A clear statement of the relationship between quality and productivity comes from my friend Dr. Yoshikasu Tsuda of Rikkyo University in Tokyo, who wrote to me as follows, dated 23 March 1980.

> I have just spent a year in the northern hemisphere, in twenty-three countries, in which I visited many industrial plants, and talked with many industrialists.
>
> In Europe and in America, people are now more interested in cost of quality and in systems of quality-audit. But in Japan, we are keeping very strong interest to improve quality by using statistical methods which you started in your very first visit to Japan. . . . When we improve quality we also improve productivity, just as you told us in 1950 would happen.

A Simple Example. Some simple figures taken from recent experience will illustrate what happens. Defective output of a certain production-line was running along at 11 percent (news to the management). A run-chart of proportion defective day by day over the previous six weeks showed good statistical control of the line as a whole. The main cause of the problem could accordingly only be ascribed to the system. This was also news to the management. The statisticians made the suggestion that possibly the people on the job, and inspectors also, did not understand well enough what kind of work is acceptable and what is not. The manager of the production-line and two supervisors went to work on the matter, and with trial and error came up in seven weeks with better definitions, with examples posted for everyone to see. A new set of data showed the proportion defective to be 5 percent. Cost, zero. Results:

Quality up
Productivity up 6%
Costs down
Profit greatly improved
Capacity of the production line increased 6%
Customer happier
Everybody happier

This gain was immediate (seven weeks); cost, zero: same work force, same burden, no investment in new machinery.

This is an example of gain in productivity accomplished by a change in the system, effected by the management, helping people to work smarter, not harder.

Reduction in Cost

Taken from a speech delivered in Rio de Janeiro, March 1981, by William E. Conway, president of the Nashua Corporation.

At Nashua, the first big success took place in March 1980: improvement of quality

TABLE 1. *Illustrating Gain in Productivity with Improved Quality*

Item	Before Improvement 11% Defective	After Improvement 5% Defective
Total cost	100	100
Spent to make good units	89	95
Spent to make defective units	11	5
Average number of good units per unit cost	.89	.95
Proportion of total cost spent to make defectives	.11	.05

and reduction of cost in the manufacture of carbonless paper.

Water-based coating that contains various chemicals is applied to a moving web of paper. If the amount of coating is right, the customer will be pleased with a good consistent mark when he uses the paper some months later. The coating-head applied approximately three pounds of dry coating to 3000 square feet of paper at a speed of approximately 1400 linear feet/minute on a web 6–8 feet wide. Technicians took samples of paper and made tests to determine the intensity of the mark. These tests were made on the sample both as it came off the coater and after it was aged in an oven to simulate use by the customer. When tests showed the intensity of the mark to be too low or too high, the operator made adjustments that would increase or decrease the amount of coating material. Frequent stops for new settings were a way of life. These stops were costly.

The engineers knew that the average weight of the coating material was too high, but did not know how to lower it without risk of putting on insufficient coating. A new coating-head, to cost $700,000, was under consideration. There would be, besides the cost of $700,000, time lost for installation, and the risk that the new head might not achieve uniformity of coating much better than the equipment in use.

In August 1979, the plant manager decided to utilize the statistical control of quality to study the operation. It was thereby found that the coating-head, if left untouched, was actually in pretty good statistical control at the desired level of 3.0 dry pounds of coating on the paper, plus or minus .4.

Elimination of various special causes, highlighted by points of control, reduced the amount of coating and still maintained good consistent quality.

The coater had by April 1980 settled down to an average of 2.8 pounds per 3000 square feet, varying from 2.4 to 3.2, thereby

saving 0.2 pounds per 3000 square feet, or $800,000 per year at present volume and cost levels.

What the operator of the coating-head had been doing, before statistical control was introduced and achieved, was to over-adjust his machine, to put on more coating, or less, reacting to tests of the paper. In doing his best, in accordance with the training and instructions given to him, he was actually doubling the variance of the coating. The control-charts, once in operation, helped him to do a much better job, with less effort. He is happy. His job is easier and more important.

All this was accomplished without making the proposed capital investment of $700,000, which might or might not have improved the process and the quality of the coated paper.

Engineering Innovation. Statistical control opened the way to engineering innovation. Without statistical control, the process was in unstable chaos, the noise of which would mask the effect of any attempt to bring improvement. Step by step they achieved:

· Improvement of the chemical content of the material used for coating, to use less and less coating.
· Improvement of the coating-head (without purchase of a new one) to achieve greater and greater uniformity of coating.

Today, only 1.0 pounds of improved coating is used per 3000 square feet of paper. This level is safe, as the variation lies only between .9 and 1.1. Reduction of a tenth of a pound means an annual reduction of $400,000 in the cost of coating.

The reader can do his own arithmetic to compute the annual reduction in cost from the starting point, namely, 3.0 pounds.

Before statistical control was achieved, the engineers had not entertained thought of improvement of coating-head. Once statistical control was achieved, it was easy to measure the effect of small changes in the chemistry of the coating and in the coating-head. The next step then became obvious—try to improve the coating and the coating-head, to use less coating with greater and greater uniformity.

Low Quality Means High Cost

A plant was plagued with a huge amount of defective product. "How many people have you on this line for rework of defects made in previous operations?" I asked the manager. He went to the blackboard and put down three people here, four there, etc.—in total, 21 percent of the work force on that line.

Defects are not free. Somebody makes them, and gets paid for making them. On the supposition that it costs as much to correct a defect as to make it in the first place, then 21 percent of his payroll and burden was being spent on rework. In practice, it usually costs more to correct a defect than to make it, so the figure 21 percent is a minimum.

Once the manager saw the magnitude of the problem, and came to the realization that he was paying out good money to make defects as well as to correct them, he found ways to help the people on the line to understand better how to do the job. The cost of rework went down from 21 percent to 9 percent in a space of two months.

Next step: reduce the proportion defective from 9 percent to 0.

From 15 percent to 40 percent of the manufacturer's costs of almost any American product that you buy today is for waste embedded in it—waste of human effort, waste of machine-time, loss of accompanying burden. No wonder that many American products are hard to sell at home or abroad.

American industry (including service-organizations) can no longer tolerate mistakes and defective material at the start nor anywhere along the line, nor equipment out of order.

New Machinery Is Not the Answer

Lag in American productivity had been attributed in editorials and in letters in the newspapers to failure to install new machinery and the latest types of automation. Such suggestions make interesting reading and still more interesting writing for people that do not understand problems of production. There is a quicker and surer way, namely, better administration of man and whatever machinery is in use today. Then, after the present problems are conquered, talk about new machinery.

The following paragraph received from a friend in a large manufacturing company will serve as illustration.

> This whole program (design and installation of new machines) has led to some unhappy experiences. All these wonderful machines performed their intended functions, on test, but when they were put into operation, they were out of business so much of the time for this and that kind of failure that our overall costs, instead of going down, went up. No one had evaluated the overall probable failure-rates and maintenance. As a result, we were continually caught with stoppages and with not enough spare parts, or with none at all; and no provision for alternate production-lines.

Comparison between American and Japanese production should take account of some important differences. Japanese manufacturers are already using their machinery to full advantage, not wasting materials, human effort, or machine-time. They have no unemployed people to draw upon for expansion. There are no unemployment agencies in Japan. The Monthly Report on the Labor Force for the United States, in contrast, shows at this writing 7 million unemployed. American industry can expand by drawing upon a supply of labor, a large part of it skilled, experienced, able, and willing to work. The Japanese manufacturer, on the other hand, can expand his production only by use of better machinery or improvement in design. He can not hire more people: there are not any. The fact is that there is only a small amount of automation in Japan. They have been sensible about it.

If I were a banker, I would not lend money for new equipment unless the company that asked for the loan could demonstrate by statistical evidence that they are using their present equipment to full realizable capacity.

Quality Control in Service Industries

Eventually, quality control will assist not only the production of goods and food (the birthplace of modern statistical theory was agriculture) but the service industries as well—hospitals, hotels, transportation, wholesale and retail establishments, perhaps even the U.S. mail. Statistical quality technology has for many years contributed to telecommunications, both in the manufacture of equipment and in service. Statistical quality technology is improving service and lowering costs in the banking business. In fact, one of the most successful applications of statistical methods on a huge scale, including sample-design and operations, is in the U.S. Census, not only in the decennial Census but in the regular monthly and quarterly surveys of people and of business, an example being the Monthly Report on the Labor Force.

It is interesting to note that some service-industries in Japan have been active in statistical methods from the start, e.g., the Japanese National Railways, Nippon Telegraph and Telephone Corporation, the Tobacco Monopoly of Japan, the Post Office. Department stores have taken up statistical quality control. Takenaka Komuten (architecture and construction) won recognition in 1979 for thoroughgoing improvement of buildings of all types, and for decrease in cost, by studying the needs of the users (in offices, hospitals, factories, hotels) and by reducing the costs of rework in drawings and in the actual construction.

WHAT TOP MANAGEMENT MUST DO

The purpose here is to explain to top management what their job is. No one in management need ask again, "What must we do?" This section serves two purposes: (1) It provides an outline of the obligations of top management. (2) It provides a yardstick by which anyone in the company may measure the performance of the management.

Paper profits, the yardstick by which stockholders and Boards of Directors often measure performance of the president, make no contribution to material living for people anywhere, nor do they improve the competitive position of a company or of American industry. Paper profits do not make bread: improvement of quality and productivity do. They make a contribution to better material living for all people, here and everywhere.

Short-term profits are not indication of good management. Anybody can pay dividends by deferring maintenance, cutting out research, or acquiring another company.

Ways of doing business with vendors and with customers that were good enough in the past must now be revised to meet new requirements of quality and productivity. Drastic revision is required.

What must top management do? As I noted at the beginning of this article, it is not enough for everyone to do his best. Everyone is already doing his best. Efforts, to be effective, must move in the right direction.

It is not enough that top management commit themselves by affirmation for life to quality and productivity. They must know what it is that they are committed to—i.e., what they must do. These obligations can not be delegated. Mere approval is not enough, nor New Year's resolutions. Failure of top management to act on any one of the fourteen points listed ahead will impair efforts on the other thirteen. Quality is everybody's job, but no one else in the company can work effectively on quality and productivity unless it is obvious that the top people are working on their obligations.

"Let me emphasize that where top management does not understand and does not get personally involved, nothing will happen." *(From a speech made by William E. Conway, president of the Nashua Corporation.)*

The 14 Points for Top Management

Here are the obligations of top management. These obligations continue forever: none of them is ever completely fulfilled.

1. *Create constancy of purpose in the company.* The next quarterly dividend is not as important as existence of the company ten, twenty, or thirty years from now.

a. Innovate. Allocate resources for long-term planning. Plans for the future call for consideration of:

- Possible new materials, new service, adaptability, probable cost.
- Method of production: possible changes in equipment.
- New skills required, and in what number?
- Training and retraining of personnel
- Training of supervisors.
- Cost of production.
- Performance in the hands of the user.
- Satisfaction of the user.

One requirement for innovation is faith that there will be a future. Innovation, the foundation of the future, can not thrive unless the top management has declared unshakable policy of quality and productivity. Until this policy can be enthroned as an institution, middle management and everyone in the company will be skeptical about the effectiveness of their best efforts.

The consumer is the most important part of the production-line. Japanese management took on a new turn in 1950 by putting the consumer first.

b. Put resources into

- research
- education

c. Put resources into maintenance of equipment, furniture, and fixtures, new aids to production in the office and in the plant.

It is a mistake to suppose that statistical quality technology applied to products and services offered at present can with certainty keep an organization solvent and ahead of competition. It is possible and in fact fairly easy for an organization to go broke making the wrong product or offering the wrong type of service, even though everyone in the organization performs with devotion, employing statistical methods and every other aid that can boost efficiency.

Innovation generates new and improved services. An example is new and different kinds of plans for savings in banks, financial service offered by credit agencies. Meals on Wheels, day care in out-patient clinics. Leasing of automobiles is an example of service that did not exist years ago. Express Mail is a new service of the U.S. post office (equivalent to what a postage stamp would accomplish in Japan or in Europe). Mailgram by Western Union is another. Intercity and intracity messenger service is a growth industry, thriving on the delinquency of the U.S. post office. Services can and do have problems of mistakes, costly correction of mistakes, and consequent impairment of productivity associated with mistakes.

2. *Learn the new philosophy.* We are in a new economic age. We can no longer live with commonly accepted levels of mistakes, defects, material not suited to the job, people on the job that do not know what the job is and are afraid to ask: handling damage; failure of management to understand the problems of the prod-

uct in use; antiquated methods of training on the job; inadequate and ineffective supervision.

Acceptance of defective materials and poor workmanship as a way of life is one of the most effective roadblocks to better quality and productivity. The Japanese faced it in 1950. Unreliable and nonuniform were kind words for the usual quality of incoming materials. Japanese management took aim at the problem and in time reduced it to a level never before achieved. American industry today faces the same problem. The road that Japanese manufacturers paved would be a good one for American management to copy.

3. *Require statistical evidence of process control along with incoming critical parts.* There is no other way for your supplier nor for you to know the quality that he is delivering, and no other way to achieve best economy and productivity. Purchasing managers must learn the statistical control of quality. They must proceed under the new philosophy: the right quality characteristics must be built in, without dependence on inspection. Statistical control of the process provides the only way for the supplier to build quality in, and the only way to provide to the purchaser evidence of uniform repeatable quality and of cost of production. There is no other way for your supplier to predict his costs.

Most purchasing managers do not know at present which of their suppliers are qualified. One of the first steps for purchasing managers to take is to learn enough about the statistical control of quality to be able to assess the qualifications of a supplier, to talk to him in statistical language. Don't expect him to carry on the conversation in French if you don't know French.

Some suppliers are already qualified and are conforming to this recommendation. Some follow their product through the purchaser's production-lines to learn what problems turn up, and to take action, so far as possible, to avoid problems in the future.

One company may have influence over hundreds of suppliers and over many other purchasers.

Vendors sometimes furnish reams of figures, such as records of adjustment, input of materials (2 kg. chromium added at 1000 h). Figures like these are as worthless to the buyer as they are to the vendor.

The manager of an important plant, which belongs to one of America's largest corporations, lamented to me that he spends most of his time defending good vendors. A typical problem runs like this. A vendor has not for years sent to him a defective item, and his price is right. Some other manufacturer underbids this vendor by a few cents and captures the business. The corporate purchasing department awards the business to him because of price. The plant manager can not take a chance, and must spend many hours and days arguing for the vendor that knows his business.

4. *The requirement of statistical evidence of process control in the purchase of critical parts will mean in most companies drastic reduction in the number of vendors that they deal with.*

Companies will have to consider the cost of having two or more vendors for the same item. A company will be lucky to find one vendor that can supply statistical evidence of quality. A second vendor, if he can not furnish statistical evidence of his quality, will have higher costs than the one that can furnish the evidence, or he will have to chisel on his quality, or go out of business. A man that does not know his costs nor whether he can repeat tomorrow today's distribution of quality is not a good business partner.

We can no longer leave quality and price to the forces of competition—not in today's requirements for uniformity and reliability. Price has no meaning without a measure of the quality being purchased. Without adequate measures of quality, business drifts to the lowest bidder, low quality and high cost being the inevitable result. American industry and the U.S. government are being rooked by rules that award business to the lowest bidder.

The purchasing managers of a company are not at fault for giving business to the lowest bidder, nor for seeking more bids in the hope of getting a still better price. This is their mandate. Only the top management can change their direction.

Purchasing managers have a new job. It will take five years for them to learn it.

Example. An American manufacturer of automobiles may today have 2500 vendors. A Japanese automobile company may have 380. Rapid and determined reduction of the number of vendors in American manufacturing is already under way.

Problem 1. You wish to purchase 1000 pieces of xbae. You make calls to two companies A and B that offer the product, and you explain the specifications. Each company submits 1000 pieces of xbae—all good, so the companies claim. You satisfy yourself, by your own inspection, that indeed all 2000 pieces meet your specifications. Which lot would you buy? Toss a coin?

Before you make a snap judgment, it might be wise to consider the fact that your specifications may not tell the whole story. There may be characteristics that are important to have in the pieces that you will buy, but not covered in your specifications. There may be other characteristics that you would wish to avoid, and your specifications may not protect you. Company A has been in the business and

provides continuing evidence of process control. There may be persuasive arguments in favor of Company A. One must remember that the distribution of the important quality-characteristics of the 1000 parts made by Company A will be more uniform than the distribution of those made by Company B. Uniformity is nearly always important. If the price offered by Company B is the lower of the two, it would be wise to inquire how this could be, as Company A's costs will be lower. Perhaps Company B can offer a bargain. He may have had a cancellation from another customer, and has the material on hand.

Problem 2. Now, we come to a totally different problem. You plan to purchase 1000 pieces of xbae every week. Your requirements in this problem point definitely to selection of Company A. The distribution of the important quality-characteristic of the xbae is predictable. It will be steady week after week. If the distribution falls within your specifications, you can eliminate inspection of incoming parts except for routine observations and comparisons for identification. You will know from the control charts that came along with the product, far better than any amount of inspection can tell you, what the distribution of quality is and what it will be.

If a tail of the distribution falls beyond your specifications, you will require 100 percent inspection.

Another advantage of statistical control is a better price and better quality than Company B can give. If Company B offers you a

**Made by
Company A**

**Made by
Company B**

1000 pieces
all meet
specifications

1000 pieces
all meet
specifications

Company A

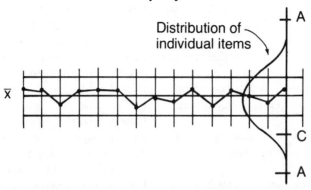

Distribution of
individual items

$\bar{x}$

$\bar{x}$-chart for a critical quality-characteristic. If the specifications are at A, A, the output of Company A requires no inspection. Problems will arise (see text) if a specification were at C.

Company B

About Company B we know nothing.

better price, it is because he knows not what his costs are. He may chisel you on quality, or he may go out of business trying to meet the price that he quoted, and leave you stranded.

You may decide, for protection, to give some of the business to Company B, just to have a second vendor to fall back on in case Company A suffers some hard luck. This is your privilege. You will have to pay the extra cost that Company B must charge you if he stays in business and delivers the quantity that you require, and you will have to inspect the material that comes from him. Whether it is wise to do business with Company B could be questioned, but you now have a rational basis for your decision.

In free enterprise you have a right to make a wrong decision, and to get beat up for it.

A request for bids usually contains a clause to say that quality may be considered along with price. That is, the award will not necessarily be given to the lowest bidder. Such a clause is meaningless without a yardstick by which to measure quality. The buyer and his purchasing manager usually lack such a yardstick. They are candidates for plunder by the lowest bidder.

A flagrant example is a request for professional help, to be awarded to the lowest bidder. Example (actual, from a government agency):

For delivery and evaluation of a course on management for quality control for supervisors . . .
An order will be issued on the basis of price.

5. Use statistical methods to find out, in any trouble spot, what are the sources of trouble. Which are local faults? Which faults belong to the system? Put responsibility where it belongs. Do not rely on judgment. Judgment always gives the wrong answer on the question of where the fault lies. Statistical methods make use of knowledge of the subject matter where it can be effective, but supplant it where it is a hazard.

Constantly improve the system. This obligation never ceases. Most people in management do not understand that the system (their responsibility) is everything not under the governance of a local group.

6. Institute modern aids to training on the job. Training must be totally reconstructed. Statistical methods must be used to learn when training is finished, and when further training would be beneficial. A man once hired and trained, and in statistical control of his own work, whether it be satisfactory or not, can do no better. Further training can not help him. If his work is not satisfactory, move him to another job, and provide better training there.

7. Improve supervision. Supervision belongs to the system, and is the responsibility of management.

- Foremen must have more time to help people on the job.
- Statistical methods are vital as aid to the foreman and to the production manager to indicate where fault lies: is it local, or is it in the system?
- The usual procedure by which the foreman calls the worker's attention to every defect, or to half of them, may be wrong—is certainly wrong in most organizations—and defeats the purpose of supervision.
- Supervision in large segments of American industry is deplorable. For example, a common practice is to look at the production

records of people on the job, supervisors, managers, and to deliberately take aim at the lowest 5 percent or the lowest 10 percent. Claims of results from this procedure are nothing but reinvention of the Hawthorne effect. The ultimate result is frustration and demoralization of the organization.

8. Drive out fear. Most people on a job, and even people in management positions, do not understand what the job is, nor what is right or wrong. Moreover, it is not clear to them how to find out. Many of them are afraid to ask questions or to report trouble. The economic loss from fear is appalling. It is necessary, for better quality and productivity, that people feel secure. *Se* comes from Latin, meaning without, *cure* means fear or care. Secure means without fear, not afraid to express ideas, not afraid to ask questions, not afraid to ask for further instructions, not afraid to report equipment out of order, nor material that is unsuited to the purpose, poor light, or other working conditions that impair quality and production.

Another related aspect of fear is inability to serve the best interests of the company through necessity to satisfy specified rules, or to satisfy a production quota, or to cut costs by some specified amount.

One common result of fear is seen in inspection. An inspector records incorrectly the result of an inspection for fear of overdrawing the quota of allowable defectives of the work force.

9. Break down barriers between departments. People in research, design, purchase of materials, sales, receipt of incoming materials, must learn about the problems encountered with various materials and specifications in production and assembly. Otherwise, there will be losses in production from necessity for rework and from attempts to use materials unsuited to the purpose. Why not spend time in the factory, see the problems, and hear about them?

I only recently saw a losing game, 40 percent defective output, the basic cause of which was that the sales department and the design department had put their heads together and come through with a style whose tolerances were beyond the economic capability of the process. This lack of coordination helps a plant to become a nonprofit organization.

In another instance, the man in charge of procurement of materials, in attendance at the seminars, declared that he has no problems with procurement, as he accepts only perfect materials. (Chuckled I to myself, "That's the way to do it.") Next day, in one of his plants, a superintendent showed to me two pieces of an item from two different suppliers, same item number, both beautifully made, both met the specifications, yet they were sufficiently different for one to be usable, the other usable only with costly rework, a heavy loss to the plant. The superintendent was charged with 20,000 of each one.

Both pieces satisfied the specifications. Both suppliers had fulfilled their contracts. The explanation lay in specifications that were incomplete and unsuited to the requirements of manufacture, approved by the man that had no problems. There was no provision for a report on material used in desperation. It seems that difficulties like this bring forth solace in such remarks as:

> This is the kind of problem that we see any day in this business.

or

> Our competitors are having the same kind of problem.

What would some people do without their competitors? Surely one responsibility of management in production is to provide help in such difficulties, and not leave a plant manager in a state of such utter hopelessness.

Purchasing managers must learn that specifications of incoming materials do not tell the whole story. What problems does the material encounter in production? It is necessary to follow a sample of materials through the whole production process to learn about the problems encountered, and onward to the consumer's problems.

10. Eliminate numerical goals, slogans, pictures, posters, urging people to increase productivity, sign their work as an autograph, etc., so often plastered everywhere in the plant. ZERO DEFECTS is an example. Posters and slogans like these never helped anyone to do a better job. Numerical goals even have a negative effect through frustration. These devices are management's lazy way out. They indicate desperation and incompetence of management. There is a better way.

11. Look carefully at work-standards. Do they take account of quality, or only numbers? Do they help anyone to do a better job? Work standards are costing the country as much loss as poor materials and mistakes.

Any day in hundreds of factories, men stand around the last hour or two of the day, waiting for the whistle to blow. They have completed their quotas for the day; they may do no more work, and they can not go home. Is this good for the competitive position of American industry? Ask these men. They are unhappy doing nothing. They would rather work.

12. Institute a massive training program for employees in simple but powerful statistical methods. Thousands of people must learn rudimentary statistical methods. One in 500 must spend the necessary ten years to become a statistician. This training will be a costly affair.

13. Institute a vigorous program for retraining people in new skills. The program should keep up with changes in model, style, materials, methods, and, if advantageous, new machinery.

14. Create a structure in top management that will push every day on the above thirteen points. Make maximum use of statistical knowledge and talent in your company. Top management will require guidance from an experienced consultant, but the consultant can not take on obligations that only the management can carry out.

Action Required

The first step is for management to understand what their job is—the fourteen points. The next step is to get into motion on them. Quality and productivity are everybody's job, but top management must lead. Until and unless top management establish constancy of purpose and make it possible for everyone in the company to work without fear for the company and not just to please someone, efforts of other people in the company, however brilliant be the fires that they start, can only be transitory.

How soon? When? A long thorny road lies ahead in American industry—ten to thirty years—to settle down to an accepted competitive position. This position may be second place, maybe fourth. Small gains will be visible within a few weeks after a company mobilizes for quality, but sweeping improvement over the whole company will take a long time, and will continue forever. Unmistakable advances will be obvious within five years, more in ten. Management must learn the new economics: likewise government regulatory agencies, and they

may require thirty years. Meanwhile, American industry will continue to suffer under the supposition that competition is the secret to better quality and service, and to lower costs.

Products that have been the backbone of American industry may in time decline to secondary importance. New products and new technology may ascend to top place. Agriculture may move up further in foreign trade.

Tangible results from each of the fourteen points will not all be visible at the same time. Perhaps the best candidate for quickest results is to supplant work-standards (No. 11) with statistical aids to the worker and to supervision. No one knows what productivity can be achieved with statistical methods that help people to accomplish more by working smarter, not harder.

A close second for quick results would be to start to drive out fear (No. 8), to help people to feel secure to find out about the job and about the product, and unafraid to report trouble with equipment and with incoming materials. Once top management takes hold in earnest, this goal might be achieved with 50 percent success, and with powerful economic results, within two or three years. Continuation of effort will bring further success.

A close third, and a winner, would be to break down barriers between departments (No. 9).

Survival of the fittest. Companies that adopt constancy of purpose for quality and productivity, and go about it with intelligence, have a chance to survive. Others have not. Charles Darwin's law of survival of the fittest, and that the unfit do not survive, holds in free enterprise as well as in nature's selections.

SECTION
V

Managing Linkages

A continuous stream of innovation requires strong functional areas that are well coordinated. Chapter 8 deals with formal and informal linking—achieving coordination across functions, divisions and/or geographic boundaries. Nadler and Tushman provide a framework and tools for achieving linkage through the formal organization. Shanklin and Ryans discuss the R&D/marketing interface, while Ghoshal and Bartlett explore various formal and informal levers for achieving integration across multinational organizations. Hatvany and Pucik focus on managing informal linkages. Hatvany and Pucik use their knowledge of Japanese organizations to probe the underlying determinants of organization culture—determinants that are broadly generalizable. Imai, Nonaka, and Takeuchi explore new product development processes in Japan, with particular emphasis on the interrelations between formal and informal practices in managing organization linkages and organization learning.

8

MANAGING LINKAGES

Strategic Linking:
Designing Formal
Coordination Mechanisms

David A. Nadler

Michael L. Tushman

Cases:

Jean Shaeffer, President of Federal Engineering, is grappling with an important decision for her engineering products firm (see Figure 1). Currently the industry leader in scientific instruments and process control products, her firm has not gotten any of the burgeoning systems business. This new market requires the production of integrated systems tailored to different user-needs. While Federal Engineering, with its divisional structure, has been able to provide first quality instruments, it has not been able to produce competitive systems. Somehow, Federal's systems have been late

and of low quality and have generated substantial customer dissatisfaction. Shaeffer's hunch is that while her firm has excellent technical talent, the distinct groups just do not work well together. There are priority difficulties, a lack of information sharing, and a lack of working together for the good of the systems business. Shaeffer's dilemma is that she must not jeopardize the instrument business while attacking the firm's problems in the systems area.

Roger Laffer, Senior Strategy Officer for Office Products Corporation, has been asked by the firm's CEO to help come up with an institutional framework for corporate technology transfer (see Figure 2). While this firm has been widely known for its technological and

---

From *Strategic Organization Design* by David Nadler and Michael L. Tushman. Copyright © 1987 by Scott, Foresman and Company. Reprinted by permission.

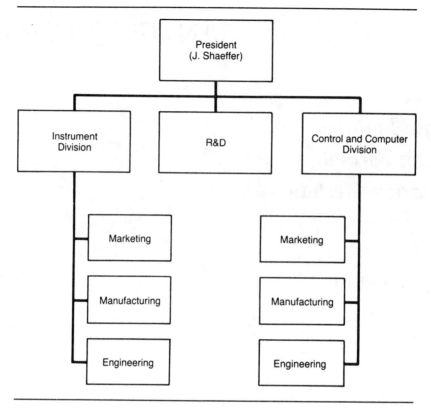

FIGURE 1.   *Federal Engineering*

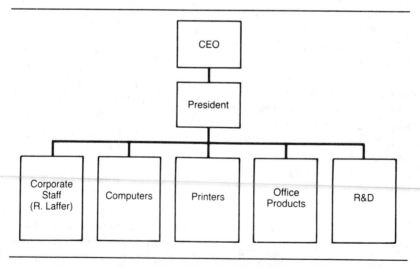

FIGURE 2.   *Office Products Company*

market excellence in computers, printers, and office products, the firm has a dismal record in capitalizing on technologies across existing business units. Laffer feels that the corporation's focus on distinct products/market niches hinders its ability to deal with opportunities that did not neatly fit into corporate categories. Laffer wonders whether a revised structure, along with top management support, could provide an institutional infrastructure to support technology transfer.

Both Shaeffer and Laffer are dealing with problems of strategic linking. Both must develop a set of formal linking mechanisms that will work to enhance, encourage, and facilitate coordination between distinct groups in their firms. Strategic linking issues follow directly from strategic grouping choices. Strategic grouping focuses resources by product, market, discipline, or geography. This grouping of resources puts some resources together *and* splits other resources. For example, a disciplinary-organized pharmaceutical laboratory focuses attention on disciplines but scatters individuals who are interested in therapeutic areas. In a product organization, functional expertise is split among product areas. Strategic linking involves choosing formal structures that link units that have been split during strategic grouping. Once strategic grouping decisions have been made, the next step is to coordinate, or link, the units so that the firm can operate as an integrated whole.

Where grouping decisions are driven by strategy considerations, strategic linking is driven by the degree of **task interdependence** between areas. Different degrees of task interdependence require different types of formal linking mechanisms. The objective is to build linking mechanisms that allow adequate information processing between groups. Linking mechanisms that are not adequate to handle

necessary information will result in poorly coordinated work. Those linking mechanisms that are more extensive than necessary will hinder information flow and result in unnecessary cost. This chapter discusses different types of work interdependence, presents a range of formal linking mechanisms, and concludes with a methodology for making linking decisions (see Figure 3).

VARIETIES OF TASK INTERDEPENDENCE

Strategic grouping results in a set of groups that are dedicated to product, markets, function, and/or discipline. Strategic grouping provides the basic architecture of the firm at each level of analysis. Strategic linking involves choosing those sets of formal linking mechanisms to coordinate the different groups so the organization functions as a whole. For example, at Federal Engineering, Jean Shaeffer must choose a set of formal linking mechanisms to link the two divisions together in service of the systems business.

Linking follows directly from grouping and, as with strategic grouping, must be accomplished at multiple levels of analysis. Our design problem involves choosing the right set of linking mechanisms to deal with: (1) work flows between distinct units, (2) the need for disciplinary- or staff-based professionals to have contact across the firm, and (3) work flows associated with emergencies, crises, or other nonroutine events.

The conceptual thread across work flow, disciplinary linkages, and work flows under crisis conditions is work-related interdependence. Managers choose linking mechanisms to deal with this source of uncertainty. The greater the task interdependence, the greater the need for coordination and joint problem solving. The more complex the degree of work/task inter-

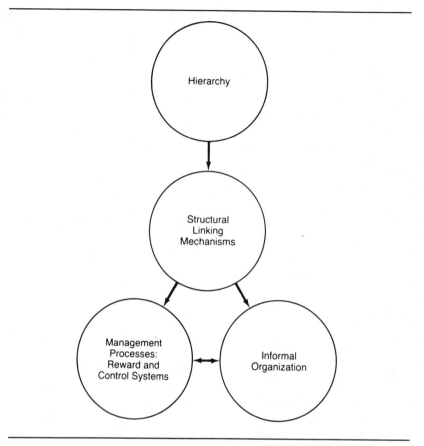

FIGURE 3.   *A Range of Formal Linking Mechanisms*

dependence, the more complex the formal linkage devices must be to handle work-related uncertainty. On the other hand, groups that are only weakly interdependent have relatively little need for coordination and joint problem solving and therefore need simple formal linking devices.

Consider branch banks located throughout a city. Each branch bank runs essentially independently of each other except for the common sharing of advertising and marketing resources. Similarly, business units within a diversified firm with completely different product/market niches are also essentially independent of each other except for those corporate resources that are shared between divisions (e.g., technology, staff). Both of these examples illustrate **pooled interdependence**. Units that operate independently but are part of the same organization and therefore share scarce resources must deal with the pooled nature of their interdependence. Units with pooled interdependence have a minimal amount of coordination and linking requirements (see Figure 4).

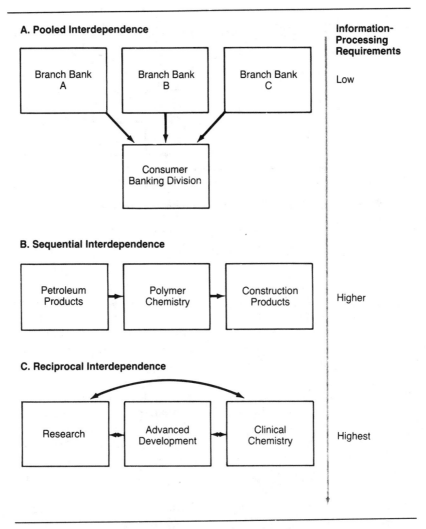

FIGURE 4.   *Forms of Interdependence*

At Olympic Oil, the Petroleum Products Division extracts petroleum from the ground and provides the raw material for the Polymer Chemistry Division (see Figure 5). This division makes a variety of products that various end-use divisions use in the production of consumer and construction products.

Similarly, in the back office of a bank, checks move through a series of groups before they exit the bank. These examples illustrate **sequential interdependence** (review Figure 4). Sequentially interdependent units must deal with a greater degree and variety of problem-solving and coordination requirements than

473

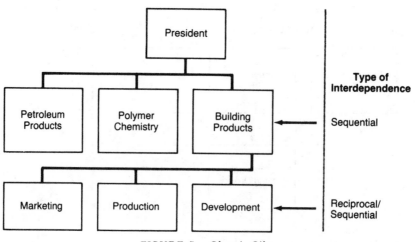

FIGURE 5.   Olympic Oil

units that have pooled interdependence. Groups that have sequential interdependence must attend to close coordination and timing so that work flows remain smooth and uninterrupted; each unit in the work flow is dependent on prior units.

At Federal Engineering, Marketing must work with R&D and with Production in the development of new products. Each functional area must be in close contact with the others to ensure the synthesis of market, technological, and production considerations. Similarly, in an advertising agency, the media, creative, and account services areas must work closely with each other in the development of ad campaigns for their clients. These examples illustrate **reciprocal interdependence**, that is, interdependence in which each group must work with each other unit in the production of a common product (review Figure 4). Reciprocal interdependence imposes substantial coordination and problem-solving requirements between units; no one unit can accomplish its task without the active contribution of each other unit.

Pooled, sequential, and reciprocal interdependence represent different degrees of work-related interdependence. Reciprocal interdependence imposes greater coordination costs and complexity than sequential, which, in turn, requires greater coordination than pooled interdependence. Beyond work flows, accentuated task interdependence can also arise in emergency, temporary, or crisis situations, in which units that normally only pool their activities must suddenly work together. For example, in our branch bank example, if one part of town suffers a black out, then the other branch banks must work together more closely to deal with the emergency. Similarly, independent product divisions that share a common technology must work together for those unique corporate ventures that attempt to combine the strengths of the two divisions.

Finally, quite apart from work flow considerations, knowledge-based staff and/or professionals must retain contact across organization boundaries or they will become overly specialized and/or lose touch with the state-of-the-art in their respective fields. The greater the rate of change in the discipline or staff areas, the greater the professionally anchored interdependence. At Warner-Lambert, for example, biologists in various areas of the corporation held monthly seminars to inform each other of current biological developments. This need for continual updating was much less critical in the more routine toxicological areas.

Whether driven by work flows, crises, or professionally anchored need for collaboration, these differing degrees of work-related interdependence impose different information processing requirements. Those units that have pooled interdependence (or in which the rate of change of the underlying knowledge base is low) have fewer coordination demands and information processing requirements than units that have reciprocal interdependence (or in which the rate of change of the underlying knowledge base is rapid). The designer's challenge is to choose the appropriate set of linking mechanisms to deal with the information-processing requirements that arise from work-related interdependence.

Finally, just as strategic grouping is relevant at multiple levels of analysis, so too is the assessment of work-related interdependence. For example, at Olympic Oil (review Figure 5), the divisions are sequentially interdependent. However, within each division, the respective functional areas must attend to reciprocal interdependence as they attempt to develop new and innovative products in their respective markets. Similarly, at Federal Engineering, not only was there reciprocal interdependence between the functional areas, but within R&D, each discipline was reciprocally interdependent in new product development efforts. Thus, the degree of task/work interdependence is not constant across organizations. The degree of work-related interdependence must be assessed at each level of analysis.

STRATEGIC LINKING: A RANGE OF LINKING MECHANISMS

Various types of formal mechanisms can be used to link, or coordinate, the efforts of organizational groups. Our objective is to choose those structural linking mechanisms that provide adequate information flows, procedures, and structures to deal with the information requirements imposed by work-related interdependence. Formal linking mechanisms that are not sufficient to handle linking requirements will result in poorly coordinated work. Those linking mechanisms that are more extensive than need be will result in unnecessary costs and overcomplexity. Structural linking mechanisms can be analyzed in terms of their ability to handle information flows and complex problem-solving requirements.[1]

Hierarchy

The most simple form of structural linking is the **hierarchy**: the formal distribution of power and authority. The hierarchy of authority follows directly from grouping decisions. For example, in a divisional structure, the divisional general managers report to the president of the firm, while functional managers report to their respective divisional general managers (review Figure 1). Coordination and linking between managers at the same level can be accomplished via their common boss. The common boss serves as an information channel, can exercise control over how much and what types of information move between groups, and can adjudicate problems that arise in his or her area.

The formal hierarchy is the simplest and one of the most pervasive formal linking mechanisms. Focused, sustained, and consistent behavior by the manager can both direct and set the stage for the effective coordination between organizational groups. The hierarchy is, however, a limited linking mechanism. Because of inherent cognitive/information-processing capacity, even modest amounts of task interdependence, exceptions, crises, or environmental uncertainty can overload the individual manager. When linking requirements begin to overload the first common supervisor (e.g., see earlier Shaeffer, Laffer examples), other formal mechanisms must be used to complement the manager as a linking mechanism.

Structural Linking

Liaison Roles.   At Federal Engineering, the development of process control systems requires the close coordination between the Instrument and Control Divisions. While some linking occurs via the hierarchy (i.e., through Shaeffer), much more intense problem solving occurs between two **liaison individuals**. John O'Connor and Phil Dinsky, respected members of the two divisions, are the point-men on the systems business. These two individuals serve as sources of information and expertise for problems and as contacts and advisors on systems work that affects their two divisions.

O'Connor and Dinsky represent formal liaison roles—formal roles that serve as information conduits and initiators of problem-solving endeavors deep in the organization. These liaison roles are responsible for enhanced information flows and coordination between units, although they rarely have authority to back up their positions. The liaison role is not usually a full-time responsibility but rather is done in conjunction with other activities (see Figure 6).

Cross-Unit Groups.   At Federal Engineering, the Air Force is a particularly important customer; Federal Engineering supplies the Air Force with a range of products and services. There are numerous Air Force complaints, however, of sloppy coordination and incomplete and/or inconsistent information among Federal Engineering professionals. To provide focused Air Force coordination between the different divisions, Shaeffer convened an Air Force integrating committee. This committee brought all key Federal Engineering actors together—a cross-unit group—to ensure a common posture toward the Air Force (see Figure 7).

The Air Force coordinating committee is

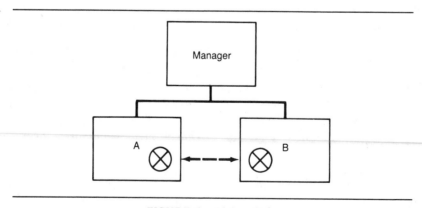

FIGURE 6.   Liaison Roles

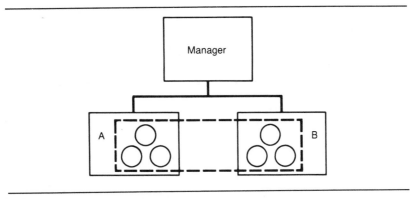

FIGURE 7. *Cross-Unit Groups*

one example of a range of possible group-based formal coordinating mechanisms. Groups made up of task-relevant representatives meet to focus on particular clients, products, markets, and/or problems. These groups can be permanent, temporary, or ad hoc. Their objective is to assure that relevant expertise comes together to deal with their joint task/problem.

In contrast with liaison roles, cross-unit groups provide a more extensive forum for information exchange, for coordination, and for the resolution of conflict between work units. Although these task forces, teams, or groups may form as need be, it may also be appropriate to design cross-unit groups into the structure if there are ongoing cross-unit projects. In a medical center, for example, a representative group of individuals from the key divisions might be responsible for establishing and adjusting guidelines and processes that affect work flows across divisions.

Integrator Role or Department. If problem-solving requirements increase and more decisions affecting multiple groups must be made at lower levels of the organization, teams and/or liaison individuals might not be sufficient.

Cross-unit groups may result in no one person feeling accountable for the total performance of the group. Conflicts sometimes arise within cross-unit groups or between liaison individuals, yet the first common boss might not have the expertise and/or time to adjudicate these differences. A solution to the need for real-time problem solving and for bringing general management's point of view to lower levels in the organization is to appoint an individual as **integrator**. This integrator role is responsible for taking a general management point of view in helping multiple-work groups accomplish a joint task, such as a specific product or project (see Figure 8).

Product, brand, geographic, and account managers are examples of formal roles created to bring a general management perspective to specialized managers who bring focused expertise, yet relatively narrow concerns to team meetings. Integrators have the formal responsibility of achieving coordination across the organization. While integrators report to senior management, they usually do not have formal authority to direct their functional and/or disciplinary colleagues. Because of this dotted-line relationship to members of their team, integrators must rely on expertise, inter-

477

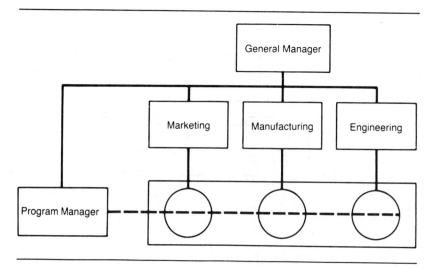

FIGURE 8.   Integrators (Project, Brand, Program, Account Managers)

personal competence, and team and conflict-resolution skills to shape the efforts of frequently recalcitrant team members.

Integrator roles must acquire functional or disciplinary resources to accomplish their work. When there are several projects, accounts, or products, each of these must compete with one another for resources. For example, in a functional organization, if there are five new product efforts, each of these must acquire scarce resources and attention from functional managers. To increase the power of the product/project organization and to help coordinate resources among products, a product development department is sometimes created (see Figure 9).

In product/project organizations, the product side of the organization has its own senior manager who reports along the same line as the functional managers. This senior manager formally represents the product side of the organization at senior levels and assists in resource allocation across projects. However, the functional organization still reports to

its functional supervisors and has a dotted-line relationship with the project/product manager. While our example has centered on project/product integrator roles, the role is quite general. It is a role to counter the consequences of strategic grouping, to achieve coordination and real-time problem solving at lower levels of complex organizations.

Matrix Structures. Some strategies require equal attention to several strategic contingencies, for example, products and markets or product and geography. Similarly, in highly uncertain environments with highly interdependent tasks, great pressures for coordination may come from both the functional and product sides of the organization. Whenever strategy requires the simultaneous maximization of several dimensions (e.g., product, market, geographic, time) and/or when information-processing requirements demand simultaneous consideration of several dimensions, integrating roles are not sufficient to handle the enormous information-processing requirements.

478

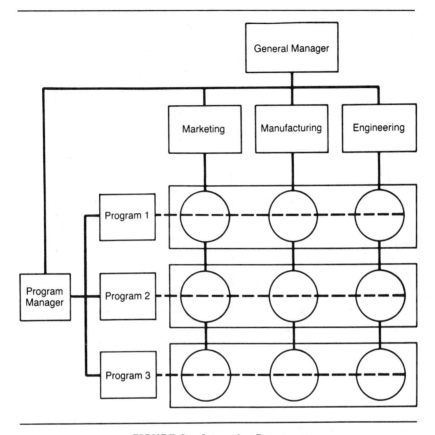

FIGURE 9. *Integrating Department*

When it is important to give equal attention to several critical contingencies and when information-processing demands are substantial, **matrix structures** are appropriate.

A matrix organization structurally improves coordination between multiple perspectives by balancing the power between dimensions of the organization and by installing systems and roles to achieve multiple objectives at once. For example, an R&D facility that wants to maximize disciplinary competence *and* product focus might invest in a matrix structure. Directors of the different laboratories would then report to both their disciplinary and product managers. The dotted-line relationship (seen in the integrator role) becomes solid; key members of the laboratory have two bosses.

Figure 10 presents a matrix organization structure. It has two chains of command. On the right side, the functional departments continue to exist. The organization still benefits from the information exchange and control provided by the grouping of people by function. On the left is another chain of command, with a product manager for each major new

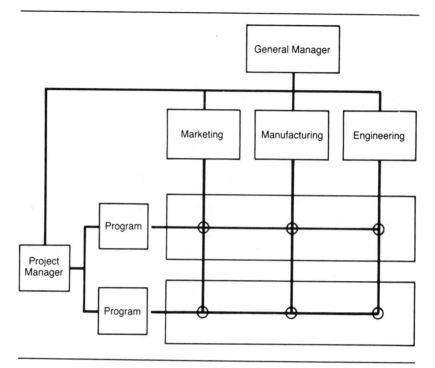

FIGURE 10.   Matrix Organization

product coordinating the activities of individuals across functional groups. Thus, those managers within each function who head product activities report to two bosses at once, a functional boss and a product boss. In this way, information is processed both within and across functional groups and coordination of different product-oriented activities is achieved.

Matrix structures are very complex. They require dual systems, roles, controls, and rewards. Systems, structures, and processes must be developed to handle both dimensions of the matrix. Further, matrix managers must deal with the difficulties of sharing a common subordinate, while the common subordinate must face off against two bosses. As seen in Figure 11, the general manager is the single boss, where each of the sides of the matrix come together. This individual must assure equal power and influence to each side of the matrix. Otherwise the organization will revert back to a single-focus organization. Below the matrix manager there is also a clear hierarchy. His or her subordinates report to one boss. The matrix is most directly felt by the matrix manager and the two matrix supervisors. It is this relatively narrow slice of the organization that really sees matrix systems, roles, procedures, and processes. This set of four roles must constantly balance the pressures and conflicts in a structure that attempts to work several strategic directions at once.

While the matrix structure is the most complex and conflictual linking mechanism, it is also the only structure that attempts to maxi-

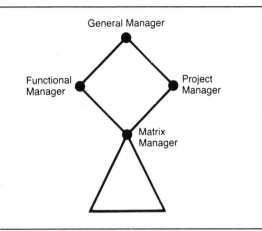

FIGURE 11.   *Matrix Organization: Another Perspective*

mize several strategically important considerations at once. Given its complexity and inherent instability, a matrix structure should be reserved for situations in which no other linking alternative is workable.[2]

Making Structural Linking Decisions

Costs and Benefits of Linking Options. Structural linking mechanisms vary by cost and resources utilized, their dependence on the informal organization, and their inherent information-processing capacities. The essence of making linking decisions is to choose those formal linking mechanisms that most effectively handle work-related interdependence. Using overly complex linking mechanisms will be too costly and inefficient, while using too simple linking mechanisms will not get the work done. The hierarchy of structural linking mechanisms can be evaluated along the following set of dimensions:

1. The cost and/or amount of resources devoted to each mechanism differs greatly.

The formal hierarchy or liaison roles require sustained attention to coordination by a few key individuals. Matrix structures, on the other hand, require dual structures, systems, and procedures. Matrix structures also require time, energy, and effort devoted to committees and teams that attend to both axes of the matrix. The more extensive the linking mechanism in terms of individuals involved, systems, and procedures, the more resources must be devoted and the more costly the linking mechanism.

2. Formal linking mechanisms also differ in their dependence on the informal organization. Where the hierarchy and some liaison roles rest firmly on the formal organization, cross-group units, integrator roles, and matrix structures depend more and more on a healthy informal organization. Those more complex linking mechanisms actually build on organization conflict. These linking mechanisms require an informal organization that can handle the ambiguity and conflict associated with substantial work-related interdependence. Indeed, without an informal organization

481

that deals openly with conflict, that has collaborative norms and values, and that can deal with the complexities of dual-boss relations, matrix organizations will not work. Thus, the more complex the formal linking mechanism, the greater the dependence on the informal system.

3. Finally, information-processing capacities of the various linking mechanisms are different. The hierarchy and liaisons are limited by individual cognitive limitations. These simple linking mechanisms deal well with simple interdependence but cannot deal with substantial uncertainty or complex work interdependence. Integrator roles, task forces, and matrix structures push decision making deep into the system and take advantage of many more resources and perspectives. These linking mechanisms allow for multiple points of view and real-time problem solving and error correction, and they are not dependent on individuals.

More complex linking mechanisms can handle more information and deal more effectively with uncertainty than can simpler linking mechanisms. Liaison roles can only relay limited amounts of informa-

tion and, while they can identify issues needing coordination, their ability to resolve conflicts is limited. Cross-unit groups can identify issues needing coordination and can involve the requisite number of individuals in inter-unit problem solving. Figure 12 compares information-processing capacity with cost and dependence on the informal organization for the set of structural linking devices.

Making Strategic Linking Decisions. Jean Shaeffer at Federal Engineering required a structure that would continue to produce top-quality scientific instruments *and* would compete more effectively for the systems business. Shaeffer decided to keep her product organization but to add a systems-business manager. Reflecting the importance of the systems business, this systems manager reported directly to Shaeffer and was fully responsible for the systems business. The systems manager was evaluated on systems business and acquired resources from relevant divisions within Federal Engineering. Divisional employees working on systems business were evaluated by their divisional managers as well as the systems manager.

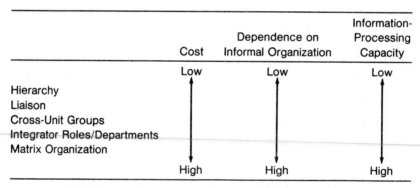

FIGURE 12.   *Consequences of Structural Linking Mechanisms*

Roger Laffer did some diagnostic work on his firm's technology transfer problem and discovered that there was no corporate vehicle to capitalize on technology transfer opportunities. While the divisions had plenty of ideas, there was no corporate instrument to evaluate and/or take action on those ideas. Laffer recommended the creation of a technology transfer board. This committee would be made up of senior technologists and divisional general managers and would be a corporate focal point for technology transfer. The committee was charged with evaluating, pushing, and funding technology transfer opportunities throughout the corporation. This technology transfer board was headed by a senior manager with both technological and market competence and was actively supported by the president and CEO.

Shaeffer and Laffer both faced problems in strategic linking. Shaeffer's organization had to deal with substantial work-related interdependence and considerable time pressure. Shaeffer's choice of a project organization with a powerful project manager reflects the demands of the reciprocal interdependence between divisions. Laffer's task force/committee solution to his firm's technology transfer problem reflects the pooled nature of the work interdependence and the weak-to-moderate perceived time pressure. Consistent with work requirements, Shaeffer chose a complex set of linking mechanisms, whereas Laffer chose a simpler, committee-based linking mechanism.

More generally, the problem in strategic linking is to choose those *sets* of formal linking mechanisms that effectively deal with work-related interdependence. The linking mechanisms discussed here are not mutually exclusive. Rather, managers must choose those sets of linking devices that are able to deal with the work-related uncertainty. For example, at Federal Engineering, Shaeffer must utilize the hierarchy, liaison individuals, cross-group teams, and project organization, all in service of the systems effort. Shaeffer added the project organization only because simpler linking mechanisms were not sufficient for the systems business. Thus, more complex formal linking mechanisms are utilized to deal with work interdependence with which more simple mechanisms are unable to cope.

Structural linking involves choosing the set of formal linking mechanisms that deals with work-related interdependence. Managers must balance the cost of more complex linking mechanisms with their increased information-processing capacity. Structural linking mechanisms must be extensive enough to handle information-processing requirements. Overly complex linking mechanisms will be costly to the organization in terms of time, money, energy, and effort. For example, a matrix organization to handle Laffer's problem would only create confusion and chaos in his technology transfer efforts. On the other hand, linking mechanisms that are not adequate to meet work, professional, or problem demands will result in poorly coordinated work. For example, if Shaeffer used a committee or task force to deal with the systems business, it is highly unlikely that the systems business would be taken seriously throughout Federal Engineering.

The choice of structural linking mechanisms should, then, be based on work-related interdependence. Those more complex linking requirements require more complex formal linking devices. Managers must choose those sets of linking mechanisms that match the information-processing demands of their unit's work interdependence. A mismatch between information-processing requirements and the choice of linking devices will be associated with relatively poor coordination and lower organization performance (see Figure 13).

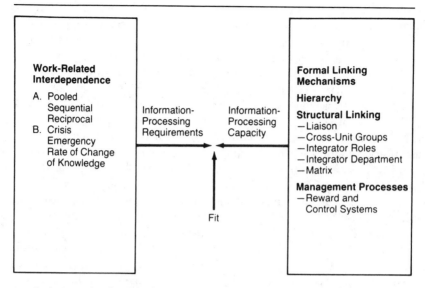

-Performance is a function of choosing that set of formal linking mechanisms to deal with work-related interdependence.

-As interdependence changes, so too should linking mechanisms change.

-Structural linking is required at multiple levels of analysis.

FIGURE 13.   *Strategic Linking: Formal Coordinating Mechanisms*

STRATEGIC LINKING IN ACTION

Formal linking mechanisms address the communication and information-processing needs that arise from grouping decisions. When grouping decisions both group together and split apart resources, linking mechanisms work to achieve coordination between interdependent units. As grouping decisions are made at several levels of analysis, so too are linking decisions. At Olympic Oil (review Figure 5), management must make linking decisions at the corporate, divisional, and functional levels. As each level deals with different degrees of work interdependence, the choices of structural linking mechanisms will differ. For example, the sequential interdependence at the corporate level can be handled via a senior team or committee; more complex linking mechanisms are required to deal with reciprocal interdependence within the division. Linking decisions, like grouping decisions, are important at each level of an organization.

Structural linking is an important managerial tool. Whereas strategic grouping decisions are single decisions made at each level of analysis, there may be a host of structural linking mechanisms within a particular unit. For example, in an R&D facility, structural linking might be accomplished via a matrix organization throughout the laboratory. Further, special task forces might be utilized to deal with the impact of new technologies on the organization, a top team might be convened to deal with a new competitive threat, and informal committees might be established to share expertise across disciplines within the laboratory. Linking, then, can be a powerful and flexible tool to deal with the different coordination requirements that exist within all organizations. Again, the choice of linking mechanism must be contingent on work-related interdependence.

Finally, as work interdependence shifts over time, so too should the choice of linking mechanisms. For example, if Laffer's technology transfer board (a simple linking mechanism) comes up with a viable candidate for internal development, the increased task interdependence between divisions will require some form of project organization to provide direction and real-time coordination efforts. Again, the choice of linking mechanisms must deal with the requisite task interdependence. As task interdependencies are not fixed, neither can our choices of structural linking mechanisms be fixed (review Figure 13).

MANAGEMENT PROCESSES: REWARD AND CONTROL SYSTEMS

Closely related to formal linking mechanisms is the design of formal reward and evaluation systems. Individuals are motivated by those factors that affect their formal as well as informal rewards. Individuals pay attention to those dimensions on which they are evaluated. As such, any formal linking mechanism must also be tied to complementary formal reward and control systems. If there is an inconsistency between structural linking mechanisms and reward patterns, coordination will suffer. For example, if a sales department is rewarded for sales volume and manufacturing is rewarded for gross margin, then these two groups are working at cross-purposes. No set of structural linking mechanisms can deal with these reward inconsistencies.

As with the choice of linking mech-

anisms, the choice of reward and control systems must be contingent on work requirements. More complex tasks require complex and subtle reward systems, while simple tasks require elementary reward systems. For example, at Federal Engineering, members of the systems team must be evaluated both for quality instruments as well as for their contribution to the systems business. These more complex reward systems must assess hard and soft criteria, both of which are critical for successful systems products.

Whatever the nature of the reward and/or control systems, they should (1) have clearly specified and operational objectives, (2) reward the total task as well as component tasks, (3) eliminate zero-sum situations, and (4) clearly link performance to valued outcomes. Bonus and incentive systems should be clearly linked to subunit performance and organizational performance. At both Federal Engineering and Laffer's office products company, the choices of structural linking mechanisms must be bolstered with formal reward and incentive systems that clearly evaluate and reward individuals for their contributions to the systems business and technology transfer.[3]

NOTES

1. Our linking ideas build on much earlier work, including J. Galbraith, *Designing Complex Organizations*, (Reading, MA: Addison-Wesley, 1973); L. Sayles and M. Chandler, *Managing Large Systems*, (New York: Harper & Row, 1971); L. Sayles, "Matrix Management: The Structure with a Future," *Organizational Dynamics* (Autumn 1976); R. Katz and T. Allen, "Project Performance and the Locus of Influence in the R&D Matrix," *Academy of Management Journal* 28 (1985) 67–87; A. Van de Ven, A. Delbecq, and R. Koenig, "Determinants of Coordination Modes Within Organizations," *American Sociological Review* 41 (1976) 322–37.

2. See S. Davis, and P. Lawrence, *Matrix,* (Reading, MA: Addison-Wesley, 1977) for an in-depth discussion of matrix organizations.

3. For much more detail on reward and control systems and their linkage to organization design, see E. Lawler, *Pay and Organization Development,* (Reading, MA: Addison-Wesley, 1981); R. Dunbar, "Designs for Organizational Control," in P. Nystrom and W. Starbuck, eds., *Handbook of Organization Design,* (New York: Oxford University Press, 1981); E. Lawler and J. Rhode, *Information and Control in Organizations,* (Santa Monica, CA: Goodyear, 1976).

Organizing for High-Tech Marketing

William L. Shanklin

John K. Ryans, Jr.

For many established companies, becoming more effective marketers is a matter of fine tuning—increasing expenditures on promotion and advertising or reorganizing the sales force. For high-technology companies, becoming effective marketers is often a matter of starting from scratch. Such companies typically derive their initial strength from the innovations provided by the research and development function. In the midst of a chaotic and fast-changing competitive environment, the importance of research and development cannot be overestimated. As the competitive environment becomes more orderly, the marketing function becomes important for high-technology companies.

The authors contend that high-technology companies can make a successful transition from being innovation-driven to being market-driven only by effectively linking the R&D and marketing efforts. Such linkages can take any number of forms, from special committees to close contact by key officials at all levels of product development, testing, research, and selling. Regardless of what specific form the linkages take, high-technology companies must be on guard to prevent animosity from creeping into the marketing-R&D relationship.

High-technology companies often have research and development capabilities superior to those of competitors, yet achieve only mediocre commercial success or fail completely. Are such companies simply in need of improved marketing skills? Not necessarily. Rather, they need to link R&D with marketing.

R&D or marketing prowess taken singularly, or even coexisting in the same organization, will not necessarily translate into financial success. Companies that appropriately link the two areas, though, can effectively anticipate, analyze, and exploit market opportunities. This article suggests how managers can forge such a linkage.

WHAT SETS HIGH TECHNOLOGY APART?

High-tech industries tend to be volatile at best, and as seen in the semiconductor, microcomputer, and robotics fields, a sorting-out of competition occurs quickly compared to less technology-intense endeavors. Several examples offer evidence of the ways revolutionary changes can occur in high technology.

Five years ago, a robotics trade show could have been held in a closet. Since then, the industry has grown to include dozens of competitors catering to a fast-growing and potentially huge market. Only recently have microcomputer companies begun to compete with one another on the basis of marketing skills and savvy. Before this industry's shakeout began, technological capability was the main success criteria; technological expertise and reputation carried Texas Instruments (TI) along for several years in home computers. Yet, once sophisticated marketing abilities became indispensable in the home computer market, companies dependent mostly on technological strengths were no match for IBM, Apple, and others. TI eventually withdrew from the home computer market in the face of sizable monetary losses. (It had suffered a similar fate some years before in digital watches.) Other market casualties in the infant home computer industry already include Panasonic, Tomy Corporation, Mattel, and Timex. Turbulence of this kind makes high-technology competition different from that in most other businesses.

One way to improve understanding of high-tech markets is to divide them into either supply-side or demand-side markets. In the former, technological progress is literally creating markets and demand. In this supply-side stage of market development, marketing strategy is highly entrepreneurial—formulated on sketchy market information and on intuition. The remarks of Akio Morita, CEO of Sony Corporation, typify the marketing philosophy of a prototype supply-side thinker:

"The newer and more innovative a product is, the more likely it is that the public might not appreciate it at the beginning. In 1950, our company marketed a tape recorder. Despite the fact that it was a great achievement and a technological innovation for us, at the time it looked like a toy to the general public. Nobody thought about recording speeches or using a tape recorder to learn languages. . . . In the case of an entirely new product, a market must be created."[1]

Supply-side markets are associated with what we refer to as innovation-driven high technology. The company's top strategic and marketing objective is to achieve profitable commercial applications for laboratory output. In this seminal phase of market development, R&D is the prime mover behind marketing's efforts. As an advertising agency executive specializing in high-technology products notes, high-tech efforts of this genre rely on a "presumptive need" rather than identification of buyers' desires.

Later, as high-tech markets mature, more normal demand-side conditions begin to prevail. In this market-driven kind of situation, R&D's task is to respond to the specific market needs identified by marketing, R&D, top management, and others. Here marketing's role is more traditional (i.e., more focused on advertising, pricing, distribution, and so on), much less entrepreneurial, and far more concerned with tight linkage between marketing and R&D. In brief, laboratory efforts are patterned after market objectives.

Today's high-tech companies must face the ongoing challenge of adapting organizationally and philosophically as their markets evolve from supply side to demand side. That evolution is usually not a smooth process. Early on, any market-share leader is in a precarious position. A continual threat exists that a giant (e.g., AT&T or Exxon) will enter into its industry or market niche. Also, a competitor's technological discovery or improvement, even one made by a start-up company, can make obsolete the leader's technology and the products or processes on which it is based. Some suggest, for example, that a tiny Findlay, Ohio-based holography start-up has outpaced two long-standing *Fortune* "500" entrants. In fact, one day's leader, perhaps an Osborne Computer, can become the next day's Chapter 11 or state-of-the-art victim.

Executives successful in an innovation-driven situation may have difficulty adjusting to a market-driven environment. These managers fail to reorient themselves so that they can compete effectively under the new conditions.

Several years ago we encountered a start-up consumer-products company that had received substantial backing from venture capitalists. This funding was based largely on the president's previous development of an eminently successful high-tech industrial product. However, he had sold the company producing the product while it was still in the innovation-driven stage of market development, and thus he had never directed a company in a competitive environment.

The president and others who ran the new consumer-products company had engineering backgrounds. They had little marketing experience; indeed, they were contemptuous of marketing activities and refused to commit any funds to explore consumer sentiment, potential market demand, and possible advertising themes. Not surprisingly, the company has yet to earn a dollar of earnings and is on the verge of bankruptcy. The technical skills and acumen that served the president so well before have been much less relevant to the different demands in a market-driven milieu.

We believe that management should not view marketing and R&D as an either-or strategic proposition. Just as there are technically trained top executives who are also good marketers, there are marketing professionals with no technological training who successfully help guide high-tech companies. A few high-technology companies, such as Apple Computer, have even looked to consumer-goods marketing professionals to provide marketing leadership.

DISTINGUISHING BETWEEN HIGH-TECH COMPANIES

"High tech" has become a buzzword to describe everything from the space shuttle to the electric frying pan. In our judgment, businesses must meet three criteria to be labeled "high technology":

1. The business requires a strong scientific-technical basis.
2. New technology can quickly make existing technology obsolete.
3. As new technologies come on stream, their applications create or revolutionize markets and demand.

For the purpose of discussing marketing-R&D linkage, the most important distinction is between what we have called market-driven and innovation-driven high technology. Market-driven high-technology companies assign R&D the task of producing innovations that meet specific market objectives. By contrast, for innovation-driven high-technology companies, what customers need or want is residual. After the R&D breakthrough is made, customer needs or wants are considered. G.D. Searle's low-calorie sweetener, Aspartame, was discovered accidentally by a lab researcher who was involved with a quite different research project. Only then was the commercial application made. We can differentiate further between market-driven and innovation-driven companies.

Market-Driven Companies

These companies fall into one of two groups:

1. **The state-of-the-art-plus group.** R&D advances move deliberately as competitors turn modifications and improvements in existing technology into incremental advantage. In the area of robotics, for in-

stance, new steps often occur from customer suggestions. As DeVilbiss Company robotics division marketing executive Timothy Bublick explains, "You develop a robot spray-gun attachment to paint for a specific customer and then it [the modification] becomes a product adaptation to the current state of the art."

If followed exclusively, such a deliberate approach could result in delays in technological breakthroughs. In the semiconductor field, the industry's ability to find ways to cram more transistors into a single chip may forestall advances in optics.

2. **The problem-solving group.** These companies do not restrict themselves to current state-of-the-art techniques. As the first high-tech marketer, Thomas Edison, observed, "First, be sure a thing is wanted or needed, then go ahead."

Seconding his view, Edward W. Ungar, head of Battelle Memorial Institute (one of the world's foremost contract R&D organizations) remarked, "In corporate R&D, most ideas for new products need to be evaluated against the test of whether or not the product will be accepted in the competitive marketplace." In applying this problem-solving orientation, Schering-Plough is coupling its discovery abilities with its marketing strengths. Schering-Plough delicately balances a heavy R&D effort with a practical view of how it will position each scientific breakthrough to meet the company's specific long-run objectives.

ing research are companies like Biogen and Genentech, along with a number of the more market-driven pharmaceutical and chemical companies. The possible applications of biotechnology are so numerous that the initial battle has been waged for basic patents.[2] While forecasters talk of developments ranging from disease-free grains to microorganisms for gobbling up oil spills, the range may be so vast that some companies may delay or ignore entirely important commercial applications.

In addition, U.S. military and space research has made possible sundry products for consumer and industrial markets. The resulting nonmilitary by-products often have little resemblance to the original applications. Already on the market, for example, in Pizza Time Theatres, are video games that use laser technology; the fruits of Singer Aerospace research are in its sewing machines and handheld power tools. In effect, market-driven research for the military and space programs is innovation-driven when it comes to consumer and industrial markets.

Market planning that explicitly recognizes and accounts for the strategic distinction between market-driven and innovation-driven research goes a long way toward yielding better corporate performance. Establishing realistic expectations early on mitigates cross fire later between R&D and marketing about "who is at fault" when a project fails to yield products that are readily marketable. Moreover, it gives solid direction to the marketing research people who seek the strongest commercial applications for a technological advance.

Innovation-Driven Companies

Many high-tech companies, laboratories, or divisions or subsidiaries of major corporations are essentially research businesses. For example, companies of all types and sizes are currently giving attention to biotechnology.

Among those pitted in basic gene-splic-

LINKAGE ROLES FOR MARKETING AND R&D

Managers of high-technology companies like to think of themselves as being market-driven. Three-fourths of the executives we surveyed

RESEARCH METHODOLOGY

We collected the data for this article via in-person and telephone interviews, questionnaires, and studies of specific marketing areas, such as personal selling, distribution channels, and advertising in high-technology companies. In total, we examined over 125 companies.

Carefully, we selected companies for their successful financial and market track records in high technology—companies of all sizes that have proved themselves in marketplace competition. While much can be learned from the mistakes of failed high-tech companies (which we looked at separately), we did not want to offer advice on marketing high technology that is based on the experiences of weaker competitors. We also wanted to avoid validity and reliability problems of studying in-depth only a small number of companies in a few industries or, alternatively, of studying a cross-section of many companies in too cursory a way.

We interviewed high-ranking executives from key companies in such industries as robotics, computers, software, medical instruments, biotechnology, military systems, and ceramics. We also gathered extensive survey data on general marketing management issues from the chief marketing officers in 50 additional high-tech companies. To get more specific information on three marketing areas, we conducted separate surveys and interviews.

By design, we studied personal selling within a single company, distribution channels within an industry, and advertising in a cross-section of high-technology industries. This research procedure allowed us to achieve a balance among company-specific, industry-specific, and industrywide perspectives. The company-specific study focused on the high-tech business-to-business sales force of a single, but premier, comprehensive computer producer-marketer. We directed the industry-specific inquiry to robotics (more than one-fourth of the major U.S. manufactures in this industry provided data). The advertising study included one-half of the top high-technology industrial advertisers and a sampling of high-tech consumer-product companies.

indicate that their new product ideas typically result from specific responses to market opportunities, rather than from R&D initiatives. (For details on our research, see the insert.)

In its 1982 annual report, German-based Nixdorf Computer stated candidly, "[Nixdorf] maintains two technology centers [Silicon Valley and Tokyo] which monitor . . . trends in fundamental research and user-oriented technology and assess their potential benefits for

Nixdorf's own R&D efforts." Thus the company uses the marketplace and competitor developments to provide its window on technology.

Another example of market specificity is Genetic Systems Corp.'s approach in writing lucid market goals guiding its R&D efforts over the short-, intermediate-, and long-term. Its short-term goals identify the need for products to diagnose infectious diseases and cancer. Intermediate goals focus on automated products to identify blood types, and its long-term goals pertain to products for treating infectious diseases and cancer.

As shown in the Exhibit, in market-driven high technology, the main direction for R&D is from marketing. R&D's reaction (dotted arrow) comes in the form of guidance on what is technically feasible and ideas from scientific circles. Formal marketing research, typical to consumer and industrial markets, is helping high-tech managers guide R&D. For example, our data show that executives extensively use traditional research techniques, such as concept testing, product prototypes, and market tests. Over 90% of the executives in our study find that concept testing is helpful in forecasting the success of potential new products.

Innovation-driven high technology offers a marked contrast, as R&D provides the stimulus and marketing officials must find applications or simply sell the product. These efforts can help create new markets by applying lab breakthroughs to largely unperceived buyer needs. A latent demand for in-home pregnancy tests may have existed for centuries, but biotechnology made these tests feasible and inexpensive. Researchers have also been successful in deriving other low-cost diagnostic tests for hepatitis, prostate cancer, and venereal disease. Yet only a few years ago the commercial potential of the underlying technology appeared poor. Certainly, few businesses would have bet that such diagnostic tests would lead to a major new medical market so soon.

Our research indicates that innovation-driven high-technology companies rely on qualitative marketing research techniques. Their managers place little stock in the mathematically based methods of marketing research that more mature companies use—methods requiring an abundance of data from a representative sampling for drawing statistical inferences.

Innovation-driven companies' reluctance to use quantitative approaches stems

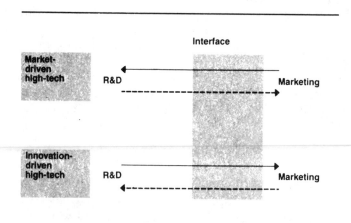

EXHIBIT.   *Conceptualizing the High-Technology Marketing-R&D Interface*

from the small amount of useful historical data. And primary data from traditional interviews of prospective buyers or users are of dubious value in answering questions about possible products or applications resulting from new technologies. In fact, some 85% of the executives in our research report using focus groups to qualify new products.

WHAT MARKETING WITH WHAT R&D?

As the organizational complexity of a company increases, the possibilities for linking marketing and R&D in profitably innovative ways become more intriguing. Our studies have led us to identify four general types of high-technology companies, according to their degree of organizational complexity:

1. The high-technology company that is not part of a larger corporation and is engaged basically in one industry (for example, Prab Robots). It may be a fledgling venture or a market leader.
2. A traditional consumer or industrial enterprise with one high-tech division or subsidiary. The corporation's primary business is not high technology, but it has diversified into high-tech via start-ups (Goodyear Tire & Rubber and Goodyear Aerospace) or acquisitions (Champion Spark Plug and DeVilbiss).
3. A high-technology division or subsidiary that is but one of several or many such units in a company. The corporation is almost exclusively (Hazeltine) or partly (United Technologies) involved in high technology.
4. Companies competing in the same industry forming a cooperative R&D venture. This arrangement prevents duplication of each other's research efforts. For instance, semiconductor and computer concerns have formed the Microelectronics Computer Research Corporation in Austin, Texas, to derive technologies for use by the member companies and for licensing the technologies to nonmembers.

In the first and second types of companies, the linkage problem is straightforward, but, in the third and fourth types, marketing and R&D appear at many places and levels in the organization. They may even be directed from corporate headquarters.

We suggest that management ask the question "what marketing with what R&D?" again and again. A high-tech division in a diversified corporation may be able to give a sister division in a different industry or in a different market the technology it needs to develop new market niches or entire markets. Marketing-R&D linkage between and among divisions as well as within divisions can stimulate ideas for new products and commercial applications. The rewards of cross-pollination of ideas among divisions come from triggering previously undiscovered market applications or correcting competitive shortcomings in product quality.

Goodyear Aerospace, the high-technology subsidiary of the Goodyear Tire & Rubber Company, has at least twice shared technological know-how from Aerospace's Defense Systems Division with its sister division, the Aircraft Wheel and Brake Division, to solve several of the latter's competitive problems. Defense Systems offers defense-oriented electronics to the U.S. military. In contrast, Aircraft sells aircraft wheels, brakes, and brake-control systems to both commercial and military aircraft manufacturers. In the early 1970s, Defense Systems transferred integrated circuit technology to Aircraft. Later, at the end of the 1970s, Defense Systems transferred digital design technology. In both instances, the technology shared had positive effects on Aircraft's market competitiveness.

An official of Corning Glass Works provided us with a better illustration of a diverse cross-divisional application. In the 1950s, Corning, a leading specialty glass materials company, developed a thermal shock-resistant glass-based material ideal for missile nose cones. It was able to withstand the heat and stress of going from ground zero to space and atmospheric re-entry, even at supersonic speeds. While this in itself was a major break-through, the late R. Lee Waterman, former CEO of Corning, saw a different and unique application for this material (trademarked PYRO-CERAM). In 1958, he introduced versatile Corning Ware cookware that could go directly from the freezer to the oven or stovetop. However, the story does not end there. Some 20 years later, another property of this original nose cone material—its transparency to radio signals—permitted an extension of its consumer application as an ideal cookware for microwave ovens.

ESTABLISHING THE INTERFACE

Marketers in high-tech businesses need to know what R&D is developing today and on the horizon, so they can realistically analyze and understand their planning constraints. The linkage in market-driven high technology comes primarily through R&D's active participation in the market planning process, especially in the objective-setting stage. In market planning, R&D can guarantee that marketing does not lose sight of R&D's vision for the product. The corollary is that marketers can offer parameters for the researchers' efforts. Through the give-and-take of setting objectives, R&D and marketing can agree on the target market, priorities, expectations, and timing.

The linkage in market-driven situations needs to be formal and carefully designed; a high-tech marketing team is a must. Face-to-face, in-person interaction and an agenda for

meetings are most productive. Marketing and R&D people should talk almost daily during the initial market planning effort for a new product or application and regularly thereafter for updating and revision. To avoid later misunderstandings, both R&D and marketing should agree on and write the goals and objectives of the marketing plan. Our research finds that R&D's actual involvement in formulating marketing strategies and tactics is best off limited to a technical-consultative role. The approach of Micom Systems, Inc. illustrates:

Micom, a microcomputer company formed in 1980, has grown from $5 million to $83 million in annual sales. For the first three years, Micom followed the all-too-usual corporate path of not linking marketing and R&D, and according to Steve Frankel, vice president of marketing and development, R&D and marketing were competing. To build the team orientation needed, the two activities merged, as Frankel explains:

"My organization, Marketing and Development, has the corporate responsibility for all marketing (product planning, marketing management, promotional input, and manuals) and development (all engineering and R&D) activities performed by the corporation. Within Marketing and Development, we have three assistant vice presidents who manage, on a product-line basis, the marketing and development activities of their respective product areas. Within each product-line organization, we have (an) individual marketing and development team. . . . At Micom, marketing and development groups work together to define new directions and solutions to the needs of their specific product areas. Once agreement is made, however, . . . the marketing activity proceeds with the classical marketing functions while the engineering team begins the engineering process.

"As the project proceeds, project status information is available to both groups, and in-house quality and functional testing, beta-test site testing, and product-user manual reviews are also joint activities. So, in summary, the marketing and development teams work together on the front and tail end of projects, and work independently from project go-ahead to the product testing and evaluation phase. Of course, if the development project does not proceed according to plan (i.e., cost or schedule) or market conditions change, we jointly reevaluate our position."

The interface needs of innovation-driven high-technology companies are different. The possibilities for applications may be less obvious or so numerous that the company must establish priorities for exploitation. Biotechnology again provides an example.

Marketable opportunities for gene-splicing seem vast, including cancer therapy, disease-free orchids, and who knows what else. Therefore, much of the early interface efforts should address such questions as what industry is the company in, what are the conceivable market opportunities, and what are the market development priorities? Once these are answered, the types of linkage guidelines appropriate for market-driven high technology apply to innovation-driven high technology as well.

Finally, consider the linkage needs of the multidivision major corporation with market-driven and innovation-driven interface demands and, perhaps, a portfolio of high-tech and low-tech strategic business units (SBUs). In this kind of organization, the marketing-R&D interface becomes a two-stage process. Within each SBU, either a market-driven or an innovation-driven interface is appropriate, but the creative linkage process should not end there. A vehicle for cross-pollination of ideas and technologies among the various SBUs is vital, but may be difficult to implement. Although company-specific recommendations are needed, possibilities might include both the establishment of a conventional corporatewide marketing-R&D committee or project team and a more experimental approach such as in-house

trade shows for the company's SBUs to demonstrate and publicize otherwise proprietary knowledge and offerings to one another. The following example of Allied Corporation illustrates a cross-pollination effort:

Top management at Allied Corporation is seeking to eliminate the historical barriers between R&D and marketing. For example, new product venture groups have formed and major laboratory facilities are consolidated at corporate headquarters. As a cornerstone, Allied has created an in-house office of science and technology, which brokers technology; it will "match 'clients' in operating companies with technology in the labs."[3]

Again, as with Goodyear Aerospace and Corning Glass, we see a multidivisional company that has not stopped with the question of how to link marketing and R&D *within* business units. Allied's technology broker will promote new marketing-R&D linkage opportunities by asking, in effect, "How about the commercial market opportunities of matching Division X's technological know-how with the needs of Division Y's marketing?"

WHO BELONGS ON THE HIGH-TECH TEAM?

The makeup of the marketing-R&D team often depends on whether the product or application is in development or being sold. In development, the company focuses on product and market planning, from conception through actual introduction. Our data note that marketing's consulting is prevalent in the research and development phase, but, contrary to some views, not to the detriment of R&D.

In successful companies, both of these functions work in tandem with each other throughout production. Additionally, the top executive in the perceptive high-tech company stays involved in every major development decision. Apple Computer's president, John Sculley, is right on target; he has mandated that Apple's top management pay close attention to product development so that more disciplined market strategy will result.

The high-tech development strategy group consists of the CEO or president, marketing, R&D, and production in roughly three-fourths of the companies we sampled. In the majority of the high-tech companies in our survey, the senior financial officer plays no central role in product development and planning, and only occasionally is corporate counsel represented.

So, with few exceptions, marketing and R&D are members of the key development decision-making team. We were not surprised, then, to see a company like McDonnell Douglas Automation Company (McAuto), whose main business is CAD/CAM time sharing, merge the formerly separate areas of marketing and R&D, enabling the company to establish a development-to-marketing line for technical products.[4]

Just as marketing is heavily involved in R&D during precommercialization, R&D participates in marketing activities during the selling stage. The high-tech marketers surveyed register strong dissent to the notion that the R&D people's involvement with a new product ends when it is turned over to marketing; 85% encourage a continuing role for R&D.

Clearly, R&D's efforts do not end once the selling begins. What form might R&D involvement on the marketing side take? Our findings show that R&D is active in a number of disparate marketing functions, including preparation of brochures and technical manuals, marketing research, and, to a lesser extent, consultation in public relations, advertising and sales promotion, and pricing. R&D's participation in trade shows is direct; over one-third of the companies we surveyed have R&D representation at trade shows. We witnessed this at the 13th ROBOTS 7 Exposition in Chicago in April 1983, where R&D people from many of the exhibitors held discussions with prospective buyers. As a corollary, our research reveals that in the robotics industry, about one-fifth of the companies involve the R&D department in selling the product.

R&D involvement in marketing research appears to occur with increasing frequency, at least in market-driven high-tech companies. For example, Masahiko Kajitani, the general manager of video planning at Matsushita Electric Industrial Company, has said that companies should avoid overdelegating marketing and consumer studies to people uninvolved in product development. General Electric sends out design engineers to obtain customer opinions on GE's electronic wares.[5] Further, many companies consider the introductory sales stage a quasi-test market and use the resulting feedback from customers to modify the product. In such instances, obviously, R&D remains involved.

Besides R&D officials, who else is on the interface team during the selling stage? The composition may vary by high-tech industry. Our inquiries reveal that applications support personnel are directly represented in the selling efforts of more than three-fourths of the robotics companies. Thus, their presence is also appropriate for the interface group. Middlemen, such as industrial distributors or agents, may also be involved in the robotics companies' linkage efforts.

Similarly, in the medical field, customer-service people are sometimes part of the marketing team and the linkage activity. Backing up the high-tech marketing team are the traditional specialized support organizations, mainly advertising agencies and marketing research firms, as well as other consultants. The advertising agency is more integrally involved in the company's total marketing efforts than in typical consumer marketing, where in-house marketers chart the course. In the opinion of

Richard Reiser, the California-based advertising executive, "True marketing—marketing strategy and positioning, not sales alone—often only occurs when the advertising agency sits down to plan a campaign."

WHEN MARKETING AND R&D GO THEIR OWN WAYS

We hear repeatedly that the lack of integration of marketing and R&D is a major obstacle to market success in high technology, and where an interface does exist, it is an inherently adversarial situation in which lab jockeys and pitchmen do not mix.

When we asked executives in a broad spectrum of high-technology companies for their opinions, more than three-fourths think that most companies make some effort to link marketing and R&D. Moreover, they see product planning as both an R&D and a marketing function. Also we find that high-technology marketing people have a predominately scientific background. Ostensibly, then, marketing and R&D managers have little reason to fail to communicate effectively with one another. Even so, we find that power plays or strained relations do frequently occur and that marketing-R&D linkage suffers in the process. A widely publicized example is the friction at the newly deregulated American Telephone & Telegraph Company. Intense infighting for control between AT&T's marketing people and its technically oriented Western Electric group has become so bitter that the company has already lost key members of its cadre of marketing executives. Turning a longtime technology-driven regulated monopoly into a market-driven competitor is a formidable chore.

Fortunately, lack of sufficient interface in most companies usually results from more benign organizational factors like geographical separation and the difficulty of melding high-tech subsidiaries with a company's traditional product offerings. But geography and product line differences are not insurmountable obstacles if a company works at linkage. TRW Inc., a diverse and geographically dispersed manufacturer headquartered in Cleveland, infused marketing concepts into its West Coast high-technology group while instilling a high-technology orientation in its traditional smokestack operations in the Midwest.

TRW's situation is not unusual. Often a company's headquarters, by design or through acquisition, is located some distance from its high-tech unit or subsidiary. For strategic reasons, the company may have placed the high-tech group in a research environment (Silicon Valley, Route 128, or Research Triangle), or an attractive site (Aspen or Austin, for example), or at some distance from headquarters' distractions. With such real or perceived obstacles, roughly half the executives we surveyed see R&D as having limited meaningful involvement in marketing planning.

What can happen if a company limits R&D participation principally to the research arena? Mark Frantz, president of the rapidly growing Frantz Medical Development Ltd., comments, "Our business is built on the fact that so many companies do not involve their inventors in the final development of the product and the marketing side of the business." Frantz's company obtains promising medical patents and, with the inventor acting as product champion, produces and markets these innovations. This example of a company taking advantage of corporate-research separation highlights the need for ongoing top management attention to the linkage problem.

We believe that marketing and R&D linkage is achievable, regardless of company size and complexity. Undoubtedly, top management's direct input is necessary if either marketing or R&D is subordinate or totally removed from the other's decision making. Top management must carefully orchestrate these first steps to ensure that the linkage has the appropriate priority. Once set, each marketing-R&D group will be concerned not only with market planning but also with offering directions for new research and applications. As we have emphasized, large multiple-division companies need to develop systems that ensure consideration of all possible applications of new technology.

NOTES

1. "Creativity in Modern Industry," *Omm,* March 1981, p. 6.
2. Tamar Lewin, "The Patent Race in Gene-Splicing," *New York Times,* August 29, 1982.
3. "Allied After Bendix: R&D is the Key," *Business Week,* December 12, 1983, p. 79.
4. Philip Maher, "CAD/CAM Vendors Plot a New Course," *Business Marketing,* April 1983, p. 66.
5. "Listening to the Voice of the Marketplace," *Business Week,* February 21, 1983, p. 90.

Innovation Processes in Multinational Corporations

Sumantra Ghoshal

Christopher A. Bartlett

INTRODUCTION

It was over twenty years ago that Vernon (1966) proposed the product cycle theory that identified the ability to innovate as the raison d'être for multinational corporations (MNCs). Over the last two decades, many new theories have been proposed to explain why MNCs exist, but innovations have continued to occupy the center stage in all the diverse and eclectic approaches (see Calvet 1981 for a brief review). The strength that allows a firm to invest and manage its affairs in many different countries is its ability to create new knowledge—to innovate—and to appropriate the benefits of such innovations in multiple locations through its own organization more effectively than through market-mediated mechanisms (Buckley and Casson 1976; Rugman 1982).

While theories of the multinational firm

highlight the importance of innovations for the existence of such organizations, the emerging phenomenon of global competition (Hout, Porter, and Rudden 1982; Hamel and Prahalad 1985) has made innovations even more important for their survival. While traditionally many MNCs could compete successfully by exploiting scale economies or arbitraging imperfections in the world's goods, labor, and capital markets, such advantages have tended to erode over time. In many industries, MNCs no longer compete primarily with numerous national companies, but with a handful of other giants who tend to be comparable in terms of size, international resource access, and worldwide market position. Under these circumstances, the ability to innovate and to exploit those innovations globally in a rapid and efficient manner has become essential for survival and perhaps the most important source of a multinational's competitive advantage.

In sharp contrast to this practical importance, the topic of innovations in MNCs has received relatively little research attention. Not one of the over 4,000 studies on the topic of innovations (for references, see Gordon et al., 1975; Kelly and Kranzberg 1978; Mohr 1982) has focused specifically on the innovation process in the setting of a multinational corporation. Similarly, in the field of management of

"Innovation Processes in Multinational Corporations" by Sumantra Ghoshal and Christopher A. Bartlett. Copyright 1987 by Sumantra Ghoshal and Christopher A. Bartlett. Reprinted by permission.

We are grateful to the Division of Research, Harvard Business School for funding the research project of which this is a preliminary and partial report. We have benefited significantly from the comments and suggestions of Louis T. Wells, Howard Stevenson, Joseph Bower, Nitin Nohria, Dominique Heau, Susan Schneider, and two anonymous reviewers of the *Strategic Management Journal.*

MNCs, past research has focussed overwhelmingly on strategy, defined implicitly as the way to enhance efficiency of current operations (see Ghoshal 1986b for a review), or structure, with most attention paid to the determinants of headquarters-subsidiary relations as opposed to their consequences. While some efforts have been made to investigate certain isolated aspects of distributed research and development in MNCs (Ronstadt 1977; Terpstra 1977), the issue of management of innovations has remained peripheral to research on the topic of management of multinational corporations.

THE STUDY

This paper is based on some of the findings of a recently concluded study of innovations in nine large MNCs, viz., Philips, GE, and Matsushita in the consumer electronics industry; L.M. Ericsson, ITT, and NEC in the telecommunications switching industry; and Unilever, Procter and Gamble, and Kao in the soaps and detergents industry. The choice of these industries and companies was based on the logic of maximum variety—the three industries represented very different requirements in terms of local responsiveness and global integration (Prahalad 1975; Porter 1986); within each industry, the selected firms were comparable in terms of size and strategic positions but, because of the differences in their national origins and administrative histories, had very significant differences in their organizational forms and processes (for descriptions and illustrations of these differences, see Bartlett and Ghoshal 1987).[1]

In each of these companies, we tried to identify as many specific cases of innovations as possible and to document their histories in the richest possible detail. To this end, 184 managers of these companies were interviewed, both at the corporate headquarters and also in their national subsidiaries in the United States, the United Kingdom, Germany, Italy, Japan, Singapore, Taiwan, Australia, and Brazil. None of the interviews lasted less than an hour and some took place during multiple meetings involving up to eight hours. We also collected and analyzed relevant internal documents relating to the histories of these innovations. This effort led to identification of thirty-eight cases of innovations for which we could reconstruct fairly extensive histories. While the descriptions possibly suffered from the well-known biases of historical reconstruction, we made all possible efforts to cross validate the stories thorugh multiple sources and eliminated from the list those cases where different respondents differed significantly in their narration of the sequence of events that led to the innovations. These thirty-eight innovation cases constitute the primary data base for this report.

FOUR DIFFERENT INNOVATION PROCESSES

The innovation process[2] is one of the most complex of all organizational processes, and any stylized representation of this complexity cannot but be guilty of oversimplification. However, past research has suggested a generic stages model, shown in Figure 1, that views the innovation process as consisting of three se-

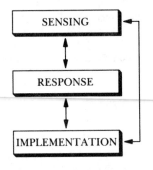

FIGURE 1. *A Model of the Innovation Process*

quential but also interacting sub-processes of sensing, response, and implementation.[3]

To innovate, a firm must sense changes that may demand adaptation or allow exploitation of an internal capability. The acquired stimuli must then be addressed through the firm's response mechanisms: technologies and products must be developed, processes must be improved or adapted, or an available capability must be converted into a functional form that satisfies a latent, emerging, or existing demand. Finally, the innovation must be exploited through efficient and effective implementation.

As suggested earlier, this is a highly simplified representation of a complex organizational process. In practice the different stages may be neither as discrete, nor as neatly sequential (Gross et al. 1971; Ginzberg and Reilly 1957). In any specific case, it may be extremely difficult to specify where the sensing process ends and the response process begins, or at what point the implementation phase may be said to have commenced. Similarly, the sequence suggested in the model, while logical, is not an invariant order of events. In reality, the process may be much more iterative, or even circular, with a high degree of interaction among all the three stages (Zaltman et al. 1973).

Despite its simplicity, the model provides a useful starting point for analyzing the administrative tasks of organizing for innovations. To innovate, a firm must develop appropriate capabilities to sense, respond, and implement. But just the capabilities are not enough; the firm must also create appropriate linkages to tie these capabilities together so that they function in an integrative manner. These two dimensions, viz., the configuration of organizational capabilities and the nature of their interlinkages provide, in Roethlisberger's (1977) terms, a "walking stick" for exploring the phenomenon of innovation-organization links.

Table 1 summarizes our analysis of the thirty-eight innovation cases in terms of this process model.[4] For each case, we identified the administrative unit or units that carried out the sensing, response, and implementation tasks. This analysis revealed four different patterns in terms of the locations where the three tasks were carried out and, hence, in terms of the interlinkages among organizational units that were required to create and implement the innovation. Each of these patterns represents a different organizational process; collectively they suggest a scheme for classification of innovation processes in multinational corporations.[5] The table represents this classification scheme, and groups the thirty-eight innovation cases according to these categories, each of which is described and illustrated in the following pages.

The Center-for-Global Innovation Process

Center-for-global innovations are those where the center, i.e., the parent company or a central facility such as the corporate R&D laboratory, creates a new product, process, or system for worldwide use.[6] Most instances of center-for-global innovations that we came across in the course of our study were technological innovations but they were spread around a wide spectrum from minor modifications to substantial reorientations (Normann 1971). Most of the cases involved no participation of the national subsidiaries except for relatively routine tasks such as marketing support or nominal assembly at the implementation stage. In some others, one or more national organizations also contributed in relatively minor ways in the sensing process, while the response task, in all cases, was entirely carried out at the center. The process by which L.M. Ericsson, the Swedish manufacturer of telecommunications switching and terminal equipments, created the AXE digital switch is one example of this innovation process.

TABLE 1.   *Innovation Processes in Multinational Corporations*

Innovation Process	Description of Process (locations where different tasks are carried out)			Number of Cases Observed
	Sensing	Response	Implementation	
1. Center-for-global	At the center (occasionally, some input may be provided by a particular national subsidiary)	Always at the center	In a number of organizational units world-wide	13
2. Local-for-local	In a particular national unit	In the same national unit	In the same national unit	11
3. Local-for-global	In a particular national unit	In the same national unit (possibly with some minor help from the center)	Initially in the national unit, subsequently in many units in the world-wide organization of the company	8
4. Global-for-global	Many organizational units, including the center and a number of national subsidiaries	Many organizational units, including the center and a number of national subsidiaries	A number of organizational units world-wide	6

Impetus for the AXE came from early sensing of both shifting market needs and emerging technological changes. The loss of an expected order from the Australian Post Office, combined with the excitement generated by the new digital switch developed by CIT-Alcatel, a small French competitor virtually unknown outside its home country, set in motion a formal review process within Ericsson's headquarters. The review resulted in a proposal for developing a radically new switching system based on new concepts and a new technology. The potential for such a product was high, but the costs and risks were also enormous. The new product was estimated to require over $50 million and about 2,000 man-years of development effort and take at least 5 years before it could be offered in the market. Even if the de-

sign turned out to be spectacular, diverting all available development resources during the intervening period could erode the company's competitive position beyond repair.

In sharp contrast to almost all the "principles of innovation" proposed by Drucker (1985), corporate managers of Ericsson decided to place their bet on the proposal for the AXE switch, as the new product came to be called. The process they adopted was not "incremental" (Quinn 1985), unless the term is so defined as to be all encompassing. A detailed, event-by-event documentation of the history of the switch by a key participant in the development process (Meurling 1985) shows little "controlled chaos" but rather the deliberateness and commitment of a programed reorientation (Normann 1971). The company provided full authority and all resources so that Ellemtel, the R&D joint venture of Ericsson and the Swedish telecommunications administration, could develop the product as quickly as possible. For over four years, the technological resources of the company were devoted exclusively to this task. The development was carried out entirely in Sweden, and by 1976, the company had the first AXE switch in operation. By 1984 the system was installed in fifty-nine countries around the world.

Not all the cases of center-for-global innovations that we documented were equally effective. NEC, for example, designed the NEAC sixty-one as a global digital switch and developed it through its traditional centralized development process. However, while the Japanese engineers at the corporate headquarters had excellent technical skills, they were not familiar with the highly sophisticated and complex software requirements of the telephone operating companies in the United States, the principal market at which the product was aimed. As a result, while the switch was appreciated for its hardware capabilities, early sales suffered because the software did not meet some specific end user needs that were significantly different from those of Japanese customers.

The Local-for-Local Innovation Process

Local-for-local innovations are those that are created and implemented by a national subsidiary entirely at the local level. In other words, the sensing, response, and implementation tasks are all carried out within the subsidiary. Most cases of such innovations that we came across tended to be market led rather than technology driven and usually involved only minor modifications of an existing technology, product, or administrative system.

The ability of its local subsidiaries to

sense and respond in innovative ways to local needs and opportunities has been an important corporate asset for Unilever. While advanced laundry detergents were not appropriate for markets like India, where much of the laundry was done in streams, a local development that allowed synthetic detergents to be compressed into solid tablet form gave the local subsidiary a product that could capture a significant share of the traditional bar soap market. Similarly, in Turkey, while the company's margarine products did not sell well, an innovative application of Unilever's expertise in edible fats allowed the company to develop a product from vegetable oils that competed with the traditional local clarified butter product, ghee.

As with center-for-global innovations, local-for-local innovations are not always as effective. In Philips, for example, the British subsidiary spent a large amount of resources to create a new TV chassis for its local market that turned out to be indistinguishable from the parent company's standard European model. As a consequence, for years Philips had to operate five instead of four television set factories in Europe.

The Local-for-Global Innovation Process

Local-for-global innovations are those which emerge as local-for-local innovations, are subsequently found to be applicable in multiple locations, and are then diffused to a number of organizational units. Thus, while the initial sensing, response, and implementation tasks are undertaken by a single subsidiary, other subsidiaries participate in the subsequent implementation process, as the innovation is diffused within the company.

Such was the case when Philips' British subsidiary reorganized the structure of its consumer electronics marketing division based on an analysis of changes in its product line and a growing concentration in its distribution channels. The traditional marketing organization which operated with a standard set of distribution, promotion, and sales policies applied uniformly to all product lines was proving to be increasing ineffective in dealing with the large-volume chains that had come to dominate the retail market. Further, Philips' undifferentiated marketing strategies were constraining efforts in the differentiated and rapidly changing markets for its diverse products. To cope with this problem, the U.K. subsidiary abolished this uniform structure for each product line and organized the marketing department into three groups; an advanced system group for dealing with the technologically sophisticated, high-margin, and image-building products like Laservision and compact disc players; a mainstay group for marketing high-volume mature products such as color TV and VCR; and a mass group for mass merchandizing of the older, declining products like portable cassette players and black-and-white TV sets.

This new organization allowed the company to differentiate the nature and intensity of marketing support it provided to different products according to their stages in the product life cycle and to engage various elements of the marketing mix—including promotion, pricing, and distribution—in a more selective and differentiated manner. Within the first year of implementation, the new organization had helped reduce aggregate selling expenses from 18 to 12 percent. During the same period, while overall market demand for consumer electronics products in the United Kingdom had fallen by 5 percent, the subsidiary's sales in this business had risen by 49 percent, including a 400 percent rise in sales to Dixons, the largest reseller chain.

Meanwhile, increasing concentration in the distribution channels and growing necessity for differentiating marketing approaches for different products became manifest Europewide. The new model of the marketing organization developed by the British subsidiary was clearly appropriate for many other subsidiaries and, despite some initial resistance, the innovation was transferred to most other national organizations.

Resistance to such transfers, however, is both widespread and strong in MNCs, and it blocked several attempted local-for-global innovations we studied. For example, management of Unilever was unable to transfer a zero phosphate detergent developed by its German subsidiary to other European locations. Insisting that its market needs were different, the French subsidiary insisted on developing its own zero-P project.

The Global-for-Global Innovation Process

Global-for-global innovations are those that are created by pooling the resources and capabilities of many different organizational units of the MNC, including the headquarters and a number of different subsidiaries, so as to arrive at a jointly developed general solution to an emerging global opportunity, instead of finding different local solutions in each environment or a central solution that is imposed on all the units. As an ideal type, this category of innovations involves participation of multiple organizational units in each of the three stages of sensing, response, and implementation. However, the key feature that distinguishes it from the other categories is that the response task is shared, instead of being carried out by a single unit. One of the best examples we observed of this mode of innovation was the way in which Proctor and Gamble developed its global liquid detergent.

Despite the success of liquid laundry detergents in the United States, all attempts to create a heavy-duty liquid detergent category in Europe failed due to different washing practices and superior performance of European powder detergents which contained levels of enzymes, bleach, and phosphates not permitted in the United States. But P&G's European scientists were convinced that they could enhance the performance of the liquid to match the local powders. After seven years of work they developed a bleach substitute, a fatty acid with water softening capabilities equivalent to phosphate, and a means to give enzymes stability in liquid form.

Meanwhile, researchers in the United States had been working on a new liquid better able to deal with the high-clay soil content in dirty clothes in the United States, and this group developed improvements in the builders, the ingredients that prevent redisposition of dirt in the wash. Also during this period, the company's International Technology Coordination Group was working with P&G scientists in Japan to develop a more robust surfactant (the ingredient that removes greasy stains), making the liquid more effective in the cold water washes that were common in Japan. Thus, the units in Europe, the United States, and Japan had each developed effective responses to its local needs, yet none of them had cooperated to share their breakthroughs.

When the company's head of R&D for Europe was promoted to the top corporate research job, one of his primary objectives was to create more coordination and cooperation among the diverse local-for-local development efforts, and the world liquid project became a test case. Plans to launch Omni, the new liquid the U.S. group had been working on, was shelved until the innovations from Europe and Japan could be incorporated. Similarly, the Japanese and the Europeans picked up on the new developments from the other laboratories. Joint effort on the part of all these groups ultimately led to the launch of Liquid Tide in the United States, Liquid Cheer in Japan, and Liquid Ariel in Europe. All these products incorporated the best of the developments created in response to European, American, and Japanese market needs.

ASSOCIATIONS BETWEEN INNOVATION PROCESSES AND ORGANIZATIONAL ATTRIBUTES

As we reviewed the key characteristics of the participating organizational components for each of the innovations listed in Table 1, four attributes, viz., (1) configuration of organizational assets and slack resources, (2) nature for inter-unit exchange relationships, (3) socialization processes, and (4) intensity of communication, appeared to have some systematic associations with the organization's ability to create innovations through the different processes we have described. These associations among the organizational factors and innovation processes are schematically represented in Figure 2, and are briefly described and illustrated in the following pages.

Configuration of Organizational Assets and Slack Resources

In some companies such as Matsushita, most key organizational assets and slack resources were centralized at the headquarters. Even though 40 percent of Matsushita's sales were made abroad, only 10 percent of its products were manufactured outside of Japan. The Japanese manufacturing facilities were also the most advanced and well-equipped plants of the company, producing almost all of its sophisticated products. R&D, similarly, was fully centralized in seven research laboratories in Japan. The center-for-global process appeared to contribute most of the significant innovations in companies with such a centralized configuration of organizational assets and resources. In Masushita, for example, *all* new consumer electronics products introduced between 1983 and 1986 were developed by the parent company in Japan and were subsequently introduced in its different foreign markets.

In companies like Philips, ITT and Unilever, on the other hand, manufacturing, mar-

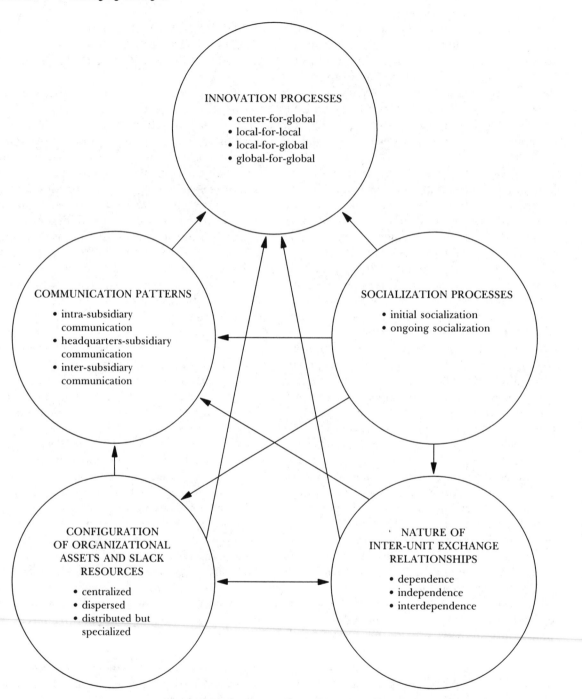

FIGURE 2.   Associations between Innovation Processes and Organizational Attributes

keting, and even R&D facilities were widely dispersed throughout the organization. The local-for-local (and, to a lesser extent, local-for-global) process contributed a significant number of innovations in these companies. The dispersal of assets and resources was perhaps at its most extreme in the telecommunications business of ITT. The company had practically no central research or manufacturing activity, and each major national subsidiary was fully intergrated and self-sufficient in its ability to develop, manufacture, and market new products. Up until the advent of digital switching, all major products including the Metaconta and Pentaconta switches were initially developed in one or the other subsidiary and were subsequently "redeveloped" by other subsidiaries, resulting in many different varieties of the same product being sold in different markets. In Philips, similarly, the list of local-for-local innovations is endless—the first stereo color TV set was developed by the Australian subsidiary, teletext TV sets were created by the British subsidiary, "smart cards" by the French subsidiary, and the programed word processing typewriter by North American Philips—to cite but a few examples.

Some of the companies we surveyed were gradually adopting a third system of asset and resource configuration. Instead of either centralization or decentralization, they were developing an interconnected network of specialized assets distributed around the world. Ericsson, NEC, and Proctor and Gamble were the most advanced in building such a system, and it is only in these companies that we saw some cases of successful global-for-global innovations (even in these companies, however, most innovations came through the other processes).

In NEC, organizational assets were traditionally centralized and most innovations were created through the center-for-global process. The NEAC 61 digital switch, for example, was developed entirely in the company's central facilities in Japan, even though the product was principally aimed at the North American market. Subsequently, however, the company developed specialized software capabilities in the United States, while hardware expertise remained at the center. Such a distribution of resources allowed NEC to approximate the global-for-global process in developing the NEAC 61E auxiliary switch: the headquarters took the lead in building the hardware while the subsidiary participated significantly in designing the software. Similarly, the global liquid project of P&G that we have described earlier in the paper was made possible and necessary because three different research units responsible for product development in Japan, Europe, and the United States had each developed specialized capabilities that the others did not possess.

Several researchers have shown now resource configuration constrains all aspects of organizational actions and interactions (Emery and Trist 1965; Aldrich 1976). In the specific context of research on organizational innovations, a number of authors have highlighted the importance of distributed slack resources for creation of innovations (March and Simon 1958; Cyert and March 1963; Mohr 1969; Mansfield 1968; Kanter 1983). It has also been argued that local slack may have opposite effects on creation of innovations and adoption of innovations created elsewhere (Wilson 1966; Sapolsky 1967). Finally, the observation that interdependent, specialized resources might lead to joint innovations is also not new to the literature (Thompson 1967; Kanter 1983). Thus, the innovation process–resource configuration associations we observed are entirely consistent with what would be predicted by existing theory, if applied to the specific context of large, multi-unit organizations.

Nature for Interunit Exchange Relationships

In most of the companies in our sample, interactions among the national subsidiaries were extremely limited and dyadic relationships between the headquarters and each of the different subsidiaries were the dominant form of interunit exchanges. In some of these companies such as Kao, the large Japanese manufacturer of soaps and detergents, these dyadic exchange relationships were based primarily on subsidiary dependence on the headquarters. National subsidiaries of this company had neither the competence nor the legitimacy to initiate any new programs or even to modify any product or administrative system developed by the parent company. For example, a particular brand of liquid shampoo that was extremely successful in Japan failed to produce desired effects when introduced in Thailand. The product, aimed to suit the sophisticated needs of the Japanese market, could not compete effectively with simpler but less expensive local products and developed only a marginal 7 percent of market share despite considerable marketing investments. However, the nature of the problem could be identified and some remedial measures taken only after marketing experts from the headquarters visited the subsidiary along with executives from Dentsu, Kao's Japanese advertising agents. The local manager acknowledged that "Japan's expertise, knowledge, and resources made it appropriate for them to make such decisions." In companies where subsidiaries had developed such highly dependent relationships with the headquarters, the center-for-global process was often the exclusive source of innovations. For example, in Kao we did not come across a single case of any other innovation process.

In some other companies, such as ITT and Philips, the subsidiaries had considerable strategic and operational autonomy, though the headquarters exercised varying degrees of administrative control through the budgeting and financial reporting systems. In these companies, where subsidiaries were relatively independent of the headquarters, local-for-local innovations were far more prevalent.

The local-for-global and global-for-global processes essentially require the involvement of multiple organizational units, including a number of different national subsidiaries, and are realizable only when inter-subsidiary exchange relationships are prevalent in the company. However, in all the cases where these innovation processes were effective, such exchange relationships among organizational units appeared to be based on reciprocal interdependence (Thompson 1967), rather than on either dependence or independence.

In Proctor and Gamble, for example, teams consisting of representatives from different national organizations in Europe (Eurobrand teams) coordinate regional strategies for different products of the company. For each product group, the team is headed by the general manager of a particular subsidiary and includes brand managers from other major subsidiaries. These teams provide one of the many mechanisms in the company that promote exchange relationships among the different national subsidiaries. Further, by ensuring that general managers from different subsidiaries head the different teams, the company creates reciprocal interdependencies in the relationships since each general manager recognizes that the level of cooperation he can expect from the brand managers of other subsidiaries in his team is dependent on the level of cooperation his brand managers extend to the other product teams that are headed by the general managers of other subsidiaries.

In contrast, ITT's attempt to develop the System 12 digital switch through a similar global-for-global process floundered because of the absence of such interdependencies. Recognizing that the technical resources required for developing the switch could not be assembled in any one location in its highly decentralized international organization, ITT management decided to adopt the global-for-global process of designing and building the switch through coordinated and joint action involving most of its major national operations.

However, conditioned by a long history of local independence, the national subsidiaries resisted joint efforts and common standards leading to constant duplication of efforts, divergence of specifications, delays and an enormous budget overrun.

The effects of centralization and decentralization on innovation has been a topic of considerable debate in the literature. While most authors have argued in favor of a negative correlation between centralization and creation of innovations and a positive correlation between centralization and internal diffusion of innovations (for a review, see Zaltman et al. 1973), it has also been suggested that the relationships may be contingent on the type of innovation as well as the extent to which information and perspectives are shared among members of the organization (Downs and Mohr, 1979). Our findings support the general view that centralization of authority and the resulting dependency relationship between the headquarters and subsidiaries facilitate diffusion of center-for-global innovations but impede creation of local-for-local innovations, while decentralization of authority and the resulting independence of subsidiaries have precisely the opposite effects.

The facilitating influence of reciprocal interdependencies on local-for-global and global-for-global innovations observed by us also finds some support in exchange theory (Emerson 1972). Such interdependencies induce mutual cooperation (March and Simon 1958) and overcome both the bureaucratic and entrepreneurial traps described by Kanter (1983) as impediments to joint problem solving on the part of different organizational members.

Before concluding our discussion on inter-unit exchange relationships, we should note that the configuration of organizational resources tend to have considerable influence on the governance of such exchange, and vice versa. Theoretically, such associations are predicted by the resource dependency perspective

(Pfeffer and Salancik 1978). When resources are centralized, as in Matsushita, Kao, and NEC, dyadic relationships of subsidiary dependence on the headquarters can be expected and are also observed. When resources are decentralized, as in Philips and Unilever, the subsidiaries exercise considerable independence and autonomy. Similarly, autonomous and resourceful subsidiaries are able to attract further resources for extention of current activities or creation of new ones, thereby establishing the reverse link between the nature of exchange relationships and the future flow of resources within the organization. Historically, location decisions for new manufacturing capacity in Philips were influenced by the relative power of different subsidiaries almost as much as by the dictates of production and distribution economies. On the other hand, Matsushita captured a very significant research facility in the United States when it acquired Motorola's TV business and senior corporate managers expected this research unit to play a major role in designing components and products for worldwide use. However, this role was inconsistent with the traditional basis of headquarters-subsidiary relationship in the company, and the capability was lost when most research engineers left in response to increasing functional control from the headquarters and the resulting loss of their local independence.

Organizational Socialization Processes

One interesting observation we made in the course of the study is that managers in both Philips and Matsushita were highly socialized into the very strong cultures of their respective organizations, though in very different ways. Further, while collectively, at the center, Matsushita managers continually sought change, individually, as expatriate managers in different national subsidiaries, they were relatively more likely to take a "custodial" stance, resisting change to centrally designed products, processes, and even routine administrative systems. Expatriate managers in Philips, in contrast, were generally more willing to champion local initiatives and thereby foster local-for-local innovations.

Van Maanen and Schein (1979) have argued that the nature of initial and ongoing socialization processes of organizations have some important influences on members' attitudes toward change. Certain socialization processes lead to custodial behavior and resistance to change, while others facilitate both content and role innovations on the part of socialized members. At least in the cases of Philips and Matsushita, the associations between socialization processes and attitude toward change proposed by these authors appear to explain the observed behavioral differences quite remarkably.

Initial post-recruitment training and subsequent career strutures are two important constituents of the organizational socialization process, and are both quite different for managers in the two companies. In Matsushita, college graduates are recruited at the center in large batches and are collectively socialized through a common training program that continues for one year or more. Managerial recruits in Philips, on the other hand, are recruited from diverse locations in small numbers, and are quickly posted to different units so as to learn on the job. While cohorts meet infrequently for some classroom sessions, initial socialization tends to be relatively more individually oriented, dependent on the new member establishing personal relationships with existing members, often his or her senior colleagues, on a one-to-one basis.

Before being assigned overseas, Matsushita managers are exposed to another strong dose of formal training. The company's formidable Overseas Training Center (OTC) prepares managers for overseas tours of duty by ensuring that they thoroughly understand Matsushita's practices and values. Expatriate managers are usually posted to a foreign location for relatively long periods, usually five to eight years, after which they return to the headquarters. They may be posted abroad once more, later in their careers, and often to the same location where they had worked earlier, though at more senior levels. In Philips, given the relatively insignificant role of the home country operations in the worldwide business of the company, a large number of managers spend a significant part of their careers abroad continuously, spending between two and three years in a number of different subsidiaries. Many of these managers retire abroad, while some return to take up top level corporate positions toward the end of their careers.

As suggested by Van Maanen and Schein, collective socialization such as in Matsushita results in relatively stronger conformity to the values that the collectivity is socialized into. Thus, when change is proposed within that collectivity (such as in the headquarters which is seen as the repository of those values), it tends to be supported. However, local changes in national subsidiaries that attempt to modify values, systems, or processes designed by that collectivity tend to be resisted. In contrast, individual socialization, as in Philips, tends to produce less homogeneity of views and greater willingness to change at local levels.

Similarly, the differences in career systems can also be expected to result in very different attitudes to local innovations. In Philips, expatriate managers follow each other into key management and technical positions in the company's national organizations around the world, they perceive themselves as a distinct subgroup within the organization and come to develop and share a distinct subculture. In Matsushita, on the other hand, there is very little interaction among the expatriate managers in different subsidiaries, and they tend to view themselves as part of the parent company on temporary assignment to foreign locations.

Consequently, Philips managers tend to identify strongly with the national organization's point of view and to serve as champions of local level changes, while Matsushita managers become more prone to implementing centrally designed products and policies.

Received theory, as well as our own observations in the nine companies, suggests that organizational socialization processes have significant influences on both the basis of internal exchange relations within organizations, and the configuration of assets and slack resources—the two other influencing variables in our model. Institutionalized processes of initial and ongoing socialization of new members lie at the core of organizational cultures and form the administrative routines that govern internal exchange behavior and exercise of choice (Ouchi 1980; Schein 1985; Nelson and Winter 1982). Shared goals and values lead to fluidity in the internal distribution of power, and to flexibility in its use (Pascale and Athos 1981; Kanter 1983). Configuration of organizational resources is a product of organizational goals and internal power structures (Pfeffer and Salancik 1978; Mintzberg 1983) and is thereby indirectly influenced by the socialization processes that affect both these determinants. Illustrations of these influences abound in the organizations we studied. The "ization" program of Unilever—a commitment to "localize" Unilever in each country as well as to "Unileverize" each local operation—is a case in point. This program—backed by extensive ongoing training activities, a planned company-wide transfer policy, and significant commitment of top management time devoted to ensure its reinforcement and salience—has led simultaneously to a gradual reconfiguration of the company's resources from being almost totally decentralized to being more specialized and interdependent, and the restructuring of interunit interactions from an internal norm of fiercely protected subsidiary autonomy to much greater sharing of influence in which subsidiary initiatives are circumscribed by a significant and substantive role of the center in setting and coordinating strategies, particularly with regard to development and introduction of new products.

Communication Patterns

Almost all studies on innovation-organization linkages have emphasized the central role of communication in facilitating organizational innovations (Allen 1977; Allen, Lee, and Tushman 1980; Burns and Stalker 1961; Kanter 1983; Rogers and Shoemaker 1971). Collectively, this body of theoretical and empirical literature has found consistent support for the proposition that communication is a prerequisite for innovations and, more specifically, that intra-organizational communication is perhaps the most important determinant of an organization's ability to create and institutionalize innovations.

Internal communication patterns in the nine companies we studied could be broadly categorised into three groups. In some companies such as ITT and Philips, internal communication within each subunit (the headquarters or individual national subsidiaries) was intense, but the level of communication among subunits (between the headquarters and each of the subsidiaries as well as among the subsidiaries themselves) was relatively low. In some others, such as Matsushita, communication links between the headquarters and most of the different subsidiaries were especially strong, but internal communication within the subsidiaries as well as communication among the subsidiaries tended to be limited. Finally, the third group consisted of companies like Procter and Gamble and L. M. Ericsson, where both internal communication within subunits and communication among subunits tended to be relatively rich and frequent. Local-for-local innovations were the most common in the first group of companies (although some of them, such as

511

Philips, could also create innovations through the local-for-global process), center-for-global innovations were dominant in the second, and only the third group could create innovations through all the four processes we have described.

By examining internal communication patterns in Philips, one can see how they support local-for-local innovations. Historically, the top management in all national subsidiaries of Philips consisted not of an individual CEO but a committee made up of the heads of the technical, commercial, and finance functions. This system of three-headed management had a long history in Philips, stemming from the functional independence of the two founding Philips brothers, one an engineer and the other a salesman, and has endured as a tradition of intensive intra-unit cross-functional communication and joint decision-making within each subsidiary.

In most subsidiaries, these integration mechanisms exist at three organizational levels. At the product management level, article teams prepare annual sales plans and budgets and develop product policies. A second tier of cross-functional coordination takes place through the group management team, which meets once a month to review results, suggest corrective actions, and resolve any inter-functional differences. The highest level cross-functional coordination and communication forum within the subsidiary is the senior management committee (SMC) consisting of the top commercial, technical, and financial managers of the subsidiary. Acting essentially as the local board, the SMC ensures overall unity of effort among the different functional groups within the local unit, and protects the legitimacy and effectiveness of the communication forums at lower levels of the organization. These multilevel cross-functional integrative mechanisms within each subsidiary lie at the heart of Philips' ability to create local innovations in its different operating environments.

If the challenge for improving the efficiency of local-for-local innovations lies in strengthening cross-functional communication within subsidiaries, the key task for enhancing the effectiveness of center-for-global innovations lies in building linkages between the headquarters and the different subsidiaries of the company. The main problem of centrally created innovations is that those developing new products or processes may not understand market needs, or that those required to implement the new product introduction are not committed to it. Matsushita overcomes these problems of center-for-global innovations by creating multilevel and multifunctional linkages between the headquarters and each of the different subsidiaries and these linkages facilitate both the communication of local market demands from the subsidiary to the center, and also central coordination and control over the subsidiary's implementation of the company's strategies and plans, including those of implementing innovations.

The communication links that connect different parts of the Matsushita organization in Japan with the video department of MESA, its U.S. subsidiary, are illustrative of headquarters-subsidiary communication systems in the company. The vice president in charge of this department of MESA has his roots in Matsushita Electric Trading Company (METC), the central organization that has overall responsibility for the company's overseas business. Although formally posted to the United States, he continues to be a member of the senior management committee of METC and spends about a third of his time in Japan. The general manager of this department had worked for fourteen years in the video product division of Matsushita Electric Industries (MEI), the central production and domestic marketing company in Japan. He maintains almost daily communication with the central product division in Japan and acts as its link to the local American market. The assistant manager in the depart-

512

ment, the most junior expatriate in the organization, links the local unit to the central factory in Japan. Having spent five years in the factory, he is and acts as its local representative and handles all day-to-day communication with factory personnel.

None of these linkages is accidental. They are deliberately created and maintained and reflect the company's desire to preserve the different perspectives and priorities of its diverse groups worldwide, and ensure that they have linkages to those in the headquarters who can represent and defend their views. Unlike in companies that try to focus headquarters-subsidiary communication through a single channel for the sake of efficiency, Matsushita's multilevel and multifunctional linkages create a broad band of communication through which each central unit involved in creating center-for-global innovations have direct access to local market information, while each local unit involved in implementing those innovations also has the opportunity to influence the innovation process.

Finally, a few companies like P&G and Ericsson are able to create organizational mechanisms that facilitate simultaneously intense intra-unit communication, extensive headquarters-subsidiary communication, and also considerable flow of information among the different subsidiaries. As a result, these companies are able to create innovations through all the four processes.

In Ericsson, for example, intra-subsidiary communication is facilitated both by a culture and tradition of open communication and, more specifically, by extensive use of ad hoc teams and special liason roles with the express mandate of facilitating intra-unit integration. Headquarters-subsidiary communication is strengthened by mechanisms such as deputing one or more senior corporate managers as members of subsidiary boards. Unlike many companies whose local boards are pro forma bodies aimed at satisfying national legal re-

quirements, Ericsson uses its local boards as legitimate forums for communicating objectives, resolving differences, and making decisions. Intersubsidiary communication is facilitated by a number of processes such as allocating global roles to subsidiaries for specific tasks (for example, Italy is the center for transmission system development, Finland for mobile telephones, and Australia for rural switches) which require them to establish communication links worldwide. However, perhaps the single factor that has the strongest effect on facilitating communication in the dispersed Ericsson organization is its long-standing policy of transferring large number of people back and forth between headquarters and subsidiaries.

Executive transfers in Ericsson differ from the more common transfer patterns in multinational corporations in both direction and intensity, as a comparison with NEC's transfer processes will demonstrate. Where NEC may transfer a new technology through one or perhaps a few key managers, Ericsson will send a team of 50 or 100 engineers and managers for a year or two; while NEC's flows are primarily from headquarters to subsidiary, Ericsson's is a balanced, two-way flow with people coming to the parent not only to learn but also to bring their expertise; and while NEC's transfers are predominantly Japanese, Ericsson's multidirectional process involves all nationalities.

Australian technicians seconded to Stockholm in the mid-1970s to bring their experience with digital switching into the development of AXE developed enduring relationships that helped in the subsequent development of a rural switch in Australia through the global-for-global process. Confidences built when an Italian team of forty spent eighteen months in Sweden to learn about electronic switching provided the basis for greater decentralization of AXE software development and a delegated responsibility for developing the switch's central transmission system through a local-for-global process.

Communication may be the final cause (Mohr 1982) that influences innovation processes in organizations, but it is itself a product of different organizational attributes such as the configuration of resources (Pfeffer 1982), internal governance systems (Kaufman 1960), and culture (Schein 1985). Our descriptions of the mechanisms that facilitate communication in some of the companies we surveyed illustrate these linkages, which are part of the model represented in Figure 2.

CONCLUSION

In this paper we have identified four organizational attributes that influence the different multinational innovation processes: configuration of organizational assets and slack resources; basis for inter-unit exchange relationships that reflect the distribution of power within the company; training, transfer, and other processes of socializing members; and the nature of intra and inter-unit communication. Each of these have been identified by earlier researchers as key factors that influence an organization's ability to innovate. Burns and Stalker (1961) emphasized the importance of decentralized authority and intraunit communication for promoting "grass-roots" innovations (local-for-local, in our terms). Lorsch and Lawrence (1965) highlighted the relevance of cross-functional integration. Quinn (1985) and Peters and Waterman (1982) illustrated the need for fluid power structures and dispersal of organizational resources. And Kanter (1983), in her description of the "integrative organization," identified each of these elements as key requirements for promoting organizational innovations. Thus, our overall findings broadly confirm those of many others who have investigated the effects of different organizational attributes on innovative capability of firms.

At the same time, our findings provide a point of departure and an avenue for incremental extention of existing theory. The source of this extention lies in our explicit focus on multi-unit organizations which necessitates simultaneous consideration of organizational attributes both within individual units and across multiple units. In contrast, most of past research has been limited to organizational subcomponents as the level of analysis, even though conclusions have sometimes been generalized at the level of the total organization. In the case of Burns and Stalker, the level of analysis is stated quite explicitly: "The twenty concerns that were subject of these studies were not all separately constituted business companies. This is why we have used concern as a generic term. . . . [Some of them] were small parts of the parent organization." The other researchers have similarly observed a district sales office of General Electric or a department in the headquarters of 3M, or a divisional data processing office of Polaroid, but not the overall organizational configuration in any one of these physically and goal-dispersed organizations. Given the possibility that there may be trade-offs between integration and differentiation at different levels of the organization (Lawrence and Lorsch 1967), findings at the level of organizational subcomponents can serve as useful hypothese but not as validated conclusions at the level of multiunit, complex organizations.

By broadening the focus to include inter-unit interactions, we have identified four different organizational processes through which innovations may be created and institutionalized in multiunit companies. It is also manifest that the different processes are facilitated by organizational attributes that are not only different but possibly also contradictory. Local-for-local innovations, for example, tend to be incremental (Quinn 1985) and unprogramed (Drucker 1985) changes that are facilitated by distributed resources and decentralized authority—attributes of the organic concern described by Burns and Stalker. Some center-for-global innovations, on the other hand, can be reorientations (Normann 1971) that are highly programmed, and they may be facilitated by precisely the opposite characteristics of centralization of organizational resources and authority. Factors that facilitate innovativeness on the part of a subunit may not be those that facilitate their adoption of innovations created elsewhere in the company, nor their participation in joint efforts. By differentiating among the processes, we have taken a step, albeit small, in the direction of a more disaggregated analysis of innovation-organization links advocated by Downs and Mohr (1979).

Our findings of the organizational attributes that facilitate each of these different innovation processes are summarized in Table 2. Given the exploratory nature of the study, these findings are, at best, grounded hypotheses that clearly require more systematic and rig-

TABLE 2. *Associations between Innovation Processes and Organizational Factors: A Summary*

Innovation Process	Configuration of Assets and Stock Resources	Socialization Processes	Nature of Inter-Unit Exchange Relationships	Communication Patterns
1. Center-for-global	Centralised at headquarters	Formal and collective initial training, transfers of few people from headquarters to subsidiaries, infrequently and for long terms	Subsidiaries dependent on headquarters	High density of communication between headquarters and subsidiaries
2. Local-for-local	Dispersed to subsidiaries	Informal and individual initial training; subsidiary-to-subsidiary transfers of an international cadre of managers	Subsidiaries independent of headquarters	High density of communication within subsidiaries
3. Local-for-global	Dispersed to subsidiaries	Informal but both collective and individual initial training; subsidiary-to-subsidiary transfers of an international cadre of managers	Subsidiaries independent of headquarters but mutually dependent on each other	High density of communication within and among subsidiaries.
4. Global-for-global	Distributed, specialized	Both collective and individual initial training, two-way transfers of large numbers of managers among headquarters and the different subsidiaries	Headquarters and subsidiaries mutually dependent on one another	High density of communication within subsidiaries, among subsidiaries, and between the headquarters and the subsidiaries

orous analysis. These hypotheses, however, have some significant consequences for theory and therefore appear to be deserving of the additional efforts that are necessary to test and validate them.[7]

At a more normative level, the complexity and diversity of technological, competitive, and market environments confronting most worldwide industries may require participating multinationals to create organizational mechanisms that would facilitate simultaneously all the innovation processes we have described. Although a few companies in our sample had begun to achieve this state on a partial and temporary basis, creating such a capability on a more general and permanent basis may be a challenge of considerable magnitude, given the potential contradictions in organizational attributes that facilitate each of these innovation processes. A more systematic study on the topic can lead to reliable suggestions on how these potential contradictions can be overcome. Based on our discussions with a number of MNC managers to whom we have presented our findings, we are convinced that such a study would be of great value to them.

NOTES

1. This study of innovations in MNCs was a part of a larger research project on management of multinational corporations which covered a number of issues other than management of innovations. The overall findings of the project are being reported in our forthcoming book, *Managing Across Borders: The Transnational Solution.*

2. The term *innovation* has been defined in many different ways. However, these definitions can be broadly classified into two categories: those that see innovation as the final event—"the idea, practice, or material artifact that has been invented or that is regarded as novel independent of its adoption or nonadoption" (Zaltman et al. 1973:7), and those who, like Myers and Marquis,

see it as a process "which proceeds from the conceptualization of a new idea to a solution of the problem and then to the actual utilization of a new item of economic or social value" (1969:1). We adopt the latter definition and, throughout the paper, use the terms innovation and innovation process interchangeably.

3. The sense-response-implement model has an extensive history in multiple fields. It is directly adopted from the unfreeze-change-refreeze framework in the field of organization development proposed by Lewin and subsequently enhanced by Bennis, Schein, Beckhard, and others. For a brief review of this literature, see Lorange et al. (1986). The same model, with different labels, has been adopted in the marketing field to describe the new product introduction process (see, for example, Urban and Hauser 1980), and by many scholars who have studied the organizational innovation process (see Zaltman et al. 1973 for a review).

4. To save space, we do not list the thirty-eight innovation cases, but interested readers can find such a list in Ghoshal (1986:a).

5. Both the spirit of this analysis and the actual methodology were inspired by the work of Bower (1980). However, given a relatively large number of cases, a more formal case clustering approach was adopted.

6. The terms global and worldwide have been used somewhat loosely in the paper to imply many different national subsidiaries or environments.

7. We have since pursued this research direction and the results tend to support the hypotheses. These findings will be reported in a forthcoming paper.

BIBLIOGRAPHY

Aldrich, H.E. "Ressource Dependency and Interorganizational Relations." *Administration and Society* 7 (1976): 419–54.

Allen, T.J. *Managing the Flow of Technology.* Cambridge, MA: MIT Press, 1977.

Allen, T.J., and S. Cohen. "Information Flow in R&D Laboratories." *Administrative Science Quarterly* 14 (1969): 12–19.

Allen, T.J.; D.M.S. Lee; and M.L. Tushman. "R&D performance as a Function of Internal Communication, Project Management, and the Nature of Work." *IEEE Transactions on Engineering Management* EM-27 (1980): 2–12.

Bartlett, C.A., and S. Ghoshal. "Managing Across Borders: New Strategic Requirements." *Sloan Management Review* (Summer 1987): 7–17.

————. *Managing Across Borders: The Transnational Solution.* Boston: Harvard Business School Press, forthcoming.

Bower, J.L. *Managing the Resource Allocation Process.* Boston: Harvard Business School Press, 1980.

Buckley, P.J., and M.C. Casson. *The Future of the Multinational Enterprise.* London: MacMillan Press, 1976.

Burns, T., and G.M. Stalker. *The Management of Innovation.* London: Tavistock, 1961.

Calvet, A.L. "A Synthesis of Foreign Direct Investment Theories and Theories of the Multinational Firm." *Journal of International Business Studies* (Spring–Summer 1981): 43–59.

Cyert, R.M., and J.G. March. *A Behavioral Theory of the Firm.* Englewood Cliffs, NJ: Prentice-Hall, 1963.

Downs, G.W., and L.B. Mohr. "Conceptual Issues in the Study of Innovation." *Administrative Science Quarterly* 21 (1976): 700–14.

————. "Toward a Theory of Innovation." *Administration and Society* 10, no. 4 (1979): 379–407.

Drucker, P.F. *Innovation and Entrepreneurship.* New York: Harper and Row, 1985.

Emerson, R.N. "Exchange Theory, Part II: Exchange Relations, Exchange Networks, and Groups as Exchange Systems." In J. Berger, M. Zelditch, and B. Anderson, (eds.), *Sociological Theories in Progress,* (vol. 2). Boston: Houghton Mifflen, 1972.

Emery, F.E., and E.L. Trist. "The Contextual Texture of Organizational Environments." *Human Relations* 18 (1965): 21–31.

Ghoshal, S. "The Innovative Multinational: A Differentiated Network of Organizational Roles and Management Processes." Ph.D. diss. Graduate School of Business Administration, Harvard University, 1986(a).

————. "Global Strategy: An Organizing Framework." Paper presented to the Annual Conference of the Academy of International Business,

London, 1986(b), forthcoming in the *Strategic Management Journal.*

Ginzberg, E. and E. Reilly. *Effective Change in Large Organizations.* New York: Columbia University Press, 1957.

Gordon, G.; J.R. Kimberley; and A. MacEachron. "Some Considerations in the Design of Problem Solving Research on the Diffusion of Medical Technology." In W.J. Abernathy, A. Sheldon, and C.K. Prahalad, eds., *The Management of Health Care.* Cambridge, MA: Ballinger, 1975.

Gross, N.; J.B. Giacquinta; and M. Berstein. *Implementing Organizational Innovations: A Sociological Analysis of Planned Educational Change.* New York: Basic Books, 1971.

Hamel, G., and C.K. Prahalad. "Do You Really Have a Global Strategy?" *Harvard Business Review* (July–August 1985): 139–148.

Hout, T.; M.E. Porter; and E. Rudden. "How Global Companies Win Out." *Harvard Business Review* (September–October 1982): 98–108.

Kanter, R.M. *The Change Masters.* New York: Simon and Schuster, 1983.

Kaufman, H. *The Forest Ranger: A Study in Administrative Behavior.* Baltimore: Johns Hopkins University Press, 1960.

Kelly, P., and M. Kranzberg. *Technological Innovations: A Critical Review of Current Knowledge.* San Francisco: San Francisco University Press, 1978.

Lawrence, P.R., and J.W. Lorsch. *Organization and Environment.* Boston: Graduate School of Business Administration, Harvard University, 1967.

Lorange, P.; M. Scott Morton; and S. Ghoshal. *Strategic Control.* St. Paul: West Publishing Co., 1986.

Lorsch, J.W., and P.A. Lawrence. "Organizing for Product Innovation." *Harvard Business Review* (January–February 1965): 109–120.

Mansfield, E. *The Economics of Technological Change.* New York: W.W. Norton, 1968.

March, J.G., and H.A. Simon. *Organizations.* New York: Wiley, 1958.

Meurling, J. *A Switch in Time.* Chicago: Telephony Publishing Corp., 1985.

Mintzberg, H. *Power in and Around Organizations.* Englewood Cliffs, NJ: Prentice Hall, 1983.

Mohr, L.B. "Determinants of Innovation in Organizations." *American Political Science Review* 63 (1969): 111–26.

————. *Explaining Organizational Behavior,* San Francisco, Jossey-Bass, 1982.

Myers, S., and D.G. Marquis. *Successful Industrial Innovations.* Washington, D.C.: National Science Foundation, NSF 69-17, 1969.

Nelson, R.R., and S.G. Winter. *An Economic Theory of Evolutionary Capabilities and Behavior.* Cambridge: Harvard University Press, 1982.

Normann, R. "Organizational Innovativeness: Product Variation and Reorientation." *Administrative Science Quarterly* 16, no. 2 (1971): 203–15.

Ouchi, W.G. "Markets, Bureaucracies, and Clans." *Administrative Science Quarterly* 25 (1980): 129–41.

Pascale, R.T., and A.G. Athos. *The Art of Japanese Management.* New York: Warner Books, 1981.

Peters, T.J., and R.H. Waterman. *In Search of Excellence.* New York: Harper and Row, 1982.

Pfeffer, J. *Power in Organizations.* Boston: Pitman, 1982.

Pfeffer, J., and G.R. Salancik. *The External Control of Organizations: A Resource Dependency Perspective.* New York: Harper and Row, 1978.

Porter, M.E. "Competition in Global Industries: A Conceptual Framework." In M.E. Porter, ed., *Competition in Global Industries.* Boston: Harvard Business School Press, 1986.

Prahalad, C.K. "The Strategic Process in a Multinational Corporation." Ph.D. diss., Graduate School of Business Administration, Harvard University, Boston, 1975.

Quinn, J.B. "Managing Innovations: Controlled Chaos." *Harvard Business Review* (May–June 1985): 73–84.

Roethlisberger, F.J. *The Elusive Phenomenon.* Boston:

Division of Research, Graduate School of Business Administration, Harvard University, 1977.

Rogers, E.M., and F.F. Shoemaker. *Communication of Innovations: A Cross-cultural Approach.* New York: Free Press, 1971.

Ronstadt, R.C. *Research and Development Abroad by U.S. Multinationals.* New York: Praeger, 1977.

Rugman, A.M. *New Theories of the Multinational Enterprise.* New York: St. Martin's Press, 1982.

Sapolsky, H.M. "Organizational Structure and Innovation." *Journal of Business* 40 (1967): 497–510.

Schien, E.H. *Organizational Culture and Leadership.* San Francisco: Jossey-Bass, 1985.

Terpstra, V. "International Product Policy: The Role of Foreign R&D." *The Columbia Journal of World Business* (Winter 1977): 24–32.

Thompson, J.D. *Organizations in Action: Social Science Bases of Administrative Theory.* New York: McGraww Hill, 1967.

Urban, G.L., and J.R. Hauser. *Design and Marketing of New Products.* Englewood Cliffs, NJ: Prentice-Hall, 1980.

Van Maanen, J., and E.H. Schien. "Toward a Theory of Organizational Socialization." In B. Shaw, ed., *Research in Organizational Behavior.* JAI Press, 1979.

Vernon, R. "International Investment and International Trade in the Product Cycle." *Quarterly Journal of Economics* (May 1966): 190–207.

Wilson, J.Q. "Innovation in Organization: Notes toward a Theory." In J.D. Thompson, ed., *Approaches to Organization Design.* Pittsburg: University of Pittsburg Press, 1966.

Zaltman, G.; R. Duncan; and J. Holbeck. *Innovations and Organizations.* New York: Wiley, 1973.

Japanese Management: Practices and Productivity

Nina Hatvany

Vladimir Pucik

Japanese management policies and practices shape a paradigm of concern for human resources—a paradigm that blends the hopes of humanistic thinkers with the pragmatism of those who must show a return on investment.

The [United States] is the most technically advanced country and the most affluent one. But capital investment alone will not make the difference. In any country the quality of products and the productivity of workers depend on management. When Detroit changes its management system we'll see more powerful American competitors.

—Hideo Sugiura, Executive Vice-President,
Honda Motor Co.

Productivity—or output per worker—is a key measure of economic health. When it increases, the economy grows in real terms and so do standards of living. When it declines, real economic growth slows or stagnates. Productivity is the result of many factors, including invest-ment in capital goods, technological innovation, and workers' motivation.*

After a number of years of sluggish productivity growth, the United States now trails most other major industrial nations in the rise in output per worker, although it still enjoys the best overall rate. This state of affairs is increasingly bemoaned by many critics in both academic and business circles. Some reasons suggested to "explain" the U.S. decline in productivity rankings include excessive government regulation, tax policies discouraging investment, increases in energy costs, uncooperative unions, and various other factors in the business environment.

Some observers, however—among them Harvard professors Robert Hayes and William J. Abernathy—put the blame squarely on American managers. They argue that U.S. firms prefer to service existing markets rather than create new ones, imitate rather than innovate, acquire existing companies rather than develop a superior product or process technology and, perhaps most important, focus on short-run returns on investment rather than long-term

Reprinted with permission of the publisher, from *Organizational Dynamics* (Spring 1981). © 1981 American Management Association, New York. All rights reserved.

*The authors would like to thank Mitsuyo Hanada, Blair McDonald, William Newman, William Ouchi, Thomas Roehl, Michael Tushman, and others for their helpful comments on earlier drafts of this paper. We are grateful to Citibank, New York, and the Japan Foundation, Tokyo, for their financial support of the work in the preparation of this paper.

growth and research and development strategy. Too many managers are setting and meeting short-term, market-driven objectives instead of adopting the appropriate time-horizon needed to plan and execute the successful product innovations needed to sustain worldwide competitiveness.

The performance of the American manufacturing sector is often contrasted with progress achieved by other industrialized countries—particularly Japan. Japan's productivity growth in manufacturing has been nearly three times the U.S. rate over the past two decades—the average annual growth rate between 1960 and 1978 was 7.8 percent. In the last five years alone, the productivity index has increased by more than 40 percent and most economists forecast similar rates for the 1980s. Such impressive results deserve careful examination.

Students of the Japanese economy generally point out that Japanese investment outlays as a proportion of gross national product are nearly twice as large as those in the United States, and this factor is backed by a high personal savings ratio and the availability of relatively cheap investment funds. Also, a massive infusion of imported technology contributed significantly to the growth of productivity in Japan. Among noneconomic factors, the Japanese political environment seems to support business needs, especially those of advanced industries. In addition, the "unique" psychological and cultural characteristics of the Japanese people are frequently cited as the key reason for Japan's success.

It is indeed a well-known fact that absenteeism in most Japanese companies is low; turnover rates are about half the U.S. figures, and employee commitment to the company is generally high. But although cultural factors are important in any context, we doubt that any peculiarities of Japanese people (if they exist) have much impact on their commitment or productivity. In fact, several recent research studies indicate that Japanese and American work-ers show little or no difference in the personality attributes related to performance. Rather, we join Robert Hayes and William Abernathy in believing that, in this context, productivity stems from the superior management systems of many Japanese firms. But the focus of our analysis is not on such areas as corporate marketing and production strategies. Instead, we will examine management practices in Japan as they affect one key company asset: human resources.

Our analysis is guided by our experience with subsidiaries of Japanese firms in the United States. Typically, these companies are staffed by a small group of Japanese managers with varying levels of autonomy relative to the company's parent. The rest of the employees are American. Although they operate in an alien culture, many of these subsidiaries are surprisingly successful. While it is often very difficult to measure the performance of newly established operations, it is no secret that production lines in several Japanese subsidiaries operate at the same productivity rate as those in Japan (for example, the Sony plant in San Diego).

This example—as well as others—serves to demonstrate that what works in Japan can often work in the United States. The techniques used by the management of Japanese subsidiaries to motivate their American workers seem to play an important part in the effort to increase productivity. Therefore, a careful examination of management practices in Japan is useful not only for a specialist interested in cross-cultural organization development, but also for the management practitioner who is losing to foreign competition even on his or her home-ground. What is it that the Japanese do better?

Our discussion attempts to answer this question by presenting a model of the Japanese management system that rests on a few elements that can be examined in different cultural settings. The model will be used to highlight the relationship between the management strategies and techniques observed in Japan and positive work outcomes, such as commitment and productivity. Our review is not intended to be exhaustive, but rather to suggest the feasibility of integrating findings from Japan with more general concepts and theories. We will therefore focus on relationships that may be verified by observations of behavior in non-Japanese, especially U.S., settings.

We propose that positive work outcomes emanate from a complex set of behavioral patterns that are not limited to any specific culture. The emphasis is on management practices as a system and on the integration of various strategies and techniques to achieve desired results. We hope thus to provide an alternative to statements—often cited but never empirically supported—that the high commitment and productivity of Japanese employees is primarily traceable to their cultural characteristics.

A MANAGEMENT SYSTEM FOCUSED ON HUMAN RESOURCES

Most managers will probably agree that management's key concern is the optimal utilization of a firm's various assets. These assets may vary—financial, technological, human, and so on. Tradeoffs are necessary because utilization of any one asset may result in underutilization of another. We propose that in most Japanese companies, *human assets are considered to be the firm's most important and profitable assets in the long run.* Although the phrase itself sounds familiar, even hollow, to many American managers and OD consultants, it is important to recognize that this management orientation is backed up by a well-integrated system of strategies and techniques that translate this abstract concept into reality.

First, long-term and secure employment is provided, which attracts employees of the

desired quality and induces them to remain with the firm. Second, a company philosophy is articulated that shows concern for employee needs and stresses cooperation and teamwork in a unique environment. Third, close attention is given both to hiring people who will fit well with the particular company's values and to integrating employees into the company at all stages of their working life. These general strategies are expressed in specific management techniques. Emphasis is placed on continuous development of employee skills; formal promotion is of secondary importance, at least during the early career stages. Employees are evaluated on a multitude of criteria—often including group performance results—rather than on individual bottomline contribution. The work is structured in such a way that it may be carried out by groups operating with a great deal of autonomy. Open communication is encouraged, supported, and rewarded. Information about pending decisions is circulated to all concerned before the decisions are actually made. Active observable concern for each and every employee is expressed by supervisory personnel (Figure 1). Each of these management practices, either alone or in combination with the others, is known to have a positive influence on commitment to the organization and its effectiveness.

We will discuss these practices as we have observed them in large and medium-size firms in Japan and in several of their subsidiaries in the United States. Although similar practices are often also in evidence in small Japanese companies, the long-term employment policies in these firms are more vulnerable to drops in economic activity and the system's impact is necessarily limited.

Strategies
Once management adopts the view that utilizing human assets is what matters most in the organization's success, several strategies have

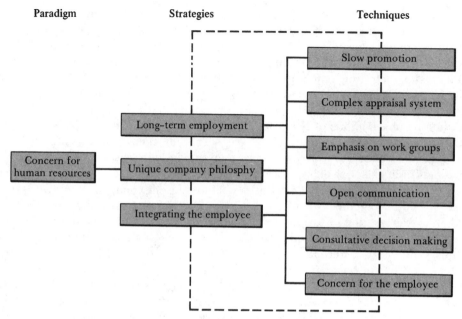

FIGURE 1. Japanese Management Paradigm

to be pursued to secure these assets in the desired quality and quantity. These strategies involve the following:

Provide Secure Employment. Although Japanese companies typically provide stable and long-term employment, many smaller firms find it difficult to do so in times of recession. The policy calls for hiring relatively unskilled employees (often directly from high schools or universities), training them on the job, promoting from within, and recognizing seniority.

The implicit guarantee of the employee's job, under all but the most severe economic circumstances, is a marked departure from conventional managerial thinking about the need to retain flexibility in work force size in order to respond efficiently to cyclical variations in demand. However, this employment system, at least as practiced in large corporations in Japan, is far from being inflexible. Several techniques can be applied to ride out recession with a minimum burden of labor cost while keeping a maximum number of regular workers on their jobs—a freeze on new hiring, solicitation of voluntary retirement sweetened by extra benefits, use of core employees to replace temporaries and subcontractors doing nonessential work, and so forth. Thus a labor force cut of approximately 10-15 percent in a short time period is not an unusual phenomenon. In addition, across-the-board salary and bonus cuts for all employees, including management, would follow if necessary.

Japanese managers believe that job security has a positive impact on morale and productivity, limits turnover and training costs, and increases the organization's cohesiveness. For that reason, they are willing to accept its temporary negative effect in a period of reduced demand. Long-term employment security is also characteristic of all the U.S. subsidiaries that we have visited. Layoffs and terminations occur extremely rarely. For example, the Kikkoman Company instituted across-the-board wage cuts in an attempt to preserve employment during the last recession. Murata instituted a four-day workweek, and at Matsushita's Quasar plant, a number of employees were shifted from their regular work to functions such as repairs, maintenance, and service. It should be noted that there are several well-known U.S. corporations—for example, IBM and Hewlett-Packard—that follow similar practices when the alternative would be layoff.

In Japanese companies, even poor performers are either retrained or transferred, instead of simply being dismissed. The plant manager in an electronics component factory explained how the company copes with personal failures: "We give a chance to improve even if there has been a big mistake. For example, the quality control manager didn't fit, so we transferred him to sales engineering and now he is doing fine."

Research on behavior in organizations suggests that the assumptions of Japanese managers and some of their U.S. colleagues about the positive impact of job security are, at least to some degree, justified. It has been shown that long tenure is positively associated with commitment to the organization, which in turn reduces turnover. High commitment in conjunction with a binding choice (employees of large firms in Japan have difficulty finding jobs of the same quality elsewhere, given the relatively stable labor market) also leads to high satisfaction, but whether this contributes to high productivity still remains to be proved. It is, however, necessary to view the policy of secure employment as a key condition for the implementation of other management strategies and techniques that have a more immediate impact on the organization's effectiveness.

Articulate a Unique Company Philosophy. A philosophy that is both articulated and carried through presents a clear picture of the organization's objectives, norms, and values—and thus helps transform commitment into productive effort. Familiarity with organizational goals gives direction to employees' actions, sets constraints on their behavior, and enhances their motivation. The understanding of shared meanings and beliefs expressed in the company philosophy binds the individual to the organization and, at the same time, stimulates the emergence of goals shared with others, as well as the stories, myths, and symbols that form the fabric of a company philosophy. William Ouchi and Raymond Price suggest that an organizational philosophy is an elegant informational device that provides a form of control at once pervasive and effective; at the same time it provides guidance for managers by presenting a basic theory of how the firm should be managed.

An explicit management philosophy of how work should be done can be found in many successful corporations in both Japan and the United States; examples in the United States include IBM, Texas Instruments, and U.S. Homes. Nevertheless, it is fair to say that the typical Japanese firm's management philosophy has a distinct flavor. It usually puts a heavy emphasis on cooperation and teamwork within a corporate "family" that is unique and distinct from that of any other firm. In return for an employee's effort, the family's commitment to the employee is translated into company determination to avoid layoffs and to provide a whole range of supplementary welfare benefits for the employee and his or her family. Naturally, without reasonable employment security, the fostering of team spirit and cooperation would be impossible. The ideal is thus to reconcile two objectives: pursuit of profits and perpetuation of the company as a group.

In a number of cases, a particular management philosophy that originated within the parent company in Japan is also being actively disseminated in its U.S. subsidiaries. Typically, claims of uniqueness range from the extent of the company's concern for employees' worklives to the quality of service to the customer. We quote from the in-house literature issued by one of the fastest growing Japanese-owned electronics component makers in California:

Management Philosophy
Our goal is to strive toward both the material and the spiritual fulfillment of all employees in the Company, and through this successful fulfillment, serve mankind in its progress and prosperity.

Management Policy
[. . .] Our purpose is to fully satisfy the needs of our customers and in return gain a just profit for ourselves. We are a family united in common bonds and singular goals. One of these bonds is the respect and support we feel for our fellow family co-workers.

Integrate Employees into the Company. The benefits of an articulated company philosophy are lost, however, if it's not properly communicated to employees or if it's not visibly supported by management's behavior. A primary function of the company's socialization effort, therefore, is to ensure that employees have understood the philosophy and seen it in action. Close attention is given to hiring people who are willing to endorse the values of the particular company and to the employees' integration into the organization at all stages of their working life. The development of cohesiveness within the firm, based on the acceptance of goals and values, is a major focus of personnel policies in many Japanese firms.

Because employees are expected to remain in the company for a major part of their careers, careful selection is necessary to ensure that they fit well into the company climate. In many U.S.-based Japanese firms also, new hires at all levels are carefully screened with this aspect in mind. As in Japan, basic criteria for hiring are moderate views and a harmonious personality, and for that reason a large proportion of new hires come from employee referrals. In general, "virgin" workforces are preferred, since they can readily be assimilated into each

company's unique environment as a community.

The intensive socialization process starts with the hiring decision and the initial training program and continues in various forms thereafter. Over time, the employee internalizes the various values and objectives of the firm, becomes increasingly committed to them, and learns the formal and informal rules and procedures, particularly through job rotation. That process usually includes two related types of job transfers. First, employees are transferred to new positions to learn additional skills in on-the-job training programs. These job changes are planned well in advance for all regular employees, including blue-collar workers. Second, transfers are part of a long-range, experience-building program through which the organization develops its future managers; such programs involve movement across departmental boundaries at any stage of an employee's career.

While employees rotate semilaterally from job to job, they become increasingly socialized into the organization, immersed in the company philosophy and culture, and bound to a set of shared goals. Even in the absence of specific internal regulations that might be too constraining in a rapidly changing environment, a well-socialized manager who has held positions in various functions and locations within the firm has a feel for the organization's needs.

Techniques

The basic management orientation and strategies that we have just discussed are closely interrelated with specific management techniques used in Japanese firms and in their subsidiaries in the United States. The whole system is composed of a set of interdependent employment practices in which the presence of one technique complements as well as influences the effectiveness of others. This interdependence, rather than a simple cause-effect relationship, is the key factor that helps maintain the organization's stability and effectiveness. Additional environmental variables may determine which of the strategies or techniques will require most attention from top management, but in their impact on the organization no single technique listed below is of prime importance.

Slow Promotion, Job Rotation, and Internal Training. All Japanese subsidiaries that we have visited have seniority-based promotion systems. At one of them, a medium-size motorcycle plant, a seniority-based promotion system has been reinstituted after an experiment with a merit-based system proved highly unpopular with employees. Training is conducted, as expected, mostly on the job, and as one textile company executive noted, career paths are flexible: "You can get involved in what you want to do." Hiring from outside into upper-level positions is rare. According to another Japanese plant manager: "We want someone who understands the management system of the company. We want to keep the employees with us; we want to keep them happy."

Although promotion is slow, early informal identification of the elite is not unusual and carefully planned lateral job transfers add substantial flexibility to job assignments. Not all jobs are equally central to the workflow in an organization, so employees—even those with the same status and salary—can be rewarded or punished by providing or withholding positions in which they could acquire the skills needed for future formal promotions.

Job rotation in U.S.-based Japanese firms seems less planned or structured than in Japan and more an ad hoc reaction to organizational needs—but in general, the emphasis on slow promotion and job rotation creates an environment in which an employee becomes a generalist rather than a specialist in a particular functional area. For the most part, however,

these general skills are unique to the organization. Several of the Japanese manufacturers that invested in the United States were forced to simplify their product technology because they were not able to recruit qualified operators versatile enough to meet their needs, and there was not enough time to train them internally.

In Japan, well-planned job rotation is the key to the success of an in-company training program that generally encompasses all the firm's employees. For some categories of highly skilled blue-collar workers, training plans for a period of up to ten years are not unusual. Off-the-job training is often included, again for managers and nonmanagers alike. However, whether such an extensive training system will be transferred to U.S. subsidiaries remains to be seen.

In addition to its impact on promotion and training, job rotation also promotes the development of informal communication networks that help in coordinating work activities across functional areas and in resolving problems speedily. This aspect of job rotation is especially important for managerial personnel. Finally, timely job rotation relieves an employee who has become unresponsive to, or bored with, the demands of his or her job.

Some observers argue that deferred promotion may frustrate highly promising, ambitious employees. However, the personnel director of a major trading company has commented: "The secret of Japanese management, if there is any, is to make everybody feel as long as possible that he is slated for the top position in the firm—thereby increasing his motivation during the most productive period of his employment." The public identification of "losers," who of course far outnumber "winners" in any hierarchical organization, is postponed in the belief that the increased output of the losers, who are striving hard to do well and still hoping to beat the odds, more than compensates for any lags in the motivation of the impatient winners. By contrast, top management in many American organizations is preoccupied with identifying rising stars as soon as possible and is less concerned about the impact on the losers' morale.

Complex Appraisal System. In addition to emphasizing the long-term perspective, Japanese companies often establish a complex appraisal system that includes not only individual performance measures tied to the bottom line, but also measures of various desirable personality traits and behaviors—such as creativity, emotional maturity, and cooperation with others as well as team performance results. In most such companies, potential, personality, and behavior, rather than current output, are the key criteria, yet the difference is often merely symbolic. Output measures may easily be "translated" into such attributes as leadership skills, technical competence, relations with others, and judgment. This approach avoids making the employee feel that the bottom line, which may be beyond his or her control, in part or in whole, is the key dimension of evaluation. Occasional mistakes, particularly those made by lower-level employees, are considered part of the learning process.

At the same time, evaluations do clearly discriminate among employees because each employee is compared with other members of an appropriate group (in age and status) and ranked accordingly. The ranking within the cohort is generally not disclosed to the employees, but of course it can be partially inferred from small salary differentials and a comparison of job assignments. At least in theory, the slow promotion system should allow for careful judgments to be made even on such subjective criteria as the personality traits of honesty and seriousness. However, the authors' observations suggest that ranking within the cohort is usually established rather early in one's career and is generally not very flexible thereafter.

Employees are not formally separated

according to their ability until later in their tenure; ambitious workers who seek immediate recognition must engage in activities that will get them noticed. Bottom-line performance is not an adequate criterion because, as noted, it is not the only focus of managerial evaluation. This situation encourages easily observable behavior, such as voluntary overtime, that appears to demonstrate willingness to exert substantial effort on the organization's behalf. The evaluation process becomes to a large degree self-selective.

Several other facets of this kind of appraisal system deserve our attention. Because evaluations are based on managerial observations during frequent, regular interactions with subordinates, the cost of such an evaluation system is relatively low. When behavior rather than bottom-line performance is the focus of evaluation, means as well as ends may be assessed. This very likely leads to a better match between the direction of employee efforts and company objectives, and it encourages a long-term perspective. Finally, since group performance is also a focus of evaluation, peer pressure on an employee to contribute his or her share to the group's performance becomes an important mechanism of performance control. Long tenure, friendship ties, and informal communication networks enable both superiors and peers to get a very clear sense of the employee's performance and potential relative to others.

Among the management techniques characteristic of large Japanese enterprises, the introduction of a complex appraisal system is probably the least visible in their U.S. subsidiaries. Most of their U.S.-based affiliates are relatively young; thus long-term evaluation of employees, the key point in personnel appraisal as practiced in Japan, is not yet practicable. Furthermore, the different expectations of American workers and managers about what constitutes a fair and equitable appraisal system might hinder acceptance of the parent company's evaluation system.

Emphasis on Work Groups. Acknowledging the enormous impact of groups on their members—both directly, through the enforcement of norms, and indirectly, by affecting the beliefs and values of members—Japanese organizations devote far greater attention to structural factors that enhance group motivation and cooperation than to the motivation of individual employees. Tasks are assigned to groups, not to individual employees, and group cohesion is stimulated by delegating responsibility to the group not only for getting the tasks performed, but also for designing the way in which they get performed. The group-based performance evaluation has already been discussed.

Similarly, in the U.S.-based Japanese firms that we have visited, the group rather than an individual forms the basic work unit for all practical purposes. Quality of work and speed of job execution are key concerns in group production meetings that are held at least monthly, and even daily in some companies. The design function, however, is not yet very well developed; many workers are still relative newcomers unfamiliar with all aspects of the advanced technology. Intergroup rivalry is also encouraged. In one capacitor company, a group on a shift that performs well consistently is rewarded regularly. Sometimes news of a highly productive group from another shift or even from the Japanese parent is passed around the shop floor to stimulate the competition.

In Japan, group autonomy is encouraged by avoiding any reliance on experts to solve operational problems. One widely used group-based technique for dealing with such problems is quality control (QC) circles. A QC circle's major task is to pinpoint and solve a particular workshop's problem. Outside experts are called in only to educate group members in the analytical tools for problem solving or to provide a specialized technical service. Otherwise, the team working on the problem operates autonomously, with additional emphasis on self-improvement activities that will help achieve group goals. Fostering motivation through direct employee participation in the work process design is a major consideration in the introduction of QC circles and similar activities to the factory floor.

Nevertheless, work-group autonomy in most work settings is bound by clearly defined limits, with the company carefully coordinating team activities by controlling the training and evaluation of members, the size of the team, and the scope and amount of production. Yet within these limits, teamwork is not only part of a company's articulated philosophy, it actually forms the basic fabric of the work process. Job rotation is encouraged both to develop each employee's skills and to fit the work group's needs.

From another perspective, the group can also assist in developing job-relevant knowledge by direct instruction and by serving as a model of appropriate behavior. The results of empirical studies suggest that structuring tasks around work groups not only may improve performance, but also may contribute to increased esteem and a sense of identity among group members. Furthermore, this process of translating organizational membership into membership in a small group seems, in general, to result in higher job satisfaction, lower absenteeism, lower turnover rates, and fewer labor disputes.

Open and Extensive Communication. Even in the Japanese-owned U.S. companies, plant managers typically spend at least two hours a day on the shop floor and are readily available for the rest of the day. Often, foremen are deliberately deprived of offices so they can be with their subordinates on the floor throughout the whole day, instructing and helping whenever necessary. The same policy applies to personnel specialists. The American personnel manager of a Japanese motorcycle plant, for example, spends between two and four hours a day on the shop floor discussing issues that concern employees. The large number of employees he is able to greet by their first name testifies to the amount of time he spends on the floor. "We have an open-door policy—but it's their door, not management's" was his explanation of the company's emphasis on the quality of face-to-face vertical communication.

Open communication is also inherent in the Japanese work setting. Open work spaces are crowded with individuals at different hierarchical levels. Even high-ranking office managers seldom have separate private offices. Partitions, cubicles, and small side rooms are used to set off special areas for conferences with visitors or small discussions among the staff. In one Japanese-owned TV plant on the West Coast, the top manager's office is next to the receptionist—open and visible to everybody who walks into the building, whether employee, supplier, or customer.

Open communication is not limited to vertical exchanges. Both the emphasis on team spirit in work groups and the network of friendships that employees develop during their long tenure in the organization encourage the extensive face-to-face communication so often reported in studies involving Japanese companies. Moreover, job rotation is instrumental in building informal lateral networks across departmental boundaries. Without these networks, the transfer of much job-related information would be impossible. These informal networks are not included in written work manuals, thus they are invisible to a newcomer; but their use as a legitimate tool to get things done is implicitly authorized by the formal control system. Communication skills and related behavior are often the focus of yearly evaluations. Frequently, foreign observers put too much emphasis on vertical ties and other hierarchical aspects of Japanese organizations. In fact, the ability to manage lateral communication is perhaps even more important to effective performance, particularly at the middle-management level.

Consultative Decision Making. Few Japanese management practices are so misunderstood by outsiders as is the decision-making process. The image is quite entrenched in Western literature on Japanese organizations: Scores of managers huddle together in endless discussion until consensus on every detail is reached, after which a symbolic document, "ringi," is passed around so they can affix their seals of approval on it. This image negates the considerable degree of decentralization for most types of decisions that is characteristic in most subsidiaries we have visited. In fact, when time is short, decisions are routinely made by the manager in charge.

Under the usual procedure for top-management decision making, a proposal is initiated by a middle manager (but often under the directive of top management). This middle manager engages in informal discussion and consultation with peers and supervisors. When all are familiar with the proposal, a formal request for a decision is made and, because of earlier discussions, is almost inevitably ratified—often in a ceremonial group meeting or through the "ringi" procedure. This implies not unanimous approval, but unanimous consent to its implementation.

This kind of decision making is not participative in the Western sense of the word, which encompasses negotiation and bargaining between a manager and subordinates. In the Japanese context, negotiations are primarily lateral between the departments concerned with the decision. Within the work group, the emphasis is on including all group members in the process of decision making, not on achieving consensus on the alternatives. Opposing parties are willing to go along, with the consolation that their viewpoint may carry the day the next time around.

However, the manager will usually not implement his or her decision "until others who will be affected have had sufficient time to offer their views, feel that they have been fairly heard, and are willing to support the decision even though they may not feel that it is the best one," according to Thomas P. Rohlen. Those outside the core of the decision-making group merely express their acknowledgement of the proposed course of action. They do not participate; they do not feel ownership of the decision. On the other hand, the early communication of the proposed changes helps reduce uncertainty in the organization. In addition, prior information on upcoming decisions gives employees an opportunity to rationalize and accept the outcomes.

Japanese managers we have interviewed often expressed the opinion that it is their American partners who insist on examining every aspect and contingency of proposed alternatives, while they themselves prefer a relatively general agreement on the direction to follow, leaving the details to be solved on the run. Accordingly, the refinement of a proposal occurs during its early implementation stage.

Although the level of face-to-face communication in Japanese organizations is relatively high, it should not be confused with participation in decision making. Most communication concerns routine tasks; thus it is not surprising that research on Japanese companies indicates no relationship between the extent of face-to-face communication and employees' perceptions of how much they participate in decision making.

Moreover, consultation with lower-ranking employees does not automatically imply that the decision process is "bottom up," as suggested by Peter Drucker and others. Especially in the case of long-term planning and strategy, the initiative comes mostly from the top. Furthermore, consultative decision making does not diminish top management's responsibility for a decision's consequences. Although the ambiguities of status and centrality may make it difficult for outsiders to pinpoint responsibility, it is actually quite clear within the organization. Heads still

roll to pay for mistakes, albeit in a somewhat more subtle manner than is customary in Western organizations: Departure to the second- or third-ranking subsidiary is the most common punishment.

Concern for the Employee. It is established practice for managers to spend a lot of time talking to employees about everyday matters. Thus they develop a feeling for employees' personal needs and problems, as well as for their performance. Obviously, gaining this intimate knowledge of each employee is easier when an employee has long tenure, but managers do consciously attempt to get to know their employees, and they place a premium on providing time to talk. The quality of relationships developed with subordinates is also an important factor on which a manager is evaluated.

Various company-sponsored cultural, athletic, and other recreational activities further deepen involvement in employees' lives. The heavy schedule of company social affairs is ostensibly voluntary, but virtually all employees participate. Typically, an annual calendar of office events might include two overnight trips, monthly Saturday afternoon recreation, and an average of six office parties—all at company expense. A great deal of drinking goes on at these events and much good fellowship is expressed among the employees.

Finally, in Japan the company allocates substantial financial resources to pay for benefits for all employees, such as a family allowance and various commuting and housing allowances. In addition, many firms provide a whole range of welfare services ranging from subsidized company housing for families and dormitories for unmarried employees, through company nurseries and company scholarships for employees' children, to mortgage loans, credit facilities, savings plans, and insurance. Thus employees often perceive a close relationship between their own welfare and the company's financial welfare. Accordingly, behavior for the company's benefit that may appear self-sacrificing is not at all so; rather, it is in the employee's own interest.

Managers in U.S.-based companies generally also voiced a desire to make life in the company a pleasant experience for their subordinates. As in Japan, managers at all levels show concern for employees by sponsoring various recreational activities or even taking them out to dinner to talk problems over. Again, continuous open communication gets special emphasis. However, company benefits are not as extensive as in Japan because of a feeling that American employees prefer rewards in the form of salary rather than the 'golden handcuff' of benefits. Furthermore, the comprehensive government welfare system in the United States apparently renders such extensive company benefits superfluous.

In summary, what we observed in many Japanese companies is an integrated system of management strategies and techniques that reinforce one another because of systemic management orientation to the quality of human resources. In addition to this system's behavioral consequences, which we have already discussed, a number of other positive outcomes have also been reported in research studies on Japanese organizations.

For example, when the company offers desirable employment conditions designed to provide job security and reduce voluntary turnover, the company benefits not only from the increased loyalty of the workforce, but also from a reduction in hiring, training, and other costs associated with turnover and replacement. Because employees enjoy job security, they do not fear technical innovation and may, in fact, welcome it—especially if it relieves them of tedious or exhausting manual tasks. In such an atmosphere, concern for long-term growth, rather than a focus on immediate profits, is also expected to flourish.

An articulated philosophy that expresses the company's family atmosphere as well as its uniqueness enables the employee to justify loyalty to the company and stimulates healthy competition with other companies. The management goals symbolized in company philosophy can give clear guidance to the employee who's trying to make the best decision in a situation that is uncertain.

Careful attention to selection and the employee's fit into the company results in a homogeneous workforce, easily able to develop the friendship ties that form the basis of information networks. The lack of conflict among functional divisions and the ability to communicate informally across divisions allow for rapid interdivisional coordination and the rapid implementation of various company goals and policies.

The other techniques we've outlined reinforce these positive outcomes. Slow promotion reinforces a long-range perspective. High earnings in this quarter need not earn an employee an immediate promotion. Less reliance on the bottom line means that an employee's capabilities and behaviors over the long term become more important in their evaluations. Groups are another vehicle by which the company's goals, norms, and values are communicated to each employee. Open communication is the most visible vehicle for demonstrating concern for employees and willingness to benefit from their experience, regardless of rank. Open communication is thus a key technique that supports consultative decision making and affects the quality of any implementation process. Finally, caring about employees' social needs encourages identification with the firm and limits the impact of personal troubles on performance.

What we have described is a system based on the understanding that in return for the employee's contribution to company growth and well-being, the profitable firm will provide a stable and secure work environment and protect the individual employee's welfare even during a period of economic slowdown.

The Transferability of Japanese Management Practices

As in Japan, a key managerial concern in all U.S.-based Japanese companies we have investigated was the quality of human resources. As one executive put it, "We adapt the organization to the people because you can't adapt people to the organization." A number of specific instances of how Japanese management techniques are being applied in the United States were previously cited. Most personnel policies we've observed were similar to those in Japan, although evaluation systems and job-rotation planning are still somewhat different, probably because of the youth of the subsidiary companies. Less institutionalized concern for employee welfare was also pointed out.

The experience of many Japanese firms that have established U.S. subsidiaries suggests that the U.S. workers are receptive to many management practices introduced by Japanese managers. During our interviews, many Japanese executives pointed out that the productivity level in their U.S. plants is on a level similar to that in Japan—and occasionally even higher. Other research data indicate that American workers in Japanese-owned plants are even more satisfied with their work conditions than are their Japanese or Japanese-American colleagues.

The relative success of U.S.-based Japanese companies in transferring their employment and management practices to cover the great majority of their U.S. workers is not surprising when we consider that a number of large U.S. corporations have created management systems that use some typical Japanese techniques. Several of these firms have an outstanding record of innovation and rapid growth. A few examples are Procter & Gamble, Hewlett-Packard, and Cummins Engine.

William Ouchi and his colleagues call these firms Theory Z organizations. Seven key characteristics of Theory Z organizations are the following:

1. Long-term employment.
2. Slow evaluation and promotion.
3. Moderately specialized careers.
4. Consensus decision making.
5. Individual responsibility.
6. Implicit, informal control (but with explicit measures).
7. Wholistic concern for the employee.

The Theory Z organization shares several features with the Japanese organization, but there are differences: In the former, responsibility is definitely individual, measures of performance are explicit, and careers are actually moderately specialized. However, Ouchi tells us little about communication patterns in these organizations, the role of the work group, and some other features important in Japanese settings.

Here's an example of a standard practice in the Theory Z organization that Ouchi studied in depth:

> [The Theory Z organization] calculated the profitability of each of its divisions, but it did not operate a strict profit center or other marketlike mechanism. Rather, decisions were frequently made by division managers who were guided by broader corporate concerns, even though their own divisional earnings may have suffered as a result.

A survey by Ouchi and Jerry Johnson showed that within the electronics industry perceived corporate prestige, managerial ability, and reported corporate earnings were all strongly positively correlated with the "Z-ness" of the organization.

It is also significant that examples of successful implementation of the Japanese system can be found even in Britain, a country notorious for labor-management conflict. In our interpretation, good labor-management relations—even the emergence of a so-called company union—is an effect, rather than a cause, of the mutually beneficial, reciprocal relationship enjoyed by the employees and the firm. Thus we see the co-existence of our management paradigm with productivity in companies in Japan, in Japanese companies in the United States and Europe, and in a number of indigenous U.S. companies. Although correlation does not imply cause, such a causal connection would be well supported by psychological theories. Douglas McGregor summarizes a great deal of research in saying: "Effective performance results when conditions are created such that the members of the organization can achieve their own goals best by directing their efforts toward the success of the enterprise."

CONCLUSION

Many cultural differences exist, of course, between people in Japan and those in Western countries. However, this should not distract our attention from the fact that human beings in all countries also have a great deal in common. In the workplace, all people value decent treatment, security, and an opportunity for emotional fulfillment. It is to the credit of Japanese managers that they have developed organizational systems that, even though far from perfect, do respond to these needs to a great extent. Also to their credit is the fact that high motivation and productivity result at the same time.

The strategies and techniques we have reviewed constitute a remarkably well-integrated system. The management practices are highly congruent with the way in which tasks are structured, with individual members' goals, and with the organization's climate. Such a fit is expected to result in a high degree of organizational effectiveness or productivity. We believe that the management paradigm of concern for human resources blends the hopes of humanistic thinkers with the pragmatism of those who

need to show a return on investment. The evidence strongly suggests that this paradigm is both desirable and feasible in Western countries and that the key elements of Japanese management practices are not unique to Japan and can be successfully transplanted to other cultures. The linkage between human needs and productivity is nothing new in Western management theory. It required the Japanese, however, to translate the idea into a successful reality.

SELECTED BIBLIOGRAPHY

Robert Hayes and William Abernathy brought the lack of U.S. innovation to public attention in their article, "Managing Our Way to Economic Decline" (*Harvard Business Review,* July-August 1980).

Thomas P. Rohlen's book, *For Harmony and Strength: Japanese White-Collar Organization in Anthropological Perspective* (University of California Press, 1974), is a captivating description of the Japanese management system as seen in a regional bank. Peter Drucker has written several articles on the system, including "What We Can Learn from Japanese Management" (*Harvard Business Review,* March-April 1971). His thoughts are extended to the United States by the empirical work of Richard Pascale, "Employment Practices and Employee Attitudes: A Study of Japanese and American Managed Firms in the U.S." (*Human Relations,* July 1978).

For further information on the Theory Z organization see "Type Z Organization: Stability in the Midst of Mobility" by William Ouchi and Alfred Jaeger (*Academy of Management Review,* April 1978), "Types of Organizational Control and Their Relationship to Emotional Well-Being" by William Ouchi and Jerry Johnson (*Administrative Science Quarterly,* Spring 1978), and "Hierarchies, Clans, and Theory Z: A New Perspective on Organization Development" by William Ouchi and Raymond Price (*Organizational Dynamics,* Autumn 1978).

Douglas McGregor explains the importance of a fit between employee and organizational goals in *The Human Side of Enterprise* (McGraw-Hill, 1960).

Managing the New Product Development Process:
How Japanese Companies
Learn and Unlearn

Ken-ichi Imai

Ikujiro Nonaka

Hirotaka Takeuchi

INTRODUCTION

In our travels through corporate America several years ago, we found America in search of itself. "What went wrong?" many would ask. The talk would be about Rip van Winkle and about the past.

In our most recent travels through corporate America, we found it ready for action. "Can we do it?" many would ask. The talk would be about the future, about renaissance, renewal, and turnaround.

We heard a number of suggestions concerning how to bring about a more competitive future for corporate America. Not surprisingly, many emphasized innovation as the key missing link. For example:

- Abernathy, Clark, and Kantrow note that a change in the nature of innovation is both a sign of dematurity in process and, in competitive terms, its most far-reaching effect.[1]

- Lawrence and Dyer believe that a readaptive process—"the process by which organizations repeatedly reconcile efficiency and innovation"—is essential to renewing American industry.[2]

- Kanter argues: "As America's economy slips further into the doldrums, innovation is beginning to be recognized as a national priority. But there is a clear and pressing need for more innovations, for we face social and economic changes of unprecedented magnitude and variety, which past practices cannot accommodate and which instead require innovative responses."[3]

We were asked, by practitioners and researchers alike, to offer them clues on how the Japanese do it. How can Japanese companies be productive and innovative at the same time? How can they support such a rapid new product development program? How can they be so flexible in seeking out new technology and, at the same time, adaptive to changing market requirements?

It is with this kind of an orientation that we embarked on our investigation of the inno-

Reprinted by permission of the Harvard Business School Press from Kim B. Clark, Robert H. Hayes and Christopher Lorenz, eds., *The Uneasy Alliance.* Copyright © 1985 by the President and Fellows of Harvard College.

vative behavior of Japanese companies with respect to new product development. Two dimensions of new product development were highlighted: (1) the speed with which new product development takes place, and (2) the flexibility with which companies adapt their development process to changes in the external environment. The basic rationale for treating speed and flexibility as the central issues of our research rests on our belief that they collectively lead to competitive advantages in the forms of increased productivity, reduced costs, improved quality, and higher market share, among others.

Methodology

To understand the dynamic process that enables certain Japanese companies to develop new products rapidly and with maximum flexibility, we selected five innovative models as our primary units of analysis. They include: (1) the FX-3500 copier made by Fuji-Xerox; (2) the City box-car made by Honda; (3) the Auto Boy lens shutter camera made by Canon (known as Sure Shot in the United States); (4) NEC's PC 8000 personal computer; and (5) Epson's MP-80 dot-matrix printer. These five models were selected with the following criteria in mind: market success, innovativeness of product features, strength of impact and visibility within the company as being a "breakthrough" development process, variety of product and process technology utilized, and spread of product categories along the product life cycle curve. See Table 1 for a more detailed description of our units of analysis.

In-depth field research was conducted with the five manufacturers mentioned above, as well as with their affiliated companies and subcontractors. The latter group was included since product development in Japan cannot be viewed solely as an intrafirm activity. As is discussed below, an interorganizational network formed between the manufacturer and its outside suppliers plays an important role toward

making speed and flexibility possible. As shown in Table 2, interviews were conducted with more than forty people from thirteen companies over a six-month period.

Descriptive Model

The key to identifying the various factors that make speed and flexibility possible is to view product development as a dynamic and continuous process of adaptation to changes in the environment. Within this framework, which is presented in Figure 1, companies adapt to changes or uncertainties in the market environment through an iterative process of variety reduction[4] and "learning-by-doing." Even a seemingly minor change in competitive behavior or consumer preference forces a manufacturer to make choices and to engage in learning, as Abernathy et al. point out:

> For whatever reasons—a sudden shift, say, in the prices of substitute products—a demand may arise among buyers for new dimensions of product performance or for a different set of trade-offs among product attributes. If this demand is sufficiently unlike the one it supersedes, producers may need to seek out new technology, to revise design concepts, to reintroduce innovation as an important element in competition, and to undertake a new round of iterative learning.[5]

A large proportion of the differences in the product development process between Japanese and U.S. manufacturers can be explained by examining how variety amplification, variety reduction, and learning actually take place.

INTRAFIRM PROCESS

This chapter consists partly of a seven-piece jigsaw puzzle. In this section, six of the pieces are identified and put into place; they all are

TABLE 1. *Description of the Units of Analysis Used in the Study*

Brand (company)	Product Description	Innovative Features	Life Cycle of Product Category	When Introduced	Approximate Peak Production Units per Year (peak year)	Share of Relevant Domestic Market
1. FX-3500 (Fuji-Xerox)	Medium-sized plain paper copier	·One-half the cost of a high-speed copier ·Speed of 40 copies/min. ·Compact size	Growth	1978	30,000 (1978)	60%
2. City (Honda)	1200cc box-car	·"Tall and short" car concept ·Large interior ·Fuel efficient	Mature	1981	125,000 (1982)	12%
3. Auto Boy (Canon)	Lens shutter camera	·Fully automatic camera with automatic winding and rewinding	Mature	1979	1,400,000 (1982)	20%
4. PC 8000 (NEC)	8-bit personal computer	·First fully packaged personal computer introduced in Japan	Introduction/ growth	1979	105,000 (1981)	45%
5. MP-80 (Epson)	Dot-matrix printer for personal computers	·Separate unit from computer ·Compact	Introduction/ growth	1980	700,000 (1981)	60%

factors supporting a speedy and flexible new product development process from the inside. The final and missing piece, which identifies the contribution of outside suppliers, is examined in the next section. Needless to say, only after all the pieces are put in place can the entire picture be appreciated.

The process of developing new products within Fuji-Xerox, Honda, Canon, NEC, and Epson itself resembles the way in which a jigsaw puzzle is put together. It requires an incremental and iterative process of what Abernathy calls "learning-by-doing" as opposed to "analytic-strategy-synthesis." The "doing" sometimes comes before the "thinking" part. Or in the words of Weick, "How can I know what I think until I see what I say?"[7]

The two processes are also similar in other respects. Variety reduction occurs in jigsaw puzzles by "building up" the pieces from scratch. It is often easier to finish a puzzle if others chip in, especially if they see things differently. Learning takes place as well, as one moves on from a seven-piece set to a more chal-

535

TABLE 2.   Interview List

Brand	Company	No. of Interviewees	Total Interview Hours
FX-3500	Fuji-Xerox[a]	7	16
	Dengen-Automation[b]	2	2
	Sanyo Seisakusho[b]	1	1
	Toritsu-Kogyo[b]	4	2
	Tesco Industrial Co.[c]	1	1
	Mitsuba Electronics[c]	1	1
City	Honda	8	12
	Masuda Manufacturing[b]	1	1
Auto Boy	Canon	6	5
PC 8000	NEC	5	7
MP-80	Epson[d]	4	7
	Shiojiri Kogyo[b]	1	1
	Standard Press[b]	1	1
		42	57

a. Joint venture between Xerox Corporation and Fuji Film.
b. Primary subcontractor.
c. Secondary subcontractor.
d. Independent subsidiary of Seiko Group.

lenging one, with more pieces, almost as a natural course of events.

The six intrafirm factors that contribute to a speedy and flexible development process are as follows:

1. Top management as catalyst
2. Self-organizing project teams
3. Overlapping development phases
4. Multilearning
5. Subtle control
6. Organizational transfer of learning

Top Management as Catalyst

Top management plays a key strategic role in new product development. It provides the initial kickoff to the development process by signaling a broad strategic direction or goal for the company. Top management rarely hands down a clearcut new product concept or a specific work plan. Rather, it intentionally leaves considerable room for discretion and local autonomy to those in charge of the development project. A certain degree of built-in ambiguity is considered healthy, especially in the early stages of development.

Top management decides on a broad strategic direction or goal by constantly monitoring the external environment—that is, competitive threats and market opportunities—and evaluating company strengths and weaknesses. Competitive threats from rival companies, for example, forced Fuji-Xerox and Canon to seek a *reactive* strategic direction:

·   Prior to 1970, Fuji-Xerox was the only plain paper copier (PPC) manufacturer in Japan. But starting with the entry of Canon in the fall of 1970, Japanese manufacturers (such as Konishiroku, Ricoh, Minolta, Copia, and Toshiba) began to make inroads into the PPC market, especially at the low end. By 1977 Ricoh, which introduced a liquid dry PPC model called "DT 1200" in 1975, had overtaken Fuji-Xerox as the market leader

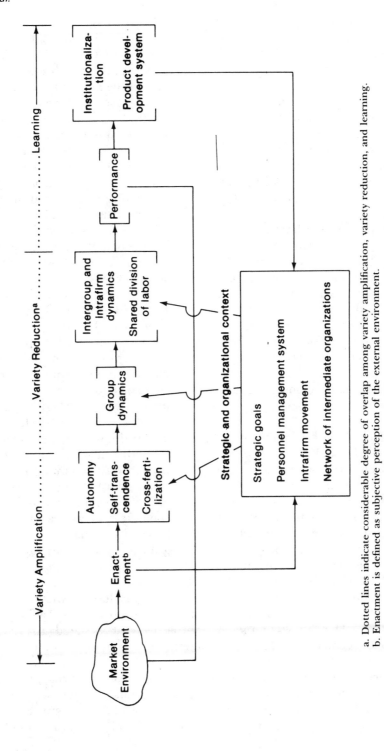

a. Dotted lines indicate considerable degree of overlap among variety amplification, variety reduction, and learning.
b. Enactment is defined as subjective perception of the external environment.

FIGURE 1.   Descriptive Model for New Product Development

in terms of units installed. Fuji-Xerox therefore launched the development of FX-3500 with a sense of urgency and a determination to regain market leadership.

- Canon was the undisputed leader in the medium-priced 35mm lens shutter camera market in the 1960s, as a result of the 1961 introduction of Canonet, a lens shutter camera with an electric exposure capability. It also became the leader in the higher-priced 35mm single-lens reflex (SLR) market with the successful 1976 introduction of the AE-1, which was the first automatic exposure SLR camera with a built-in microprocessor. But while Canon was concentrating its effort on the SLR market in the mid-1970s, Konishiroku overtook it in lens shutter cameras, increasing its share of the Japanese market to over 40 percent by introducing two new Konica models, one with a built-in flash and the second with an automatic focus. It was this serious competitive threat from Konica that prompted Canon to redirect its development efforts behind Auto Boy.

An assessment of market opportunities and company strengths and weaknesses led Honda, NEC, and Epson to pursue more of a *proactive* strategic direction:

- Honda's top management felt a sense of crisis as its best-selling lines—Civic and Accord—were beginning to lose their appeal to the youth market. The City development project was initiated in 1978, just as the postwar generation (under thirty-three years at the time) began to outnumber the prewar generation. The City was targeted toward the youth segment and developed by a young project team, whose average age was only twenty-seven. As is described in greater detail below, this team was given full autonomy to develop "the kind of car that we, the young, would like to drive."

- NEC's strength in microprocessors led to the eventual development of its personal computer PC 8000. NEC began to mass produce microprocessors to reduce costs and, at the same time, broaden their application base. Mr. Watanabe, head of the PC 8000 development team, was originally asked by top management to create a market for microprocessors and to sell them by the bundle. He visited Silicon Valley and saw the successful launching of personal computer prototypes in the United States, which confirmed the large market potential of microprocessor applications within personal computers. As a result, TK 80, a training kit for hobbyists and the predecessor to PC 8000, was introduced in 1976.

- Ever since its establishment in 1961, Epson's history has been marked by what appears to be almost an obsession with creating a market for "products of the future." Epson, which was originally a parts manufacturing plant for Seiko, was the first company in Japan to enter the miniprinter market with its EP-101 just as electronic calculators were taking off in the late 1960s. Epson also developed an electronic printer (MP-80) just when the personal computer market was beginning to boom. And it introduced a letter-sized computer (HC-20) in 1982, opening up a market for hand-held computers. Top management constantly keeps its watchful eye on tomorrow's growth opportunities and directs its development people to think the unthinkable.

Regardless of whether the strategic direction or goal is determined reactively or proactively, it is stated in rather nonspecific terms. For example, Canon's top management directed the Auto Boy team "to think of something new that will surpass all preceding competitive brands." Honda's top management told the City team "to create a radically different concept of what a car should be like." Top management at Fuji-Xerox instructed the FX-3500 team "to come up with a product head and shoulders above others." These directions or goals are intentionally left vague, to give the development team maximum latitude toward creative problem solving.

But, at the same time, top management is not at all hesitant about setting very challenging parameters. Canon's Auto Boy team, for example, was given a free hand to develop an auto-focus camera as long as it was done "on its own." Unlike all other front-runners, who licensed the auto-focusing technology from Honeywell, the challenge was to develop the new product using Canon's original core technology. Similarly, Honda's top management asked for a radically different concept within the constraints that City be a "resource-saving, energy-efficient, mass-oriented automobile." Fuji-Xerox's FX-3500 team was given two years to come up with a new product that could be produced at half the cost of the high-end line and still be equipped with similar performance standards. As a point of reference, it took five years for Fuji-Xerox to develop an earlier domestic model (FX-2200) and over four years for Xerox Corporation in the United States to develop a model comparable to the FX-3500 at that time.

Top management implants a certain degree of tension within the project team by giving it a wide degree of freedom in carrying out a project of great strategic importance to the company and by setting very challenging parameters. This creation of tension, if managed properly, helps to cultivate a "must-do" attitude and a sense of cohesion among members of the crisis-solving project team. Examples of how tension is created are described by top management in the following manner:

- Mr. Kawamoto, vice president of Honda in charge of development, remarked: "At times, management needs to do something drastic like setting the objective, giving the team full responsibility, and keeping its mouth shut. It's like putting the team members on the second floor, removing the ladder, and telling them to jump, or else. I believe creativity is born by pushing people against the wall and pressuring them almost to the extreme."

- Mr. Kobayashi, president of Fuji-Xerox, noted: "I kept on rejecting the proposals repeatedly for about half a year. In retrospect, I'm amazed how persistent I was about sending them back. Engineers can think up all kinds of reasons why something is impossible to do. But I was able to resist giving in because everyone in the company shared an acute sense of crisis."

Self-organizing Project Teams

A new product development team, consisting of members with diverse backgrounds and temperaments, is hand picked by top management and is given a free hand to create something new. Given unconditional backing from the top, this team begins to operate like a corporate entrepreneur and engage in strategic initiatives that go beyond the current corporate domain. Members of this team often risk their reputation and sometimes their career to carry out their role as change agents for the organization at large.

Within the context of evolutionary theory, such a group is said to possess a self-reproductive capability. Several evolutionary theorists use the word "self-organization" to refer to a group capable of creating its own dynamic orderliness.[8] A recent study by Burgelman found that a new venture group within a diversified firm in the United States takes on a self-organizing character.[9] Another study by Nonaka has shown that Japanese companies with a self-organizing characteristic tend to have higher performance records than others.[10]

The creation and, more importantly, the

propagation of this kind of self-organizing product development team within Japanese companies represents a rare opportunity for the organization at large to break away from the built-in rigidity and hierarchy of day-to-day operations. It is quite difficult for a highly structured and seniority-based organization to mobilize itself for change, especially under noncrisis conditions. The effort collapses somewhere in the hierarchy. A new product development team is better suited to serve as a motor for corporate change because of its visibility ("we've been hand picked"), its legitimate power ("we have the unconditional support from the top to create something new"), and its sense of mission ("we're working to solve a crisis situation").

To become self-organizing, a group needs three qualifications. First of all, it has to be completely autonomous. Our case studies support this condition. For example:

· A Honda City design engineer recalled: "It is incredible how the company called in young engineers like ourselves to design a new car and gave us the freedom to do it our way." Mr. Kawashima, then the president of Honda, promised at the outset that he would not intervene with the City project. "Yes, we've given them freedom," commented Mr. Kawamoto, vice president in charge of development, "but we've also transferred a strong sense of responsibility to them."

· Mr. Watanabe, who headed the PC 8000 project for NEC, recalled: "We were given the go-ahead from top management to proceed with the project provided that we would work all by ourselves in developing the product and be responsible to manufacture, sell, and service the product on our own as well."

Second, given this autonomy, a group comes up with extremely challenging goals on

its own and tries to keep elevating these goals. It does not seem to be content with incremental improvements alone and is in constant pursuit of a quantum leap. We observed this tendency toward "self-transcendence," or the creative overcoming of the status quo, in all of the development projects we analyzed. For example:

· Epson typifies a company with a never-ending quest for approaching the limit. Its corporate target is to have the next-generation model developed by the time the first-generation model is introduced on the market. There is an unwritten rule that the next-generation model be at least 40 percent better than the existing model.

· The self-motivating factor behind Canon's Auto Boy development team was to improve on the company's achievement with its AE-1 in the single-lens reflex camera market. "You weren't even considered human at that time if you weren't somehow associated with the AE-1," said one of the Auto Boy team members, half jokingly. "We wanted to challenge that legacy." The end result was a fully automatic lens shutter camera with an improved auto-focusing technology that became the top-selling brand immediately after introduction, despite a two-year lag in market entry.

· Both Fuji-Xerox and Honda challenged the status quo. Fuji-Xerox overcame the preconceived notion within the company that a new product normally be developed through conversion engineering (of a U.S. model) and undertook a self-development program. Similarly, Honda overcame what appeared to be an unshakable market preference for a "long and low" car and developed a "short and tall" City.

Third, a self-organizing group is usually composed of members of diverse functional specializations, thought processes, or behavior patterns. The total becomes much more than the sum of its parts when these members assemble and begin to interact with each other. Variety is amplified and new ideas are generated as a result. We found this phenomenon, which can be termed cross-fertilization, to be widespread among the companies interviewed. Cross-fertilization can be seen in the following examples:

· Honda's City development team members included representatives from product development (D), production engineering (E), and sales (S), as shown in Figure 2.

Interactions across these functional boundaries were substantial. Throughout the City development project, which took three years, more than 2,000 visits from E to D and back were recorded by team members and other employees involved in the project. The two physical locations (Suzuka for E and Wako for D) were 300 miles apart, or five hours by train.

· Fuji-Xerox's FX-3500 was also developed by a multifunctional team, consisting of managers from planning, design, production, sales, distribution, and evaluation.

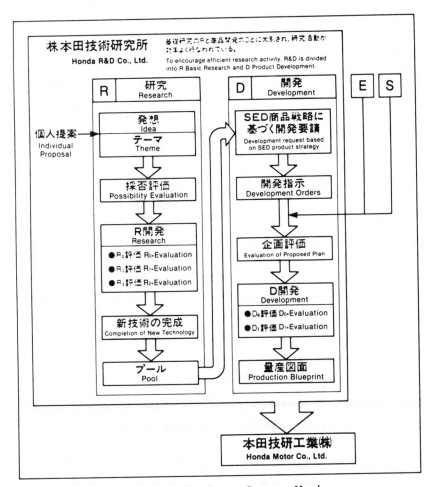

FIGURE 2. *Development System at Honda*

Unlike City, these managers were physically located in one large room where open communication and sharing of information took place continuously.

Figure 3 is a listing of the functional backgrounds of the key team members for the five development projects we analyzed.

Overlapping Development Phases

The group dynamics of the self-organizing team strongly influence the manner in which the development project proceeds. The autonomous, self-transcendent, cross-fertilizing nature of the group produces a unique set of dynamics. For example:

- Cohesion is promoted as team members face some challenging goals. The broad nature of the goals also helps to alleviate detailed differences.

- Ambiguity is tolerated, given the diverse backgrounds of the group.
- Overspecification is avoided, since it may impair creativity.
- Sharing of information is encouraged so as to become better acquainted with the realities of the market.
- Decision making is intentionally delayed to extract as much up-to-date information as possible from the marketplace and technical communities.
- Sharing of responsibility is accepted as the group embarks on a risk-taking mission.

These dynamics help to explain why phase management in the Japanese companies we investigated tends to be holistic and overlapping rather than analytical and sequential. Variety reduction is delayed as long as possible in these companies as the self-organizing team engages itself in the search for information— from both the marketplace and the technical

Functional Background[b] Company	R&D	Production	Sales	Planning	Service	Quality control	Others	Total
Fuji-Xerox	5	4	1	4	1	1	1	17
Honda	18	6	4	-	1	1		30
Canon	12	10	-	-	-	2	4	28
NEC	5	-	2	2	2		-	11
Epson	10	10	8	-	-		-	28

a. Numbers indicate number of key team members.
b. Designates the assignment prior to joining the new product development team.

FIGURE 3.   Functional Background of the Key New Product Development Team Members[a]

communities—and in an iterative process of experimenting even at the very late phases of the development process. As mentioned earlier, variety reduction under the analytical/sequential approach is conducted more systematically, and as early as possible.

A simplified illustration of the overlapping nature of phase management is depicted in Figure 4. The sequential approach, labeled Type A, is typified by the NASA-type phased program planning (PPP) system adopted by a number of U.S. companies. Under this system, a new product development project moves through different phases—for example, concept, feasibility, definition, design, and production—in a logical, step-by-step fashion. The project proceeds to the next phase only after all the requirements are satisfied, thereby minimizing risk. But, at the same time, a bottleneck in one phase can slow down the timetable of the entire development process. The overlapping approach is represented by Type B, in which the overlapping occurs only at the interface of adjacent phases, and Type C, in which the overlapping extends across several phases. The product development process at Fuji-Xerox and Honda is closer to Type C than to Type B.

The overlapping approach has its merits and demerits as well. The more obvious merits include: faster speed of development, increased flexibility, and sharing of information. It also leads to a more subtle, but equally important, set of merits dealing with human resources management. Among others, it helps to do the following:

- Foster the more strategic point of view of a generalist
- Enhance shared responsibility and cooperation
- Stimulate involvement and commitment
- Sharpen a problem-solving orientation
- Encourage initiative taking
- Develop diversified skills
- Create grounds for peer recognition
- Increase sensitivity of everyone involved to changes in market conditions

On the other hand, the burden of managing the process increases exponentially. By necessity, the overlapping approach amplifies ambiguity, tension, and conflict within the group. The burden of coordinating the intake and dissemination of information rises as well, as does the responsibility for management to carry out on-the-job training on an ad hoc and intensive basis.

The loose coupling of phases also makes division of labor, in the strict sense of the word, ineffective. Division of labor works well in a Type A system where the tasks to be accomplished in each phase are clearly delineated and defined. Each project member knows his or her responsibility, seeks depth of knowledge in a specialized area, and is evaluated on an individ-

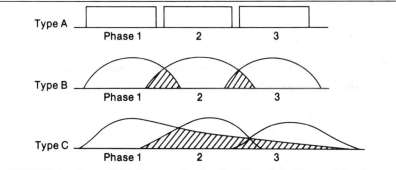

FIGURE 4. *Sequential (A) versus Overlapping (B and C) Phases of Development*

ual basis. But such segmentalism, to use Kanter's terminology,[11] works against the grain of a loosely coupled system (that is, Type B and Type C) where the norm is to reach out across functional boundaries as well as across different phases. Project members are expected to interact with each other extensively, to share everything from risk, responsibility, information, to decision making, and to acquire breadth of knowledge and skills. Under a loosely coupled system, then, the tasks can only be accomplished through what we call "shared division of labor." This shared division of labor takes place not only within the company but also with outside suppliers. We witnessed varying degrees of coupling or overlapping among the Japanese companies:

• Fuji-Xerox revised the PPP system, which it inherited from the parent company, in two respects. First, it reduced the number of phases required to develop FX-3500 from six to four by redefining some of the phases and aggregating them differently. Second, a linear, sequential system was changed into what Mr. Kobayashi referred to as a "sashimi" system. Sashimi—or sliced raw fish—is served in Japan by tilting the slices and placing them on a plate with one piece overlapping another. (See Figure 5 for a detailed graphic representation.) Such a system requires extensive social interactions on the part of all those involved in the project, as well as the existence of a cooperative network with suppliers. On this latter

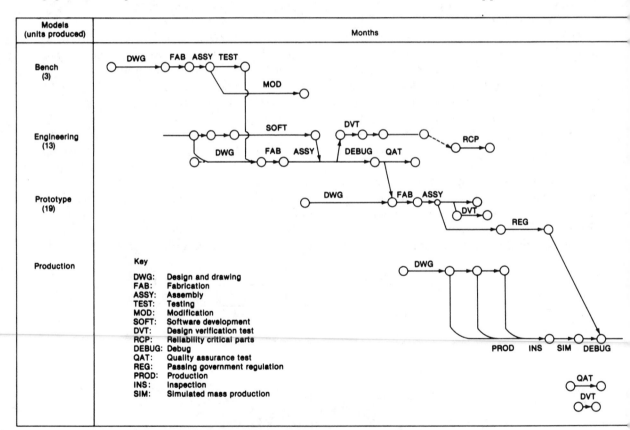

FIGURE 5.   Overlapping Nature of the Development Schedule at Fuji-Xerox

point, a project member in charge of design commented as follows: "We ask our suppliers to come to our factory and start working together with us as early in the development process as possible. The suppliers also don't mind our visiting their plants. This kind of mutual exchange and openness about information works to enhance flexibility. Early participation on the part of the supplier enables them to understand where they are positioned within the entire process. Furthermore, by working with us on a regular basis, they learn how to bring in precisely what we are looking for, even if we only show them a rough sketch. When we reach this point, our designers can simply concentrate on work requiring creative thinking." As a result of these efforts, Fuji-Xerox was able to shorten the development time from thirty-eight months on a similar prior model to twenty-nine months for FX-3500.

· Canon experienced several conflict situations as the Auto Boy development proceeded with an extensive degree of overlapping. One project member recalled: "Someone from development thinks that if one out of 100 is good, that's a clear sign for going ahead. Someone from production thinks that if one out of 100 is not good, we've got to start all over. That gap has to be narrowed. But both sides have absolutely no question in their minds that the conflict can be resolved." Conflict induces high levels of social interaction, which in turn gives rise to creative solutions, according to the same project member: "The design people keep a watchful eye on the entire project to make sure that their original design becomes converted into a truly good product at the very end. The production people, on the other hand, try to come to grips with what the designer had in mind by asking themselves, 'Why did he do it this way?' Creative solutions are

born by each side intruding on the turf of the other." This mutual "reaching out" enabled Canon to remain flexible even under intense conflict situations.

· Honda's City team adopted what we decided to call a "rugby" approach toward product development. Mr. Watanabe explained: "I always tell my team members that our work cannot be done on the basis of a relay. In a relay someone says, 'My job is done, now you take it from here.' But that's not right. Everyone has to run the entire distance. Like in rugby, every member of the team runs together, tosses the ball left and right, and dashes toward the goal." The important point to remember here is that critical problems occur most frequently at relay points within the sequential approach. The "rugby" approach smoothes out the process by involving everyone in the development project. Individual initiative is also a prerequisite, argued Mr. Kawamoto: "If each and every one of us does his or her job well, then we basically won't need a structure." In fact, we found that Honda's project members deviated freely from the step-by-step structure for product development, shown in Figure 2.

Multilearning

The five Japanese companies in our study have another commonality that became increasingly clearer as our study progressed. They possess an almost fanatical devotion to learning—both within organizational membership and with outside members of the interorganizational network. To them, learning is something that takes place continuously in a highly adaptive and interactive manner. This discovery is neither new nor original. Lawrence and Dyer, for example, point this out as follows: "Japanese management does not now need to be convinced, if it ever did, that their organizations are learning and social systems as well as production systems."[12]

What is somewhat new and original is the discovery that learning plays a key role in enabling companies to achieve speed and flexibility within the new product development process. Continuous interactions with outside information sources, for instance, allow them to respond rapidly to changing market preferences. The iterative process of trial-and-error or learning-by-doing gives considerable degrees of freedom in responding to outside challenges or to challenging goals emanating from within. The constant encouragement to acquire diversified knowledge and skills also helps to create a versatile team capable of solving a wide array of problems in a relatively short period of time.

Broadly speaking, learning manifests itself within the organization along two dimensions: across multiple levels, and across multiple functions. We decided to refer collectively to these two types of intrafirm learning—that is, multilevel and multifunctional—as "multilearning."

We witnessed Japanese companies promoting learning at three levels—individual, group, and companywide. Examples of each are described below:

- Epson encouraged learning at the individual level to develop as many generalists within the company as possible. Mr. Aizawa, executive vice president in charge of R&D, stressed this point when he said: "I have been telling my development staff members that they need to be both an engineer and a marketer in order to be promoted within our firm. Even in an engineering company like ours you can't get on top without the ability to foresee future developments in the market."
- Honda fostered learning within the production group by establishing a special corner within the factory where the rank-and-

file workers could experiment with work simplification ideas during normal working hours, using tools and materials provided by the company. Referred to as the "handmade automation" program, it has been instrumental in elevating the skill levels of workers in the manufacturing group and in positioning automation within the minds of the workers as a positive force that could make their work simpler and safer.
- Fuji-Xerox utilized the TQC (Total Quality Control) movement on a companywide basis as a means of learning how to bring about a more creative and speedy new product development process. The switch from the PPP system to the "sashimi" system came about as a result of the TQC movement.

We also witnessed Japanese companies treating "learning in breadth," or learning across functional lines, as the cornerstone of their human resources management program. Several examples of this very pervasive practice include the following:

- All of the project members who developed Epson's first miniprinter (EP-101) were mechanical engineers. They knew very little about electronics at the start of the project. In fact, Mr. Aizawa, also a mechanical engineer by training, returned to his university and studied electrical engineering for two years as a special researcher while trying to serve as the project leader of the EP-101 team. All project members were well versed in electronics by the time EP-101 was completed.
- A small group of sales engineers from the Electronic Devices Division, who originally sold microprocessors, were responsible for developing NEC's PC 8000. They acquired much of the necessary know-how by (1) putting TK 80, a computer kit, on the market two years prior to the PC 8000 intro-

duction; (2) setting up an NEC service center called BIT-IN in the middle of Akihabara, a consumer electronics center in Tokyo, soon after introducing TK 80; (3) stationing themselves for about a year, even on weekends, at BIT-IN; and (4) interacting with hobbyists who frequented BIT-IN, which almost became a "club," and extracting as much useful information from them as possible.

- Mr. Nakamura, who had recently become a department manager of the assembly line operation at Honda's motorcycle plant in Suzuka, said that the recent lateral move took him by surprise. He admitted: "After all, I was in the painting operation for twenty-two years. I guess management wanted some change. But for me, it meant learning everything all over again from scratch."

Honda also has a so-called "practical training" program in which all department managers, like Mr. Nakamura, are asked to select a functional area in which they have never worked before and to spend one week every two years "getting their hands dirty." NEC enhances mobility across functional lines by transferring technical people from its R&D center to its divisions. As Mr. Miya, director of R&D, noted: "When a researcher starts producing results, the division comes to us and says, 'Give us that person.' Our current president started out in R&D and was transferred to the division." NEC's rotation plan, shown in Figure 6, calls for a transfer of more than half of the newly recruited researchers (holding a master's degree) from R&D to the divisions at the end of about ten years and more than 80 percent after twenty years. Canon also utilizes similar programs of employee exchange and job rotation to encourage its employees to become "U-shaped" individuals—that is, individuals with a broad base of skills and knowledge.

Other personnel policies commonly found in Japanese companies—such as long-term employment and group evaluation—also foster multifunctional learning. Long-term employment makes the retention of valuable lessons from the past possible. Since people do not leave the company, members of a project team can easily seek out past "stars" and extract words of wisdom from them. Group evaluation enhances interactions within the group and fosters sharing of skills and know-how among each other.

From management's point of view, a new product development project offers an ideal springboard for creating a group of employees with broad skills and knowledge and an organizational climate conducive to bringing about change. But from the team members' point of view, multilearning requires an extraordinary amount of effort and dedication. Of course, it helps to have the rails already laid out (in the form of personnel policies that facilitate multilearning). It also helps to be working with a self-organizing team, a crisis situation, and an overlapping system of development. But management did not happen to find these supporting factors by chance. Rather, management made them happen.

Subtle Control

Some checks are needed to prevent looseness, ambiguity, tension, or conflict from getting out of control. These are manifested within the five companies we studied in subtle forms of control, rather than in more formal or systematic forms. Consistent with the self-organizing nature of the team, the emphasis is on self-control and on control through peer pressure or "control by love." More specifically, management uses selection of team members, openness in working environment, sharing of information, group-oriented evaluation, and sharing of values as a mechanism for implanting subtle control within the product development process. Each of these mechanisms is discussed below.

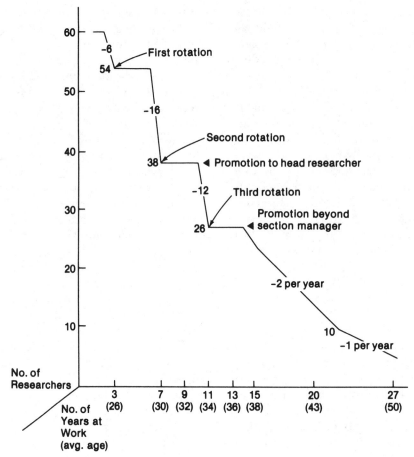

a. Job rotation of newly hired graduates with a master's degree.

FIGURE 6.   *Job Rotation Plan of R&D Researchers at NEC*[a]

First, management implants the seeds of control by selecting the right people onto the project team, constantly monitoring the balance in team membership, and adding or deleting specific members if deemed necessary. For example:

Honda handpicked team members mostly in their twenties to develop the youth-oriented City. As Mr. Kawamoto noted: "It's our responsibility to assign the right individuals to the appropriate positions. We also need to monitor the project closely and transfuse new blood midstream into the project. An elderly or conservative member may be added to the team when the balance shifts too dangerously toward a radically new approach. Or an engineer with a different technical background may be added when the project appears to have hit a stalemate."[13]

Mr. Yoshino, who supervised the FX-3500 project, recalled: "When the design manager was assigned, I told him to give me the names of people he'd like to have. It wasn't 100 percent, but we were able to uproot the people we wanted with a probability of about 90 percent. And if we thought that someone was not living up to our expectations, I'd send the word out to have that person replaced midstream in the process."

Second, subtle control is exercised in the form of an open and visible working environment. For example:

- Honda enhances visibility by holding meetings in a large room with glass walls. "We can see what other people are up to," commented a City team member. This philosophy is also reflected at the executive level, where all the top executives have desks in one large room and hold meetings (among themselves and with those reporting to them) around three round tables situated in the center of that large room.
- Fuji-Xerox also espouses this so-called "large-room" system. Mr. Suzuki, a project member for FX-3500, gave the rationale for this system as follows: "When all the team members are located in one large room, someone's information becomes yours even without trying. You then start thinking in terms of what's best or second best for the group at large and not only about where you stand. If everyone understands the other person's position, then each of us is more willing to give in or at least try to talk to each other. Initiatives emerge as a result."

Third, management implants seeds of subtle control by encouraging team members to extract as much information from the field—from customers and competitors—and, more importantly, to bounce them off other members. This sharing of information helps to keep everyone up to date, build cohesion within the group, and act as a source of peer pressure:

- Fuji-Xerox encourages its design people to go out into the field to talk with users and suppliers during slack periods. A member of the design team commented: "We go out into the field to listen to the voice of our customers, study our competitors' models, or search for any useful research findings. As a result, we are in a better position to finalize the product concept as we see fit and to respond to what a planner has to say about quality, cost, or delivery time. We try to digest what we sensed in the field and have it reflected in the design. This exposure helps in a lot of ways. For example, a designer may be tempted to take the easy way out at times, but he may reflect on what the customer had to say and try to find some way of meeting that requirement. Even if we get into an argument, we can go back to what we saw and heard in the field and build a consensus around our common experience. As you can see, we're constantly trying to find ways of feeding back the demands from the field into the design of the product."
- Sharing information about competition also helps to push the project to a higher level. As the same member of the Fuji-Xerox design team observed: "The development process in Japan is characterized by a daily influx of information about competition. This information exerts pressure which, in turn, acts as a driving force for us to do better. I wonder if this kind of pressure we feel in Japan—i.e., pressure every day by every member of the team—is felt in the U.S. as well."

Fourth, the Japanese evaluation system, which is based on group rather than individual performance, serves as another form of subtle control. Such a system encourages the formation of a self-organizing team and fosters multilearning among the team members. It also helps to build trust and cohesion, on the one hand, and peer pressure, on the other.

Lastly, management exerts subtle control by establishing some overriding values shared by everyone in the organization. These values or basic beliefs give people within the organization a sense of how to behave and what they ought to be doing. The shared value at NEC is "C & C," which stands for "Computer and Communication." NEC, which started out as a communications company, has made major inroads into the computer business under this slogan. The shared value at Canon is to become a "superior company." It is now in the second phase of the "superior company" plan, whose core theme is the strengthening of its R&D capability. The shared value at Epson, which is "thinking the unthinkable," sets a very aggressive pattern for product development, as exemplified by the unwritten "40 percent improvement" rule.

Organizational Transfer of Learning

We noted earlier that all those involved in the development project are engaged in a constant process of learning, across both levels and functions. The know-how accumulated at the individual level is transferred to other divisions or to subsequent projects within the organization and becomes institutionalized over time. Lawrence and Dyer describe this process as follows:

> It is true . . . that members of an organization cannot only learn as individuals but can transmit their learning to others, can codify it and embody it in the standard procedures of the organization. In this limited sense, the organization can be said to learn. When certain organizational arrangements are in place, an organization will foster the learning of its members and take the follow-up steps that convert that learning into standard practice. Then it is functioning as a learning system, generating innovations.[14]

We observed the process of institutionalization most vividly within Fuji-Xerox and Canon:

- The FX-3500 project was instrumental in making self-development the standard practice within Fuji-Xerox. Prior to this project, new product development was synonymous with conversion engineering, in which a basic U.S. model was converted into a Japanese one through minor engineering modifications. As mentioned earlier, self-development was accompanied by major process improvements, including the condensing of phases from six to four.

- Canon established a model for new product development through the AE-1 project and refined it in the Auto Boy project. Prior to these projects, Canon did not have a standardized format for new product development. One former member of the Auto Boy project team recalled: "When we were developing Auto Boy, we used to meet once a month or so to exchange notes on individual subprojects in progress, and once in three months or so to discuss the project from a larger perspective. We didn't know it at the time, but the pattern became institutionalized into monthly and quarterly progress reviews later." The know-how accumulated in these two camera projects was transferred to the Business Machines Group when it developed PC-10, a personal or microcopier introduced in 1982. Project members for the PC-10 sought out previous leaders of the AE-1 and Auto Boy projects to extract as many live lessons as possible.

The importance of having a role model within the organization is emphasized by Mr. Kawamoto of Honda as well: "Leave the organization somewhat untidy and leave sufficient room for self-growth to emerge. Everyone will be able to see that one team is doing something outstanding. There's no need for everyone to be doing it, but one outstanding example will set the standards for others." But since these teams are dissolved after the product is developed, whatever learning takes place is carried over to the next generation through individual team members. Mr. Kawamoto continued: "If the factory is up and running and the early-period claims are resolved, we dismantle the project team, leaving only a few people to follow through. We only have a limited number of very able people, so we turn them loose to another key project immediately."

Besides passing down "words of wisdom" from the past and establishing standard practices within the organization, we also noticed a simultaneous attempt on the part of Japanese companies to "unlearn" the lessons of the past and to engage themselves in a continuous process of what Schumpeter called "creative destruction." This process of unlearning helps to prevent the development process from becoming too rigid. It also acts as the springboard for making further incremental improvements. For example:

- Fuji-Xerox has been reviewing and refining its "sashimi" approach toward product development. Compared with the days of the FX-3500, the manpower required to develop a comparable new product today has been reduced to about one-half and the product development cycle to twenty-four months. Among other things, it allowed suppliers to participate from the early stages of development and eliminated the prototype production phase from the development process.
- Mr. Aizawa of Epson described the continuous nature of the development process as follows: "We have this constant fear that we're going to be left behind. That's why we want to have a product in test production when a new product is being introduced in the market." Epson always approaches a new product idea from two opposing points of view. One idea is pitted squarely against another even when developing the next generation model of a successful product already on the market. This approach opens the door for unlearning to

take place and helps to maximize flexibility within the development process.

- Honda had to unlearn the lessons from the past to develop a totally new concept of cars. Mr. Watanabe recalled: "At one point in time, we had a choice between a modified version of Civic or a totally different 'short and tall' car. We opted for the latter, knowing full well the risks involved."

We also observed that much of the unlearning is triggered when a new crisis from the market environment confronts the organization. What used to work in the past is no longer valid, given the changes in the external environment. To adapt to these changes, the challenge is to retain some of the useful learning accumulated from the past and, at the same time, throw away that portion of learning which is no longer applicable. In this regard, we agree with Levitt, who says that "translation of knowledge into results is almost surely a matter of 'tinkering' and, more importantly, a matter of management."[15]

INTERORGANIZATIONAL NETWORK

Six intrafirm factors that lead to speedy and flexible new product development processes in Japan have now been examined. To use a mundane analogy, we have just completed looking at the inside of the house we are interested in buying. To stop here is like deciding to buy a house without even looking at the outside. This section of the paper examines the impact of the outside—that is, outside suppliers and the interorganizational network surrounding our five companies—on new product development. Our study shows that interorganizational factors make just as important a contribution as intrafirm factors in speeding up the new product development process, as well as in making it more flexible.

What do we mean by an interorganizational network? The concept may still be somewhat foreign to many U.S. executives, although Schonberger observes the following: "In the past few years U.S. executives have been exposed to news stories on the Japanese success phenomenon, including stories on the Japanese tendency . . . to rely heavily on extensive networks of suppliers."[16]

Most of the large Japanese companies in the machinery business, including our five sample firms, utilize an interorganizational network similar to the one presented in Figure 7. The overall framework consists of three separate networks:

A. Affiliated network: consists of affiliated companies that supply raw materials and parts or serve as sales organizations. Honda, for example purchases parts from outside vendors, such as Nippon Denso, but also buys from its affiliated companies such as Keihin Seiki and Yachiyo Kogyo.

B. Supplier network: consists of small- and medium-sized companies that manufacture and process parts. These companies are also starting to take the initiative of developing and test-producing their own parts.

C. R&D network: consists of research institutions (either company-, government-, or university-owned) with a cooperative relationship on R&D.

Of these three, we will focus our attention on the supplier network in this chapter, since it has the most direct impact on the speed and flexibility of new product development.

To cite a specific example on how a supplier can speed up product development, consider the progress made by two vendors of Fuji-Xerox in shortening the delivery time for some parts, as shown in Table 3. Both vendors have been able to reduce the delivery time at least by one-half between 1978 and 1983.

To illustrate how flexibility can be achieved, consider how the supplier network is structured for Fuji-Xerox (see Figure 8). Toritsu-Kogyo, one of its primary subcontractors, operates a factory of its own and utilizes seventy-seven secondary subcontractors. A large number of these subcontractors are located within walking distance from Toritsu-Kogyo. Toritsu-Kogyo is a fairly large subcontractor, by Japanese standards, with sales of about 7 million and fifty employees, while most of the secondary subcontractors have fewer than ten employees. A clear-cut division of

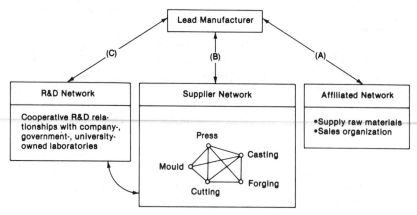

FIGURE 7.   *Interorganizational Network*

TABLE 3. *Time Required to Deliver Parts to Fuji-Xerox*

	Model/Year			
Vendor/Part	Model 3500 1976	Model 4800 1978	Model 4370 1981	1983
Dengen Automation, Inc.				
Bench model	10 mos.	6 mos.	—	3–4 mos.
Feasibility model	4 mos.	4 mos.	2 mos.	2 mos.
Sanyo Seisakusho, Inc.				
Specific part (A)		1.5 mos.	1 mo.	3 wks.
General part (B)		4 wks.	3 wks.	2 wks.

labor sets in as each secondary subcontractor tries to differentiate itself from the rest according to two criteria. The first differentiation occurs on the basis of the seventeen skills shown in Figure 8. The second differentiation takes place on the basis of products handled. The six secondary subcontractors in pressing, for example, may each handle a different-sized product.

Because of this division of labor, each secondary subcontractor acquires a high level of technological skill in one specialized area, as well as a high level of competence in problem-solving. Thus, secondary subcontractors can respond very quickly to special requests made by either Toritsu-Kogyo or Fuji-Xerox, and can adapt very effectively to changes in the environment.

Fuji-Xerox can consider these secondary subcontractors as highly specialized and skilled task forces possessing up-to-date information who can be called into the product development process when necessary. All seventy-seven of these task forces can be mobilized very quickly since they are physically located in one general area. They can be mobilized in any combination or in any sequence as Toritsu-Kogyo selects and coordinates the most appropriate secondary subcontractors for the job at hand.

An understanding of how the network is structured is a prerequisite for studying the dynamic process by which speed and flexibility are achieved within the supplier network. The general conceptual framework for analyzing the intrafirm process presented earlier (Figure 1) applies, to a large extent, to the interorganizational process as well. Suppliers also adapt to changes and uncertainties in the market environment through an iterative process of variety reduction and learning-by-doing. The similarities as well as some differences are discussed below to further understanding of how a supplier network can contribute to new product development.

Self-organizing Network

Various supplier networks emerged in Japan during the postwar economic boom as a natural response to the expansion in the market size for goods and services. One similarity of these interorganizational networks to the intrafirm development teams is the self-organizing manner in which they both emerged. For one thing, supplier networks are made up of autonomous firms that gathered around the lead manufacturer's plants, most of which were located in Tokyo because of its development in the postwar period as the focal market for most goods and services. So it was only natural for the suppliers to locate in Tokyo as well, since that proximity would bring several mutual benefits, such as shorter delivery time, lower inventory

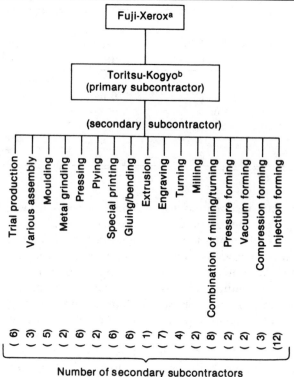

a. Has other primary subcontractors.
b. Serves as subcontractor for other manufacturers.

FIGURE 8.   Interorganizational Network for Fuji-Xerox

carrying costs, lower transportation costs, and improved communication.[17]

The number of these Tokyo-based suppliers grew rapidly, as expert machinists with an entrepreneurial flair left their positions in larger companies to start up their own ventures. Each supplier, equipped with a unique set of skills, would initially enter the market in a small way by finding a particular niche. As these ventures grew, some of the more talented employees would create spin-offs by developing new skills or mastering a new technology. Since these employees were usually treated like members of the extended family, their former

bosses viewed these spin-offs as a natural course of events, like a child leaving home to become independent. Some would even go as far as lending financial support.

These spin-offs are also viewed positively by the lead manufacturer. They generally raise the overall level of skills and technology within the network as each start-up company enters the market with some sort of incremental improvement. As the number of narrowly focused firms proliferates, they also give the lead manufacturer a greater degree of freedom in selecting a subcontractor.

In addition to being autonomous, these

small firms possess a strong motivation to constantly upgrade their skills or technology, thereby satisfying the second qualification for self-organization (self-transcendence). Most of the suppliers pride themselves on running a flexible operation capable of producing special-order or experimental items at short notice. But each time the lead manufacturer makes very tough demands—shorter delivery time or higher technical content—the suppliers have to overcome the status quo and reach out for something higher.

The third qualification for cross-fertilization is also evident as suppliers with differentiated functional skills, technological backgrounds, and behavior patterns are assembled in one general location to form a loosely coupled network. Although these suppliers are autonomous units, we observed substantial degrees of social interaction, open communication, and information exchange taking place among them.

Shared Division of Labor

Self-organization fosters variety amplification as the number of start-ups and spin-offs proliferates within the process. Variety reduction takes over as these autonomous units form a loosely coupled network and begin to exercise shared division of labor. Recall how members of the development team reached out across functional boundaries, interacted with each other, and shared risk, responsibility, and information among themselves. A similar phenomenon occurs among members of the supplier network as well.

Each supplier knows that it cannot survive on its own. Its survival is very much a function of how well it can coexist with others within the network. To coexist, each supplier has to think of the common good and behave like a team player. But coexistence alone does not ensure growth. Each member has to sharpen its own skills, and at the same time, mutually sup-

port the continuity of the network at large. Mutual support is enhanced through tightly knit interactions among the members on a day-to-day basis to exchange relevant business information and ideas. Continuity is maintained by establishing a system of shared division of labor among the network members.

The Fuji-Xerox network, presented in Figure 8, shows how seventy-seven secondary subcontractors are organized to perform seventeen separate functions for Toritsu-Kogyo. A clear-cut division of labor is in place as each of the seventy-seven specializes either on the basis of function performed or product type handled. The Toritsu-Kogyo example may suggest that the network is composed only of subcontractors with very specialized skills. But a separate group of subcontractors who handle standardized processing is also in existence. Such nonspecialized subcontractors take part in the division of labor by working on evenings or holidays to speed up the overall process.

The shared nature of division of labor can be graphically represented by taking the static organization chart shown in Figure 8 and overlaying horizontal lines interconnecting the seventy-seven subcontractors. In other words, these subcontractors are mutually dependent and are in constant interaction with each other. A considerable amount of reaching out takes place as they try to share risk, responsibility, and information. For the entire network to be effective (or to put it differently, for each subcontractor to survive), the subcontractors need to run together like a rugby team, maintaining cohesiveness and balance.

A company's new product development project benefits from the existence of an interrelated group of specialized suppliers on the outside. It is in a position, for example, to invite specific suppliers to join the project team in the early phases of the program to develop or test produce some parts. In the case of Fuji-Xerox, 90 percent of the parts used during test production are manufactured outside. In the later phases of the project, including mass production, it can rely on the primary subcontractor to select and control the right number of qualified secondary subcontractors. This delegation of authority to someone outside adds another level of flexibility to the new product development process. All of the Japanese companies in our study rely heavily on outside suppliers to produce parts at the mass production stage. The percentage of parts produced outside for our units of analysis (that is, FX-3500, City, and so forth) were roughly 90 percent for Fuji-Xerox, between 70–75 percent for Honda, about 65 percent for Canon, over 70 percent for NEC, and 70 percent for Epson.

Learning

We noted earlier that learning at the intrafirm level played a key role in enabling companies to achieve speed and flexibility within product development. The same conclusion can be drawn about learning at the interorganizational level, although the nature of learning is quite different at the two levels. In the former, learning (or multilearning as we called it) has a strong human resources management orientation, whereas learning at the latter has a straightforward economics orientation.

We know that learning takes place within the production process in the form of lower costs as the production volume increases over a given period of time. This so-called "learning curve effect" takes place, albeit to a lesser extent, among subcontractors as well. We observed a learning curve among them, both for lot production and for line production, indicated by line A in Figure 9.

Can we expect a similar learning curve effect during trial production? In general, the answer is no, since requests for a bench model or a prototype come a little at a time and trial production volume is low. But subcontractors within the supplier network are an exception. They are able to realize a learning curve effect even for bench models.

This exception is made possible through what we call "learning in arrangement." Although numerous companies order bench models and their specifications call for different shapes and sizes, many of the orders require the same production technology and skill. So the basic difference boils down to the materials used and the arrangement of work flow. Thus, if learning can take place in how to arrange the work flow effectively, some savings in cost and time are possible. Even if different types of bench models are being produced, the production process can be made to run continuously without having to halt every time a new bench model is introduced.

A supplier network facilitates learning in

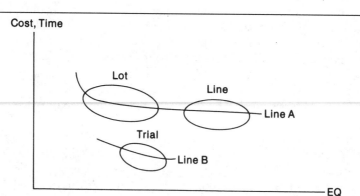

FIGURE 9.   *Learning Curve Effect among Subcontractors*

arrangement for the following reasons. First, several suppliers that specialize in prototype production exist within the network. In the supplier network for Fuji-Xerox, six such subcontractors were present (see Figure 8). Second, since each of these six specialized further according to the types of products handled, members of the network knew which type of prototype to send to whom. Third, this specialization allows the subcontractors to receive the same kinds of orders from numerous sources, thereby giving them some advantage in volume. Fourth, the geographic concentration of these specialized subcontractors in or near Tokyo prompts referrals from other companies outside the network.

Lower costs and time—achieved through learning at an early phase of product development—have a positive impact on the overall process. The manufacturer is induced to experiment with a wide variety of prototypes, thereby keeping flexibility at the maximum. At the same time, it speeds up the development process.

Information Exchange

As noted earlier, members of the supplier network try to build mutual support by actively interacting with each other and exchanging as much useful information as possible. These exchanges help them to keep abreast of the most recent developments in the marketplace and the technical community. They also expedite the development process, as suppliers can have vital parts ready just in time for the manufacturer.

Information exchange takes place both laterally and vertically. Lateral exchange in the case of Toritsu-Kogyo occurs among its seventy-seven secondary subcontractors. These subcontractors share several commonalities that facilitate the flow of information. First, they all work for the same primary subcontractor. Second, their factories are located within

walking distance from each other in downtown Tokyo. Third, most of the owners live there as well, which means that they can be in touch with each other even after working hours. Fourth, almost all of the owners are machinists by training and, therefore, share similar backgrounds. Fifth, they tend to share a trait typical of downtown Tokyo when communicating with each other—to use few words and be as straightforward as possible.

Lateral flow of information is also intensified as a result of the "weak tie" nature of the way in which the secondary subcontractors are organized. Previous research has shown that a weak tie makes a faster and broader exchange of information possible.[18] Those who emit the information feel free to say whatever they please on a wide variety of topics; they feel it is the responsibility of those receiving the information to screen out and digest what they have heard. Contained in this kind of free-wheeling exchange is "leading-edge" information on what is happening in the marketplace and the technical community, as well as hints on how to improve existing products or what new products to develop in the future.

Vertical information exchange occurs within the three levels of hierarchy—the lead manufacturer, primary subcontractor, and secondary subcontractor. The "strong tie" nature of the organization of the vertical network gives rise to a more orderly and planned exchange of information. For example, both top-down flow and bottom-up flow of information are funneled through the primary subcontractor, which plays an important role as the link between the lead manufacturer and the secondary subcontractor.

But the tie is not so strong as to prevent direct exchange of information from taking place. A division or department manager of the lead manufacturer may visit a secondary subcontractor on occasion to learn as much from "the man on the spot" as possible. The reverse also takes place when a secondary subcontractor visits the lead manufacturer to participate in a new product development project or to discuss specific ideas on quality improvement. The most frequent exchange, of course, is undertaken between the engineers from both sides who can talk freely about technical matters without having to go through middle managers or the salespeople. This direct linkage not only saves time but makes the subcontractor far more a part of the new product development team than an outside vendor.

Reciprocity

Information is only part of what members of the network give and take. Many of the business transactions that take place within the network are also based on a give-and-take relationship. Two such transactions are highlighted here.

First, the lead manufacturer may sometimes push its subcontractors very hard to have some part delivered by a certain deadline but compensate them well when the task is accomplished. For example:

· Fuji-Xerox decided to change the basic design of a certain part midway in the development process and made an extremely tough demand on one of its subcontractors about when the delivery should be made. The subcontractor complied with this "utterly insane" request by working at night and completing the assignment on time.

Fuji-Xerox reciprocated later by paying the subcontractor handsomely. Mr. Kawamoto of Honda summed up this reciprocal relationship when he said, "We're buying time with money."

Second, where market conditions are unfavorable, a lead manufacturer may sometimes make a quite unreasonable demand on the subcontractor about the purchase price of a certain part. It tries to make up for whatever opportunity loss the subcontractor may have incurred by giving it very attractive margins when conditions turn favorable at a later date.

There is no guarantee that the lead manufacturer will return the favor in the future. Written contracts are unheard of. But if the lead manufacturer delivers, a trusting relationship begins to develop. In the long run, this kind of relationship leads the subcontractors as a group to accept the lead manufacturer as the legitimate leader and establish a strong cooperative system in support of it. A set of "shared network norms" is established over time, laying out a basic understanding of how business should be conducted within the network. Such a norm may tolerate an unreasonable demand made by the lead manufacturer during times of competitive crisis.

Subcontractors are willing to make sacrifices in the short run because they understand that their own survival is largely dependent on how well the leader performs in the market. They also know that certain new product development projects are a matter of life or death for the leader and, therefore, go out of their way to help. "Unreasonable demands are easier to swallow during wartime than peacetime," commented one subcontractor.

CONCLUSION

We embarked on this study by observing the dynamic process by which new products are developed within five Japanese companies. We then tried to put some order to what we observed by identifying seven commonalities in the process. This section compares and contrasts what has been learned from these Japanese companies with what is known about the new product development process in the United States.

Our conclusions are speculative at best since we have not conducted a comparable in-depth interview of U.S. companies. But our involvement in a recent comparative study of management practices in the United States and Japan[19] has given us a basic understanding of how U.S. companies approach the development process. Our exposure to corporate America and to leading U.S. researchers in this field has given us further insights. With this limitation in mind, we highlight three major differences between the United States and Japan.

First of all, the process itself is viewed differently. In general, new product development within U.S. companies is viewed more as an analytical and systematic process. Many companies utilize a sequential approach involving some variation of the phased program planning system. Variety reduction within such a system takes place in a topdown manner, with as many uncertainties as possible removed upstream in the process.

In contrast, Japanese companies view new product development as a trial-and-error process (learning-by-doing) and resort to a considerably looser format of phase management. Top management keeps goals purposely broad and tolerates ambiguity to encourage an iterative process of information-seeking and solution-seeking to emerge in a bottom-up manner. Variety reduction is conducted on a more ad hoc basis and is more spread out across the overlapping phases. This allows a more flexible response to last-minute feedback from the marketplace.

Second, a different kind of learning takes place. In the United States, product development is undertaken by a highly competent and innovative group of specialists. Most of the learning is done by an elite group of technical people, largely on an individual basis, within a narrow area of specialization. Thus an accumulation of knowledge based on "learning in depth" takes place.

In contrast, product development in Japan is often undertaken by a team of nonexperts who are encouraged to become general-

ists by interacting with each other throughout the development process. The development project may be headed by a nontechnical person, as in the case of the FX-3500 made by Fuji-Xerox, or by a sales engineer from a different division, as in the case of NEC's PC 8000. Everyone participating in the development process is engaged in learning, even outside suppliers. Learning also takes place across all phases of management and across functional boundaries. It is this kind of "learning in breadth" that supports the dynamic process of product development among Japanese companies. This learning emanating from the development process, in turn, serves as the trigger to set total organizational learning in motion. In this sense, new product development is the particular device that fosters corporatewide learning.

Third, the organizational impact is different. In the United States, new product development is an important strategic tool that enables companies to adapt to changes in the external environment by taking advantage of market opportunities or responding to competitive threats. As such, it plays a central role in strategy formulation.

In addition to its impact on strategy, product development in Japan takes on another important role. It serves as a change agent for reshaping corporate culture. Product development breaks down the hierarchy or rigidity normally associated with Japanese organizations—such as the seniority system or lifetime employment. Management gives the development team very broad but challenging goals as well as full autonomy to come up with something new. It uproots a competent middle manager from the hierarchical organizational structure and assigns innovative young talents to the team. Management gives unconditional support and legitimizes these unconventional moves by declaring a state of emergency or crisis. As mentioned earlier, organizational members and outside suppliers are willing to "swallow" more during times of war than during peacetime. The multilearning that takes place during wartime often breaks down the traditional ways of doing things. These changes are institutionalized within the entire organization until another crisis situation forces it to unlearn the lessons of the past. In a sense, management entrusts to the new product development team the mission to bring about an iterative process of learning and unlearning within the entire organization.

Caveats

Some final words of caution are in order. First, the Japanese approach toward product development has some built-in limitations, among which are the following:

1. It requires an extraordinary effort on the part of all project members during the entire span of the development process. Monthly overtime of 100 hours during the peak and 60 hours during the rest of the time is not uncommon.
2. It may not apply to breakthrough projects requiring a truly revolutionary innovation, as in the areas of biotechnology or chemistry.
3. It may not apply to mammoth projects, as in the aerospace business, where extensive face-to-face interactions are limited by the sheer scale of the project.
4. It may not apply within organizations where new product development is carried out by a genius at the top who makes the invention and hands down a well-defined set of specifications for people below to follow.

Second, we can expect our findings to be relevant in the short run; but given the evolutionary nature of the new product development process in Japan, they may soon be outdated. Extensive reliance on CAD/CAM, for example, may have far-reaching implications for how product development is managed within the organization. Externally, recent developments in telecommunications may make an interorganizational network based on geographical proximity obsolete. In fact, a recent study by Imai has already documented the rise of a telecommunications-based network in Japan.[20]

Finally, we must recognize that generalizations are misleading. For lack of better terminology, we labeled what we observed in the five Japanese companies as the "Japanese approach." Such a generalization may not necessarily apply when the sample size is enlarged, although our instincts tell us otherwise. We have also generalized about U.S. companies, describing their approach as analytical and sequential. For some companies, such a description may be very far from reality. For example, we see a flexible approach in place among a few U.S. companies that have established such systems as internal corporate ventures,[21] product champions,[22] skunkworks,[23] and others. The difference between the United States and Japan, therefore, may not be so much a "difference of kind" but more a "difference of degree."

NOTES

1. William J. Abernathy, Kim B. Clark, and Alan M. Kantrow, *Industrial Renaissance* (New York: Basic Books, Inc., 1983), 107.
2. Paul R. Lawrence and Davis Dyer, *Renewing American Industry* (New York: The Free Press, 1983), 8.
3. Rosabeth Moss Kanter, *The Change Masters* (New York: Simon and Schuster, 1983), 19.
4. Variety is defined as the number of distinguishable elements, as used in cybernetics. A company reduces the number of options available by making choices on an iterative basis.
5. Abernathy et al., *Industrial Renaissance*, 27.
6. William J. Abernathy, abstract to "The Anatomy of the Product Development Cycle: An

Historical Perspective," Colloquium on Productivity and Technology, Harvard Business School, March 1984, 27–29.

7. Karl E. Weick, *The Social Psychology of Organizing* (Reading, Mass.: Addison-Wesley, 1979), 133.

8. See Eric Jantsch, "Unifying Principles of Evolution," in *The Evolutionary Vision*, ed. E. Jantsch (Boulder, Colorado: Westview Press, 1981); Devendra Sahal, "A Unified Theory of Self-Organization," *Journal of Cybernetics* 9 (1979): 127–42.

9. Robert A. Burgelman, "A Model of Internal Corporate Venturing in the Diversified Major Firm," *Administrative Science Quarterly*, June 1983, 223–44.

10. Ikujiro Nonaka, "Evolutionary Strategy and Corporate Culture" (in Japanese), *Soshiki Kagaku* 17, 3 (1983): 47–58.

11. Kanter says segmentalism obstructs innovation and change in the following manner: "Segmentalism sets in when people are never given the chance to think beyond the limits of their job, to see it in a larger context, to contribute what they know from doing it to the search for even better ways. The hardening of organization arteries represented by segmentalism occurs when job definitions become prison walls and when the people in the more constrained jobs become viewed as a different and lesser breed." Kanter, *Change Masters*, 180–81.

12. Lawrence and Dyer, *Industry*, 262.

13. A similar finding is reported by Quinn who responds in the following manner to the question, How do executives cross-sectionally coordinate the various interacting subsystems in the decision dynamic? "In addition to selecting people with the technical skills most likely to be relevant over the time horizon of the strategy, most top executives tried consciously to team different management styles together: a 'tough'

manager with a 'people-oriented' manager, a conceptualizer with an implementer, an entrepreneur with a controller, and so on." James Brian Quinn, *Strategies for Change: Logical Incrementalism* (Homewood, Ill.: Richard D. Irwin, 1980), 138–39.

14. Lawrence and Dyer, *Industry*, 263.

15. Theodore Levitt, *The Marketing Imagination* (New York: The Free Press, 1983), 164.

16. Richard J. Schonberger, *Japanese Manufacturing Techniques* (New York: The Free Press, 1982), 175.

17. This supplier proximity to lead manufacturer plants also took place in the United States in the heavy-industry region around the Great Lakes and Ohio. According to Schonberger (ibid., 173), "Independent machine shops and foundries abound, locating near to buyer plants that make farm machinery, autos, machine tools, and so forth."

18. Mark S. Granovetter, "The Strength of Weak Ties," *American Journal of Sociology* 78 (1973): 1360–80; Everett M. Rogers, *Diffusion of Innovations*, 3d ed. (New York: The Free Press, 1983), Chapter 8, 293–303.

19. Tadao Kagono, Ikujiro Nonaka, Kiyonori Sakakibara, and Akihiro Okumura, *An Evolutionary View of Organizational Adaptation* (Tokyo: Nippon Keizai Shinbun, 1983). An English translation is forthcoming from North-Holland.

20. Ken-ichi Imai, *Japanese Industrial Society* (Tokyo: Chikuma Shobo, 1983), in Japanese.

21. Burgelman, "Internal Corporate Venturing."

22. Thomas J. Peters and Robert H. Waterman, Jr., *In Search of Excellence* (New York: Harper & Row 1982).

23. Thomas J. Peters, "The Mythology of Innovation, or a Skunkworks Tale Part II," *The Stanford Magazine*, Fall 1983, 11–19.

SECTION
VI

Venturing and
Organization Learning

Prior sections have focused on the business unit. We have explored building functional strength and cross-functional integration driven by business unit strategy. As a product class matures, the dominant form of innovation becomes ever more incremental. While incremental product and/or process innovation may be profitable in the short run, the firm must also develop the capacity for major product or process innovation to ensure long-term success.

Major innovation requires expertise, competence, and systems outside of the firm's main focus. Corporate venturing is a mechanism that permits incremental innovation to coexist with discontinuous innovation. Venturing facilitates organization learning as new sources of expertise and resources are forged both within and outside of the business unit. Chapter 9 examines, then, both internal and external corporate venturing as mechanisms to enhance organization learning. Maidique argues that organizational processes that make for early success are fundamentally different from those needed to innovate over time. Maidique focuses on key roles that seem to be required as the organization evolves. Burgelman presents a framework for thinking about corporate venturing and describes the managerial and organizational issues involved in effectively managing internal corporate venturing. Roberts and Berry discuss alternative modes of corporate venturing. Through a rich case study, they illustrate how their framework can be applied to help organizations select optional entry strategies. Finally, Teece builds on the industrial organization and technology literatures in exploring alternative mechanisms firms can use to capture benefit from innovation. Teece discusses strengths and limitations of internal integration, licensing, contracting, and joint ventures as levers to enhance profiting from innovation.

9

VENTURING AND ORGANIZATION LEARNING

Entrepreneurs, Champions, and Technological Innovation

Modesto A. Maidique

Successful radical innovation requires a special combination of entrepreneurial, managerial, and technological roles within a firm. As the firm grows and changes, these roles also change, and they tend to be performed by different people in different ways. The author draws conclusions from several cases of radical technological innovation to support these hypotheses.

> *There is plenty of reason to suppose that individual talents count for a good deal more than the firm as an organization.* Kenneth J. Arrow[1]

At all stages of development of the firm, highly enthusiastic and committed individuals who are willing to take risks play an important role in technological innovation. In the initial stages of

Reprinted from "Entrepreneurs, Champions, and Technological Innovation" by Modesto A. Maidique, *Sloan Management Review,* Winter, 1980, pp. 59–76, by permission of the publisher. Copyright 1980 by the Sloan Management Review Association. All rights reserved.

the technological firm's development, these entrepreneurial individuals are the force that moves the firm forward. In later stages, they absorb the risks of radical innovation, that is, of those innovations that restructure the current business or create new businesses. As Ed Roberts has pointed out, "In the large firm as well as in the foundling enterprise, the entrepreneur is the central figure in successful technological innovation."[2]

Successful innovation, however, requires a special combination of entrepreneurial, managerial, and technological roles. Furthermore, the characteristics of this network[3] of roles is a function of the stage of development of the firm. In this article, the theory of entrepreneurial roles—what we know about entrepreneurship and technological innovation—is combined with Scott's theory of corporate development to generate several hypotheses regarding the evolution of entrepreneurial roles, as the firm evolves from a small firm to a large, diversified firm.[4]

This paper makes three principal arguments: first, that the entrepreneurial role is essential for radical technological innovation, but that it manifests itself differently depending on the firm's stage of development; second, that radical technological innovation, to be successful, requires top management participation in the entrepreneurial network; third, that in addition to the independent entrepreneur and the product champion, there is an important intermediate entrepreneurial role especially prominent in diversified firms; that is, the executive champion.[5]

Radical technological innovation, at any stage of development, can be viewed as requiring a recreation of the original entrepreneurial network—the merging together of the roles that the original entrepreneurial team performed: business definition, sponsorship, technical definition, and technical communication. As a business grows or becomes more diverse, the original entrepreneurial network becomes fragmented: these critical roles are decoupled, and a conservative bias is often introduced into subsequent innovations. Business definition becomes separated from technical definition by administrative systems, organizational hierarchy, and market and technological diversity. The business readjusts more slowly, or not at all, to technological discontinuities. Nonetheless, this separation is natural and necessary. It is impossible for the original entrepreneur to be closely coupled to an increasingly larger number of more diverse innovations. He or she must, however, seek out and complete the network of entrepreneurship for that handful of radical innovations that will have significant impact on the future of the firm.

This article is organized into five sections. First, the literature on entrepreneurial roles and corporate development is briefly summarized, and we identify gaps in the literature addressed later in the article. In the following three sections, the evolution of the network of entrepreneurial roles is examined by analyzing radical innovations in three different contexts: the small, integrated, and diversified technological firm.[6] In the last section some hypotheses are drawn from this analysis regarding the task of top management in the technology-based firm.

ENTREPRENEURSHIP AND CORPORATE DEVELOPMENT: A BRIEF REVIEW OF THE LITERATURE

A rich literature exists on heroic, independent, technological entrepreneurs such as Thomas Watson, Jr., Henry Ford, and Edwin Land. During the last two decades, however, a new literature on entrepreneurship has developed that emphasizes the role of individuals within the firm who also exhibit entrepreneurial characteristics. But this development in the literature has identified a plethora of new, and often confusing, internal entrepreneurial roles that make interpretation of the new literature difficult.

The literature on entrepreneurship is principally concerned with static behavior. To date, little attention has been paid to the evolution of the entrepreneurial role. On the other hand, a substantial literature on corporate development does exist. This literature, however, makes only a general reference to the entrepreneur and the evolution of his or her role.

In the remainder of this section, I briefly summarize those aspects of these literatures that are relevant to this work. I conclude that the deficiencies in the literatures, which are discussed in detail below, have resulted in gaps that I address in the subsequent sections.

Entrepreneurs and Champions

The significance of the role of the entrepreneur has been recognized for at least two centuries. Schumpeter credits J. B. Say, an early nineteenth-century French economist, for being the first to recognize that the entrepreneur in a capitalist society is "the pivot on which everything turns."[7] According to Schumpeter, the entrepreneur

> reforms or revolutionizes the pattern of production by exploiting an invention or, more generally, an untried technological possibility for producing a new commodity or producing an old one in a new way, by opening up a new source of supply of materials or a new outlet for products by reorganizing an industry. . . .[8]

On the other hand, the critical role that "champions" of technological change play within industrial organizations has been recognized only during the last two decades. In a seminal study of radical military innovations, Schon observed that certain committed individuals, champions, played the key role in successful innovations. Schon lists some fifteen major inventions of the twentieth century, such as the jet engine and the gyrocompass, in which individuals played a major role. In his studies, Schon found that "the new idea either finds a champion or *dies.*" To Schon, the "product champions" are critical, for

> no ordinary involvement with a new idea provides the energy required to cope with the indifference and resistance that major technical change provokes. . . . champions of new inventions . . . display persistence and courage of heroic quality.[9]

Schon's analysis led him to four basic conclusions:

1. At the outset, the new idea encounters sharp resistance.[10]
2. Overcoming this resistance requires vigorous promotion.
3. Proponents of the idea work primarily through the *informal* rather than the formal organization.
4. Typically, *one person* emerges as champion of the idea.

The product champion served as a catalyst for the development of a literature on internal entrepreneurial roles. In the decade following Schon's work, several new entrepreneurial (and related) roles and new names for old roles appeared in the innovation literature, such as "business innovators,"[11] "internal entrepreneurs,"[12] "sponsors,"[13] "change agents,"[14] "Maxwell demons or mutation selectors,"[15] "technical and manager champions,"[16] and "administrative entrepreneurs."[17]

Collins and Moore[18] found very strong similarities between traditional or "independent" entrepreneurs and certain managers within the firm who operated like Schon's product champions: they called them "administrative entrepreneurs." Collins and Moore developed a psychological profile of the 150 independent and administrative entrepreneurs in their study and concluded that "the entrepreneurial personality, in short, is characterized by an unwillingness to submit to authority, an inability to work with it, and a consequent need to escape from it."[19] Some potential entrepreneurs find ways, at least temporarily, to satisfy their psychological needs by pursuing—sometimes in unorthodox ways—high risk projects *within* the organization; others finally break away and create a new structure.

In two recent studies, Duchesneau[20] and Olsen[21] obtained statistical data on the presence of champions in the footwear and textile industries respectively. Duchesneau interviewed senior managers in sixty-nine footwear firms and found that about two-thirds of them recognized the presence of "product champions" in their firms. In his study of twelve textile firms, Olsen found a "strong correlation" between early adoption of innovations and the existence of an identifiable champion. Olsen found identifiable champions in eight of the twelve firms in his study.[22]

SAPPHO

One of the first studies that attempted both to quantify the product champion function and to break it down into subroles was the SAPPHO[23] study. In phase I of the SAPPHO project, twenty-nine pairs of successful and unsuccessful innovations were studied. Of the forty-three pairs of innovations in SAPPHO II, twenty-two were in the chemical industry and twenty-one in scientific instruments.[24] The forty-three pairs were compared along 122 dimensions; fifteen of which were found (on an aggregate basis) to have statistical significance higher than .1 percent (as determined by the binomial test). Another nine variables had statistical significance higher than 1 percent. The SAPPHO investigators used multivariate analysis to extract from these twenty-four variables five underlying areas of difference between successful and unsuccessful innovations:[25]

1. Strength of management and characteristics of managers.
2. Marketing performance.
3. Understanding of user needs.
4. Research and development (R & D) efficiency of development.
5. Communications.

To study the first of these factors—the role of key managers and technologists—the SAPPHO investigators defined four categories of key individuals:

1. *Technical innovator.* The individual who made the "major contribution on the technical side" to the development and/or design of the innovation.
2. *Business innovator.* The individual within the managerial structure who was responsible for the "overall progress" of the project.
3. *Product champion.* Any individual who made a decisive contribution to the innovation by "actively and enthusiastically promoting its progress through critical stages."
4. *Chief executive.* The "head of the executive structure" of the innovating organization, but not necessarily the managing director or chief executive officer.

The forty-three pairs of innovations were then tested to determine how—if at all—the presence of such key individuals explained the success of the innovation. The results for the five most significant parameters are summarized in Table 1. The striking feature of this data was that the individual that emerged as the principal factor was *not* the product champion, but the "business innovator." In particular, the business innovator's power, respectability, status, and experience were important. However, the role of the product champion was also shown to discriminate significantly for success.

Thus, the SAPPHO study provides systematic evidence in favor of the champion hypothesis. The study also indicates that besides commitment and enthusiasm, the power and status of the sponsoring executive also play an important role in determining the success of an innovation. Often this latter role is played by someone other than the "product champion"; that is (using the SAPPHO terminology), the "business innovator."

The significance of the SAPPHO data is

TABLE 1.

Variable*	S	N**	F	%S†	Binomial Test††
1. The executive in charge of the successful innovation has more power.	20	19	4	75%	7.7×10^{-4}
2. The executive in charge of success has more responsibility.	18	20	4	72%	1.3×10^{-3}
3. The executive in charge of success has more diverse experience.	20	18	5	73%	2×10^{-3}
4. The executive in charge of success has more enthusiasm.	14	27	2	71%	2×10^{-3}
5. The executive in charge of success has higher status.	18	21	4	72%	2×210^{-3}

Notes:

*S = variable discriminated for success; F = variable discriminated for failure; N = variable did not discriminate in either direction, or insufficient data were available to form the comparison.

**The N grouping presents a data interpretation problem since it's not clear how this group broke down between the "insufficient data" category or situations where the variable had a "neutral" effect or none at all.

†The %S was calculated by assuming that the N group was equally split between the insufficient data group and the neutral grouping. The "insufficient data" half was discarded and the remainder was split equally between the S and F groups. The "%S" was then calculated as a percentage of the new total S + F population.

††Calculated by SAPPHO group.

made clearer by using Bower's model for the resource allocation process.[26] He proposed a three stage model for resource allocation in a large firm: top management provides a business and structural *context* for decision making; within this context, middle management selects the projects that they support; higher level sponsorship, *impetus,* is required for successful completion of funding. At the root of the resource allocation process are the specialists and lower level managers who give *definition* to the projects.

Using Bower's terms, the technical innovator provides technical *definition,* the business innovator provides *sponsorship* or *impetus,* and the chief executive provides business definition or *context.* [27] Viewed from this vantage point, the role of the product champion is to serve as a catalyst for increased sponsorship or impetus.

Similarly, the innovation process can also be viewed from the perspective of the key roles defined by Roberts. He expands the set of critical functions to include a technical information role and a project management role (see Tables 2 and 3 and Figure 1). The project manager anticipates the need for sponsorship by planning for the requirement of the innovation. The gatekeeper, a function characterized originally by Allen, acts as a clearing house for technical information for the technologists in the firm. In a subsequent paper, Roberts suggests that marketing and manufacturing gatekeepers can usually be identified and that they also play important roles in the innovation process.[28]

TABLE 2.   *Key Roles and Functions According to Different Investigators**

Function	Bower (1972)	Schwartz (1973)	SAPPHO (1974)	Kusiatin (1976)	Roberts** (1968, 1972, 1977, 1978)	This Article†
Business, Structure & Definition	Context	Context	Chief Executive	—	—	Business Definition, Technological Entrepreneur
Sponsorship	Impetus	Impetus	Business Innovator, Product Champion	Manager Champion, Technical Champion	Sponsor, Product Champion, Project Manager, Internal Entrepreneur	Executive Champion, Product Champion, Sponsorship, Technological Entrepreneur
Technical Definition	Definition	Technical Definition, Factoring	Technical Innovator	Technical Champion	Creative Scientist	Technical Definition, Technologist
Technical Information	—	—	—	Technical Champion	Gatekeeper	—
Market Information	—	—	—	—	Market Gatekeeper	—

Notes:
*Roles and definitions vary from researcher to researcher and are only roughly comparable.
**See also Rhoades, Roberts, and Fusfeld, 1978 (reference 28).
†See list of definitions in Table 3.

STAGES OF DEVELOPMENT OF THE FIRM

While the literature on entrepreneurs and champions is primarily concerned with static behavior, a substantial literature exists on corporate development. However, this literature is concerned with overall description of the stages of corporate development, not with the specific issue of the entrepreneurial team. Nonetheless, the corporate development literature does provide a framework for analyzing the development of entrepreneurial networks.

Galbraith proposed that the three stages of company evolution are small, medium, and large.[29] Chandler, on the other hand, sug-

gested that the three basic stages are small, integrated, diversified; Scott, Salter, and others concur.[30]

Based on his study of the histories of about seventy large U.S. firms, Chandler proposed that, in most cases, firms follow a typical sequence of development.[31] After the initial entrepreneurial stage, the firm becomes vertically integrated and managed by a centralized functional organization. Normally, this stage is followed by diversification, which is managed by a decentralized divisional organization.

This basic idea was further developed and extended by Scott while he was working with Christensen and McArthur. A set of propositions delineating the various stages of

TABLE 3.   *Definitions of Key Roles*

Technological Entrepreneur	The organizer of a technological venture who exercises control of the venture (typically by owning a substantial percentage of the equity) and assumes the risks of the business. Usually he is the chief executive officer.
Product Champion	A member of an organization who creates, defines, or adopts an idea for a new technological innovation and who is willing to risk his or her position and prestige to make possible the innovation's successful implementation.
Executive Champion	An executive in a technological firm who has direct or indirect influence over the resource allocation process and who uses this power to channel resources to a new technological innovation, thereby absorbing most, but usually not all, the risk of the project.
Technical Definition	The basic performance requirements and associated specifications that characterize a proposal for a new technological innovation.
Sponsorship	The actions by which executives channel resources to innovative projects that they have chosen to support.
Business Definition	A description of the business within which a firm chooses to compete and of the overall administrative practices that the firm will follow in that business.

corporate development resulted from a cooperative effort by Scott and Salter (see Table 4).[32]

Wrigley further elaborated on the three original stages and proposed three subclasses of diversified businesses: dominant business diversified, related business diversified, and unrelated business diversified.[33] Scott's classes, as

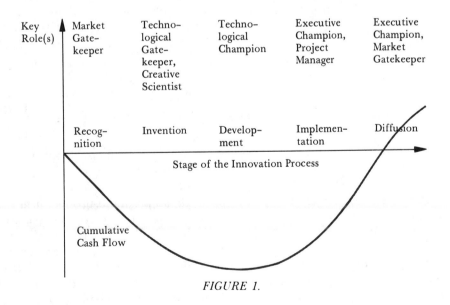

FIGURE 1.

TABLE 4.   The Three Stages of Organizational Development

Company Characteristics	Stage I: Small	Stage II: Integrated	Stage III: Diversified
Product Line	Single product or single line	Single product line	Multiple product lines
Distribution	One channel or set of channels	One set of channels	Multiple channels
Organization Structure	Little or no formal structure; "one-man show"	Specialization based on function	Specialization based on product-market relationships
R & D Organization	Not institutionalized; guided by owner-manager	Increasingly institutionalized search for product or process improvements	Institutionalized search for new products as well as for improvements
Performance Measurement	By personal contact and subjective criteria	Increasingly impersonal, using technical and/or cost criteria	Increasingly impersonal, using market criteria (return on investment and market share)
Rewards	Unsystematic and often paternalistic	Increasingly systematic, with emphasis on stability and service	Increasingly systematic, with variability related to performance
Control System	Personal control of both strategic and operating decisions	Personal control of strategic decisions, with increasing delegation of operating decisions through policy	Delegation of product-market decisions within existing businesses, with indirect control based on analysis of "results"
Strategic Choices	Needs of owner versus needs of company	Degree of integration; market-share objective; breadth of product line	Entry and exit from industries; allocation of resources by industry; rate of growth

modified by Wrigley, include the following (also see Table 4):

- *Small (or entrepreneurial):* single product or single product line company, with little formal structure, controlled by owner-manager.
- *Integrated:* single product line firm with vertically integrated manufacturing and specialized functional organizations. Owner-

manager retains control of strategic decisions. Most operating decisions are delegated through policy.

Diversified: multi-product firm with formalized managerial systems that are evaluated by objective criteria, such as rate of growth and returns on investment. Product-market decisions within existing businesses are delegated. Within the diversified group, there are three subcategories:

1. *Dominant business firms* that derive 70–95 percent of their sales from a single business[34] or a vertically integrated chain of businesses (e.g., General Motors, IBM, Xerox, U.S. Steel).
2. *Related business firms* that diversify into related areas where no single business accounts for more than 70 percent of sales (e.g., DuPont, Eastman Kodak, General Electric).
3. *Unrelated business firms* that have diversified without necessarily relating new business to old, and where no single business accounts for as much as 70 percent of sales (e.g., Litton, North American, Rockwell, and Textron).

Abernathy and Utterback have studied the process of corporate evolution using the "productive unit."[35] An economist might call the productive unit a simple firm: Abernathy and Utterback use the term to describe a related line of products, the manufacturing process and the overhead structure required to develop, make, and market the products. Their work has significantly extended the literature on corporate evolution, but their focus has been on the evolution of the process segment rather than on managerial characteristics.

Abernathy and Utterback visualized new created manufacturing processes as initially having some slack: procedures are not fixed, job design and material flow are informal and flexible. In short, the process is *fluid*. At the other extreme, once the process has been "perfected" by accumulation of experience, jobs become standardized, manual procedures are automated, and rigid specifications are instituted. The process becomes, in Abernathy and Utterback's terms, *specific*. A transition phase joins these two external conditions.

In this model (Figure 2), product and process innovation proceed at different rates. Product innovation is highest in the fluid phase and declines monotonically through the transition and specific phases—the slack is managed out. Process innovation appears slowly and rises as the accumulated flow of products through the processing stages increases the opportunity—and managerial pressures—to formalize the process and reduce costs.

A product or family of products emerges

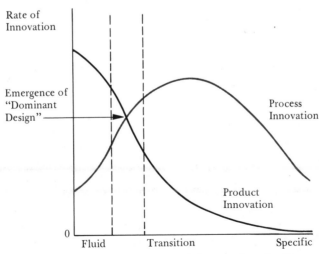

FIGURE 2. Stage of Development of the Manufacturing Process

from the initial crops of new products and attains wide market acceptance and high volume. These products, called "dominant designs," make it possible to increase the rate at which the manufacturing process becomes systematized. As standardization increases, process innovation declines (see Figure 2).

Although we might expect some major transformations in managerial relationships as a process follows the path described above, Abernathy and Utterback do not discuss these transformations. Yet, in an earlier paper, Abernathy and Townsend acknowledged (following Bright) that "management has a critical role in causing the process to evolve and in readying it for technological innovation."[36] These transformations in managerial relationships were not studied in the subsequent paper, which used the extensive Myers and Marquis[37] data base (567 innovations, 120 firms, 5 industries), partly because the data base did not contain data on managerial issues.[38] Thus, regarding the evolution of managerial relationships in innovations, Abernathy and Utterback do not go beyond referring to the broad "organic" and "mechanistic" descriptions of organizational relationships suggested by Burns and Stalker.[39]

In the next three sections, we use the three major corporate evolution contexts defined by Scott to examine entrepreneurial roles in radical innovations. We start with the small, or entrepreneurial, technological firm.

Stage I: The Small Technological Firm

The technological entrepreneur, in addition to defining the firm's business, plays (and enjoys playing) the dual role of sponsor and definition agent. Technological entrepreneurs often intervene (sometimes excessively) in the definition phase of innovations.

At Redactron Office Automation, a start-up firm specializing in automatic editing components, founder and President Evelyn Berezin[40] explained her part in the defining of her company's first product:

> When we reached the stage of spelling out detailed specifications, it became apparent that the engineers simply hated to commit things to writing. So I wrote out all the detailed specifications myself, usually dictating them while I was commuting back and forth to the plant. That's how the system got designed.[41]

And it wasn't by whim that Berezin decided on the specifications. Prior to the definition phase, she had spent weeks in the field observing people who used IBM automatic editing equipment. Berezin considered closeness to the market imperative:

> If you don't go there yourself but rely instead on what other people tell you, the information is likely to be distorted or the impressions incomplete. When you're designing a new product, you simply cannot afford to have layers of people between you and the eventual product users.[42]

Clearly, Berezin believed that the power and vision of the CEO should be brought into intimate contact with both technical experts and customers.

A similar pattern becomes evident from studying Henry Kloss, the founder of Advent Corporation. For Kloss, a classic technological entrepreneur, Advent was his third technology venture. He was a founder of Acoustic Research (AR). From there, he moved on to start KLH (both AR and KLH make high-quality consumer audio products). One of Kloss's principal reasons for starting Advent was his desire to retain control—he sold his shares of AR "under duress, and his share of KLH Corporation with mixed feelings."

Like Berezin, Kloss was familiar with the market and the most advanced technologies. He directed Video Beam, Advent's large screen television project, from start to finish. According to Kloss, "I am responsible for the concept of the product and the initial demonstration that the idea is feasible. . . ." He delineated personal contributions, such as

> choosing the optimal size of tubes; deciding the critical tolerances in the final product, making tradeoff choices between costs and qualities; working with consultants in the tube-making business and making the best judgment out of their recommendations as to production processes, specifications and material tolerances. . . .[43]

Kloss laid out the manufacturing and assembly plant when the time came to make the product. He was then the chief executive officer of Advent, which then had annual sales of $16 million.

The Kloss-Berezin phenomenon can occur in other countries. Efraim "Efi" Irazi, an Israeli electrical engineer educated at M.I.T., founded Sci-Tex in 1968. The company makes electro-optical systems, and it now has annual sales of over $10 million. Sci-Tex had an appointed administrator of R & D, but "Efi felt that would not preclude him from being involved, from continuing to participate very actively in the problem solution process which had become his own lifeblood."[44]

In all three cases (Advent, Redactron, and Sci-Tex), the entrepreneur helped define new products while retaining control as CEO. For the entrepreneur sees the company as a giant erector set and retains the right to play with any of the pieces. For Kloss, the ideal situation would be that of Edwin Land, chairman and technical director of billion-dollar Polaroid. Land reputedly has access to any level of R & D while functioning as CEO.

Stage II: The Integrated Firm

Thus what the entrepreneur created passed inexorably beyond the scope of his authority. . . . What the entrepreneur created, only a group of men sharing specialized information could ultimately operate. John Kenneth Galbraith[45]

If the entrepreneur succeeds, he or she creates a dilemma. Growth means more products, more people, more managers—a transition from a small to an integrated, technological firm. Continued growth requires changes in the entrepreneur's role. Technological progress and organizational complexity act as dual forces on the entrepreneur to cause him or her to give up technical definition and sponsorship of most new projects.

Robert Noyce, the coinventor of the integrated circuit and cofounder and chairman of Intel, puts it this way:

> Maybe you can do good technical work for ten years, if you work hard at it, but after that the younger guys are better prepared. It's a question of technical obsolescence, if you will.[46]

Like Noyce, Ken Olsen, chairman of Digital Equipment Corporation (DEC), has moved away from technical definition, though he, like Henry Kloss, is reputedly a "shirt-sleeved engineering type."[47] After a recession, Olsen explained in an interview: "I let the engineers do the designing; my concern was to keep the team together."[48] Olsen is vitally concerned with maintaining open communications to insure that the best proposals come to his attention. He meets regularly with an engineering committee comprised of about twenty engineers from all levels of DEC. Olsen sets the agenda and periodically disbands and reconstitutes the committee to maintain a fresh flow of ideas. He sees his role as that of a catalyst, or a "devil's advocate": he expects that the best solutions to technical problems will be developed by his technicians.

Often Olsen's role is mostly that of a sponsor. One DEC manager had been championing a project whose approval had become mired in red tape. Olsen sat in on a meeting in which this man's difficulties emerged. Olsen asked about the project and wondered out loud why such a promising idea was finding such little support. The manager describes what followed: "Suddenly the barriers to my project came down. What normally might have taken a year or more to complete became a six-month project."[49] Sponsorship had replaced technical definition.

Such sponsorship of product champions is critically important for continued innovation. Schwartz studied innovation in two major technology-based firms (one was DEC) and concluded that middle managers served as integrators between technical specialists and top management.[50] The middle managers decided which proposals would be submitted for approval, negotiated terms of support for a proposal, and selected the criteria for program evaluation. According to Schwartz, these functions "tended to add a conservative bias to proposals. As a result, the innovations studied were incremental rather than radical."[51] The technological or product champion seeks to break through these barriers, but his or her attempts meet with little success without adequate sponsorship.

The development of the float glass process at Pilkington Brothers, an integrated British glass manufacturer, is a classic case of technological championing. A few years after joining the firm, Alistair Pilkington, a distant relative of the founding family, conceived a radically new way of making plate glass one evening while he was washing the family dishes. That evening was the beginning of a risky crusade to develop the float glass process, during which, according to Pilkington, "chaps were literally taken off on stretchers

from heat exhaustion, yet [they] came back for more."[52]

Developing this process was a big financial gamble for Pilkington Brothers. After the fundamentals of the process had been proved in the lab, over 100,000 tons of unsalable glass costing $3.6 million were produced in the pilot plant. Month after month, Alistair Pilkington faced the firm's directors with a new request for $280,000 of operating funds and with promises of progress on the project. For a company with net profits of about $400,000 per month, this was a major risk; yet Pilkington continued to get approval until salable glass was finally produced in 1958. Alistair Pilkington succeeded by persisting in his role as technological champion despite continued setbacks and high risks.[53] His credit, however, must clearly be shared with Harry Pilkington, the entrepreneurial chairman of the board, who absorbed the risks of young Alistair's innovation and who made it possible for the company later to reap $250 million in licensing fees from its competitors. The Pilkington story is often given as a classic example of the product champion. In fact, it is a classic example of the entrepreneur and the champion working in unison—the simplest entrepreneurial network. The champion proposes, the entrepreneur disposes.

Both at DEC and at Pilkington Brothers, the entrepreneur's role had evolved from technical definition to sponsorship. But the entrepreneur was still the primary impetus for new projects, especially large ones. Other people, like the young engineering manager at DEC and Alistair Pilkington, had assumed the role of technical definition and had become technological champions for their projects. Olsen and the senior Pilkington found it easy to continue to act as sponsors for key projects in their firms. This was possible partly because the technologies of their integrated, single product line firms could be grasped by top management more easily than could the technologies of a diversified firm.

Stage III: The Diversified Firm

Diversified firms that have a dominant business (one that accounts for over 70 percent of sales) operate that business as if they were integrated, single product line firms. Such dominant businesses are usually controlled by top management through a functional structure, while the remainder of the businesses are managed through product divisions.[54] In these firms, the relationship between entrepreneurs and champions in the dominant business resembles that which we found in integrated, single product line firms like DEC and Pilkington. The entrepreneur gives up definition of products in the dominant business, but he keeps tight reins on sponsorship, particularly for major products. IBM in the late 1960s is a case in point.

Outsiders see IBM as a reflection of its products—a model of orderliness and rationality. To many, IBM is the epitome of the modern corporation: technologically powerful and highly innovative, but predictable and smoothly managed. Yet in 1964, after months of chaotic infighting, Tom Watson, Jr. (then IBM's CEO) made the extraordinarily risky decision to commit IBM to a revolutionary new line of computers, the System 360. The program's projected cost, $5 billion, exceeded the total assets (or, for that matter, the annual sales) of IBM that year. The System 360 was revolutionary in three major ways:

1. It depended heavily on microcircuitry (technology now commonplace, but then in its infancy);
2. The series was comprised of six basic computers designed so that users could scale up from one machine to another without having to rewrite existing programs;
3. The six models (30, 40, 50, 60, 62, 70) were to be made available simultaneously.

Bob Evans, the line manager who acted as technological champion for the new computers, explained: "We called this project 'you bet your company.' "[55]

According to one IBM executive, Watson had grown up with the IBM computer business and therefore "was able to use the informal organization to obtain the knowledge necessary to make the right decisions."[56] Watson once invited Fred Brooks, who at that time was the 360 project design manager, and other technical experts to his ski lodge in Vermont for a detailed discussion of the critical programming compatability issue. Watson also relied on T. Vincent Learson, a group executive whom Watson had tapped in 1954, to head IBM's entry into computers. Learson, a "tough and forceful personality," was "impatient with staff reports and committees," and he tended "to operate outside the conventional chain of command."[57] Learson was known to go directly to lower level management when he needed information.

The decision to go ahead with the 360 system shook IBM to its core. Sweeping organizational changes were instituted (three reorganizations over a six-year period), technically oriented executives diluted some of the traditional power of the marketing staff, IBM World Trade stopped trying to develop its own computers. IBM shifted from being simply an assembler of computer components to making its own components. In the process it became for a time the world's largest manufacturer of integrated circuits.

The success of IBM's System 360 is now legendary. It became the dominant design in business computers, and Learson was promoted to president of IBM shortly after its introduction. But in 1966 it was far from clear that Watson's gamble would be successful, and IBM's management agreed that "no meaningful figure could be put on the gamble." But a decade later it was clear that Watson had bet his company—and won.

Similarly to the senior Pilkington and Olsen, Watson and Learson communicated directly with the technical experts, and they relied on informal information networks to supplement the information that they were able to obtain directly. Middle managers, like Brooks, Evans, and Alistair Pilkington, championed the new products. But top management made the major resource allocation decisions, and they had the insight to make winning decisions because they were dealing with single product line businesses, with which they had grown up.

When diversified firms enter fields with which the dominant entrepreneur is not intimately familiar, the process changes. Now a new kind of champion emerges, bridging the gap between the entrepreneur and the technological champion. Within the scope of their authority and responsibility, these "executive champions" are modeled after the independent entrepreneurs discussed above. Sometimes, they are simply the original independent entrepreneur in a new corporate context, in which their power is circumscribed. In other cases, they grow out of the roots of the existing corporate culture. Examples of each of these situations are discussed below.

In 1975 Redactron, facing financial difficulties, agreed to merge with the Burroughs Corporation. Though Evelyn Berezin kept her title of president of Redactron, she became a Burroughs employee in 1976. In an interview shortly after joining Burroughs, she explained her philosophy of management:

> In a small company the real cost is not in the operating expenses, or in buying some new pieces of gear, or hiring a new person—it's time. Reducing your development or commercialization time is worth virtually whatever you have to pay.

Redactron and three other companies in the Burroughs structure operated as relatively autonomous businesses within Burroughs, which was then a $2 billion corporation with 50,000 employees and fifty plants. At Burroughs, Berezin's role had changed considerably, partly as a result of her continued quest to gain precious time. Now she was far more concerned with resource allocation than with product definition. She explained her new role:

> Burroughs's product management is involved in the resource allocation process, thus I am constantly involved in funding decisions. I have told my managers "you go do it, I'll get the funds." But there are limitations: For instance, I don't have authority to increase the engineering budget, but I can determine how programs are carried out. Secondly, all raises go through Detroit. I can't unilaterally increase people's salaries, but my people know that I will champion them.

Berezin cited a specific example:

> We had developed a new peripheral device which was very sophisticated yet inexpensive. It was a new technology. It was, in a word, "gorgeous." Yet corporate engineering said no, they were working on something else and this project could not be supported. The tooling that was required was on the order of $50,000 and had to be approved by corporate. My capital budget, however, had been approved at the beginning of the year, and I could only approve new items below $5,000.
>
> What did I do? I took the device to top management. What happened? The project was approved. I had to fight for four months, but I won the fight. In a good large company, if you fight—long enough—you win.

Evelyn Berezin, similarly to Tom Watson, Jr., was sponsoring a member of her organization who had proposed a promising new project. The key difference was that the circumscription of her authority required that she seek higher level sponsorship to implement the project. Berezin was an "executive champion" several times removed from the detailed technical definition but without the entrepreneurial clout to be the ultimate sponsor. Thus, in this example, we find a new kind of champion—more senior in the managerial structure than the technological champion, who, in fact, is often a sponsor for the latter within the limits of his or her responsibility.

In 1974 just after Ronald Peterson had returned to Grumman Aerospace from a year as a Sloan Fellow at M.I.T., he was tapped by the president of Grumman Aerospace to head an exploratory group on advanced energy systems. His immediate concern was with resource allocation, and in particular, with the authority that would go with his new position. Peterson, who had done his M.I.T. thesis on innovation, felt that unless he reported directly to Joseph G. Gavin, Jr., the chief executive of Grumman Aerospace, the probability of success for his project would be reduced. "I've learned enough about new ventures," explained Peterson, "to know they don't work unless they get top level attention." Soon after, Peterson became manager of the Energy Program Department, and he reported directly to Gavin.[58]

After studying the alternatives open to Grumman, Peterson decided not to follow in the footsteps of the other aerospace firms, which were bidding on proposals for esoteric new technologies, such as giant wind machines, oceanic platforms, and orbiting power stations.[59] Peterson decided instead to shun government proposals and to concentrate first on low technology solar hot water and space heating installations. He knew that in order to achieve his goals he needed the freedom to operate the business differently and independently from the traditional management style of the parent firm.

In 1977 when Joseph G. Gavin, Jr. was appointed president of the Grumman Corporation, the parent company of Grumman Aerospace, he appointed Peterson general manager of the Grumman Advanced Energy Systems Division. "I've been fortunate that Gavin has had a long-term interest in energy and since the beginning has played the sponsor role on this project." Peterson explained why the new division was established:

> One of the main reasons for establishing this division was not to be hampered with aerospace procedures, salaries, and infrastructure . . .
>
> In this position I have a lot of authority . . . I can sign for anything up to a million dollars on the annual plan without any counter signature. I have the power to raise salaries and hire and fire people. In effect, I am the president of a virtually independent corporation.[60]

Peterson had become, within the limits approved by Gavin, an executive champion, who was willing and able to risk within the limits of his substantial authority, Grumman's resources—on projects proposed by technological champions in his organization.

CONCLUSION: EVOLUTION OF THE ENTREPRENEURIAL NETWORK

The relatively small number of cases and industries studied, the incompleteness of some of the secondary data, and the complexity of the processes examined limit us to hypothesis generation. But the data suggest the outline of a framework to explain how the relationship between the network of entrepreneurial roles and radical innovation changes with stage of process development and increasing organizational complexity. The proposed framework is illustrated in Figure 3.

The small firm is easiest to analyze. Here, there is usually one business unit and one source of managerial sponsorship for technological projects—the entrepreneur. However, as in Efi Erazi's case, problem solving is part of the technological entrepreneur's lifeblood.

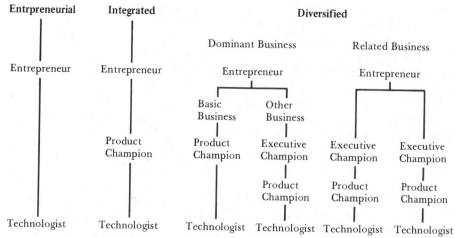

FIGURE 3.   *The Evolution of the Entrepreneurial Network*

Thus, most technological entrepreneurs continue to hold on to most—or all—of the reins of product definition, often until it is too late: Henry Kloss neglected financial and other duties of corporate management until he eventually lost control of his firm. Involvement in product definition without upsetting the firm's organizational hierarchy is not difficult for entrepreneurs in small firms. Most of the technical people are generally old colleagues of the entrepreneur or were hired directly by him or her. But even the most capable entrepreneur has limits. One of the most brilliant entrepreneurs I've met thought himself impervious to these limitations when he moved into his plant, cot and all. Within a few months, after a divorce and a nervous breakdown, he too realized that he had a limit.

Unless the entrepreneur recognizes that at some point—two or three projects for some, perhaps a dozen for others—he or she must retreat from the product definition role, the firm is likely to fail. Berezin, as is evident from our second encounter with her, had begun such a transition.

The integrated technology firm, such as DEC and Pilkington Brothers, illustrates the entrepreneur's transition from the definition role. In these firms, the CEO was technically knowledgeable and still played a pivotal role in the innovation process—for radical innovations. Harry Pilkington and Olsen did not define the new products, but they used their technical expertise and their organization's informal channels to gather information from their technical people. Although Pilkington and Olsen were managing firms several times smaller than Watson's IBM, all three chief executive officers viewed providing impetus to at least the radical technical innovations as a key part of their role. Nonetheless, there is a limit to how many projects a chief executive can sponsor. Olsen, for instance, could afford to be very visible in his firm's resource allocation process. Watson, as CEO of a $4 billion firm, could only be involved in those projects that might reshape the entire future of his firm—like the System 360. Similarly to the small company entrepreneurs, Watson and Olsen had excellent informal information networks that made available to them vitally important information for their decision making without destroying the firm's organizational structure. And that is precisely what they did: they made the major technological decisions and absorbed risks. Sponsorship replaced definition.

Increasing organizational complexity requires that others also sponsor innovations. The CEO, or the overall top management team, should be the sponsor for radical changes in large organizations; yet many other important projects also need sponsors.[61] Ronald Peterson argued that Joseph G. Gavin, Jr. had been the "sponsor" for his energy group. It was clear, however, that by the time Grumman Advanced Energy Systems had become a division, Peterson, not Gavin, had become the sponsor for most new projects. Gavin primarily helped to set overall business direction. In short, Peterson and Berezin functioned as executive champions who, *within their business units,* behaved similarly to Olsen, Watson, and Pilkington, Sr.

The inherent disorder produced by direct interaction of the central sources of sponsorship with the defining or proposing agents is a unifying thread in all the cases studied. By promoting informality in communications, interacting with first-line technical people, reviewing proposals outside of the "conservative bias" of middle management, and finally *deciding,* the entrepreneurial heads of the executive structures studied helped to perpetuate the fluidness, slack, and disorder that Abernathy and Utterback found in the highly innovative initial stages of process development.

In a study of the evolution of the structure of the firm,[62] Abernathy and Utterback suggest that the "normal" direction of transition in a business unit, and more generally, in a firm, is toward a more rigid process that implies more product homogeneity, increased automation, and more bureaucratic management. In this phase, both product and process innovation begin to approach zero (see Figure 2).

However, Abernathy and Utterback do not propose that the "normal" direction of process and organizational evolution is inexorable. They leave the door open for reversals. Utterback has argued that major innovation, usually originating outside the existing industry, is the key catalyst for a reversal. New ventures—larger firms entering a new business—are credited with the lion's share of the innovations that create important threats for the present competitors.[63] In a later paper, Abernathy and Utterback concluded that government sponsorship, through either purchases or direct regulation, can also have a major impact on innovation in an industry.[64] The recent downsizing of automobiles is a case in point.

Some firms, however, do revolutionize themselves from within. We argue here that when this happens it is the consequence of entrepreneurial sponsorship. This lever for change is not often used. Established firms generally respond to technological invasions by perfecting their present technology. The history of the fountain pen, steam locomotive, and razor blade industries and their response to ballpoint pens, diesel locomotives, and electric razors are illustrative examples.[65]

When the sources of sponsorship are buffered, major changes either do not come about, or, alternatively, occur through a spin-off. Soon after the System 360 had been introduced at IBM, Gene Amdahl, a brilliant designer who had been part of its creation, proposed a new generation of computers based on advanced integrated circuits. After failing to get sponsorship for his proposals, Amdahl quit IBM in frustration and set out to develop his own computer in 1971. The result was the Amdahl 470, introduced in 1975, that was 1.4 to 1.8 times faster than the IBM 370/168 while costing 8 to 12 percent less. Amdahl's sales, four years after the introduction of the 470, were running at an annualized rate of over $400 million.[66] Although IBM later introduced a system with comparable performance, the attitude toward technological champions seems to have changed significantly since the Watson days. Frank Cary, IBM's present chairman, says, "From now on, change will be evolutionary rather than revolutionary."[67]

In summary, managing radical technological change is a fundamental element of top management's task in the technology-based firm. To succeed in coping with new waves of technology, the chief executive must, in Ted Levitt's words, "attack the problem of change."[68] He or she should develop an environment where risk taking by executive champions and product champions will lead to new ventures and products. Most importantly, if the top manager is to thwart the drift toward rigidity and organizational inertia that usually accompanies success, he or she must be personally involved in the entrepreneurial networks that lead to radical technological change, for only the chief executive can make decisions, provide the resources, and absorb the risks necessary for such change.

REFERENCES

1. See K. J. Arrow in *The Rational Direction of Inventive Activity: Economic and Social Factors*, ed. R. Nelson (Princeton: Princeton University Press, 1962), p. 624.

2. See E. B. Roberts, "Entrepreneurship and Technology," in *The Factors in the Transfer of Technology*, ed. W. H. Gruber and D. G. Marquis (Cambridge, MA: MIT Press, 1969), p. 259.

3. I am indebted to Anil Gupta for first suggesting the use of the word network to describe this system of communication and political links.

4. The stages of corporate development used in this article follow the work of B. R. Scott. See for instance: B. R. Scott, "The New Industrial State, Old Myths and New Realities," *Harvard Business Review*, March-April 1973, pp. 133–148; B. R. Scott, "Stages of Corporate Development 1, 2" (Case Clearing House, Harvard Business School, the President and Fellows of Harvard College, 1971).

5. For a definition, see Table 3.

6. See terms for small, integrated, and diversified as defined later in this article.

7. See J. A. Schumpeter, *History of Economic Analysis* (England: Oxford University Press, 1954), p. 554.

8. See J. A. Schumpeter, *Capitalism, Socialism, and Democracy* (New York: Harper & Row, 1975), p. 132.

9. See D. A. Schon, "Champions for Radical New Inventions," *Harvard Business Review*, March-April 1963, p. 84.

10. One of the first to suggest means of managing this resistance to change was P. R. Lawrence, "How to Deal with Resistance to Change," *Harvard Business Review*, January-February 1954. However, in 1969 in a disarmingly candid retrospective commentary on his *Harvard Business Review* classic, Professor Lawrence wrote, "There is . . . an implication in the article that the social and human costs of change can be largely avoided by thoughtful management effort. Today I am less sanguine about this."

11. See R. Rothwell, C. Freeman, A. Horlsey, V. T. P. Jervis, A. B. Robertson and J. Townsend, "SAPPHO Updated—Project SAPPHO Phase II," *Research Policy* 3 (1974): 258–291.

12. See E. B. Roberts and A. B. Frohman, "Internal Entrepreneurship: Strategy for Growth," *The Business Quarterly*, Spring 1972, pp. 71–78.

13. See E. B. Roberts, "Generating Effective Corporate Innovation," *Technology Review*, October-November 1977, pp. 27–33.

14. See L. Grossman, *The Change Agent* (New York: AMACOM, 1974).

15. See J. A. Morton, *Organizing for Innovation* (New York: McGraw-Hill, 1971), p. 95.

16. See I. Kusiatin, "The Process and Capacity for Diversification through Internal Development" (D.B.A. diss., Harvard Graduate School of Business Administration, April 1976).

17. See O. Collins and D. G. Moore, *The Organization Makers* (New York: Appleton-Century-Croft, 1970).

18. Ibid.

19. See O. F. Collins, D. G. Moore, and D. B. Umwalla, *The Enterprising Man* (Board of Trustees, Michigan State University, 1964).

20. See T. D. Duchesneau and J. B. Lafond, "Characteristics of Users and Nonusers of an Innovation: The Role of Economical Organizational Factors" (Paper presented at the Annual Convention of the Eastern Economic Association, Hartford, Connecticut, April 1977). In addition, private correspondence of November 30,

1977 provided expanded details on research data.

21. See R. P. Olsen, "Equipment Supplier—Producer Relationships and Process Innovation in the Textile Industry" (Harvard University Graduate School of Business Administration, November 17, 1975).

22. Ibid. Olsen's champions generally were each identified with a series of innovations.

23. See Rothwell, Freeman, Horlsey, Jervis, Robertson, and Townsend (1974).

24. Ibid.

25. The five key areas identified by the SAPPHO group are all interdisciplinary in character. The first is at the boundary of research and development and organizational behavior; the fourth and fifth are at the boundary of research and development and administrative systems; the second and the third are at the interface between marketing and R&D. Two of these areas have been studied by M.I.T. investigators, Allen (communications) and von Hippel (user needs), who have broadly confirmed and significantly extended the SAPPHO results. A substantial literature also exists in the area of marketing management and project management and related techniques for maximizing the efficiency of R&D. Managerial characteristics, particularly those of the technological entrepreneur, have been studied extensively by another M.I.T. scholar, Roberts, whose conclusions regarding entrepreneurship are also broadly consistent with the SAPPHO group. See also: Roberts and Frohman (Spring 1972); Roberts (October-November 1977); T.A. Allen, *Managing the Flow of Technology: Technology Transfer and the Dissemination of Technological Information within the R&D Organization* (Cambridge, MA: MIT Press, 1977). R.S. Rosenbloom and F.W. Wolek, *Technology and Information Transfer* (Boston: Division of Research, Graduate School of Business Administration, Harvard University, 1977); E. von Hippel, "The Dominant Role of Users in the Scientific Instrument Innovation Process," *Research Policy* 5 (1976): 212–239.

26. See J.L. Bower, *Managing the Resource Allocation Process* (Homewood, IL: Richard Irwin, 1972).

27. In this article, sponsorship and business definition will be used in preference to impetus and context. Bower's context also includes the administrative system, while business definition, as it is used here, is a narrower term that includes only the strategic part of the context.

28. See: Roberts (October-November 1977); "What Do We Really Know about Managing R&D" (A talk with Ed Roberts), *Research Management,* November TK, 1978; R. Rhoades, E.B. Roberts, and A.R. Fusfeld, "A Correlation of R&D Laboratory Performance with Critical Function Analysis," *Research Management* 9, October 1978, pp. 13–17.

29. See J. K. Galbraith, *The New Industrial State* (Boston: Houghton Mifflin, 1967).

30. See: A.D. Chandler, *Strategy and Structure* (Cambridge, MA: MIT Press, 1962); B.R. Scott, *An Open Model of the Firm* (D.B.A. diss., Graduate School of Business Administration, Harvard University, 1962); J.H. McArthur and B.R. Scott, *Industrial Planning in France* (Boston: Division of Research, Harvard Business School, 1969); M. Salter, *Stages of Corporate Development: Implications for Management Control* (Mimeo, 1967); L. Wrigley, *Diversification and Divisional Autonomy* (D.B.A. diss., Graduate School of Business Administration, Harvard University, 1970).

31. See Chandler (1962).

32. See Scott (1971).

33. See Wrigley (1970).

34. A single business is defined here as one that manufactures a single product, a line of products with variations in size and style, or a closely related set of products linked by technology or market structure.

35. See W. J. Abernathy and J. Utterback, "A Dynamic Model of Process and Product Innovation," *Omega* 3 (1975): 639–656.

36. See W. J. Abernathy and P. L. Townsend, "Technology, Productivity, and Process Change," in *Technological Forecasting and Social Change* 7 (New York: American Elsevier Publishing, 1975), pp. 379–396.

37. See S. Myers and D.G. Marquis, *Successful Technological Innovations* (Washington, D.C.: National Science Foundation, NSF 69–17).

38. Dr. James Utterback—personal communication.

39. See T. Burns and G. M. Stalker, *The Management of Innovation* (London: Tavistock Publications, 1966).

40. See Rimbruster Office Automation, Inc. (Intercollegiate Case Clearing House, 4–674–009, Rev. 10.76), p. 9. An updated case of this firm reveals that Rimbruster was a disguised name for Redactron (see Redactron Corporation, Intercollegiate Case Clearing House, 1–276–163) and Redactron's president, Evelyn Berezin).

41. Ibid., p. 3.

42. Ibid., p. 8.

43. For background information on Henry Kloss, see "Advent Corporation (C)" (Intercollegiate Case Clearing House, 9–674–027). The quotation is from "Advent Corporation (D)" (Intercollegiate Case Clearing House, 9–676-053), p.3.

44. See Sci-Tex (Intercollegiate Case Clearing House, 1–678–009), p. 10.

45. See J. K. Galbraith, *The New Industrial State* (Boston: Houghton Mifflin, 1967), pp. 88–89.

46. See G. Bylinsky, *The Innovation Millionaires* (New York: Charles Scribner's Sons, 1976), p. 161.

47. See R. Adams, "Do You Sincerely Want to Be a Millionaire?" *Boston Magazine*, November 1972, p. 45.

48. See J. S. Schwartz, "The Decision to Innovate" (D.B.A. diss., Harvard University Graduate School of Business Administration, 1973), p. 107.

49. Interview with Digital Equipment Corporation executive.

50. See: Schwartz (1973); J.W. Lorsch and P. J. Lawrence, "Organizing for Product Innovation," *Harvard Business Review*, January-February 1965.

51. See Schwartz (1973), p. 111.

52. See "Pilkington Float Glass (A)" (International Case Clearing House, Harvard Business School, 9–670–069).

53. Alistair Pilkington is now chairman of Pilkington Brothers, the first person outside the direct lineage of the founder to hold that position. See J. B. Quinn, "Technological Innovation, Entrepreneurship, and Strategy," *Sloan Management Review*, Spring 1979, pp. 19–30.

54. See L. Wrigley, "Divisional Autonomy and Diversification" (D.B.A. diss., Harvard Business School, 1970).

55. Bob Evans, as quoted by T. A. Wise, "IBM's $5,000,000 Gamble," *Fortune*, September 1966.

56. Interview with IBM executive.

57. See: Wise (September 1966); T.A. Wise, "The Rocky Road to the Marketplace," *Fortune*, October 1966.

58. See M.A. Maidique and J. Ince, " 'The Grumman Corporation' and 'Grumman Energy Systems' " (Intercollegiate Case Clearing House, 1979).

59. See *Business Week*, 27 June 1977.

60. Shortly after his appointment as general manager, Peterson was promoted to president of Grumman Energy Systems, Inc., a wholly owned subsidiary of the Grumman Corporation.

61. See Roberts (October-November 1977).

62. See W. J. Abernathy and J. M. Utterback, "Innovation and the Evolving Structure of the Firm" (Harvard University Graduate School of Business Administration, Working Paper 75018, June 1975).

63. See J. M. Utterback, "Management of Technology" (Center for Policy Alternatives, M.I.T., February 28, 1978).

64. See W. J. Abernathy and J. M. Utterback, "Patterns of Industrial Innovation," *Technology Review*, June-July 1978.

65. See A.C. Cooper and D. Schendel, "Strategic Responses to Technological Threats," *Business Horizons*, February 1976.

66. See B. Uttal, "Gene Amdahl Takes Aim at IBM," *Fortune*, September 1977. For a case study of a data processing center (Hughes) that saw performance/price ratio in favor of the Amdahl 470 over the IBM 370, see J. B. Woods, "Converting from 370 to 470," *Datamation*, July 1977. The sales rate information is from Amdahl 1978 quarterly reports.

67. Frank Cary, as quoted by Uttal (September 1977).

68. See T. Levitt, *Marketing for Business Growth* (New York: McGraw-Hill, 1974), p. 148.

Managing the Internal Corporate Venturing Process

Robert A. Burgelman

The strategic management of internal corporate venturing (ICV) presents a major challenge for many large established firms. The author's conceptualization of ICV suggests that vicious circles and managerial dilemmas typically emerge in the development of new ventures. These problems are exacerbated by the indeterminateness of the strategic context for ICV in the corporation, and by perverse selective pressures exerted by its structural context. This article presents four major recommendations for improving the effectiveness of a firm's ICV strategy. Ed.

Many large established firms currently seem to be trying hard to improve their capacity for managing internal entrepreneurship and new ventures. Companies like Du Pont and General Electric have appointed CEOs with a deep understanding of the innovation process.[1] IBM has generated much interest with its concept of "independent business units."[2] To head its new ventures division, Allied Corporation has attracted the person who ran 3M's new ven-

Reprinted from "Managing the Internal Corporate Venturing Process" by Robert A. Burgelman, *Sloan Management Review*, Winter 1984. pp. 33–48, by permission of the publisher. Copyright 1984 by the Sloan Management Review Association. All rights reserved.

The author gratefully acknowledges the support received from the Strategic Management Program of Stanford University's Graduate School of Business and the helpful comments made by Leonard R. Sayles, Steven C. Wheelwright, and an anonymous reviewer on an earlier version of this paper. Parts of the paper were presented at the Third Strategic Management Society Conference in Paris in October 1983.

tures group for many years.[3] These are only some of the better publicized cases.

Most managers in large established firms will probably agree that internal corporate venturing (ICV) is an important avenue for corporate growth and diversification. However, they will also probably observe that it is a hazardous one, and will be ready to give examples of new ventures (and managerial careers) gone for naught.

Systematic research suggests that such apprehension is not unfounded. In a large sample study of firms attempting to diversify through internal development, Ralph Biggadike found that it takes on the average about eight years for a venture to reach profitability, and about ten to twelve years before its ROI equals that of mainstream business activities.[4] He concludes his study with the caveat that new business development is "not an activity for the impatient or for the fainthearted." Norman

Fast did a study of firms that had created a separate new venture division to facilitate internally developed ventures.[5] He found that the position of such new venture divisions was precarious. Many of these were short-lived, and most others suffered rather dramatic changes as a result of often erratic changes in the corporate strategy and/or in their political position. An overview of earlier studies on new ventures is provided by Eric von Hippel, who observed a great diversity of new venture practices.[6] He also identified some key factors associated with the success and failure of new ventures, but did not document how the ICV *process* takes shape.

The purpose of this article is to shed additional light on some of the more deep-rooted problems inherent in the ICV *process* and to suggest recommendations for making a firm's ICV strategy more effective. This article presents a new model capable of capturing the intricacies of managerial activities involved in the ICV process. This model provides a fairly complete picture of the organizational dynamics of the ICV process. By using this new tool, we can identify and discuss key problems and their interrelationships and then suggest some ideas for alleviating, if not eliminating, these deeprooted problems.

A NEW MODEL OF THE ICV PROCESS

The hazards facing internal corporate ventures are similar in many ways to those confronting new businesses developed by external entrepreneurs. Not surprisingly, the ICV process has typically been conceptualized in terms of a "stages model" which describes the evolution and organization development of a venture as a *separate* new business. Such a model emphasizes the *sequential* aspects of the development process, and focuses on problems within the various stages and on issues pertaining to the transitions between stages. For example, Jay Galbraith has recently proposed a model of new venture development that encompasses five generic stages: (1) proof of principle, prototype; (2) model shop; (3) start-up volume production; (4) natural growth; (5) strategic maneuvering.[7] He discusses the different requirements of these five stages in terms of tasks, people, rewards, processes, structures, and leadership.

Such a stages model is useful for helping managers to organize their experiences and to anticipate problems of fledgling businesses. However, it does not really address the problems of growing a new business *in a corporate context.* Many difficult problems generated and encountered by ICV result from the fact that related strategic activities take place at multiple levels of corporate management. These must be considered *simultaneously* as well as sequentially in order to understand the special problems associated with ICV.

A Process Model of ICV

The work of Joseph Bower and his associates has laid the foundation for a "process model" approach which depicts the simultaneous as well as sequential managerial activities involved in strategic decision making in large complex firms.[8] Recently, I have proposed an extension of this approach that has generated a new model of the ICV process.[9] This new model is based on the findings of an in-depth study of the complete development process of six ICV projects in the context of the new venture division of one large diversified firm. These ICV projects purported to develop new businesses based on new technologies, and constituted radical innovation efforts from the corporation's viewpoint. The Appendix provides a brief description of the methodology used in the study. Figure 1 shows the new model of ICV.

Figure 1 shows the *core* processes of an ICV project and the *overlaying* processes (the

Levels	Core Processes		Overlaying Processes	
	Definition	Impetus	Strategic Context	Structural Context
Corporate Management	Monitoring	Authorizing	Rationalizing	Structuring
NVD Management	Coaching Stewardship	Strategic Building	Organizational Championing Delineating *Selecting*	Negotiating
Group Leader Venture Leader	Technical & Need Linking *Product Championing*	Strategic Forcing	Gatekeeping Idea Generating Bootlegging	Questioning

☐ = Key Activities

Source: Reprinted from "A Process Model of Internal Corporate Venturing in the Diversified Major Firm" by R.A. Burgelman, published in the *Administrative Science Quarterly*, vol. 28. no. 2. June 1983 by permission of the *Administrative Science Quarterly*. Copyright 1983 Cornell University.

FIGURE 1. *Key and Peripheral Activities in a Process Model of ICV*

corporate context) in which the core processes take shape. The core processes of ICV comprise the activities through which a new business becomes defined (definition process) and its development gains momentum in the corporation (impetus process). The overlaying processes comprise the activities through which the current corporate strategy is extended to accommodate the new business thrusts resulting from ICV (strategic context determination), and the activities involved in establishing the administrative mechanisms to implement corporate strategy (structural context determination).

The model shows how each of the processes is constituted by activities of managers at different levels in the organization. Some of these activities were found to be more important for the ICV process than others. These key activities are indicated by the shaded areas. They represent new concepts which are useful to provide a more complete description of the complexities of the ICV process. Because they allow us to refer to these complexities in a concise way, they also serve to keep the discussion manageable. The process model shown in Fig-

ure 1 is *descriptive*. It does *not* suggest that the pattern of activities is optimal from a managerial viewpoint. In fact, many of the problems discussed below result from this particular pattern that the ICV process seems to take on naturally.

MAJOR PROBLEMS IN THE ICV PROCESS

The process model provides a framework for elucidating four important problem areas observed in my study:

- Vicious circles in the definition process,
- Managerial dilemmas in the impetus process,
- Indeterminateness of the strategic context of ICV development, and
- Perverse selective pressures exerted by the structural context on ICV development.

Table 1 serves as a road map for discussing each of these problem areas.

TABLE 1.   Major Problems in the ICV Process

| Levels | Core Processes | | Overlaying Processes | |
	Definition	Impetus	Strategic Context	Structural Context
Corporate Management	Top management lacks capacity to assess merits of specific new venture proposals for corporate development.	Top management relies on purely quantitative growth results to continue support for new venture.	Top management considers ICV as insurance against mainstream business going badly. ICV objectives are ambiguous & shift erratically.	Top management relies on reactive structural changes to deal with problems related to ICV.
NVD Management	Middle-level managers in corporate R&D are not capable of coaching ICV project initiators.	Middle-level managers in new business development find it difficult to balance strategic building efforts with efforts to coach venture managers.	Middle-level managers struggle to delineate boundaries of new business field. They spend significant amounts of time on political activities to maintain corporate support.	Middle-level managers struggle with unanticipated structural impediments to new venture activities. There is little incentive for star performers to engage in ICV activities.
Group Leader Venture Leader	Project initiators cannot convincingly demonstrate in advance that resources will be used effectively. They need to engage in scavenging to get resources.	Venture managers find it difficult to balance strategic forcing efforts with efforts to develop administrative framework of emerging ventures.	Project initiators do not have clear idea of which kind of ICV projects will be viable in corporate context. Bootlegging is necessary to get new idea tested.	Venture managers do not have clear idea of what type of performance will be rewarded, except fast growth.

Vicious Circles in the Definition Process

The ICV projects in my study typically started with opportunistic search activities at the group leader level (first-level supervisor) in the firm's research function. Technical linking activities led to the assembling of external and/or internal pieces of technological knowledge to create solutions for new or known, but unsolved, technical problems. Need linking activities involved the matching of new technical solutions to new or poorly served market needs. Both types of linking activities took place in an iterative fashion. Initiators of ICV projects perceived their initiatives to fall outside of the current strategy of the firm, but felt that there was a good chance they would be included in future strategic development if they proved to be successful.

At the outset, however, project initiators

typically encountered resistance and found it difficult to obtain resources from their managers to demonstrate the feasibility of their project. Hence, the emergence of *vicious circles:* resources could be obtained if technical feasibility was demonstrated, but such a demonstration required resources. Similar problems arose with efforts to demonstrate commercial feasibility. Even when a technically demonstrated product, process, or system existed, corporate management was often reluctant to start commercialization efforts, because they were unsure about the firm's capabilities to do this effectively.

Product championing activities, which have been well documented in the literature, served to break through these vicious circles.[10] Using bootlegging and scavenging tactics, the successful product champion was able to provide positive information which reassured middle-level management and provided them with a basis for claiming support for ICV projects in their formal plans. As the product initiator of a medical equipment venture explained:

> When we proposed to sell the ANA product by our own selling force, there was a lot of resistance, out of ignorance. Management did numerous studies, had outside consultants on which they spent tens of thousands of dollars; they looked at ZYZ Company for a possible partnership. Management was just very unsure about its marketing capability. I proposed to have a test marketing phase with twenty to twenty-five installations in the field. We built our own service group; we pulled ourselves up by the "bootstrap." I guess we had more guts than sense.

Why Does the Problem of Vicious Circles Exist?
The process model provides some insight about this by showing the connection between the activities of the different levels of management involved in the definition process (see Table 1). Operational-level managers typically struggled to conceptualize their somewhat nebulous (at least to outsiders) business ideas,

which made communication with management difficult. Their proposals often went against conventional corporate wisdom. They could not clearly specify the development path of their projects, and they could not demonstrate in advance that the resources needed would be used effectively in uncharted domains.

Middle-level managers in corporate R&D (where new ventures usually originated) were most concerned about maintaining the integrity of the R&D work environment, which is quite different from a business-oriented work environment. They were comfortable with managing relatively slow-moving exploratory research projects and well-defined development projects. However, they were reluctant to commit significant amounts of resources (especially people) to suddenly fast-moving areas of new development activity that fell outside of the scope of their current plans and that did not yet have demonstrated technical and commercial feasibility. In fact, the middle-level manager often seemed to encourage, not just tolerate, the sub-rosa activities of a project's champion. As one such manager said, "I encourage them to do 'bootleg' research; tell them to come back [for support] when they have results."

At the corporate level, managers seemed to have a highly reliable frame of reference to evaluate business strategies and resource allocation proposals pertaining to the main lines of the corporation's business. However, their capacity to deal with substantive issues of new business opportunities was limited, and their expectations concerning what could be accomplished in a short time frame were often somewhat unrealistic. Also, ICV proposals competed for scarce top management time. Their relatively small size combined with the relative difficulty in assessing their merit made it at the outset seem uneconomical for top management to allocate much time to them.

The process model shows the lack of articulation between the activities of different levels of management; this may, to a large extent, account for the vicious circles encountered in the definition process.

Managerial Dilemmas in the Impetus Process

Successful efforts at product championing demonstrated that the technical and commercial potential of a new product, process, or system was sufficient to result in a sizable new business. This, in turn, allowed an ICV project to receive "venture" status: to become a quasi-independent, embryonic new business organization with its own budget and general manager. From then on, continued impetus for its development seemed entirely dependent on achieving fast growth in order to convince top management that it could grow to a $50 to $100 million business within a five- to ten-year period.

My findings suggest that this created a *dilemmatic* situation for the venture manager: maximizing growth with the one product, process, or system available versus building the functional capabilities of the embryonic business organization. Similarly, the middle-level manager was confronted with a *dilemmatic* situation: focusing on expanding the scope of the new business versus spending time coaching the (often recalcitrant) venture manager.

Ironically, my study indicates that new product development was likely to be a major problem of new ventures.[11] Lacking the carefully evolved relationships between R&D, engineering, marketing, and manufacturing typical for the mainstream operations of the firm, the venture's new product development schedules tended to be delayed and completed products often showed serious flaws. This was exacerbated by the tendency of the venture's emerging R&D group to isolate itself from the corporate R&D department, partly in order to establish its own "identity." A related, and somewhat disturbing, finding was that new venture managers seemed to become the victims of

their own success at maintaining impetus for the venture's development. Here are some examples from my study:

- In an environmental systems venture, the perceived need to grow very fast led to premature emphasis on commercialization. Instead of working on the technical improvement of the new system, the venture's resources were wasted on (very costly) remedial work on systems already sold. After a quick rise, stagnation set in and the venture collapsed.
- In a medical equipment venture, growth with one new system was very fast and could be sustained. However, after about five years, the new products needed for sustaining the growth rate turned out to be flawed. As one manager in the venture commented: "Every ounce of effort with Dr. S. [the venture manager] was spent on the short run. There was no strategizing. New product development was delayed, was put to corporate R&D. Every year we had doubled in size, but things never got any simpler."

In both cases the venture manager was eventually removed.

How Do These Dilemmas Arise? The process model shows how the strategic situation at each level of management in the impetus process is different, with fast growth being the only shared interest (see Table 1).

At the venture manager level, continued impetus depended on strategic forcing efforts: attaining a significant sales volume and market share position centered on the original product, process, or system within a limited time horizon.[12] To implement a strategy of fast growth, the venture manager attracted generalists who could cover a number of different functional areas reasonably well. Efficiency considerations became increasingly important with the growth of the venture organization and with competitive pressures due to product matura-tion. New functional managers were brought in to replace the generalists. They emphasized the development of routines, standard operating procedures, and the establishment of an ad-ministrative framework for the venture. This, however, was timeconsuming and detracted from the all-out efforts to grow fast. Growth concerns tended to win out, and organization building was more or less purposefully neg-lected.

While the venture manager created a "beachhead" for the new business, the middle-level manager engaged in strategic building ef-forts to sustain the impetus process. Such ef-forts involved the conceptualization of a master strategy for the broader new field within which the venture could fit. They also involved the integration of projects existed elsewhere in the corporation, and/or of small firms that could be acquired with the burgeoning venture. These efforts became increasingly important as the strategic forcing activities of the venture manager reached their limit, and major discon-tinuities in new product development put more stress on the middle-level manager to find sup-plementary products elsewhere to help main-tain the growth rate. At the same time, the ad-ministrative problems created by the strategic forcing efforts increasingly required the atten-tion of the venture manager's manager. Given the overwhelming importance of growth, how-ever, the coaching activities and organization building were more or less purposefully neg-lected.

The decision by corporate-level man-agement to authorize further resource alloca-tions to a new venture was, to a large extent, dependent on the credibility of the managers involved. Credibility, in turn, depended pri-marily on the quantitative results produced. Corporate management seemed to have some-what unrealistic expectations about new ven-tures. They sent strong signals concerning the importance of making an impact on the overall corporate position soon. This, not surprisingly,

reinforced the emphasis to achieve growth on the part of the middle and operational levels of management. One manager in a very successful venture said, "Even in the face of the extraordi-nary growth rate of the ME venture, the ques-tions corporate management raised when they came here concerned our impact on the overall position of GAMMA, rather than the perform-ance of the venture per se."

Indeterminateness of the Strategic Context of ICV

The problems encountered in the core process of ICV are more readily understood when ex-amining the nature of the overlaying processes (the corporate context) within which ICV pro-jects took shape. My findings indicated a high level of indeterminateness in the strategic con-text of ICV. Strategic guidance on the part of top management was limited to declaring cor-porate interest in broadly defined fields like "health" or "energy." Also, there seemed to be a tendency for severe oscillations in top man-agement's interest in ICV—a "now we do it, now we don't" approach. It looked very much as if new ventures were viewed by top manage-ment as insurance against mainstream business going badly, rather than as a corporate objec-tive per se.[13] As one experienced middle-level manager said:

> They are going into new areas because they are not sure that we will be able to stay in the cur-rent mainstream businesses. That is also the rea-son why the time of maturity of a new venture is never right. If current business goes OK, then it is always too early, but when current business is not going too well, then we will just jump into anything!

In other words, corporate management's inter-est in new ventures seemed to be activated pri-marily by the expectation of a relatively poor performance record with mainstream business activities—a legacy most top managers want to avoid. Treating ICV as "insurance" against such an undesirable situation, however, implies the unrealistic assumption that new ventures can be developed at will within a relatively short time frame, and plays down the importance of crafting a corporate development strategy in substantive terms. Lacking an understanding of substantive issues and problems in particular new venture developments, top management is likely to become disenchanted when progress is slower than desired. Perhaps not surprisingly, venture managers in my study seemed to prefer less rather than more top management attention until the strategic context of their activities was more clearly defined.

Why This Indeterminateness? In determining the strategic context (even more than the impetus process), the strategies of the various levels of management showed a lack of articulation with each other. The process model in Table 1 allows us to depict this.

Corporate management's objectives concerning ICV seemed to be ambiguous. Top management did not really know which specific new businesses they wanted until those businesses had taken some concrete form and size, and decisions had to be made about whether to integrate them into the corporate portfolio through a process of retroactive rationalizing. Top management's actual (as opposed to declared) time horizon was typically limited to three to five years, even though new ventures take between eight and twelve years on the average to become mature and profitable.

Middle managers were aware that they had to take advantage of the short-term windows for corporate acceptance. They struggled with delineating the boundaries of a new business field. They were aware that it was only through their strategic building efforts and the concomitant articulation of a master strategy for the ongoing venture initiatives that the new business fields could be concretely delineated and possible new strategic directions determined. This indeterminateness of the strategic

context of ICV required middle-level managers to engage in organizational championing activities.[14] Such activities were of a political nature and time-consuming. As one venture manager explained, these activities required an "upward" orientation which is very different from the venture manager's substantive and downward (hands-on) orientation. One person who had been general manager of the new venture division said: "It is always difficult to get endorsement from the management committee for ventures which require significant amounts of resources but where they cannot clearly see what is going to be done with these resources. It is a matter of proportion on the one hand, but it is also a matter of educating the management committee which is very difficult to do."

The middle-level manager also had to spend time working out frictions with the operating system that were created when the strategies of the venture and mainstream businesses interfered with each other. The need for these activities further reduced the amount of time and effort the middle-level manager spent coaching the venture manager.

At the operational level, managers engaged in opportunistic search activities which led to the definition of ICV projects in new areas. These activities were basically independent of the current strategy of the firm. The rate at which mutant ideas were pursued seemed to depend on the amount of slack resources available at the operational level. Many of these autonomous efforts were started as "bootlegged" projects.

Perverse Selective Pressures of the Structural Context

Previous research indicates that reaching high market share fast has survival value for new ventures.[15] Hence, the efforts to grow fast which I found pervading the core processes correspond to the managers' correct assessment of the external strategic situation.

My study, however, suggests that the

firm's structural context exerted *perverse selective pressures* to grow fast which exacerbated the external ones. This seemed in part due to the incompleteness of the structural context in relation to the special nature of the ICV process. Establishing a separate new venture division was useful for nurturing and developing new businesses that fell outside of the current corporate strategy. (It was also convenient to have a separate "address" for projects that were "misfits" or "orphans" in the operating divisions.) However, because the managerial work involved in these was very different from that of the mainstream business, the corporate measurement and reward systems were not adequate, and yet they remained in effect, mostly, in the new venture division (NVD).

Another part of the structural context problems resulted from the widely shared perception that the position of the NVD was precarious.[16] This, in turn, created an "it's now or never" attitude on the part of the participants in the NVD, adding to the pressures to grow fast.

How Do These Selective Pressures Arise? Table 1 shows the situation at each of the management levels. Corporate management did not seem to have a clear purpose or strong commitment to new ventures. It seemed that when ICV activities expanded beyond a level that corporate management found opportune to support in light of their assessment of the prospects of mainstream business activities, changes were effected in the structural context to "consolidate" ICV activities. These changes seemed reactive and indicative of the lack of a clear strategy for diversification in the firm. One high-level manager charged with making a number of changes in which the NVD would operate in the future said:

> To be frank, I don't feel corporate management has a clear idea. Recently, we had a meeting with the management committee, and there are now new directives. Basically, it de-emphasizes

diversification for the moment. The emphasis is on consolidation, with the recognition that diversification will be important in the future. . . . The point is that we will not continue in four or five different areas anymore.

At the middle level, the incompleteness of the structural context also manifested itself in the lack of integration between the ICV activities and the mainstream businesses. Middle-level managers of the new venture division experienced resistance from managers in the operating divisions when their activities had the potential to overlap. Ad hoc negotiations and reliance on political savvy substituted for long-term, joint optimization arrangements.[17] This also created the perception that there was not much to gain for middle-level star performers by participating directly in the ICV activities. In addition, the lack of adequate reward systems also made middle-level managers reluctant to remove venture managers in trouble. One middle-level manager talked about the case of a venture manager in trouble who had grown a project from zero to about $30 million in a few years:

> When the business reached, say, $10 million, they should have talked to him; have given him a free trip around the world, $50 thousand, and six months off; and then have persuaded him to take on a new assignment. But that's not the way it happened. For almost two years, we knew that there were problems, but no one would touch the problem until it was too late and he had put himself in a real bind. We lost some good people during this period, and we lost an entrepreneur.

At the operational level, managers felt that the only reward available was to become general manager of a sizeable new business in the corporate structure. This "lure of the big office" affected the way in which they searched for new opportunities. One high-level manager observed, "People are looking around to find a program to latch onto, and that could be developed into a demonstration plan. Business research always stops after one week, so to speak."

MAKING THE ICV STRATEGY WORK BETTER

Having identified major problem areas with the help of the process model, we can now propose recommendations for improving the strategic management of ICV. They serve to alleviate, if not eliminate, the problems by making the corporate context more hospitable to ICV. This could allow management to focus more effectively on the problems inherent in the core processes.

There are four "themes" for the recommendations which correspond to the four major problem areas already discussed:

· Facilitating the definition process,
· Moderating the impetus process,
· Elaborating the strategic context of ICV, and
· Refining the structural context of ICV.

Each of these themes encompasses more specific action items for management. Table 2 summarizes the various recommendations and their expected effects on the ICV process.

Facilitating the Definition Process

Timely assessment of the true potential of an ICV project remains a difficult problem. This follows from the very nature of such projects: the many uncertainties around the technical and marketing aspects of the new business, and the fact that each case is significantly different from all others. These factors make it difficult to develop standardized evaluation procedures and development programs, without screening to death truly innovative projects.

Managing the definition process effectively poses serious challenges for middle-level managers in the corporate R&D department. They must facilitate the integration of technical and business perspectives, and they must maintain a life line to the technology developed in corporate R&D as the project takes off. As stated earlier, the need for product championing efforts, if excessive, may cut that life line early on and lead to severe discontinuities in new product development after the project has reached the venture stage. The middle-level manager's efforts must facilitate both the product championing efforts and the continued development of the technology base by putting the former in perspective and by making sure that the interface between R&D and businesspeople works smoothly.

Facilitating the Integration of R&D-Business Perspectives.   To facilitate the integration of technical and business perspectives, the middle manager must understand the operating logic of both groups, and must avoid getting bogged down in technical details yet have sufficient technical depth to be respected by the R&D people. Such managers must be able to motivate the R&D people to collaborate with the businesspeople toward the formulation of business objectives against which progress can be measured. Formulating adequate business objectives is especially important if corporate management becomes more actively involved in ICV and develops a greater capacity to evaluate the fit of new projects with the corporate development strategy.

Middle-level managers in R&D must be capable of facilitating give-and-take between the two groups in a process of mutual adjustment toward the common goal of advancing the progress of the new business project. It is crucial to create mutual respect between technical and businesspeople. If the R&D manager shows respect for the contribution of the businesspeople, this is likely to affect the attitudes

594

TABLE 2. *Recommendations for Making ICV Strategy Work Better*

Levels	Core Processes		Overlaying Processes	
	Definition	Impetus	Strategic Context	Structural Context
Corporate Management	ICV proposals are evaluated in light of corporate development strategy. Conscious efforts are made to avoid subjecting them to conventional corporate wisdom.	New venture progress is evaluated in substantive terms by top managers who have experience in organizational championing.	A process is in place for developing long-term corporate development strategy. This strategy takes shape as result of ongoing interactive learning process involving top & middle levels of management.	Managers with successful ICV experience are appointed to top management. Top management is rewarded financially & symbolically for long-term corporate development success.
NVD Management	Middle-level managers in corporate R&D are selected who have both technical depth & business knowledge necessary to determine minimum amount of resources for project, & who can coach star players.	Middle-level managers are responsible for use & development of venture managers as scarce resources of corporation, & they facilitate intrafirm project transfers if new business strategy warrants it.	Substantive interaction between corporate & middle-level management leads to clarifying merits of new business field in light of corporate development strategy.	Star performers at middle level are attracted to ICV activities. Collaboration of mainstream middle level with ICV activities is rewarded. Integrating mechanisms can easily be mobilized.
Group Leader Venture Leader	Project initiators are encouraged to integrate technical & business perspectives. They are provided access to resources. Project initiators can be rewarded by means other than promotion to venture manager.	Venture managers are responsible for developing functional capabilities of emerging venture organizations, & for codification of what has been learned in terms of required functional capabilities while pursuing new business opportunity.	Slack resources determine level of emergence of mutant ideas. Existence of substantive corporate development strategy provides preliminary self-selection of mutant ideas.	A wide array of venture structures & supporting measurement & reward systems clarifies expected performance for ICV personnel.

of the other R&D people. Efforts will probably be better integrated if regular meetings are held with both groups to evaluate, as peers, the contribution of different team members.

The Middle Manager as Coach.   Such meetings also provide a vehicle to better coach the product champion, who is really the motor of the ICV project in this stage of development. There are some similarities between this role and that of the star player on a sports team. Product champions are often viewed in either/ or terms: either they can do their thing and chances are the project will succeed (although there may be discontinuities and not fully exploited ancillary opportunities), or we harness them but they will not play.

A more balanced approach is for the middle-level manager to use a process that recognizes the product champion as the star player, but that, at times, challenges him or her to maintain breadth by having to respond to queries:

- How is the team benefiting more from this particular action than from others that the team may think to be important?
- How will the continuity of the team's efforts be preserved?
- What will the next step be?

To support this approach, the middle manager should be able to reward team members differently. This, of course, refers back to the determination of the structural context, and reemphasizes the importance of recognizing, at the corporate level, that different reward systems are necessary for different business activities.

Moderating the Impetus Process

The recommendations for improving the corporate context (the overlaying processes) of ICV, have implications for the way in which the impetus process is allowed to take shape. Corporate management should expect the middle-

level managers to think and act as corporate strategists and the operational-level managers to view themselves as organization builders.

The Middle-Level Managers as Corporate Strategists.   Strategy making in new ventures depends, to a very great extent, on the middle-level managers. Because new ventures often intersect with multiple parts of mainstream businesses, middle managers learn what the corporate capabilities and skills—and shortcomings—are, and they learn to articulate new strategies and build new businesses based on new combinations of these capabilities and skills. This, in turn, also creates possibilities to enhance the realization of new operational synergies existing in the firm. Middle-level managers can thus serve as crucial integrating and technology transfer mechanisms in the corporation, and corporate management should expect them to perform this role as they develop a strategy for a new venture.

The Venture Managers as Organization Builders.   Pursuing fast growth and the administrative development of the venture simultaneously is a major challenge during the impetus process. This challenge, which exists for any start-up business, is especially treacherous for one in the context of an established firm. This is because managers in ICV typically have less control over the selection of key venture personnel, yet, at the same time, have access to a variety of corporate resources. There seems to be less pressure on the venture manager and the middle-level manager to show progress in building the organization than there is to show growth.

The recommendations concerning measurement and reward systems should encourage the venture manager to balance the two concerns better. The venture manager should have leeway in hiring and firing decisions, but should also be held responsible for the development of new functional capabilities and the administrative framework of the venture. This would reduce the probability of major discontinuities in new product development mentioned earlier. In addition, it would provide the corporation with codified know-how and information which can be transferred to other parts of the firm or to other new ventures, even if the one from which it is derived ultimately fails as a business. Know-how and information, as well as sales and profit, become important outputs of the ICV process.

Often the product champion or venture manager will not have the required capabilities to achieve these additional objectives. The availability of compensatory rewards and of avenues for recycling the product champion or venture manager would make it possible for middle management to better tackle deteriorating managerial conditions in the new business organization. Furthermore, the availability of a competent replacement (after systematic corporate search) may induce the product champion or venture manager to relinquish his or her position, rather than see the venture go under.

Elaborating the Strategic Context of ICV

Determining the strategic context of ICV is a subtle and somewhat elusive process involving corporate and middle-level managers. More effort should be spent on developing a long-term corporate development strategy explicitly encompassing ICV. At the same time measures should be taken to increase corporate management's capacity to assess venture strategies in substantive terms as well as in terms of projected quantitative results.

The Need for a Corporate Development Strategy. Top management should recognize that ICV is an important source of strategic renewal for the firm and that it is unlikely to work well if treated as insurance against poor mainstream business prospects. ICV should, therefore, be considered an integral and continuous part of the strategy-making process. To dampen the oscillations in corporate support for ICV, top management should create a process for developing an explicit long-term (ten to twelve years) strategy for corporate development, supported by a resource generation and allocation strategy. Both should be based on ongoing efforts to determine the remaining growth opportunities in the current mainstream businesses and the resource levels necessary to exploit them. Given the corporate objectives of growth and profitability, a resource pool should be reserved for activities outside the mainstream business. This pool should not be affected by short-term fluctuations in current mainstream activities. The existence of this pool of "slack" (or perhaps better, "uncommitted") resources would allow top management to affect the rate at which new venture initiatives will emerge (if not their particular content). This approach reflects a broader concept of strategy making than maintaining corporate R&D at a certain percentage of sales.

Substantive Assessment of Venture Strategies. To more effectively determine the strategic context of ICV and to reduce the political emphasis in organizational championing activities, top management should increase their capacity to make substantive assessments of the merits of new ventures for corporate development. Top management should learn to assess better the strategic importance of ICV projects to corporate development and their degree of relatedness to core corporate capabilities. One way to achieve this capacity is to include in top management people with significant experience in new business development. In addition, top management should require middle-level organizational champions to explain how a new field of business would further the corporate development objectives in substantive rather than purely numerical terms and how they expect to create value from the corporate viewpoint with a new business field. Operational-level managers would then be able to assess better which of the possible directions their envisaged projects could take and would be more likely to receive corporate support.

Refining the Structural Context of ICV

Refining the structural context requires corporate management to use the new venture division design in a more deliberate fashion, and to complement the organization design effort with supporting measurement and reward systems.

More Deliberate Use of the New Venture Division Design. Corporate management should develop greater flexibility in structuring the relationships between new venture projects and the corporation. In some instances, greater efforts would seem to be in order to integrate new venture projects directly into the mainstream businesses, rather than transferring them to the NVD because of lack of support in the operating division where they originated. In other cases, projects should be developed using external venture arrangements. Where and how a new venture project is developed should depend on top management's assessment of its strategic importance for the firm, and of the degree to which the required capabilities are related to the firm's core capabilities. Such assessments should be easier to implement by having a wide range of available structures for venture-corporation relationships.[18]

Also, the NVD is a mechanism for decoupling the activities of new ventures and those of mainstream businesses. However, because this decoupling usually cannot be perfect, integrative mechanisms should be established to deal constructively with conflicts that will unavoidably and unpredictably arise. One such mechanism is a "steering committee" involving managers from operating divisions and the NVD.

Finally, top management should facilitate greater acceptance of differences between the management processes of the NVD and the mainstream businesses. This may lead to more careful personnel assignment policies and to greater flexibility in hiring and firing policies in the NVD to reflect the special needs of emerging businesses.

Measurement and Reward Systems in Support of ICV. Perhaps the most difficult aspect concerns how to provide incentives for top management to seriously and continuously support ICV as part of corporate strategy making. Corporate history writing might be an effective mechanism to achieve this. This would involve the careful tracing and periodical publication (e.g., a special section in annual reports) of decisions whose positive or negative results may become clear only after ten or more years. Corporate leaders (like political ones) would, presumably, make efforts to preserve their position in corporate history.[19] Another mechanism is to attract "top performers" from the mainstream businesses of the corporation to ICV activities. To do this, at least a few spots on the top management team should always be filled with managers who have had significant experience in new business development. This will also eliminate the perception that NVD participants are not part of the real world and, therefore, have little chance to advance in the corporation as a result of ICV experience.

The measurement and reward systems should be used to alleviate some of the more destructive consequences of the necessary emphasis on fast growth in venture development. This would mean, for instance, rewarding accomplishments in the areas of problem finding, problem solving, and know-how development. Success in developing the administrative aspects of the emerging venture organization should also be included, as well as effectiveness in managing the interfaces with the operating division.

At the operational level where some managerial failures are virtually unavoidable, top management should create a reasonably foolproof safety net. Product champions at this level should not have to feel that running the business is the only possible reward for getting it started. Systematic search for and screening of potential venture managers should make it easier to provide a successor for the product champion in *time*. Avenues for recycling product champions and venture managers should be developed and/or their reentry into the mainstream businesses facilitated.

Finally, more flexible systems for measuring and rewarding performance should accompany the greater flexibility in structuring the venture-corporate relations mentioned earlier. This would mean greater reliance on negotiation processes between the firm and its entrepreneurial actors. In general, the higher the degree of relatedness (the more dependent the new venture is on the firm's resources) and the lower the expected strategic importance for corporate development, the lower the rewards the internal entrepreneurs would be able to negotiate. As the venture evolves, milestone points could be agreed upon to revise the negotiations. To make such processes symmetrical (and more acceptable to the nonentrepreneurial participants in the organization), the internal entrepreneurs should be required to substitute negotiated for regular membership awards and benefits.[20]

CONCLUSION: NO PANACEAS

This article proposes that managers can make ICV strategy work better if they increase their capacity to conceptualize the managerial activities involved in ICV in process model terms. This is because the process model approach allows the managers involved to think through how their strategic situation relates to the strategic situation of managers at different levels who are simultaneously involved in the process. Understanding the interplay of these different strategic situations allows managers to see the relationships between problems which otherwise remain unanticipated and seemingly disparate. This may help them perform better as *individual* strategists while also enhancing the *corporate* strategy-making process.

Of course, by focusing on the embedded, nested problems and internal organizational dynamics of ICV strategy making, this article has not addressed other important problems. I believe, however, that the vicious circles, managerial dilemmas, indeterminateness of the strategic context, and perverse selective pressures of the structural context are problem areas that have received the least systematic attention.

The recommendations (based on this viewpoint) should result in a somewhat better use of the individual entrepreneurial resources of the corporation and, therefore, in an improvement of the corporate entrepreneurial capability. Yet, the implication is not that this process can or should become a planned one, or that the discontinuities associated with entrepreneurial activity can be avoided. ICV is likely to remain an uncomfortable process for the large complex organization. This is because ICV upsets carefully evolved routines and planning mechanisms, threatens the internal equilibrium of interests, and requires revising a firm's self-image. The success of radical innovations, however, is ultimately dependent on whether they can become institutionalized. This may pose the most important challenge for managers of large established firms in the eighties.

APPENDIX

A Field Study of ICV

A qualitative method was chosen as the best way to arrive at an encompassing view of the ICV process.

Research Setting.   The research was carried out in one large, diversified, U.S.-based, high technology firm, which I shall refer to as GAMMA. GAMMA had traditionally produced and sold various commodities in large volume, but it had also tried to diversify through the internal development of new products, processes, and systems in order to get closer to the final user or consumer and to catch a greater portion of the total value added in the chain from raw materials to end products. During the sixties, diversification efforts were carried out within existing operating divisions, but in the early seventies, the company established a separate new venture division (NVD).

Data Collection.   Data were obtained on the functioning of the NVD. The charters of its various departments, the job descriptions of the major positions in the division, the reporting relationships and mechanisms of coordination, and the reward system were studied. Data were also obtained on the relationships between the NVD and the rest of the corporation. In particular, the collaboration between the corporate R&D department and divisional R&D groups was studied. Finally, data were also obtained on the role of the NVD in the implementation of the corporate strategy of unrelated diversification. These data describe the historical evolution of the structural context of ICV development at GAMMA before and during the research period.

The bulk of the data was collected by studying the six major ICV projects in progress at GAMMA at the time of the research. These ranged from a case where the business objectives were still being defined to one where the venture had reached a sales volume of $35 million.

In addition to the participants in the six ICV projects, I interviewed NVD administrators, people from several operating divisions, and one person from corporate management. A total of sixty-one people were interviewed. The interviews were unstructured and took from 1½ to 4½ hours. Tape recordings were not made, but the interviewer took notes in shorthand. The interviewer usually began with an open-ended invitation to discuss work-related activities and then directed the interview toward three major aspects of the ICV development process: the evolution over time of a project, the involvement of different functional groups in the development process, and the involvement of different hierarchical levels in the development process. Respondents were asked to link particular statements they made to statements of other respondents on the same issues or problems and to give examples where appropriate. After completing an interview, the interviewer made a typewritten copy of the conversation. About 435 legal-size pages of typewritten field notes resulted from these interviews.

The research also involved the study of documents. As it might be expected, the ICV project participants relied little on written procedures in their day-to-day working relationships with other participants. One key set of documents, however, was the written corporate long-range plans concerning the NVD and each of the ICV projects. These official descriptions of the evolution of each project between 1973 and 1977 were compared with the interview data.

Finally, occasional behavioral observations were made, for example, when other people would call or stop by during an interview or in informal discussions during lunch at the research site. These observations, though not systematic, led to the formulation of new questions for further interviews.

Conceptualization.   The field notes were used to write a case history for each of the venture projects which put together the data obtained from all participants on each of the three major aspects of venture development. The comparative analysis of the six ICV cases allowed the construction of a stages model that described the sequence of stages and their key activities. The process model resulted from combining the analysis at the project level with the data obtained at the corporate level.

REFERENCES

1. See "DuPont: Seeking a Future in Biosciences," *Business Week,* 24 November 1980, pp. 86–98; "General Electric: The Financial Wizards Switch Back to Technology," *Business Week,* 16 March 1981, pp. 110–114.
2. See "Meet the New Lean, Mean IBM," *Fortune,* 13 June 1983, pp. 68–82.
3. "Allied after Bendix: R&D Is the Key," *Business Week,* 12 December 1983, pp. 76–86.
4. See R. Biggadike, "The Risky Business of Diversification," *Harvard Business Review,* May–June 1979, p. 111.
5. See N.D. Fast, "The Future of Industrial New Venture Departments," *Industrial Marketing Management* (1979): 264–273.
6. See E. von Hippel, "Successful and Failing Internal Corporate Ventures: An Empirical Analysis," *Industrial Marketing Management* (1977): 163–174. Some of the diversity found by von Hippel, however, may be due to a somewhat unclear distinction between new product development and new business development.
7. See J. R. Galbraith, "The Stages of Growth," *Journal of Business Strategy* 4 (1983) pp. 70–79.
8. See J. L. Bower, *Managing the Resource Allocation Process* (Boston: Graduate School of Business Administration, Harvard University, 1970).
9. See R. A. Burgelman, "Managing Innovating Systems: A Study of the Process of Internal Corporate Venturing" (unpublished doctoral dissertation, Columbia University, 1980); R. A. Burgelman, "A Process Model of Internal Corporate Venturing in the Diversified Major Firm," *Administrative Science Quarterly,* June 1983, pp. 223–244.
10. See D. A. Schon, "Champions for Radical New Inventions," *Harvard Business Review,* March–April 1963, pp. 77–86; E. B. Roberts, "Generating Effective Corporate Innovation," *Technology Review,* October-November 1977, pp. 27–33.
11. One of the key problems encountered by Exxon Enterprises was precisely the existence of these new product development problems in the entrepreneurial ventures (Qyx, Quip, and Vydec) it had acquired and was trying to integrate. See "What's Wrong at Exxon Enterprises," *Business Week,* 24 August 1981, p. 87.
12. The need for strategic forcing is consistent with findings suggesting that attaining large market share fast at the cost of early profitability is critical for venture survival. See Biggadike (May–June 1979).
13. See Entrepreneurial activity used as insurance against environmental turbulence was first documented by R. A. Peterson and D. G. Berger, "Entrepreneurship in Organizations: Evidence from the Popular Music Industry," *Administrative Science Quarterly* 16 (1971): 97–106; R. A. Burgelman, "Corporate Entrepreneurship and Strategic Management: Insights from a Process Study," *Management Science* 29 (1983): 1649–1664.
14. The importance of the middle-level manager in ICV was already recognized by E. von Hippel (1977). The role of a "manager champion" or "executive champion" has also been discussed by: I. Kusiatin, "The Process and Capacity for Diversification through Internal Development" (unpublished doctoral dissertation, Harvard University, 1976); M. A. Maidique, "Entrepreneurs, Champions, and Technological Innovation," *Sloan Management Review,* Winter 1980, pp. 59–76.
15. See Biggadike (May–June 1979).
16. See Fast (1979).
17. These frictions are discussed in more detail in R. A. Burgelman, "Managing the New Venture Division: Research Findings and Implications for Strategic Management," *Strategic Management Journal,* in press.
18. An overview of different forms of corporate venturing is provided in E. B. Roberts, "New Ventures for Corporate Growth," *Harvard Business Review,* July–August 1980, pp. 132–142. A design framework is suggested in R. A. Burgelman, "Designs for Corporate Entrepreneur-

ship in Established Firms," *California Manage-*
ment Review, in press.

19. Some firms seem to have developed the posi-
    tion of corporate historian. See "Historians
    Discover the Pitfalls of Doing the Story of a
    Firm," *Wall Street Journal,* 27 December 1983.
    Without underestimating the difficulties such a
    position is likely to hold, one can imagine the
    possibility of structuring it in such a way that
    the relevant data would be recorded. Another
    instance, possibly a board-appointed commit-

tee, could periodically interpret these data
along the lines suggested.

20. Some companies have developed innovative
    types of arrangements to structure their rela-
    tionships with internal entrepreneurs. Other
    companies have established procedures to help
    would-be entrepreneurs with their decision to
    stay with the company or to spin off. Control
    Data Corporation, for example, has established
    an "Employee Entrepreneurial Advisory Of-
    fice."

Entering New Businesses:
Selecting Strategies for Success

Edward B. Roberts

Charles A. Berry

Selective use of the alternative strategies available for entering new businesses is a key issue for diversifying corporations. The approaches include internal development, acquisition, licensing, joint ventures, and minority venture capital investments. Using the existing literature, the authors devise a matrix of company "familiarity" with relevant market and technological experiences. Through a case study of a successful diversified technological firm, they demonstrate how the familiarity matrix can be applied to help a company select optimum entry strategies. Ed.

Entry into new product-markets, which represents diversification for the existing firm, may provide an important source of future growth and profitability. Typically, such new businesses are initiated with low market share in high growth markets and require large cash inflows to finance growth. In addition, many new product-market entries fail, draining additional cash resources and incurring high opportunity costs to the firm. Two basic strategic questions are thus posed: (1) Which product-markets should a corporation enter? and (2) How

should the company enter these product-markets to avoid failure and maximize gain?

Although these questions are fundamentally different, they should not be answered independently of one another. Entering a new business may be achieved by a variety of mechanisms, such as internal development, acquisition, joint ventures, and minority investments of venture capital. As Roberts indicates, each of these mechanisms makes different demands upon the corporation.[1] Some, such as internal development, require a high level of commitment and involvement. Others, such as venture capital investment, require much lower levels of involvement. What are the relative benefits and costs of each of these entry mechanisms? When should each be used?

This article attempts to analyze and answer these questions, first by proposing a framework for considering entry issues, second

Reprinted from "Entering New Businesses: Selecting Strategies for Success" by Edward B. Roberts and Charles A. Berry, *Sloan Management Review*, Spring 1985, pp. 3–17, by permission of the publisher. Copyright 1985 by the Sloan Management Review Association. All rights reserved.

The authors express their deep appreciation to an anonymous reviewer who assisted in clarifying a number of aspects of this article.

by a review of relevant literature, third by application of this literature to the creation of a matrix that suggests optimum entry strategies, and finally by a test of the matrix through a case analysis of business development decisions by a successful diversified corporation.

ENTRY STRATEGY: A NEW SELECTION FRAMEWORK

New business development may address new markets, new products, or both. In addition, these new areas may be ones that are familiar or unfamiliar to a company. Let us first define "newness" and "familiarity":

- Newness of a Technology or Service: the degree to which that technology or service has not formerly been embodied within the products of the company.
- Newness of a Market: the degree to which the products of the company have not formerly been targeted at that particular market.
- Familiarity with a Technology: the degree to which knowledge of the technology exists within the company, but is not necessarily embodied in the products.

- Familiarity with a Market: the degree to which the characteristics and business patterns of a market are understood within the company, but not necessarily as a result of participation in the market.

If the businesses in which a company presently competes are its *base* businesses, then market factors associated with the new business area may be characterized as *base, new familiar,* or *new unfamiliar*. Here, "market factors" refers not only to particular characteristics of the market and the participating competitors, but also includes the appropriate pattern of doing business that may lead to competitive advantage. Two alternative patterns are performance/premium price and lowest cost producer. Similarly, the technologies or service embodied in the product for the new business area may be characterized on the same basis. Figure 1 illustrates some tests that may be used to distinguish between "base" and "new" areas. Table 1 lists questions that may be used to distinguish between familiar and unfamiliar technologies. (Equivalent tests may be applied to services.) Questions to distinguish between familiar and unfamiliar markets are given in Table 2.

The application of these tests to any new business development opportunity enables it to be located conceptually on a 3 × 3 technology/

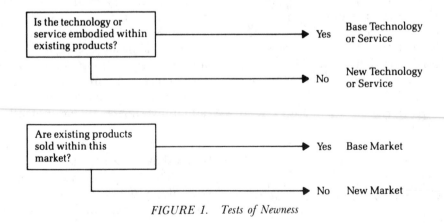

FIGURE 1.   Tests of Newness

TABLE 1. *Tests of Technological Familiarity*

	Decreasing Familiarity

1. Is the technological capability used within the corporation without being embodied in products, e.g., required for component manufacture (incorporated in processes rather than products)?
2. Do the main features of the new technology relate to or overlap with existing corporate technological skills or knowledge, e.g., coating of optical lenses and aluminizing semiconductor substrates?
3. Do technological skills or knowledge exist within the corporation without being embodied in products or processes, e.g., at a central R&D facility?
4. Has the technology been systematically monitored from within the corporation in anticipation of future utilization, e.g., by a technology assessment group?
5. Is relevant and reliable advice available from external consultants?

TABLE 2. *Tests of Market Familiarity*

	Decreasing Familiarity

1. Do the main features of the new market relate to or overlap existing product markets, e.g., base and new products are both consumer products?
2. Does the company presently participate in the market as a buyer (relevant to backward integration strategies)?
3. Has the market been monitored systematically from within the corporation with a view to future entry?
4. Does knowledge of the market exist within the corporation without direct participation in the market, e.g., as a result of previous experience of credible staff?
5. Is relevant and reliable advice available from external consultants?

market *familiarity matrix* as illustrated in Figure 2. The nine sectors of this matrix may be grouped into three regions, with the three sectors that comprise any one region possessing broadly similar levels of familiarity.

LITERATURE REVIEW: ALTERNATIVE STRATEGIES

Extensive writings have focused on new business development and the various mechanisms by which it may be achieved. Much of this literature concentrates on diversification, the most demanding approach to new business development, in which both the product and market dimensions of the business area may be new to a company. Our review of the literature supports and provides details for the framework shown in Figure 2, finding that familiarity of a company with the technology and market being addressed is the critical variable that explains much of the success or failure in new business development approaches.

Rumelt's pioneering 1974 study of diversification analyzed company performance against a measure of the relatedness of the various businesses forming the company.[2] Rumelt identified nine types of diversified companies, clustered into three categories: dominant business companies, related business compa-

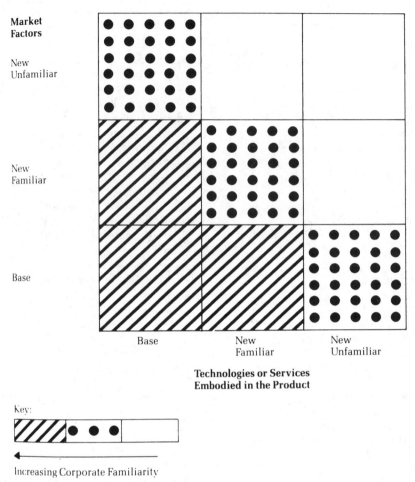

Market Factors

New Unfamiliar

New Familiar

Base

Base   New Familiar   New Unfamiliar

Technologies or Services Embodied in the Product

Key:

Increasing Corporate Familiarity

FIGURE 2.   The Familiarity Matrix

nies, and unrelated business companies. From extensive analysis Rumelt concluded that related business companies outperformed the averages on five accounting-based performance measures over the period 1949 to 1969.

Rumelt recently updated his analysis to include Fortune 500 companies' performances through 1974 and drew similar conclusions: the related constrained group of companies was the most profitable, building on single strengths or resources associated with their original businesses.[3] Rumelt, as well as Chris-

tensen and Montgomery, also found, however, that the performance in part reflected effects of concentrations in certain categories of industrial market clusters.[4] While some (e.g., Bettis and Hall)[5] have questioned Rumelt's earlier conclusions, still others (e.g., Holzmann, Copeland, and Hayya)[6] have supported the findings of lower returns by unrelated business firms and highest profitability for the related constrained group of firms.

Peters supports Rumelt's general conclusions on the superior performance of related business companies.[7] In his study of thirty-seven "well-managed" organizations, he found that they had all been able to define their strengths and build upon them: they had not moved into potentially attractive new business areas that required skills that they did not possess. In their recent book Peters and Waterman termed this "sticking to the knitting."[8]

Even in small high technology firms similar effects can be noted. Recent research by Meyer and Roberts on ten such firms revealed that the most successful firms in terms of growth had concentrated on one key technological area and introduced product enhancements related to that area.[9] In contrast, the poorest performers had tackled "unrelated" new technologies in attempts to enter new product-market areas.

The research discussed above indicates that in order to ensure highest performance, new business development should be constrained within areas related to a company's base business—a very limiting constraint. However, no account was taken of how new businesses were in fact entered and the effect that the entry mechanism had on subsequent corporate performance. As summarized in Table 3, the literature identifies a wide range of approaches that are available for entering new business areas, and highlights various advantages and disadvantages.

TABLE 3. *Entry Mechanisms: Advantages and Disadvantages*

New Business Development Mechanisms	Major Advantages	Major Disadvantages
Internal Developments	Use existing resources	Time lag to break even tends to be long (on average eight years) Unfamiliarity with new markets may lead to errors
Acquisitions	Rapid market entry	New business area may be unfamiliar to parent
Licensing	Rapid access to proven technology Reduced financial exposure	Not a substitute for internal technical competence Not proprietary technology Dependent upon licensor
Internal Ventures	Use existing resources May enable a company to hold a talented entrepreneur	Mixed record of success Corporation's internal climate often unsuitable
Joint Ventures or Alliances	Technological/marketing unions can exploit small/large company synergies Distribute risk	Potential for conflict between partners
Venture Capital and Nurturing	Can provide window on new technology or market	Unlikely alone to be a major stimulus of corporate growth
Educational Acquisitions	Provide window and initial staff	Higher initial financial commitment than venture capital Risk of departure of entrepreneurs

New Business Development Mechanisms

Internal Developments.   Companies have traditionally approached new business development via two routes: internal development or acquisition. Internal development exploits internal resources as a basis for establishing a business new to the company. Biggadike studied Fortune 500 companies that had used this approach in corporate diversification.[10] He found that, typically, eight years were needed to generate a positive return on investment, and performance did not match that of a mature business until a period of ten to twelve years had elapsed. However, Weiss asserts that this need not be the case.[11] He compared the performance of internal corporate developments with comparable businesses newly started by individuals and found that the new independent businesses reached profitability in half the time of corporate effort—approximately four years versus eight years. Although Weiss attributes this to the more ambitious targets established by independent operations, indeed the opposite may be true. Large companies' overhead allocation charges or their attempts at large-scale entry or objectives that preclude early profitability may be more correct explanations for the delayed profitability of these ventures.

Miller indicates that forcing established attitudes and procedures upon a new business may severely handicap it, and suggests that success may not come until the technology has been adapted, new facilities established, or familiarity with the new markets developed.[12] Miller stresses this last factor as very important. Gilmore and Coddington also believe that lack of familiarity with new markets often leads to major errors.[13]

Acquisitions.   In contrast to internal development, acquisition can take weeks rather than years to execute. This approach may be attractive not only because of its speed, but because it offers a much lower initial cost of entry into a new business or industry. Salter and Weinhold point out that this is particularly true if the key parameters for success in the new business field are intangibles, such as patents, product image, or R&D skills, which may be difficult to duplicate via internal developments within reasonable costs and time scales.[14]

Miller believes that a diversifying company cannot step in immediately after acquisition to manage a business it knows nothing about.[15] It must set up a communication system that will permit it to understand the new business gradually. Before this understanding develops, incompatibility may exist between the managerial judgment appropriate for the parent and that required for the new subsidiary.

Licensing.   Acquiring technology through licensing represents an alternative to acquiring a complete company. J.P. Killing has pointed out that licensing avoids the risks of product development by exploiting the experience of firms who have already developed and marketed the product.[16]

Internal Ventures.   Roberts indicates that many corporations are now adopting new venture strategies in order to meet ambitious plans for diversification and growth.[17] Internal ventures share some similarities with internal development, which has already been discussed. In this venture strategy, a firm attempts to enter different markets or develop substantially different products from those of its existing base business by setting up a separate entity within the existing corporate body. Overall the strategy has had a mixed record, but some companies such as 3M have exploited it in the past with considerable success. This was due in large part to their ability to harness and nurture entrepreneurial behavior within the corporation. More recently, IBM's Independent Business Units (especially its PC venture) and Du Pont's new electronic materials division demonstrate the effectiveness of internal ventures for market expansion and/or diversification. Burgelman has suggested that corporations need to "develop greater flexibility between new venture projects and the corporation," using external as well as internal ventures.[18]

The difficulty in successfully diversifying via internal ventures is not a new one. Citing Chandler,[19] Morecroft comments on Du Pont's failure in moving from explosive powders to varnishes and paints in 1917:

> [C]ompeting firms, though much smaller and therefore lacking large economies of scale and production, were nonetheless profitable. . . . Their sole advantage lay in the fact that they specialized in the manufacture, distribution, and sale of varnishes and paints. This focus provided them with clearer responsibilities and clearer standards for administering sales and distribution.[20]

Joint Ventures or Alliances. Despite the great potential for conflict, many companies successfully diversify and grow via joint ventures. As Killing points out, when projects get larger, technology more expensive, and the cost of failure too large to be borne alone, joint venturing becomes increasingly important.[21] Shifts in national policy in the United States are now encouraging the formation of several large research-based joint ventures involving many companies. But the traditional forms of joint ventures, involving creation of third corporations, seem to have limited life and/or growth potential.

Hlavacek et al. and Roberts[22] believe one class of joint venture to be of particular interest—"new style" joint ventures in which large and small companies join forces to create a new entry in the market place. In these efforts of "mutual pursuit," usually without the formality of a joint venture company, the small company provides the technology, the large company provides marketing capability, and

the venture is synergistic for both parties. Recent articles have indicated how these large company/small company "alliances," frequently forged through the creative use of corporate venture capital, are growing in strategic importance.[23]

Venture Capital and Nurturing. The venture strategy that permits some degree of entry, but the lowest level of required corporate commitment, is that associated with external venture capital investment. Major corporations have exploited this approach in order to become involved with the growth and development of small companies as investors, participants, or even eventual acquirers. Roberts points out that this approach was popular as early as the mid-to-late 1960s with many large corporations, such as Du Pont, Exxon, Ford, General Electric, Singer, and Union Carbide.[24] Their motivation was the so-called "window on technology," the opportunity to secure closeness to and possibly later entry into new technologies by making minority investments in young and growing high-technology enterprises. However, few companies in the 1960s were able to make this approach alone an important stimulus of corporate growth or profitability. Despite this, ever increasing numbers of companies today are experimenting with venture capital, and many are showing important financial and informational benefits.

Studies carried out by Greenthal and Larson[25] show that venture capital investments can indeed provide satisfactory and perhaps highly attractive returns, if they are properly managed, although Hardymon et al.[26] essentially disagree. Rind distinguishes between direct venture investments and investment into pooled funds of venture capital partnerships.[27] He points out that although direct venture investments can be carried out from within a corporation by appropriate planning and organization, difficulties are often encountered because of a lack of appropriately skilled people, contradictory rationales between the investee company and parent, legal problems, and an inadequate time horizon. Investment in a partnership may remove some of these problems, but if the investor's motives are something other than simply maximizing financial return, it may be important to select a partnership concentrating investments in areas of interest. Increasingly, corporations are trying to use pooled funds to provide the "windows" on new technologies and new markets that are more readily afforded by direct investment, but special linkages with the investment fund managers are needed to implement a "window" strategy. Fast cites 3M and Corning as companies that have invested as limited partners in venture capital partnerships.[28] This involvement in business development financing can keep the company in touch with new technologies and emerging industries as well as provide the guidance and understanding of the venture development process necessary for more effective internal corporate venturing.

In situations where the investing company provides managerial assistance to the recipient of the venture capital, the strategy is classed "venture nurturing" rather than pure venture capital. This seems to be a more sensible entry toward diversification objectives as opposed to a simple provision of funds, but it also needs to be tied to other company diversification efforts.

Educational Acquisitions. Although not discussed in the management literature, targeted small acquisitions can fulfill a role similar to that of a venture capital minority investment and, in some circumstances, may offer significant advantages. In an acquisition of this type, the acquiring firm immediately obtains people familiar with the new business area, whereas in a minority investment, the parent relies upon its existing staff to build familiarity by interacting with the investee. Acquisitions for educational purposes may therefore represent a faster route to familiarity than the venture capital "window" approach. Staff acquired in this manner may even be used by the parent as a basis for redirecting a corporation's primary product-market thrust. Harris Corporation (formerly Harris-Intertype) entered the computer and communication systems industry using precisely this mechanism; it acquired internal skills and knowledge through its acquisition of Radiation Dynamics, Inc. Procter & Gamble recently demonstrated similar behavior in citing its acquisition of the Tender Leaf Tea brand as "an initial learning opportunity in a growing category of the beverage business."[29]

One potential drawback in this educational acquisition approach is that it usually requires a higher level of financial commitment than minority investment and therefore increases risk. In addition, it is necessary to ensure that key people do not leave soon after the acquisition as a result of the removal of entrepreneurial incentives. A carefully designed acquisition deal may be necessary to ensure that incentives remain. When Xerox acquired Versatec, for example, the founder and key employees were given the opportunity to double their "sellout" price by meeting performance targets over the next five years.

Though not without controversy, major previous research work on large U.S. corporations indicated that the highest performers had diversified to some extent but had constrained the development of new business within areas related to the company's base business. The range of mechanisms employed for entering new businesses, previously displayed in Table 3, is divided in Figure 3 into three regions, each requiring a different level of corporate involvement and commitment. No one mechanism is ideal for all new business development. It may be possible, therefore, that selective use of entry mechanisms can yield substantial benefits over concentration on one particular approach. If this presumption is valid, then careful strat-

Increasing Corporate Involvement Required

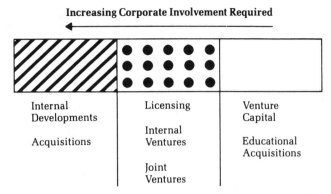

FIGURE 3. Spectrum of Entry Strategies

egy selection can reduce the risk associated with new business development in unrelated areas.

DETERMINING OPTIMUM ENTRY STRATEGIES

How can the entry strategies of Figure 3 be combined with the conceptual framework of Figure 2? Which entry strategies are appropriate in the various regions of the familiarity matrix? The literature provides some useful guides.

In his discussion of the management problems of diversification, Miller proposes that acquisitive diversifiers are frequently required to participate in the strategic and operating decisions of the new subsidiary before they are properly oriented towards the new business.[30] In this situation the parent is "unfamiliar" with the new business area. It is logical to conclude that if the new business is unfamiliar *after* acquisition, it must also have been unfamiliar *before* acquisition. How then could the parent have carried out comprehensive screening of the new company before executing the acquisition? In a situation in which familiarity was low or absent, preacquisition screening most probably overlooked many fac-

tors, turning the acquisition into something of a gamble from a business portfolio standpoint. Similar arguments can be applied to internal development in unfamiliar areas and Gilmore and Coddington specifically stress the dangers associated with entry into unfamiliar markets.[31]

This leads to the rather logical conclusion that entry strategies requiring high corporate involvement should be reserved for new businesses with familiar market and technological characteristics. Similarly, entry mechanisms requiring low corporate input seem best for unfamiliar sectors. A recent discussion meeting with a number of chief executive officers suggested that, at most, 50 percent of major U.S. corporations practice even this simple advice.

The three sections of the Entry Strategy Spectrum in Figure 3 can now be aligned with the three regions of the familiarity matrix in Figure 2. Let us analyze this alignment for each region of the matrix, with particular regard for the main factors identified in the literature.

Region 1: Base/Familiar Sectors

Within the base/familiar sector combinations illustrated in Figure 4, a corporation is fully equipped to undertake all aspects of new business development. Consequently, the full range of entry strategies may be considered, including internal development, joint ventur-

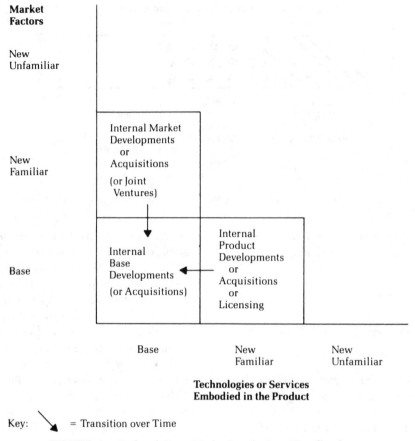

Key: ⬊ = Transition over Time

FIGURE 4.   *Preferred Entry Mechanisms in Base/Familiar Sectors*

ing, licensing, acquisition, or minority investment of venture capital. However, although all of these are valid from a corporate familiarity standpoint, other factors suggest what may be the optimum entry approach.

The potential of conflict between partners may reduce the appeal of a joint venture, and minority investments offer little benefit since the investee would do nothing that could not be done internally.

The most attractive entry mechanisms in these sectors probably include internal development, licensing, and acquisition. Internal development may be appropriate in each of these

sectors, since the required expertise already exists within the corporation. Licensing may be a useful alternative in the base market/new familiar technology sector since it offers fast access to proven products. Acquisition may be attractive in each sector but, as indicated by Shanklin, may be infeasible for some companies in the base/base sector as a result of antitrust legislation.[32] For example, although IBM was permitted to acquire ROLM Corporation, the Justice Department did require that IBM divest ROLM's MIL-SPECS Division because of concern for concentration in the area of military computers.

It may therefore be concluded that in these base/familiar sectors, the optimum entry strategy range may be limited to internal development, licensing, and acquisition as illustrated in Figure 4. In all cases a new business developed in each of these sectors is immediately required to fulfill a conventional sales/profit role within the corporate business portfolio.

Finally, since new businesses within the base market/new familiar technology and new familiar market/base technology sectors immediately enter the portfolio of ongoing business activities, they transfer rapidly into the base/base sector. These expected transitions are illustrated by the arrows in Figure 4.

Region 2: Familiar/Unfamiliar Sectors

Figure 5 illustrates the sectors of lowest familiarity from a corporate standpoint. It has already been proposed that a company is potentially competent to carry out totally appropriate analyses only on those new business opportunities that lie within its own sphere of familiarity. Large-scale entry decisions outside this sphere are liable to miss important characteristics of

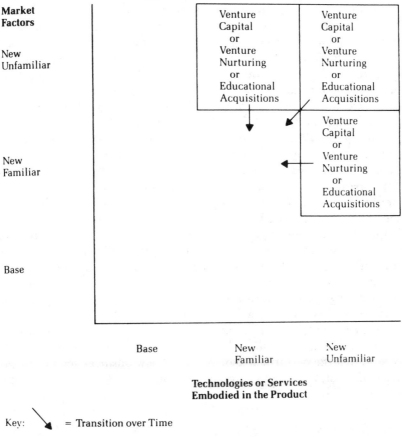

FIGURE 5. *Preferred Entry Mechanisms in Familiar/Unfamiliar Sectors*

the technology or market, reducing the probability of success. This situation frequently generates unhappy and costly surprises. Furthermore, if the unfamiliar parent attempts to exert strong influence on the new business, the probability of success will be reduced still further.

These factors suggest that a two-stage approach may be best when a company desires to enter unfamiliar new business areas. The first stage should be devoted to building corporate familiarity with the new area. Once this has been achieved, the parent is then in a position to decide whether to allocate more substantial resources to the opportunity and, if appropriate, select a mechanism for developing the business.

As indicated earlier, venture capital provides one possible vehicle for building corporate familiarity with an unfamiliar area. With active nurturing of a venture capital minority investment, the corporation can monitor, firsthand, new technologies and markets. If the investment is to prove worthwhile, it is essential for the investee to be totally familiar with the technology/market being monitored by the investor. The technology and market must therefore be the investee's base business. Over time active involvement with the new investment can help the investor move into a more familiar market/technology region, as illustrated in Figure 5, from which the parent can exercise appropriate judgment on the commitment of more substantial resources.

Similarly, educational acquisitions of small young firms may provide a more transparent window on a new technology or market, and even on the initial key employees, who can assist the transition toward higher familiarity. It is important, however, that the performance of acquisitions of this type be measured according to criteria different from those used to assess the "portfolio" acquisitions discussed earlier. These educational acquisitions should be measured initially on their ability to provide increased corporate familiarity with a new tech-

nology or market, and not on their ability to perform immediately a conventional business unit role of sales and profits contributions.

Region 3: Marginal Sectors

The marginal sectors of the matrix are the two base/new unfamiliar combinations plus the new familiar market/new familiar technology area, as illustrated in Figure 6. In each of the base/new unfamiliar sectors, the company has a strong familiarity with either markets or technologies, but is totally unfamiliar with the other dimension of the new business. In these situations joint venturing may be very attractive to the company and prospective partners can see that the company may have something to offer. However, in the new familiar technology/market region the company's base business strengths do not communicate obvious familiarity with that new technology or market. Hence, prospective partners may not perceive that a joint venture relationship would yield any benefit for them.

In the base market/new unfamiliar technology sector the "new style" joint venture or alliance seems appropriate. The large firm provides the marketing channels and the small company provides the technological capability, forming a union that can result in a very powerful team.[33] The complement of this situation may be equally attractive in the new unfamiliar market/base technology sector, although small companies less frequently have strong marketing/distribution capabilities to offer a larger ally.

The various forms of joint ventures such as these not only provide a means of fast entry into a new business sector, but offer increased corporate familiarity over time, as illustrated in Figure 6. Consequently, although a joint venture may be the optimum entry mechanism into the new business area, future development of that business may be best achieved by internal development or acquisition as discussed in the

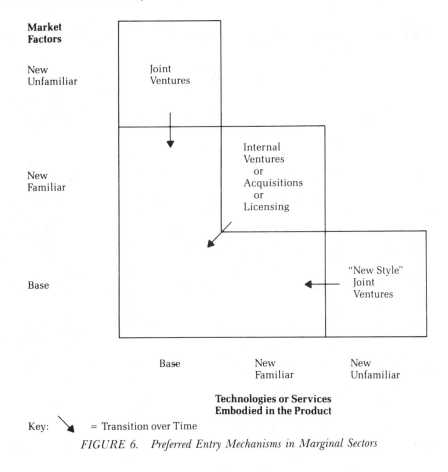

Market Factors

New Unfamiliar — Joint Ventures

New Familiar — Internal Ventures or Acquisitions or Licensing

Base — "New Style" Joint Ventures

Base   New Familiar   New Unfamiliar

Technologies or Services Embodied in the Product

Key: = Transition over Time

FIGURE 6. *Preferred Entry Mechanisms in Marginal Sectors*

earlier base/familiar sectors section of this article.

In the new familiar market/new familiar technology sector, the company may be in an ideal spot to undertake an internal venture. Alternatively, licensing may provide a useful means of obtaining rapid access to a proven product embodying the new technology. Minority investments can also succeed in this sector but, since familiarity already exists, a higher level of corporate involvement and control may be justifiable.

Acquisitions may be potentially attractive in all marginal sectors. However, in the base/new unfamiliar areas this is dangerous since the company's lack of familiarity with the technology or market prevents it from carrying out comprehensive screening of candidates. In contrast, the region of new familiar market/ new familiar technologies does provide adequate familiarity to ensure that screening of candidates covers most significant factors. In this instance an acquisitive strategy is reasonable.

Sector Integration: Optimum Entry Strategies

The above discussion has proposed optimum entry strategies for attractive new business opportunities based on their position in the familiarity matrix. Figure 7 integrates these propos-

615

Market Factors

	Base	New Familiar	New Unfamiliar
New Unfamiliar	Joint Ventures	Venture Capital or Venture Nurturing or Educational Acquisitions	Venture Capital or Venture Nurturing or Educational Acquisitions
New Familiar	Internal Market Developments or Acquisitions (or Joint Ventures)	Internal Ventures or Acquisitions or Licensing	Venture Capital or Venture Nurturing or Educational Acquisitions
Base	Internal Base Developments (or Acquisitions)	Internal Product Developments or Acquisitions or Licensing	"New Style" Joint Ventures

Technologies or Services Embodied in the Product

FIGURE 7.   *Optimum Entry Strategies*

als to form a tool for selecting entry strategies based on corporate familiarity.

TESTING THE PROPOSALS

In testing the proposed entry strategies, Berry studied fourteen new business development episodes that had been undertaken within one highly successful diversified technological corporation.[34] These episodes were all initiated within the period 1971 to 1977, thus representing relatively recent activity while still ensuring that sufficient time had elapsed for performance to be measurable.

The sample comprised six internal developments (three successful, three unsuccessful); six acquisitions (three successful, three incompatible); and two successful minority investments of venture capital. These were analyzed in order to identify factors that differentiated successful from unsuccessful episodes, measured in terms of meeting very high corporate standards of growth, profitability, and return on investment. Failures had not achieved these standards and had been discontinued or divested. The scatter of these episodes on the familiarity matrix is illustrated in Figure 8. Internal developments are represented by symbols A to F, acquisitions by symbols G to L, with symbols M and N showing the location of the minority investments.

The distribution of success and failure on the matrix gives support to the entry strategy proposals that have been made in this arti-

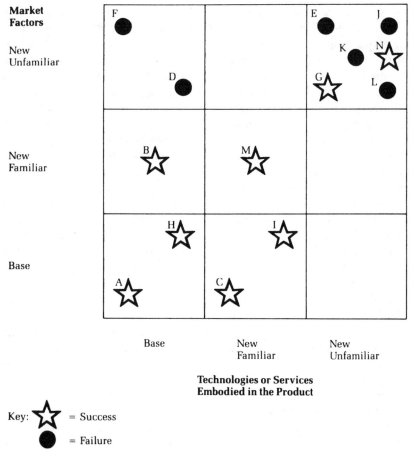

Key: ☆ = Success

● = Failure

FIGURE 8. *Episode Scatter on the Familiarity Matrix*

cle. All high corporate involvement mechanisms (internal development and regular "portfolio" acquisitions) in familiar sectors were successful. However, in unfamiliar areas, only one of this category of entry mechanism, acquisition G, succeeded. This acquisition was a thirty-year-old private company with about 1,000 employees, producing components for the electronics and computer industries. Company G was believed to offer opportunities for high growth although it was unrelated to any area of the parent's existing business. The deal was completed after an unusually long period of two years of candidate evaluation carried out from within the parent company. The only constraint imposed upon Company G following acquisition was the parent's planning and control system, and in fact the acquired company was highly receptive to the introduction of this system. This indicated that Company G was not tightly integrated with the parent and that any constraints imposed did not severely disrupt the established operating procedures of the company.

All factors surrounding the acquisition of Company G—its size, growth market, low level of constraints, and low disruption by the parent—suggest that Company G might have continued to be successful even if it had not been acquired. Representatives of the parent agreed that this might be the case, although they pointed out that the levels of performance obtained following acquisition might not have occurred if Company G had remained independent. Hence, if an acquired company is big enough to stand alone and is *not* tightly integrated with the parent, its degree of operational success is probably independently determined.

It is important to point out that despite the success that occurred in instance G, an acquisition of this type in unfamiliar areas must carry some degree of risk. The parent is liable to overlook many subtle details while screening candidates. Furthermore, when an established company is acquired and continues to operate with a high degree of independence, identification of synergy becomes difficult. Synergy must exist in any acquisitive development if economic value is to be created by the move.[35] Consequently, an acquisition of this type not only carries risk, but may also be of questionable benefit to shareholders, especially if a high price was required because of an earlier good performance record.

The other success in an unfamiliar area, episode N, is a minority investment of venture capital. By the very nature of minority investments, corporate involvement is limited to a low level. Although some influence may be exerted via participation on the board of directors of the investee, again, the investee is not tightly bound to the parent. Consequently, the success of the investee tends to be determined to a large extent by its own actions.

Detailed examination of episodes G and N suggests a good reason for the subject companies' success despite their location in unfamiliar sectors—the companies did not require significant input from the unfamiliar parent in the decision-making process. This suggests that new business development success rates in unfamiliar areas may be increased by limiting corporate input with the decision-making process to low levels until corporate familiarity with the new area has developed.

These experiences support the entry proposals already outlined in this article.

Some companies have already adopted entry strategies that seem to fit the proposals of this article, and Monsanto represents one of the best examples. Monsanto is now committed to significant corporate venturing in the emerging field of biotechnology. Its first involvement in this field was achieved with the aid of its venture capital partnership Innoven, which invested in several small biotechnology firms, including Genentech. During this phase Monsanto interacted closely with the investees, inviting them in-house to give seminars to senior management on their biotechnology research and opportunities. Once some internal familiarity with the emerging field had developed, Monsanto decided to commit substantial resources to internal research-based ventures. Monsanto used venture capital to move from an unfamiliar region to an area of more familiar technology and market, and is currently continuing those venture capital activities to seek new opportunities in Europe. Joint ventures with Harvard Medical School and Washington University of St. Louis are further enhancing its familiarity with biotechnology, while producing technologies that Monsanto hopes to market. Contract research leading to licenses from small companies, primarily those in which it holds minority investments, is another strategy Monsanto is employing. Although the outcome is far from determined, Monsanto seems to be effectively entering biotechnology by moving from top right to bottom left across the familiarity matrix of Figure 7.

CONCLUSION

A spectrum of entry strategies was presented in this article, ranging from those that require high corporate involvement, such as internal development or acquisition, to those that require only low involvement, such as venture capital. These were incorporated into a new conceptual framework designed to assist in selecting entry strategies into potentially attractive new business areas. The framework concentrates on the concept of a corporation's "familiarity" with the technology and market aspects of a new business area, and a matrix was used to relate familiarity to optimum entry strategy.

In this concept, no one strategy is ideal for all new business development situations. Within familiar sectors virtually any strategy may be adopted, and internal development or acquisition is probably most appropriate. However, in unfamiliar areas these two high involvement approaches are very risky and greater familiarity should be built *before* they are attempted. Minority investments and small targeted educational acquisitions form ideal vehicles for building familiarity and are therefore the preferred entry strategies in unfamiliar sectors.

Early in this article, research results from the literature were outlined which indicated that in order to ensure highest performance, new business development should be constrained within areas related to a company's base business. However, this research did not account for alternative entry mechanisms. This article proposes that a multifaceted approach, encompassing internal developments, acquisitions, joint ventures, and venture capital minority investments, can make available a much broader range of business development opportunities, at lower risk than would otherwise be possible.

REFERENCES

1. See E.B. Roberts, "New Ventures for Corporate Growth," *Harvard Business Review*, July–August 1980, pp. 134–142.

2. See R.P. Rumelt, *Strategy, Structure and Economic Performance* (Harvard Business School, Division of Research, 1974).

3. See R.P. Rumelt, "Diversification Strategy and Profitability," *Strategic Management Journal* 3 (1982): 359–369.

4. See H.R. Christensen and C.A. Montgomery, "Corporate Economic Performance: Diversification Strategy versus Market Structure," *Strategic Management Journal* 2 (1981): 327–344.

5. See R.A. Bettis and W.K. Hall, "Risks and Industry Effects in Large Diversified Firms," *Academy of Management Proceedings '81*, pp. 17–20.

6. See O.J. Holzmann, R.M. Copeland, and J. Hayya, "Income Measures of Conglomerate Performance," *Quarterly Review of Economics and Business* 15 (1975): 67–77.

7. See T. Peters, "Putting Excellence into Management," *Business Week*, 21 July 1980, pp. 196–205.

8. See T.J. Peters and R.H. Waterman, *In Search of Excellence* (New York: Harper and Row, 1982).

9. See M.H. Meyer and E.B. Roberts, "New Product Strategy in Small High Technology Firms: A Pilot Study" (MIT Sloan School of Management Working Paper #1428-1-84, May 1984).

10. See H.R. Biggadike, "The Risky Business of Diversification," *Harvard Business Review*, May–June 1979, pp. 103–111.

11. See L.A. Weiss, "Start-Up Businesses: A Comparison of Performances," *Sloan Management Review*, Fall 1981, pp. 37–53.

12. See S.S. Miller, *The Management Problems of Diversification* (New York: John Wiley & Sons, 1963).

13. See J.S. Gilmore and D.C. Coddington, "Diversification Guides for Defense Firms," *Harvard Business Review*, May–June 1966, pp. 133–159.

14. See M.S. Salter and W.A. Weinhold, "Diversification via Acquisition: Creating Value," *Harvard Business Review*, July-August 1978, pp. 166–176.

15. See Miller (1963).

16. See J.P. Killing, "Diversification through Licensing," *R&D Management*, June 1978, pp. 159–163.

17. See Roberts (1980).

18. See R.A. Burgelman, "Managing the Internal Corporate Venturing Process," *Sloan Management Review,* Winter 1984, pp. 33–48.

19. See A.D. Chandler, *Strategy and Structure* (Cambridge, MA: MIT Press, 1962).

20. See. J.D.W. Morecroft, "The Feedback Viewpoint in Business Strategy for the 1980s" (MIT Sloan School of Management, Systems Dynamics Memorandum D-3560, April 1984).

21. See J.P. Killing, "How to Make A Global Joint Venture Work," *Harvard Business Review,* May–June 1982, pp. 120–127.

22. See J.D. Hlavacek, B.H. Dovey, and J.J. Biondo, "Tie Small Business Technology to Marketing Power," *Harvard Business Review,* January-February 1977; Roberts (1980).

23. See *Business Week,* "Acquiring the Expertise but Not the Company," 25 June 1984, pp. 142B–142F; *Inc.,* "The Age of Alliances," February 1984, pp. 68–69.

24. See Roberts (1980).

25. R.P. Greenthal and J.A. Larson, "Venturing into Venture Capital," *Business Horizons,* September–October 1982, pp. 18–23.

26. See G.F. Hardymon, M.J. Denvino, and M.S. Salter, "When Corporate Venture Capital Doesn't Work," *Harvard Business Review,* May–June 1983, pp. 114–120.

27. See K.W. Rind, "The Role of Venture Capital in Corporate Development," *Strategic Management Journal* 2 (1981): 169–180.

28. See N.D. Fast, "Pitfalls of Corporate Venturing," *Research Management,* March 1981, pp. 21–24.

29. See Procter & Gamble Company, *1983 Annual Report* (Cincinnati, OH: 1984), p. 5.

30. See Miller (1963).

31. See Gilmore and Coddington (1966).

32. See W.L. Shanklin, "Strategic Business Planning: Yesterday, Today and Tomorrow," *Business Horizons,* October 1979, pp. 7–14.

33. See *Business Week* (25 June 1984); Hlavacek, Dovey, and Biondo (1977); *Inc.* (February 1984).

34. C.A. Berry, "New Business Development in a Diversified Technological Corporation" (MIT Sloan School of Management/Engineering School Master of Science Thesis, 1983).

35. See Salter and Weinhold (July–August 1978).

Profiting from Technological Innovation: Implications for Integration, Collaboration, Licensing and Public Policy

David J. Teece

It is quite common for innovators—those firms that are first to commercialize a new product or process in the market—to lament the fact that competitors and imitators have profited more from the innovation than the firm first to commercialize it! Because it is often held that being first to market is a source of strategic advantage, the clear existence and persistence of this phenomenon may appear perplexing if not troubling. The aim of this chapter is to explain why a fast second or even a slow third might outperform the innovator. The message is par-

"Profiting from Technological Innovation: Implications for Integration, Collaboration, Licensing and Public Policy" by David J. Teece, from *The Competitive Challenge: Strategies for Industrial Innovation and Renewal*, edited by David J. Teece. Copyright 1987 by Center for Research in Management, School of Business Administration, University of California, Berkeley. Reprinted by permission of Ballinger Publishing Company.

I thank Raphael Amit, Harvey Brooks, Chris Chapin, Therese Flaherty, Richard Gilbert, Heather Haveman, Mel Horwitch, David Hulbert, Carl Jacobsen, Michael Porter, Gary Pisano, Richard Rumelt, Raymond Vernon, and Sidney Winter for helpful discussions relating to the subject matter of this chapter. Three anonymous referees also provided valuable criticisms. I gratefully acknowledge the financial support of the National Science Foundation under grant no. SRS-8410556 to the Center for Research in Management, University of California, Berkeley. Earlier versions of this chapter were presented at a National Academy of Engineering Symposium titled "World Technologies and National Sovereignty," February 1986, and at a conference on innovation at the University of Venice, March 1986.

ticularly pertinent to those science-and engineering-driven companies that labor under the mistaken illusion that developing new products that meet customer needs will ensure fabulous success. It may possibly do so for the product but not for the innovator.

In this chapter, a framework is offered that identifies the factors that determine who wins from innovation: the firm that is first to market, follower firms, or firms that have related capabilities that the innovator needs. The follower firms may or may not be imitators in the narrow sense of the term, although they sometimes are. The framework appears to have utility for explaining the share of the profits from innovation accruing to the innovator compared to its followers and suppliers (see Figure 1), as well as for explaining a variety of interfirm activities such as joint ventures, coproduction agreements, cross-distribution arrangements, and technology licensing. Implications for strategic management, public policy, and international trade and investment are also discussed.

THE PHENOMENON

Figure 2 presents a simplified taxonomy of the possible outcomes from innovation. Quadrant

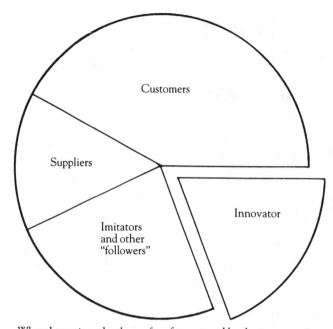

What determines the share of profits captured by the innovator?

FIGURE 1.   Explaining the Distribution of the Profits from Innovation

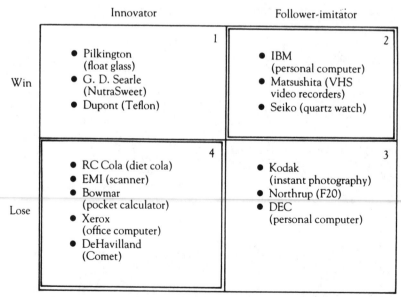

FIGURE 2.   Taxonomy of Outcomes from the Innovation Process

1 represents positive outcomes for the innovator. A first-to-market advantage is translated into a sustained competitive advantage that either creates a new earnings stream or enhances an existing one. Quadrant 4 and its corollary quadrant 2 are the ones that are the focus of this paper.

The EMI CAT Scanner is a classic case of the phenomenon to be investigated.[1] By the early 1970s, the U.K. firm Electrical Musical Industries (EMI) Ltd. was in a variety of product lines including phonographic records, movies, and advanced electronics. EMI had developed high resolution televisions in the 1930s, pioneered airborne radar during World War II, and developed the U.K.'s first all-solid-state computers in 1952.

In the late 1960s Godfrey Hounsfield, an EMI senior research engineer engaged in pattern recognition research that resulted in his displaying a scan of a pig's brain. Subsequent clinical work established that computerized axial tomography (CAT) was viable for generating cross-sectional views of the human body, the greatest advance in radiology since the discovery of x-rays in 1895.

Although EMI was initially successful with its CAT scanner, within six years of its introduction into the United States in 1973 the company had lost market leadership and by the eighth year had dropped out of the CT scanner business. Other companies successfully dominated the market, although they were late entrants and are still profiting in the business today.

Other examples include RC Cola, a small beverage company that was the first to introduce cola in a can and the first to introduce diet cola. Both Coca Cola and Pepsi followed almost immediately and deprived RC of any significant advantage from its innovation. Bowmar, which introduced the pocket calculator, was not able to withstand competition from Texas Instruments, Hewlett Packard, and others and went out of business. Xerox failed to succeed with its entry into the office computer business, even though Apple succeeded with the MacIntosh, which contained many of Xerox's key product ideas, such as the mouse and icons. The de Havilland Comet saga has some of the same features. The Comet I jet was introduced into the commercial-airline business two years or so before Boeing introduced the 707, but de Havilland failed to capitalize on its substantial early advantage. MITS introduced the first personal computer, the Altair, experienced a burst of sales, then slid quietly into oblivion.

If there are innovators who lose, there must be follower imitators who win. A classic example is IBM with its PC, a great success since the time it was introduced in 1981. Neither the architecture nor components embedded in the IBM PC were considered advanced when introduced, nor was the way the technology was packaged a significant departure from then-current practice. Yet the IBM PC was fabulously successful and established MS-DOS as the leading operating system for 16-bit PCs. By the end of 1984, IBM had shipped over 500,000 PCs, and many considered that it had irreversibly eclipsed Apple in the PC industry.

PROFITING FROM INNOVATION: BASIC BUILDING BLOCKS

In order to develop a coherent framework within which to explain the distribution of outcomes illustrated in Figure 2, three fundamental building blocks must first be put in place: the appropriability regime, the dominant design paradigm, and complementary assets.

Regimes of Appropriability
A regime of appropriability refers to the environmental factors, excluding firm and market structure, that govern an innovator's ability to

capture the profits generated by an innovation. The most important dimensions of such a regime are the nature of the technology and the efficacy of legal mechanisms of protection (Figure 3).

It has long been known that patents do not work in practice as they do in theory. Rarely, if ever, do patents confer perfect appropriability although they do afford considerable protection on new chemical products and rather simple mechanical inventions. Many patents can be "invented around" at modest costs. They are especially ineffective at protecting process innovations. Often patents provide little protection because the legal requirements for upholding their validity or for proving their infringement are high.

In some industries, particularly where the innovation is embedded in processes, trade secrets are a viable alternative to patents. Trade-secret protection is possible, however, only if a firm can put its product before the public and still keep the underlying technology secret. Usually only chemical formulas and industrial-commercial processes (such as cosmetics and recipes) can be protected as trade secrets after they are out.

The degree to which knowledge is tacit or codified also affects ease of imitation. Codified knowledge is easier to transmit and receive and is more exposed to industrial espionage and the like. Tacit knowledge by definition is difficult to articulate, so transfer is hard unless those who possess the know-how in question can demonstrate it to others (Teece 1981). Survey research indicates that methods of appropriability vary markedly across industries and probably within industries as well (Levin *et al.* 1984).

- Legal instruments
  - Patents
  - Copyrights
  - Trade secrets
- Nature of technology
  - Product
  - Process
  - Tacit
  - Codified

FIGURE 3.   Appropriability Regime: Key Dimensions

The property-rights environment within which a firm operates can thus be classified according to the nature of the technology and the efficacy of the legal system to assign and protect intellectual property. Although a gross simplification, a dichotomy can be drawn between environments in which the appropriability regime is tight (technology is relatively easy to protect) and weak (technology is almost impossible to protect). Examples of the former include the formula for Coca Cola syrup; an example of the latter would be the Simplex algorithm in linear programming.

The Dominant Design Paradigm

It is commonly recognized that there are two stages in the evolutionary development of a given branch of a science: the preparadigmatic stage, when there is no single generally accepted conceptual treatment of the phenomenon in a field of study, and the paradigmatic stage, which begins when a body of theory appears to have passed the canons of scientific acceptability. The emergence of a dominant paradigm signals scientific maturity and the acceptance of agreed on standards by which what has been referred to as normal scientific research can proceed. These standards remain in force unless or until the paradigm is overturned. Revolutionary science is what overturns normal science, as when Copernicus's theories of astronomy overturned Ptolemy's in the seventeenth century.

Abernathy and Utterback (1978) and Dosi (1982) have provided a treatment of the technological evolution of an industry that appears to parallel Kuhnian notions of scientific evolution (see Kuhn 1970). In the early stages of industry development, product designs are fluid, manufacturing processes are loosely and adaptively organized, and generalized capital is used in production. Competition among firms manifests itself in competition among designs that are markedly different from each other.

This might be called the preparadigmatic stage of an industry.

At some point in time and after considerable trial and error in the marketplace, one design or a narrow class of designs begins to emerge as the more promising. Such a design must be able to meet a whole set of user needs in a relatively complete fashion. The Model T Ford, the IBM 360, and the Douglas DC-3 are examples of dominant designs in the automobile, computer, and aircraft industry, respectively.

Once a dominant design emerges, competition shifts to price and away from design. Competitive success then shifts to a whole new set of variables. Scale and learning become much more important, and specialized capital gets deployed as incumbents seek to lower unit costs through exploiting economies of scale and learning. Reduced uncertainty over product design provides an opportunity to amortize specialized long-lived investments.

Innovation is not necessarily halted once the dominant design emerges; as Clark (1985) points out, it can occur lower down in the design hierarchy. For instance, a *V* cylinder configuration emerged in automobile engine blocks during the 1930s with the Ford V-8 engine. Niches were quickly found for it. Moreover, once the product design stabilizes, there is likely to be a surge of process innovation as producers attempt to lower production costs for the new product (see Figure 4).

The Abernathy-Utterback framework does not characterize all industries. It seems more suited to mass markets where consumer tastes are relatively homogeneous. It would appear to be less characteristic of small-niche markets where the absence of scale and learning economies attaches much less of a penalty to multiple designs. In these instances, generalized equipment will be employed in production.

The existence of a dominant design watershed is of great significance to the distribution of profits between innovator and follower. The innovator may have been responsible for the fundamental scientific breakthroughs as well as the basic design of the new product. However, if imitation is relatively easy, imitators may enter the fray, modifying the product in important ways yet relying on the fundamental designs pioneered by the innovator. When the game of musical chairs stops and a domi-

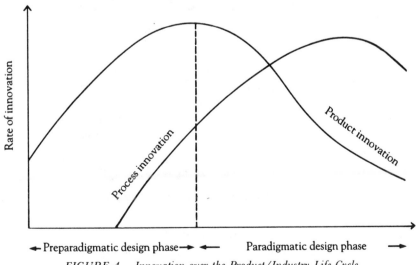

FIGURE 4. Innovation over the Product/Industry Life Cycle

nant design emerges, the innovator may well end up positioned disadvantageously relative to a follower. Hence, when imitation is possible and occurs coupled with design modification before the emergence of a dominant design, followers have a good chance of having their modified product anointed as the industry standard, often to the great disadvantage of the innovator.

Complementary Assets

Let the unit of analysis be an innovation. An innovation consists of certain technical knowledge about how to do things better than with the existing state of the art. Assume that the know-how in question is partly codified and partly tacit. For such know-how to generate profits, it must be sold or utilized in some fashion in the market.

In almost all cases, the successful commercialization of an innovation requires that the know-how in question be utilized in conjunction with other capabilities or assets. Services such as marketing, competitive manufacturing, and after-sales support are almost always needed. These services are often obtained from complementary assets that are specialized. For example, the commercialization of a new drug is likely to require the dissemination of information over a specialized information channel. In some cases, as when the innovation is systemic, the complementary assets may be other parts of a system. For instance, computer hardware typically requires specialized software, both for the operating system and for applications. Even when an innovation is autonomous, as with plug-compatible components, certain complementary capabilities or assets will be needed for successful commercialization. Figure 5 summarizes this schematically.

Whether the assets required for least-cost production and distribution are specialized to the innovation turns out to be important in the development presented below. Accord-

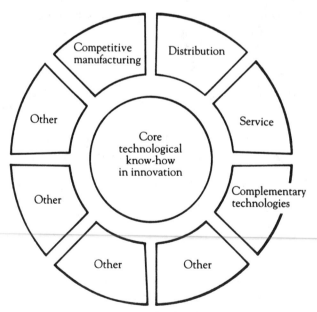

FIGURE 5.   *Complementary Assets Needed to Commercialize an Innovation*

ingly, the nature of complementary assets is explained in some detail. Figure 6 differentiates among complementary assets that are generic, specialized, and cospecialized.

Generic assets are general-purpose assets that do not need to be tailored to the innovation in question. Specialized assets are those with unilateral dependence between the innovation and the complementary asset. Cospecialized assets are those with bilateral dependence. For instance, specialized repair facilities were needed to support the introduction of the rotary engine by Mazda. These assets are cospecialized because of the mutual dependence of the innovation on the repair facility. Containerization similarly required the deployment of some cospecialized assets in ocean shipping and terminals. However, the dependence of trucking on containerized shipping was less than that of containerized shipping on trucking as trucks can convert from containers to flat beds at low cost. An example of a generic asset would be the manufacturing facilities needed to make running shoes. Generalized equipment can be employed in the main, exceptions being the molds for the soles.

IMPLICATIONS FOR PROFITABILITY

These three concepts can now be related in a way that will shed light on the imitation process and the distribution of profits between innovator and follower. We begin by examining tight appropriability regimes.

Tight Appropriability Regimes

In those few instances where the innovator has an ironclad patent or copyright protection or

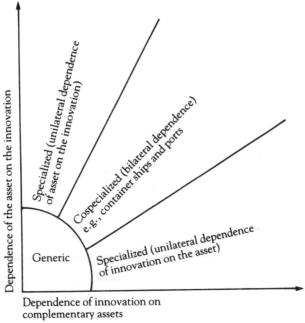

FIGURE 6. *Complementary Assets: Generic, Specialized, and Cospecialized*

where the nature of the product is such that trade secrets effectively deny imitators access to the relevant knowledge, the innovator is almost assured of translating its innovation into market value for some period of time. Even if the innovator does not possess the desirable endowment of complementary costs, ironclad protection of intellectual property will afford the innovator the time to access these assets. If these assets are generic, contractual relation may well suffice; and the innovator may simply license its technology. Specialized R&D firms are viable in such an environment. Universal Oil Products, an R&D firm developing refining processes for the petroleum industry, was one such case in point. If, however, the complementary assets are specialized or cospecialized, contractual relationships are exposed to hazards because one or both parties will have to commit capital to certain irreversible investments that will be valueless if the relationship between innovator and licensee breaks down. Accordingly, the innovator may find it prudent to expand its boundaries by integrating into specialized and cospecialized assets. Fortunately, the factors that make for difficult imitation will enable the innovator to build or acquire those complementary assets without competing with innovators for their control.

Competition from imitators is muted in this type of regime, which sometimes characterizes the petrochemical industry. In this industry, the protection offered by patents is fairly easily enforced. One factor assisting the licensee in this regard is that most petrochemical processes are designed around a specific variety of catalysts which can be kept proprietory. An agreement not to analyze the catalyst can be extracted from licensees, affording extra protection. However, even if such requirements are violated by licensees, the innovator is still well positioned because the most important properties of a catalyst are related to its physical structure, and the process for generating this structure cannot be deduced from structural analysis alone. Every reaction technology a company acquires is thus accompanied by an ongoing dependence on the innovating company for the catalyst appropriate to the plant design. Failure to comply with various elements of the licensing contract can thus result in a cutoff in the supply of the catalyst and possibly in facility closure.

Similarly, if the innovator comes to market in the preparadigmatic phase with a sound product concept but the wrong design, a tight appropriability regime will afford the innovator the time needed to perform the trials needed to get the design right. As discussed earlier, the best initial design concepts often turn out to be hopelessly wrong, but if the innovator possesses an impenetrable thicket of patents or has technology that is simply difficult to copy, then the market may well afford the innovator the necessary time to ascertain the right design before being eclipsed by imitators.

Weak Appropriability

Tight appropriability is the exception rather than the rule. Accordingly, innovators must turn to business strategy if they are to keep imitator followers at bay. The nature of the competitive process will vary according to whether the industry is in the paradigmatic or the preparadigmatic phase.

Preparadigmatic Phase. In the preparadigmatic phase, the innovator must be careful to let the basic design float until sufficient evidence has accumulated that a design has been delivered that is likely to become the industry standard. In some industries, there may be little opportunity for product modification. In microelectronics, for example, designs become locked in when the circuitry is chosen. Product modification is limited to debugging and software modification. An innovator must begin the design process anew if the product does not fit the market well. In some respects, however, selecting designs is dictated by the need to meet certain compatibility standards so that new hardware can interface with existing applications software. In one sense, therefore, the design issue for the microprocessor industry today is relatively straightforward: deliver greater power and speed while meeting the computer industry standards of the existing software base. However, from time to time, windows of opportunity emerge for the introduction of entirely new families of microprocessors that will define a new industry and software standard. In these instances, basic design parameters are less well defined and can be permitted to float until market acceptance is apparent.

The early history of the automobile industry exemplifies exceedingly well the importance for subsequent success of selecting the right design in the preparadigmatic stages. None of the early producers of steam cars survived the early shakeout when the closed-body internal-combustion-engine automobile emerged as the dominant design. The steam car, nevertheless, had numerous early virtues such as reliability, which the internal-combustion-engine autos could not deliver.

The British fiasco with the Comet I is also instructive. De Havilland had picked an early design with both technical and commercial flaws. By moving into production, de Havilland was hobbled by significant irreversibilities and loss of reputation to such a degree that it was unable to convert to the Boeing design, which subsequently emerged as dominant. It was not able to occupy even second place, which went instead to Douglas.

As a general principle, it appears that innovators in weak appropriability regimes need to be coupled intimately to the market so that user needs can fully influence designs. When multiple parallel and sequential prototyping is feasible, it has clear advantages. Generally such an approach is prohibitively costly. When development costs for a large commercial aircraft exceed $1 billion, variations on a theme are all that is possible.

Hence, the probability that an innovator—defined here as a firm that is first to commercialize a new product design concept—will enter the paradigmatic phase possessing the dominant design is problematic. The lower the cost of prototyping, the higher the relative probabilities and the more tightly coupled the firm is to the market. The latter is a function of organizational design and can be influenced by managerial choices. The cost is embedded in the technology and cannot be influenced except in minor ways by managerial decisions. In industries with large developmental and prototyping costs—and hence significant irreversibilities—where innovation of the product concept is easy, one would expect that the probability that the innovator would emerge as the winner or among the winners at the end of the preparadigmatic stage is low.

Paradigmatic Stage. In the preparadigmatic phase, complementary assets do not loom large. Rivalry is focused on trying to identify the design that will be dominant. Production volumes are low, and there is little to be gained in deploying specialized assets as scale economies are unavailable and price is not a principal competitive factor. However, as the leading design or designs begin to be revealed by the market, volumes increase and opportunities for economies of scale will induce firms to begin gearing up for mass production by acquiring specialized tooling and equipment and possibly specialized distribution as well. Because these investments involve significant irreversibilities, producers are likely to proceed with caution. Islands of specialized capital will begin to appear in an industry that otherwise features a sea of general-purpose manufacturing equipment.

However, as the terms of competition begin to change and prices become increasingly important, access to complementary assets becomes absolutely critical. Because the core technology is easy to imitate, by assumption, commercial success swings on the terms and conditions on which the required complementary assets can be accessed.

It is at this point that specialized and cospecialized assets become critically important. Generalized equipment and skills, almost by definition, are always available in an industry, and even if they are not, they do not involve significant irreversibilities. Accordingly, firms have easy access to this type of capital, and even if there is insufficient capacity available in the relevant assets, it can be put in place easily as it involves few risks. Specialized assets, on the other hand, involve significant irreversibilities and cannot be easily accessed by contract as the risks are significant for the party making the dedicated investment. The firms that control the cospecialized assets, such as distribution channels, specialized manufacturing capacity, and so forth are clearly advantageously positioned relative to an innovator. Indeed, in rare instances where incumbent firms possess an airtight monopoly over specialized assets and the innovator is in a regime of weak appropriability, all of the profits to the innovation could conceivably inure to the firms possessing the specialized assets, who should be able to get the upperhand.

Even when the innovator is not confronted by situations where competitors or potential competitors control key assets, the innovator may still be disadvantaged. For instance, the technology embedded in cardiac pacemakers was easy to imitate, and so competitive outcomes quickly came to be determined by who had easiest access to the complementary assets, in this case specialized marketing. A similar situation has recently arisen in the United States with respect to personal computers. As an industry participant (Norman 1986: 438) recently observed:

> There are a huge number of computer manufacturers, companies that make peripherals (e.g., printers, hard disk drives, floppy disk drives), and software companies. They are all trying to get marketing distributors because they cannot afford to call on all of the U.S. companies directly. They need to go through retail distribution channels, such as Businessland, in order to reach the marketplace. The problem today, however, is that many of these companies are not able to get shelf space and thus are having a very difficult time marketing their products. The point of distribution is where the profit and the power are in the marketplace today.

CHANNEL STRATEGY ISSUES

The above analysis indicates how access to complementary assets, such as manufacturing and distribution, on competitive teams is critical if the innovator is to avoid handing over the lion's share of the profits to imitators, and/or to the owners of the complementary assets that are specialized or cospecialized to the innovation. It is now necessary to delve more deeply into the appropriate control structure that the innovator ideally ought to establish over these critical assets.

There are a myriad of possible channels that could be employed. At one extreme, the innovator could integrate into all of the necessary complementary assets, as illustrated in Figure 7, or into just some of them, as illustrated in Figure 8. Complete integration (Figure 7) is likely to be unnecessary as well as prohibitively expensive. It is well to recognize that the variety of assets and competences that need to be accessed is likely to be quite large even for only modestly complex technologies. To produce a personal computer, for instance, a company needs access to expertise in semiconductor technology, display technology, disk-drive technology, networking technology, keyboard technology, and several others. No company by itself can keep pace in all of these areas.

At the other extreme, the innovator

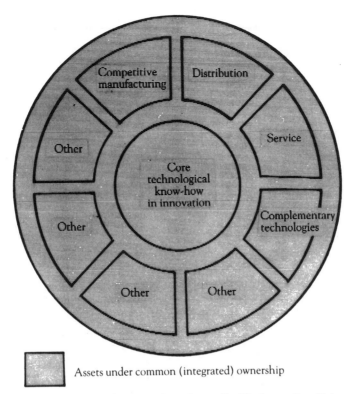

Assets under common (integrated) ownership

FIGURE 7.   *Complementary Assets Internalized for Innovation: Hypothetical Case #1 (Innovator Integrated into All Complementary Assets)*

could attempt to access these assets through straightforward contractual relationships (such as component supply contracts, fabrication contracts, and service contracts). In many instances, such contracts may suffice, although sometimes exposing the innovator to various hazards and dependencies that it may well wish to avoid. Between the fully integrated and fully contractual extremes, there are a myriad of intermediate forms and channels available. An analysis of the properties of the two extreme forms is presented below. A brief synopsis of mixed modes then follows.

Contractual Modes

The advantages of a contractual solution—whereby the innovator signs a contract, such as a license, with independent suppliers, manufacturers, or distributors—are obvious. The innovator will not have to make the upfront capital expenditures needed to build or buy the assets in question, thus reducing risks as well as cash requirements.

Contracting, rather than integrating, is likely to be the optimal strategy when the innovator's appropriability regime is tight and the complementary assets are available in competitive supply (that is, there is adequate capacity and a choice of sources).

Both conditions apply in petrochemicals—for instance, so that an innovator does not need to be integrated to be successful. Consider first the appropriability regime. As discussed earlier, the protection offered by patents is fairly easily enforced, particularly for

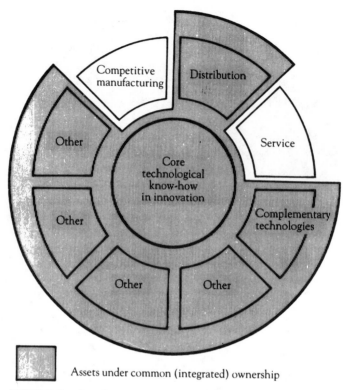

FIGURE 8.   Complementary Assets Internalized for Innovation: Hypothetical Case #2 (Innovator Subcontracts for Manufacturing and Service)

process technology, in the petrochemical industry. Given the advantageous feedstock prices available in hydrocarbon rich petrochemical exporters and the appropriability regime characteristic of this industry, there is no incentive or advantage in owning the complementary assets (production facilities) as they are not typically highly specialized to the innovation. Union Carbide appears to realize this and has recently adjusted its strategy accordingly. Essentially, Carbide is placing its existing technology into a new subsidiary, Engineering and Hydrocarbons Service. The company is engaging in licensing and offers engineering, construction, and management services to customers who want to take their feedstocks and integrate them forward into petrochemicals.

But Carbide itself appears to be backing away from an integration strategy.

Chemical and petrochemical product innovations are not quite so easy to protect, thus raising new challenges to innovating firms in the developed nations as they attempt to shift out of commodity petrochemicals. There are already numerous examples of new products that made it to the marketplace, filled a customer need, but, because of imitation, never generated competitive returns to the innovator. For example, in the 1960s, Dow decided to start manufacturing rigid polyurethane foam. However, it was imitated very quickly by numerous small firms that had lower costs.[2] The absence of low-cost manufacturing capability left Dow vulnerable.

Contractual relationships can bring added credibility to the innovator, especially if the innovator is relatively unknown when the contractual partner is established and viable. Indeed, arm's-length contracting that embodies more than a simple buy/sell agreement is becoming so common and is so multifaceted that the term *strategic partnering* has been devised to describe it. Even large companies such as IBM are now engaging in it. For IBM, partnering buys access to new technologies enabling the company to "learn things we couldn't have learned without many years of trial and error."[3] IBM's arrangement with Microsoft to use the latter's MS-DOS operating-system software on the IBM PC facilitated the timely introduction of IBM's personal computer into the market.

Smaller, less integrated companies are often eager to sign on with established companies because of name recognition and reputation spillovers. For instance, Cipher Data Products, Inc. contracted with IBM to develop a low-priced version of IBM's 3480 .5-inch streaming cartridge drive, which is likely to become the industry standard. As Cipher management points out, "one of the biggest advantages to dealing with IBM is that, once you've created a product that meets the high quality standards necessary to sell into the IBM world, you can sell into any arena."[4] Similarly, IBM's contract with Microsoft "meant instant credibility" to Microsoft (McKenna 1985: 94).

It is most important to recognize, however, that strategic (contractual) partnering, which is currently very fashionable, is exposed to certain hazards, particularly for the innovator, when the innovator is trying to use contracts to access specialized capabilities. First, it may be difficult to induce suppliers to make costly irreversible commitments that depend for their success on the success of the innovation. To expect suppliers, manufacturers, and distributors to do so is to invite them to take risks along with the innovator. The problem that this poses for the innovator is similar to the problems associated with attracting venture capital. The innovator must persuade its prospective partner that the risk is a good one. The situation is one open to opportunistic abuses on both sides. The innovator has incentives to overstate the value of the innovation, while the supplier has incentives to run with the technology should the innovation be a success.

Instances of both parties making irreversible capital commitments nevertheless exist. Apple's Laserwriter—a high-resolution laser printer that allows PC users to produce near-typeset-quality text and art-department graphics—is a case in point. Apple persuaded Canon to participate in the development of the Laserwriter by providing subsystems from its copiers—but only after Apple contracted to pay for a certain number of copier engines and cases. In short, Apple accepted a good deal of the financial risk in order to induce Canon to assist in the development and production of the Laserwriter. The arrangement appears to have been prudent, yet there were clearly hazards for both sides. It is difficult to write, execute, and enforce complex development contracts, particularly when the design of the new product is still floating. Apple was exposed to the risk that its co-innovator Canon would fail to deliver, and Canon was exposed to the risk that the Apple design and marketing effort would not succeed. Still, Apple's alternatives may have been rather limited, inasmuch as it did not command the requisite technology to go it alone.

In short, the current euphoria over strategic partnering may be partially misplaced. The advantages are being stressed (for example, McKenna 1985) without a balanced presentation of costs and risks. Briefly, there is the risk that the partner will not perform according to the innovator's perception of what the contract requires; there is the added danger that the partner may imitate the innovator's technology and attempt to compete with the innovator. This latter possibility is particularly acute if the provider of the complementary asset is uniquely situated with respect to the complementary asset in question and has the capacity to imitate the technology that the innovator is unable to protect. The innovator will then find that it has created a competitor who is better positioned than the innovator to take advantage of the market opportunity at hand. *Business Week* (1986: 57–59) has expressed concerns along these lines in its discussion of the "hollow corporation."[5]

It is important to bear in mind, however, that contractual or partnering strategies in certain cases are ideal. If the innovator's technology is well protected, and if what the partner has to provide is a generic capacity available from many potential partners, then the innovator will be able to maintain the upper hand while avoiding the costs of duplicating downstream capacity. Even if the partner fails to perform, adequate alternatives exist (by assumption, the partner's capacities are commonly available) so that the innovator's efforts to commercialize its technology successfully ought to proceed profitably.

Integration Modes

Integration, which by definition involves ownership, is distinguished from pure contractual modes in that it typically facilitates incentive alignment and control. If an innovator owns rather than rents the complementary assets needed to commercialize, then it is in a position to capture spillover benefits stemming from increased demand for the complementary assets caused by the innovation.

Indeed, an innovator might be in the position, at least before its innovation is announced, to buy up capacity in the complementary assets, possibly to its great subsequent advantage. If futures markets exist, simply taking forward positions in the complementary assets might suffice to capture much of the spillovers.

Even after the innovation is announced, the innovator may still be able to build or buy complementary capacities at competitive prices if the innovation has ironclad legal protection (that is, if the innovation is in a tight appropriability regime). However, if the innovation is not tightly protected and once out is easy to imitate, then securing control of complementary capacities is likely to be the key success factor, particularly if those capacities are in fixed supply—so-called bottlenecks. Distribution and specialized manufacturing competences often become bottlenecks.

As a practical matter, however, an innovator may not have the time to acquire or build the complementary assets that ideally it would like to control. This is particularly true when imitation is so easy that timing becomes critical. Additionally, the innovator may simply not have the financial resources to proceed. The implications of timing and cash constraints are summarized in Figure 9.

Accordingly, in weak appropriability regimes, innovators need to rank complementary assets as to their importance. If the complementary assets are critical, ownership is warranted although if the firm is cash constrained, a minority position may well represent a sensible tradeoff.

Needless to say, when imitation is easy, strategic moves to build or buy complementary assets that are specialized must occur with due reference to the moves of competitors. There is no point in moving to build a specialized asset, for instance, if one's imitators can do it more quickly and cheaply.

It is self-evident that if the innovator is already a large enterprise with many of the relevant complementary assets under its control, integration is not likely to be the issue that it might otherwise be, because the innovating firm will already control many of the relevant specialized and cospecialized assets. However,

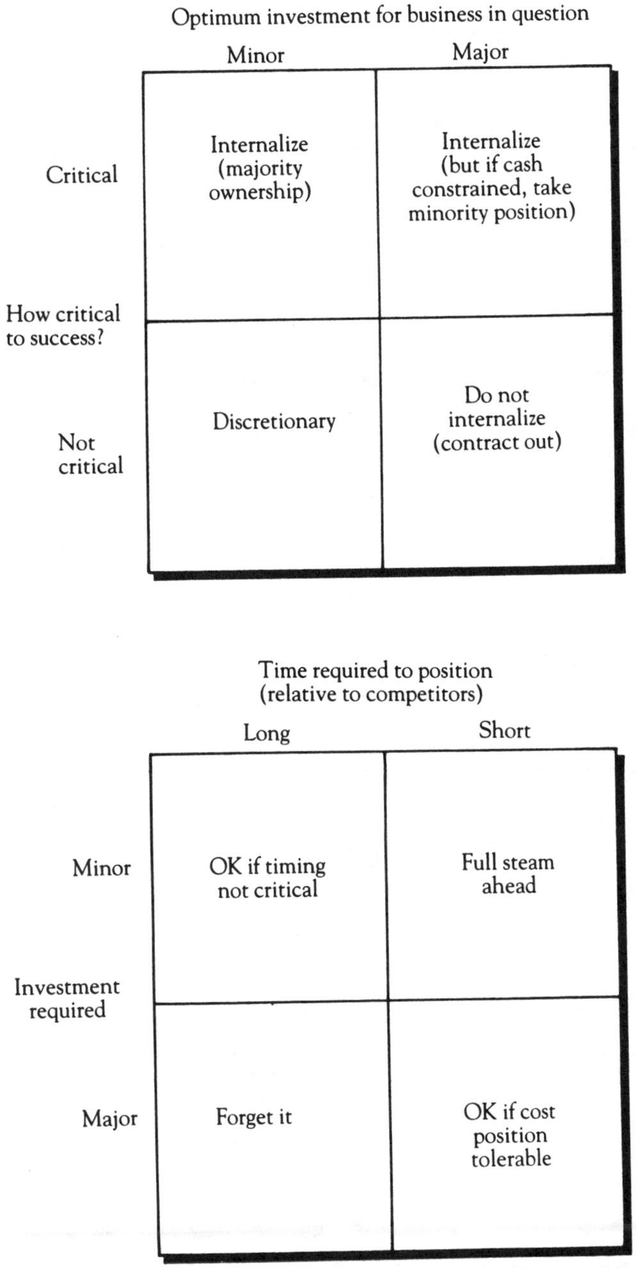

Optimum investment for business in question

	Minor	Major
Critical	Internalize (majority ownership)	Internalize (but if cash constrained, take minority position)
Not critical	Discretionary	Do not internalize (contract out)

How critical to success?

Time required to position
(relative to competitors)

	Long	Short
Minor	OK if timing not critical	Full steam ahead
Major	Forget it	OK if cost position tolerable

Investment required

FIGURE 9.   *Specialized Complementary Assets and Weak Appropriability: Integration Calculus*

in industries experiencing rapid technological change, technologies advance so rapidly that it is unlikely that a single company will have the full range of expertise needed to bring advanced products to market in a timely and cost-effective fashion. Hence, the integration issue is not just a small-firm issue.

Integration versus Contract Strategies: An Analytic Summary

Figure 10 summarizes some of the relevant considerations in the form of a decision flow chart. It indicates that a profit-seeking innovator, confronted by weak intellectual property

protection and the need to access specialized complementary assets or capabilities, is forced to expand its activities through integration if it is to prevail over imitators. Put differently, innovators who develop new products that possess poor intellectual property protection but that require specialized complementary capacities are more likely to parlay their technology into a commercial advantage rather than see it prevail in the hands of imitators.

Figure 10 makes apparent that the difficult strategic decisions arise in situations where the appropriability regime is weak and where specialized assets are critical to profitable commercialization. These situations, which

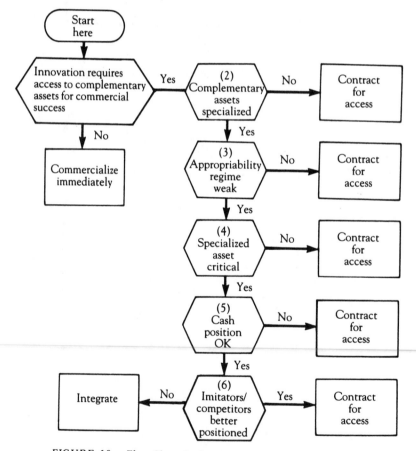

FIGURE 10.   Flow Chart for Integration versus Contract Design

in reality are very common, require that a fine-grained competitor analysis be part of the innovator's strategic assessment of its opportunities and threats. This is carried a step further in Figure 11, which looks only at situations where commercialization requires certain specialized capabilities. It indicates the appropriate strategies for the innovators and predicts the outcomes to be expected for the various players.

Three classes of players are of interest: innovators, imitators, and the owners of cospecialized assets (such as distributors). All three can potentially benefit or lose from the innovation process. The latter can potentially benefit from the additional business that the innovation may direct in the asset owner's direction. Should the asset turn out to be a bottleneck with respect to commercializing the innovation, the owner of the bottleneck facilities is obviously in a position to extract profits from the innovator or imitators.

The vertical axis in Figure 11 measures how those who possess the technology (the innovator or possibly its imitators) are positioned *vis-à-vis* those firms that possess required specialized assets. The horizontal axis measures the tightness of the appropriability regime, tight regimes being evidenced by ironclad legal protection coupled with technology that is simply difficult to copy; weak regimes offer little in the way of legal protection and the essence of the technology, once released, is transparent to the imitator. Weak regimes are further subdivided according to how the innovator and imitators are positioned *vis-à-vis* each other. This is likely to be a function of factors such as lead time and prior positioning in the requisite complementary assets.

Figure 11 makes it apparent that even when firms pursue the optimal strategy, other industry participants may take the jackpot. This possibility is unlikely when the intellectual property in question is tightly protected. The only serious threat to the innovator is where a specialized complementary asset is completely locked up, a possibility recognized in cell 4. This can rarely be done without the cooperation of government. But it frequently occurs, as when a foreign government closes off access to a foreign market, forcing the innovators to license to foreign firms with the effect of government cartelizing of the potential licensees. With weak intellectual-property protection, however, it is quite clear that the innovator will often lose out to imitators or asset holders even when the innovator is pursuing the appropriate strategy (cell 6). Clearly, incorrect strategies can compound problems. For instance, if innovators integrate when they should contract, a heavy commitment of resources will be incurred for little if any strategic benefit, thereby exposing the innovator to even greater losses than would otherwise be the case. On the other hand, if an innovator tries to contract for the supply of a critical capability when it should build the capability itself, it may well find it has nurtured an imitator better able to serve the market than the innovator itself.

Mixed Modes

The real world rarely provides extreme or pure cases. Decisions to integrate or license involve tradeoffs, compromises, and mixed approaches. It is not surprising, therefore, that the real world is characterized by mixed modes of organization involving judicious blends of integration and contracting. Sometimes mixed modes represent transitional phases. For instance, because of the convergence of computer and telecommunication technology, firms in each industry are discovering that they often lack the requisite technical capabilities in the other. Because the technological interdependence of the two requires collaboration among those who design different parts of the system, intense cross-boundary coordination and information flows are required. When separate enterprises are involved, agreement must be reached on complex protocol issues among

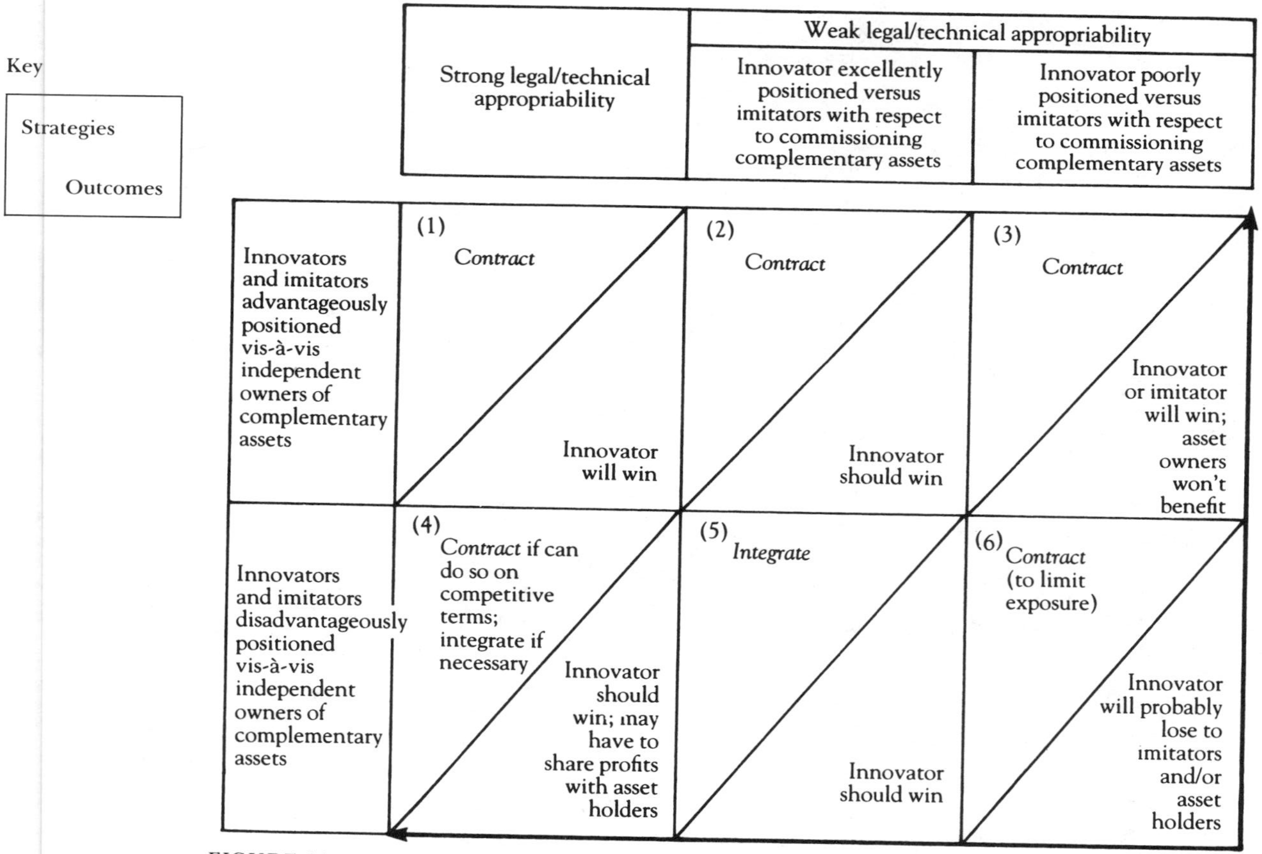

FIGURE 11.   Contract and Integration Strategies and Outcomes for Innovators: Specialized Asset Case

parties who see their interests differently. Contractual difficulties can be anticipated because the selection of common technical protocols among the parties will often be followed by transaction-specific investments in hardware and software. There is little doubt that this was the motivation behind IBM's purchase of 15 percent of PBX manufacturer Rolm in 1983, a position that was expanded to 100 percent in 1984. IBM's stake in Intel, which began with a 12 percent purchase in 1982, is probably not a transitional phase leading to 100 percent purchase because both companies realize that the two corporate cultures are not very compatible, and IBM may not be so impressed with Intel's technology as it once was.

The CAT Scanner, the IBM PC, and NutraSweet: Insights from the Framework

EMI's failure to reap significant returns from the CAT scanner can be explained in large measure by reference to the concepts developed above. The scanner that EMI developed was of a technical sophistication much higher than would normally be found in a hospital, requiring a high level of training support and servicing. EMI had none of these capabilities, could not easily contract for them, and was slow to realize their importance. It most probably could have formed a partnership with a company like Siemens to access the requisite capabilities. Its failure to do so was a strategic error compounded by the very limited intellectual property protection that the law afforded the scanner. Although subsequent court decisions have upheld some of EMI's patent claims, once the product was in the market it could be reverse-engineered and its essential features copied. Two competitors, GE and Technicare, already possessed the complementary capabilities that the scanner required, and they were also technologically capable. In addition, both were experienced marketers of medical equipment and had reputations for quality, reliability, and service. GE and Technicare were thus able to commit their R&D resources to developing a competitive scanner, borrowing ideas from EMI's scanner, which they undoubtedly had access to through cooperative hospitals, and improving on it where they could while they rushed to market. GE began taking orders in 1976 and soon after made inroads on EMI. In 1977, concern for rising health-care costs caused the Carter administration to introduce certificate-of-need regulation, which required HEW's approval of expenditures on big-ticket items like CAT scanners. This severely cut the size of the available market.

By 1978 EMI had lost market-share leadership to Technicare, which was in turn quickly overtaken by GE. In October 1979 Godfrey Hounsfield of EMI shared the Nobel prize for invention of the CAT scanner. Despite this honor and the public recognition of its role in bringing this medical breakthrough to the world, the collapse of its scanner business forced EMI in the same year into the arms of a rescuer, Thorn Electrical Industries, Ltd. GE subsequently acquired what was EMI's scanner business from Thorn for what amounted to a pittance ("GE Gobbles a Rival in CAT Scanners" 1980). Though royalties continued to flow to EMI, the company had failed to capture the lion's share of the profits generated by the innovation it had pioneered and successfully commercialized.

If EMI illustrates how a company with outstanding technology and an excellent product can fail to profit from innovation while the imitators succeed, the story of the IBM PC indicates how a new product representing a very modest technological advance can yield remarkable returns to the developer.

The IBM PC, introduced in 1981, was a success despite the fact that the architecture was ordinary and the components standard. Philip Estridge's design team in Boca Raton, Florida, decided to use existing technology to produce a solid, reliable micro rather than state of the art. With a one-year mandate to develop a PC, Estridge's team could do little else.

However, the IBM PC did use what at the time was a new 16-bit microprocessor (the Intel 8088) and a new disk operating system (DOS) adapted for IBM by Microsoft. Other than the microprocessor and the operating system, the IBM PC incorporated existing micro standards and used off-the-shelf parts from outside vendors. IBM did write its own BIOS (Basic Input/Output System), which is embedded in ROM, but this was a relatively straightforward programming exercise.

The key to the PC's success was not the technology. It was the set of complementary assets that IBM either had or quickly assembled around the PC. In order to expand the market for PCs, there was a clear need for an expandable, flexible microcomputer system with extensive applications software. IBM could have based its PC system on its own patented hardware and copyrighted software. Such an approach would cause complementary products to be cospecialized, forcing IBM to develop peripherals and a comprehensive library of software in a very short time. Instead, IBM adopted what might be called an induced contractual approach. By adopting an open-system architecture, as Apple had done, and by making the operating-system information publicly available, a spectacular output of third-part software was induced. IBM estimated that by mid-1983, at least 3,000 hardware and software products were available for the PC (Gens and Christiansen 1983: 88). Put differently, IBM pulled together the complementary assets, particularly software, which success required, without even using contracts, let alone integration. This was despite the fact that the software developers were creating assets that in part cospecialized the IBM PC, at least in the first instance.

A number of special factors made this seem a reasonable risk to the software writers.

A critical one was IBM's name and commitment to the project. The reputation behind the letters *I, B, M* is perhaps the greatest cospecialized asset the company possesses. The name implied that the product would be marketed and serviced in the IBM tradition. It guaranteed that PC-DOS would become an industry standard, so that the software business would not be dependent solely on IBM because emulators were sure to enter. It guaranteed access to retail distribution outlets on competitive terms. The consequence was that IBM was able to take a product that represented at best a modest technological accomplishment and turn it into a fabulous commercial success. The case demonstrates the role that complementary assets play in determining outcomes.

The spectacular success and profitability of G.D. Searle's NutraSweet is an uncommon story that is also consistent with the above framework. In 1982 Searle reported combined sales of $74 million for NutraSweet and its tabletop version, Equal. In 1983 this surged to $336 million. In 1985 NutraSweet sales exceeded $700 million (*Monsanto Annual Report* 1985) and Equal had captured 50 percent of the U.S. sugar-substitute market and was number one in five other countries.

NutraSweet, which is Searle's trade name for aspartame, has achieved rapid acceptance in each of its FDA-approved categories because of its good taste and ability to substitute directly for sugar in many applications. However, Searle's earnings from NutraSweet and the absence of a strategic challenge can be traced in part to Searle's clever strategy.

It appears that Searle has managed to establish an exceptionally tight appropriability regime around NutraSweet—one that may well continue for some time after the patent has expired. No competitor appears successfully to have invented around the Searle patent and commercialized an alternative, no doubt in part because the FDA approval process would have to begin anew for an imitator who was not violating Searle's patents. A competitor who tried to replicate the aspartame molecule with minor modification to circumvent the patent would probably be forced to replicate the hundreds of tests and experiments that proved aspartame's safety. Without patent protection, FDA approval would provide no shield against imitators that would come to market with an identical chemical and that could establish to the FDA that it was the same compound that had already been approved. Without FDA approval, on the other hand, the patent protection would be worthless, for the product would not be sold for human consumption.

Searle has aggressively pushed to strengthen its patent protection. The company was granted U.S. patent protection in 1970. It has also obtained patent protection in Japan, Canada, Australia, United Kingdom, France, West Germany, and a number of other countries. However, most of these patents carry a seventeen-year life. Because the product was approved for human consumption only in 1982, the seventeen-year patent life was effectively reduced to five. Recognizing the obvious importance of its patent, Searle pressed for and obtained special legislation in November 1984 extending the patent protection on aspartame for another five years. The United Kingdom provided a similar extension. In almost every other nation, however, 1987 will mark the expiration of the patent.

When the patent expires, however, Searle will still have several valuable assets to help keep imitators at bay. Searle has gone to great lengths to create and promulgate the use of its NutraSweet name and a distinctive *Swirl* logo on all goods licensed to use the ingredient. The company has also developed the *Equal* trade name for a tabletop version of the sweetener. Trademark law in the United States provides protection against unfair competition in branded products as long as the owner of the mark continues to use it. Both the NutraSweet and Equal trademarks will become essential assets when the patents on aspartame expire. Searle may well have convinced consumers that the only real form of sweetener is NutraSweet/Equal. Consumers know most other artificial sweeteners by their generic names-saccharin and cyclamates.

Clearly, Searle is trying to build a position in complementary assets to prepare for the competition that will surely arise. Searle's joint venture with Ajinomoto ensures them access to that company's many years of experience in the production of biochemical agents. Much of this knowledge is associated with techniques for distillation and synthesization of the delicate hydrocarbon compounds that are the ingredients in NutraSweet and is therefore more tacit than codified. Searle has begun to put these techniques to use in its own $160 million Georgia production facility. It can be expected that Searle will use trade secrets to the maximum to keep this know-how proprietary.

By the time its patent expires, Searle's extensive research into production techniques for L-phenylalanine and its eight years of experience in the Georgia plant should give it significant cost advantage over potential aspartame competitors. Trade-secret protection, unlike patents, has no fixed lifetime and may well sustain Searle's position for years to come.

Searle, moreover, has wisely avoided renewing contracts with suppliers when they have expired. Had Searle subcontracted manufacturing for NutraSweet, it would have created a manufacturer who would have been in a position to enter the aspartame market itself or to team up with a marketer of artificial sweeteners. By keeping manufacturing in-house and by developing a valuable trade name, Searle has a good chance of protecting its market position from dramatic inroads once patents expire. Clearly, Searle seems to be astutely aware of the importance of maintaining a tight appropriability regime and using cospecialized assets strategically.

IMPLICATIONS FOR R&D STRATEGY, INDUSTRY STRUCTURE, AND TRADE POLICY

Allocating R&D Resources

The analysis so far assumes that the firm has developed an innovation for which a market exists. It indicates the strategies that the firm must follow to maximize its share of industry profits relative to imitators and other competitors. There is no guarantee of success even if optimal strategies are followed.

The innovator can improve its total return to R&D, however, by adjusting its R&D investment portfolio to maximize the probability that technological discoveries will emerge that either are easy to protect with existing intellectual property law or require for commercialization cospecialized assets already within the firm's repertoire of capabilities. Put differently, if an innovating firm does not target its R&D resources toward new products and processes that it can commercialize advantageously relative to potential imitators or followers, it is unlikely to profit from its investment in R&D. In this sense, a firm's history—and the assets it already has in place—ought to condition its R&D investment decisions. Clearly, an innovating firm with considerable assets already in place is free to strike out in new directions as long as in doing so it is aware of the kinds of capabilities required to commercialize the innovation successfully. It is therefore rather clear that the R&D investment decision cannot be divorced from the strategic analysis of markets and industries and the firm's position within them.

Small Firm versus Large Firm Comparisons

Business commentators often remark that many small entrepreneurial firms that generate new, commercially valuable technology fail while large multinational firms, often with less meritorious records with respect to innovation, survive and prosper. One set of reasons for this phenomenon is now clear. Large firms are more likely to possess the relevant specialized and cospecialized assets within their boundaries at the time of new-product introduction. They can therefore do a better job of milking their technology, however meager, to maximum advantage. Small domestic firms are less likely to have the relevant specialized and cospecialized assets within their boundaries and so either will have to incur the expense of trying to build them or will have to try to develop coalitions with competitors/owners of the specialized assets.

Regimes of Appropriability and Industry Structure

In industries where legal methods of protection are effective or where new products are just hard to copy, the strategic necessity for innovating firms to integrate into cospecialized assets would appear to be less compelling than in industries where legal protection is weak. In cases where legal protection is weak or nonexistent, the control of cospecialized assets will be needed for long-run survival.

In this regard, it is instructive to examine the U.S. drug industry (Temin 1979). Beginning in the 1940s, the U.S. Patent Office began, for the first time, to grant patents on certain natural substances that involved difficult extraction procedures. Thus, in 1948 Merck received a patent on Streptomycin, which was a natural substance. However, it was not the extraction process but the drug itself that received the patent. Hence, patents were important to the drug industry in terms of what could be patented (drugs), but they did not prevent imitation (Temin 1979: 436). Sometimes just changing one molecule will enable a company to come up with a different substance that does not violate the patent. Had patents been more all inclusive—and I am not suggesting they should be—licensing would have been an effective mechanism for Merck to extract profits from its innovation. As it turns out, the emergence of close substitutes—coupled with FDA regulation that had the *de facto* effect of reducing the elasticity of demand for drugs—placed high rewards on a product differentiation strategy. This required extensive marketing, including a salesforce that could directly contact doctors, who were the purchasers of drugs through their ability to create prescriptions.[6] The result was exclusive production (that is, the earlier industry practice of licensing was dropped) and forward integration into marketing (the relevant cospecialized asset).

Generally, if legal protection of the innovator's profits is secure, innovating firms can select their boundaries based simply on their ability to identify user needs and respond to those through research and development. The weaker the legal methods of protection, the greater the incentive to integrate into the relevant cospecialized assets. Hence, as industries in which legal protection is weak begin to mature, integration into innovation-specific cospecialized assets will occur. Often this will take the form of backward, forward, and lateral integration. (Conglomerate integration is not part of this phenomenon.) For example, IBM's purchase of Rolm can be seen as a response to the impact of technological change on the identity of the cospecialized assets relevant to IBM's future growth.

Industry Maturity, New Entry, and History

As technologically progressive industries mature and a greater proportion of the relevant cospecialized assets are brought in under the corporate umbrellas of incumbents, new entry becomes increasingly difficult. Moreover, when it does occur, it is more likely to involve coalition formation very early on. Incumbents will for sure own the cospecialized assets, and new entrants will find it necessary to forge links with them. Here lies the explanation for the sudden surge in strategic partnering now occurring internationally, and particularly in the computer and telecommunications industry. Note that it should not be interpreted in anticompetitive terms. Given existing industry structure, coalitions ought to be seen not as attempts to stifle competition but as mechanisms for lowering entry requirements for innovators.

In industries where there has occurred technological change of a kind that required deployment of specialized or cospecialized assets at the time, a configuration of firm boundaries may well have arisen that no longer has compelling efficiencies. Considerations that once dictated integration may no longer hold, yet there may not be strong forces leading to divestiture. Hence, existing firm boundaries in some industries—especially those where the technological trajectory and attendant specialized asset requirements have changed—may be rather fragile. In short, history matters in terms of understanding the structure of the modern business enterprise. Existing firm boundaries cannot always be assumed to have obvious rationales in terms of today's requirements.

The Importance of Manufacturing to International Competitiveness

Practically all forms of technological know-how must be embedded in goods and services to yield value to the consumer. An important policy issue for the innovating nation is whether the identity of the firms and nations performing this function matter.

In a world of tight appropriability and zero transactions cost—the world of neoclassical trade theory—it is a matter of indifference whether an innovating firm has an in-house manufacturing capability, domestic or foreign. It can simply engage in arm's-length contracting (patent licensing, know-how licensing, coproduction, and so forth) for the sale of the output of the activity in which it has a comparative advantage (in this case R&D) and will maximize returns by specializing in what it does best.

However, in a regime of weak appropriability, especially where the requisite manufacturing assets are specialized to the innovation, which is often the case, participation in manufacturing may be necessary if an innovator is to appropriate the rents from its innovation. Hence, if an innovator's manufacturing costs are higher than those of its imitators, the innovator may well end up ceding the lion's share of profits to the imitator.

In a weak appropriability regime, low-cost imitator-manufacturers may end up capturing all of the profits from innovation. In a weak appropriability regime where specialized manufacturing capabilities are required to produce new products, an innovator with a manufacturing disadvantage may find that its advantage at early-stage research and development will have no commercial value. This will eventually cripple the innovator, unless it is assisted by governmental processes. For example, it appears that one of the reasons why U.S. color television manufacturers did not capture the lion's share of the profits from the innovation, for which RCA was primarily responsible, was that RCA and its U.S. licensees were not competitive at manufacturing. In this context, concerns that the decline of manufacturing threatens the entire economy appear to be well founded.

A related implication is that as the technology gap closes, the basis of competition in an industry will shift to the cospecialized assets. This appears to be what is happening in microprocessors. Intel is no longer ahead technologically. As Gordon Moore, CEO of Intel points out ("Institutionalizing the Revolution" 1986: 35),

> Take the top 10 [semiconductor] companies in the world . . . and it is hard to tell at any time who is ahead of whom. . . . It is clear that we have to be pretty damn close to the Japanese from a manufacturing standpoint to compete.

It is not just that strength in one area is necessary to compensate for weakness in another. As technology becomes more public and less proprietary through easier imitation, strength in manufacturing and other capabilities is necessary to derive advantage from whatever technological advantages an innovator may possess.

Put differently, the notion that the United States can adopt a designer role in international commerce while letting independent firms in other countries such as Japan, Korea, Taiwan, or Mexico do the manufacturing is unlikely to be viable as a long-run strategy because profits will accrue primarily to the low-cost manufacturers (by providing a larger sales base over which they can exploit their special skills). Where imitation is easy, and even where it is not, there are obvious problems in transacting in the market for know-how—problems that are described in more detail elsewhere (Teece 1981). In particular, there are difficulties in pricing an intangible asset whose true performance features are difficult to ascertain *ex ante.*

The trend in international business towards what Miles and Snow (1986) call *dynamic networks*—characterized by vertical disintegration and contracting—thus ought to be viewed with concern. *(Business Week* 1986 has referred to the same phenomenon as the *hollow corporation.)* Dynamic networks may not so much reflect innovative organizational forms as the disassembly of the modern corporation because of deterioration in national capacities, in manufacturing particularly, that are complementary to technological innovation. Dynamic networks may therefore signal not so much the rejuvenation of U.S. enterprise as its piecemeal demise.

How Trade and Investment Barriers Can Affect Innovators' Profits

In regimes of weak appropriability, governments can move to shift the distribution of the gains from innovation away from foreign innovators and toward domestic firms by denying innovators ownership of specialized assets. The foreign firm, which by assumption is an innovator, will be left with the option of selling its intangible assets in the market for know-how if both trade and investment are foreclosed by government policy. This option may appear better than the alternative (no remuneration at all from the market in question). Licensing may then appear profitable but only because access to the complementary assets is blocked by government.

Thus when an innovating firm generating profits needs to access complementary assets abroad, host governments, by limiting access, can sometimes milk the innovators for a share of the profits, particularly that portion that originates from sales in the host country. However, the ability of host governments to do so depends importantly on how critical the host country's assets are to the innovator. If the cost and infrastructure characteristics of the host country are such that it is the world's lowest-cost manufacturing site and if domestic industry is competitive, then by acting as a *de facto* monopsonist, the host-country government ought to be able to adjust the terms of access to the complementary assets so as to appropriate a greater share of the profits generated by the innovation.[7] If, on the other hand, the host country offers no unique complementary assets, except access to its own market, restrictive practices by the government will only redistribute profits with respect to domestic rather than worldwide sales.

Implications for the International Distribution of the Benefits from Innovation

The above analysis indicates that innovators who do not have access to the relevant specialized and cospecialized assets may end up ceding profits to imitators and other competitors or simply to the owners of the specialized or cospecialized assets. Even when the specialized assets are possessed by the innovating firm, they may be located abroad. Under these circumstances, foreign factors of production are likely to benefit from the research and development activities. There is little doubt, for instance, that the inability of many U.S. multinationals to sustain competitive manufacturing in the United States is resulting in declining returns to U.S. labor. Stockholders and top management probably do as well, if not better, when a multinational accesses cospecialized assets in the firm's foreign subsidiaries; however, if there is unemployment in the factors of production supporting the specialized and cospecialized assets in question, then the foreign factors of production will benefit from innovation originating beyond national borders. This speaks to the importance to innovating nations of maintaining competence and competitiveness in the assets that complement technological innovation, manufacturing being a case in point. It also speaks to the importance to innovating nations of enhancing the protection afforded worldwide to intellectual property.

However, it must be recognized that there are inherent limits to the legal protection of intellectual property and that business and national strategy are therefore likely to be critical factors in determining how the gains from innovation are shared worldwide. By making the correct strategic decision, innovating firms can move to protect the interests of stockholders; however, to ensure that domestic rather than foreign cospecialized assets capture the lion's share of the externalities spilling over to complementary assets, the supporting infrastructure for those complementary assets must not be allowed to decay. In short, if a nation has prowess at innovation, then in the absence of ironclad protection for intellectual property, it must maintain well-developed complementary assets if it is to capture the spillover benefits from innovation.

Conclusion

This analysis has attempted to synthesize from recent research in industrial organization and strategic management a framework within which to analyze the distribution of the profits from innovation. The framework indicates that the boundaries of the firm are an important strategic variable for innovating firms. The ownership of complementary assets, particularly when they are specialized or cospecialized, helps establish who wins and who loses from innovation. Imitators can often outperform innovators if they are better positioned with respect to critical complementary assets. Hence, public policy aimed at promoting innovation must focus not only on R&D but also on complementary assets as well as the underlying infrastructure. If government decides to stimulate innovation, it would seem important to clear away barriers that would impede the development of complementary assets that tend to be specialized or cospecialized to innovation. To fail to do so would cause an unnecessarily large portion of the profits from innovation to flow to imitators and other competitors. When these firms lie beyond national borders, there are obvious implications for the international distribution of income.

When applied to world markets, results similar to those obtained from the new trade theory are suggested by the framework. In particular, tariffs and other restrictions on trade can in some cases injure innovating firms while simultaneously benefiting protected firms when they are imitators. However, the propositions suggested by the framework are particularized to appropriability regimes, suggesting that economywide conclusions will be illusive. The policy conclusions derivable for commodity petrochemicals, for instance, are likely to be different from those that would be arrived at for semiconductors.

The approach also suggests that the product life cycle model of international trade will play itself out very differently in different industries and markets, in part according to appropriability regimes and the nature of the assets that need to be employed to convert a technological success into a commercial one.

Whatever its limitations, the approach establishes that it is not so much the structure of markets but the structure of firms, particularly the scope of their boundaries, coupled with national policies with respect to the development of complementary assets, that determines the distribution of the profits among innovators and imitator/followers.

NOTES

1. The EMI story is summarized in Martin (1984).
2. Robert D. Kennedy, executive vice president of Union Carbide, quoted in *Chemical Week.*
3. Comment attributed to Peter Olson III, IBM's director of business development, as reported in "The Strategy behind IBM's Strategic Alliances" (1985: 126).
4. Comment attributed to Norman Farquhar, Cipher's vice president for strategic development, as reported in *Electronic Business* (1985: 128).
5. *Business Week* uses the term to describe a corporation that lacks in-house manufacturing capability.
6. In the period before FDA regulation, all drugs other than narcotics were available over the counter. Because the end user could purchase drugs directly, sales were price sensitive. Once prescriptions were required, this price sensitivity collapsed; the doctors not only did not have to pay for the drugs, but in most cases they were unaware of the prices of the drugs that they were prescribing.
7. If the host-country market structure is monopolistic in the first instance, private actors might be able to achieve the same benefit. What government can do is to force collusion of domestic enterprises to their mutual benefit.

REFERENCES

"A Bad Aftertaste." 1985. *Business Week.* (July 15).

Abernathy, W.J., and J.M. Utterback. 1978. "Patterns of Industrial Innovation." *Technology Review* 80(7) (June/July): 40–47.

Clark, Kim B., 1985. "The Interaction of Design Hierarchies and Market Concepts in Technological Evolution." *Research Policy* 14: 235–251.

Dosi, G. 1982. "Technological Paradigms and Technological Trajectories." *Research Policy* 11(3): 147–62.

"GE Gobbles a Rival in CT Scanners." 1980. *Business Week.* (May 19): p. 47.

Gens, F., and C. Christiansen. 1983. "Could 1,000,-000 IBM PC Users Be Wrong?" *Byte* (November): 88.

"The Hollow Corporation." 1986. *Business Week* (March 3): 57–59.

"Institutionalizing the Revolution." 1986. *Forbes* (June 16): 35.

Kuhn, Thomas. 1970. *The Structure of Scientific Revolutions,* 2d ed. Chicago: University of Chicago Press.

Levin, R., A. Klevorick, N. Nelson, and S. Winter. 1984. "Survey Research on R&D Appropriability and Technological Opportunity." Unpublished manuscript, Yale University.

Martin, Michael. 1984. *Managing Technical Innovation and Entrepreneurship.* Reston, Va.: Reston.

McKenna, Regis. 1985. "Market Positioning in High Technology." *California Management Review* 27(3) (Spring): 82–108.

Miles, R.E., and C.C. Snow. 1986. "Network Organizations: New Concepts for New Forms." *California Management Review* 28(3) (Spring): 62–73.

Monsanto Annual Report. 1985. St. Louis, Mo.

Norman, David A. 1986. "Impact of Entrepreneurship and Innovations on the Distribution of Personal Computers." In *The Positive Sum Strategy,* edited by R. Landau and N. Rosenberg, pp. 437–39. Washington, D.C.: National Academy Press.

"The Strategy behind IBM's Strategic Alliances." 1985. *Electronic Business* 11(19) (October 1): 126–29.

Teece, D.J. 1981. "The Market for Know-How and the Efficient International Transfer of Technology." *Annals of the American Academy of Political and Social Science* 458 (November): 81–96.

Temin, P. 1979. "Technology, Regulation, and Market Structure in the Modern Pharmaceutical Industry." *Bell Journal of Economics* 10(2) (Autumn): 429–46.

Williamson, O.E. 1975. *Markets and Hierarchies.* New York: Free Press.

SECTION
VII

Governmental Influence
on Innovation

Innovation has become an increasingly global phenomenon. As industries become ever more competitive, and as technologies change rapidly and are quickly transferred across national boundaries, action by governments and industry associations have a heightened impact on the performance of firms and industries. Chapter 10 presents several articles that reflect the increasing importance of national technology and/or innovation policy. Based on American and German examples, Sabel, Herrigel, Kazis, and Deeg discuss alternative forms of "pre-competitive competition" including government-firm, firm-firm, and firm-labor linkages necessary to sustain organization flexibility and innovation over time. Nelson and Langlois review the American experience to date on industrial innovation policy, while Hannay reviews the impacts of innovation policy in several U.S. industries.

10

GOVERNMENTAL INFLUENCE ON INNOVATION

How to Keep Mature Industries Innovative

Charles F. Sabel

Gary Herrigel

Richard Kazis

Richard Deeg

Much of the debate about how to promote economic growth in the United States assumes that products, industries, and even whole economies pass through inevitable lifecycles. The core idea is that leaders innovate and laggards emulate. A new product, in this view, is born in the most advanced markets. At first it is manufactured in small batches by skilled workers using flexible production machinery. In time it is mass-produced by semi-skilled workers using machinery designed expressly for making that product. Finally, the product becomes so standard and the manufacturing process so routine that firms in less-developed countries with low labor costs can begin competing with the innovator. By then, the leading markets have come up with

new products, and the cycle starts again. According to a similar logic, countries first master agriculture, then manufacturing, then the provision of sophisticated services, all the while ceding routine industries and whole economic sectors to less-developed competitors.

The stories of two Massachusetts industries are often taken as examples of this natural industrial history. First was the rise and decline of textile manufacturing and the textile-machine industry, which supplied mills with equipment such as looms. Then came the rise of what is loosely called high tech—principally mini-computers, as well as semiconductors, test equipment, and products that incorporate or complement them. The state's programs for transferring technology from universities to industry, providing start-up firms with venture capital, and training workers for jobs in high

Reprinted with permission from *Technology Review*, Copyright 1987.

tech are often cited as models of how a government that has learned the lesson of the product-lifecycle theory can foster a permanently innovative economy.

There is something wrong with the idea that all industries or even manufacturing economies have but a few decades in the sun. Consider Baden-Württemberg, a state in southwest Germany whose prosperity rivals that of Massachusetts. Baden-Württemberg makes its fortune through many of the traditional industries Massachusetts has abandoned—the machine-tool, special-machine, automobile, and automotive-parts industries. In particular, this German state is among the world leaders in textile-machinery production. Although Baden-Württemberg has only a small high-tech sector in the sense familiar to Americans, firms are rapidly developing high-tech products, such as industrial lasers for production machinery, that could eventually compete with the most sophisticated creations of America's youngest sectors.

To understand the current prosperity of the two states and the implications for public policy, it is necessary to look beyond the theory of product lifecycle to the precise organization of industry. One way to do this is to contrast the historical decline of the textile-machinery industry in Massachusetts with its rise in Baden-Württemberg. The same mechanisms that explain America's loss of leadership in textile machinery explain the dwindling competitiveness in other industries such as machine tools and special machines. These mechanisms also shed surprising light on the vulnerability of high-tech industry in this country.

THE RISE AND FALL OF A TRADITIONAL INDUSTRY

The decline of the textile-machinery industry can now be described in considerable detail. Changes in this industry took place over a 25-year period—slow motion by contemporary standards. And enough time has gone by so that the participants can take a contemplative, if not detached, view of their past. The story is disconcerting precisely because, despite its historical remoteness, it undeniably parallels the difficulties of so many of today's industries.

The first firms making looms and equipment for preparing yarn grew out of machine shops in the textile mills that Boston entrepreneurs founded in the early nineteenth century. By the Civil War the most successful of these firms had become independent and often large companies. Throughout the late nineteenth century they experimented with a variety of strategies to regulate competition and thereby minimize the sharp fluctuations in earnings characteristic of firms producing capital goods. They tried price-fixing agreements and horizontal combinations, in which different firms specialized in particular products so that they could market a full line of equipment jointly. But the strategy that won out was merger.

Mergers produced a highly concentrated industry. In 1890 there had been twelve major loom producers in the United States; by 1930 only two remained. Draper made automatic looms for the coarse, simple fabrics that constituted the bulk of the American market. Crompton and Knowles made more flexible box looms for fancier goods. Two other firms—the Whitin Machine Works and the Saco-Lowell Shops—split the market for yarn-preparation equipment. Through the 1960s, these four firms dominated their branches of the U.S. textile-machinery industry. They faced negligible competition for the U.S. market and held significant shares in some foreign markets as well.

The key to the firms' success—and eventually, we believe, to their downfall—was their relationship with their customers, the textile mills. By the late nineteenth century, equipment makers had gained a tight, often controlling grip on the mills. Mills were dependent on them for service and technical advice, and in some cases for capital as well. Machine builders regularly accepted stock in new Southern textile mills as payment. These relations sheltered machine makers from competition. Once a mill became dependent, it was unlikely to turn to a competitor for new machines. Particular textile firms were known for decades as "Whitin" mills or "Saco-Lowell" mills.

This stability, buffeted of course by periodic swings in the demand for textiles, allowed equipment makers to standardize their products, apply mass-production principles to further reduce manufacturing costs, and thus tighten their hold on mills. Draper may have gone the farthest down this path. It invested in facilities designed expressly for producing a few types of looms. According to one machinery executive, mill owners were told: "Here is what we produce. How many do you want?" The company also integrated production vertically, producing almost all the needed materials and parts. It owned 150,000 acres of forest to guarantee the supply of wood for shuttles and harnesses. The foundry was such an exemplar of advanced automation that General Motors regularly sent production engineers to inspect it. Asked what the company did not produce for itself, a former president could think only of small supplies of dogwood that were needed for special components.

This manufacturing strategy reduced machine builders' ability to respond to shifts in demand. Lack of skilled workers also lessened flexibility in manufacturing. Mass-production techniques required that an increasing percentage of unskilled workers perform repetitive tasks. Because of the reduced need for skilled labor, firms allowed the vocational-training system to atrophy. Even in the late nineteenth century American producers had found it difficult to copy British yarn-preparation machinery because they could not match the precision work of British hand-fitters.

Machine makers' rigidity encouraged furtive technical experimentation by mill owners who were not content with the available product lines. They modified standard machines to increase efficiency or achieve new effects and kept the results of their tinkering to themselves. This cut the machine makers off from an invaluable source of new ideas. For example, the modern air-jet loom was anticipated in the 1920s in the Chini Silk Mill, where technicians secretly produced a novelty silk on a Draper loom modified so that the shuttle could be propelled back and forth by compressed air.

In the 1950s and 1960s Southern mills grew very large as compared with even the largest machine makers, and the balance of forces began to change. Mills such as Milliken, Burlington, and J.P. Stevens scanned the world for equipment better suited to their needs. U.S. mill owners started attending International Textile Machinery Exhibitions in 1959. They were surprised to see a wide range of sophisticated machinery at these European trade shows, and they met producers who were willing to accommodate customer specifications. Not everything the Europeans made was better than its American counterpart, and many superior products—say, for making particularly delicate cloth—were of no use to U.S. mills. But enough of the European innovations were applicable to create a new relationship between U.S. mills and foreign capital-goods producers.

By the late 1960s, intense international competition in textiles led to rapid shifts in fabric production. Hence new kinds of textile machines were in demand. The European innovations became decisive. By the early 1980s most of the advanced products in the industry—sophisticated air-jet looms, highly automated and precise spinning equipment—were either unavailable from American producers or available only at prohibitive prices.

Why did the Americans fail to break free of the old mass-production system? Several apparently plausible answers turn out to be wrong. Money, for example, was not a problem.

The firms were cashrich, in part simply because they continued to earn good revenues from selling replacement parts and equipment. Nor was indifference to foreign achievements a problem. Thoughtful managers in the industry—and we spoke with many—had a clear sense of the technological threat they faced. In some cases firms even considered licensing designs that later took away a large portion of their business. For instance, Draper considered licensing the Sulzer air-driven, shuttleless loom.

The American textile-machine industry was paralyzed by the rigidities of the concentrated, vertically integrated, mass-production system. Foreign competitors took the technological lead not in a single dramatic dash, but step by almost imperceptible step. Each isolated European refinement in machine design looked inconsequential by itself. Few innovations promised large increases in productivity, even in the unlikely case that they could be made by the largely unskilled work force in the U.S. textile-machine factories and used by the still less skilled mill workers. In a few cases innovation did promise productivity breakthroughs—for example, with the shuttleless loom. But American manufacturers thought that improvements based on their own design traditions would accomplish the same thing.

Decision by decision, the Americans' calculations are difficult to fault. Most of the potential breakthroughs could theoretically have been achieved by alternative means. But by systematically screening out refinements that would have taxed and thus developed engineering and manufacturing capacities, the U.S. firms left their competition in unchallenged possession of a growing stock of new ideas. Some small fraction of these ideas were eventually incorporated into machine designs compatible with the around-the-clock operations and unskilled work force of the U.S. textile mills. Unfortunately, these were designs U.S. machine makers could not match.

INNOVATION IN BADEN-WÜRTTEMBERG

The German textile-machine industry, like the American one, grew up wherever textile producers existed. Saxony focused on equipment for worsteds, Monchengladbach on finishing machines, and Württemberg on knitting machines. But unlike the Americans, the Germans were not able to concentrate demand on standardized products.

The chief reason was competition from powerful British textile-machine companies with decades of experience and intimidating reputations. Almost from the first, German textile-machine manufacturers focused on specialized products that complemented British equipment or produced yarn or cloth impossible to make with standard machines.

In some instances the state learned to assist machine builders. For example, the Kingdom of Saxony subsidized the development of cotton-yarn and yarn-equipment industries during Napoleon's continental blockade, when Germany was protected from British competition. The end of the blockade thwarted these efforts. But a second program to create an indigenous industry succeeded. Saxony subsidized the production of worsted yarn and yarn-making equipment, a sector in which the British were not competitive, and the state became a world center for woolen textiles and worsted-spinning technologies.

Because German textile producers faced British competition in standard cotton items, they had to specialize in this area, too. Their constant search for new products created a demand for new spinning, weaving, knitting, and finishing technologies. German machine builders started to see themselves as custom producers whose survival depended on accommodating shifting desires.

Textile-machine makers came to view their industry as an association of specialists, each with unmatched expertise and flexibility in a particular phase or type of production. The system recalled the horizontal combinations of nineteenth-century New England machine makers that jointly marketed a single line in the South. However, unlike U.S. firms, German firms did not fuse into a few corporations dominating broad segments of the market. On the contrary, in the 1920s the trade association helped stabilize the status quo. Firms continued dividing the market according to their areas of specialization. while achieving economies of scale through joint marketing and research.

These arrangements were called specialization cartels or more often finishing associations to distinguish them from price-fixing combinations. Their function was to regulate competition by guaranteeing each firm that its market niche would not be invaded by other association members during downturns. Without such assurance, few firms would have run the risk of extreme specialization. The most famous association, Unionmatex, was established in 1918 by 14 builders of carding, combing, spinning, and cloth-finishing machinery.

The legal system encouraged the development of specialization cartels but probably did not determine it. Though such cartels were illegal during the formative period of the textile-machine industry in the United States, American executives have discreetly admitted that they were skilled in avoiding legal restrictions on the kinds of cooperation they found advantageous. Conversely, while cartels were legal in Germany, German firms would not have made use of their legal opportunities unless they had had economic motives for doing so.

The textile-machinery producers' trade association and its eight sub-associations had become indispensable to the German industry by the middle of the 1920s. One sub-association, the Union of Spinning Machine and Loom Manufacturers, pooled advertising expenses and established offices to represent members in foreign markets. This organization acted as a forum for setting industrial standards and fostered cooperation between the industry and its customers.

Research institutes were another important part of the structure upon which firms relied. Many institutes were established jointly by machine makers, textile manufacturers, and local governments. For example, in 1855 a group of Württemberg industrialists and state officials aided textile production by founding a technical institute for research and training in Reutlingen. A broad movement eventually led to public vocational services that provided small and medium-sized firms with craft workers.

The division of labor between the individual firm and the industry was self-reinforcing. The more specialized each firm became, the more it depended on the success of products complementing its own. Firms became more interested in exchanging information with related producers. They also began to further the well-being of the industry as a whole by supporting broad institutions—vocational schools, research institutes, and marketing agencies. The industry was not simply the sum of autonomous production units, but rather a set of institutions that made the survival of individual companies possible.

This system put pressure on firms to innovate, then helped them translate their innovative ideas into reality. Because firms could not diversify to reduce losses during downturns, they had to improve or customize their product lines. Progress in one phase of production created bottlenecks in others, and stimulated complementary innovations. Firms succeeded in producing these innovations because they could draw on skilled workers, well-equipped research institutes, and the advice of customers and machine makers in related fields. The kinds of incremental changes that were ruled out in the Massachusetts system stimulated self-renewal in the German model.

BADEN-WÜRTTEMBERG TODAY

As world markets for textile machinery became more volatile in the 1970s, these principles were extended in two ways. One was the spread of subcontracting. In part this was a matter of necessity. Firms faced with precipitous shifts in demand had to reduce their fixed costs by selling machinery, laying off workers, and purchasing parts from subcontractors. At least three major producers in Baden-Württemberg—Sulzer Morat, Terrot, and Zinser—are currently taking this approach.

More fundamentally, the move to subcontracting reflects a redefinition of strategy. As development costs rise with increasingly rapid technological progress and product changes, firms begin to share the additional expenses with subcontractors. The firms concentrate their expertise in coordinating the design and assembling the full product, and in advancing a few key technologies. They develop complementary technologies in collaboration with selected subcontractors.

This trend goes hand in hand with the creation of a production network cutting across industries. When subcontractors provide common processes or products to a variety of industries, firms do not fear that information passed to subcontractors will wind up with the competition. Firms benefit from the subcontractors' experiences with customers in other industries. At the same time, diversified subcontractors are hedged against slumps in any one industry.

No one in Baden-Württemberg could direct us to statistics confirming the spread of such inter-industry supplier networks, but virtually everyone was familiar with the phenomenon. Consider the story of Robert Bosch, GmbH, one of the world's most sophisticated automotive-parts manufacturers. The firm is extraordinarily knowledgeable as both a subcontractor to others and an organizer of its own subcontracting system.

For Bosch, creating a supplier network has been explicit corporate policy since 1970. The firm establishes long-term relations with promising subcontractors and supports them. This support ranges from giving general advice on new technologies—acting, the firms' managers said, like a technical university—to collaborating in producing single parts. The company insists that subcontractors do no more than 20 percent of their business with Bosch. This means that Bosch can reduce orders to them without endangering their existence—to which its own is tied. It also means that Bosch can learn from what they are doing for other customers. Bosch's largest subcontractors, we found, apply this model of organization to their own suppliers. Those suppliers sometimes impose the same rules on companies working for them. The result is that even shops with as few as 50 employees share in this network of subcontracting and information exchange.

An example of the spread of this model to the textile-machinery industry is the firm Kern & Liebers of Schramberg-Sulgen. Typical of the emerging class of large, diversified suppliers, Kern & Liebers began manufacturing thin stamped-metal parts for the watch industry in the Black Forest in the nineteenth century. It used its experience in stamping technology to manufacture similar parts for the knitting-machine industry. Today the firm has mastered a series of complementary technologies, including flangeless stamping (which eliminates the need to remove burrs from the workpiece), continuously stressed springs (which must meet particularly stringent quality standards), and plastic lamination for embedding small parts into sub-assemblies. Kern & Liebers supplies consumer electronics firms, special-machinery producers, textile-machine makers, and the automobile industry (including Bosch). It has begun to apply the Bosch model of organization to its own suppliers.

Because of the expansion of subcontracting, many consulting firms have emerged to offer services that the new small and medium-sized producers cannot provide for themselves. Between 1976 and 1984, the number of firms engaged in "technical consulting and planning" in Baden-Württemberg rose 40 percent from 8,662 to 12,130. Their revenue more than doubled from 2,240,593 deutsche marks (DM) to 4,935,535. Software houses providing customized programs seem to have grown similarly.

An expansion of public-sector services to industry has paralleled the expansion of services by the private sector. Income from established state-run technical consulting services increased from 4.9 million DM in 1983 to 8.4 million DM in 1985. Revenues from a new, related consulting service increased by almost five times from 1 million DM in 1984, when it was founded, to 4.9 million DM in 1985.

There has been a corresponding effort to increase the skill level and hence the flexibility of the work force. A coalition of industrialists, state politicians, and vocational-education teachers has sponsored remarkable improvements in an already excellent vocational and technical training system. Vocational high schools *(Berufsschulen)* once provided elementary instruction for apprentices. Now they are teaching the skills that have been taught to technicians and engineering students in polytechnics *(Fachhochschulen)*. Meanwhile, polytechnics are beginning to do the kind of research and teaching formerly reserved for technical universities. The traditional blue-collar educational system has become so good that it attracts students who have met the requirements for university study. The boundary between working-class and middle-class education is breaking down, just as the drive for flexibility inside the factories is blurring the distinction between white-collar workers who conceptualize tasks and blue-collar workers who execute them.

HIGH TECH AND HIGH FINANCE

Our sketch of traditional industries in Baden-Württemberg and Massachusetts shows how different organizational principles have produced strikingly different outcomes. These same principles explain why separate high-tech and high-finance sectors emerged in only one of the two economies.

As its name suggests, high tech grew up outside what are called in Massachusetts the "mature" industries. The rigid structure of traditional industries in Massachusetts impeded them in developing or rapidly absorbing radical innovations such as computer-based technologies. Ironically, Massachusetts, one of the homelands of high tech, makes little use of these breakthroughs in the rest of its industrial economy, while Baden-Württemberg, lost beyond the horizon of invention, turns them to profitable purpose.

It is hardly novel to observe that Massachusetts high tech arose in a world of its own. Key innovations in computer development were made in university labs financed by government contracts, usually for armaments. New ideas were commercialized by researchers in collaboration with financiers and business people chosen through personal ties. The first customers were often research labs or firms that were themselves studying or developing high tech. The prototypical example is the PDP, a minicomputer inspired by Kenneth Olson's work at M.I.T., commercialized by his Digital Equipment Corp. (DEC), and sold to engineers in M.I.T. labs.

In their internal organization and their relation to other high-tech producers, companies like DEC have more in common with Baden-Württemberg industries than with traditional industries in Massachusetts. Work in large high-tech firms is organized in semi-autonomous teams that recall the federated craft units of the German machine makers. Firms with complementary products tend to collaborate closely for long periods. Universities act as surrogate industrial associations, facilitating the exchange of information. This exchange eventually allows competing firms to set joint standards.

However, there are important limits to the analogy. Relations with suppliers are less stable in the United States because as firms grow, they tend to hire managers from established corporations. These managers usually have little experience in operating a Bosch-model subcontracting system. And since universities are not industrial associations, they do not have the authority to coordinate specialization among firms. Finally, production work is usually regarded as a mere appendage when firms are doing essentially custom jobs for a few clients. Hence such work is rarely integrated into the team model. If it is organized at all, it is organized according to the principles of old-line mass-production industry.

Furthermore, most of the flexibility has been devoted to designing standard products. Until recently, firms such as DEC, Prime, and Data General dedicated themselves to building faster and cheaper machines for customers who were expected to handle any necessary customization themselves. This was a natural enough strategy. Many scientific and engineering users knew just what they wanted to do with the machines. As for industrial customers, computer producers could have only the vaguest idea what applications they might desire. The reason was that American firms, as exemplified in the textile-machinery case, were slow to integrate new developments into production. Put another way, the PDP was like the Draper loom, a sophisticated, general-purpose product that sold itself. For a long time DEC knew even less about the varied needs of customers than Draper did.

Now competition in the minicomputer industry is changing. Many companies can or will soon be able to make high-performance "boxes." Firms will have to survive by cutting production costs and customizing products to suit particular needs. This puts the American manufacturers at a double disadvantage. First, even the largest firms have traditionally not paid much attention to the organization of production. They will have to decide whether to extend the flexible work procedures in engineering departments to manufacturing areas. Second, greater customization will be difficult, not simply because firms will be forced to overcome their ignorance about their customers' industries. The customers themselves represent an obstacle, since they know so little about their own markets.

No doubt companies like DEC will survive. It has already cut manufacturing costs through automation, introduced sophisticated systems for changing standard models into semi-custom systems, and moved through joint ventures to provide software for applications in particular industries. DEC has also extended its contacts with innovative manufacturers by entering a joint venture with COMAU, the FIAT machine-tool subsidiary in Turin. Like the best German machine-tool makers, COMAU collaborates closely with manufacturers that purchase its machine tools in many countries. DEC can expect such collaboration to keep it abreast of changing demands for factory automation. Furthermore, according to marketing specialists in the computer industry, DEC is likely to ally with business consulting firms to help address its customers' confusion about their markets. The consulting firms will help DEC's customers choose a business strategy and the computing equipment to go with it.

In Baden-Württemberg and West Germany more generally, the condition of high tech is in many ways the reverse of what it is in Massachusetts. There is little high-tech industry—almost none in Baden-Württemberg except for subsidiaries of U.S. multinationals. But there are many high-tech products such as machines that incorporate sophisticated computer controls. These products are a natural result of the firms' drive to produce higher-performance, increasingly flexible equipment. Stoll was the first company in the world to manufacture a computer-controlled flat-knitting machine, the type used for producing fashionable knit-wear. Traub and Index, once producers of traditional lathes, have both become world leaders in computer-controlled lathe and machining-center technology.

Even the few high-tech firms that do exist in West Germany follow the tendency of the country's mature industries toward specialization. For example, Nixdorf, a minicomputer producer, does not try to compete by producing the most sophisticated machines. Instead, it specializes in adapting its equipment to the needs of individual customers. Nixdorf's strategy resembles that of the early German textile-machine makers. Just as these firms found the core of their market occupied by well-designed British machinery, so Nixdorf found the core of its market occupied by Massachusetts firms. In both cases the answer was to make a virtue of necessity and go into the customization business.

Some firms in traditional industries have also begun to produce high-tech products as part of their specialization strategy. For example, Trumpf, a maker of computer-controlled machines for cutting sheet metal, found a California supplier of industrial lasers unwilling to customize a component. Trumpf designed a laser that meets its requirements, and it will sell this product in competition with its former American supplier.

Financial services in Massachusetts are, like high tech, a world apart. Most of the banks, insurance companies, and mutual funds that dominate Boston's financial district simply aggregate funds from the region and nation and invest them in world markets. Boston's financial institutions are not the organizational core of the region's economy or important actors in world money markets. Their major contribution to the regional economy has been to supply venture capital to high-tech start-ups.

The financial-services sector in Baden-Württemberg is small by comparison with Bos-

ton's and much more tightly integrated into the rest of the economy. Many local firms are financed in part by cooperative banks and community savings and loan institutions. Larger companies engaged in extensive international operations usually use a large Frankfurt bank but retain their connection to the local institution with which they grew up. The boards of directors of the cooperative banks are typically comprised of owners or directors of the leading firms in the district. This system guarantees that judgments on granting credit or loaning funds to fuse distressed firms will be made by people dedicated to advancing the interests of the community and long familiar with the companies and personalities concerned.

Just as Baden-Württemberg has high-tech products without high-tech industry, so it has a supple system for providing capital to local firms without a "modern" financial-services sector. It is hard to escape the feeling that in both cases, apparent backwardness obscures resilient modernity.

SEEKING A MORE FLEXIBLE ECONOMY

In challenging the product-lifecycle theory and the policies derived from it, this account raises two obvious questions. First, can the institutions required for a flexible, specialized economy be created where they do not already exist? And if so, what role can government play in creating them?

We believe that the emergence of economies such as Baden-Württemberg's is not simply a matter of historical luck. We are convinced that government can, and probably must, help form them. Similarly structured economies can be found in a number of regions, including the Danish province of Jutland, the Swedish region of Småland, the French province of Rhône-Alpes, and what is now called the Third Italy (roughly the area defined by Venice, Florence, and Ancona). Additionally, these economies have developed in urban centers such as Turin and, at least incipiently, Los Angeles and Route 128 in Massachusetts. There proved to be many paths from pre-industrial, "traditional" societies to "modern" industrial economies, and by the same token, the diverse experience of these areas suggests that there are many, often unexpected ways to achieve a robustly specialized economy.

Furthermore, in many of these regions, the state has a long tradition of supporting the growth of specialized industry. Baden-Württemberg is a case in point. In 1848 the Kingdom of Württemberg established a Central Office for Manufacture and Commerce (*Zentralstelle für Gewerbe und Handel*). This office in turn created a bank for small and medium-sized firms, a technology consulting service, a service for demonstrating prototype machines, an export-assistance program, and an information-gathering program, which sent technicians and managers abroad to learn about new technologies. Later the state assumed responsibility for the vocational and technical education system.

Many of these programs are at first glance indistinguishable from programs in place or under discussion in Massachusetts and other U.S. states. The key to their success, and ultimately to the durable prosperity of Baden-Württemberg, was the way their operation meshed with and reinforced the organized specialization of local firms.

Here is the decisive lesson in the success of "mature" industry in Baden-Württemberg. By themselves, policies to foster innovation or entrepreneurship in general, or to promote a shift of economic activity to high tech or high finance, are insufficient. All efforts will probably come to naught unless the government simultaneously encourages or at least, as in Baden-Württemberg, allows industry to reorganize in a manner that encourages innovative specialization.

Basic American ways of thinking must change, too. The trade associations, specialization cartels, and cooperative banks that help institutionalize flexibility in Baden-Württemberg strike many of us as collusive. We are used to the notion that the only way to encourage innovation is to remove obstacles to competition, including private agreements by firms to limit their freedom of action.

Recently, economists, public officials, and business managers have gingerly begun to concede that the idea of competition as unlimited freedom can be a barrier to innovation. Through joint ventures and participation in collective research efforts, firms in many industries are learning that cooperation can be crucial in developing profitable new ideas. States such as Michigan and Massachusetts have instituted programs aimed at revitalizing the automobile-parts, cutting-tool, and apparel industries. These programs are helping the state governments understand how to foster the necessary cooperation among firms, and between management and labor.

Moreover, what is today called "precompetitive" cooperation has precedents in American business tradition. In the late nineteenth and early twentieth centuries, for example, Justice Louis Brandeis represented a movement of small and medium-sized firms that sought legal authority to form just the sort of associations characteristic of industry in Baden-Württemberg.

Habits of thought are hard to break, and forgotten, failed movements are a poor source of inspiration in confused times. If our story about Massachusetts and Baden-Württemberg went to the heart of the matter, America is not losing its industrial base because it has priced itself out of world markets through an overvalued dollar or high wages. America's failure is due to the very success with which it applied its concepts of production efficiency and market competition. To regain what the country has lost, government, industry, and labor will have to redefine both concepts.

Industrial Innovation Policy: Lessons from American History

Richard R. Nelson

Richard N. Langlois

Summary. The historical interrelations of government support of R & D and technical change in seven major American industries point to three types of policy that have been successful in the past: (i) government R & D support for technologies in which the government has a strong and direct procurement interest; (ii) decentralized systems of government-supported research in the "generic" area between the basic and the applied; and (iii) a decentralized system of clientele-oriented support for applied R & D. A fourth type of policy, under which the government attempts to "pick winners" in commercial applied R & D, has been a clear-cut failure.

Government involvement in the research and development (R & D) process has a long history in this country. As is too often the case, this rich experience has seldom been consulted in policy debates over government programs to stimulate industrial innovation.

This article is an attempt to identify some of the lessons of past federal R & D policy. It summarizes the conclusions of a study, recently completed at the Center for Science and Technology Policy at New York University, of how such policies have shaped technological change in seven major American industries—semiconductors, computers, aircraft, phar-maceuticals, agriculture, residential construction, and automobiles (1). What makes this study unique is its explicitly comparative, cross-industry focus: it was predicated on the hypothesis—amply supported in the resulting case studies—that the kinds of government programs that have shown themselves feasible and effective vary greatly among industrial sectors, depending upon the nature of the governmental involvement and the nature of competition in the industry.

The selection of industries for study was made with an eye toward obtaining a sample with a broad spectrum of characteristics: industries with fragmented as well as with concentrated structures and industries subject to much government intervention and to relatively little. The design of the study was also informed by a desire to attract recognized

"Industrial Innovation Policy: Lessons from American History" by Richard R. Nelson and Richard N. Langlois, from *Science*, 18 February 1983, Volume 219, pp. 814–818. Copyright 1983 by the American Association for the Advancement of Science. Reprinted by permission.

scholars knowledgeable about the technology in each industry: the prior interests and areas of expertise of these scholars were therefore a factor in the selection of industries. The study was then carried out as a cooperative research effort (2).

THE UNRAVELING CONSENSUS

In treating the questions of innovation policy as warranting detailed empirical exploration, we were acknowledging, reluctantly, that the general theoretical analyses and statistical observations of economists provide only limited and incomplete guidance for policy. We are not alone in this perception; the most significant aspect of the recent economic literature on innovation is its progressive inconclusiveness about the appropriate role for government.

It was not always that way. Economic research a decade or more ago had settled on two closely related sets of propositions about industrial innovation. The first of these was that technological change is an important source of productivity growth and, simultaneously, that R & D expenditure is a principal determinant of technological advance. The implication drawn from this was that R & D spending is a kind of "control variable" through which one would affect macroeconomic productivity.

The second set of arguments derived from theoretical rather than statistical work. Economists during the 1950's and '60's developed models in which private firms possessed an inherent tendency to "underinvest" in R & D. There were several reasons for this unhappy circumstance: (i) R & D creates knowledge, and knowledge is a "public good" in the sense that the firm cannot fully appropriate to itself the benefits of the knowledge it creates (3); (ii) the payoff to R & D is uncertain, and risk-averse firms will therefore wish to do less of it than a

(risk-neutral) society would prefer (4); and (iii) the fragmented structure of certain industries militates against sufficient R & D spending, the firms being too small to undertake certain kinds of projects (5).

Taken together, these sets of arguments strongly supported the notion that the government's role lay in correcting a global problem of inadequate private R & D. But over the last decade the consensus surrounding this conclusion has essentially come undone.

First, and perhaps most important, the experience of the 1970's cast doubt on the presumed tight link between a nation's overall R & D spending and its rate of productivity growth. Although it is true that high rates of R & D spending attended the rapid growth of productivity in the United States, Western Europe, and Japan during the 1950's and '60's, spending for R & D continued to be high in most countries during the ubiquitous slowdown in productivity growth after 1973. Only in the United States and France was the slowdown presaged by a decrease in R & D expenditures, and in both those countries the decrease was almost exclusively in defense and space rather than in civilian areas. Moreover, recent studies of the differences in productivity growth among countries suggests that, even in the 1950's and '60's, the countries with the highest ratio of R & D spending to gross national product—the United States and Britain—had among the lowest rates of productivity, growth (6).

Part of the message seems to be that it matters where a country is relative to the frontier of technology and productivity. Countries not on the frontier can "play catch up" fairly easily without much R & D spending so long as their rates of physical investment are high. Countries nearer the frontier have to work harder for each percent increase in productivity. This begins to suggest that it is not necessarily to a country's great advantage to be alone on the frontier. There is evidence that, as before World War II, the United States is again benefiting from technological ideas developed elsewhere; and there may be much to recommend a world in which many countries share the technological frontier and therefore share a common economic environment in an interdependent way *(7)*.

At the theoretical level, many economists have also begun to believe that the relation between competition and innovative behavior is more than a matter of some tendency to "underinvest" (or even to "overinvest") in R & D. The relationships of innovation to information "externalities," to risk, and to market structure are increasingly seen to be subtle and complex *(8, 9)*.

In the first place, the implications for total R & D spending of imperfect appropriability are now understood to be less clear-cut than they once seemed. Economists are coming to realize that, in a world of patents and industrial secrecy, firms in some instances have an incentive to engage in duplicative or near-duplicative R & D in an effort to copy a rival's technology or "invent around" its patents. This at once calls into question the idea that firms necessarily engage in too little R & D; more important, it begins to focus our attention not on the level of R & D but on the types of R & D projects the industry engages in.

The economist's view of the "uncertainty" issues had taken a similar turn. Rather than focusing on the amount of R & D that uncertainty is likely to draw forth, economists are now recognizing that what uncertainty really demands is the exploration of a diverse set of approaches. This way of looking at things suggests that heavy commitments to any one approach are dangerous in the early stages of development of a technology and should be avoided until the uncertainties (both market uncertainties and technological ones) are significantly reduced.

What that suggests in turn is that the focus of the policy issue should shift from R & D levels to the portfolio of R & D projects an industry tends to generate. This entails turning our attention toward the incentive effects of policies and institutional structures and toward considerations of access, secrecy, and information flow. In practice, some kinds of R & D projects will tend to be "underfunded" and some to be "overfunded"; and a simple R & D subsidy is not the sort of policy such a situation demands.

THE CASE STUDY APPROACH

Although many economists are exploring new approaches to the theory of innovation and technical change, analyzing the effects of government policy on industrial innovation must still be seen as largely an empirical problem—which policies have worked, which have not, and why. This was the premise behind our seven industry case studies. One conclusion, which we develop below, is that what the government can do effectively differs from industry to industry. There are nevertheless certain features common to technological change in all industries that should be kept clearly in mind in designing government policies.

One theme that unites the history of technological change is the pervasiveness of uncertainty. Although it takes a form in (say) the pharmaceutical industry very different from that in the commercial aircraft industry, uncertainty seems nevertheless to be endemic. A quick reading of the case studies is enough to dash any supposition that technological change is somehow a cleanly plannable activity. In fact, it is an activity characterized as much by false starts, missed opportunities, and lucky breaks as by brilliant insights and clever strategic decisions. Only in hindsight does the right approach seem obvious; before the fact, it is far from clear which of a bewildering array of options will prove most fruitful or even feasible. Strange as it now seems to us, aviation experts were once divided on the relative merits of the turboprop and turbojet engines as power plants for the aircraft of the future; and the computer industry was by no means unanimous that transistors—or, later, integrated circuits—were to be the technology of the future. Policy must recognize uncertainty as a fact of life, and must not try to repress it or analyze it away.

A second and related universal theme is the importance of detailed knowledge of the technology, of its strengths and weaknesses, and of user needs in guiding the innovation process. In all the case studies, either the producer-provider or the user-demander played a major role in generating and screening technological advances. Whenever major innovation was attempted without access to their knowledge, the results were disastrous. This fact imposes severe constraints on what government can do effectively.

The implications become clear if we consider four general kinds of government support for civilian R & D: (i) programs attendant on government procurement or some other well-defined public objective; (ii) programs to support research on "generic" technologies—research in the gray area between basic scientific research and applied R & D; (iii) programs to support applied R & D in the service of well-defined clientele demands; (iv) programs that insist on "picking winners" in commercial applied R & D. Programs of each sort show up in several guises in the industry case studies.

PROCUREMENT-RELATED R & D

In three of the industries studied—aviation, computers, and semiconductors—the government was heavily involved as a user-demander of the technology. This kind of government involvement has two important policy implications.

The first has to do with the ability of the government to guide R & D effectively. In cases of government procurement for defense, space, or similar clearly defined public projects, the government is itself the user-demander. It thus has knowledge of its own needs and, usually, at least a modicum of expertise in the technology it proposes to use. Motivation and knowledge line up fairly well in such circumstances, and the government is frequently able to sponsor effective R & D on the relevant technology. To the extent that the technology can be easily transferred to commercial application, the result is the well-known "spillover" into civilian technology.

Second, a public belief in the legitimacy of the government's primary mission—defense, for example—smooths the political waters for any related program of government R & D support. In the semiconductor industry the Department of Defense and the National Aeronautics and Space Administration (NASA) played a crucial role. Both the transistor itself and the integrated circuit were first developed with private funds; but, in the latter case at least, military and space demand was certainly an important motivation, and once those innovations were clearly identified, government support of product and process engineering helped speed their advance. Similarly, the government funded much of the early research, and provided much of the early market, for the digital computer. Defense procurement and in recent years heavy government R & D funding have played a major role in the evolution of aircraft technology.

It is important to recognize that the efficacy of government procurement-related R & D depends on the knowledge-advantage that comes from the government's position as user and on the political legitimacy of its mission as justified on grounds other than spillover benefits. The conclusion thus does not extend to government procurement projects, the justification of which is the spillover itself or in which the procurement is intended to make a market for the technology. [The supersonic transport (SST) project remains the best case in point.]

Moreover, our case studies suggest that the potential for the generation of spillover by procurement-related government R & D support may be limited to the early stages of a technology's development, when government and civilian demands are not yet specialized. As a technology matures, the requirements of the government and the private sector normally diverge. This means not only that spillover diminishes but also that military and commercial R & D increasingly compete for resources. In the mature phases of a technology's development, spillover may be as much to the military from the commercial sector as the other way round.

GENERIC TECHNOLOGIES

When there is no recognized public-sector demand for a technology, the government's ability to fund R & D effectively and to guide the development of that technology is more limited. The government does not then have natural access to the sorts of information necessary to guide allocation, and may in fact be blocked from getting the information. Furthermore, the legitimizing effect of a public sector purpose is not there to protect a support program from strong political opposition.

Nonetheless, these problems may be attenuated if the government restricts its attentions to areas, such as so-called generic technology, that are a step or two removed from specific commercial application. The reason is that, at this "directed basic" level of research, the knowledge involved has a large public component: much of it is the sort of nonpatentable and nonspecific knowledge—broad design concepts, properties of materials, and testing concepts—that is generally shared among scientists and does not pose a strong threat to proprietary interests.

In a sense, such generic work falls in between the sorts of work that an academic researcher, pursuing fashionable questions within the bounds of a standard scientific field, would tackle and the kinds of result-oriented research that would interest most corporate R & D laboratories. Of course, some companies do support generic research, and the findings are very often treated as public rather than as proprietary. In many instances, the funding for such research comes at least in part from governmental sources. In either case, the keys to success seem to be, first, involving the relevant scientific and technical communities in the allocation process and, second, recognizing that research ought to be influenced both by the purely scientific disciplines and by those interested in applications; indeed, a tension between the pure and the applied is generally salutary.

Our case studies provide examples of generic research, some associated with government procurement (in aviation, computers, semiconductors) and some more commercially oriented work (certain aspects of agricultural and pharmaceutical science).

The agricultural sciences, viewed as a generic research system, seem to have defined and filled their niche appropriately. Such work fits in between the academic basic sciences (like chemistry and biology) and the applied R & D carried out in private firms and in the experiment stations (like the development of new seeds or fertilizers). Interests on both sides of the line pull and tug to influence the kinds of research that are done as well as to monitor its quality and efficacy. The biomedical research community is another example. Research here too is pulled by applied interests (the physicians) and tugged by scientists in the more basic fields. Interestingly, both the agricultural

665

and the biomedical sciences typically reside in university settings, but in separate professional schools rather than main-line departments.

Another similarity between agricultural science and biomedical research is the way funding allocation is carried out. Both disciplines take the majority of their support from the government, but the funding agencies keep their distance, allowing the allocation machinery to be manipulated by the research communities themselves.

The National Advisory Committee on Aeronautics (NACA), the forerunner of NASA, is another example of a generic research system. Here the setting was a freestanding organization, not a university. But NACA's research concerns were certainly generic—broad-gauged aviation problems rather than specific designs—and the relevant engineering societies played a significant role in guiding and monitoring work at NACA. After World War II, the military increasingly assigned to private contractors the sort of work NACA had carried out, a trend that both reflected and abetted the divergence of military from civilian aircraft technology.

Because generic research poses a diffuse rather than a visible threat to established competitive positions, it may be possible to mount such a program successfully in any industry. But the size of the gray area between the basic and the applied may vary greatly among industries. Of course, the extent of this generic range may itself be influenced by the presence of a government program: the public financing of R & D often proves contagious, luring business scientists into a wider communications network and increasing the public flavor of private work. This is certainly desirable to the extent that it does not diminish the private incentive or ability to seize upon new ideas and develop them for market.

The manner in which a generic program develops may also be of critical importance. The aborted Cooperative Automotive Research Program (CARP) and the Cooperative Generic Technologies Program (COGENT) of the last administration seem to fit the description of generic systems, yet neither attracted the enthusiasm of the industries it proposed to aid, perhaps because the initiative and the design of the programs came strictly from Washington, with little participation by industry.

CLIENTELE-ORIENTED APPLIED R & D

The most important characteristic of government support for basic and generic research is that it does not require government administrators to make decisions that involve considerations of profitability and commercial potential. A basic or generic research program seeks to advance scientific and technical knowledge, and the decisions involved require primarily scientific and technical knowledge. While acquiring this sort of knowledge is not a trivial matter, it can at least conceivably be marshaled by program administrators, especially since the relevant scientific community can often be enlisted to help guide allocation. Moreover, basic and generic research, which involves exploring widely applicable technological options, seldom poses a concentrated threat to proprietary interests.

When we move closer to the level of applied research, however, the problems of government involvement multiply. The knowledge involved is both specific and idiosyncratic in form and may be proprietary in character. This instantly puts the government administrator at an informational disadvantage vis-à-vis firms that have no incentive, and sometimes no ability, to transmit what they know to Washington. Moreover, it is difficult to maintain a political constituency for a program that poses visible threats to established competitive positions.

Now, there is one much-discussed example of a strikingly successful government program of applied R & D: the agricultural research system. As noted above, much of the research supported in this program has been basic and generic, but a sizable proportion has been extremely applied in character, focusing on particular objectives like better seed varieties or more effective pest control. The interesting question is: what special conditions have made this applied R & D program feasible and productive?

A crucial feature of the agricultural industry is that it is largely atomistic in form. The competition among farms is something near to the "perfect competition" described in economics textbooks rather than the more rivalrous kind that characterizes most manufacturing industries. For this reason, fellow competitors are seen as inherently less threatening in farming than in most other industries; technological knowledge is therefore far less proprietary, and there is a public cast to the results of even very applied R & D.

The federal-state system of agricultural experiment stations evolved in a way that took advantage of the market structure in agriculture, marshaling the support of the farmers and giving them an important position in the evaluation and selection of projects. Coupled with the regional nature of agricultural technology, this led to a system in which farmers see it as advantageous to them to advance even very specific technologies as quickly as possible. As a model for the administration of a government-supported applied R & D program, the agricultural system is quite instructive. It is highly decentralized, and specific resource allocation decisions are made at state and county levels. Those decisions respond with some sensitivity to the demands of two constituencies: farmers (given voice through state legislatures) and the agricultural science community.

In the language of the social scientist, we might call this a "captured" system, in much the same sense that transportation, communication, and other industries are said to have captured their regulators. Capture of this sort is not very often congruent with the general interest of consumers; but in the case of agriculture the system seems to have evolved in a salutary fashion.

The residential construction industry is also relatively atomistic in structure, and has therefore long been seen as conformable to the agricultural model. Yet several government efforts to spur housing R & D have not worked; more accurately, the housing industry beat back or cut back the government attempts to mount such programs.

There are probably many reasons for the nontransferability of the agricultural model to housing. Building is somewhat more rivalrous in character than farming, at least at the level of materials suppliers. More important, the atomistic home builder is very likely more conscious of a threat from housing innovation than is (or was) his atomistic counterpart in farming. Although agricultural innovation did in the long run lead to the demise of the small farmer by increasing the scale of farming operations, each small farmer could nonetheless see the benefit to him (in the short and medium run, at least) of improved farming methods. The builder may well be more aware that any exploitation of scale economies brought about by innovation would very likely redound to his disadvantage, posing a clear threat to the system of small local firms in which he operates. Another factor is that building codes—a very old form of "new social regulation"—are intractable and entrenched. And there is not the background of good basic science in housing that there is in agriculture.

Beyond that, however, it may have been crucial that the agricultural research system evolved slowly over time and was not constructed de novo or centrally designed. It may even be that its success derives critically from the particular path the system's development followed (for example, its growth out of what was essentially a training program for farmers) and particular historical circumstances (such as the characteristics of the 19th-century industry) that are not easily replicated.

PICKING WINNERS

Which brings us to the final approach to government R & D support—"picking winners" in commercial competition. Here the historical record seems, for a change, unequivocal. Unequivocally negative.

The SST project and Operation Breakthrough were two examples touched upon in our case studies. In both cases, the government did not attempt to create a framework in which scientific and user interests could guide allocation: rather, the federal agencies attempted to insert themselves directly into the business of developing particular technologies for a commercial market in which they had little or no procurement interest.

In the case of Operation Breakthrough, the Department of Housing and Urban Development had been neither a major builder nor a buyer of nonsubsidized housing. It had neither the technical expertise nor the market experience that commercial success demands. Similarly, the government was not in the business of making or buying commercial airplanes; those who were, the commercial aircraft companies, showed little interest in such a plane until the prospects for high subsidies appeared. Very few of the housing designs that came out of Operation Breakthrough have since had any commercial value; and the lesson from the British-French Concorde experience is that we are lucky we went no further than we did with plans for an American SST.

The lesson here is not specific to these cases; it is not that these particular government agencies lacked some necessary expertise that could in principle have been remedied by hiring a larger or better cadre of experts. The lesson is a general one, about the location of knowledge and the mechanism of its transmission in the R & D system.

European experience testifies to the generality of the lesson *(10)*. In many cases, government attempts to enter the business of commercial applied R & D led to (i) duplicating private efforts or (ii) subsidizing those efforts and thereby replacing private with public funds or (iii) investing in designs the private sector had long abandoned as unpromising. There is certainly an argument that the government can be more forward looking than a private firm, supporting projects that are unpromising today but may be promising tomorrow. But the most effective way to perform such a next-generation function is not by competing in the commercial marketplace but through research of a more generic sort.

LESSONS FOR POLICY

The conclusions of a comparative historical analysis can only be qualitative and judgmental. But perhaps the lesson of economic theory and political practice during the last couple of decades is precisely the importance of this sort of empirical analysis.

Our central conclusion might be summed up in one word: complexity. The wide diversity of technological and institutional details, of knowledge structures and incentive structures, among American industries recommends against an industrial policy to boost "industrial innovation" in some global sense in the hope of affecting macroeconomic problems. Broad-brush measures like tax policy, antitrust policy, and patent policy, which affect each of the various industries in a very different way, should be assessed on their own merits and should not be viewed as "control variables" to stimulate innovation.

We do not propose a general rationale or justification for active government support of R & D. Applying the lessons of history to create programs that are both politically viable and socially desirable is no straightforward task. But the historical experience we have examined reveals three approaches that have worked in the past: support associated with government procurement or some other well-defined public sector objective; support of defined nonproprietary research, with allocation funds guided by the appropriate scientific community; and provision of an institutional structure that allows potential users to guide the allocation of applied R & D funds.

A fourth kind of policy, whereby government officials themselves try to identify projects that will be winners in a commercial market competition, is always seductive, but the evidence, from our studies and others, suggests that such strategy is to be avoided.

REFERENCES AND NOTES

1. R. R. Nelson, Ed., *Government and Technical Change: A Cross-Industry Analysis* (Pergamon, New York, 1982). The authors responsible for the industry studies are: agriculture. Robert Evenson, Yale University, automobiles, Lawrence J. White, New York University; commercial aviation, David Mowery and Nathan Rosenberg, Stanford University; computers, Barbara Katz, New York University, and Almarin Phillips, University of Pennsylvania; housing, John Quigley, University of California, Berkeley; pharmaceuticals, Henry Grabowski and John Vernon, Duke University; and semiconductors, Richard Levin, Yale University. The study was supported by grant NSG7636 from the National Aeronautics and Space Administration.

2. Each author (or pair of authors) was responsible for an individual chapter, but the group agreed on a common format for the cases, and it met several times (sometimes with other scholars and representatives from government and industry) to discuss and integrate the chapters. R. Nelson was responsible for a synthetic "cross-cutting" chapter, which was in turn discussed and commented upon by the group.

3. The most influential of these models was by Kenneth Arrow [K. Arrow, "The allocation of resources to invention," in *The Rate and Direction of Inventive Activity: Economic and Social Factors*, R. R. Nelson, Ed. (Princeton Univ. Press, Princeton, N.J., 1962)]. But see also H. Demsetz, "Information and efficiency: Another view," *J. Law Econ.* **12**, 1 (1969).

4. K. Arrow and R. Lind, "Uncertainty and the evaluation of public investment." *Am. Econ. Rev.* **60** (No. 3), 364 (1970).

5. The notion that "competitive"—that is, atomistic—firms innovate less than larger, "oligopolistic" firms has been called the Schumpeterian hypothesis, since those who read the work of Joseph Schumpeter through a certain set of spectacles discerned an argument of this sort in his *Capitalism, Socialism, and Democracy* (Harper, New York, 1942). The argument was stripped down to a more elemental form and popularized by John Kenneth Galbraith, especially in his *The New Industrial State* (Houghton-Mifflin, Boston, 1967).

6. Organization for Economic Cooperation and Development, *Technical Change and Economic Progress* (Organization for Economic Cooperation and Development, Paris, 1980). It may be significant that both the United States and Britain devoted an unusually high fraction of R & D to defense. But even when differences in defense R & D spending are accounted for, inter-country differentials in productivity growth seem better correlated with such matters as physical investment than with overall R & D spending.

7. N. Rosenberg, "U.S. technological leadership and foreign competition: 'De te fabula narratur?'" (photocopy, Stanford University, November 1981).

8. R. R. Nelson and S. Winter, "In search of a useful theory of innovation." *Res. Policy* **6**. 36 (1977).

9. R. R. Nelson, "Research on productivity growth and productivity differences: Dead ends and new departures." *J. Econ. Lit.* **19**. 1029 (1981).

10. See generally K. Pavitt and W. Walker, "Government policies toward industrial innovation: A review." *Res. Policy* **5** (No. 1) (1976).

Technology and Trade:
A Study of U.S. Competitiveness
in Seven Industries

N. Bruce Hannay

In the aggregate, the U.S. economy is less dependent on foreign trade than many other nations, but more and more U.S. industries are finding that they must compete internationally to survive. Despite the emerging competitive situation, international trade and relationships with other economies simply are not yet accorded the same importance in the United States as in Japan and Western Europe. It is essential that public policy take into account the international implications of any new initiatives, not only in policy formulation but also in administrative practice.

For some years after World War II, the leadership of the United States in the development and application of technology, and in world trade that stemmed from it, was unchallenged. At least in part, this resulted from the circumstances that left us as the only major industrialized country with a manufacturing capability that was intact at the end of the war. But, as other countries rebuilt their economies over the next two decades, they employed the best available technology in modern, efficient production facilities. With active participation by their governments, their industries developed along lines that not only took care of national needs, but also gave them an advantage in world trade competition in selected market areas. The United States actively encouraged and supported this rebuilding with both financial and technical assistance.

By the early 1970s, doubts about our competitiveness began to emerge, and they have been expressed with increasing urgency since then. Our productivity growth rate was significantly lower than that of other countries, and it was declining. Our industrial plant was aging. Our historically favorable trade balance first diminished, then turned negative. Our trade in technology-intensive products began to reflect a shrinking share of world markets. It was evident that we were not using technology as effectively as we might, or perhaps as well as

"Technology and Trade: A Study of U.S. Competitiveness in Seven Industries" by N. Bruce Hannay, from *The Competitive Status of U.S. Industry—An Overview,* edited by Lowell W. Steele and N. Bruce Hannay. Reprinted by permission of National Academy of Engineering.

some other countries were using it. Our innovative capacity was, for the first time, in doubt. A large part of our R&D was for defense and space, and without commercial objectives. Key industries, like steel, autos, and consumer electronics, were in deep trouble. Countries we had been assisting were becoming, to an ever greater extent, successful competitors in world trade.

The unanswered question then, and to a certain degree today also, was whether this trend reflected only an inevitable closing of the gap, as the war-damaged economies of Europe and Japan recovered, or whether it was a sign of inherent weakness in our own system, weakness that would eventually undermine our preeminent world position and even bring a loss of leadership. There were those who thought they saw such weakness in our educational system, or in industry's management practices, or in government policy and the relationship between the public and private sectors. Compounding the difficulty in understanding the implication of trade shifts has been the persistent imbalance in exchange rates.

When these concerns first emerged, the National Academy of Engineering (NAE) undertook what turned out to be a series of studies that examined central issues relating to technical and international economic and trade issues.[1-4] The studies were conducted by a committee, established in cooperation with the National Research Council, of experts from industry, academia, labor and government— scientists, engineers, economists, business and financial experts, labor representatives, and government specialists.

The first major study of the committee was a broad examination of the relationship among technology, trade, and U.S. competitiveness.[2] The purpose of the study was to reach an understanding of the issues and to determine priorities for the committee's future work. The main conclusion was that our national performance with respect to technological innovation, productivity improvement, and competitiveness in world trade was primarily determined by the health of the domestic economy and the constraints put on it, rather than by events outside the United States. Based on this examination the committee subsequently studied the effects of federal tax policy, regulation, and antitrust policy on technological innovation and recommended modification of those policies in ways that would be likely to increase our rate of technological change.[3] In certain areas, policies have since moved in ways consistent with those recommendations, to some extent at least.

The conclusions reached in these studies tended to be generalizations. Even though federal policies have a unique impact on each industrial sector, very little research had been done to disaggregate industry in the analysis of governmental policies in the areas of technology and trade. This led the committee to embark on industry-by-industry sectoral studies, choosing seven industries that represented a broad spectrum of characteristics. Each of the sectors contributes significantly to employment and to the GNP; the industries range from high to low technology, from rapidly growing to mature, from capital-intensive to those that are not, from industrial to consumer products, and from industries dominated by large firms to those with many small companies. The industries selected were automobiles, electronics, pharmaceuticals, machine tools, fibers and textiles, steel, and commercial aircraft.[4]

The committee's belief in undertaking these studies was that by understanding specific industrial sectors better, public policies could be more effectively formulated. The study program was designed to identify global shifts in production and in trade, to relate shifts in international comparative industrial advantage to technological and other factors, and to assess the probable impact of public policy options on

671

the rate and nature of technological change and on the international competitiveness of the U.S. industry.

There follows a very brief review of some of the results of these sectoral studies. First, specific findings are presented, by sector, then some similarities and differences among the sectors are noted, and finally some general conclusions are drawn.

FINDINGS OF INDUSTRY STUDIES

Automobiles

At the time of the study, the automobile industry had recently undergone major changes, and there was great uncertainty as to the outcome of those changes. The U.S. automobile companies faced severe foreign competition at a time when they had to deal with restructuring their products in the face of great market uncertainty. Huge amounts of capital were required at a time when future profits were in doubt. It was unclear whether the prevailing situation was temporary, whether maturation of the industry was forcing a long-term shift to lower-cost foreign manufacture, or whether new technology and production practices would fundamentally alter the industry.

For some years prior to the mid-1970s, technological change had been incremental in the industry. Key competitive factors in the mass market were cost, styling, and a strong dealer network. Over the course of the last several years, however, the industry had once again introduced new technological concepts, including, for example, downsizing, new materials, electronic controls, and engine design.

The imposition of government regulations regarding safety, pollution, and energy efficiency also had a major impact on the indus-

try. These mandates claimed both resources and management attention at a time of competitive upheaval, high interest rates, inflation, and a sluggish economic growth.

The study also analyzed foreign competition in detail. In the early 1980s the Japanese automobile firms had an estimated landed cost advantage of $750 to $1,500 per small vehicle. This advantage reflected differences in labor rates, materials costs, and productivity, as well as the effect of exchange rates. Despite the popular image of Japanese superiority in advanced technology, the Japanese advantage was found to lie rather in management—management of technology, of operations, and of the work force—and in culture—attitudes toward work and both individual and corporate responsibility. Other major factors that contributed to the Japanese advantage were the reliability of parts suppliers' delivery schedules, elimination of downtime, drastic reduction of setup and rework time, and a job structure and workplace environment that placed responsibility for quality and output on the workers.

The Japanese situation contrasts starkly with the adversarial labor-management relationship in the U.S. industry. Planning and control of work have been performed by staff groups organizationally remote from the workplace, and supervision has emphasized the meeting of demanding production goals. This system does not inspire loyalty or commitment, and it fails to take advantage of the knowledge and experience of the work force.

The efficacy of available policy options in the automobile industry is strongly influenced by the scenario selected for future events. If, for example, the industry is mature, production is likely to continue to diffuse around the world; the U.S. industry is likely to be much smaller and to emphasize specialty manufacture, with less value added. Without permanent trade barriers, cost disadvantages on standard models would be unlikely to be reversible. If the difficulties are seen as transient, while the industry restructures its product line and manufacturing capacity, then temporary protective measures help. If a new period of technological innovation and performance-oriented competition is emerging, then U.S. management has an opportunity to re-establish leadership. Even so, the U.S. share of value added is likely to decline, especially on standard models.

Electronics

The United States was the unchallenged leader in the electronics industry for some years after World War II and has maintained its leadership in much, but not all, of the industry for four decades. A conspicuous loss of leadership to Japan has occurred in consumer electronics, and the United States now faces a major challenge from Japan in other areas of electronics.

The development of the industry in the United States differs from that in the rest of the world. Except for defense and space electronics, the government has had little involvement (although the Defense Department's VHSIC program is expected to have significant commercial fallout in semiconductors). The industry has been characterized by a few dominant, innovative, giant firms in telecommunications and in computers, and by many entrepreneurial firms that have excelled at innovating new products and developing new markets in semiconductors, components, subsystems and systems, and, more recently, in telecommunications and computers. In these smaller firms, the level of vertical integration is low. The consumer electronics industry, before it lost its markets to Japan, was dominated by old-line radio manufacturers.

In contrast to the U.S. situation, the involvement of foreign governments in supporting and guiding the industry is common. Electronics is regarded as vital to future economic growth, national security, and even a self-image of leadership or equality in the industrialized world. Most of the manufacture comes from large, broadly based, highly integrated companies.

Japan has advanced dramatically in electronics. Government support is extensive—for very large scale integration (VLSI), pattern recognition, artificial intelligence, and fifth-generation computers and supercomputers. The financial structure of Japanese companies and the financial environment in which they operate are very different from those of U.S. companies—in recent years the cost of capital has been little more than half the U.S. rate.

In the United States, R&D and capital costs in electronics are very high, because the technology is changing extremely rapidly. Volume must be sufficient to generate the needed resources for investing in technology, added capacity, and new equipment. Another industry problem is the serious shortage of the electronic engineers, computer scientists and engineers, software programmers, and technicians needed to maintain a strong competitive position.

The U.S. position in four key industry segments—semiconductors, computers, telecommunications, and consumer electronics—is summarized below:

1. In semiconductors, the United States retains a lead but is under serious challenge by the Japanese. Japanese trade and investment barriers severely restrict imports, but U.S. firms supply more than half of Europe's total needs. The U.S. industry is changing. New entrants are inhibited by high start-up costs. Systems manufacturers are integrating forward. Major foreign investments are being made in U.S. firms to acquire technology and market share. U.S. strength results from aggressive technology development and a strong equipment

industry. Japanese focus on narrow, high-volume markets, such as 16K and 64K RAMs, has enabled heavy penetration in those markets, but the United States is ahead in microprocessors and custom circuits. In the future its lead will depend on its success in resolving capital and human resource problems, in maintaining its present leadership in basic research and innovation, and in removing trade and investment barriers.

2. In computers, the United States retains a powerful position in mainframes, minicomputers, and microcomputers, in software, and in distribution and service. In hardware and standardized high-volume manufacture, it faces a severe challenge from Japan.

3. In telecommunications, government intervention plays a crucial role. The United States has a strong position in switching and transmission. The Japanese are challenging in optical transmission. Competition in digital technology is severe. The structure of the U.S. industry is undergoing profound change, the consequences of which are not yet clear. The ability of the restructured industry to continue to lead the world in basic advances in telecommunications science and in new telecommunications technology remains to be seen. Interestingly, both Japan and the United Kingdom are privatizing their telephone monopolies.

4. In consumer electronics, since the mid-1950s the United States has fallen from a position of dominance in market share and in pioneer technology and all but ceded position to Japan. Japanese firms have been aggressive in adopting integrated circuitry and in developing manufacturing techniques that cost and improve quality. A combination of long-term commitment to consumer markets and aggressive application of technology, aided by long-term

availability of capital, a well-trained work force, a protected home market, favorable exchange rates, and willingness to use discriminatory pricing, have propelled Japan to a position of world dominance.

Any consideration of possible steps to strengthen the competitive position of the U.S. electronics industry should address four issues: long-range research, capital formation, human resources, and international trade policies. The management of this industry has a record of innovative, flexible approaches to problems. It should be encouraged to continue its experimentation in cooperative programs, joint ventures, and the like. Antitrust policy must recognize the imperative of evolving to meet world competition. Tax and depreciation policies that recognize the large and rapid obsolescence of equipment are of critical importance.

Strengthening the academic base that produces needed skills, from technician to Ph.D. scientist and engineer, warrants high priority. Government support of basic research and increased cooperative industry-university programs should both be strengthened.

The high leverage associated with electronics leadership has led virtually all developed countries to undertake programs to foster a domestic industry. Aggressive pursuit of multilateral trade liberalization must receive persistent, high-priority attention.

Steel

The study of the steel industry concentrated on the integrated producers, who constitute 80 percent of capacity and face the greatest competitive difficulties.

The importance of the steel industry to the economy and national security is universally accepted, but it is far from clear what part of our needs should be supplied internally. The industry is no longer technologically progressive. Of 28 process advances currently under development, only 2—direct reduction and continuous casting, both well established technologically—are expected to achieve significant adoption in 5 years, and only 5 others are projected to be adopted in 20 years. It appears that capital limitations and the projected rate of return on the investment, rather than the proprietary nature of the technology, are the problems. In recent years some of the principal changes in process technology have been the result of investment in plant that utilizes technology developed many years earlier but not adopted previously. New alloys are introduced more rapidly than new production processes because the plant investment is much lower. For this reason, the specialty steel industry has fared better than the large producers.

Leadership in technology does not assure economic success, and technology by itself cannot solve the steel problem. In addition to the pricing and capacity policies of foreign competitors, such factors as labor productivity, the cost of raw materials, energy, and labor, and plant location in relation to markets play a powerful role.

In terms of delivered cost, which is the important criterion, the study estimated that most of the scrap-based producers and many current-practice integrated steel plants in the interior of the United States should be able to meet the full-cost delivered price of Japanese competitors. Ten percent or more of domestic capacity is estimated to be nonviable and a candidate for shutdown.

Long-term estimates of capacity and consumption suggest an overcapacity problem for many years. Developing and Eastern bloc countries account for most of the additions to capacity. Thus, the domestic industry can expect to face increased pressure from imports and worldwide potential for overproduction that will lead to lower prices. The problem of chronic overcapacity results from a number of reinforcing circumstances: (1) foreign government investment in capacity, irrespective of demand; (2) foreign subsidies that increase output and reduce the rate of plant closures; (3) protectionism in domestic markets; and (4) growth in the use of steel that lags the growth in GNP.

Any attempt to revitalize the industry must balance a number of complex and often contradictory factors: determining the minimum domestic capacity needed for national security, achieving the inevitable restructuring while protecting the interests of affected workers and communities, providing U.S. consumers with access to the lowest-cost steel available worldwide, ensuring free and fair global trade in steel, and recognizing the aspirations of developing countries. Irrespective of the policy changes implemented, the industry will encounter some permanent shrinkage and represent a declining fraction of world capacity, and no measures will make all of those involved—steel producers, steel workers, and consumers—better off.

Fibers, Textiles, and Apparel

Developing countries regard manufacture of fabrics and apparel as an important source of employment and export earnings. They have extensive government programs to encourage investment and growth, to establish favorable terms for exports, and to restrict imports. Conversely, U.S. policies have been designed to slow the decrease in employment resulting from import competition.

Each of the three segments of this textile complex—fibers and yarn, textiles, and apparel—stands in a somewhat different competitive position. Fibers and yarn are produced primarily by large, powerful concerns that are able to finance investment in technology development and new equipment. The United States enjoys a favorable position in both the technology and trade of these products. The industry is capital-intensive, and its technology diffuses rapidly. The technology for this industry is developed by the fiber producers, who emphasize new fibers first and then reductions in cost. Economies of scale and aggressive R&D have enabled the U.S. firms to maintain competitive leadership, and this leadership can be sustained.

Technological advance in textiles is concentrated in the equipment manufacturers; little R&D is conducted by the fabric and apparel producers. Advances in textile equipment diffuse rapidly around the world and enable the developing countries to upgrade their operations. The U.S. position in textile machinery has weakened dramatically. Imports represented 9 percent of shipments in 1963 and 50 percent in 1980, with West Germany and Switzerland accounting for over 60 percent. Many key technical advances are now being made overseas.

Technological advances in fabrics have emphasized improved productivity. Many of the major advances have been made abroad, but U.S. firms have adopted them, along with other international competitors. The United States now has a clear technological lead in nonwovens, but that lead is expected to narrow. In general, the United States has a strong favorable balance of trade, but the picture is highly variable.

Computer technology has had a substantial impact on apparel manufacture, but the industry is still labor-intensive; economies of scale are not an important driving force. Smaller firms have been particularly hard hit by changes in technology and competition. Most have lacked the expertise, capital, and vision to take advantage of foreign market opportunities and to establish lower-cost foreign manufacturing facilities. As a consequence, these firms have been under severe competitive and financial pressure, and many have disappeared.

Japanese firms have responded to changing international competition by following the shifting of comparative advantage to developing countries. By a combination of establishing local facilities and partnerships, licensing, loans, and intricate purchasing and selling arrangements, the Japanese have played an active role in the emerging textile complex in the Asian-Pacific area.

The textile complex in the United States faces a shortage of technical workers and managers at all levels in comparison with what is needed to sustain a strong competitive position. The levels of compensation and limited attractiveness of careers in the industry contribute to the problem. Aggressive pursuit of technological developments and improved competence in international business are critical to success.

The need for developing countries to increase their exports and the high future growth in developing-country markets create political and diplomatic dilemmas that complicate any attempts to change the international framework of trade.

Trade in textiles is subject to destabilizing surges. Mechanisms and resources for more rapid response to sudden changes could be helpful. Even though tariff barriers are substantial, nontariff barriers, such as customs clearance, inspection, and local-content requirements are often greater deterrents and more difficult to identify. Present restrictions on offshore processing of some stages of manufacture reduce the flexibility of producers to achieve lowest cost and thus diminish the U.S. international competitiveness.

Machine Tools

The very competitive machine tool industry is highly fragmented with many small, independent, family-owned firms. The industry is relatively small. Nevertheless, it is of key strategic importance both to national security and to international economic competitiveness. Continued improvements in productivity are critically dependent on a healthy, technologically advanced machine tool industry. The availability of sophisticated but inexpensive new electronic devices, especially microprocessors and sensors, is opening up major opportunities for automation of production equipment.

A number of indications of the declining health of the U.S. industry have emerged. The U.S. share of world exports dropped from 23 percent in 1964 to 7 percent in 1980, while imports increased from 4 percent to 24 percent of domestic consumption. The United States now has a negative balance of trade in machine tools, and it is worsening. The major problem facing the industry is the traditional one of extreme cyclicality. The severe swings in volume reduce the investment attractiveness of the industry and lead to undercapitalization. This, in turn, severely impedes the upgrading of facilities and introduction of new technology. The same conditions have led to a persistent inability of the industry to attract skilled manpower at all levels—tool and die makers, industrial engineers, software programmers, and general management. Employment uncertainty has deterred entry. College courses pertaining to manufacturing technology generally have not been highly regarded by students, and careers have had a low appeal. There has been little government funding of manufacturing research. Only very recently has attention in the United States been given to these circumstances, which are in sharp contrast to the situation in both Europe and Japan.

The reduction in exports takes on added significance because export sales can help alleviate the extreme swings of the domestic market. Export sales also provide additional revenue to help defray the cost of developing and introducing new products and new manufacturing technology.

In general, the technology of American machine tools as products is roughly comparable to that of other nations, although American products are regarded as behind competition in the use of electronic controls and the associated software. Also, U.S. manufacturing technology employing machine tools is behind in moving to flexible, computer-integrated manufacture and in applying numerical controls, both of which were first introduced in the United States. Given the decline in market share and unattractive financial performance, the prospects are uncertain that the industry will exploit technology to the extent necessary to attain a competitive edge.

Small, family-owned U.S. firms are poorly equipped to pursue international sales. Moreover, the loan criteria of the Export-Import Bank focus on transactions that are much larger than typical machine tool sales.

The long-term viability of the U.S. industry will be strongly influenced by the growth and vitality of the U.S. economy, the level and stability of interest rates, and the development of a coherent export policy. Measures aimed more directly at supporting exports by small business would be especially useful to the industry, as would increased attention to development of applicable human resource skills and of advanced manufacturing technology. Changes in management with respect to pursuit of exports, investment in new technology, human resource development, and closer ties with customers are also called for.

Pharmaceuticals

The profitability, excellent growth, and dramatic technical advances of the U.S. pharmaceutical industry have tended to obscure the pronounced deterioration in relative performance of U.S. pharmaceutical firms vis-à-vis their foreign counterparts. In part, this unnoticed deterioration results from the long time lag—as much as 20 years—between decisions to invest in discovery of new drugs and a perceptible impact of any drugs discovered on the sales and profitability of the firm. In addition, the general and very rapid advance in the basic sciences of human health is generating sales growth worldwide, and this makes the U.S. industry appear to be growing, innovative, and profitable.

Thus, the relative performance of pharmaceutical firms vis-à-vis other sections of the U.S. economy looks favorable despite the relative decline internationally.

As evidence of our deteriorating relative position, in roughly the past two decades the U.S. share of world pharmaceutical R&D expenditures has fallen by one-half, as have the number of new U.S.-owned drugs entering clinical trial, the U.S. share of world production, and the U.S. share of world exports. Foreign firms now market their innovations directly in the United States. For their part, U.S. firms have invested widely in other countries and this is affecting the planning and conduct of their R&D.

The principal determinant of competitive success is the ability to introduce a continuing stream of commercially successful new products through technological innovation. However, the regulatory costs and delays imposed on U.S.-based R&D are significantly higher than elsewhere. The costs associated with the development and introduction of new drugs have become so large that access to the sales volume available from worldwide markets has become a critical consideration in determining competitive viability. Thus, the deterioration of the U.S. share of world markets is cause for concern. Small firms are in an especially precarious position because they lack the financial resources to develop new drugs and clear them through the regulatory agency.

Changes in government regulations and in the regulatory climate with respect to R&D, introduction of new therapeutic agents, export of experimental drugs, and acceptance of foreign data could have high leverage on competitive position. The process is at present subject to intense political pressure, requires massive amounts of documentation, and tends to delay clinical trial, even under carefully controlled conditions. Reforms that clarify and expedite the Food and Drug Administration's (FDA's) new drug-approval process, by providing a more significant role for experts from outside the FDA and by encouraging a more productive dialogue with industry, could significantly reduce the cost and time required to introduce new drugs. Also, the U.S. effort to deter economic concentration can limit the merging of firms that are not large enough themselves to be viable in global competition.

The lengthy time required to obtain FDA approval eliminates nearly half of the intended 17-year protection granted by a patent. This led the NAE committee to recommend restoring patent life consumed by regulatory review to increase the incentive for innovation. Very recently, the government has taken a step of this kind.

Civil Aviation Manufacturing

The civil aviation industry, including both manufacturers and the commercial airlines, is in the midst of profound change. Some features of the change result from domestic actions and circumstances, for example, economic deregulation of air transport, while others result from external trends and events, such as the emerging foreign competition in commercial transports, civil helicopters, and business aircraft.

No other industry plays as crucial a role as aviation in national security, national economic health, and foreign trade. Civil aircraft manufacture provides both the base load for key design and production teams and a huge (15,000 firms) production infrastructure in a high state of readiness for national defense. Export of aircraft continues to be the largest single source of revenue from trade in manufactured goods (and second only to agriculture overall). After dominating world markets since the end of World War II, the U.S. aircraft industry now faces a significantly more challenging competitive environment. Among the factors worthy of special notice are the following:

1. Due to a combination of deep recession and economic deregulation, the financial performance of domestic airlines has deteriorated drastically, and continuation of the airlines' traditional role in launching new aircraft is uncertain.

2. Aircraft manufacture is recognized as an attractive industry worldwide. After decades of persistence the Europeans, through Airbus Industries, have demonstrated commercial success. The Japanese have targeted aircraft as a growth industry of the future. Many smaller countries are mounting programs in helicopters and small aircraft. These foreign competitors enjoy a special supportive relationship with their governments that gives them access to sources of financing for developing, production, and sale of aircraft that are not available to a private firm in the United States.

3. Air travel in the United States is projected to grow less rapidly than in foreign markets. Thus, export sales and product planning for export markets will become increasingly important.

4. Countries are demanding a participative role in manufacture as the price of entry into their markets. The manufacturers seek to spread risks and to develop additional capital. Thus, aircraft manufacture is becoming increasingly internationalized.

5. Because of the industry's close connection with national security, the U.S. government plays a determining role in controlling aircraft exports. The task of balancing national security and commercial interests is becoming increasingly complex and controversial.

6. The technology underlying the design and manufacture of aircraft and engines offers major opportunities for improved performance, economy, and reliability. The United States has leadership or parity in all the key technologies. However, the margin of leadership has narrowed, and competitors have the capability to equal or even surpass us if U.S. effort loses momentum either in R&D or in its application to new aircraft. Since trade in aircraft is dominated by foreign governmental actions that apply economic and social criteria not possible for a privately owned company, trade negotiations become central to competitive success. In the competitive environment that is emerging, the traditional U.S. approach of seeking to create discipline in the rules of international trade faces serious handicaps, unless it is pursued more vigorously with respect to (a) negotiation of agreements that prohibit trade-distorting practices, (b) inclusion in the agreements of all countries competing in aircraft markets for all classes of planes, and (c) provision of adequate response mechanisms and deterrents to violators.

In considering policy initiatives for the future, the following areas warrant special attention:

·   Ensuring pursuit of U.S. interests in trade agreements and in mechanisms for timely, effective response to predatory practices.
·   Modifying lending practices of the Export-Import Bank to ensure that its terms and conditions are adequate to meet the behavior of competitors.
·   Preserving momentum in research and technology development.
·   Examining more broadly the trade-offs between technology transfer and the impact of export restrictions on the U.S. competitive position.
·   Ensuring maximum synergy between national defense and commercial interests in the development, design, and production of aircraft.

SECTORAL SIMILARITIES AND DIFFERENCES

What similarities and differences appear among these various industries? What lessons can be drawn with respect to public policy, management practices, and academic programs and priorities?

Need for a Global Perspective

The most dramatic common theme that emerges is that, despite the disparate nature of these various industries, all must now be termed world-scale industries. They must be managed in that context, and public policy must reflect the reality of growing and more pervasive international competition. For some industries, escalating costs of R&D, combined with burgeoning demands for large quantities of capital to obtain modern facilities, necessitate tapping global markets in order to generate the needed sales volume. For others, decisions regarding capacity expansion and future demand must be made in a global context; otherwise, serious errors—either in creating overcapacity or in lacking capacity to serve growth—are almost inevitable. Thus, even though, in the aggregate, the U.S. economy is less dependent on foreign trade than many other nations, more and more U.S. industries are finding that they must compete internationally in order to survive. Moreover, in most of the industries studied, foreign markets will be growing more rapidly than domestic ones.

Despite the emerging competitive situation, international trade and relationships with other economies simply are not yet accorded the same importance in the United States as in Japan and Western Europe. The sheer size and vitality of the U.S. market, combined with the size and richness of the land mass of the United States, make foreign trade and relations with other countries seem remote and relatively unimportant.

The low level of proficiency in foreign languages and the limited knowledge of foreign cultures and customs provided by our educational system are an unmistakable indicator of the limited importance attached to international trade. Government policies—antitrust, regulatory, tax, trade, and many others—mostly reflect concern with domestic issues, and there is little regard for their effect on U.S. competitiveness in international markets.

This same situation is mirrored in U.S. executive development programs, which usually put little weight on international experience, and in the U.S. approach to product planning. Most U.S. firms develop products for the American market and then offer them, more or less as an afterthought, for export. Consumer products reflect American tastes and standards of living. Industrial products are built to U.S. standards and reflect U.S. trade-offs among the costs of labor, capital, and energy. In contrast, the Japanese work diligently and remarkably effectively to achieve congruence between the requirements of domestic and export markets.

The recurring hostility between government and industry on market matters, and the bureaucratic tangles that ensnarl licensing, certification, approvals, and so on, also reflect the low importance attached to trade, as do the limited resources made available to support trade negotiations, to administer customs regulations, and to collect and analyze trade and economic data and information on foreign technology.

Control of technology transfer, while legitimate for national security, has not been consistent and imposes delays that call into question the reliability of U.S. shippers. In addition, the control is sometimes imposed without sufficient perspective on the availability to foreign customers of alternatives that could negate the results the United States seeks and without adequate consideration of the negative impact on the competitive status of U.S. firms.

The value-added tax (VAT) widely used in Europe has a built-in bias that supports exports as compared with our corporate income tax. In the United States continual and extensive education and persuasion are required to preserve critical financial supports, such as DISC and Export-Import Bank loans, while, in contrast, foreign government representatives are frequently virtual partners in the negotiations for large transactions and provide visible evidence of their government's support for the transaction.

Inconsistency of Policies, Institutions, and Priorities

A second major common theme was the lack of coherence and mutual reinforcement among policies and institutions and the lack of consistency in setting priorities that one generally finds in the United States. This contrasts with the situation in Japan and, to some extent, in West Germany. In those countries monetary, fiscal, export, and tax policies, the educational system, capital markets, and industrial management and labor relations have a consistency and coherence of purpose that we lack. This is not to suggest that we should adopt their ways, as the pluralism of our society and our institutions and a government based on checks and balances have both served us well. Nevertheless, it is imperative for us to scrutinize our own strengths and weaknesses with a realistic eye in the light of both the growing importance of international trade and the strengthened competition we face.

Small Firms Handicapped

Another sectoral similarity is the handicap of small firms in pursuing international sales. Some important industries, such as machine tools and textiles, are characterized by small firms, and in electronics small firms are prominent. The foreign-language deficiency noted above is one impediment. Inadequate knowledge of foreign markets and foreign business and legal practices is another. Many banks, especially those outside the major coastal cities, have no experience in international finance to aid local businessmen. The NAE studies pointed out that appropriate help for small business is thought to be lacking at government agencies. The priorities and lending practices of the Export-Import Bank in the past have been directed very heavily toward large transactions, which virtually exclude smaller firms; fortunately, that situation is now changing.

High Cost of Capital

The cost and availability of capital for U.S. companies (particularly in electronics and steel), as compared with foreign competitors, and the projected rate of return on investments increasingly threaten the ability of important U.S. industries to invest the capital necessary to remain competitive in international markets. In part, the difference in the cost of capital reflects special foreign government tax programs, as well as direct subsidies, to foster exports. In part, it also includes general economic considerations, such as the rate of saving, interest rates, taxes, rate of inflation, and monetary policy. But the problem of projected rates of return goes farther. It involves problems of highly cyclical industries (such as steel and machine tools), inadequate rate of return in critical industries, short time frames for evaluating investment by U.S. managers and U.S. investors, and volatility in flows of capital. This last subject is particularly important because of its impact on the time horizon for planning investment—the pursuit of higher return on investment, in principle, leads to greater efficiency in the allocation of resources, but it may lead to shortened time horizons and risk-aversion in investment decisions.

Role of Developing Countries

The developing countries are becoming increasingly important in the competitive equation. In some industries—steel, and fibers and textiles—their impact is evident through increases in capacity. In other industries—autos, pharmaceuticals, and aircraft—their impact is a combination of the growing importance of their markets, due to the more rapid growth of these less mature economies, and their insistence on domestic content as the price of market access.

Shortage of Trained Personnel

In several sectors—electronics, machine tools, and textiles—there is a widespread shortage of trained people at various levels, from shop floor to management. The shortage applies to specific technical skills in electronics, computers, software, and machine design. Two other broad personnel categories of special importance are people trained in international business, with direct experience in foreign commerce, and people trained in sophisticated manufacturing. The latter reflects the low status of manufacturing in the United States, in education and as a career, a situation that is in striking contrast to that in Japan, West Germany, and elsewhere.

The shortage of specific technical skills is particularly acute in electronics. Our production of electronic and other engineers is relatively much lower than that of our major international competitors, especially Japan and West Germany. Moreover, the shortage of faculty in electronic engineering and computer science, resulting from the competition with industry for these specialists, is serious. The escalating cost of modern equipment and the high proportion of foreign students are also important elements in our inability to provide sufficient numbers of well-educated professionals in these fields.

This problem appears not to be severe in the automobile and steel industries, because they have been undergoing major retrenchment, or in pharmaceuticals, because of the massive government support of university research in the life sciences. On the other hand, both machine tools and textiles suffer because they are not viewed as glamorous, high-growth industries. In aircraft the principal concerns involve ameliorating the effects of extreme cyclicality and holding skilled design and production teams together during troughs in volume.

Role of New Technology

The prospects for the development of new technology are very bright in most of the industries studied. This is particularly true of electronics, aircraft, pharmaceuticals, and machine tools, and the opportunity is there in autos and in fibers and textiles. Only in steel was there any real doubt about the possibility that new technology could produce a comparative advantage for U.S. industry. These observations point to the importance of strengthening our national capability in the development of new technology through measures ranging from the reinforcement of our science and engineering base in the universities to tax incentives to industry for R&D and for investment in new plants.

Some additional common themes were noted with respect to conditions needed for maintenance of technological leadership. The close tie between technological leadership and financial performance, including the ability to obtain capital, was noted particularly in electronics, pharmaceuticals, machine tools, and textiles. Similarly, the requirement for large, well-funded R&D programs for maintenance of the technological leadership needed to achieve and sustain a strong competitive advantage was evident in the industries—electronics, pharmaceuticals, and aircraft—that are experiencing the most rapid technological progress. The competitive leverage obtainable from technology is also very important in machine tools, but in steel, autos, and textiles it is diluted by such factors as labor and energy costs, the cost of raw materials, and government trade policies. In no case was technological leadership, by itself, an adequate basis for competitive success. Adequate technology is a necessary but not a sufficient basis for success.

Other Common Themes

Several other common themes were apparent. managerial skills and practices were highlighted as critical factors in automobiles, machine tools, and textiles. Deficiencies were noted in the U.S. capability to pursue international business and to achieve high productivity and high quality in mass-production industries. In machine tools the tradition of independence in pursuing technological development and relatively limited interactions with customers were noted as special management problems.

Three industries, automobiles, textiles, and steel, are projected to experience permanent decline from earlier levels of output, irrespective of public policy. In these cases, policy initiatives should include consideration of needed restructuring and ameliorating the disruption caused by the transition to a new, sustainable level of operations.

POLICY IMPLICATIONS AND CONCLUSIONS

The original premise on which the study of the seven industries was based was borne out by the studies themselves; namely, no two industries are alike in their patterns of technological development, in the problems they must solve in order to remain healthy and competitive in international markets, and in the public policies that would be most helpful to them in achieving their ends. Despite the differences among the industries, however, it is generally the case that important concerns go beyond the bounds of a single industry. Thus, there is ample opportunity for policy actions that would have broad, if not universal, effects on industry. Several conclusions with respect to policy actions of this kind follow.

The first general conclusion drawn from the studies is that government policy must be based on a substantially more informed view of the characteristics, needs, and prospects of individual industries than it has been to date. This is not to say that government policy should amount to nothing more than an accumulation of responses to perceived or claimed needs of every industry, but rather that it should be an enlightened policy in the sense that it recognizes that no single action can meet all needs, no simple "fixes" exist, and not all sectors can be equally satisfied by any policy. The studies do demonstrate that there are abundant opportunities for policy changes that would benefit important segments of industry, the U.S. economy, and the U.S. position in international technology competition. Some examples of such policy actions are changes in regulation in pharmaceuticals, support for exports from small machine-tool manufacturing firms, and steps to lower the cost of capital for the steel and electronics industries. In any efforts to strengthen government policies and actions, three overriding requirements must remain paramount: (1) the need for consistency over time in our approach to issues, (2) the need for persistent, visibly high-priority attention to international trade negotiations and the monitoring of the behavior of foreign competitors, including foreign governments (a particularly troublesome problem in aircraft), and (3) the need to establish a mutually reinforcing set of policies and actions relating to trade negotiations, monetary and fiscal policy, encouragement of capital formation, export support instruments, education, restructuring of industry, and so on.

Clearly, this implies a need for a continuing, coordinated review and awareness of technology and trade issues at a high enough level in the government that effective action can be taken. It is essential that public policy increasingly take into account the international implications, as well as the domestic effects, of any new initiatives, not only in policy formulation but also in administrative practice. While our system of government and the limits of our understanding of the dynamics of the economy certainly do not permit us to adopt a fully articulated "industrial policy" in the foreseeable future, a higher degree of coordination among the many separate policies and policymakers of our government is clearly called for. This same high degree of coordination is needed to provide the knowledge and information base to support policy formulation and administration.

A closely related general conclusion is that there must be continuing attention at the highest levels of government to the basic contributors to the country's economic health—education, science and technology, and a climate that is conducive to the industrial application of new technology being conspicuous among them. Thus, we must address such shortages as those in manufacturing engineering and in the supply of electronics engineers and computer scientists. We must give greater attention to the basic health of science and engineering in our universities. And, we must adopt policies that encourage capital formation and investment in new technology in the private sector.

Another general conclusion from the studies is that there is a challenge to U.S. management and labor. We have no monopoly on managerial competence, foresight, or competitive drive. Exogenous factors, such as a giant dynamic market, plentiful national resources, an educated, industrious work force, political stability, and private enterprise, have contributed historically to managerial success. They may also have delayed recognition of managerial weaknesses. These exogenous

strengths are no longer the dominant force they once were. Increased attention on the part of both large and small companies to world markets, to foreign competition, to foreign policies on trade and on technology, and to foreign managerial innovation is becoming critical to survival. As a corollary, increased public support for education in foreign languages and foreign cultures, as well as more rigorous standards for public literacy in science and technology, are other dimensions of the change that is needed. Unless the public comes to recognize the vital role that international trade plays in the nation's economic health and in the competitive viability of our own industries, the sustained support that is required for progress in other areas is unlikely to emerge.

The challenge to labor is to show that American labor can make contributions to productivity and to product quality that match those of our principal foreign competitors, especially the Japanese. On the positive side, U.S. labor has generally been more flexible in the acceptance of technological advances than have its counterparts in many European countries. As the pace of technological change increases, labor and management will need to work together creatively to develop mechanisms for ameliorating the disruptions brought about by technological change. Both management and labor will have to accept job retraining as a way of life.

A fourth general conclusion is that the government needs to give more sustained attention to the problems faced by small companies. Tax policy has alternately favored and discouraged venture companies—at present, it mostly favors them, but it is far from clear that this will last. Small companies often need special help from the government in competing in foreign markets, and for the most part this help has not been forthcoming. Small companies, individually, generally lack the expertise and experience to deal successfully with foreign regulations, procedures, and market systems and could benefit greatly from government assistance in those areas. Until recently the financial assistance available from the Export-Import Bank has been unavailable to smaller companies or for smaller transactions in all companies.

Finally, what would the committee now say about the concern that started it on its studies nearly a decade ago? Was it inevitable that U.S. preeminence in technology and trade would erode as Japan and Europe rebuilt their economies, or were there basic weaknesses in the U.S. system that were primarily responsible for our apparent loss of momentum?

As is usually the case, there are elements of truth in the opposite viewpoints. Certainly Europe and Japan found it possible to take advantage of existing technology and to use that as an important lever for the rapid economic growth they experienced and the gains they made relative to us after World War II. The readjustment process seems to have about run its course, and as far as Europe is concerned we may now be in balance. But Japan has gone beyond this and has emerged as our real economic competitor, and the outcome is not at all clear.

At the same time, we are no longer complacent. The very fact that we have recognized our previous inattention to our economic vitality has led us to at least some remedial policies and actions, although probably not enough. There are ample signs that we have not lost our ability to innovate, our productivity has turned up, and we are becoming more competitive. However, we have not yet fully met the Japanese challenge, and we will not until we give more serious national attention to issues of international trade and to its dependence on technology. Not only in the so-called high technology industries, but in the others as well, we need a continuing flow of new technology if we are to remain the world's economic leader. Technologists and economists are in essential agreement on the issues and on the directions in which the United States must move—the challenge to them is to make their voices heard.

NOTES

1. The studies were funded primarily by the National Science Foundation: portions were also sponsored by the U.S. Department of Commerce and the National Aeronautics and Space Administration. See notes 2, 3, and 4 for titles of specific studies.

2. National Research Council and National Academy of Engineering, Committee on Technology and International Economic and Trade Issues, *Technology, Trade, and the U.S. Economy* (Washington, D.C.: National Academy of Sciences, 1978).

3. National Research Council and National Academy of Engineering. Committee on Technology and International Economic and Trade Issues, *The Impact of Regulation on Industrial Innovation* (1979); *The Impact of Tax and Financial Regulatory Policies on Industrial Innovation* (1980): *Antitrust, Uncertainty, and Technological Innovation* (1980) (Washington, D.C.: National Academy of Sciences).

4. National Academy of Engineering, Committee on Technology and International Economic and Trade Issues, *The Competitive Status of the U.S. Auto Industry* (1982); *The Competitive Status of the U.S. Machine Tool Industry* (1983); *The Competitive Status of the U.S. Pharmaceutical Industry* (1983); *The Competitive Status of the U.S. Fibers, Textiles, and Apparel Complex* (1983); *The Competitive Status of the U.S. Electronics Industry* (1984); *The Competitive Status of the U.S. Civil Aviation Manufacturing Industry* (1985); *The Competitive Status of the U.S. Steel Industry* (1985). (Washington, D.C.: National Academy Press, 1982–1985).

SECTION
VIII

Executive Leadership and Management of Innovation and Change

Innovation always involves disruption. As innovation disturbs the status quo and is often associated with uncertainty, managing innovation involves managing conflict and the politics of stability versus change. The senior management team must mediate between inherent internal forces for stability versus external forces for change. This last chapter focuses on the management team's role in setting direction, creating a vision, and building organizations capable of both short-term efficiency and long-term innovation. Maidique and Hayes report on the characteristics of effective versus ineffective organizations, providing several ideas in managing the dual requirements for innovation and stability. Tushman, Newman and Romanelli provide evidence that highly effective organizations evolve through long periods of incremental change punctuated by discontinuous changes throughout the firm. Where low performing firms initiate system-wide changes in response to crisis, high performing organizations initiate sweeping changes to take advantage of environmental opportunities (often technological opportunities). As organization change is always a part of the innovation/adaptation process, Nadler and Tushman focus on managing strategic change, discussing generic factors that inhibit organization change and, in turn, specific actions in the management of politics, organization control, and individual resistance to change. Quinn's article also focuses on managing strategic change. Where the prior readings emphasize discontinuous change, Quinn argues for purposeful "muddling." Our final article again raises the issue of organization history and precedent. Managing innovation and change takes place in the context of a firm's unique history; managers must manage for today and tomorrow in the context of a firm's past. Smith and Wright discuss the need for innovation and change at ALCOA—which might conflict with ALCOA's distinguished past. Again, the managerial dilemma is to be able to build on an organization's traditions even as innovation and change take the firm to an uncertain future.

11

EXECUTIVE LEADERSHIP AND THE MANAGEMENT OF INNOVATION AND CHANGE

The Art of High-Technology Management

Modesto A. Maidique

Robert H. Hayes

The authors argue that, contrary to popular opinion, U.S. firms need not look overseas for models of successfully managed companies. Instead, many U.S. companies can benefit from using well-managed American high-tech firms as their guides. Through their studies of a wide range of high-technology firms, the authors identified those characteristics they believe make a company successful, and grouped them into six themes. Analysis of their findings has led them to conclude that well-managed companies have found ways to resolve a critical dilemma—the ability to manage the conflict between continuity and rapid change. Ed.

Over the past fifteen years, the world's perception of the competence of U.S. companies in managing technology has come full circle. In 1967, a Frenchman, J.-J. Servan-Schreiber, expressed with alarm in his book, *The American Challenge,* that U.S. technology was far ahead of the rest of the industrialized world.[1] This "technology gap," he argued, was continually widening because of the superior ability of Americans to organize and manage technological development.

Today, the situation is perceived to have changed drastically. The concern now is that

Reprinted from "The Art of High-Technology Management" by Modesto A. Maidique and Robert H. Hayes, *Sloan Management Review,* Winter 1984, pp. 17–31, by permission of the publisher. Copyright 1984 by the Sloan Management Review Association. All rights reserved.

the gap is reversing: the onslaught of Japanese and/or European challenges is threatening America's technological leadership. Even such informed Americans as Dr. Simon Ramo express great concern: In his book, *America's Technology Slip*, Dr. Ramo notes the apparent inability of U.S. companies to compete technologically with their foreign counterparts.[2] Moreover, in the best seller *The Art of Japanese Management*, the authors use as a basis of comparison two technology-based firms: Matsushita (Japanese) and ITT (American).[3] Here, the Japanese firm is depicted as a model for managers, while the management practices of the U.S. firm are sharply criticized.

Nevertheless, a number of U.S. companies appear to be fending off these foreign challenges successfully. These firms are repeatedly included on lists of "America's best-managed companies." Many of them are competitors in the R&D intensive industries, a sector of our economy that has come under particular criticism. Ironically, some of them have even served as models for highly successful Japanese and European high-tech firms.

For example, of the forty-three companies that Peters and Waterman, Jr., judged to be "excellent" in *In Search of Excellence*, almost half were classified as "high technology," or as containing a substantial high-technology component.[4] Similarly, of the five U.S. organizations that William Ouchi described as best prepared to meet the Japanese challenge, three (IBM, Hewlett-Packard, and Kodak) were high-technology companies.[5] Indeed, high-technology corporations are among the most admired firms in America. In a Fortune study that ranked the corporate reputation of the 200 largest U.S. corporations, IBM and Hewlett-Packard (HP) ranked first and second, respectively.[6] And of the top ten firms, nine complete in such high-technology fields as pharmaceuticals, precision instruments, communications, office equipment, computers, jet engines, and electronics.

The above studies reinforce our own findings, which have led us to conclude that U.S. high-technology firms that seek to improve their management practices to succeed against foreign competitors need not look overseas. The firms mentioned above are not unique. On the contrary, they are representative of scores of well-managed small and large U.S. technology-based firms. Moreover, the management practices they have adopted are widely applicable. Thus, perhaps the key to stimulating innovation in our country is not to adopt the managerial practices of the Europeans or the Japanese, but to adapt some of the policies of our *own* successful high-technology firms.

The Study

Over the past two decades, we have been privileged to work with a host of small and large high-technology firms as participants, advisors, and researchers. We and our assistants interviewed formally and informally over 250 executives, including over 30 CEOs, from a wide cross section of high-tech industries—biotechnology, semiconductors, computers, pharmaceuticals, and aerospace. About 100 of these executives were interviewed in 1983 as part of a large-scale study of product innovation in the electronics industry (which was conducted by one of this article's authors and his colleagues).[7] Our research has been guided by a fundamental question: what are the strategies, policies, practices, and decisions that result in successful management of high-technology enterprises? One of our principal findings was that no company has a monopoly on managerial excellence. Even the best run companies make big mistakes, and many smaller, lesser regarded companies are surprisingly sophisticated about the factors that mediate between success and failure.

It also became apparent from our interviews that the driving force behind the successes of many of these companies was strong leadership. All companies need leaders and visionaries, of course, but leadership is particularly essential when the future is blurry and when the world is changing rapidly. Although few high-tech firms can succeed for long without strong leaders, leadership itself is not the subject of this article. Rather, we accept it as given and seek to understand what strategies and management practices can *reinforce* strong leadership.

The companies we studied were of different sizes ($10 million to $30 billion in sales); their technologies were at different stages of maturity; their industry growth rates and product mixes were different; and their managers ranged widely in age. But they all had the same unifying thread: a rapid rate of change in the technological base of their products. This common thread, rapid technological change, implies novel products and functions and thus usually rapid growth. But even when growth is slow or moderate, the destruction of the old capital base by new technology results in the need for rapid redeployment of resources to cope with new product designs and new manufacturing processes. Thus, the two dominant characteristics of the high-technology organizations that we focused on were growth and change.

In part because of this split focus (growth and change), the companies we studied often appeared to display contradictory behavior over time. Despite these differences, in important respects, they were remarkably similar because they all confronted the same two-headed dilemma: how to unleash the creativity that promotes growth and change without being fragmented by it, and how to control innovation without stifling it. In dealing with this concern, they tended to adopt strikingly similar managerial approaches.

The Paradox: Continuity and Chaos

When we grouped our findings into general themes of success, a significant paradox gradually emerged—which is a product of the unique challenge that high-technology firms face. Some of the behavioral patterns that these companies displayed seemed to favor promoting disorder and informality, while others would have us conclude that it was consistency, continuity, integration, and order that were the keys to success. As we grappled with this apparent paradox, we came to realize that continued success in a high-technology environment requires periodic shifts between chaos and continuity.[8] Our originally static framework, therefore, was gradually replaced by a dynamic framework within whose ebbs and flows lay the secrets of success.

SIX THEMES OF SUCCESS

The six themes that we grouped our findings into were: (1) business focus; (2) adaptability; (3) organizational cohesion; (4) entrepreneurial culture; (5) sense of integrity; and (6) "hands-on" top management. No one firm exhibits excellence in every one of these categories at any one time, nor are the less successful firms totally lacking in all. Nonetheless, outstanding high-technology firms tend to score high in most of the six categories, while less successful ones usually score low in several.[9]

1. Business Focus

Even a superficial analysis of the most successful high-technology firms leads one to conclude that they are highly focused. With few exceptions, the leaders in high-technology fields, such as computers, aerospace, semiconductors, biotechnology, chemicals, pharmaceuticals, electronic instruments, and duplicating machines, realize the great bulk of their sales either from a single product line or from a closely related set of product lines.[10] For example, IBM, Boeing, Intel, and Genentech confine themselves almost entirely to computer products, commercial aircraft, integrated circuits,

691

and genetic engineering, respectively. Similarly, four-fifths of Kodak's and Xerox's sales come from photographic products and duplicating machines, respectively. In general, the smaller the company, the more highly focused it is. Tandon concentrates on disk drives; Tandem on high-reliability computers; Analog Devices on linear integrated circuits; and Cullinet on software products.

Closely Related Products.   This extraordinary concentration does not stop with the dominant product line. When the company grows and establishes a secondary product line, it is usually closely related to the first. Hewlett-Packard, for instance, has two product families, each of which accounts for about half of its sales. Both families—electronic instruments and data processors—are focused on the same technical, scientific, and process control markets. IBM also makes two closely related product lines—data processors (approximately 80 percent of sales) and office equipment—both of which emphasize the business market.

Companies that took the opposite path have not fared well. Two of yesterday's technological leaders, ITT and RCA, have paid dearly for diversifying away from their strengths. Today, both firms are trying to divest many of what were once highly touted acquisitions. As David Packard, chairman of the board of Hewlett-Packard, once observed, "No company ever died from starvation, but many have died from indigestion."[11]

A communications firm that became the world's largest conglomerate, ITT began to slip in the early 1970s after an acquisition wave orchestrated by Harold Geneen. When Geneen retired in 1977, his successors attempted to redress ITT's lackluster performance through a far-reaching divestment program.[12] So far, forty companies and other assets worth over $1 billion have been sold off—and ITT watchers believe the program is just getting started. Some analysts believe that

ITT will ultimately be restructured into three groups, with the communications/electronics group and engineered products (home of ITT semiconductors) forming the core of a "new" ITT.

RCA experienced a similar fate to ITT. When RCA's architect and longtime chairman, General David Sarnoff, retired in 1966, RCA was internationally respected for its pioneering work in television, electronic components, communications, and radar. But by 1980, the three CEOs who followed Sarnoff had turned a technological leader into a conglomerate with flat sales, declining earnings, and a $2.9 billion debt. This disappointing performance led RCA's new CEO, Thorton F. Bradshaw, to decide to return RCA to its high-technology origins.[13] Bradshaw's strategy is to now concentrate on RCA's traditional strengths—communications and entertainment—by divesting its other businesses.

Focused R&D.   Another policy that strengthens the focus of leading high-technology firms is concentrating R&D on one or two areas. Such a strategy enables these businesses to dominate the research, particularly the more risky, leading edge explorations. By spending a higher proportion of their sales dollars on R&D than their competitors do, or through their sheer size (as in the case of IBM, Kodak, and Xerox), such companies maintain their technological leadership. It is not unusual for a leading firm's R&D investment to be one and a half to two times the industry's average as a percent of sales (8–15 percent) and several times more than any individual competitor on an absolute basis.[14]

Moreover, their commitment to R&D is both enduring and consistent. It is maintained through slack periods and recessions because it is believed to be in the best, longterm interest of the stockholders. As the CEO of Analog Devices, a leading linear integrated circuit manufacturer, explained in a quarterly report which noted that profits had declined 30 percent, "We are sharply constraining the growth of fixed expenses, but we do not feel it is in the best interest of shareholders to cut back further on product development . . . in order to relieve short-term pressure on earnings."[15] Similarly, when sales, as a result of a recession, flattened and profit margins plummeted at Intel, its management invested a record-breaking $130 million in R&D, and another $150 million in plant and equipment.[16]

Consistent Priorities. Still, another way that a company demonstrates a strong business focus is through a set of priorities and a pattern of behavior that is continually reinforced by top management: for example, planned manufacturing improvement at Texas Instruments (TI); customer service at IBM; the concept of the entrepreneurial product champion at 3M; and the new products at HP. Belief in the competitive effectiveness of their chosen theme runs deep in each of these companies.

A business focus that is maintained over extended periods of time has fundamental consequences. By concentrating on what it does well, a company develops an intimate knowledge of its markets, competitors, technologies, employees, and of the future needs and opportunities of its customers.[17] The Stanford Innovation Project recently completed a three-year study of 224 U.S. high-technology products (half of which were successes, half of which were failures) and concluded that a continuous, in-depth, informal interaction with leading customers throughout the product development process was the principal factor behind successful new products. In short, this coupling is the cornerstone of effective high-technology progress. Such an interaction is greatly facilitated by the longstanding and close customer relationships that are fostered by concentrating on closely related product-market choices.[18] "Customer needs," explains Tom Jones, chairman of Northrop Corpora-

tion, "must be understood *way ahead of time*" (authors' emphasis).[19]

2. Adaptability

Successful firms balance a well-defined business focus with the willingness, and the will, to undertake major and rapid change when necessary. Concentration, in short, does not mean stagnation. Immobility is the most dangerous behavioral pattern a high-technology firm can develop: technology can change rapidly, and with it the markets and customers served. Therefore, a high-technology firm must be able to track and exploit the rapid shifts and twists in market boundaries as they are redefined by new technological, market, and competitive developments.

The cost of strategic stagnation can be great, as General Radio (GR) found out. Once the proud leader of the electronic instruments business, GR almost single-handedly created many sectors of the market. Its engineering excellence and its progressive human relations policies were models for the industry. But when its founder, Melville Eastham, retired in 1950, GR's strategy ossified. In the next two decades, the company failed to take advantage of two major opportunities for growth that were closely related to the company's strengths: microwave instruments and minicomputers. Meanwhile, its traditional product line withered away. Now all that remains of GR's once dominant instruments line, which is less than 10 percent of sales, is a small assembly area where a handful of technicians assemble batches of the old instruments.

It wasn't until William Thurston, in the wake of mounting losses, assumed the Presidency at the end of 1972 that GR began to refocus its engineering creativity and couple it to its new marketing strategies. Using the failure of the old policies as his mandate, Thurston deemphasized the aging product lines, focused GR's attention on automated test equipment, balanced its traditional engineering excellence with an increased sensitivity to market needs, and gave the firm a new name—GenRad. Since then, GenRad has resumed rapid growth and has won a leadership position in the automatic test equipment market.

The GenRad story is a classic example of a firm making a strategic change because it perceived that its existing strategy was not working. But even successful high-technology firms sometimes feel the need to be rejuvenated periodically to avoid technological stagnation. In the mid-1960s, for example, IBM appeared to have little reason for major change. The company had a near monopoly in the computer mainframe industry. Its two principal products—the 1401 at the low end of the market and the 7090 at the high end—accounted for over two-thirds of its industry's sales. Yet, in one move the company obsoleted both product lines (as well as others) and redefined the rules of competition for decades to come by simultaneously introducing six compatible models of the "System 360," based on proprietary hybrid integrated circuits.[21]

During the same period, GM, whose dominance of the U.S. auto industry approached IBM's dominance of the computer mainframe industry, stoutly resisted such a rejuvenation. Instead, it became more and more centralized and inflexible. Yet, GM was also once a high-technology company. In its early days when Alfred P. Sloan ran the company, engines were viewed as high-technology products. One day, Charles F. Kettering told Sloan he believed the high efficiency of the diesel engine could be engineered into a compact power plant. Sloan's response was: "Very well—we are now in the diesel engine business. You tell us how the engine should run, and I will . . . capitalize the program."[22] Two years later, Kettering achieved a major breakthrough in diesel technology. This paved the way for a revolution in the railroad industry and led to GM's preeminence in the diesel locomotive markets.

Organizational Flexibility. To undertake such wrenching shifts in direction requires both agility and daring. Organizational agility seems to be associated with organizational flexibility—frequent realignments of people and responsibilities as the firm attempts to maintain its balance on shifting competitive sands. The daring and the willingness to take "you bet your company" kind of risks is a product of both the inner confidence of its members and a powerful top management—one that either has effective shareholder control or the full support of its board.

3. Organizational Cohesion

The key to success for a high-tech firm is not simply periodic renewal. There must also be cooperation in the translation of new ideas into new products and processes. As Ken Fisher, the architect of Prime Computer's extraordinary growth, puts it, "If you have the driving function, the most important success factor is the ability to integrate. It's also the most difficult part of the task."[23]

To succeed, the energy and creativity of the whole organization must be tapped. Anything that restricts the flow of ideas, or undermines the trust, respect, and sense of a commonality of purpose among individuals is a potential danger. This is why high-tech firms fight so vigorously against the usual organizational accoutrements of seniority, rank, and functional specialization. Little attention is given to organizational charts: often they don't exist.

Younger people in a rapidly evolving technological field are often as good—and sometimes even better—a source of new ideas as are older ones. In some high-tech firms, in fact, the notion of a "halflife of knowledge" is used; that is, the amount of time that has to elapse before half of what one knows is obsolete. In semiconductor engineering, for example, it is estimated that the halflife of a newly minted Ph.D. is about seven years. Therefore, any practice that relegates younger engineers to secondary, nonpartnership roles is considered counterproductive.

Similarly, product design, marketing, and manufacturing personnel must collaborate in a common cause rather than compete with one another, as happens in many organizations. Any policies that appear to elevate one of these functions above the others—either in prestige or in rewards—can poison the atmosphere for collaboration and cooperation.

A source of division, and one which distracts the attention of people from the needs of the firm to their own aggrandizement, are the executive "perks" that are found in many mature organizations: pretentious job titles, separate dining rooms and restrooms for executives, larger and more luxurious offices (often separated in some way from the rest of the organization), and even separate or reserved places in the company parking lot all tend to establish "distance" between managers and doers and substitute artificial goals for the crucial real ones of creating successful new products and customers. The appearance of an executive dining room, in fact, is one of the clearest danger signals.

Good Communication. One way to combat the development of such distance is by making top executives more visible and accessible. IBM, for instance, has an open-door policy that encourages managers at different levels of the organization to talk to department heads and vice-presidents. According to senior IBM executives, it was not unusual for a project manager to drop in and talk to Frank Cary (IBM's chairman) or John Opel (IBM's president) until Cary's recent retirement. Likewise, an office with transparent walls and no door, such as that of John Young, CEO at HP, encourages communication. In fact, open-style offices are common in many high-tech firms.

A regular feature of 3M's management process is the monthly Technical Forum where technical staff members from the firm exchange views on their respective projects. This emphasis on communication is not restricted to internal operations. Such a firm supports and often sponsors industry-wide technical conferences, sabbaticals for staff members, and cooperative projects with technical universities.

Technical Forums serve to compensate partially for the loss of visibility that technologists usually experience when an organization becomes more complex and when production, marketing, and finance staffs swell. So does the concept of the dual-career ladder that is used in most of these firms; that is, a job hierarchy through which technical personnel can attain the status, compensation, and recognition that is accorded to a division general manager or a corporate vice-president. By using this strategy, companies try to retain the spirit of the early days of the industry when scientists played a dominant role, often even serving as members of the board of directors.[24]

Again, a strategic business focus contributes to organizational cohesion. Managers of firms that have a strong theme/culture and that concentrate on closely related markets and technologies generally display a sophisticated understanding of their businesses. Someone who understands where the firm is going and why is more likely to be willing to subordinate the interests of his or her own unit or function in the interest of promoting the common goal.

Job Rotation. A policy of conscious job rotation also facilitates this sense of communality. In the small firm, everyone is involved in everyone else's job: specialization tends to creep in as size increases and boundary lines between functions appear. If left unchecked, these boundaries can become rigid and impermeable. Rotating managers in temporary assignments across these boundaries helps keep the lines fluid and informal, however. When a new process is developed at TI, for example, the process developers are sent to the production unit where the process will be implemented. They are allowed to return to their usual posts only after that unit's operations manager is convinced that the process is working properly.

Integration of Roles.   Other ways that high-tech companies try to prevent organizational, and particularly hierarchical, barriers from rising is through multidisciplinary project teams, "special venture groups," and matrixlike organizational structures. Such structures, which require functional specialists and product/market managers to interact in a variety of relatively short-term problem-solving assignments, both inject a certain ambiguity into organizational relationships and require each individual to play a variety of organizational roles.

For example, AT&T uses a combination of organizational and physical mechanisms to promote integration. The Advanced Development sections of Bell Labs are physically located on the sites of the Western Electric plants. This location creates an organizational bond between Development and Bell's basic research and an equally important spatial bond between Development and the manufacturing engineering groups at the plants. In this way, communication is encouraged among Development and the other two groups.[25]

Long-term Employment.   Long-term employment and intensive training are also important integrative mechanisms. Managers and technologists are more likely to develop satisfactory working relationships if they know they will be harnessed to each other for a good part of their working lives. Moreover, their loyalty and commitment to the firm is increased if they know the firm is continuously investing in upgrading their capabilities.

At Tandem, technologists regularly train administrators on the performance and function of the firm's products and, in turn, administrators train the technologists on per-

sonnel policies and financial operations.[26] Such a firm also tends to select college graduates who have excellent academic records, which suggest self-discipline and stability, and then encourages them to stay with the firm for most, if not all, of their careers.

4. Entrepreneurial Culture

While continuously striving to pull the organization together, successful high-tech firms also display fierce activism in promoting internal agents of change. Indeed, it has long been recognized that one of the most important characteristics of a successful high-technology firm is an entrepreneurial culture.[27]

Indeed, the ease with which small entrepreneurial firms innovate has always inspired a mixture of puzzlement and jealousy in larger firms. When new ventures and small firms fail, they usually do so because of capital shortages and managerial errors.[28] Nonetheless, time and again they develop remarkably innovative products, processes, and services with a speed and efficiency that baffle the managers of large companies. The success of the Apple II, which created a new industry, and Genentech's genetically engineered insulin are of this genre. The explanation for a small entrepreneurial firm's innovativeness is straightforward, yet it is difficult for a large firm to replicate its spirit.

Entrepreneurial Characteristics.   First, the small firm is typically blessed with excellent communication. Its technical people are in continuous contact (and oftentimes in cramped quarters). They have lunch together, and they call each other outside of working hours. Thus, they come to understand and appreciate the difficulties and challenges facing one another. Sometimes they will change jobs or double up to break a critical bottleneck; often the same person plays multiple roles. This overlapping of responsibilities results in a second blessing: a dissolving of the classic organizational barriers that are major impediments to the innovating process. Third, key decisions can be made immediately by the people who first recognize a problem, not later by top management or by someone who barely understands the issue. Fourth, the concentration of power in the leader/entrepreneurs makes it possible to deploy the firm's resources very rapidly. Lastly, the small firm has access to multiple funding channels, from the family dentist to a formal public offering. In contrast, the manager of an R&D project in a large firm has effectively only one source, the "corporate bank."

Small Divisions. In order to recreate the entrepreneurial climate of the small firm, successful large high-technology firms often employ a variety of organizational devices and personnel policies. First, they divide and subdivide. Hewlett-Packard, for example, is subdivided into fifty divisions: the company has a policy of splitting divisions soon after they exceed 1,000 employees. Texas Instruments is subdivided into over thirty divisions and 250 "tactical action programs." Until recently, 3M's business was split into forty divisions. Although these divisions sometimes reach $100 million or more in sales, by Fortune 500 standards they are still relatively small companies.

Variety of Funding Channels. Second, such high-tech firms employ a variety of funding channels to encourage risk taking. At Texas Instruments managers have three distinct options in funding a new R&D project. If their proposal is rejected by the centralized Strategic Planning (OST) System because it is not expected to yield acceptable economic gains, they can seek a "Wild Hare Grant." The Wild Hare program was instituted by Patrick Haggerty, while he was TI's chairman, to insure that good ideas with long-term potential were not systematically turned down. Alternatively, if the project is outside the mainstream of the OST System managers or engineers can contact one of dozens of

individuals who hold "IDEA" grant purse strings and who can authorize up to $25,000 for prototype development. It was an IDEA grant that resulted in TI's highly successful "Speak and Spell" learning aid.

3M managers also have three choices: they can request funds from (1) their own division, (2) corporate R&D, or (3) the new ventures division.[29] This willingness to allow a variety of funding channels has an important consequence: it encourages the pursuit of alternative technological approaches, particularly during the early stages of a technology's development, when no one can be sure of the best course to follow.

IBM, for instance, has found that rebellion can be good business. Arthur K. Watson, the founder's son and a longtime senior manager, once described the way the disk memory, a core element of modern computers, was developed:

> [It was] not the logical outcome of a decision made by IBM management; [Because of budget difficulties] it was developed in one of our laboratories as a bootleg project. A handful of men . . . broke the rules. They risked their jobs to work on a project they believed in.[30]

At Northrop the head of aircraft design usually has at any one time several projects in progress without the awareness of top management. A lot can happen before the decision reaches even a couple of levels below the chairman. "We like it that way," explains Northrop Chairman Tom Jones.[31]

Tolerance of Failure. Moreover, the successful high-technology firms tend to be very tolerant of technological failure. "At HP," Bob Hungate, general manager of the Medical Supplies Division, explains, "it's understood that when you try something new you will sometimes fail."[32] Similarly, at 3M, those who fail to turn their pet project into a commercial success almost always get another chance. Richard Frankel, the president of the Kevex Corporation, a $20 million instrument manufacturer, puts it this way, "You need to encourage people to make mistakes. You have to let them fly in spite of aerodynamic limitations."[33]

Opportunity to Pursue Outside Projects.   Finally, these firms provide ample time to pursue speculative projects. Typically, as much as 20 percent of a productive scientist's or engineer's time is "unprogrammed," during which he or she is free to pursue interests that may not lie in the mainstream of the firm. IBM Technical Fellows are given up to five years to work on projects of their own choosing, from high speed memories to astronomy.

5. Sense of Integrity

While committed to individualism and entrepreneurship, at the same time successful high-tech firms tend to exhibit a commitment to long-term relationships. The firms view themselves as part of an enduring community that includes employees, stockholders, customers, suppliers, and local communities: their objective is to maintain stable associations with all of these interest groups.

Although these firms have clearcut business objectives, such as growth, profits, and market share, they consider them subordinate to higher order ethical values. Honesty, fairness, an openness—that is, integrity—are not to be sacrificed for short-term gain. Such companies don't knowingly promise what they can't deliver to customers, stockholders, or employees. They don't misrepresent company plans and performance. They tend to be tough but forthright competitors. As Herb Dwight—president of Spectra-Physics, one of the world's leading laser manufacturers—says, "The managers that succeed here go *out of their way* to be ethical."[34] And Alexander d'Arbeloff, cofounder and president of Teradyne, states

bluntly, "Integrity comes first. If you don't have that, nothing else matters."[35]

These policies may seem utopian, even puritanical, but in a high-tech firm they also make good business sense. Technological change can be dazzlingly rapid; therefore, uncertainty is high, risks are difficult to assess, and market opportunities and profits are hard to predict. It is almost impossible to get a complex product into production, for example, without solid trust between functions, between workers and managers, and between managers and stockholders (who must be willing to see the company through the possible dips in sales growth and earnings that often accompany major technological shifts). Without integrity the risks multiply and the probability of failure (in an already difficult enterprise) rises unacceptable. In such a context, Ray Stata, cofounder of the Massachusetts High Technology Council, states categorically, "You need an environment of mutual trust."[36]

This commitment to ethical values must start at the top, otherwise it is ineffective. Most of the CEOs we interviewed consider it to be a cardinal dimension of their role. As Bernie Gordon, president of Analogic explains, "The things that make leaders are their philosophy, ethics, and psychology."[37] Nowhere is this dimension more important than in dealing with the company's employees. Paul Rizzo, IBM's vice chairman, puts it this way, "At IBM we have a fundamental respect for the individual . . . people must be free to disagree and to be heard. Then, even if they lose, you can still marshall them behind you."[38]

Self-understanding.   This sense of integrity manifests itself in a second, not unrelated, way—self-understanding. The pride, almost arrogance, of these firms in their ability to compete in their chosen fields is tempered by a surprising acknowledgment of their limitations. One has only to read Hewlett-Packard's corporate objectives or interview one of its top managers to sense this extraordinary blend of strength and humility. Successful high-tech companies are able to reconcile their "dream" with what they can realistically achieve. This is one of the reasons why they are extremely reticent to diversify into unknown territories.

6. "Hands-on" Top Management

Notwithstanding their deep sense of respect and trust for individuals, CEOs of successful high-technology firms are usually actively involved in the innovation process to such an extent that they are sometimes accused of meddling. Tom McAvoy, Corning's president, sifts through hundreds of project proposals each year trying to identify those that can have a "significant strategic impact on the company"—the potential to restructure the company's business. Not surprisingly, most of these projects deal with new technologies. For one or two of the most salient ones, he adopts the role of "field general": he frequently visits the line operations, receives direct updates from those working on the project, and assures himself that the required resources are being provided.[39]

Such direct involvement of the top executive at Corning sounds more characteristic of vibrant entrepreneurial firms, such as Tandon, Activision, and Seagate, but Corning is far from unique. Similar patterns can be identified in many larger high-technology firms. Milt Greenberg, president of GCA, a $180 million semiconductor process equipment manufacturer, stated: "Sometimes you just have to short-circuit the organization to achieve major change."[40] Tom Watson, Jr. (IBM's chairman) and Vince Learson (IBM's president) were doing just that when they met with programmers and designers and other executives in Watson's ski cabin in Vermont to finalize software design concepts for the System 360—at a point in time when IBM was already a $4 billion firm."[41]

Good high-tech managers not only understand how organizations, and in particular engineers, work, they understand the fundamentals of their technology and can interact directly with their people about it. This does not imply that it is necessary for the senior managers of such firms to be technologists (although they usually are in the early stages of growth): neither Watson nor Learson were technical people. What appears to be more important is the ability to ask lots of questions, even "dumb" questions, and dogged patience in order to understand in-depth such core questions as: (1) how the technology works; (2) its limits, as well as its potential (together with the limits and potential of competitors' technologies); (3) what these various technologies require in terms of technical and economic resources; (4) the direction and speed of change; and (5) the available technological options, their cost, probability of failure, and potential benefits if they prove successful.

This depth of understanding is difficult enough to achieve for one set of related technologies and markets; it is virtually impossible for one person to master many different sets. This is another reason why business focus appears to be so important in high-tech firms. It matters little if one or more perceptive scientists or technologists forsees the impact of new technologies on the firm's markets, if its top management doesn't internalize these risks and make the major changes in organization and resource allocation that are usually necessitated by a technological transition.

THE PARADOX OF HIGH-TECHNOLOGY MANAGEMENT

The six themes around which we arranged our findings can be organized into two, apparently paradoxical groupings: business focus, organizational cohesion, and a sense of integrity fall into one group; adaptability, entrepreneurial culture, and hands-on management fall into the other group. On the one hand, business focus, organizational cohesion, and integrity imply stability and conservatism. On the other hand, adaptability, entrepreneurial culture, and hands-on top management are synonymous with rapid, sometimes precipitous change. The fundamental tension is between order and disorder. Half of the success factors pull in one direction; the other half tug the other way.

This paradox has frustrated many academicians who seek to identify rational processes and stable cause-effect relationships in high-tech firms and managers. Such relationships are not easily observable unless a certain constancy exists. But in most high-tech firms, the only constant is continual change. As one insightful student of the innovation process phrased it, "Advanced technology requires the collaboration of diverse professions and organizations, often with ambiguous or highly interdependent jurisdictions. In such situations, many of our highly touted rational management techniques break down."[42] One recent researcher, however, proposed a new model of the firm that attempts to rationalize the conflict between stability and change by splitting the strategic process into two loops, one that extends the past, the other that periodically attempts to break with it.[43]

Established organizations are, by their very nature, innovation resisting. By defining jobs and responsibilities and arranging them in serial reporting relationships, organizations encourage the performance of a restricted set of tasks in a programmed, predictable way. Not only do formal organizations resist innovation, they often act in ways that stamp it out. Overcoming such behavior—which is analogous to the way the human body mobilizes antibodies to attack foreign cells—is, therefore, a core job of high-tech management.

The Paradoxical Challenge. High-tech firms deal with this challenge in different ways. Texas Instruments, long renowned for the complex, interdependent matrix structure it used in managing dozens of product-customer centers (PCCs), recently consolidated groups of PCCs and made them into more autonomous units. "The manager of a PCC controls the resources and operations for his entire family . . . in the simplest terms, the PCC manager is to be an entrepreneur," explained Fred Bucy, TI's president.[44]

Meanwhile, a different trend is evident at 3M, where entrepreneurs have been given a free rein for decades. A recent major reorganization was designed to arrest snowballing diversity by concentrating its sprawling structure of autonomous divisions into four market groups. "We were becoming too fragmented," explained Vincent Ruane, vice-president of 3M's electronics division.[45]

Similarly, HP recently reorganized into five groups, each with its own strategic responsibilities. Although this simply changes some of its reporting relationships, it does give HP, for the first time, a means for integrating product and market development across generally autonomous units.[46]

These reorganizations do not mean that organizational integration is dead at Texas Instruments, or that 3M's and HP's entrepreneurial cultures are being dismantled. They signify first, that these firms recognize that both (organizational integration and entrepreneurial cultures) are important, and second, that periodic change is required for environmental adaptability. These three firms are demonstrating remarkable adaptability by reorganizing from a position of relative strength—not, as is far more common, in response to financial difficulties. As Lewis Lehr, 3M's president explained, "We can change now because we're not in trouble."[47]

Such reversals are essentially anti-bureaucratic, in the same spirit as Mao's admonition to "let a hundred flowers blossom and a hundred schools of thought contend."[48] At IBM, in 1963, Tom Watson, Jr., temporarily abolished the corporate management committee in an attempt to push decisions downward and thus facilitate the changes necessary for IBM's great leap forward to the System 360.[49] Disorder, slack, and ambiguity are necessary for innovation, since they provide the porosity that facilitates entrepreneurial behavior—just as do geographically separated, relatively autonomous organizational subunits.

But the corporate management committee is alive and well at IBM today. As it should be. The process of innovation, once begun, is both self-perpetuating and potentially self-destructive: although the top managers of high-tech firms must sometimes espouse organizational disorder, for the most part they must preserve order.

Winnowing Old Products. Not all new product ideas can be pursued. As Charles Ames, former president of Reliance Electric, states, "An enthusiastic inventor is a menace to practical businessmen."[50] Older products, upon which the current success of the firm was built, at some point have to be abandoned: just as the long-term success of the firm requires the planting and nurturing of new products, it also requires the conscious, even ruthless, pruning of other products so that the resources they consume can be used elsewhere.

This attitude demands hard-nosed managers who are continually managing the functional and divisional interfaces of their firms. They cannot be swayed by nostalgia, or by the fear of disappointing the many committed people who are involved in the development and production of discontinued products. They must also overcome the natural resistance of their subordinates, and even their peers, who often have a vested interest in the products that brought them early personal success in the organization.

Yet, firms also need a certain amount of continuity because major change often emerges from the accretion of a number of smaller, less visible improvements. Studies of petroleum refining, rayon, and rail transportation, for example, show that half or more of the productivity gains ultimately achieved within these technologies were the result of the accumulation of minor improvements.[51] Indeed, most engineers, managers, technologists, and manufacturing and marketing specialists work on what Thomas Kuhn might have called "normal innovation,"[52] the little steps that improve or extend existing product lines and processes.

Managing Ambivalently. The successful high-technology firm, then, must be managed ambivalently. A steady commitment to order and organization will produce one color Model T Fords. Continuous revolution will bar incremental productivity gains. Many companies have found that alternating periods of relaxation and control appear to meet this dual need. Surprisingly, such ambiguity does not necessarily lead to frustration and discontent.[53] In fact, interspersing periods of tension, action, and excitement with periods of reflection, evaluation, and revitalization is the same sort of irregular rhythm that characterizes many favorite pastimes—including sailing, which has been described as "long periods of total boredom punctuated with moments of stark terror."

Knowing when and where to change from one stance to the other, and having the power to make the shift, is the core of the art of high-technology management. James E. Webb, administrator of the National Aeronautics and Space Administration during the successful Apollo ("man on the moon") program, recalled that "we were required to fly our administrative machine in a turbulent environment, and . . . a certain level of *organizational instability was essential if NASA was not to lose control*" (authors' emphasis).[54]

In summary, the central dilemma of the

high-technology firm is that it must succeed in managing two conflicting trends: continuity and rapid change. There are two ways to resolve this dilemma. One is an old idea: managing different parts of the firm differently—some business units for innovation, others for efficiency.

A second way—a way which we believe is more powerful and pervasive—is to manage differently at different times in the evolutionary cycle of the firm. The successful high-technology firm *alternates* periods of consolidation and continuity with sharp reorientations that can lead to dramatic changes in the firm's strategies, structure, controls, and distribution of power, followed by a period of consolidation.[55] Thomas Jefferson knew this secret when he wrote 200 years ago, "A little revolution now and then is a good thing."[56]

REFERENCES

1. See J.-J. Servan-Schreiber, *The American Challenge* (New York: Atheneum Publishers, 1968).

2. See S. Ramo, *America's Technology Slip* (New York: John Wiley & Sons, 1980).

3. See R. Pascale and A. Athos, *The Art of Japanese Management* (New York: Simon & Schuster, 1981).

4. See T. J. Peters and R. H. Waterman, Jr., *In Search of Excellence* (New York: Harper and Row, 1982). For purposes of this article, the high-technology industries are defined as those which spend more than 3 percent of sales on R&D. These industries, though otherwise quite different, are all characterized by a rapid rate of change in their products and technologies. Only five U.S. industries meet this criterion: (1) chemicals and pharmaceuticals; (2) machinery (especially computers and office machines); (3) electrical equipment and communications; (4) professional and scientific instruments; and (5) aircraft and missiles. See National Science Foundation, *Science Resources Studies Highlights*, NSF81-331, December 31, 1981, p. 2.

5. See W. Ouchi, *Theory Z: How American Management Can Meet the Japanese Challenge* (New York: John Wiley & Sons, 1980).

6. See C. E. Makin, "Ranking Corporate Reputations," *Fortune*, 10 January 1983, pp. 34–44. Corporate reputation was subdivided into eight attributes: quality of management, quality of products and services, innovativeness, long-term investment value, financial soundness, ability to develop and help talented people, community and environmental responsibility, and use of corporate assets.

7. See M. A. Maidique and B. J. Zirger, "Stanford Innovation Project: A Study of Successful and Unsuccessful Product Innovation in High-Technology Firms," *IEEE Transactions on Engineering Management*, in press; M. A. Maidique, "The Stanford Innovation Project: A Comparative Study of Success and Failure in High-Technology Product Innovation," *Management of Technological Innovation Conference Proceedings* (Worcester Polytechnic Institute, 1983).

8. A similar conclusion was reached by Romanelli and Tushman in their study of leadership in the minicomputer industry, which found that successful companies alternated long periods of continuity and inertia with rapid reorientations. See E. Romanelli and M. Tushman, "Executive Leadership and Organizational Outcomes: An Evolutionary Perspective," *Management of Technological Innovation Conference Proceedings* (Worcester Polytechnic Institute, 1983).

9. One of the authors in this article has employed this framework as a diagnostic tool in audits of high-technology firms. The firm is evaluated along these six dimensions on a 0–10 scale by members of corporate and divisional management, working individually. The results are then used as inputs for conducting a strategic review of the firm.

10. General Electric evidently has also recognized the value of such concentration. In 1979, Reginald Jones, then GE's CEO, broke up the firm into six independent sectors led by "sector executives." See R. Vancil and P. C. Browne, "General Electric Consumer products and Services Sector" (Boston, MA: Harvard Business School Case Services 2-179-070).

11. Personal communication with David Packard, Stanford University, March 4, 1982.

12. After only eighteen months as Geneen's successor as president, Lyman Hamilton was summarily dismissed by Geneen for reversing Geneen's way of doing business. See G. Colvin, "The Re-Geneening of ITT," *Fortune*, 11 January 1982, pp. 34–39.

13. See "RCA: Still Another Master," *Business Week*, 17 August 1981, pp. 80–86.

14. See "R&D Scoreboard," *Business Week*, 6 July 1981, pp. 60–75.

15. See R. Stata, Analog Devices *Quarterly Report*, 1st Quarter, 1981.

16. See "Why They Are Jumping Ship at Intel," *Business Week*, 14 February 1983, p. 107; M. Chase, "Problem-Plagued Intel Bets on New Products, IBM's Financial Help," *Wall Street Journal*, 4 February 1983.

17. These SAPPHO findings are generally consistent with the results of the Stanford Innovation Project, a major comparative study of U.S. high-technology innovation. See M. A. Maidique, "The Stanford Innovation Project: A Comparative Study of Success and Failure in High Technology Product Innovation," *Management of Technology Conference Proceedings* (Worcester Polytechnic Institute, 1983).

18. See Maidique and Zirger (in press). Several other authors have reached similar conclusions. See, for example, Peters and Waterman (1982).

19. Pesonal communication with Tom Jones, chairman of the board, Northrop corporation, May 1982.

20. See W. R. Thurston, "The Revitalization of GenRad," *Sloan Management Review*, Summer 1981, pp. 53–57.

21. See T. Wise, "IBM's 5 Billion Dollar Gamble," *Fortune*, September 1966; "A Rocky Road to the Marketplace," *Fortune*, October 1966.

22. See A. P. Sloan, *My Years with General Motors* (New York: Anchor Books, 1972), p. 401.

23. Personal communication with Ken Fisher, 1980. Mr. Fisher was president and CEO of Prime Computer from 1975 to 1981.

24. At Genentech, Cetus, Biogen, and Collaborative Research, four of the leading biotechnology firms, a top scientist is also a member of the board of directors.

25. See, for example, J. A. Morton, *Organizing for Innovation* (New York: McGraw-Hill, 1971).

26. Jimmy Treybig, president of Tandem Computer, Stanford Executive Institute Presentation, August 1982.

27. See D. A. Schon, *Technology and Change* (New York: Dell Publishing, 1967); Peters and Waterman (1982).

28. See S. Myers and E. F. Sweezy, "Why Innovations Fail," *Technology Review*, March–April 1978, pp. 40–46.

29. See *Texas Instruments* (A), 9-476-122, Harvard Business School case; *Texas Instruments Shows U.S. Business How to Survive in the 1980's*, 3-579-092, Harvard Business School case; Texas Instruments *"Speak and Spell Product,"* 9-679-089, revised 7/79, Harvard Business School case.

30. Arthur K. Watson, Address to the Eighth International Congress of Accountants, New York City, September 24, 1962, as quoted by D. A. Shon, "Champions for Radical New Inventions," *Harvard Business Review*, March–April 1963, p. 85.

31. Personal communication with Tom Jones, chairman of the board, Northrop Corporation, May 1982.

32. Personal communication with Bob Hungate, general manager, Medical Supplies Division, Hewlett-Packard, 1980.

33. Personal communication with Richard Frankel, president, Kevex Corporation, April 1983.

34. Personal communication with Herb Dwight, president and CEO, Spectra-Physics, 1982.

35. Personal communication with Alexander d'Arbeloff, cofounder and president of Teradyne, 1983.

36. Personal communication with Ray Stata, president and CEO, Analog Devices, 1980.

37. Personal communication with Bernie Gordon, president and CEO, Analogic, 1982.

38. Personal communication with Paul Rizzo, 1980.

39. Personal communication with Tom McAvoy, president of Corning Glass, 1979.

40. Personal communication with Milt Greenberg, president of GCA, 1980.

41. See Wise (September 1966).

42. See L. R. Sayles and M. K. Chandler, *Managing Large Systems: Organizations for the Future* (New York: Harper and Row, 1971).

43. See R. A. Burgelman, "A Model of the Interaction of Strategic Behavior, Corporate Context and the Concept of Corporate Strategy," *Academy of Management Review* (1983): 61–70.

44. See S. Zipper, "T1 Unscrambling Matrix Management to Cope with Gridlock in Major Profit Centers;" *Electronic News,* 26 April 1982, p. 1.

45. See M. Barnfather, "Can 3M Find Happiness in the 1980's?" *Forbes,* 11 March 1982, pp. 113–116.

46. See R. Hill, "Does a 'Hands Off' Company Now Need a 'Hands On' Style?' *International Management,* July 1983, p. 35.

47. See Barnfather (March 11, 1982).

48. *Quotations from Chariman Mao Tse Tung,* ed. S. R. Schram (Bantam Books, 1967), p. 174.

49. See D. G. Marquis, "Ways of Organizing Projects," *Innovation,* August 1969, pp. 26–33; T. Levitt, *Marketing for Business Growth* (New York: McGraw-Hill, 1974), in particular, ch. 7.

50. Charles Ames, former CEO of Reliance Electric, as quoted in "Exxon's $600-million Mistake," *Fortune,* 19 October 1981.

51. See, for example, W. J. Abernathy and J. M. Utterback, "Patterns of Industrial Innovation," *Technology Review,* June–July 1978, pp. 40–47.

52. See T. Kuhn, *The Structure of Scientific Revolutions,* 2d ed. (Chicago, IL: University of Chicago Press, 1967).

53. After reviewing an early draft of this article, Ray Stata wrote, "The articulation of dynamic balance, of ying and yang, . . . served as a reminder to me that there isn't one way forever, but a constant adaption to the needs and circumstances of the moment." Ray Stata, president, Analog Devices, letter of 29 November 1982.

54. Quoted in "Some Contributions of James E. Webb to the Theory and Practice of Management," a presentation by Elmer B. Staats before the annual meeting of the Academy of Management on 11 August 1978.

55. See Romanelli and Tushman (1983).

56. See J. Bartlett, *Bartlett's Familiar Quotations,* 14th ed. (Boston, MA: Little, Brown), p. 471B.

Convergence and Upheaval: Managing the Unsteady Pace of Organizational Evolution

Michael L. Tushman

William H. Newman

Elaine Romanelli

A snug fit of external opportunity, company strategy, and internal structure is a hallmark of successful companies. The real test of executive leadership, however, is in maintaining this alignment in the face of changing competitive conditions.

Consider the Polaroid or Caterpillar corporations. Both firms virtually dominated their respective industries for decades, only to be caught off guard by major environmental changes. The same strategic and organizational factors which were so effective for decades became the seeds of complacency and organization decline.

Recent studies of companies over long periods show that the most successful firms maintain a workable equilibrium for several years (or decades), but are also able to initiate

and carry out sharp, widespread changes (referred to here as reorientations) when their environments shift. Such upheaval may bring renewed vigor to the enterprise. Less successful firms, on the other hand, get stuck in a particular pattern. The leaders of these firms either do not see the need for reorientation or they are unable to carry through the necessary frame-breaking changes. While not all reorientations succeed, those organizations which do not initiate reorientations as environments shift underperform.

This article focuses on reasons why for long periods most companies make only incremental changes, and why they then need to make painful, discontinuous, system-wide shifts. We are particularly concerned with the role of executive leadership in managing this pattern of convergence punctuated by upheaval. Here are four examples of the convergence/upheaval pattern:

- Founded in 1915 by a set of engineers from MIT, the General Radio Company was established to highly innovative and high-

© 1986 by the Regents of the University of California. Reprinted from the *California Management Review*, Vol. 29, No. 1. By permission of The Regents.

The authors thank Donald Hambrick and Kathy Harrigan for insightful comments and the Center for Strategy Research and the Center for Research on Innovation and Entrepreneurship at the Graduate School of Business, Columbia University, for financial support.

705

quality (but expensive) electronic test equipment. Over the years, General Radio developed a consistent organization to accomplish its mission. It hired only the brightest young engineers, built a loose functional organization dominated by the engineering department, and developed a "General Radio culture" (for example, no conflict, management by consensus, slow growth). General Radio's strategy and associated structures, systems, and people were very successful. By World War II, General Radio was the largest test-equipment firm in the United States.

After World War II, however, increasing technology and cost-based competition began to erode General Radio's market share. While management made numerous incremental changes, General Radio remained fundamentally the same organization. In the late 1960s, when CEO Don Sinclair initiated strategic changes, he left the firm's structure and systems intact. This effort at doing new things with established systems and procedures was less than successful. By 1972, the firm incurred its first loss.

In the face of this sustained performance decline, Bill Thurston (a long-time General Radio executive) was made President. Thurston initiated system-wide changes. General Radio adopted a more marketing-oriented strategy. Its product line was cut from 20 different lines to 3; much more emphasis was given to product-line management, sales, and marketing. Resources were diverted from engineering to revitalize sales, marketing, and production. During 1973, the firm moved to a matrix structure, increased its emphasis on controls and systems, and went outside for a set of executives to help Thurston run this revised General Radio. To perhaps more formally symbolize these changes and

the sharp move away from the "old" General Radio, the firm's name was changed to GenRad. By 1984, GenRad's sales exploded to over $200 million (vs. $44 million in 1972).

After 60 years of convergent change around a constant strategy, Thurston and his colleagues (many new to the firm) made discontinuous system-wide changes in strategy, structure, people, and processes. While traumatic, these changes were implemented over a two-year period and led to a dramatic turnaround in GenRad's performance.

Prime Computer was founded in 1971 by a group of individuals who left Honeywell. Prime's initial strategy was to produce a high-quality/high-price minicomputer based on semiconductor memory. These founders built an engineering-dominated, loosely structured firm which sold to OEMs and through distributors. This configuration of strategy, structure, people, and processes was very successful. By 1974, Prime turned its first profit; by 1975, its sales were more than $11 million.

In the midst of this success, Prime's board of directors brought in Ken Fisher to reorient the organization. Fisher and a whole new group of executives hired from Honeywell initiated a set of discontinuous changes throughout Prime during 1975–1976. Prime now sold a full range of minicomputers and computer systems to OEMs and end-users. To accomplish this shift in strategy, Prime adopted a more complex functional structure, with a marked increase in resources to sales and marketing. The shift in resources away from engineering was so great that Bill Poduska, Prime's head of engineering, left to form Apollo Computer. Between 1975–1981, Fisher and his colleagues consolidated and incrementally adapted structure, systems, and processes to better accomplish the new strategy. During this convergent period, Prime grew dramatically to over $260 million by 1981.

In 1981, again in the midst of this continuing sequence of increased volume and profits, Prime's board again initiated an upheaval. Fisher and his direct reports left Prime (some of whom founded Encore Computer), while Joe Henson and a set of executives from IBM initiated wholesale changes throughout the organization. The firm diversified into robotics, CAD/CAM, and office systems; adopted a divisional structure; developed a more market-driven orientation; and increased controls and systems. It remains to be seen how this "new" Prime will fare. Prime must be seen, then, not as a 14-year-old firm, but as three very different organizations, each of which was managed by a different set of executives. Unlike General Radio, Prime initiated these discontinuities during periods of great success.

· The Operating Group at Citibank prior to 1970 had been a service-oriented function for the end-user areas of the bank. The Operating Group hired high school graduates who remained in the "back-office" for their entire careers. Structure, controls, and systems were loose, while the informal organization valued service, responsiveness to client needs, and slow, steady work habits. While these patterns were successful enough, increased demand and heightened customer expectations led to ever decreasing performance during the late 1960s.

In the face of severe performance decline, John Reed was promoted to head the Operating Group. Reed recruited several executives with production backgrounds, and with this new top team he initiated system-wide changes. Reed's vision was to transform the Operating Group from a *service-*oriented back office to a *factory* producing high-quality products. Consistent with this new mission, Reed and his colleagues initiated sweeping changes in strategy, structure, work flows, controls, and culture. These changes were initiated concurrently throughout the back office, with very little participation, over the course of a few months. While all the empirical performance measures improved substantially, these changes also generated substantial stress and anxiety within Reed's group.

· For 20 years, Alpha Corporation was among the leaders in the industrial fastener industry. Its reliability, low cost, and good technical service were important strengths. However, as Alpha's segment of the industry matured, its profits declined. Belt-tightening helped but was not enough. Finally, a new CEO presided over a sweeping restructuring: cutting the product line, closing a plant, trimming overhead; then focusing on computer parts which call for very close tolerances, CAD/CAM tooling, and cooperation with customers on design efforts. After four rough years, Alpha appears to have found a new niche where convergence will again be warranted.

These four short examples illustrate periods of incremental change, or convergence, punctuated by discontinuous changes throughout the organization. Discontinuous or "frame-breaking" change involves simultaneous and sharp shifts in strategy, power, structure, and controls. Each example illustrates the role of executive leadership in initiating and implementing discontinuous change. Where General Radio, Citibank's Operating Group, and Alpha initiated system-wide changes only after sustained performance decline, Prime proactively initiated system-wide changes to take advantage of competitive/technological conditions. These patterns in organization evolution are not unique. Upheaval, sooner or later, follows convergence if a company is to survive; only a farsighted minority of firms initiate upheaval prior to incurring performance declines.

The task of managing incremental change, or convergence, differs sharply from managing frame-breaking change. Incremental change is compatible with the existing structure of a company and is reinforced over a period of years. In contrast, frame-breaking change is abrupt, painful to participants, and often resisted by the old guard. Forging these new strategy-structure-people-process consistencies and laying the basis for the next period of incremental change calls for distinctive skills.

Because the future health, and even survival, of a company or business unit is at stake, we need to take a closer look at the nature and consequences of convergent change and of differences imposed by frame-breaking change. We need to explore when and why these painful and risky revolutions interrupt previously successful patterns, and whether these discontinuities can be avoided and/or initiated prior to crisis. Finally, we need to examine what managers can and should do to guide their organizations through periods of convergence and upheaval over time.

THE RESEARCH BASE

The research which sparks this article is based on abundant company histories and case studies. The more complete case studies have tracked individual firms' evolution and various crises in great detail (e.g., Chandler's seminal study of strategy and structure at Du Pont, General Motors, Standard Oil, and Sears[1]). More recent studies have dealt systematically with whole sets of companies and trace their experience over long periods of time.

A series of studies by researchers at McGill University covered over 40 well-known firms in diverse industries for at least 20 years per firm (e.g., Miller and Friesen[2]). Another research program conducted by researchers at Columbia, Duke, and Cornell Universities is tracking the history of large samples of companies in the minicomputer, cement, airlines, and glass industries. This research program builds on earlier work (e.g., Greiner[3]) and finds that most successful firms evolve through long periods of convergence punctuated by frame-breaking change.

The following discussion is based on the history of companies in many different industries, different countries, both large and small organizations, and organizations in various stages of their product class's life-cycle. We are dealing with a widespread phenomenon—not just a few dramatic sequences. Our research strongly suggests that the convergence/upheaval pattern occurs within departments (e.g., Citibank's Operating Group), at the business-unit level (e.g., Prime or General Radio), and at the corporate level of analysis (e.g., the Singer, Chrysler, or Harris Corporations). The problem of managing both convergent periods and upheaval is not just for the CEO, but necessarily involves general managers as well as functional managers.

PATTERNS IN ORGANIZATIONAL EVOLUTION: CONVERGENCE AND UPHEAVAL

Building on Strength: Periods of Convergence

Successful companies wisely stick to what works well. At General Radio between 1915 and 1950, the loose functional structure, committee management system, internal promotion practices, control with engineering, and the high-quality, premium-price, engineering mentality all worked together to provide a highly congruent system. These internally consistent patterns in strategy, structure, people, and processes served General Radio for over 35 years.

Similarly, the Alpha Corporation's customer driven, low-cost strategy was accomplished by strength in engineering and production and ever more detailed structures and systems which evaluated cost, quality, and new product development. These strengths were epitomized in Alpha's chief engineer and president. The chief engineer had a remarkable talent for helping customers find new uses for industrial fasteners. He relished solving such problems, while at the same time designing fasteners that could be easily manufactured. The president excelled at production—producing dependable, low-cost fasteners. The pair were role models which set a pattern which served Alpha well for 15 years.

As the company grew, the chief engineer hired kindred customer-oriented application engineers. With the help of innovative users, they developed new products, leaving more routine problem-solving and incremental change to the sales and production departments. The president relied on a hands-on manufacturing manager and delegated financial matters to a competent treasurer-controller. Note how well the organization reinforced Alpha's strategy and how the key people fit the organization. There was an excellent fit between strategy and structure. The informal structure also fit well—communications were open, the simple mission of the company was widely endorsed, and routines were well understood.

As the General Radio and Alpha examples suggest, convergence starts out with an effective dovetailing of strategy, structure, people, and processes. For other strategies or in other industries, the particular formal and in-

formal systems might be very different, but still a winning combination. The formal system includes decisions about grouping and linking resources as well as planning and control systems, rewards and evaluation procedures, and human resource management systems. The informal system includes core values, beliefs, norms, communication patterns, and actual decision-making and conflict resolution patterns. It is the whole fabric of structure, systems, people, and processes which must be suited to company strategy.[4]

As the fit between strategy, structure, people, and processes is never perfect, convergence is an ongoing process characterized by incremental change. Over time, in all companies studied, two types of converging changes were common: fine-tuning and incremental adaptations.

- *Converging Change: Fine-Tuning*—Even with good strategy-structure-process fits, well-run companies seek even better ways of exploiting (and defending) their missions. Such effort typically deals with one or more of the following:
- *Refining* policies, methods, and procedures.
- Creating *specialized units and linking mechanisms* to permit increased volume and increased attention to unit quality and cost.
- *Developing personnel* especially suited to the present strategy—through improved selection and training, and tailoring reward systems to match strategic thrusts.
- Fostering individual and group *commitments* to the company mission and to the excellence of one's own department.
- Promoting *confidence* in the accepted norms, beliefs, and myths.
- *Clarifying* established roles, power, status, dependencies, and allocation mechanism.

The fine-tuning fills out and elaborates the consistencies between strategy, structure, people, and processes. These incremental changes lead to an ever more interconnected (and therefore more stable) social system. Convergent periods fit the happy, stick-with-a-winner situations romanticized by Peters and Waterman.[5]

Converging Change: Incremental Adjustments to Environmental Shifts—In addition to fine-tuning changes, minor shifts in the environment will call for some organizational response. Even the most conservative of organizations expect, even welcome, small changes which do not make too many waves.

A popular expression is that almost any organization can tolerate a "ten-percent change." At any one time, only a few changes are being made; but these changes are still compatible with the prevailing structures, systems, and processes. Examples of such adjustments are an expansion in sales territory, a shift in emphasis among products in the product line, or improved processing technology in production.

The usual process of making changes of this sort is well known: wide acceptance of the need for change, openness to possible alternatives, objective examination of the pros and cons of each plausible alternative, participation of those directly affected in the preceding analysis, a market test or pilot operation where feasible, time to learn the new activities, established role models, known rewards for positive success, evaluation, and refinement.

The role of executive leadership during convergent periods is to reemphasize mission and core values and to delegate incremental decisions to middle-level managers. Note that the uncertainty created for people affected by such changes is well within tolerable limits. Opportunity is provided to anticipate and learn what is new, while most features of the structure remain unchanged.

The overall system adapts, but it is not transformed.

Converging Change: Some Consequences—For those companies whose strategies fit environmental conditions, convergence brings about better and better effectiveness. Incremental change is relatively easy to implement and ever more optimizes the consistencies between strategy, structure, people, and processes. At AT&T, for example, the period between 1913 and 1980 was one of ever more incremental change to further bolster the "Ma Bell" culture, systems, and structure all in service of developing the telephone network.

Convergent periods are, however, a double-edged sword. As organizations grow and become more successful, they develop internal forces for stability. Organization structures and systems become so interlinked that they only allow compatible changes. Further, over time, employees develop habits, patterned behaviors begin to take on values (e.g., "service is good"), and employees develop a sense of competence in knowing how to get work done within the system. These self-reinforcing patterns of behavior, norms, and values contribute to increased organizational momentum and complacency and, over time, to a sense of organizational history. This organizational history—epitomized by common stories, heroes, and standards—specifies "how we work here" and "what we hold important here."

This organizational momentum is profoundly functional as long as the organization's strategy is appropriate. The Ma Bell and General Radio culture, structure, and systems—and associated internal momentum—were critical to each organization's success. However, if (and when) strategy must change, this momentum cuts the other way. Organizational history is a source of tradition, precedent, and pride which are, in turn, anchors to the past. A proud history often restricts vigilant problem solving and may be a source of resistance to change.

When faced with environmental threat, organizations with strong momentum

· may not register the threat due to organization complacency and/or stunted external vigilance (e.g., the automobile or steel industries), or

· if the threat is recognized, the response is frequently heightened conformity to the status quo and/or increased commitment to "what we do best."

For example, the response of dominant firms to technological threat is frequently increased commitment to the obsolete technology (e.g., telegraph/telephone; vacuum tube/transistor; core/semiconductor memory). A paradoxical result of long periods of success may be heightened organizational complacency, decreased organizational flexibility, and a stunted ability to learn.

Converging change is a double-edged sword. Those very social and technical consistencies which are key sources of success may also be the seeds of failure if environments change. The longer the convergent period, the greater these internal forces for stability. This momentum seems to be particularly accentuated in those most successful firms in a product class (for example, Polaroid, Caterpillar, or U.S. Steel), in historically regulated organizations (for example, AT&T, GTE, or financial service firms), or in organizations that have been traditionally shielded from competition (for example, universities, not-for-profit organizations, government agencies and/or services).

On Frame-Breaking Change

Forces Leading to Frame-Breaking Change—What, then, leads to frame-breaking change? Why defy tradition? Simply stated, frame-breaking change occurs in response to or, better yet, in anticipation of major environmental changes—changes which require more than incremental adjustments. The need for discontinuous change springs from one or a combination of the following:

· *Industry Discontinuities*—Sharp changes in legal, political, or technological conditions shift the basis of competition within industries. *Deregulation* has dramatically transformed the financial services and airlines industries. *Substitute product technologies* (such as jet engines, electronic typing, microprocessors) or *substitute process technologies* (such as the planar process in semiconductors or float-glass in glass manufacture) may transform the bases of competition within industries. Similarly, the emergence of industry standards, or *dominant designs* (such as the DC-3, IBM 360, or PDP-8) signal a shift in competition away from product innovation and towards increased process innovation. Finally, *major economic changes* (e.g., oil crises) and *legal shifts* (e.g., patent protection in biotechnology or trade/regulator barriers in pharmaceuticals or cigarettes) also directly affect bases of competition.

· *Product-Life-Cycle Shifts*—Over the course of a product class life-cycle, different strategies are appropriate. In the emergence phase of a product class, competition is based on product innovation and performance, where in the maturity stage, competition centers on cost, volume, and efficiency. Shifts in patterns of demand alter key factors for success. For example, the demand and nature of competition for minicomputers, cellular telephones, wide-body aircraft, and bowling alley equipment was transformed as these products gained acceptance and their product classes evolved. Powerful international competition may compound these forces.

· *Internal Company Dynamics*—Entwined with these external forces are breaking points

711

within the firm. Sheer size may require a basically new management design. For example, few inventor-entrepreneurs can tolerate the formality that is linked with large volume; even Digital Equipment Company apparently has outgrown the informality so cherished by Kenneth Olsen. Key people die. Family investors may become more concerned with their inheritance taxes than with company development. Revised corporate portfolio strategy may sharply alter the role and resources assigned to business units or functional areas. Such pressures especially when coupled with external changes, may trigger frame-breaking change.

Scope of Frame-Breaking Change—Frame-breaking change is driven by shifts in business strategy. As strategy shifts so too must structure people, and organizational processes. Quite unlike convergent change, frame-breaking reforms involve discontinuous changes throughout the organization. These bursts of change do not reinforce the existing system and are implemented rapidly. For example, the system-wide changes at Prime and General Radio were implemented over 18–24-month periods, where as changes in Citibank's Operating Group were implemented in less than five months. Frame-breaking changes are revolutionary changes *of* the system as opposed to incremental changes *in* the system.

The following features are usually involved in frame-breaking change:

- *Reformed Mission and Core Values*—A strategy shift involves a new definition of company mission. Entering or withdrawing from an industry may be involved; at least the way the company expects to be outstanding is altered. The revamped AT&T is a conspicuous example. Success on its new course calls for a strategy based on competition, aggressiveness, and responsiveness, as well

as a revised set of core values about how the firm competes and what it holds as important. Similarly, the initial shift at Prime reflected a strategic shift away from technology and towards sales and marketing. Core values also were aggressively reshaped by Ken Fisher to complement Prime's new strategy.

- *Altered Power and Status*—Frame-breaking change always alters the distribution of power. Some groups lose in the shift while others gain. For example, at Prime and General Radio, the engineering functions lost power, resources, and prestige as the marketing and sales functions gained. These dramatically altered power distributions reflect shifts in bases of competition and resource allocation. A new strategy must be backed up with a shift in the balance of power and status.

- *Reorganization*—A new strategy requires a modification in structure, systems, and procedures. As strategic requirements shift, so too must the choice of organization form. A new direction calls for added activity in some areas and less in others. Changes in structure and systems are means to ensure that this reallocation of effort takes place. New structures and revised roles deliberately break business-as-usual behavior.

- *Revised Interaction Patterns*—The way people in the organization work together has to adapt during frame-breaking change. As strategy is different, new procedures, work flows, communication networks, and decision-making patterns must be established. With these changes in work flows and procedures must also come revised norms, informal decision-making/conflict-resolution procedures, and informal roles.

- *New Executives*—Frame-breaking change also involves new executives, usually brought in from outside the organization (or business unit) and placed in key managerial positions. Commitment to the new mission, energy to overcome prevailing inertia, and freedom from prior obligations are all needed to refocus the organization. A few exceptional members of the old guard may attempt to make this shift, but habits and expectations of their associations are difficult to break. New executives are most likely to provide both the necessary drive and an enhanced set of skills more appropriate for the new strategy. While the overall number of executive changes is usually relatively small, these new executives have substantial symbolic and substantive effects on the organization. For example, frame-breaking changes at Prime, General Radio, Citibank, and Alpha Corporation were all spearheaded by a relatively small set of new executives from outside the company or group.

Why All at Once?—Frame-breaking change is revolutionary in that the shifts reshape the entire nature of the organization. Those more effective examples of frame-breaking change were implemented rapidly (e.g., Citibank, Prime, Alpha). It appears that a piecemeal approach to frame-breaking changes gets bogged down in politics, individual resistance to change, and organizational inertia (e.g., Sinclair's attempts to reshape General Radio). Frame-breaking change requires discontinuous shifts in strategy, structure, people, and processes concurrently—or at least in a short period of time. Reasons for rapid, simultaneous implementation include:

- *Synergy* within the new structure can be a powerful aid. New executives with a fresh mission, working in a redesigned organization with revised norms and values, backed up with power and status, provide strong reinforcement. The pieces of the revitalized organization pull together, as opposed to piecemeal change where one part of the

new organization is out of synch with the old organization.

- *Pockets of resistance* have a chance to grow and develop when frame-breaking change is implemented slowly. The new mission, shifts in organization, and other frame-breaking changes upset the comfortable routines and precedent. Resistance to such fundamental change is natural. If frame-breaking change is implemented slowly, then individuals have a greater opportunity to undermine the changes and organizational inertia works to further stifle fundamental change.

- Typically, there is a *pent-up need for change.* During convergent periods, basic adjustments are postponed. Boat-rocking is discouraged. Once constraints are relaxed, a variety of desirable improvements press for attention. the exhilaration and momentum of a fresh effort (and new team) make difficult moves more acceptable. Change is in fashion.

- Frame-breaking change is an inherently *risky and uncertain venture.* The longer the implementation period, the greater the period of uncertainty and instability. The most effective frame-breaking changes initiate the new strategy, structure, processes, and systems rapidly and begin the next period of stability and convergent change. The sooner fundamental uncertainty is removed, the better the chances of organizational survival and growth. While the pacing of change is important, the overall time to implement frame-breaking change will be contingent on the size and age of the organization.

Patterns in Organization Evolution—This historical approach to organization evolution focuses on convergent periods punctuated by reorientation—discontinuous, organization-wide upheavals. The most effective firms take advantage of relatively long convergent periods.

These periods of incremental change build on and take advantage of organization inertia. Frame-breaking change is quite dysfunctional if the organization is successful and the environment is stable. If, however, the organization is performing poorly and/or if the environment changes substantially, frame-breaking change is the only way to realign the organization with its competitive environment. Not all reorientations will be successful (e.g., People Express' expansion and up-scale moves in 1985–86). However, inaction in the face of performance crisis and/or environmental shifts is a certain recipe for failure.

Because reorientations are so disruptive and fraught with uncertainty, the more rapidly they are implemented, the more quickly the organization can reap the benefits of the following convergent period. High-performing firms initiate reorientations when environmental conditions shift and implement these reorientations rapidly (e.g., Prime and Citibank). Low-performing organizations either do not reorient or reorient all the time as they root around to find an effective alignment with environmental conditions.

This metamorphic approach to organization evolution underscores the role of history and precedent as future convergent periods are all constrained and shaped by prior convergent periods. Further, this approach to organization evolution highlights the role of executive leadership in managing convergent periods *and* in initiating and implementing frame-breaking change.

EXECUTIVE LEADERSHIP AND ORGANIZATION EVOLUTION

Executive leadership plays a key role in reinforcing system-wide momentum during convergent periods and in initiating and implementing bursts of change that characterize

strategic reorientations. The nature of the leadership task differs sharply during these contrasting periods of organization evolution.

During convergent periods, the executive team focuses on *maintaining* congruence and fit within the organization. Because strategy, structure, processes, and systems are fundamentally sound, the myriad of incremental substantive decisions can be delegated to middle-level management, where direct expertise and information resides. The key role for executive leadership during convergent periods is to reemphasize strategy, mission, and core values and to keep a vigilant eye on external opportunities and/or threats.

Frame-breaking change, however, requires direct executive involvement in all aspects of the change. Given the enormity of the change and inherent internal forces for stability, executive leadership must be involved in the specification of strategy, structure, people, and organizational processes *and* in the development of implementation plans. During frame-breaking change, executive leadership is directly involved in *reorienting* their organizations. Direct personal involvement of senior management seems to be critical to implement these system-wide changes (e.g., Reed at Citibank or Iacocca at Chrysler). Tentative change does not seem to be effective (e.g., Don Sinclair at General Radio).

Frame-breaking change triggers resistance to change from multiple sourceschange must overcome several generic hurdles, including:

· Individual opposition, rooted in either anxiety or personal commitment to the status quo, is likely to generate substantial individual resistance to change.
· Political coalitions opposing the upheaval may be quickly formed within the organization. During converging periods a political equilibrium is reached. Frame-breaking upsets this equilibrium; powerful individuals

and/or groups who see their status threatened will join in resistance.

- Control is difficult during the transition. The systems, roles, and responsibilities of the former organization are in suspension; the new rules of the game—and the rewards—have not yet been clarified.
- External constituents—suppliers, customers, regulatory agencies, local communities, and the like—often prefer continuation of existing relationships rather than uncertain moves in the future.

Whereas convergent change can be delegated, frame-breaking change requires strong, direct leadership from the top as to where the organization is going and how it is to get there. Executive leadership must be directly involved in: motivating constructive behavior, shaping political dynamics, managing control during the transition period, and managing external constituencies. The executive team must direct the content of frame-breaking change *and* provide the energy, vision, and resources to support, and be role models for, the new order. Brilliant ideas for new strategies, structures, and processes will not be effective unless they are coupled with thorough implementation plans actively managed by the executive team.[6]

When to Launch an Upheaval. The most effective executives in our studies foresaw the need for major change. They recognized the external threats and opportunities, and took bold steps to deal with them. For example, a set of minicomputer companies (Prime, Rolm, Datapoint, Data General, among others) risked short-run success to take advantage of new opportunities created by technological and market changes. Indeed, by acting before being forced to do so, they had more time to plan their transitions.[7]

Such visionary executive teams are the exceptions. Most frame-breaking change is postponed until a financial crisis forces drastic action. The momentum, and frequently the success, of convergent periods breeds reluctance to change. This commitment to the status quo, and insensitivity to environmental shocks, is evident in both the Columbia and the McGill studies. It is not until financial crisis shouts its warning that most companies begin their transformation.

The difference in timing between pioneers and reluctant reactors is largely determined by executive leadership. The pioneering moves, in advance of crisis, are usually initiated by executives within the company. They are the exceptional persons who combine the vision, courage, and power to transform an organization. In contrast, the impetus for a tardy break usually comes from outside stakeholders; they eventually put strong pressure on existing executives—or bring in new executives—to make fundamental shifts.

Who Manages the Transformation. Directing a frame-breaking upheaval successfully calls for unusual talent and energy. The new mission must be defined, technology selected, resources acquired, policies revised, values changed, organization restructured, people reassured, inspiration provided, and an array of informal relationships shaped. Executives already on the spot will probably know most about the specific situation, but they may lack the talent, energy, and commitment to carry through an internal revolution.

As seen in the Citibank, Prime, and Alpha examples, most frame-breaking upheavals are managed by executives brought in from outside the company. The Columbia research program finds that externally recruited executives are more than three times more likely to initiate frame-breaking change than existing executive teams. Frame-breaking change was coupled with CEO succession in more than 80 percent of the cases. Further, when frame-breaking change was combined with executive succession, company performance was significantly higher than when former executives stayed in place. In only 6 of 40 cases we studied did a current CEO initiate and implement multiple frame-breaking changes. In each of these six cases, the existing CEO made major changes in his/her direct reports, and this revitalized top team initiated and implemented frame-breaking changes (e.g., Thurston's actions at General Radio).[8]

Executive succession seems to be a powerful tool in managing frame-breaking change. There are several reasons why a fresh set of executives are typically used in company transformations. The new executive team brings different skills and a fresh perspective. Often they arrive with a strong belief in the new mission. Moreover, they are unfettered by prior commitments linked to the status quo; instead, this new top team symbolizes the need for change. Excitement of a new challenge adds to the energy devoted to it.

We should note that many of the executives who could not, or would not, implement frame-breaking change went on to be quite successful in other organizations—for example, Ken Fisher at Encore Computer and Bill Podusk at Apollo Computer. The stimulation of a fresh start and of jobs matched to personal competence applies to individuals as well as to organizations.

Although typical patterns for the when and who of frame-breaking change are clear—wait for a financial crisis and then bring in an outsider, along with a revised executive team, to revamp the company—this is clearly less than satisfactory for a particular organization. Clearly, some companies benefit from transforming themselves before a crisis forces them to do so, and a few exceptional executives have the vision and drive to reorient a business which they nurtured during its preceding period of convergence. The vital tasks are to manage incremental change during convergent periods; to have the vision to initiate and implement frame-breaking change prior to the competition; and to mobilize an executive team which can initiate and implement both kinds of change.

CONCLUSION

Our analysis of the way companies evolve over long periods of time indicates that the most effective firms have relatively long periods of convergence giving support to a basic strategy, but such periods are punctuated by upheavals—concurrent and discontinuous changes which reshape the entire organization.

Managers should anticipate that when environments change sharply:

- Frame-breaking change cannot be avoided. These discontinuous organizational changes will either be made proactively or initiated under crisis/turnaround condition.
- Discontinuous changes need to be made in strategy, structure, people, and processes concurrently. Tentative change runs the risk of being smothered by individual, group, and organizational inertia.
- Frame-breaking change requires direct executive involvement in all aspects of the change, usually bolstered with new executives from outside the organization.
- There are no patterns in the sequence of frame-breaking changes, and not all strategies will be effective. Strategy and, in turn, structure, systems, and processes must meet industry-specific competitive issues.

Finally, our historical analysis of organizations highlights the following issues for executive leadership:

- Need to manage for balance, consistency, or fit during convergent period.

- Need to be vigilant for environmental shifts in order to anticipate the need for frame-breaking change.
- Need to effectively manage incremental as well as frame-breaking change.
- Need to build (or rebuild) a top team to help initiate and implement frame-breaking change.
- Need to develop core values which can be used as an anchor as organizations evolve through frame-breaking changes (e.g., IBM, Hewlett-Packard).
- Need to develop and use organizational history as a way to infuse pride in an organization's past and for its future.
- Need to bolster technical, social, and conceptual skills with visionary skills. Visionary skills add energy, direction, and excitement so critical during frame-breaking change.

Effectiveness over changing competitive conditions requires that executives manage fundamentally different kinds of organizations and different kinds of change. The data are consistent across diverse industries and countries, an executive team's ability to proactively initiate and implement frame-breaking change *and* to manage convergent change seem to be important factors which discriminate between organizational renewal and greatness versus complacency and eventual decline.

REFERENCES

1. A. Chandler, *Strategy and Structure* (Cambridge, MA: MIT Press, 1962).
2. D. Miller and P. Friesen, *Organizations: A Quantum View* (Englewood Cliffs, NJ: Prentice-Hall, 1984).
3. L. Greiner, "Evolution and Revolution as Organizations Grow," *Harvard Business Review* (July/August 1972), pp. 37–46.
4. D. Nadler and M. Tushman, *Strategic Organization Design* (Homewood, IL: Scott Foresman, 1988).
5. T. Peters and R. Waterman, *In Search of Excellence* (New York, NY: Harper and Row, 1982).
6. Nadler and Tushman, op. cit.
7. For a discussion of preemptive strategies, see I. MacMillan, "Delays in Competitors' Responses to New Banking Products," *Journal of Business Strategy*, 4 (1984): 58–65.
8. M. Tushman and B. Virany, "Changing Characteristics of Executive Teams in an Emerging Industry," *Journal of Business Venturing* (1986).

Concepts for the Management of Organizational Change

David A. Nadler

Bringing about major change in a large and complex organization is a difficult task. Policies, procedures, and structures need to be altered. Individuals and groups have to be motivated to continue to perform in the face of major turbulence. People are presented with the fact that the "old ways," which include familiar tasks, jobs, procedures, and structures are no longer applicable. Political behavior frequently becomes more active and more intense. It is not surprising, therefore, that the process of effectively implementing organizational change has long been a topic that both managers and researchers have pondered. While there is still much that is not understood about change in complex organizations, the experiences and research of recent years do provide some guidance to those concerned with implementing major changes in organizations.

This paper is designed to provide some useful concepts to aid in understanding the dynamics of change and to help in the planning and managing of major organizational changes. The paper is organized into several sections. We will start with a brief discussion of a model of organizational behavior. This discussion is necessary since it is difficult to think about changing organizations without some notion of

why they work the way they do in the first place. Second, we will define what we mean by organizational change and identify criteria for the effective management of change. Third, we will discuss some of the basic problems of implementing change. In the last section, we will list some specific methods and tools for effective implementation of organizational changes.

A VIEW OF ORGANIZATIONS

There are many different ways of thinking about organizations and the patterns of behavior that occur within them. During the past two decades, there has emerged a view of organizations as complex open social systems (Katz & Kahn, 1966), mechanisms which take input from the larger environment and subject that input to various transformation processes that result in output.

As systems, organizations are seen as composed of interdependent parts. Change in one element of the system will result in changes in other parts of the system. Similarly, organizations have the property of equilibrium; the system will generate energy to move towards a state of balance. Finally, as open systems, organizations need to maintain favorable transactions of input and output with the environment in order to survive over time.

While the systems perspective is useful, systems theory by itself may be too abstract a concept to be a usable tool for managers. Thus, a number of organizational theorists have attempted to develop more pragmatic theories or models based on the system paradigm. There are a number of such models currently in use. One of these will be employed here.

The particular approach, called a *Congruence Model of Organizational Behavior* (Nadler & Tushman, 1977; 1979) is based on the general systems model. In this framework, the major inputs to the system of organizational behavior are the *environment* which provides constraints, demands and opportunities, the *resources* available to the organization, and the *history* of the organization. A fourth input, and perhaps the most crucial, is the organization's *strategy*. Strategy is the set of key decisions about the match of the organization's resources to the opportunities, constraints, and demands in the environment within the context of history.

The output of the system is, in general, the effectiveness of the organization's performance, consistent with the goals of strategy. Specifically, the output includes *organizational performance,* as well as *group performance* and *individual behavior and affect* which, of course, contribute to organizational performance.

The basic framework thus views the organization as being the mechanism that takes inputs (strategy and resources in the context of history and environment) and transforms them into outputs (patterns of individual, group, and organizational behavior). This view is portrayed in figure 1.

The major focus of organizational analysis is therefore the transformation process. The model conceives of the organization as being composed of four major components. The first component is the task of the organization, or the work to be done and its critical characteristics. The second component is composed of the *individuals* who are to perform organizational tasks. The third component includes all of the *formal organizational arrangements,* including various structures, processes, systems, etc. which are designed to motivate and facilitate individuals in the performance of organizational tasks. Finally, there is a set of *informal organizational arrangements,* which are usually neither planned nor written, but which tend to emerge over time. These include patterns of communication, power and influence, values and norms,

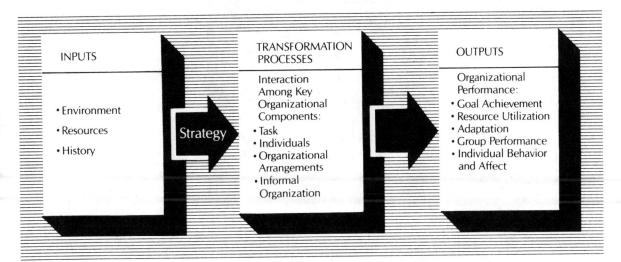

FIGURE 1. *The Systems Model Applied to Organizational Behavior*

etc. which characterize how an organization actually functions.

How do these four components (task, individuals, organizational arrangements, and the informal organization) relate to one another? The relationship among components is the basic dynamic of the model. Each component can be thought of as having a relationship with each other component. Between each pair, then, we can think of a relative degree of consistency, congruence, or "fit." For example, if we look at the type of work to be done (task) and the nature of the people available to do the work (individuals) we could make a statement about the congruence between the two by seeing whether the demands of the work are consistent with the skills and abilities of the individuals. At the same time we would compare the rewards that the work provides to the needs

and desires of the individuals. By looking at these factors, we would be able to assess how congruent the nature of the task was with the nature of the individuals in the system. In fact, we could look at the question of congruence among all the components, or in terms of all six of the possible relationships among them (see figure 2 below). The basic hypothesis of the model is therefore that *organizations will be most effective when their major components are congruent with each other.* To the extent that organizations face problems of effectiveness due to management and organizational factors, these problems will stem from poor fit, or lack of congruence, among organizational components.

This approach to organizations is thus a contingency approach. There is not one best organization design, or style of management, or method of working. Rather, different pat-

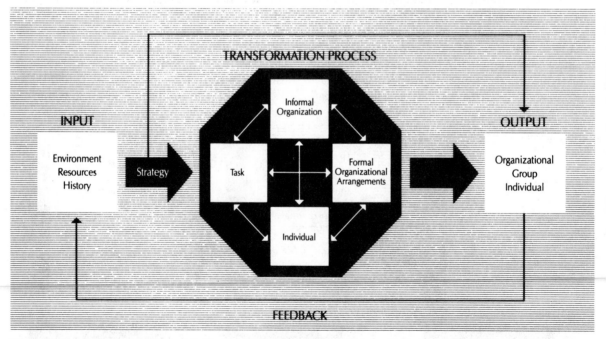

Source: David A. Nadler and Michael L. Tushman, "A Congruence Model for Diagnosing Organizational Behavior" from *Organizational Psychology: A Book of Readings* (3rd Edition) by D. Kolb, I. Rubin, and J. McIntyre. Englewood Cliffs, NJ: Prentice-Hall, 1979.

FIGURE 2.   A Congruence Model of Organizational Behavior

terns of organization and management will be most appropriate in different situations. The model recognizes the fact that individuals, tasks, strategies, and environments may differ greatly from organization to organization.

THE TASK OF IMPLEMENTING CHANGE

Having briefly presented some concepts that underlie our thinking about organizations, the question of change can now be addressed. Managers are frequently concerned about implementing organizational changes. Often changes in the environment necessitate organizational change. For example, factors related to competition, technology, or regulation, shift and thus necessitate changes in organizational

strategy. If a new strategy is to be executed, then the organization and its various subunits (departments, groups, divisions, etc.) must perform tasks that may be different from those previously performed. Building on the organizational model presented above, this means that modification may need to be made in organizational arrangements, individuals and the informal organization.

Typically, implementing a change involves moving an organization to some desired future state. As illustrated in figure 3 below, we can think of changes in terms of transitions (Beckhard & Harris, 1977). At any point in time, the organization exists in a current state (A). The current state describes how the organization functions prior to the change. The future state (B) describes how the organization should be functioning in the future. It is the state that ideally would exist after the change.

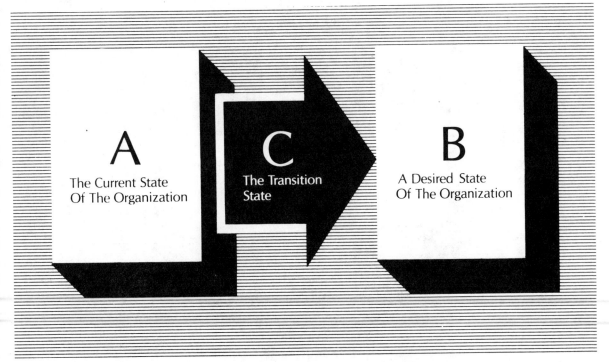

Source: Adapted from Beckhard & Harris, 1977

FIGURE 3.   *Organizational Change as a Transition State*

The period between A and B can be thought of as the transition state (C). In its most general terms, then, the effective management of change involves developing an understanding of the current state (A), developing an image of a desired future state (B), and moving the organization from A through a transition period to B (Beckhard & Harris, 1977).

Major transitions usually occur in response to changes in the nature of organizational inputs or outputs. Most significant changes are in response to or in anticipation of environmental or strategic shifts, or problems of performance. In terms of the congruence model, a change occurs when managers determine that the configuration of the components in the current state is not effective and the organization must be reshaped. Often this means a rethinking and redefining of the organization's task followed by changes in other components to support that new task (see figure 4).

What constitutes effective management of these changes? There are several criteria to consider. Building on the transition framework presented above, organizational change is effectively managed when:

1. The organization is moved from the current state to the future state.
2. The functioning of the organization in the future state meets expectations; i.e., it works as planned.
3. The transition is accomplished without undue cost to the organization.

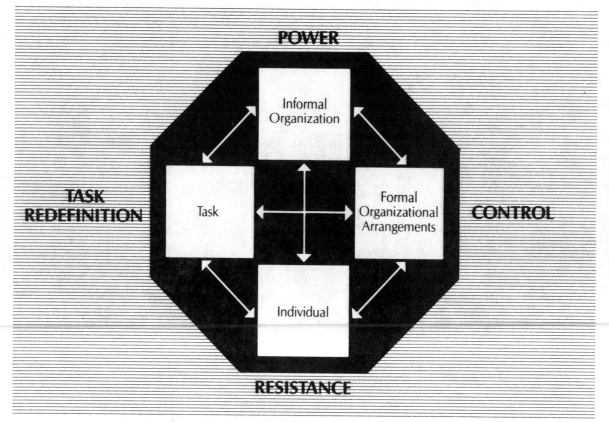

FIGURE 4.   Problems of Change in Relation to the Components of the Organizational Model

722

4. The transition is accomplished without undue cost to individual organizational members.

Of course, not every organizational change can be expected to meet these criteria, but such standards provide a target for planning change. The question is how to manage the way in which the change is implemented so as to maximize the chances that the change will be effective. Experience has shown that the way that a change is implemented can influence the effectiveness of the transition as much as the content of that change.

PROBLEMS IN IMPLEMENTING CHANGE

Experience and research have shown that the process of creating change is more difficult than it might seem. It is tempting to think of an organization as a large machine where parts can be replaced at will. On the contrary, the task of changing the behavior of organizations, groups and individuals has turned out to be a difficult and often frustrating endeavor.

Using the organizational model presented above, we can envision how organizations, as systems, are resistant to change. The forces of equilibrium tend to work to cancel out many changes. Changing one component of an organization may reduce its congruence with other components. As this happens, energy develops in the organization to limit, encapsulate, or revise the change.

The first issue in many changes is to diagnose the current system to identify the source of problems (or opportunities for improvement). In a large organization, this frequently leads to a rethinking of strategy, and a redefinition of the organization's task or work. For example, AT&T examines the environment and determines that it needs to change the primary orientation of its strategy, and thus, its task from service towards marketing.

The analysis of strategy and redefinition of task is an important step in changing an organization. On the other hand, many of the most troublesome problems of changing organizations occur not in the strategic/task shift, but in the implementation of the organizational transition to support the change in the nature of the strategy and the work. More specifically, any major organizational change presents three major problems which must be dealt with.

First is the problem of *resistance* to change (Watson, 1969; Zaltman & Duncan, 1977). Any individual faced with a change in the organization in which he/she works may be resistant for a variety of reasons. People have need for a certain degree of stability or security; change presents unknowns which cause anxiety. In addition, a change that is imposed on an individual reduces his/her sense of autonomy or self-control. Furthermore, people typically develop patterns for coping with or managing the current structure and situation. Change means that they will have to find new ways of managing their own environments—ways that might not be as successful as those currently used. In addition, those who have power in the current situation may resist change because it threatens that power. They have a vested interest in the status quo. Finally, individuals may resist change for ideological reasons; they truly believe that the way things are done currently is better than the proposed change. Whatever the source, individual resistance to change must be overcome for implementation of a change to be successful.

A second problem is that of organizational *control*. Change disrupts the normal course of events within an organization. It thus disrupts and undermines existing systems of management control, particularly those developed as part of the formal organizational arrangements. Change may make those systems irrelevant and/or inappropriate. As a result, during a change, it may become easy to lose control of the organization. As goals, structures, and people shift, it becomes difficult to monitor performance and make corrections as in normal control processes.

A related problem is that most formal organizational arrangements are designed for stable states, not transition states. Managers become fixated on the future state (B) and assume that all that is needed is to design the most effective organizational arrangements for the future. They think of change from A to B as simply a mechanical or procedural detail. The problems created by the lack of concern for the transition state are compounded by the inherent uniqueness of it. In most situations, the management systems and structures developed to manage A or B are simply not appropriate or adequate for the management of C. They are steady state management systems, designed to run organizations already in place, rather than transitional management systems.

The third problem is *power*. Any organization is a political system made up of different individuals, groups, and coalitions competing for power (Tushman, 1977; Salancik & Pfeffer, 1977).

Political behavior is thus a natural and expected feature of organizations. This occurs in both states A and B. In state C (transition), however, these dynamics become even more intense as the old order is dismantled and a new order emerges. This happens because any significant change poses the possibility of upsetting or modifying the balance of power among groups. The uncertainty created by change creates ambiguity, which in turn tends to increase the probability of political activity (Thompson & Tuden, 1959). Individuals and groups may take action based on their perception of how the change will affect their relative power position in the organization. They will try to influence where they will sit in the organization that emerges from the transition, and will be concerned about how the conflict of the transition period will affect the balance of power in the

future state. Finally, individuals and groups may engage in political action because of their ideological position on the change—it may be inconsistent with their shared values or image of the organization (Pettigrew, 1972).

In some sense, each of these problems is related primarily to one of the components of the organization (see figure 4). Resistance relates to the individual component, getting people to change their behavior. Control concerns the design of appropriate organizational arrangements for the transition period. Power relates to the reactions of the informal organization to change. Therefore, if a change is to be effective, all three problems—resistance, control, and power—must be addressed.

GUIDELINES FOR IMPLEMENTING CHANGE

The three basic problems that are inherent in change each lead to a general implication for the management of change (see figure 5 below).

The implication of the resistance problem is the need to *motivate* changes in behavior by individuals. This involves overcoming the natural resistance to change that emerges, and getting individuals to behave in ways consistent with both the short-run goals of change and the long-run organizational strategy.

The implication of the control problem is the need to *manage the transition*. Organizational arrangements must be designed and used to ensure that control is maintained during and after the transition. They must be specifically appropriate to the transition period rather than to the current or future state.

Finally, the implication of the power issue is the need to *shape the political dynamics* of change so that power centers develop that support the change, rather than block it (Pettigrew, 1975).

Each of these general implications suggests specific actions that can be taken to im-

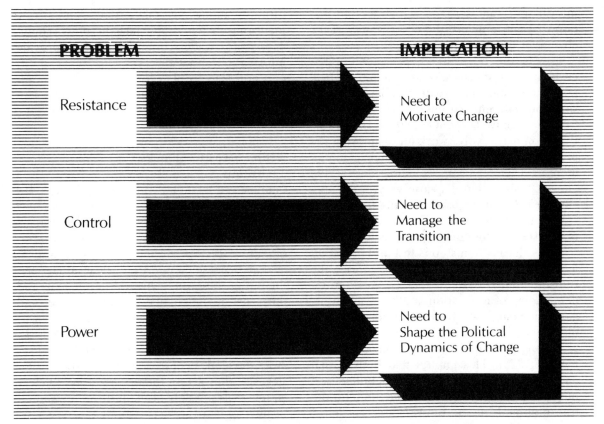

FIGURE 5. *Problems of Change and Implications for Change Management*

prove the chances of achieving an effective change. A number of action steps can be identified for each of the three implications.

Action Steps to Motivate Change

The first action step is to *identify and surface dissatisfaction with the current state.* As long as people are satisfied with the current state, they will not be motivated to change; people need to be "unfrozen" out of their inertia in order to be receptive to change (Lewin, 1947; Bennis et al, 1973). The greater the pain and dissatisfaction with the current state, the greater the motivation to change and the less the resistance to change. As a consequence, the management of change may require the creation of pain and dissatisfaction with the status quo. Dissatisfac-

tion most commonly results from information concerning some aspect of organizational performance which is different from either desired or expected performance. Discrepancies can therefore be used to create dissatisfaction. As a result, data can be an important tool to initiate a process of change (Nadler, 1977).

The second action step is to build in *participation* in the change. One of the most consistent findings in the research on change is that participation in the change tends to reduce resistance, build ownerships of the change, and thus motivate people to make the change work (Coch & French, 1948; Vroom, 1964; Kotter & Schlesinger, 1979). Participation also facilitates the communication of information about what the change will be and why it has come about. Participation may also lead to obtaining new

725

information from those participating, information that may enhance the effectiveness of the change or the future state.

On the other hand, participation has costs since it involves relinquishing control, takes time, and may create conflict. For each situation, different degrees of participation may be most effective (Vroom & Yetton, 1973). Participation may involve work on diagnosing the present situation, in planning change, in implementing change, or in combinations of the above. Participation may also vary in the specific devices that are used, ranging from large-scale data collection to sensing groups, to questionnaires, to cross unit committees, etc.

A third action step is to build in *rewards* for the behavior that is desired both during the transition state and in the future state. Our understanding of motivation and behavior in organizations suggests that people will tend to be motivated to behave in ways that they perceive as leading to desired outcomes (Vroom, 1964; Lawler, 1973). This implies that both formal and informal rewards must be identified and tied to the behavior that is needed, both for the transition and for the future state. The most frequent problem is that organizations expect individuals to behave in certain ways (particularly in a transition) while rewarding them for other conflicting behaviors (Kerr, 1975). In particular, rewards such as bonuses, pay systems, promotion, recognition, job assignment, and status symbols all need to be carefully examined during major organizational changes and restructured to support the direction of the transition.

Finally, people need to be provided with the *time and opportunity to disengage from the present state.* Change frequently creates feelings of loss, not unlike a death. People need to mourn for the old system or familiar way of doing things. This frequently is manifested in the emergence of stories or myths about the "good old days," even when those days weren't so good. The process of dealing with a loss and going

through mourning takes time, and those managing change should take this into account. This factor underscores the need to provide information about the problems of the status quo and also to plan for enough time in advance of a change to allow people to deal with the loss and prepare for it.

Action Steps to Manage the Transition

One of the first and most critical steps for managing the transition state is to *develop and communicate a clear image of the future.* (Beckhard & Harris, 1977). Resistance and confusion frequently develop during an organizational change because people are unclear about what the future state will be like. Thus the goals and purposes of the change become blurred, and individual expectancies get formed on the basis of information that is frequently erroneous. In the absence of a clear image of the future, rumors develop, people design their own fantasies, and they act on them. Therefore, as clear an image as possible of the future state should be developed to serve as a guideline, target, or goal. In particular, a written statement or description of the future state may be of value in clarifying the image. Similarly, it is important to communicate information to those involved in the change, including what the future state will be like, how the transition will come about, why the change is being implemented, and how individuals will be affected by the change. This communication can be accomplished in a variety of ways, ranging from written communications to small group meetings, large briefing sessions, video-taped presentations, etc.

A second action step for managing the transition involves the use of *multiple and consistent leverage points.* If, building on the model presented above, an organization is made up of components which are interdependent, then the successful alteration of organizational behavior patterns must involve the use of multiple

726

leverage points, or modifications in the larger set of components which shape the behavior of the organization and the people in it (Nadler & Tichy, 1980). Structural change, task change, change in the social environment, as well as changes in individuals themselves are all needed to bring about significant and lasting changes in the patterns of organizational behavior. Changes that are targeted at individuals and social relations (such as training, group interventions, etc.) tend to fade out quickly with few lasting effects when done in isolation (Porter, Lawler & Hackman, 1975). On the other hand, task and structural changes alone, while powerful and enduring, frequently produce unintended and dysfunctional consequences (see, for example, literature on control systems; e.g., Lawler & Rhode, 1976). Change which is in the direction intended and which is lasting therefore requires the use of multiple leverage points to modify more than a single component. Similarly, the changes have to be structured so that they are consistent; the training of individuals, for example, should dovetail with new job descriptions, rewards systems, or reporting relationships. In the absence of consistency, changes run the risk of creating new "poor fits" among organizational components. The result is either an abortive change, or decreases in organizational performance.

The third action step involves a number of different activities. *Organizational arrangements for the transition* need to be explicitly considered, designed, and used. As mentioned earlier, the organizational arrangements that function in either the present or future state are typically steady state designs, rather than designs for use in managing the transition state. The whole issue of developing structures to manage the transition has been discussed in depth elsewhere (see Beckhard & Harris, 1977), but a number of the most important elements should be mentioned here. In particular, the following organizational arrangements are important for managing the change:

A. A Transition Manager. Someone should be designated as the manager of the organization for the transition state. This person may be a member of management, a chief executive, or someone else, but frequently it is difficult for one person to manage the current state, prepare to manage the future state, and simultaneously manage the transition. This person should have the power and authority needed to make the transition happen, and should be appropriately linked to the steady state managers, particularly the future state manager.

B. Resources for the Transition. Major transitions involve potentially large risks for organizations. Given this, they are worth doing well and it is worth providing the needed resources to make them happen effectively. Resources such as personnel, dollars, training expertise, consultative expertise, etc. must be provided for the transition manager.

C. Transition Plan. A transition is a movement from one state to another. To have that occur effectively, and to measure and control performance, a plan is needed with benchmarks, standards of performance, and similar features. Implicit in such a plan is a specification of the responsibilities of key individuals and groups.

D. Transition Management Structures. Frequently it is difficult for a hierarchy to manage the process of change itself. As a result, it may be necessary to develop other structures or use other devices outside the regular organizational structure during the transition management period. Special task forces, pilot projects, experimental units, etc. need to be designed and employed for this period (see again Beckhard & Harris, 1977 for a discussion of these different devices).

The final action step for transition management involves developing *feedback mechanisms* to provide transition managers with information on the effectiveness of the transition

727

and provide data on areas which require additional attention or action. There is a huge amount of anecdotal data about senior managers ordering changes and assuming those changes were made, only to find out to their horror that the change never occurred. Such a situation develops because managers lack feedback devices to tell them whether actions have been effective or not. During stable periods, effective managers tend to develop various ways of eliciting feedback. During the transition state, however, these mechanisms often break down due to the turbulence of the change, or because of the natural inclination not to provide "bad news." Thus, it becomes important for transition managers to develop multiple, redundant, and sensitive mechanisms for generating feedback about the transition. Devices such as surveys, sensing groups, consultant interviews, etc. as well as informal communication channels need to be developed and used during this period.

Action Steps for Shaping the Political Dynamics of Change

If an organization is a political system composed of different groups competing for power, then the most obvious action step involves *ensuring or developing the support of key power groups.* For a change to occur successfully, a critical mass of power groups has to be assembled and mobilized in support of the change. Those groups that may oppose the change have to in some way be compensated for or have their effects neutralized. Not all power groups have to be intimately involved in the change. Some may support the change on ideological grounds, while others may support the change because it enhances their own power position. With other groups, they will have to be included in the planning of the change so that their participation will motivate them, or co-opt them (Selznick, 1949). Still others may have to be dealt with by bargaining or negotiations. The main point is that the key groups who may

be affected by the change need to be identified, and strategies for building support among a necessary portion of those groups need to be developed and carried out (Sayles, 1979).

A major factor affecting the political terrain of an organization is the behavior of key and powerful leaders. Thus a second major action step involves *using leader behavior to generate energy in support of the change.* Leaders can mobilize groups, generate energy, provide models, manipulate major rewards, and do many other things which can affect the dynamics of the informal organization. Sets of leaders working in coordination can have a tremendously powerful impact on the informal organization. Thus leaders need to think about using their own behavior to generate energy (see House, 1976 on charismatic leadership) as well as to build on the support and behavior of other leaders (both formal and informal) within the organization.

The third action step involves *using symbols and language to create energy* (Peters, 1978; Pfeffer, 1980). By providing a language to describe the change and symbols that have emotional impact, it is possible to create new power centers or to bring together power centers under a common banner. Language is also important in defining an ambiguous reality. If, for example, a change is declared a success then it may become a success in the perception of others.

Finally, there is the need to *build in stability.* Organizations and individuals can only stand so much uncertainty and turbulence. An overload of uncertainty may create dysfunctional effects, as people may begin to panic, engage in extreme defensive behavior, and become irrationally resistant to any new change proposed. The increase of anxiety created by constant change thus has its costs. One way of dealing with this is to provide some sources of stability (structures, people, physical locations, etc. that stay the same) that serve as "anchors" for people to hold onto and provide a means for definition of the self in the midst of turbulence. While too many anchors can encourage resistance, it is important to provide some stability. More importantly, it is necessary to communicate the stability. People may not take comfort from something that is stable if they are unsure of its stability. Thus those aspects of the organization that will not change during a transition period need to be identified and communicated to organization members.

SUMMARY

This paper has attempted to identify some of the problems and issues of bringing about changes in complex organizations. At the same time, a number of general and specific action steps have been suggested. To understand how to change organizational behavior, we need a tool to understand how it occurs in the first place. The model used here (Nadler & Tushman, 1977; 1979) suggests that any change will encounter three general problems: resistance, control, and power. The general implication is the need to motivate change, manage the transition, and shape the political dynamics of change. For each of these three general implications, a number of specific action steps have been identified (see figure 6 below).

Obviously, each of these action steps will be more or less critical (and more or less feasible) in different situations. Thus students of

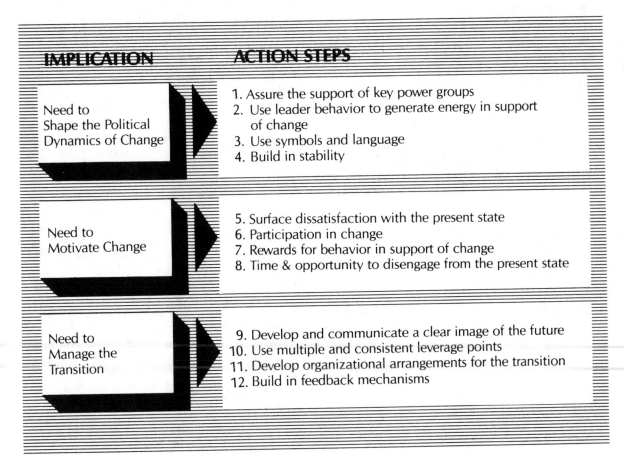

IMPLICATION	ACTION STEPS
Need to Shape the Political Dynamics of Change	1. Assure the support of key power groups 2. Use leader behavior to generate energy in support of change 3. Use symbols and language 4. Build in stability
Need to Motivate Change	5. Surface dissatisfaction with the present state 6. Participation in change 7. Rewards for behavior in support of change 8. Time & opportunity to disengage from the present state
Need to Manage the Transition	9. Develop and communicate a clear image of the future 10. Use multiple and consistent leverage points 11. Develop organizational arrangements for the transition 12. Build in feedback mechanisms

FIGURE 6. *Implications for Change Management and Related Action Steps*

organization and managers alike need to be diagnostic in their approach to the problems of managing change. Each situation, while reflecting general patterns, has unique characteristics, based on its own differences of individuals, history, and situation. Thus specific variants of the action steps need to be developed for specific situations. To do this, managers need diagnostic models to understand problems, as well as guidelines for implementing changes, as presented here. Together, these two types of tools can be powerful aids in building and maintaining effective organizations.

REFERENCES

Beckhard, R. & Harris, R. *Organizational transitions.* Reading, Massachusetts: Addison-Wesley, 1977.

Bennis, W.G., Berlew, D.E., Schein, E.H. & Steele, F.I. *Interpersonal dynamics: Essays and readings on human interaction.* Homewood, Ill.: Dorsey Press, 1973.

Coch, L. & French, J.R.P., Jr. Overcoming resistance to change. *Human Relations,* 1948, **11,** 512–532.

House, Robert J. A 1976 theory of charismatic leadership (mimeo). Faculty of Management Studies, University of Toronto, 1976.

Katz, D. & Kahn, R.L. *The social psychology of organizations.* New York: John Wiley & Sons, 1966.

Kerr, S. On the folly of rewarding A while hoping for B. *Academy of Management Journal,* December, 1975, 769–783.

Kotter, J.P. & Schlesinger, L.A. Choosing strategies for change. Harvard Business Review, 1979 (March-April), 106–114.

Lawler, E.E. *Motivation in work organizations.* Belmont, California: Wadsworth Publishing Co., 1973.

Lawler, E.E. & Rhode, J.G. *Information and control in organizations.* Santa Monica, California: Goodyear, 1976.

Lewin, K. Frontiers in group dynamics. *Human Relations,* 1947, **1,** 5–41.

Nadler, D.A. & Tushman, M.L. A congruence model for diagnosing organizational behavior. In D.A. Nadler, M.L. Tushman & N.G. Hatvany (eds.). *Approaches to Managing Organizational Behavior: Models, readings, and cases.* Boston: Little, Brown, 1981.

Nadler, D.A. *Feedback and organization development: Using data based methods.* Reading, Massachusetts: Addison-Wesley, 1977.

Nadler, D.A. & Tushman, M.L. A congruence model for diagnosing organizational behavior. In D. Kolb, I. Rubin, & J. McIntyre. *Organizational Psychology: A book of readings.* (3rd edition), Englewood Cliffs, N.J.: Prentice-Hall, 1979.

Nadler, D.A. & Tichy, N.M. The limitations of traditional intervention technology in health care organizations. In N. Margulies & J. Adams (eds.) *Organization development in health care organizations.* Reading, Mass: Addison-Wesley, 1980.

Peters, T.J. Symbols, patterns, and settings: An optimistic case for getting things done. *Organizational Dynamics,* 1978 (Autumn), 3–23.

Pettigrew, A. *The politics of organizational decision-making.* London: Tavistock Press, 1972.

Pettigrew, A. Towards a political theory of organizational intervention. *Human Relations,* 1978, **28,** 191–208.

Pfeffer, J. Management as symbolic action: The creation and maintainance of organizational paradigms. In L.L. Cummings & B.M. Staw (eds.) *Research in organizational behavior* (Vol. 3), JAI Press, 1980.

Porter, L.W., Lawler, E.E. & Hackman, J.R. *Behavior in organizations,* New York: McGraw-Hill, 1975.

Salancik, G.R. & Pfeffer, J. Who gets power and how they hold on to it: A strategic-contingency model of power. *Organizational Dynamics,* 1977 (Winter), 3–21.

Sayles, L.R. *Leadership: what effective managers really do and how they do it.* McGraw-Hill, 1979.

Selznick, P. *TVA and the Grass Roots.* Berkley: University of California Press, 1949.

Thompson, J.D. & Tuden, A. Strategies, structures and processes of organizational decision. In J.D. Thompson et al (eds.). *Comparative studies in administration.* Pittsburgh: University of Pittsburgh Press, 1959.

Tushman, M.L. A political approach to organizations: a review and rationale. *Academy of Management Review,* 1977, **2,** 206–216.

Vroom, V.H. *Work and motivation.* New York: Wiley, 1964.

Vroom, V.H. and Yetton, P.W. Leadership and decision making. Pittsburgh: University of Pittsburgh Press, 1973.

Watson, G. Resistance to change. In W.G. Bennis, K.F. Benne & R. Chin (eds.) *The planning of change.* New York: Holt, Rinehart, Winston, 1969.

Zaltman, G. & Duncan, R. *Strategies for planned change.* New York: John Wiley, 1977.

IMPLEMENTATION ANALYSIS GUIDE

	IMPLEMENTATION PRACTICES	RATING	COMMENTS/EXPLANATION
SHAPING POLITICAL DYNAMICS	1. Getting the support of key power groups.		
	2. Using leader behavior to support the direction of change.		
	3. Using symbols and language.		
	4. Building in stability.		
MOTIVATING CHANGE	5. Creating dissatisfaction with the status quo.		
	6. Participation in planning and/or implementing change.		
	7. Rewarding needed behavior in transition and future states.		
	8. Providing time and opportunity to disengage from current state.		
MANAGING THE TRANSITION	9. Developing and communicating a clear image of the future state.		
	10. Using multiple and consistent leverage points.		
	11. Using transition management structures.		
	12. Building in feedback and evaluation of the transition.		

Rating is an assessment of the general quality of action in each implementation practice area. Scale for ratings is — — — — — — — —

5 = Very good
4 = Good
3 = Fair
2 = Poor
1 = Very Poor

SUMMARY RATINGS

Motivation	Transition	Political	Overall

Managing Strategic Change

James Brian Quinn

In two articles published previously in the SMR, the author described the process of "logical incrementalism" for strategic planning and how it is used effectively in several large corporations. This third and final article in the series analyzes this approach to management—a sort of purposeful "muddling"—in greater detail, delineating the steps which successful managers generally follow in inaugurating and executing stategic change. Ed.

"Just as bad money has always driven out good, so the talented general manager—the person who makes a company go—is being overwhelmed by a flood of so-called 'professionals,' textbook executives more interested in the form of management than the content, more concerned about defining and categorizing and quantifying the job, than in getting it done. . . . They have created false expectations and wasted untold man-hours by making a religion of formal long-range planning."[1] H. E. Wrapp, *New York Times.*

Two previous articles have tried to demonstrate why executives managing strategic change in large organizations should not—and do not—follow highly formalized textbook approaches in long-range planning, goal generation, and strategy formulation.[2] Instead, they artfully blend formal analysis, behavioral techniques, and power politics to bring about cohesive, step-by-step movement toward ends

which initially are broadly conceived, but which are then constantly refined and reshaped as new information appears.[3] Their integrating methodology can best be described as "logical incrementalism."

But is this truly a process in itself, capable of being managed? Or does it simply amount to applied intuition? Are there some conceptual structures, principles, or paradigms that are generally useful? Wrapp, Normann, Braybrooke, Lindblom, and Bennis have provided some macrostructures incorporating many important elements they have observed in strategic change situations.[4] These studies and other contributions cited in this article offer important insights into the management of change in large organizations. But my data suggest that top managers in such enterprises develop their major strategies through processes which neither these studies nor more formal approaches to planning adequately explain. Managers *consciously* and *proactively* move forward *incrementally:*

·   To improve the quality of information utilized in corporate strategic decisions.

- To cope with the varying lead times, pacing parameters, and sequencing needs of the "subsystems" through which such decisions tend to be made.
- To deal with the personal resistance and political pressures any important strategic change encounters.
- To build the organizational awareness, understanding, and psychological commitment needed for effective implementation.
- To decrease the uncertainty surrounding such decisions by allowing for interactive learning between the enterprise and its various impinging environments.
- To improve the quality of the strategic decisions themselves by (1) systematically involving those with most specific knowledge, (2) obtaining the participation of those who must carry out the decisions, and (3) avoiding premature momenta or closure which could lead the decision in improper directions.

How does one manage the complex incremental processes which can achieve these goals? The earlier articles structured certain key elements;[5] these will not be repeated here. The following is perhaps the most articulate short statement on how executives proactively manage incrementalism in the development of corporate strategies:

Typically you start with general concerns, vaguely felt. Next you roll an issue around in your mind till you think you have a conclusion that makes sense for the company. You then go out and sort of post the idea without being too wedded to its details. You then start hearing the arguments pro and con, and some very good refinements of the idea usually emerge. Then you pull the idea in and put some resources together to study it so it can be put forward as more of a formal presentation. You wait for "stimuli occurrences" or "crises," and launch pieces of the idea to help in these situations. But they lead toward your ultimate aim. You know where you want to get. You'd like to get there in

six months. But it may take three years, or you may not get there. And when you do get there, you don't know whether it was originally your own idea—or somebody else had reached the same conclusion before you and just got you on board for it. You never know. The president would follow the same basic process, but he could drive it much faster than an executive lower in the organization.[6]

Because of differences in organizational form, management style, or the content of individual decisions, no single paradigm can hold for all strategic decisions.[7] However, very complex strategic decisions in my sample of large organizations tended to evoke certain kinds of broad process steps. These are briefly outlined below. While these process steps occur generally in the order presented, stages are by no means orderly or discrete. Executives do consciously manage individual steps proactively, but it is doubtful that any one person guides a major strategic change sequentially through all the steps. Developing most strategies requires numerous loops back to earlier stages as unexpected issues or new data dictate. Or decision times can become compressed and require short-circuiting leaps forward as crises occur.[8] Nevertheless, certain patterns are clearly dominant in the successful management of strategic change in large organizations.

CREATING AWARENESS AND COMMITMENT— INCREMENTALLY

Although many of the sample companies had elaborate formal environmental scanning procedures, most major strategic issues first emerged in vague or undefined terms, such as "organizational overlap," "product proliferation," "excessive exposure in one market," or "lack of focus and motivation."[9] Some appeared as "inconsistencies" in internal action patterns or "anomalies" between the enterprise's current posture and some perception of its future environment.[10] Early signals may come from anywhere and may be difficult to distinguish from the background "noise" of ordinary communications. Crises, of course, announce themselves with strident urgency in operations control systems. But, if organizations wait until signals reach amplitudes high enough to be sensed by formal measurement systems, smooth, efficient transitions may be impossible.[11]

Need Sensing: Leading the Formal Information System

Effective change managers actively develop informal networks to get objective information—from other staff and line executives, workers, customers, board members, suppliers, politicians, technologists, educators, outside professionals, government groups, and so on—to sense possible needs for change. They purposely use these networks to short-circuit all the careful screens[12] their organizations build up to "tell the top only what it wants to hear." For example:

> Peter McColough, chairman and CEO of Xerox, was active in many high-level political and charitable activities—from treasurer of the Democratic National Committee to chairman of the Urban League. In addition, he said, "I've tried to decentralize decision making. If something bothers me, I don't rely on reports or what other executives may want to tell me. I'll go down very deep into the organization, to certain issues and people, so I'll have a feeling for what they think." He refused to let his life be run by letters and memos. "Because I came up by that route, I know what a salesman can say. I also know that before I see [memos] they go through fifteen hands, and I know what that can do to them."[13]

To avoid undercutting intermediate managers, such bypassing has to be limited to information gathering, with no implication that orders or approvals are given to lower levels. Properly handled, this practice actually improves formal communications and motivational systems as well. Line managers are less tempted to screen information and lower levels are flattered to be able "to talk to the very top." Since people sift signals about threats and opportunities through perceptual screens defined by their own values, careful executives make sure their sensing networks include people who look at the world very differently than do those in the enterprise's dominating culture. Effective executives consciously seek options and threat signals beyond the *status quo.* "If I'm not two to three years ahead of my organization, I'm not doing my job" was a common comment of such executives in the sample.

Amplifying Understanding and Awareness

In some cases executives quickly perceive the broad dimensions of needed change. But they still may seek amplifying data, wider executive understanding of issues, or greater organizational support before initiating action. Far from accepting the first satisfactory (satisficing) solution—as some have suggested they do—successful managers seem to consciously generate and consider a broad array of alternatives.[14] Why? They want to stimulate and choose from the most creative solutions offered by the best minds in their organizations. They wish to have colleagues knowledgeable enough about issues to help them think through all the ramifications. They seek data and arguments sufficiently strong to dislodge preconceived ideas or blindly followed past practices. They do not want to be the prime supporters of losing ideas or to have their organizations slavishly adopt "the boss's solution." Nor do they want—through announcing decisions too early—to prematurely threaten existing power centers which could kill any changes aborning.

Even when executives do not have in mind specific solutions to emerging problems, they can still proactively guide actions in intuitively desired directions—by defining what issues staffs should investigate, by selecting principal investigators, and by controlling reporting processes. They can selectively "tap the collective wit" of their organizations, generating more awareness of critical issues and forcing initial thinking down to lower levels to achieve greater involvement. Yet they can also avoid irreconcilable opposition, emotional overcommitment,[15] or organizational momenta beyond their control by regarding all proposals as "strictly advisory" at this early stage.

As issues are clarified and options are narrowed, executives may systematically alert ever wider audiences. They may first "shop" key ideas among trusted colleagues to test responses. Then they may commission a few studies to illuminate emerging alternatives, contingencies, or opportunities. But key players might still not be ready to change their past action patterns or even be able to investigate options creatively. Only when persuasive data are in hand and enough people are alerted and "on board" to make a particular solution work, might key executives finally commit themselves to it. Building awareness, concern, and interest to attention-getting levels is often a vital—and slowly achieved—step in the process of managing basic changes. For example:

> In the early 1970s there was still a glut in world oil supplies. Nevertheless, analysts in the General Motors Chief Economist's Office began to project a developing U.S. dependency on foreign oil and the likelihood of higher future oil prices. These concerns led the board in 1972 to create an ad hoc energy task force headed by David C. Collier, then treasurer, later head of GM of Canada and then of the Buick Division. Collier's group included people from manufacturing, research, design, finance, industry-government relations, and the economics staff. After six months of research, in May of 1973 the task force went to the board with three conclusions: (1) there was a developing energy prob-lem, (2) the government had no particular plan to deal with it, (3) energy costs would have a profound effect on GM's business. Collier's report created a good deal of discussion around the company in the ensuing months. "We were trying to get other people to think about the issue," said Richard C. Gerstenberg, then chairman of GM.[16]

Changing Symbols: Building Credibility

As awareness of the need for change grows, managers often want to signal the organization that certain types of changes are coming, even if specific solutions are not in hand. Knowing they cannot communicate directly with the thousands who would carry out the strategy, some executives purposely undertake highly visible actions which wordlessly convey complex messages that could never be communicated as well—or as credibly—in verbal terms.[17] Some use symbolic moves to preview or verify intended changes in direction. At other times, such moves confirm the intention of top management to back a thrust already partially begun—as Mr. McColough's relocation of Xerox headquarters to Connecticut (away from the company's Rochester reprographics base) underscored that company's developing commitment to product diversification, organizational decentralization, and international operations. Organizations often need such symbolic moves—or decisions they regard as symbolic—to build credibility behind a new strategy. Without such actions even forceful verbiage might be interpreted as mere rhetoric. For example:

In GM's downsizing decision engineers said that one of top management's early decisions affected the credibility of the whole weight-reduction program. "Initially, we proposed a program using a lot of aluminum and substitute materials to meet the new 'mass' targets. But this would have meant a very high cost and would have strained the suppliers aluminum capacity. However, when we presented this program to management, they said, 'Okay, if necessary, we'll do it.' They didn't back down. We began to understand then that they were dead serious. Feeling that the company would spend the money was critical to the success of the entire mass reduction effort."[18]

Legitimizing New Viewpoints

Often before reaching specific strategic decisions, it is necessary to legitimize new options which have been acknowledged as possibilities, but which still entail an undue aura of uncertainty or concern. Because of their familiarity, older options are usually perceived as having lower risks (or potential costs) than newer alternatives. Therefore, the managers seeking change often consciously create forums and allow slack time for their organizations to talk through threatening issues, work out the implications of new solutions, or gain an improved information bank that will permit new options to be evaluated objectively in comparison with more familiar alternatives.[19] In many cases, strategic concepts which are at first strongly resisted, gain acceptance and support simply by the passage of time, if executives do not exacerbate hostility by pushing them too fast from the top. For example:

When Joe Wilson thought Haloid Corporation should change its name to include Xerox, he first submitted a memorandum asking colleagues what they thought of the idea. They rejected it. Wilson then explained his concerns more fully, and his executives rejected the idea again. Finally Wilson formed a committee headed by Sol Linowitz who had thought a separate Xerox subsidiary might be the best solution. As this committee deliberated, negotiations were under way with the Rank Organization and the term Rank-Xerox was

commonly heard and Haloid-Xerox no longer seemed so strange. "And so," according to John Dessauer, "six-month delay having diluted most opposition, we of the committee agreed that the change to Haloid-Xerox might in the long run produce sound advantages."[20]

Many top executives consciously plan for such "gestation periods" and often find that the strategic concept itself is made more effective by the resulting feedback.

Tactical Shifts and Partial Solutions

At this stage in the process guiding executives might share a fairly clear vision of the general directions for movement. But rarely does a total new corporate posture emerge full grown—like Minerva from the brow of Jupiter—from any one source. Instead, early resolutions are likely to be partial, tentative, or experimental.[21] Beginning moves often appear as mere tactical adjustments in the enterprise's existing posture. As such, they encounter little opposition, yet each partial solution adds momentum in new directions. Guiding executives try carefully to maintain the enterprise's ongoing strengths while shifting its total posture incrementally—at the margin—toward new needs. Such executives themselves might not yet perceive the full nature of the strategic shifts they have begun. They can still experiment with partial new approaches and learn without risking the viability of the total enterprise. Their broad early steps can still legitimately lead to a variety of different success scenarios. Yet logic might dictate that they wait before committing themselves to a total new strategy.[22] As events unfurl, solutions to several interrelated problems might well flow together in a not-yet-perceived synthesis. For example:

In the early 1970s at General Motors there was a distinct awareness of a developing fuel economy ethic. General Motors executives said, "Our conclusions were really at the conversational level—that the big car trend was at an end. But we were not at all sure sufficient numbers of large car buyers were ready to move to dramatically lighter cars." Nevertheless, GM did start concept studies that resulted in the Cadillac Seville.

When the oil crisis hit in fall 1973, the company responded in further increments, at first merely increasing production of its existing small car lines. Then as the crisis deepened, it added another partial solution, the subcompact "T car"—the Chevette—and accelerated the Seville's development cycle. Next, as fuel economy appeared more saleable, executives set an initial target of removing 400 pounds from B-C bodies by 1977. As fuel economy pressures persisted and engineering feasibilities offered greater confidence, this target was increased to 800–1000 pounds (three mpg). No step by itself shifted the company's total strategic posture until the full downsizing of all lines was announced. But each partial solution built confidence and commitment toward a new direction.

Broadening Political Support

Often these broad emerging strategic thrusts need expanded political support and understanding to achieve sufficient momentum to survive.[23] Committees, task forces, and retreats tend to be favored mechanisms for accomplishing this. If carefully managed, these do not become the "garbage cans" of emerging ideas, as some observers have noted.[24] By selecting the committee's chairman, membership, timing, and agenda, guiding executives can largely influence and predict a desired outcome, and can force other executives toward a consensus. Such groups can be balanced to educate, evaluate, neutralize, or overwhelm opponents. They can be used to legitimize new options or to generate broad cohesion among diverse thrusts, or they can be narrowly focused to

build momentum. Guiding executives can constantly maintain complete control over these "advisory processes" through their various influences and veto potentials. For example:

IBM's Chairman Watson and Executive Vice President Learson had become concerned over what to do about: third generation computer technology, a proliferation of designs from various divisions, increasing costs of developing software, internal competition among their lines, and the needed breadth of line for the new computer applications they began to foresee. Step by step, they oversaw the killing of the company's huge Stretch computer line (uneconomic), a proposed 8000 series of computers (incompatible software), and the prototype English Scamp Computer (duplicative). They then initiated a series of "strategic dialogues" with divisional executives to define a new strategy. But none came into place because of the parochial nature of divisional viewpoints.

Learson, therefore, set up the SPREAD Committee, representing every major segment of the company. Its twelve members included the most likely opponent of an integrated line (Haanstra), the people who had earlier suggested the 8000 and Scamp designs, and Learson's handpicked lieutenant (Evans). When progress became "hellishly slow," Haanstra was removed as chairman and Evans took over. Eventually the committee came forth with an integrating proposal for a single, compatible line of computers to blanket and open up the market for both scientific and business applications, with "standard interface" for peripheral equipment. At an all-day meeting of the fifty top executives of the company, the report was not received with enthusiasm, but there were no compelling objections. So Learson blessed the silence as consensus saying, "OK, we'll do it"—i.e., go ahead with a major development program.[25]

In addition to facilitating smooth implementation, many managers reported that in-

738

teractive consensus building processes also improve the quality of the strategic decisions themselves and help achieve positive and innovative assistance when things otherwise could go wrong.

Overcoming Opposition: "Zones of Indifference" and "No Lose" Situations

Executives of basically healthy companies in the sample realized that any attempt to introduce a new strategy would have to deal with the support its predecessor had. Barring a major crisis, a frontal attack on an old strategy could be regarded as an attack on those who espoused it—perhaps properly—and brought the enterprise to its present levels of success. There often exists a variety of legitimate views on what could and should be done in the new circumstances that a company faces. And wise executives do not want to alienate people who would otherwise be supporters. Consequently, they try to get key people behind their concepts whenever possible, to co-opt or neutralize serious opposition if necessary, or to find "zones of indifference" where the proposition would not be disastrously opposed.[26] Most of all they seek "no lose" situations which will motivate all the important players toward a common goal. For example:

When James McFarland took over at General Mills from his power base in the Grocery Products Division, another serious contender for the top spot had been Louis B. "Bo" Polk, a very bright, aggressive young man who headed the corporation's acquisition-diversification program. Both traditional lines and acquisitions groups wanted support for their activities and had high-level supporters. McFarland's corporate-wide "goodness to greatness" conferences (described in earlier articles) first obtained broad agreement on growth goals and criteria for all areas.

Out of this and the related acquisition proposal process came two thrusts: (1) to expand—internally and through acquisitions—in food-related sectors and (2) to acquire new growth centers based on General Mill's marketing skills. Although there was no formal statement, there was a strong feeling that the majority of resources should be used in food-related areas. But neither group was foreclosed, and no one could suggest the new management was vindictive. As it turned out, over the next five years about $450 million was invested in new businesses, and the majority were not closely related to foods.

But such tactics do not always work. Successful executives surveyed tended to honor legitimate differences in viewpoints and noted that initial opponents often shaped new strategies in more effective directions and became supporters as new information became available. But strong-minded executives sometimes disagreed to the point where they had to be moved or stimulated to leave; timing could dictate very firm top-level decisions at key junctures. Barring crises, however, disciplinary steps usually occurred incrementally as individual executives' attitudes and competencies emerged vis-à-vis a new strategy.

Structuring Flexibility: Buffers, Slacks, and Activists

Typically there are too many uncertainties in the total environment for managers to program or control all the events involved in effecting a major change in strategic direction. Logic dictates, therefore, that managers purposely design flexibility into their organizations and have resources ready to deploy incrementally as events demand. Planned flexibility requires: (1) proactive horizon scanning to identify the general nature and potential impact of opportunities and threats the firm is most likely to encounter, (2) creating sufficient resource buffers—or slacks—to respond effectively as events actually unfurl, (3) developing and positioning "credible activists" with a psychological commitment to move quickly and flexibly to exploit specific opportunities as they occur, and (4) shortening decision lines from such people (and key operating managers) to the top for the most rapid system response. These—rather than pre-capsuled (and shelved) programs to respond to stimuli which never quite occur as expected—are the keys to real contingency planning.

The concept of resource buffers requires special amplification. Quick access to resources is needed to cushion the impact of random events, to offset opponents' sudden attacks, or to build momentum for new strategic shifts. Some examples will indicate the form these buffers take.

For critical purchased items, General Motors maintained at least three suppliers, each with sufficient capacity to expand production should one of the others encounter a catastrophe. Thus, the company had expandable capacity with no fixed investment. Exxon set up its Exploration Group to purposely undertake the higher risks and longer-term investments necessary to search for oil in new areas, and thus to reduce the potential impact on Exxon if there were sudden unpredictable changes in the availability of Middle East oil. Instead of hoarding cash, Pillsbury and General Mills sold off unprofitable businesses and cleaned up their financial statements to improve their access to external capital sources for acquisitions. Such access in essence provided the protection of a cash buffer without its investment. IBM's large R&D facility and its project team approach to development assured that it had a pool of people it could quickly shift among various projects to exploit interesting new technologies.

When such flexible response patterns are designed into the enterprise's strategy, it is proactively ready to move on those thrusts—acquisitions, innovations, or resource explorations—which require incrementalism.

Systematic Waiting and Trial Concepts

The prepared strategist may have to wait for events, as Roosevelt awaited a trauma like Pearl Harbor. The availability of desired acquisitions or real estate might depend on a death, divorce, fiscal crisis, management change, or an erratic stock market break.[27] Technological advances may have to await new knowledge, inventions, or lucky accidents. Despite otherwise complete preparations, a planned market entry might not be wise until new legislation, trade agreements, or competitive shake-outs occur. Organizational moves have to be timed to retirements, promotions, management failures, and so on. Very often the specific strategy adopted depends on the timing or sequence of such random events.[28] For example:

Although Continental Group's top executives had thoroughly discussed and investigated energy, natural resources, and insurance as possible "fourth legs" for the company, the major acquisition possibilities were so different that the strategic choice depended on the fit of particular candidates—e.g., Peabody Coal or Richmond Insurance—within these possible industries. The choice of one industry would have precluded the others. The sequence in which firms became available affected the final choice, and that choice itself greatly influenced the whole strategic posture of the company.

In many of the cases studied, strategists proactively launched trial concepts—Mr. McColough's "architecture of information" (Xerox), Mr. Spoor's "Super Box" (Pillsbury)—in order to generate options and concrete proposals. Usually these "trial balloons" were phrased in very broad terms. Without making a commitment to any specific solution, the executive can activate the organization's creative abilities. This approach keeps the manager's own options open until substantive alternatives can be evaluated against each other and against concrete current realities. It prevents practical line managers from rejecting a stategic shift, as they might if forced to compare a "paper option" against well-defined current needs. Such trial concepts give cohesion to the new strategy while enabling the company to take maximum advantage of the psychological and informational benefits of incrementalism.

SOLIDIFYING PROGRESS—INCREMENTALLY

As events move forward, executives can more clearly perceive the specific directions in which their organizations should—and realistically can—move. They can seek more aggressive movement and commitment to their new perceptions, without undermining important ongoing activities or creating unnecessary reactions to their purposes. Until this point, new strategic goals might remain broad, relatively unrefined, or even unstated except as philosophic concepts. More specific dimensions might be incrementally announced as key pieces of information fall into place, specific unanswered issues approach resolution, or significant resources have to be formally committed.

Creating Pockets of Commitment

Early in this stage, guiding executives may need to actively implant support in the organization for new thrusts. They may encourage an array of exploratory projects for each of several possible options. Initial projects can be kept small, partial, or ad hoc, neither forming a comprehensive program nor seeming to be integrated into a cohesive strategy. Executives often pro-

vide stimulating goals, a proper climate for imaginative proposals, and flexible resource support, rather than being personally identified with specific projects. In this way they can achieve organizational involvement and early commitment without focusing attention on any one solution too soon or losing personal credibility if it fails.

Once under way, project teams on the more successful programs in the sample became ever more committed to their particular areas of exploration. They became pockets of support for new strategies deep within the organization. Yet, if necessary, top managers could delay until the last moment their final decisions blending individual projects into a total strategy. Thus, they were able to obtain the best possible match among the company's technical abilities, its psychological commitments, and its changing market needs. By making final choices more effectively—as late as possible with better data, more conscientiously investigated options, and the expert critiques competitive projects allowed—these executives actually increased technical and market efficiencies of their enterprises, despite the apparent added costs of parallel efforts.[29]

In order to maintain their own objectivity and future flexibility, some executives choose to keep their own political profiles low as they build a new consensus. If they seem committed to a strategy too soon, they might discourage others from pursuing key issues which should be raised.[30] By stimulating detailed investigations several levels down, top executives can seem detached yet still shape both progress and ultimate outcomes—by reviewing interim results and specifying the timing, format, and forums for the release of data. When reports come forward, these executives can stand above the battle and review proposals objectively, without being personally on the defensive for having committed themselves to a particular solution too soon. From this position they can more easily orchestrate a high-level

741

consensus on a new strategic thrust. As an added benefit, negative decisions on proposals often come from a group consensus that top executives can simply confirm to lower levels, thereby preserving their personal veto for more crucial moments. In many well-made decisions people at all levels contribute to the generation, amplification, and interpretation of options and information to the extent that it is often difficult to say who really makes the decision.[31]

Focusing the Organization

In spite of their apparent detachment, top executives do focus their organizations on developing strategies at critical points in the process. While adhering to the rhetoric of specific goal setting, most executives are careful not to state new goals in concrete terms before they have built a consensus among key players. They fear that they will prematurely centralize the organization, preempt interesting options, provide a common focus for otherwise fragmented opposition, or cause the organization to act prematurely to carry out a specified commitment. Guiding executives may quietly shape the many alternatives flowing upward by using what Wrapp refers to as "a hidden hand." Through their information networks they can encourage concepts they favor, let weakly supported options die through inaction, and establish hurdles or tests for strongly supported ideas with which they do not agree but which they do not wish to oppose openly.

Since opportunities for such focusing generally develop unexpectedly, the timing of key moves is often unpredictable. A crisis, a rash of reassignments, a reorganization, or a key appointment may allow an executive to focus attention on particular thrusts, add momentum to some, and perhaps quietly phase out others.[32] Most managers surveyed seemed well aware of the notion that "if there are no other options, mine wins." Without being Ma-

chiavellian, they did not want misdirected options to gain strong political momentum and later have to be terminated in an open bloodbath. They also did not want to send false signals that stimulated other segments of their organizations to make proposals in undesirable directions. They sensed very clearly that the patterns in which proposals are approved or denied will inevitably be perceived by lower echelons as precedents for developing future goals or policies.[33]

Managing Coalitions

Power interactions among key players are important at this stage of solidifying progress. Each player has a different level of power determined by his or her information base, organizational position, and personal credibility.[33] Executives legitimately perceive problems or opportunities differently because of their particular values, experiences, and vantage points. They will promote the solutions they perceive as the best compromise for the total enterprise, for themselves, and for their particular units. In an organization with dispersed power, the key figure is the one who can manage coalitions.[34] Since no one player has all the power, regardless of that individual's skill or position, the action that occurs over time might differ greatly from the intentions of any of the players.[35] Top executives try to sense whether support exists among important parties for specific aspects of an issue and try to get partial decisions and momenta going for those aspects. As "comfort levels" or political pressures within the top group rise in favor of specific decisions, the guiding executive might, within his or her concept of a more complete solution, seek—among the various features of different proposals—a balance that the most influential and credible parties can actively support. The result tends to be a stream of partial decisions on limited strategic issues made by constantly changing coalitions of the critical power centers.[36] These decisions steadily evolve toward a broader consensus, acceptable to both the top executive and some "dominant coalition" among these centers.

As a partial consensus emerges, top executives might crystallize issues by stating some broad goals in more specific terms for internal consumption. Finally, when sufficient general acceptance exists and the timing is right, the goals may begin to appear in more public announcements. For example:

As General Mills divested several of its major divisions in the early 1960s, its annual reports began to refer to these as deliberate moves "to concentrate on the company's strengths" and "to intensify General Mills's efforts in the convenience foods field." Such statements could not have been made until many of the actual divestitures were completed, and a sufficient consensus existed among the top executives to support the new corporate concept.

Formalizing Commitment by Empowering Champions

As each major strategic thrust comes into focus, top executives try to ensure that some individual or group feels responsible for its goals. If the thrust will project the enterprise in entirely new directions, executives often want more than mere accountability for its success—they want real commitment.[37] A significantly new major thrust, concept, product, or problem solution frequently needs the nurturing hand of someone who genuinely identifies with it and whose future depends on its success. For example:

Once the divestiture program at General Mills was sufficiently under way, General Rawlings selected young "Bo" Polk to head up an acquisition program to use the cash generated. In this role Polk had nothing to lose. With strong senior management in the remaining consumer products divisions, the ambitious Polk would have had a long road to the top there. In acquisitions, he provided a small political target, only a $50,000 budget in a $500 million company. Yet he had high visibility and could build his own power base, if he were successful. With direct access to and the support of Rawlings, he would be protected through his early ventures. All he had to do was make sure his first few acquisitions were successful. As subsequent acquisitions succeeded, his power base could feed on itself—satisfying both Polk's ego needs and the company's strategic goals.

In some cases, top executives have to wait for champions to appear before committing resources to risky new strategies. They may immediately assign accountability for less dramatic plans by converting them into new missions for ongoing groups.

From this point on, the strategy process is familiar. The organization's formal structure has to be adjusted to support the strategy.[38] Commitment to the most important new thrusts has to be confirmed in formal plans. Detailed budgets, programs, controls, and reward systems have to reflect all planned strategic thrusts. Finally, the guiding executive has to see that recruiting and staffing plans are aligned with the new goals and that—when the situation permits—supporters and persistent opponents of intended new thrusts are assigned to appropriate positions.

Continuing the Dynamics by Eroding Consensus

The major strategic changes studied tended to take many years to accomplish. The process was continuous, often without any clear beginning or end.[39] The decision process constantly molded and modified management's concerns and concepts. Radical crusades became the new conventional wisdom, and over time totally new issues emerged. Participants or observers were often not aware of exactly when a particular decision had been made[40] or when a subsequent consensus was created to supersede or modify it; the process of strategic change was continuous and dynamic. Several GM executives described the frequently imperceptible[41] way in which many strategic decisions evolved:

> We use an iterative process to make a series of tentative decisions on the way we think the market will go. As we get more data we modify these continuously. It is often difficult to say who decided something and when—or even who originated a decision. . . . Strategy really evolves as a series of incremental steps. . . . I frequently don't know when a decision is made in General Motors. I don't remember being in a committee meeting when things came to a vote. Usually someone will simply summarize a developing position. Everyone else either nods or states his particular terms of consensus.

A major strategic change in Xerox was characterized this way:

> How was the overall organization decision made? I've often heard it said that after talking with a lot of people and having trouble with a number of decisions which were pending, Archie McCardell really reached his own conclusion and got Peter McColough's backing on it. But it really didn't happen quite that way. It was an absolutely evolutionary approach. It was a growing feeling. A number of people felt we ought to be moving toward some kind of matrix organization. We have always been a pretty democratic type of organization. In our culture you can't come down with mandates or ultimatums from the top on major changes like this. You almost have to work these things through and let them grow and evolve, keep them on the table so people are thinking about them and talking about them.

Once the organization arrives at its new consensus, the guiding executive has to move immediately to insure that this new position does not become inflexible. In trying to build commitment to a new concept, individual executives often surround themselves with people who see the world in the same way. Such people can rapidly become systematic screens against other views. Effective executives therefore purposely continue the change process, constantly introducing new faces and stimuli at the top. They consciously begin to erode the very strategic thrusts they may have just created—a very difficult, but essential, psychological task.

INTEGRATION OF PROCESSES AND OF INTERESTS

In the large enterprises observed, strategy formulation was a continuously evolving analytical-political consensus process with neither a finite beginning nor a definite end. It generally followed the sequence described above. Yet the total process was anything but linear. It was a groping, cyclical process that often circled back on itself, with frequent interruptions and delays. Pfiffner aptly describes the process of strategy formation as being "like fermentation in biochemistry, rather than an industrial assembly line."[42]

Such incremental management processes are not abrogations of good management practice. Nor are they Machiavellian or consciously manipulative maneuvers. Instead, they represent an adaptation to the practical psychological and informational problems of getting a constantly changing group of people with diverse talents and interests to move *together* effectively in a continually dynamic environment. Much of the impelling force behind logical incrementalism comes from a desire to tap the talents and psychological drives of the whole organization, to create cohesion, and to generate identity with the emerging strategy. The remainder of that force results from the interactive nature of the random factors and lead times affecting the independent subsystems that compose any total strategy.

An Incremental—Not Piecemeal—Process

The total pattern of action, though highly incremental, is not piecemeal in well-managed organizations. It requires constant, conscious reassessment of the total organization, its capacities, and its needs as related to surrounding environments. It requires continual attempts by top managers to integrate these actions into an understandable, cohesive whole. How do top managers themselves describe the process? Mr. Estes, president of General Motors, said:

> We try to give them the broad concepts we are trying to achieve. We operate through questioning and fact gathering. Strategy is a state of mind you go through. When you think about a little problem, your mind begins to think how it will affect all the different elements in the total situation. Once you have had all the jobs you need to qualify for this position, you can see the problem from a variety of viewpoints. But you don't try to ram your conclusions down people's throats. You try to persuade people what has to be done and provide confidence and leadership for them.

Formal-Analytical Techniques. At each stage of strategy development, effective executives constantly try to visualize the new patterns that might exist among the emerging strategies of various subsystems. As each subsystem strategy becomes more apparent, both its executive team and top-level groups try to project its implications for the total enterprise and to stimulate queries, support, and feedback from those involved in related strategies. Perceptive top executives see that the various teams generating subsystem strategies have overlapping members. They require periodic updates and reviews before higher echelon groups that can bring a total corporate view to bear. They use formal planning processes to interrelate and evaluate the resources required, benefits sought, and risks undertaken vis-à-vis other elements of the enterprise's overall strategy. Some use scenario techniques to help visualize potential impacts and relationships. Others utilize complex forecasting models to better understand the basic interactions among subsystems, the total enterprise, and the environment. Still others use specialized staffs, "devil's advocates," or "contention teams" to make sure that all important aspects of their strategies receive a thorough evaluation.

Power-Behavioral Aspects: Coalition Management. All of the formal methodologies help, but the real integration of all the components in an enterprise's total strategy eventually takes place only in the minds of high-level executives. Each executive may legitimately perceive the intended balance of goals and thrusts differently. Some of these differences may be openly expressed as issues to be resolved when new information becomes available. Some differences may remain unstated—hidden agendas to emerge at later dates. Others may be masked by accepting so broad a statement of intention that many different views are included in a seeming consensus, when a more specific statement might be divisive. Nevertheless, effective strategies do achieve a level of understanding and consensus sufficient to focus action.

Top executives deliberately manage the incremental processes within each subsystem to create the basis for consensus. They also manage the coalitions that lie at the heart of most controlled strategy developments.[43] They recognize that they are at the confluence of innumerable pressures—from stockholders, environmentalists, government bodies, customers, suppliers, distributors, producing units, marketing groups, technologists, unions, special issue activists, individual employees, ambitious executives, and so on—and that knowledgeable people of goodwill can easily disagree on proper actions. In response to changing pressures and coalitions among these groups, the top management team constantly forms and reforms its own coalitions on various decisions.[44]

Most major strategic moves tend to assist some interests—and executives' careers—at the expense of others. Consequently, each set of interests serves as a check on the others and thus helps maintain the breadth and balance of strategy.[45] To avoid significant errors some managers try to ensure that all important groups have representation at or access to the top.[46] The guiding executive group may continuously adjust the number, power, or proximity of such access points in order to maintain a desired balance and focus.[47] These delicate adjustments require constant negotiations and implied bargains within the leadership group. Balancing the forces that different interests exert on key decisions is perhaps the ultimate control top executives have in guiding and coordinating the formulation of their companies' strategies.[48]

Establishing, Measuring, and Rewarding Key Thrusts

Few executives or management teams can keep all the dimensions of a complex evolving strategy in mind as they deal with the continuous flux of urgent issues. Consequently, effective strategic managers seek to identify a few central themes that can help to draw diverse efforts together in a common cause.[49] Once identified, these themes help to maintain focus and consistency in the strategy. They make it easier to discuss and monitor proposed strategic thrusts. Ideally, these themes can be developed into a matrix of programs and goals, cutting across formal divisional lines and dominating the selection and ranking of projects within divisions. This matrix can, in turn, serve as the basis for performance measurement, control, and re-

ward systems that ensure the intended strategy is properly implemented.

Unfortunately, few companies in the sample were able to implement such a complex planning and control system without creating undue rigidities. But all did utilize logical incrementalism to bring cohesion to the formal-analytical and power-behavioral processes needed to create effective strategies. Most used some approximation of the process sequence described above to form their strategies at both subsystem and overall corporate levels. A final summary example demonstrates how deliberate incrementalism can integrate the key elements in more traditional approaches to strategy formulation.

In the late 1970s a major nation's largest bank named as its new president and CEO a man with a long and successful career, largely in domestic operating positions. The bank's chairman had been a familiar figure on the international stage and was due to retire in three to five years. The new CEO, with the help of a few trusted colleagues, his chief planner, and a consultant, first tried to answer the questions: "If I look ahead seven to eight years to my retirement as CEO, what would I like to leave behind as the hallmarks of my leadership? What accomplishments would define my era as having been successful?" He chose the following as goals:

1. To be the country's number one bank in profitability and size without sacrificing the quality of its assets or liabilities.
2. To be recognized as a major international bank.
3. To improve substantially the public image and employee perceptions of the bank.
4. To maintain progressive policies that prevent unionization.
5. To be viewed as a professional, well-managed bank with strong, planned management continuity.

6. To be clearly identified as the country's most professional corporate finance bank, with a strong base within the country but with foreign and domestic operations growing in balance.
7. To have women in top management and to achieve full utilization of the bank's female employees.
8. To have a tighter, smaller headquarters and a more rationalized, decentralized corporate structure.

The CEO brought back to the corporate offices the head of his overseas divisions to be COO and to be a member of the Executive Committee, which ran the company's affairs. The CEO discussed his personal views concerning the bank's future with this Committee and also with several of his group VPs. Then, to arrive at a cohesive set of corporate goals, the Executive Committee investigated the bank's existing strengths and weaknesses (again with the assistance of consultants) and extrapolated its existing growth trends seven to eight years into the future. According to the results of this exercise, the bank's foreseeable growth would require that:

1. The bank's whole structure be reoriented to make it a much stronger force in international banking.
2. The bank decentralize operations much more than it ever had.
3. The bank find or develop at least 100 new top-level specialists and general managers within a few years.
4. The bank reorganize around a "four bank" principle (international, commercial, investment, and retail banks) with entirely new linkages forged among these units.
5. These linkages and much of the bank's new international thrust be built on its expertise in certain industries, which were the primary basis of its parent country's international trade.

6. The bank's profitability be improved across the board, especially in its diverse retail banking units.

To develop more detailed data for specific actions and to further develop consensus around needed moves, the CEO commissioned two consulting studies: one on the future of the bank's home country and the other on changing trade patterns and relationships worldwide. As these studies became available, the CEO allowed an ever wider circle of top executives to critique the studies' findings and to share their insights. Finally, the CEO and the Executive Committee were willing to draw up and agree to a statement of ten broad goals (parallel to the CEO's original goals but enriched in flavor and detail). By then, some steps were already under way to implement specific goals (e.g., the four bank concept). But the CEO wanted further participation of his line officers in the formulation of the goals and in the strategic thrusts they represented across the whole bank. By now eighteen months had gone by, but there was widespread consensus within the top management group on major goals and directions.

The CEO then organized an international conference of some forty top officers of the bank and had a background document prepared for this meeting containing: (1) the broad goals agreed upon, (2) the ten major thrusts that the Executive Committee thought were necessary to meet these goals, (3) the key elements needed to back up each thrust, and (4) a summary of the national and economic analyses the thrusts were based upon. The forty executives had two full days to critique, question, improve, and clarify the ideas in this document. Small work groups of line executives reported their findings and concerns directly to the Executive Committee. At the end of the meeting, the Executive Committee tabled one of the major thrusts for further study, agreed to refined wording for some of the bank's broad goals, and modified details of the major thrusts in line with expressed concerns.

The CEO announced that within three months each line officer would be expected to submit his own statement of how his unit would contribute to the major goals and thrusts agreed on. Once these unit goals were discussed and negotiated with the appropriate top executive group, the line officers would develop specific budgetary and nonbudgetary programs showing precisely how their units would carry out each of the major thrusts in the strategy. The COO was asked to develop measures both for all key elements of each unit's fiscal performance and for performance against each agreed upon strategic thrust within each unit. As these plans came into place, it became clear that the old organization had to be aligned behind these new thrusts. The CEO had to substantially redefine the COO's job, deal with some crucial internal political pressures, and place the next generation of top managers in the line positions supporting each major thrust. The total process from concept formulation to implementation of the control system was to span three to four years, with new goals and thrusts emerging flexibly as external events and opportunities developed.

CONCLUSION

In recent years, there has been an increasingly loud chorus of discontent about corporate strategic planning. Many managers are concerned that despite elaborate strategic planning systems, costly staffs for planning, and major commitments of their own time, their most elaborately analyzed strategies never get implemented. These executives and their companies generally have fallen into the trap of thinking about strategy formulation and implementation as separate, sequential processes. They rely on the awesome rationality of their formally derived strategies and the inherent power of their positions to cause their organizations to respond. When this does not occur, they become bewildered, if not frustrated and angry. Instead, successful managers in the companies observed acted logically and incrementally to improve the quality of information used in key decisions; to overcome the personal and political pressures resisting change; to deal with the varying lead times and sequencing problems in critical decisions; and to build the organizational awareness, understanding, and psychological commitment essential to effective strategies. By the time the strategies began to crystallize, pieces of them were already being implemented. Through the very processes they used to formulate their strategies, these executives had built sufficient organizational momentum and identity with the strategies to make them flow toward flexible and successful implementation.

REFERENCES

1. See H. E. Wrapp, "A Plague of Professional Managers," *New York Times*, 8 April 1979.
2. This is the third in a series of articles based upon my study of ten major corporations' processes for achieving significant strategic change. The other two articles in the series are: J. B. Quinn, "Strategic Goals: Process and Politics," *Sloan Management Review*, Fall 1977, pp. 21–37; J. B. Quinn, "Strategic Change: 'Logical Incrementalism,'" *Sloan Management Review*, Fall 1978, pp. 7–21. The whole study has been published as a book entitled *Strategies for Change: Logical Incrementalism* (Homewood, IL: Dow Jones-Irwin, 1981). All findings purposely deal only with strategic changes in large organizations.
3. See R. M. Cyert and J. G. March, *A Behavioral Theory of the Firm* (Englewood Cliffs, NJ: Prentice-Hall, 1963), p. 123. Note this learning-feedback-adaptiveness of goals and feasible alternatives over time as organizational learning.

4. See: H. E. Wrapp, "Good Managers Don't Make Policy Decisions," *Harvard Business Review,* September-October 1967, pp. 91–99; R. Normann, *Management for Growth,* trans. N. Adler (New York: John Wiley & Sons, 1977); D. Braybrooke and C. E. Lindblom, *A Strategy of Decision: Policy Evaluation as a Social Process* (New York: Free Press of Glencoe, 1963); C. E. Lindblom, *The Policy-Making Process* (Englewood Cliffs, NJ: Prentice-Hall, 1968); W. G. Bennis, *Changing Organizations: Essays on the Development and Evolution of Human Organizations* (New York: McGraw-Hill, 1966).

5. See respectively: Quinn (Fall 1977); Quinn (Fall 1978).

6. See J. B. Quinn, *Xerox Corporation (B)* (copyrighted case, Amos Tuck School of Business Administration, Dartmouth College, Hanover, NH, 1979).

7. See O. G. Brim, D. Glass et al., *Personality and Decision Processes: Studies in the Social Psychology of Thinking* (Stanford, CA: Stanford University Press, 1962).

8. Crises did occur at some stage in almost all the strategies investigated. However, the study was concerned with the attempt to manage strategic change in an ordinary way. While executives had to deal with precipitating events in this process, crisis management was not—and should not be—the focus of effective strategic management.

9. For some formal approaches and philosophies for environmental scanning, see: W. D. Guth, "Formulating Organizational Objectives and Strategy: A Systematic Approach," *Journal of Business Policy* (Autumn 1971): 24–31; F. J. Aguilar, *Scanning the Business Environment* (New York: Macmillan Co., 1967). For confirmation of the early vagueness and ambiguity in problem form and identification, see H. Mintzberg, D. Raisinghani, and A. Théorêt, "The Structure of 'Unstructured' Decision Processes," *Administrative Science Quarterly* (June 1976): 246–275.

10. For a discussion on various types of "misfits" between the organization and its environment as a basis for problem identification, see Normann (1977), p. 19.

11. For suggestions on why organizations engage in "problem search" patterns, see R. M. Cyert, H. A. Simon, and D. B. Trow, "Observation of a Business Decision," *The Journal of Business* (October 1956): 237–248; For the problems of timing in transitions, see L. R. Sayles, *Managerial Behavior: Administration in Complex Organizations* (New York: McGraw-Hill, 1964).

12. For a classic view of how these screens operate, see C. Argyris, "Double Loop Learning in Organizations," *Harvard Business Review,* September-October 1977, pp. 115–125.

13. See Quinn (copyrighted case, 1979).

14. Cyert and March (1963) suggest that executives choose from a number of satisfactory solutions; later observers suggest they choose the first truly satisfactory solution discovered.

15. See F. F. Gilmore, "Overcoming the Perils of Advocacy in Corporate Planning," *California Management Review* (Spring 1973): 127–137.

16. See J. B. Quinn, *General Motors Corporation: The Downsizing Decision* (copyrighted case, Amos Tuck School of Business Administration, Dartmouth College, Hanover, NH, 1978).

17. See E. Rhenman, *Organization Theory for Long-Range Planning* (New York: John Wiley & Sons, 1973), p. 63. Here author notes a similar phenomenon.

18. See Quinn (copyrighted case, 1978).

19. See R. M. Cyert, W. R. Dill, and J. G. March, "The Role of Expectations in Business Decision Making," *Administrative Science Quarterly* (December 1958): 307–340. The authors point out the perils of top management advocacy because existing polities may unconsciously bias information to support views they value.

20. See J. H. Dessauer, *My Years with Xerox: The Billions Nobody Wanted* (Garden City, NY: Doubleday, 1971).

21. See: H. Mintzberg, *The Nature of Managerial Work* (New York: Harper & Row, 1973). Note that this "vision" is not necessarily the beginning point of the process. Instead it emerges as new data and viewpoints interact; Normann (1977).

22. See Mintzberg, Raisinghani, and Théorêt (June 1976). Here the authors liken the process to a decision tree where decisions at each node become more narrow, with failure at any node allowing recycling back to the broader tree trunk.

23. Wrapp (September-October 1967) notes that a conditioning process that may stretch over months or years is necessary in order to prepare the organization for radical departures from what it is already striving to attain.

24. See J. G. March, J. P. Olsen, S. Christensen et al., *Ambiguity and Choice in Organizations* (Bergen, Norway: Universitetsforlaget, 1976).

25. See: T. A. Wise, "I.B.M.'s $5 Billion Gamble," *Fortune,* September 1966, pp. 118–124; T. A. Wise, "The Rocky Road to the Marketplace (Part II: I.B.M.'s $5 Billion Gamble)," *Fortune,* October 1966, pp. 138–152.

26. For an excellent overview of the processes of co-optation and neutralization, see Sayles (1964); For perhaps the first reference to the concept of the "zone of indifference," see C. I. Barnard, *The Functions of the Executive* (Cambridge, MA: Harvard University Press, 1938); The following two sources note the need of executives for coalition behavior to reduce the organizational conflict resulting from differing interests and goal preferences in large organizations: Cyert and March (1963); J. G. March, "Business Decision Making," in *Readings in Managerial Psychology,* H. J. Leavitt and L. R. Pondy, eds. (Chicago: University of Chicago, 1964).

27. Cyert and March (1963) also note that not only do organizations seek alternatives but that "alternatives seek organizations" (as when finders, scientists, bankers, etc., bring in new solutions).

28. See March, Olsen, Christensen et al. (1976).

29. Much of the rationale for this approach is contained in J. B. Quinn, "Technological Innovation, Entrepreneurship, and Strategy," *Sloan Management Review,* Spring 1979, pp. 19–30.

30. See C. Argyris, "Interpersonal Barriers to Decision Making," *Harvard Business Review,* March-April 1966, pp. 84–97. The author notes that when the president introduced major decisions from the top, discussion was "less than open" and commitment was "less than complete," although executives might assure the president to the contrary.

31. See March (1964).

32. The process tends to be one of eliminating the less feasible rather than of determining a target or objectives. The process typically reduces the number of alternatives through successive limited comparisons to a point where understood analytical techniques can apply and the organization structure can function to make a choice. See Cyert and March (1963).

33. For more detailed relationships between authority and power, see H. C. Metcalf and L. Urwick, eds., *Dynamic Administration: The Collected Papers of Mary Parker Follett* (New York: Harper & Brothers, 1941); A. Zaleznik, "Power and Politics in Organizational Life," *Harvard Business Review,* May-June 1970, pp. 47–60.

34. See J. D. Thompson, "The Control of Complex Organizations," in *Organizations in Action* (New York: McGraw-Hill, 1967).

35. See G. T. Allison, *Essence of Decision: Explaining the Cuban Missile Crisis* (Boston: Little, Brown and Company, 1971).

36. See C. E. Lindblom, "The Science of 'Muddling Through,'" *Public Administration Review* (Spring 1959): 79–88. The author notes that the relative weights individuals give to values and the intensity of their feelings will vary sequentially from decision to decision, hence the dominant coalition itself varies with each decision somewhat.

37. Zaleznik (May-June 1970) notes that confusing compliance with commitment is one of the most common and difficult problems of strategic implementation. He notes that often organizational commitment may override personal interest if the former is developed carefully.

38. See A. D. Chandler, *Strategy and Structure: Chapters in the History of the Industrial Enterprise* (Cambridge, MA: MIT Press, 1962).

39. See K. J. Cohen and R. M. Cyert, "Strategy: Formulation, Implementation, and Monitoring," *The Journal of Business* (July 1973): 349–367.

40. March (1964) notes that major decisions are "processes of gradual commitment."

41. Sayles (1964) notes that such decisions are a "flow process" with no one person ever really making the decisions.

42. See J. M. Pfiffner, "Administrative Rationality," *Public Administration Review* (Summer 1960): 125–132.

43. See R. James, "Corporate Strategy and Change — The Management of People" (monograph, The University of Chicago, 1978). The author does an excellent job of pulling together the threads of coalition management at top organizational levels.

44. See Cyert and March (1963), p. 115.

45. Lindblom (Spring 1959) notes that every interest has a "watchdog" and that purposely allowing these watchdogs to participate in and influence decisions creates consensus decisions that all can live with. Similar conscious access to the top for different interests can now be found in corporate structures.

46. See Zaleznik (May-June 1970).

47. For an excellent view of the bargaining processes involved in coalition management, see Sayles (1964), pp. 207–217.

48. For suggestions on why the central power figure in decentralized organizations must be the person who manages its dominant coalition, the size of which will depend on the issues involved, and the number of areas in which the organizations must rely on judgmental decisions, see Thompson (1967).

49. Wrapp (September-October 1967) notes the futility of a top manager trying to push a full package of goals.

Alcoa Goes Back to the Future

George David Smith

John E. Wright

In 1983, the Aluminum Company of America was the world's leading producer of aluminum and aluminum products. It was also a company in crisis. Earnings were down, energy costs were up, capacity was in excess, and demand forecasts were gloomy. Alcoa had already begun its first major staff reductions since the Great Depression. The depressed economic and business environment was placing great pressure on the company, not simply to cut costs but to reassess the very nature and mission of its business.

In that year's annual report, W.H. Krome George, Alcoa's chairman and CEO, said the "golden years" of growth for primary aluminum had passed. Casting aside the "generally optimistic" tone of previous annual reports, George offered a sober assessment of the "economic and political winds" and "their probable future direction." Alcoa, he concluded, had to face up to a fundamental climate change; it was time for a discussion of adaptation to "a world quite different from the one we have known."

In this regard, Alcoa was not unlike other contemporary American corporations operating in capital-intensive basic industries. But a unique historical tradition had imbued the company with its own ways of coping with problems, its own ways of thinking and behaving, its own values, its own corporate personality—in other words, its own culture. All this was a legacy that contained strengths as well as constraints on the company's ability to move into the future.

As discussion over the future of Alcoa's business proceeded, the constraints of culture loomed ever larger. To understand them, it was not enough to rely on employee attitude surveys or analyses of current pressures, policy, and practice. The existing culture was, after all, an evolutionary accretion of past events and decisions, and had to be understood as a dynamic problem in a stream of time.

Thus, as new strategies began to unfold, Alcoa retained a small group of business and technology historians to provide an objective analysis of the company's internal development. The team of consultants had four critical areas of concern: strategy, structure, the management of personnel, and the management of technology. After studying the history of the industry, the historians spent four months researching company records (many of which extended back to the company's founding) and interviewing dozens of current and former employees.

The interview subjects cut across Alcoa's

"Alcoa Goes Back to the Future" by George David Smith and John E. Wright, from *Across the Board*, September 1986, Volume 23, The Conference Board. Copyright 1986 by George David Smith and John E. Wright. Reprinted by permission.

George David Smith was a consultant, and John E. Wright was the Alcoa manager responsible for overseeing the project described in this article.

functional operations and product lines; some had experience with the company dating back as far as World War I. The historians also talked with customers, union officials, and even local officials in communities where the company operated. When, after four months of intensive work, the historians closeted themselves in a hotel room to discuss their findings, they were able to piece together a composite portrait of the company that took into account the dependence of the present on the past and the dynamics of change over time. Archival records provided information that could not be gleaned from the interviews and often served to correct faulty memories. The records also helped identify the origins and contexts of events that the historians found to have particular relevance.

Alcoa does not lack a sense of history. To the contrary, tradition is important. Company folklore is bound to a tradition of striving, struggle, growth, and success, celebrating Alcoa's achievement in pioneering a new metal into common use through the establishment of a great monopoly enterprise.

Familiar stories had reached mythic proportions—Charles Martin Hall's obsessive pursuit of an aluminum reduction process in a woodshed, Arthur Vining Davis's Napoleonic conquest of the primary aluminum market, the inspired progress of aluminum science in Alcoa's research laboratories, the titanic struggle of the 1937 antitrust case, the tragic fall of the American monopoly in World War II, and the triumphant resurgence of the company as leader of the world industry.

Founded in 1888 as an experimental plant on a $20,000 investment, Alcoa was nearly a century old. Into the early 1950s, though it had become one of the largest industrials in America, it remained substantially an owner-managed firm whose top executives had been around a long time. A. V. Davis, who retired in 1957, had been the effective head of the company since the turn of the century. He and his associates had developed an innovative,

entrepreneurial enterprise, eschewing bureaucratic modes of decision-making and adapting flexibly to changes in the economic, regulatory, and labor environments.

Alcoa was a model of large-scale vertical integration. By World War I, its founders had extended complete control over all the inputs to aluminum production. They had bought control of bauxite reserves in Arkansas, Georgia, and Surinam, and had built a large refinery in East St. Louis, Illinois. The company had built its own hydro-power facilities to drive its electricity-intensive reduction process, and it even owned rail lines and barges to haul its raw materials. Alcoa thus became a fully self-sufficient enterprise "upstream," with its own mining, refining, and power operations supporting its main business of smelting aluminum.

Because aluminum was a new and unfamiliar material, Alcoa had also integrated "downstream" into fabrication and distribution, to push newly found applications of aluminum into markets dominated by ferrous metals, copper, nickel, and wood. To support its far-flung operations, Alcoa developed centrally controlled R&D, accounting, and marketing capabilities, as well as elaborate support systems in engineering, transportation, and construction. Over time, vertical integration of operations "from mine to metal" gave rise to (and was reinforced by) attitudes that preferred making to buying, and that placed heavy emphasis on internal innovation.

Vertical integration supported Alcoa's growing operations in aluminum smelting so efficiently that, even after its controlling patents had expired, no one could muster both the capital and technical know-how necessary to enter the U.S. market as a viable competitor.

World War II and a landmark antitrust decision brought Alcoa's monopoly to an end. The demand for aluminum in military aircraft and supplies required a doubling of the nation's smelting capacity between 1939 and 1943, and this could be achieved only with a rapid infusion of Government funds. Alcoa built virtually all and operated most of the Government refining, smelting, and fabricating facilities, although two other companies, Olin and Reynolds, were granted contracts to conduct some refining and smelting operations.

Then, in 1945, a Federal Appeals Court ruled that Alcoa had an illegal monopoly. Despite Alcoa's history of success in warding off antitrust charges, and despite the absence of compelling proof of specific wrongdoing, Judge Learned Hand found that Alcoa's success in capturing more than 90 percent of the American primary aluminum market was, in and of itself, a violation of the Sherman Antitrust Act.

This novel interpretation of antitrust law stunned Alcoa executives, who were forced to acquiesce in a de facto remedy whereby almost all commercially viable Government aluminum facilities were sold off to new competitors at deep discounts.

The postwar aluminum industry was characterized by oligopolistic competition among four major integrated producers: Alcoa, Reynolds, Kaiser, and Alcan. Under pressure from the Justice Department and Congress, Alcoa's share of primary capacity in the United States eroded from well over 90 percent to close to 30 percent by 1960. But the company thrived amid growing demand for aluminum products in an expanding consumer society.

To compete, Alcoa relied on its storehouse of technological expertise, remaining self-reliant in its tightly integrated structure while pursuing a vigorously entrepreneurial approach to new product development. Over time, administration of the growing business was further centralized by tightening the Pittsburgh central office's control over the coordination of Alcoa's operations at home and, after 1958, overseas.

By the late 1960s, aluminum, still a "young" material, had penetrated a wide array of major industrial and consumer markets. Alcoa's center of gravity had always been its production of primary aluminum, but to convert that into sales, markets had to be taken away from other materials. For example, in the early part of the century, the company took market share from copper in electrical cable by lowering the costs of production and by constant attention to technological improvements that gave aluminum cable both economic and qualitative advantages.

A more recent example is packaging. In 1960, 100 percent of the beverage cans in the United States were made of tin-plate steel; by 1980, about 90 percent were aluminum. The light metal found major markets in residential and commercial building products, ground transportation, machinery, and aircraft—in each case by displacing other materials through lower cost and/or technical superiority. Alcoa made most of its money by linking the production of primary aluminum to specialized semifinished products made from advanced aluminum alloys, while leaving the vast majority of the more common finished aluminum products to others.

Meanwhile, changes in the environment began to challenge some basic assumptions about the business. By 1970, it had become apparent that a long-term growth rate in demand for aluminum roughly twice that of the GNP would not continue, and that the power of the North American oligopoly to "administer" prices and output was diminishing. Competition from a dozen producers combined with rising energy and labor costs to pressure the supply side.

As the tide that had lifted an increasing number of boats in the industry began to recede, Alcoa shifted its emphasis toward products and markets where its technological strengths could be applied to such sophisticated, high-margin products as aerospace components and lithographic sheet. At the same time, the company poured resources into improving and expanding its primary production. This strategy proved successful. In 1979 Alcoa enjoyed its most profitable year on volume since its record year of 1973.

Then the depression of the early 1980s struck, and Alcoa (its prosperity always sensitive to economic cycles) was hit hard. This shock was exacerbated by the gathering forces of foreign competition and domestic rivalry from new synthetic materials. The rise of subsidized third-world smelters and the trading of aluminum ingot as a commodity on the London Metals Exchange (since 1978) finally confirmed that rationally administered prices and controlled capacity of aluminum ingot in the world market were relics of a bygone era.

In this radically altered environment, it became less desirable to expand American production of primary aluminum. If the company were not to become just another American corporate dinosaur, a reconsideration of its strategy was imperative.

First, management considered the growth of the company's basic business. It was clear that the American, Japanese, and European markets for aluminum would continue to be important, but long-term growth would increasingly depend on industrialized nations with developing economies. Alcoa thus accelerated its timetable for pursuing such markets in Latin America and the Eastern Pacific Rim. Because it is neither politically nor economically feasible to export to most developing economies, Alcoa has pressed the expansion of primary capacity in Brazil and Australia, greatly increasing its commitment to overseas operations.

Domestically, Alcoa's strategy required more sophisticated and innovative thinking. At stake was the very identity of the company. In management meetings, a strong case was made for expanding the business more broadly into nonaluminum materials, such as other metals, polymers, and ceramics. Such products, it was

hoped, would offer new prospects for growth and help offset the cyclicality of the core aluminum business.

New strategies would require new structures. It was already becoming clear that Alcoa would have to develop a decentralized organization to produce for a widening array of discrete markets. Management would have to devolve decision-making authority upon more and varied business units while simultaneously increasing the corporation's ability to control costs, increase productivity, and improve the flow of information in a more complex and competitive world. Technologically, Alcoa would have to extend its fundamental research efforts in new areas while accelerating the computerization and automation of its production systems. All of this would affect existing bureaucratic arrangements, personnel requirements, and labor relations.

The historians cast Alcoa's immediate concerns in longer-term perspective. They argued that much of the fear about the potential obsolescence of aluminum was rooted more in the situational anxiety of economic recession than in historical realities. This implied not too hasty a retreat from the company's core business. At the same time, the company's strong self-image as a primary aluminum producer—which arose from Alcoa's origins and great historical success as a smelter—seemed excessive, considering that the most profitable parts of the business had moved well downstream into such fabricated products as can sheet and aerospace alloys. The company's overwhelming self-reliance on its own technological and material resources in a number of areas—a legacy of the corporation's patterns of intensive vertical integration—seemed out of proportion to the realities of the evolving business-unit management structure.

In October 1983, an interim report on the company's history was presented to Alcoa's senior management and staff. Conclusions about "the salient characteristics of the corporate culture" were presented in categories defined as "enduring cultural strengths" and "embedded cultural constraints." The former were defined as "habits of thought and behavior and structural characteristics that have served and will continue to serve the company well, if they are not undermined by change." Embedded constraints were defined as "those deeply held assumptions or interpretations of experience that affect management judgments, patterns of behavior, and relationships in the corporation." The constraints were defined not as weaknesses, but as historically rooted "aspects of the corporate culture that will, consciously or not, condition thinking about the future."

A key observation was that Alcoa had evolved as a highly paternalistic company with a remarkable continuity in management over the years, and that even with more rapid turnover at the top in recent years, the company continued to provide a high sense of security and stability among employees at all levels of the enterprise. Alcoa was historically (in comparison with other large corporations) a strongly people-oriented company, with an unusually good labor history (despite management preoccupations with labor as a "problem"), and a remarkably flexible bureaucracy. These qualities—stability, concern for people, and flexibility—promoted unusually intense loyalty, which was identified as a crucial strength in times of change.

Identifying "embedded constraints" was a more subtle problem, because their implications were harder to uncover and explain. Insights in this area illuminated many of the hidden, unconscious elements of contemporary management problems. For example, the company's recent push into international markets was a belated development, and the delay was tied directly to a nearly forgotten decision. Few modern Alcoans were aware that from World War I through the 1920s, their company had developed extensive operations in 11 foreign

countries—in bauxite mining, refining, smelting, fabrications, and sales. Then, in 1928, the company consolidated its operations (with the exception of its bauxite holdings in Surinam), and placed them under the control of its Canadian subsidiary, which it then spun off as Aluminium Limited (now Alcan).

The reasons were varied. Alcoa's executives concluded that they could not competently manage both domestic and foreign sales, especially given the gathering forces of nationalism and the resulting barriers to trade in Europe. Also, Alcoa could not legally participate in European cartels because of American antitrust laws. A more personal consideration was A.V. Davis's desire to find a position for his younger brother, Edward, who was in competition with Roy Hunt (son of the company's principal founder) for Alcoa's presidency, which the elder Davis was about to vacate. Edward became head of Aluminum Limited, owned by the same small group of shareholders that controlled Alcoa.

Aluminium Limited then served as the foreign arm of the joint stockholders, while Alcoa served the greater U.S. market. The arrangement was ideal, of course, until the common control of the two companies was stripped away by a 1950 antitrust decree. When, in the 1960s, Alcoa once again moved into international operations, it had some difficulty. Having been so long isolated from international business, the company's management had developed a strong domestic mind set, which inhibited its ability to grant its new overseas operations the importance they deserved. Charles Parry, Alcoa's current CEO, reflected on the lesson of this particular history before an assembly of the Pittsburgh business community in June 1985. It was an attitude about an event in the company's history, and not anything inherent in the nature of its abilities, technology, or markets, that subsequently kept Alcoa isolated in the U.S. market for so many years, Parry said. "World War II demands for aluminum and the outstanding growth in consumption during postwar years combined to reinforce . . . a mind set that long outlasted the context of its creation."

Another striking, though more subtle, effect of a forgotten historical process was uncovered in the company's management of research. Alcoa established a formal research program in 1919, which by World War II had attracted a team of eminent scientists and engineers working at the cutting edge of metallurgical and electrochemical research. Alcoa Laboratories developed a proud tradition as a research institution.

But some specific decisions made in the 1950s about staffing and managing research and development had steered the company away from its emphasis on fundamental research. There were good reasons for this shift at the time. The rise of domestic competition had forced the company to pay more attention to short-term engineering. In the 1950s and '60s, Alcoa Laboratories, believing that its fundamental knowledge was adequate for years to come, became more driven by the near-term needs of sales and operating efforts.

While many benefits came from this shift in emphasis, Alcoa Laboratories lost much of its ability to manage fundamental research, even for basic processes that might be essential to reducing costs and achieving competitive advantages over the long term. One exceptional long-term project designed to replace the Hall smelting process is a case in point. The effort to develop a new Alcoa Smelting Process (ASP) was written off in 1985 after more than two decades of concerted development that had reached as far as full-scale production.

The magnitude of the failure reflected attempts to pursue long-term revolutionary goals in a short-term incremental environment. In 1981, Allen Russell, a former head of Alcoa Laboratories, found that the ASP project was suffering from weaknesses in Alcoa's knowledge base, which had been depleted; excessive secrecy, which prevented calling in appropriate outside consultants; and time pressure, which may have forced too rapid a movement of the process into the manufacturing phase before fundamental problems were resolved. The historians recognized these problems as vestiges of research management that had been structured for the years following the loss of Alcoa's monopoly. As ASP moved through the pilot phase in the late 1970s, no one was sufficiently aware of the disjunction between self-image and behavior, one rooted in a pre-World War II tradition, the other in a postwar structure.

The significance of this history was grasped immediately by Peter Bridenbaugh, Alcoa's vice president for research, who by 1983 had been charged with reforming Alcoa Laboratories. "While it was relatively easy to identify the strengths and weaknesses in our organization," he says, "understanding how they came into being proved much more difficult. Insights into past changes in Alcoa's conception of what its research and development laboratory should be gave us a crucial perspective on and understanding of the roots of our current problems." In particular, says Bridenbaugh, "I began to appreciate the penalty we were paying for changes that had occurred in the years following the war."

Once again, in light of historical realities, Alcoa Laboratories had to alter its hiring and training procedures, its research management, its funding patterns, and its incentive systems.

The question remains whether Alcoa can manage two approaches to R&D, one geared to incremental innovation in a mature industry, the other to the building of new knowledge on wider fronts of technology. Alcoa's future includes not only new aluminum alloys, but other metals and materials, including synthetic materials that can be used with aluminum-based products. Alcoa's R&D will have to lead in some areas while relying on external technology in others; it will have to continue incremental innovation while making selective commitments to long-term, fundamental problems.

All the managerial and organizational issues of such a multipronged research strategy have yet to be resolved, but the company has undertaken reforms, not the least of which has shifted the emphasis of basic research away from smelting and refining processes into a wider range of technology.

The historians drew other conclusions about the impact of corporate history on present-day management goals. One powerful constraint on Alcoa's attempt to decentralize was the company's tradition of self-sufficiency. Another lay in the very origins of the enterprise as a commercial venture to exploit Hall's smelting process. The company's self-image as a primary producer remained strong, even dominant, as the enterprise broadened its product and technology lines. "At times," says Parry, "our proud tradition as an aluminum smelter became something of a constraint on our ability to respond to changing circumstances."

By the late 1970s, many managers outside the smelting orbit came to feel that their businesses had not received enough nurturing—that the corporate center of gravity had not expanded enough to give higher-value applications their full chance to grow and compete. The company's highly centralized decision-making only exacerbated the problem. The corporate planners agreed. "We came to understand that individual business did not have sufficient control over their resources and output decisions," says one. "Decentralization was the key."

Decentralization ran against the trend of Alcoa's history. It became, according to one insider, "a highly emotional debate" in top management circles, and tended to fall out along generational lines. Managers in Alcoa's vertically integrated operations would have to get used to dealing more and more in external markets. Managers used to unitary authority and managers with ties to traditionally powerful entities (such as Alcoa's highly centralized engineering department) would have to accept a greater diffusion of power and responsibility.

The concept of decentralization had won the day by late 1979, but decisive movement toward a new structure came only after pressure was applied by the tough economic environment of the early 1980s. Decentralization has proceeded apace, along with an increased willingness to buy materials and technology when it makes sense to do so. Parry is careful to note that a better understanding of the company's history has not mandated this or other strategic developments, which, after all, must be driven by present problems and expectations for the future. What history has done is to make clearer the questions management had to ask about its business before committing it to change.

There remains at least one unresolved question. Alcoa has placed great stock in what it considers to be its greatest strength: the loyalty of its personnel. Over the years the company has had few written policies, relying instead on informal methods of dealing with peers, superiors, and subordinates—the legacy of a much smaller, more closely held company. Open-door executive offices, walk-around plant managers, and the expectation of lifelong Alcoa careers all contributed to the sense of "family," one of the most important words in the company lexicon.

For overlapping generations, the corporate family has been a powerful ideal in what could have easily evolved into a more impersonal, bureaucratic corporate culture. It is an ideal that has bound employees to the firm, that has kept turnover low and productivity high. And it has tied the most remote peripheries of the enterprise to the central office in Pittsburgh through alternating periods of decentralized and centralized decision-making.

But morale and loyalty are tested in turbulent times. When the United Steelworkers and the Aluminum Brick and Glass Workers went on strike at 15 Alcoa plants on May 31 of this year, it was the first extended walkout by both the company's major unions. In early July, both unions agreed to contracts that would reduce labor costs by 95 cents an hour on average. (Two plants remained struck over local issues.)

The strike was also unusual in that Alcoa salaried personnel kept all the plants running throughout. Almost 600 employees, from clerks to scientists to executives, came from company headquarters, sales offices, research laboratories, and other facilities to perform operating, maintenance, and housekeeping chores.

There was a considerable measure of good will. Years of relatively harmonious labor relations (compared with those of other basic industries) and more recent efforts to promote participatory working relationships on the plant floor had done much to minimize the level of anger inherent in the dispute. Strikers did not interfere with the comings and goings of their salaried counterparts, and at several locations managers and strikers cheerfully fraternized. One plant manager went out of his way to provide shade for the picket line; others sent coffee and doughnuts. Strikers at a vulnerable smelter actually went in to fix an electrical transformer lest production become crippled by the costly "freezing" of aluminum pot lines.

Still, there was the reality of the clash between institutional powers. Forces of history had wrought fundamental changes in the economy and the industry, skewing the structure of the relationship between labor and management, which had been fairly stable since World War II. Worldwide competitive pressures were forcing Alcoa to cut costs in ways that made unprecedented demands on both workers and managers. Relatively centralized unions found that they had to negotiate with an increasingly decentralized corporation.

Two problems hover in the aftermath. The first is institutional: Can unions demanding uniform compensation packages continue to deal effectively with decentralized corporations whose costs are increasingly varied across products and markets? The second relates to the corporate culture: Will the very fact of the strike diminish the underlying sense of loyalty that has historically underpinned labor-management relations at Alcoa?

Even within management, the decentralization at Alcoa has upset familiar relationships and threatened employees' sense of security. There is some risk that Alcoa's sense of family will wither in the present restructuring of the company, which has called for dislocations of personnel, reductions in force, and redefinitions of roles and tasks. Whether the company can preserve its tradition of corporate loyalty remains to be seen, but careful maintenance of this "powerful asset," as Parry calls it is vital to Alcoa's plans.

Index